NEW GCSE MATHS
Edexcel Linear

Fully supports the 2010 GCSE Specification

Brian Speed • Keith Gordon • Kevin Evans • Trevor Senior • Chris Pearce

CONTENTS

INTRODUCTION

Welcome to Collins New GCSE Maths for Edexcel Linear Higher Book 2.

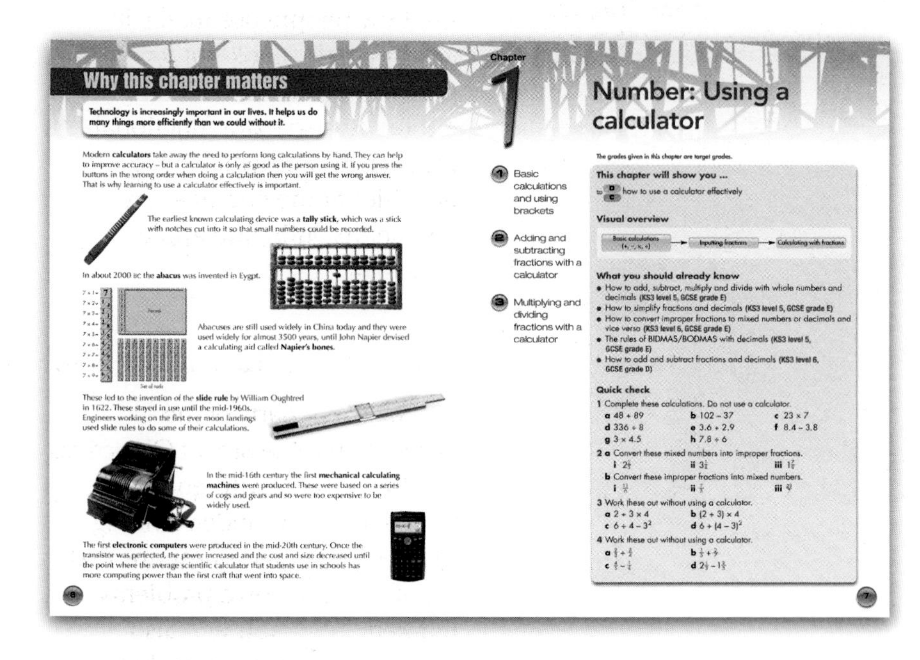

Why this chapter matters

Technology is increasingly important in our lives. It helps us do many things more efficiently than we could without it.

Modern **calculators** take away the need to perform long calculations by hand. They can help to improve accuracy – but a calculator is only as good as the person using it. If you press the buttons in the wrong order when doing a calculation then you will get the wrong answer. That is why learning to use a calculator effectively is important.

The earliest known calculating device was a **tally stick**, which was a stick with notches cut into it so that small numbers could be recorded.

In about 2000 BC the **abacus** was invented in Eygpt.

Abacuses are still used widely in China today and they were used widely for almost 3500 years, until John Napier devised a calculating aid called **Napier's bones**.

These led to the invention of the **slide rule** by William Oughtred in 1622. These stayed in use until the mid-1960s. Engineers working on the first ever moon landings used slide rules to do some of their calculations.

In the mid-16th century the first **mechanical calculating machines** were produced. These were based on a series of cogs and gears and so were too expensive to be widely used.

The first **electronic computers** were produced in the mid-20th century. Once the transistor was perfected, the power increased and the cost and size decreased until the point where the average scientific calculator that students use in schools has more computing power than the first craft that went into space.

6

Chapter 1

Number: Using a calculator

1 Basic calculations and using brackets

2 Adding and subtracting fractions with a calculator

3 Multiplying and dividing fractions with a calculator

The grades given in this chapter are target grades.

This chapter will show you ...

D to C how to use a calculator effectively

Visual overview

Basic calculations (+, −, ×, ÷) → Inputting fractions → Calculating with fractions

What you should already know

- How to add, subtract, multiply and divide with whole numbers and decimals (KS3 level 5, GCSE grade E)
- How to simplify fractions and decimals (KS3 level 5, GCSE grade E)
- How to convert improper fractions to mixed numbers or decimals and vice versa (KS3 level 6, GCSE grade E)
- The rules of BIDMAS/BODMAS with decimals (KS3 level 5, GCSE grade E)
- How to add and subtract fractions and decimals (KS3 level 6, GCSE grade D)

Quick check

1 Complete these calculations. Do not use a calculator.
a $48 + 89$ b $102 - 37$ c 23×7
d $336 \div 8$ e $3.6 + 2.9$ f $8.4 - 3.8$
g 3×4.5 h $7.8 \div 6$

2 a Convert these mixed numbers into improper fractions.
b Convert these improper fractions into mixed numbers.

3 Work these out without using a calculator.
a $2 + 3 \times 4$ b $(2 + 3) \times 4$
c $6 + 4 - 3^2$ d $6 + (4 - 3)^2$

4 Work these out without using a calculator.

7

Why this chapter matters

Find out why each chapter is important through the history of maths, seeing how maths links to other subjects and cultures, and how maths is related to real life.

Chapter overviews

Look ahead to see what maths you will be doing and how you can build on what you already know.

Colour-coded grades

Know what target grade you are working at and track your progress with the colour-coded grade panels at the side of the page.

Use of calculators

Questions where you must or could use your calculator are marked with icon.

Explanations involving calculators are based on *CASIO fx–83ES*.

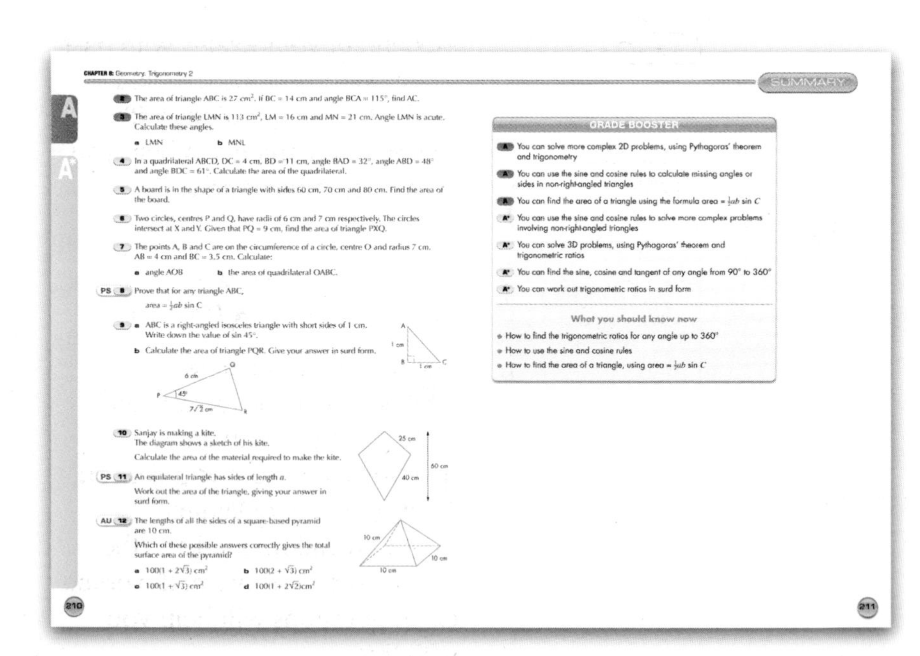

CHAPTER 8: Geometry: Trigonometry 2

SUMMARY

2 The area of triangle ABC is 27 cm². If BC = 14 cm and angle BCA = 115°, find AC.

3 The area of triangle LMN is 113 cm², LM = 16 cm and MN = 21 cm. Angle LMN is acute. Calculate these angles.
a LMN b MNL

4 In a quadrilateral ABCD, DC = 4 cm, BD = 11 cm, angle BAD = 32°, angle ABD = 48° and angle BDC = 61°. Calculate the area of the quadrilateral.

5 A board is in the shape of a triangle with sides 60 cm, 70 cm and 80 cm. Find the area of the board.

6 Two circles, centres P and Q, have radii of 6 cm and 7 cm respectively. The circles intersect at X and Y. Given that PQ = 9 cm, find the area of triangle PXQ.

7 The points A, B and C are on the circumference of a circle, centre O and radius 7 cm. AB = 4 cm and BC = 3.5 cm. Calculate:
a angle AOB b the area of quadrilateral OABC.

PS 8 Prove that for any triangle ABC,
area = $\frac{1}{2}ab \sin C$

9 a ABC is a right-angled isosceles triangle with short sides of 1 cm. Write down the value of sin 45°.
b Calculate the area of triangle PQR. Give your answer in surd form.

10 Sanjay is making a kite. The diagram shows a sketch of his kite.
Calculate the area of the material required to make the kite.

PS 11 An equilateral triangle has sides of length a.
Work out the area of the triangle, giving your answer in surd form.

AU 12 The lengths of all the sides of a square-based pyramid are 10 cm.
Which of these possible answers correctly gives the total surface area of the pyramid?
a $100(1 + 2\sqrt{3})$ cm² b $100(2 + \sqrt{3})$ cm²
c $100(1 + \sqrt{3})$ cm² d $100(1 + 2\sqrt{2})$ cm²

210

GRADE BOOSTER

A You can solve more complex 2D problems, using Pythagoras' theorem and trigonometry
A You can use the sine and cosine rules to calculate missing angles or sides in non-right-angled triangles
A You can find the area of a triangle using the formula area = $\frac{1}{2}ab \sin C$
A* You can use the sine and cosine rules to solve more complex problems involving non-right-angled triangles
A* You can solve 3D problems, using Pythagoras' theorem and trigonometric ratios
A* You can find the sine, cosine and tangent of any angle from 90° to 360°
A* You can work out trigonometric ratios in surd form

What you should know now

- How to find the trigonometric ratios for any angle up to 360°
- How to use the sine and cosine rules
- How to find the area of a triangle, using area = $\frac{1}{2}ab \sin C$

211

Grade booster

Review what you have learnt and how to get to the next grade with the Grade booster at the end of each chapter.

Worked examples

Understand the topic before you start the exercise by reading the examples in blue boxes. These take you through questions step by step.

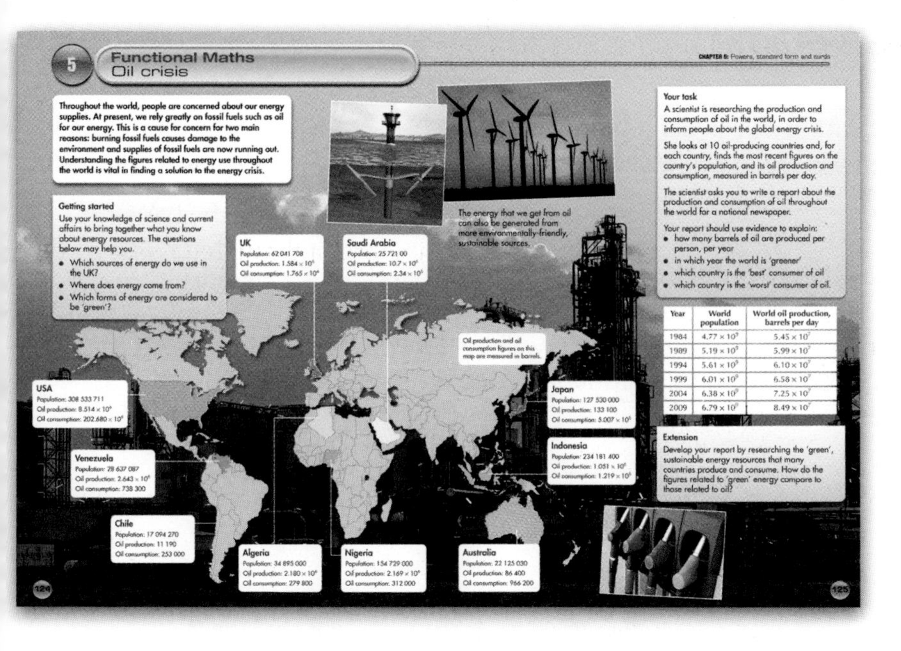

5 Functional Maths
Oil crisis

Throughout the world, people are concerned about our energy supplies. At present, we rely greatly on fossil fuels such as oil for our energy. This is a cause for concern for two main reasons: burning fossil fuels causes damage to the environment and supplies of fossil fuels are now running out. Understanding the figures related to energy use throughout the world is vital in finding a solution to the energy crisis.

Getting started

Use your knowledge of science and current affairs to bring together what you know about energy resources. The questions below may help you.

- Which sources of energy do we use in the UK?
- Where does energy come from?
- Which forms of energy are considered to be 'green'?

The energy that we get from oil can also be generated from more environmentally-friendly, sustainable sources.

UK Population: 62 041 708; Oil production: 1.584×10^6; Oil consumption: 1.765×10^6

Saudi Arabia Population: 25 721 00; Oil production: 10.7×10^6; Oil consumption: 2.34×10^6

USA Population: 308 533 711; Oil production: 8.514×10^6; Oil consumption: 202.680×10^5

Venezuela Population: 28 637 087; Oil production: 2.643×10^6; Oil consumption: 738 300

Chile Population: 17 094 270; Oil production: 11 190; Oil consumption: 253 000

Algeria Population: 34 895 000; Oil production: 2.180×10^6; Oil consumption: 279 800

Nigeria Population: 154 729 000; Oil production: 2.169×10^6; Oil consumption: 312 000

Australia Population: 22 125 030; Oil production: 86 400; Oil consumption: 966 200

Japan Population: 127 530 000; Oil production: 133 100; Oil consumption: 5.007×10^6

Indonesia Population: 234 181 400; Oil production: 1.051×10^6; Oil consumption: 1.219×10^6

Oil production and oil consumption figures on this map are measured in barrels.

CHAPTER 5: Powers, standard form and surds

Your task

A scientist is researching the production and consumption of oil in the world, in order to inform people about the global energy crisis.

She looks at 10 oil-producing countries and, for each country, finds the most recent figures on the country's population, and its oil production and consumption, measured in barrels per day.

The scientist asks you to write a report about the production and consumption of oil throughout the world for a national newspaper.

Your report should use evidence to explain:

- how many barrels of oil are produced per person, per year
- in which year the world is 'greener'
- which country is the 'best' consumer of oil
- which country is the 'worst' consumer of oil.

Year	World population	World oil production, barrels per day
1984	4.77×10^9	5.45×10^7
1989	5.19×10^9	5.99×10^7
1994	5.61×10^9	6.10×10^7
1999	6.01×10^9	6.58×10^7
2004	6.38×10^9	7.25×10^7
2009	6.79×10^9	8.49×10^7

Extension

Develop your report by researching the 'green', sustainable energy resources that many countries produce and consume. How do the figures related to 'green' energy compare to those related to oil?

124 125

Functional maths

Practise functional maths skills to see how people use maths in everyday life. Look out for practice questions marked **FM**. There are also extra functional-maths and problem-solving activities at the end of every chapter to build and apply your skills.

New Assessment Objectives

Practise new parts of the curriculum (Assessment Objectives AO2 and AO3) with questions that assess your understanding marked **AU** and questions that test if you can solve problems marked **PS**. You will also practise some questions that involve several steps and where you have to choose which method to use; these also test AO2. There are also plenty of straightforward questions (AO1) that test if you can do the maths.

Exam practice

Prepare for your exams with past exam questions and detailed worked exam questions with examiner comments to help you score maximum marks.

Quality of Written Communication (QWC)

Practise using accurate mathematical vocabulary and writing logical answers to questions to ensure you get your QWC (Quality of Written Communication) marks in the exams. The Glossary and worked exam questions will help you with this.

Why this chapter matters

Technology is increasingly important in our lives. It helps us do many things more efficiently than we could without it.

Modern **calculators** take away the need to perform long calculations by hand. They can help to improve accuracy – but a calculator is only as good as the person using it. If you press the buttons in the wrong order when doing a calculation then you will get the wrong answer. That is why learning to use a calculator effectively is important.

The earliest known calculating device was a **tally stick**, which was a stick with notches cut into it so that small numbers could be recorded.

In about 2000 BC the **abacus** was invented in Eygpt.

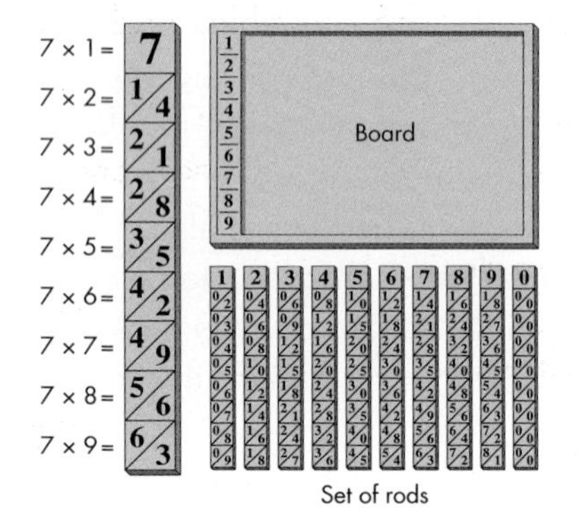

Set of rods

Abacuses are still used widely in China today and they were used widely for almost 3500 years, until John Napier devised a calculating aid called **Napier's bones**.

These led to the invention of the **slide rule** by William Oughtred in 1622. These stayed in use until the mid-1960s. Engineers working on the first ever moon landings used slide rules to do some of their calculations.

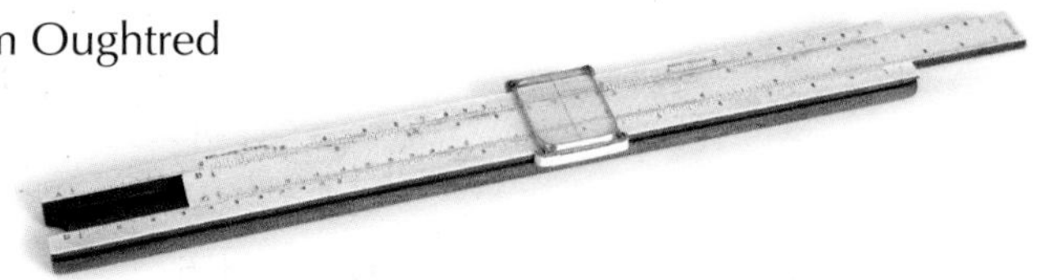

In the mid-16th century the first **mechanical calculating machines** were produced. These were based on a series of cogs and gears and so were too expensive to be widely used.

The first **electronic computers** were produced in the mid-20th century. Once the transistor was perfected, the power increased and the cost and size decreased until the point where the average scientific calculator that students use in schools has more computing power than the first craft that went into space.

Chapter

1 Number: Using a calculator

Basic calculations and using brackets

Adding and subtracting fractions with a calculator

3 Multiplying and dividing fractions with a calculator

The grades given in this chapter are target grades.

This chapter will show you …

to D C how to use a calculator effectively

Visual overview

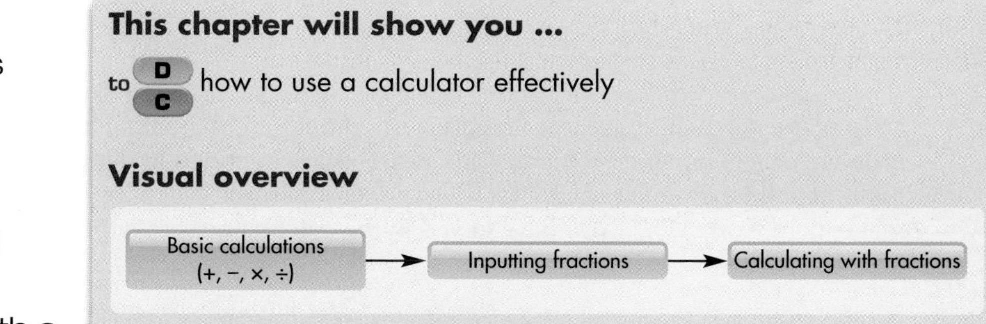

What you should already know

- How to add, subtract, multiply and divide with whole numbers and decimals **(KS3 level 5, GCSE grade E)**
- How to simplify fractions and decimals **(KS3 level 5, GCSE grade E)**
- How to convert improper fractions to mixed numbers or decimals and vice versa **(KS3 level 6, GCSE grade E)**
- The rules of BIDMAS/BODMAS with decimals **(KS3 level 5, GCSE grade E)**
- How to add and subtract fractions and decimals **(KS3 level 6, GCSE grade D)**

Quick check

1 Complete these calculations. Do not use a calculator.

a $48 + 89$ **b** $102 - 37$ **c** 23×7

d $336 \div 8$ **e** $3.6 + 2.9$ **f** $8.4 - 3.8$

g 3×4.5 **h** $7.8 \div 6$

2 a Convert these mixed numbers into improper fractions.

i $2\frac{2}{5}$ **ii** $3\frac{1}{4}$ **iii** $1\frac{7}{9}$

b Convert these improper fractions into mixed numbers.

i $\frac{11}{6}$ **ii** $\frac{7}{3}$ **iii** $\frac{23}{7}$

3 Work these out without using a calculator.

a $2 + 3 \times 4$ **b** $(2 + 3) \times 4$

c $6 + 4 - 3^2$ **d** $6 + (4 - 3)^2$

4 Work these out without using a calculator.

a $\frac{2}{3} + \frac{3}{4}$ **b** $\frac{1}{5} + \frac{2}{7}$

c $\frac{4}{5} - \frac{1}{4}$ **d** $2\frac{1}{3} - 1\frac{2}{5}$

1.1 Basic calculations and using brackets

This section will show you how to:

- use some of the important keys, including the bracket keys, to do calculations on a calculator

Key words
brackets
equals
function key
key
shift key

Most of the calculations in this unit are carried out to find the final answer of an algebraic problem or a geometric problem. The examples are intended to demonstrate how to use some of the **function keys** on the calculator. Remember that some functions will need the **shift key** SHIFT to make them work. When you have **keyed** in the calculation, press the **equals** key = to give the answer.

Some calculators display answers to fraction calculations as fractions. There is always a key to convert this to a decimal. In examinations, an answer given as a fraction or a decimal will always be acceptable unless the question asks you to round to a given accuracy.

Most scientific calculators can be set up to display the answers in the format you want.

EXAMPLE 1

These three angles are on a straight line.

To find the size of angle **a**, subtract the angles 68° and 49° from 180°.

49°
a
68°

You will learn more about angles in Chapters 3 and 8.

You can do the calculation in two ways.

180 – 68 – 49 or 180 – (68 + 49)

Try keying each calculation into your calculator.

180 – 68 – 49

1 8 0 – 6 8 – 4 9 =

The display will show 63.

180 – (68 + 49)

1 8 0 – (6 8 + 4 9) =

Again, the display should show 63.

It is important that you can do this both ways.

You must use the correct calculation or use **brackets** to combine parts of the calculation.

A common error is to work out 180 – 68 + 49, which will give the wrong answer.

EXAMPLE 2

Work out the area of this trapezium, where $a = 12.3$, $b = 16.8$ and $h = 2.4$.

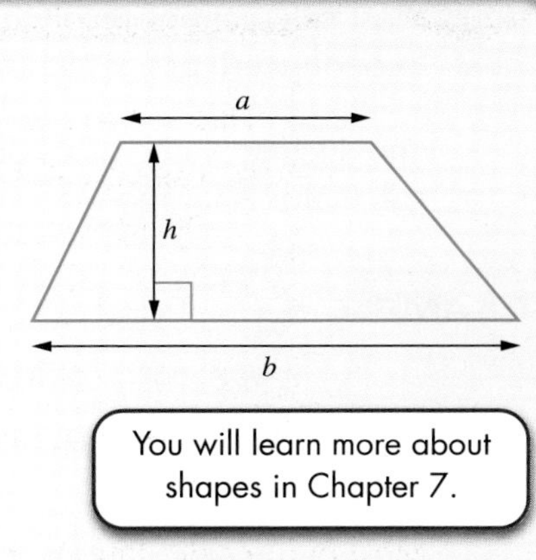

To work out the area of the trapezium, you use the formula:

$A = \frac{1}{2}(a + b)h$

Remember, you should always substitute into a formula before working it out.

$A = \frac{1}{2}(12.3 + 16.8) \times 2.4$

You will learn more about shapes in Chapter 7.

Between the brackets and the numbers at each end there is an assumed multiplication sign, so the calculation is:

$\frac{1}{2} \times (12.3 + 16.8) \times 2.4$

Be careful, $\frac{1}{2}$ can be keyed in lots of different ways:

- As a division

  1 ÷ 2 =

  The display should show 0.5.

- Using the fraction key and the arrows

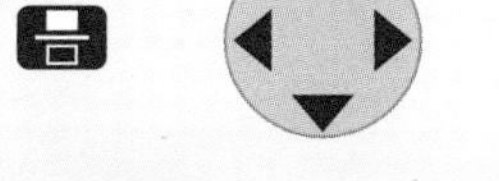

  The display should show $\frac{1}{2}$.

Keying in the full calculation, using the fraction key:

[fraction key] 1 ▼ 2 ▶ × (1 2 • 3 + 1 6 • 8) × 2 • 4 =

The display should show 34.92 or $\frac{873}{25}$.

Your calculator has a power key $x^{\square}$ and a cube key x^3.

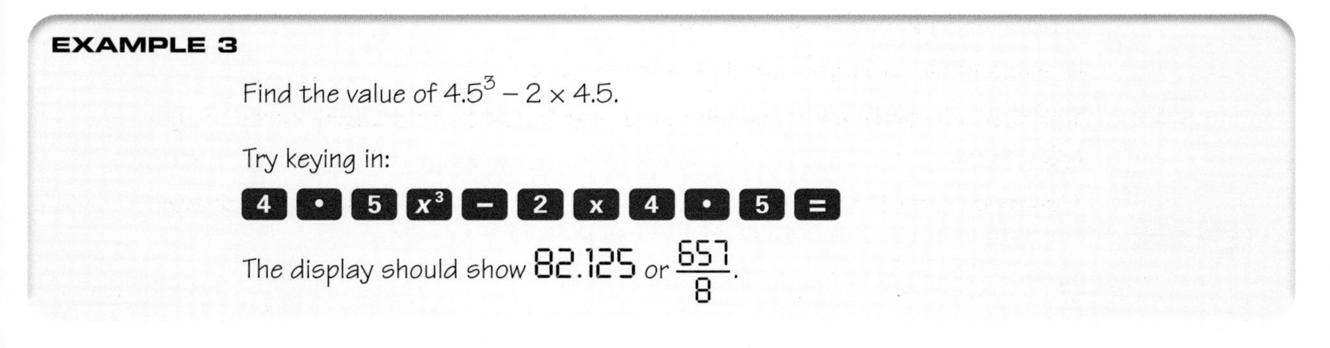

EXAMPLE 3

Find the value of $4.5^3 - 2 \times 4.5$.

Try keying in:

4 • 5 x^3 − 2 × 4 • 5 =

The display should show 82.125 or $\frac{657}{8}$.

Most calculations involving circles will involve the number π (pronounced 'pi'), which has its own calculator button π.

You will learn more about circles in Chapter 4.

The decimal value of π goes on for ever. It has an approximate value of 3.14 but the value in a calculator is far more accurate and may be displayed as 3.1415926535 or π.

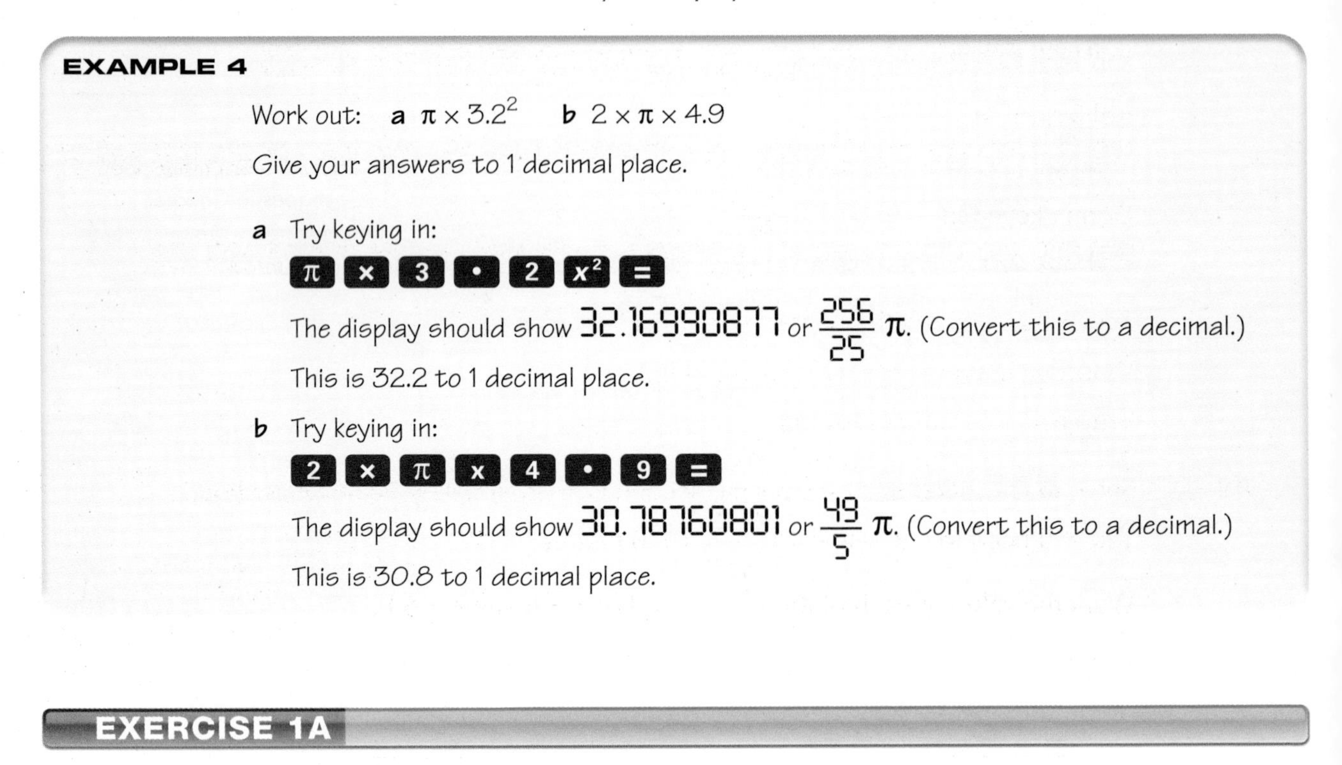

EXAMPLE 4

Work out: **a** $\pi \times 3.2^2$ **b** $2 \times \pi \times 4.9$

Give your answers to 1 decimal place.

a Try keying in:

π × 3 • 2 x^2 =

The display should show 32.16990877 or $\frac{256}{25}\pi$. (Convert this to a decimal.)

This is 32.2 to 1 decimal place.

b Try keying in:

2 × π × 4 • 9 =

The display should show 30.78760801 or $\frac{49}{5}\pi$. (Convert this to a decimal.)

This is 30.8 to 1 decimal place.

EXERCISE 1A

Use your calculator to work out the following.

Try to key in the calculation in as one continuous set, without writing down any intermediate values.

D

1 Work these out.

a $(10 - 2) \times 180 \div 10$

b $180 - (360 \div 5)$

2 Work these out.

a $\frac{1}{2} \times (4.6 + 6.8) \times 2.2$

b $\frac{1}{2} \times (2.3 + 9.9) \times 4.5$

3 Work out the following and give your answers to 1 decimal place.

a $\pi \times 8.5$ **b** $2 \times \pi \times 3.9$ **c** $\pi \times 6.8^2$ **d** $\pi \times 0.7^2$

FM 4 At Sovereign garage, Jon bought 21 litres of petrol for £21.52.

At the Bridge garage he paid £15.41 for 15 litres.

At which garage is the petrol cheaper?

D

AU **5** A teacher asked her class to work out $\frac{2.3 + 8.9}{3.8 - 1.7}$.

Abby keyed in:

(2 • 3 + 8 • 9) ÷ 3 • 8 – 1 • 7 =

Bobby keyed in:

2 • 3 + 8 • 9 ÷ 3 • 8 – 1 • 7 =

Col keyed in:

(2 • 3 + 8 • 9) ÷ (3 • 8 – 1 • 7) =

Donna keyed in:

2 • 3 + 8 • 9 ÷ (3 • 8 – 1 • 7) =

They each rounded their answers to 3 decimal places.

Work out the answer each of them found.

Who had the correct answer?

PS **6** Show that a speed of 31 metres per second is approximately 70 miles per hour.

You will need to know that 1 mile ≈ 1610 metres.

7 Work the value of each of these, if $a = 3.4$, $b = 5.6$ and $c = 8.8$.

a abc **b** $2(ab + ac + bc)$

C

8 Work out the following giving your answer to 2 decimal places.

a $\sqrt{(3.2^2 - 1.6^2)}$ **b** $\sqrt{(4.8^2 + 3.6^2)}$

9 Work these out.

a $7.8^3 + 3 \times 7.8$ **b** $5.45^3 - 2 \times 5.45 - 40$

1.2 Adding and subtracting fractions with a calculator

This section will show you how to:

- use a calculator to add and subtract fractions

Key words

fraction
improper fraction
key
mixed number
proper fraction
shift key

In this lesson, questions requiring calculation of **fractions** are set in a context linked to other topics, such as algebra or geometry.

You will recall that a fraction with the numerator bigger than the denominator is an **improper fraction** or a *top-heavy fraction*.

You will also recall that a **mixed number** is made up of a whole number and a **proper fraction**.

For example:

$\frac{14}{5} = 2\frac{4}{5}$ and $3\frac{2}{7} = \frac{23}{7}$

Using a calculator with improper fractions

Check that your calculator has a fraction key. Remember, for some functions, you may need to use the **shift key** SHIFT.

To **key** in a fraction, press [fraction key].

Input the fraction so that it looks like this:

$\frac{9}{5}$ or 9⌟5

Now press the equals key [=] so that the fraction displays in the answer part of the screen.

Pressing shift and the key S⇔D will convert the fraction to a mixed number.

1⌟4⌟5

This is the mixed number $1\frac{4}{5}$.

Pressing the equals sign again will convert the mixed number back to an improper fraction.

- Can you see a way of converting an improper fraction to a mixed number without using a calculator?
- Test your idea. Then use your calculator to check it.

Using a calculator to convert mixed numbers to improper fractions

To input a mixed number, press the shift key first and then the fraction key [fraction key].

Pressing the equals sign will convert the mixed number to an improper fraction.

- Now key in at least 10 improper fractions and convert them to mixed numbers.
- Remember to press the equals sign to change the mixed numbers back to improper fractions.
- Now input at least 10 mixed numbers and convert them to improper fractions.
- Look at your results. Can you see a way of converting a mixed number to an improper fraction without using a calculator?
- Test your idea. Then use your calculator to check it.

EXAMPLE 5

A water tank is half full. One-third of the capacity of the full tank is poured out.

What fraction of the tank is now full of water?

The calculation is $\frac{1}{2} - \frac{1}{3}$.

Keying in the calculation gives:

[▭/▭] [1] ▼ [2] ▶ [−] [▭/▭] [1] ▼ [3] ▶ [=]

The display should show $\frac{1}{6}$.

The tank is now one-sixth full of water.

EXAMPLE 6

Work out the perimeter of a rectangle $1\frac{1}{2}$ cm long and $3\frac{2}{3}$ cm wide.

To work out the perimeter of this rectangle, you can use the formula:

$$P = 2l + 2w$$

where $l = 1\frac{1}{2}$ cm and $w = 3\frac{2}{3}$ cm.

$$P = 2 \times 1\frac{1}{2} + 2 \times 3\frac{2}{3}$$

Keying in the calculation gives:

[2] [×] [SHIFT] [▭/▭] [1] ▶ [1] ▼ [2] ▶ [+] [2] [×] [SHIFT] [▭/▭] [3] ▶ [2] ▼ [3] ▶ [=]

The display should show $10\frac{1}{3}$.

So the perimeter is $10\frac{1}{3}$ cm.

EXERCISE 1B

1 Use your calculator to work these out. Give your answers as mixed numbers.

Try to key in the calculation as one continuous set, without writing down any intermediate values.

a $4\frac{3}{4} + 1\frac{4}{5}$ **b** $3\frac{5}{6} + 4\frac{7}{10}$ **c** $7\frac{4}{5} + 8\frac{9}{20}$ **d** $9\frac{3}{8} + 2\frac{9}{25}$

e $6\frac{7}{20} + 1\frac{3}{16}$ **f** $2\frac{5}{8} + 3\frac{9}{16} + 5\frac{3}{5}$ **g** $6\frac{9}{20} - 3\frac{1}{12}$ **h** $4\frac{3}{4} - 2\frac{7}{48}$

i $8\frac{11}{32} - 5\frac{1}{6}$ **j** $12\frac{4}{5} + 3\frac{9}{16} - 8\frac{2}{3}$ **k** $9\frac{7}{16} + 5\frac{3}{8} - 7\frac{1}{20}$ **l** $10\frac{3}{4} + 6\frac{2}{9} - 12\frac{3}{11}$

2 A water tank is three-quarters full. Two-thirds of a full tank is poured out.

What fraction of the tank is now full of water?

D

3

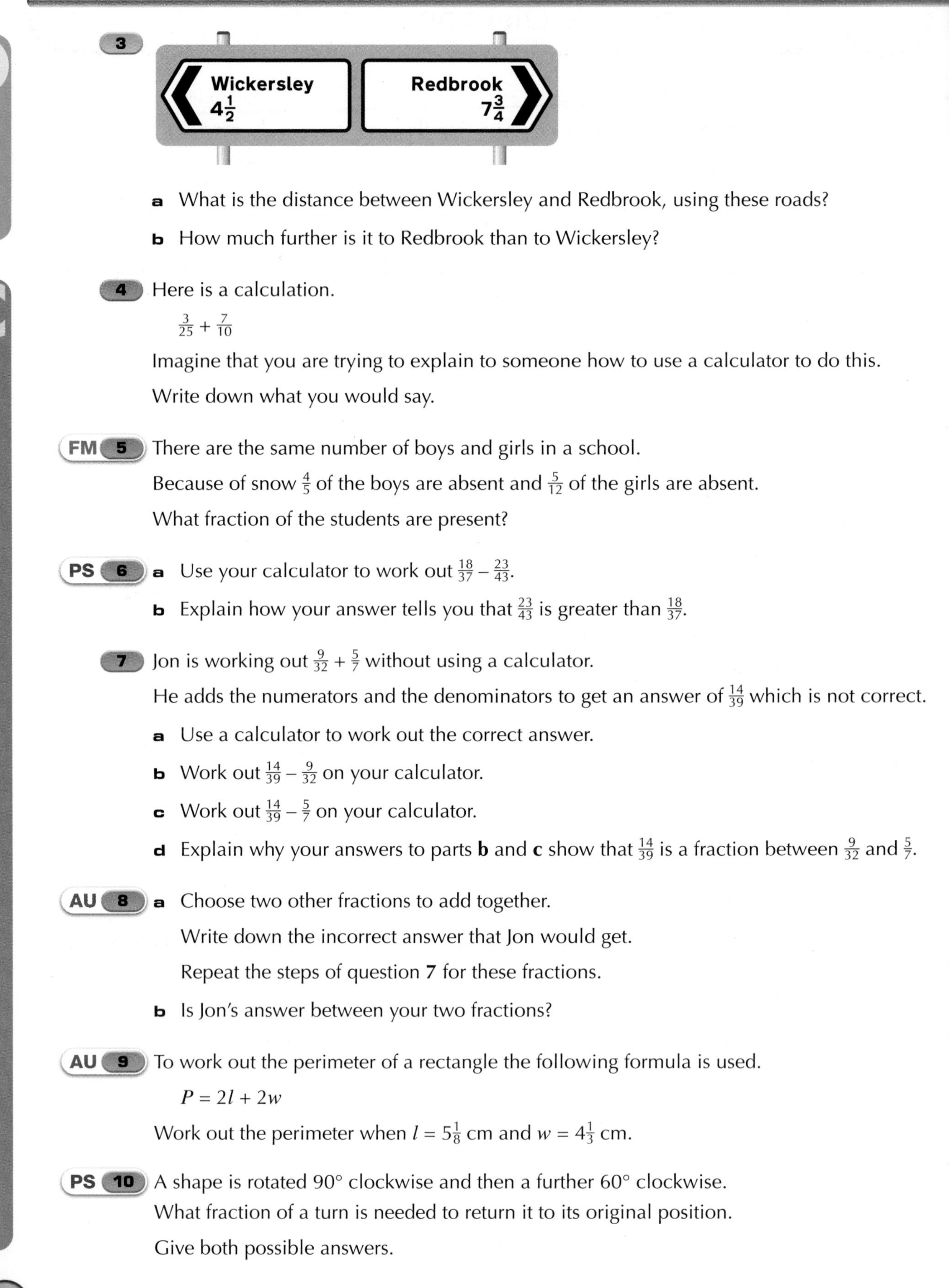

a What is the distance between Wickersley and Redbrook, using these roads?

b How much further is it to Redbrook than to Wickersley?

4 Here is a calculation.

$\frac{3}{25} + \frac{7}{10}$

Imagine that you are trying to explain to someone how to use a calculator to do this.

Write down what you would say.

FM 5 There are the same number of boys and girls in a school.

Because of snow $\frac{4}{5}$ of the boys are absent and $\frac{5}{12}$ of the girls are absent.

What fraction of the students are present?

PS 6 **a** Use your calculator to work out $\frac{18}{37} - \frac{23}{43}$.

b Explain how your answer tells you that $\frac{23}{43}$ is greater than $\frac{18}{37}$.

7 Jon is working out $\frac{9}{32} + \frac{5}{7}$ without using a calculator.

He adds the numerators and the denominators to get an answer of $\frac{14}{39}$ which is not correct.

a Use a calculator to work out the correct answer.

b Work out $\frac{14}{39} - \frac{9}{32}$ on your calculator.

c Work out $\frac{14}{39} - \frac{5}{7}$ on your calculator.

d Explain why your answers to parts **b** and **c** show that $\frac{14}{39}$ is a fraction between $\frac{9}{32}$ and $\frac{5}{7}$.

AU 8 **a** Choose two other fractions to add together.

Write down the incorrect answer that Jon would get.

Repeat the steps of question **7** for these fractions.

b Is Jon's answer between your two fractions?

AU 9 To work out the perimeter of a rectangle the following formula is used.

$P = 2l + 2w$

Work out the perimeter when $l = 5\frac{1}{8}$ cm and $w = 4\frac{1}{3}$ cm.

PS 10 A shape is rotated 90° clockwise and then a further 60° clockwise.

What fraction of a turn is needed to return it to its original position.

Give both possible answers.

1.3 Multiplying and dividing fractions with a calculator

This section will show you how to:
- use a calculator to multiply and divide fractions

Key words
fraction
key
shift key

In this lesson, questions requiring calculation of **fractions** will be set in a context linked to other topics such as algebra or geometry. Remember, for some functions, you may need to use the **shift key** SHIFT.

EXAMPLE 7

Work out the area of a rectangle of length $3\frac{1}{2}$ m and width $2\frac{2}{3}$ m.

The formula for the area of a rectangle is:

area = length × width

Keying in the calculation, where length = $3\frac{1}{2}$ and width = $2\frac{2}{3}$ gives:

The display should show $9\frac{1}{3}$.

The area is $9\frac{1}{3}$ cm^2.

EXAMPLE 8

Work out the average speed of a bus that travels $20\frac{1}{4}$ miles in $\frac{3}{4}$ hour.

The formula for the average speed is:

$$\text{average speed} = \frac{\text{distance}}{\text{time}}$$

Use this formula to work the average speed of the bus, where distance is $20\frac{1}{4}$ and time is $\frac{3}{4}$.

Keying in the calculation gives:

SHIFT ⊟ 2 0 ▶ 1 ▼ 4 ÷ ⊟ 3 ▼ 4 ▶ =

The display should show 27.

The average speed is 27 mph.

EXERCISE 1C

D

1 Use your calculator to work these out. Give your answers as fractions.

Try to key in the calculation as one continuous set, without writing down any intermediate values.

a $\frac{3}{4} \times \frac{4}{5}$ **b** $\frac{5}{6} \times \frac{7}{10}$ **c** $\frac{4}{5} \times \frac{9}{20}$ **d** $\frac{3}{8} \times \frac{9}{25}$

e $\frac{7}{20} \times \frac{3}{16}$ **f** $\frac{5}{8} \times \frac{9}{16} \times \frac{3}{5}$ **g** $\frac{9}{20} \div \frac{1}{12}$ **h** $\frac{3}{4} \div \frac{7}{48}$

i $\frac{11}{32} \div \frac{1}{6}$ **j** $\frac{4}{5} \times \frac{9}{16} \div \frac{2}{3}$ **k** $\frac{7}{16} \times \frac{3}{8} \div \frac{1}{20}$ **l** $\frac{3}{4} \times \frac{2}{9} \div \frac{3}{11}$

2 The formula for the area of a rectangle is: area = length × width

Use this formula to work the area of a rectangle of length $\frac{2}{3}$ m and width $\frac{1}{4}$ m.

3 Some steps are each $\frac{1}{5}$ m high. How many steps are needed to climb 3 m?

AU **4** **a** Use your calculator to work out $\frac{3}{4} \times \frac{9}{16}$ **b** Write down the answer to $\frac{9}{4} \times \frac{3}{16}$

AU **5** **a** Use your calculator to work out $\frac{2}{3} \div \frac{5}{6}$ **b** Use your calculator to work out $\frac{2}{3} \times \frac{6}{5}$

c Use your calculator to work out $\frac{4}{7} \div \frac{3}{4}$ **d** Write down the answer to $\frac{4}{7} \times \frac{4}{3}$

6 Use your calculator to work these out. Give your answers as mixed numbers.

Try to key in the calculation as one continuous set, without writing down any intermediate values.

a $4\frac{3}{4} \times 1\frac{4}{5}$ **b** $3\frac{5}{6} \times 4\frac{7}{10}$ **c** $7\frac{4}{5} \times 8\frac{9}{20}$ **d** $9\frac{3}{8} \times 2\frac{9}{25}$

e $6\frac{7}{20} \times 1\frac{3}{16}$ **f** $2\frac{5}{8} \times 3\frac{9}{16} \times 5\frac{3}{5}$ **g** $6\frac{9}{20} \div 3\frac{1}{12}$ **h** $4\frac{3}{4} \div 2\frac{7}{48}$

i $8\frac{11}{32} \div 5\frac{1}{6}$ **j** $12\frac{4}{5} \times 3\frac{9}{16} \div 8\frac{2}{3}$ **k** $9\frac{7}{16} \times 5\frac{3}{8} \div 7\frac{1}{20}$ **l** $10\frac{3}{4} \times 6\frac{2}{9} \div 12\frac{3}{11}$

C

7 The formula for the area of a rectangle is: area = length × width

Use this formula to work the area of a rectangle of length $5\frac{2}{3}$ mand width $3\frac{1}{4}$ m.

8 The volume of a cuboid is $26\frac{3}{4}$ cm^3. It is cut into eight equal pieces.

Work out the volume of one of the pieces.

9 The formula for the distance travelled is: distance = average speed × time taken

Work out how far a car travelling at an average speed of $36\frac{1}{4}$ mph will travel in $2\frac{1}{2}$ hours.

10 Glasses are filled from litre bottles of water.
Each glass holds $\frac{1}{2}$ pint. 1 litre = $1\frac{3}{4}$ pints

How many litre bottles are needed to fill 10 glasses?

PS FM **11** The ribbon on a roll is $3\frac{1}{2}$ m long. Joe wants to cut pieces of ribbon that are each $\frac{1}{6}$ m long.

He needs 50 pieces.

How many rolls will he need?

GRADE BOOSTER

- **D** You can use BIDMAS or BODMAS to carry out operations in the correct order
- **D** You can use a calculator to add, subtract, multiply and divide fractions
- **C** You can use a calculator to add, subtract, multiply and divide mixed numbers

What you should know now

- How to use a calculator effectively, including the brackets and fraction keys

EXAMINATION QUESTIONS

1 Use your calculator to work out the exact value of

$$\frac{15.6}{1.18 + 2.07}$$ *(2 marks)*

Edexcel, March 2008, Paper 10 (Calculator), Question 4

2 Use your calculator to work out

$$(3.4 + 2.1)^2 \times 5.7$$

Write down all the figures on your calculator display. *(2 marks)*

Edexcel, June 2008, Paper 10 (Calculator), Question 1

3 Use your calculator to work out

$$\frac{22.4 \times 14.5}{8.5 + 3.2}$$

Write down all the figures on your calculator display. *(2 marks)*

Edexcel, June 2008, Paper 15 (Calculator), Question 3

4 **a** Use your calculator to work out

$$\frac{1000}{7.3^2 - 16.3}$$

Write down all the figures on your calculator display. *(2 marks)*

b Write your answer to part a correct to 1 decimal place. *(1 mark)*

Edexcel, June 2008, Paper 11 Section A (Calculator), Question 2

5 Use your calculator to work out

$$\sqrt{12.63 + 18^2}$$

Write down all the figures on your calculator display. *(2 marks)*

Edexcel, November 2008, Paper 10 (Calculator), Question 1

6 Use a calculator to work out

$$\sqrt{\frac{21.6 \times 15.8}{3.8}}$$

Write down all the figures on your calculator display. *(2 marks)*

Edexcel, November 2008, Paper 15 (Calculator), Question 2

7 **a** Use your calculator to work out $\frac{26.4 + 8.2}{\sqrt{5.76}}$ as a decimal.

Write down all the figures on your calculator display. *(2 marks)*

b Write your answer to part **a** correct to 2 decimal places. *(1 mark)*

Edexcel, March 2009, Paper 10 (Calculator), Question 3

8 Use your calculator to work out $\frac{\sqrt{13.2 - 6.8}}{3.25 + 4.9}$

Write down all the figures on your calculator display. *(2 marks)*

Edexcel, June 2007, Paper 11 Section A (Calculator), Question 5

C D

Worked Examination Questions

AU **1** The perimeter of a rectangle is $32\frac{1}{2}$ cm.
Work out a pair of possible values for the length and the width of the rectangle.

1 Perimeter is 2 × length + 2 × width
Length + width = $32\frac{1}{2} \div 2$

This gets 1 mark for the method.

Length + width = $16\frac{1}{4}$ cm

This gets a mark for an accurate calculation.

Possible length and width are:
Length = 10 cm
Width = $6\frac{1}{4}$ cm

Any two values with a sum of $16\frac{1}{4}$ would score the final mark.

Total: 3 marks

FM **2** A driver is travelling 200 miles.
He sets off at 10 am.
He stops for a 20-minute break.
His average speed when travelling is $42\frac{1}{2}$ mph.
He wants to arrive before 3 pm.
Is he successful?

2 Time travelling = distance ÷ average speed
= $200 \div 42\frac{1}{2}$

This gets 1 mark for method.

= $4\frac{12}{17}$ or 4.7058...

20 minutes = $\frac{1}{3}$ hour or 0.33

This is the next step and gets 1 mark for method.

$4\frac{12}{17} + \frac{1}{3}$ or 4.7058... + 0.33...
= $5\frac{2}{51}$ hours 5.039 hours

This gets 1 mark for accuracy.

10 am to 3 pm is 5 hours so he arrives after 3 pm

A statement giving the correct conclusion from correct working would get 1 mark for quality of written communication (QWC).

Total: 4 marks

1

Functional Maths
Setting up your own business

You have been asked by your Business Studies teacher to set up a jewellery stall selling beaded jewellery at an upcoming Young Enterprise fair. There will be 50 stalls at this fair (many of which will be selling jewellery) and it is expected that there will be 500 attendees.

You will be competing against every other stall to sell your products to the attendees, either as one-off purchases or as bulk orders. In order to be successful in this you must carefully plan the design, cost and price of your jewellery, to ensure that people will buy your products and that you make a profit.

Getting started

Answer these questions to begin thinking about how beads can be used to make a piece of jewellery.

1 How many 6 mm beads are needed to make a bracelet?

2 How many 8 mm beads are needed to make an anklet?

3 How many 10 mm beads are needed to make a short necklace?

4 How many 10 mm beads are needed to make a long necklace?

5 You are asked to make a bracelet with beads of two different lengths. You decide to use 6 mm red beads and 8 mm blue beads. How many would you need if you used them alternately?

How to make beaded jewellery

Beads are sold in different sizes and wire is sold in different thicknesses, called the gauge. To make a piece of jewellery the beads are threaded onto the wire.

Step 1 Choose a gauge of wire and cut the length required.

Step 2 Put a fastening on one end.

Step 3 Thread on beads of different sizes in a pattern.

Step 4 Put a fastener on the other end.

Your jewellery is now complete.

Cost of materials

Here are the costs of the raw materials that you will need to make your jewellery.

6 mm beads	10p each
8 mm beads	12p each
10 mm beads	15p each
24-gauge wire	10p per centimetre
20-gauge wire	8p per centimetre
Fasteners for both ends: 30p per item of jewellery	

Advice

For bracelets and anklets use 20-gauge wire.

For necklaces use 24-gauge wire.

Beads are available in three lengths: 6 mm, 8 mm and 10 mm.

Beads are available in three colours, green, blue and red.

Standard lengths for bracelets and necklaces	
Bracelet	17 cm
Anklet	23 cm
Short necklace	39 cm
Long necklace	46 cm

Your task

With a partner, draw up a business plan for your jewellery stall, to ensure you produce high quality beaded jewellery that will turn a good profit. In your plan, you should include:

- an outline of who you expect to buy your jewellery (your 'target market')
- a design for at least one set of jewellery that will appeal to your target market
- a list of all the materials you will need
- the cost of your designs
- a fair price at which to sell your jewellery
- a discounting plan for bulk orders, or if you must reduce your prices on the day
- an expected profit.

Use all the information given on these pages to create your business plan.

Be sure to justify your plan, using appropriate mathematics and describing the calculations that you have done.

Present your business plan as a report to the Young Enterprise committee.

Why this chapter matters

About 70% of the Earth's surface is covered with water. What area of water is that? Just how big is the Earth's surface? And, what is the volume of air in the atmosphere above the Earth's surface?

The Earth is roughly a sphere and the circumference of the Earth is approximately 40 000 kilometres but how does that help to find the surface area?

Consider something on a smaller scale, such as a football. How much material is needed to make a football? The diameter of a football is about 70 cm but can you use that fact to find the area of the surface?

Volumes are difficult too. How big is the Earth or the moon? Or, how much air is there inside a football? Gas is often stored at refineries in spherical containers but how can you calculate the amount of gas a spherical container will hold?

Volume calculations apply to all sorts of shapes. If you think about a volcano, you should be able to imagine how to find the surface area but if you were a geologist you would need to know the amount of rock that might be thrown out of the volcano by an eruption. In short, you would need to know the volume of the volcano. How can you find it?

Fortunately we have the tools to answer these questions. In Book 1, you learned how to find the surface areas and volumes of prisms. In this chapter, you will be looking at more complex shapes and extending your skills in calculating volumes.

Chapter

2 Geometry: Volume

The grades given in this chapter are target grades.

This chapter will show you ...

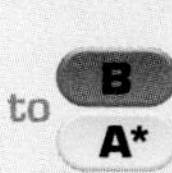

to B–A* how to calculate the volume of a pyramid

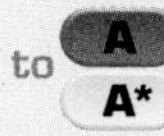

to A–A* how to calculate the volume and surface area of a cone and a sphere

Visual overview

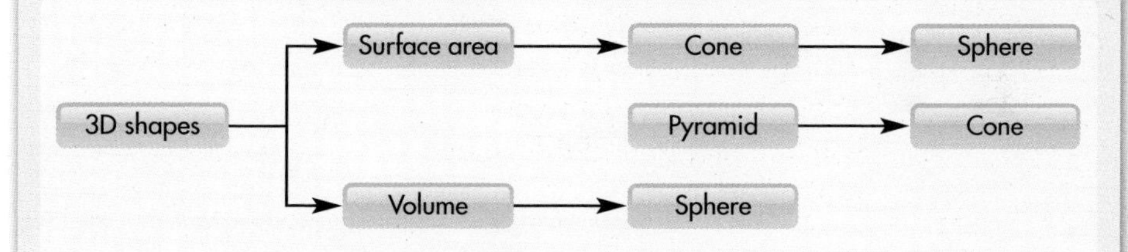

What you should already know

- The area of a rectangle is given by area = length × width or $A = lw$ **(KS3 level 5, GCSE grade F)**
- The area of a triangle is given by area = $\frac{1}{2}$ × base × height or $A = \frac{1}{2}bh$ **(KS3 level 6, GCSE grade D)**
- The area of a parallelogram is given by area = base × height or $A = bh$ **(KS3 level 6, GCSE grade E)**
- The circumference of a circle is given by $C = \pi d$, where d is the diameter of the circle **(KS3 level 6, GCSE grade D)**
- The area of a circle is given by $A = \pi r^2$, where r is the radius of the circle **(KS3 level 6, GCSE grade D)**
- The volume of a cuboid is given by volume = length × width × height or $V = lwh$ **(KS3 level 6, GCSE grade E)**
- The common metric units to measure area, volume and capacity are shown in this table **(KS3 level 6, GCSE grade E)**

Area	Volume	Capacity
100 mm^2 = 1 cm^2	1000 mm^3 = 1 cm^3	1000 cm^3 = 1 litre
10 000 cm^2 = 1 m^2	1 000 000 cm^3 = 1 m^3	1 m^3 = 1000 litres

continued

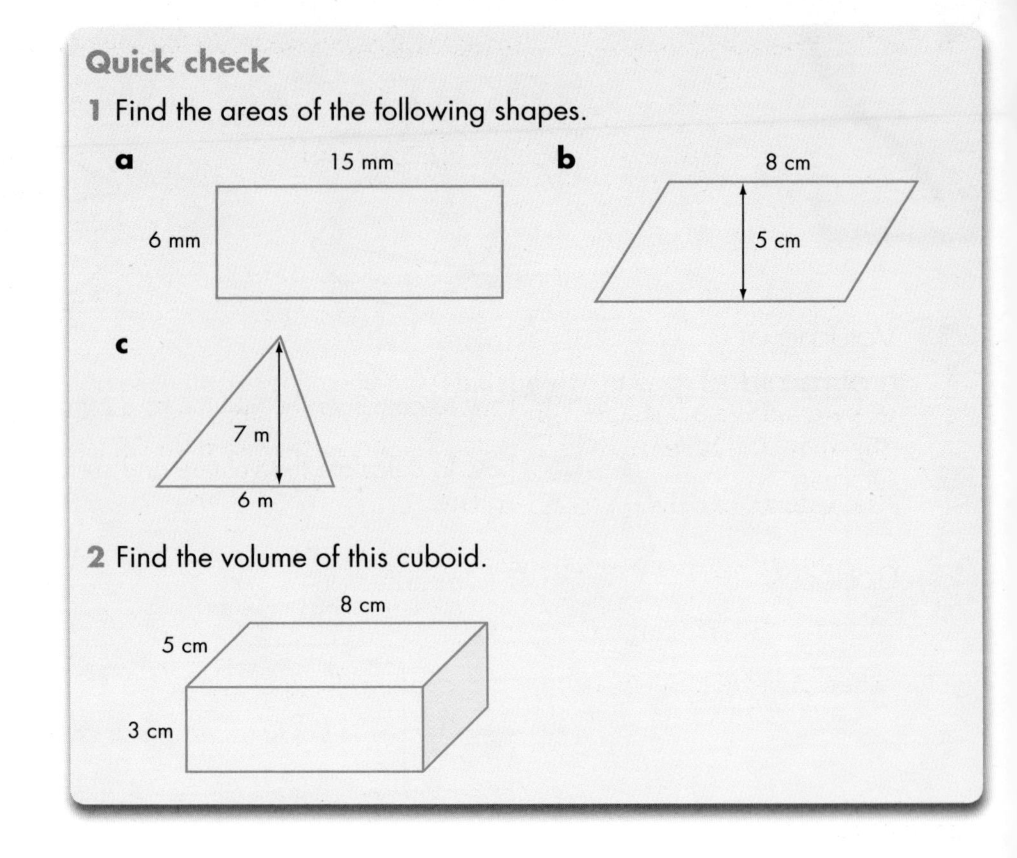
Quick check
1 Find the areas of the following shapes.
a
15 mm
6 mm
b
8 cm
5 cm
c
7 m
6 m
2 Find the volume of this cuboid.
8 cm
5 cm
3 cm

2.1 Volume of a pyramid

This section will show you how to:
- calculate the volume of a pyramid

Key words
apex
frustum
pyramid
volume

A **pyramid** is a 3D shape with a base from which triangular faces rise to a common vertex, called the **apex**. The base can be any polygon, but is usually a triangle, a rectangle or a square.

The **volume** of a pyramid is given by:

volume = $\frac{1}{3}$ × base area × vertical height

$$V = \tfrac{1}{3}Ah$$

where A is the base area and h is the vertical height.

EXAMPLE 1

Calculate the volume of the pyramid on the right.

Base area = 5 × 4 = 20 cm²

Volume = $\frac{1}{3}$ × 20 × 6 = 40 cm³

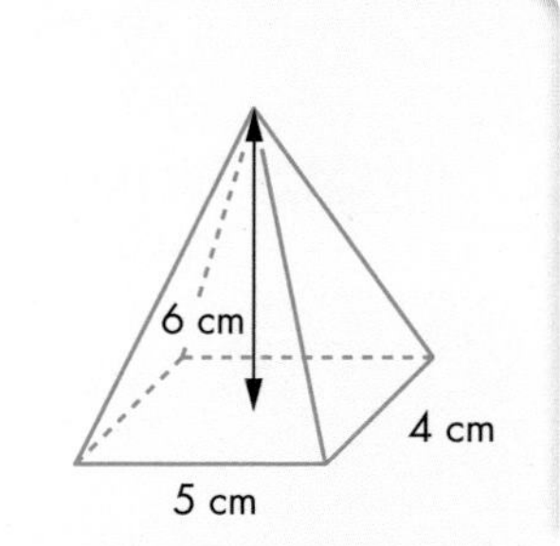

EXAMPLE 2

A pyramid, with a square base of side 8 cm, has a volume of 320 cm³. What is the vertical height of the pyramid?

Let h be the vertical height of the pyramid. Then,

volume = $\frac{1}{3} \times 64 \times h = 320$ cm³

$\frac{64h}{3} = 320$ cm³

$h = \frac{960}{64}$ cm

$h = 15$ cm

EXERCISE 2A

B

1 Calculate the volume of each of these pyramids, all with rectangular bases.

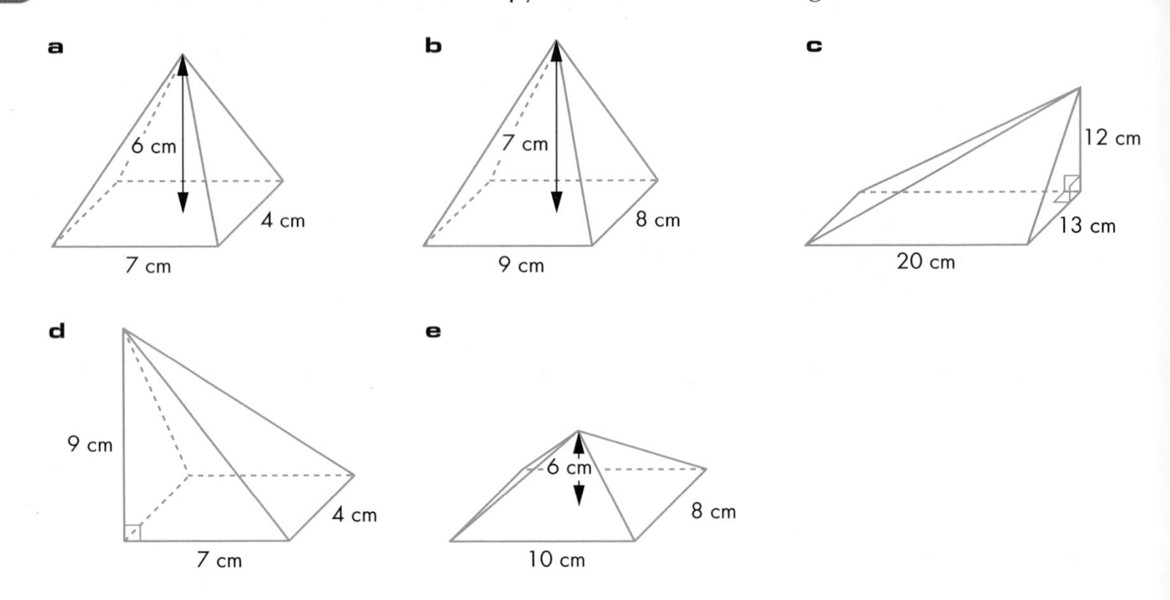

2 Calculate the volume of a pyramid having a square base of side 9 cm and a vertical height of 10 cm.

AU **3** Suppose you have six pyramids which have a height that is half the side of the square base.

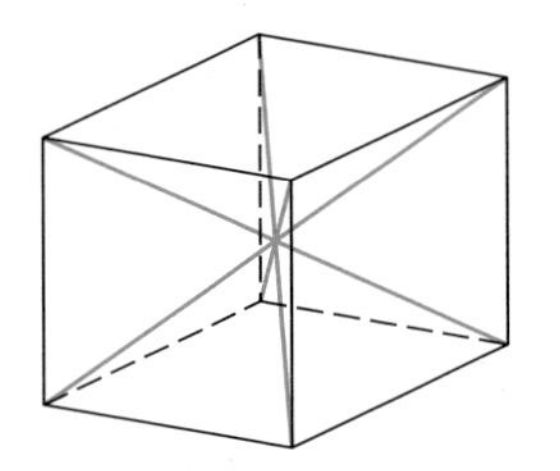

a Explain how they can fit together to make a cube.

b How does this show that the formula for the volume of a pyramid is correct?

A

4 The glass pyramid outside the Louvre Museum in Paris was built in the 1980s. It is 20.6 m tall and the base is a square of side 35 m. The design was very controversial.

Suppose that instead of a pyramid, the building was a conventional shape with the same square base, a flat roof and the same volume.

How high would it have been?

A

5 Calculate the volume of each of these shapes.

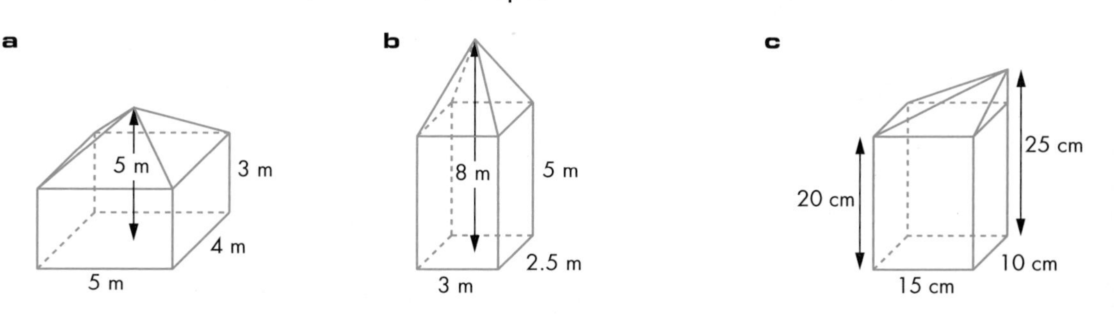

AU 6 What is the mass of a solid pyramid having a square base of side 4 cm, a height of 3 cm and a density of 13 g/cm^3? (1 cm^3 has a mass of 13 g.)

PS 7 A crystal is in the form of two square-based pyramids joined at their bases (see diagram).

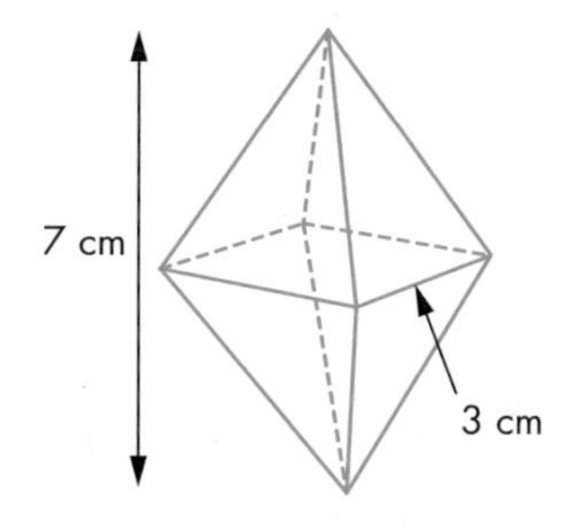

The crystal has a mass of 31.5 g.

What is the mass of 1 cm^3 of the substance?

8 A pyramid has a square base of side 6.4 cm.
Its volume is 81.3 cm^3.

Calculate the height of the pyramid.

PS 9 A pyramid has the same volume as a cube of side 10.0 cm.
The height of the pyramid is the same as the side of the square base.

Calculate the height of the pyramid.

A*

10 The pyramid in the diagram has its top 5 cm cut off as shown.
The shape which is left is called a **frustum**. Calculate the volume of the frustum.

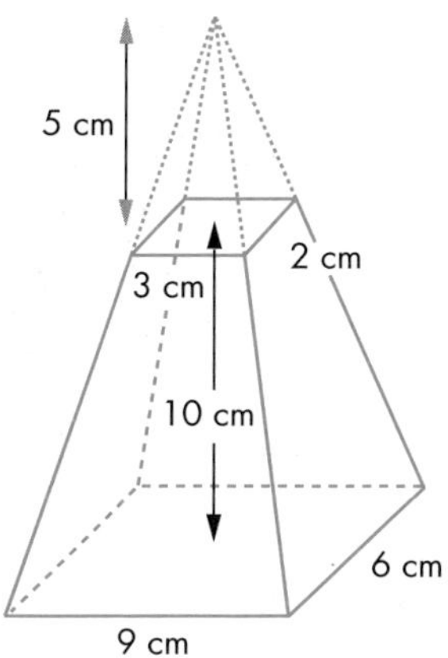

2.2 Cones

This section will show you how to:
- calculate the volume and surface area of a cone

Key words
slant height
surface area
vertical height
volume

A cone can be treated as a pyramid with a circular base. Therefore, the formula for the **volume** of a cone is the same as that for a pyramid.

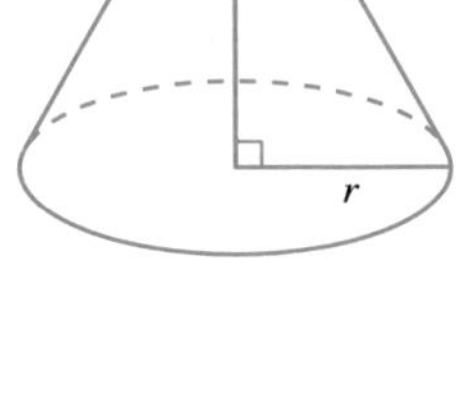

volume = $\frac{1}{3}$ × base area × vertical height

$$V = \tfrac{1}{3}\pi r^2 h$$

where r is the radius of the base and h is the **vertical height** of the cone.

The curved **surface area** of a cone is given by:

curved surface area = π × radius × slant height

$$S = \pi r l$$

where l is the **slant height** of the cone.

So the total surface area of a cone is given by the curved surface area plus the area of its circular base.

$$A = \pi r l + \pi r^2$$

EXAMPLE 3

For the cone in the diagram, calculate:

i its volume

ii its total surface area.

Give your answers in terms of π.

i The volume is given by $V = \frac{1}{3}\pi r^2 h$

$= \frac{1}{3} \times \pi \times 36 \times 8 = 96\pi \text{ cm}^3$

ii The total surface area is given by $A = \pi r l + \pi r^2$

$= \pi \times 6 \times 10 + \pi \times 36 = 96\pi \text{ cm}^2$

EXERCISE 2B

1 For each cone, calculate:

i its volume **ii** its total surface area.

Give your answers to 3 significant figures.

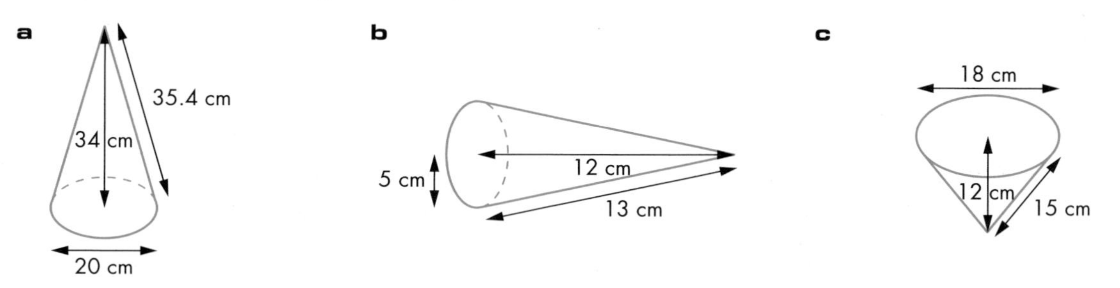

2 A solid cone, base radius 6 cm and vertical height 8 cm, is made of metal whose density is 3.1 g/cm^3. Find the mass of the cone.

3 Find the total surface area of a cone whose base radius is 3 cm and slant height is 5 cm. Give your answer in terms of π.

4 Calculate the volume of each of these shapes. Give your answers in terms of π.

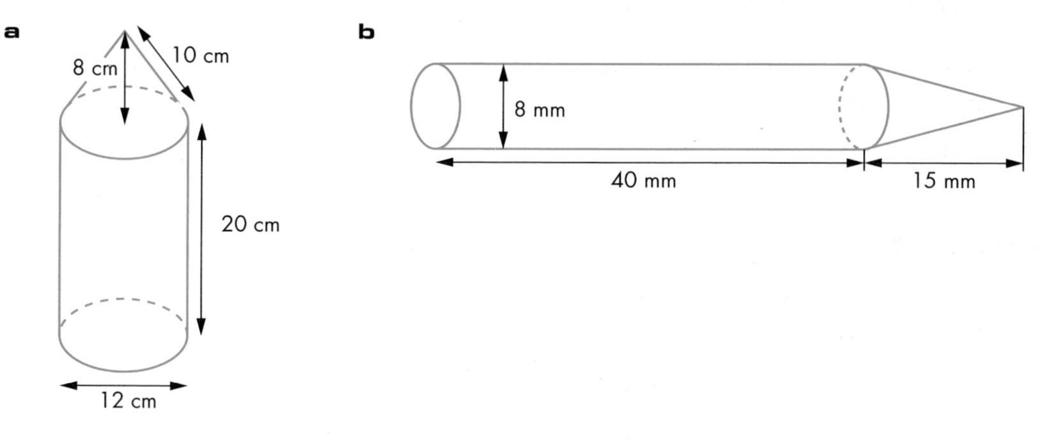

5 You could work with a partner on this question.

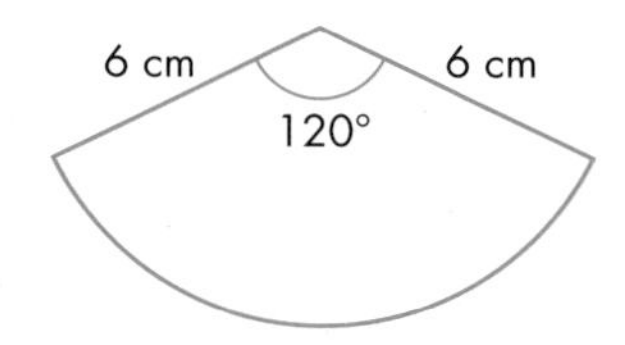

A sector of a circle, as in the diagram, can be made into a cone (without a base) by sticking the two straight edges together.

a What would be the diameter of the base of the cone in this case?

b What is the diameter if the angle is changed to 180°?

c Investigate other angles.

A

6 A cone has the dimensions shown in the diagram.

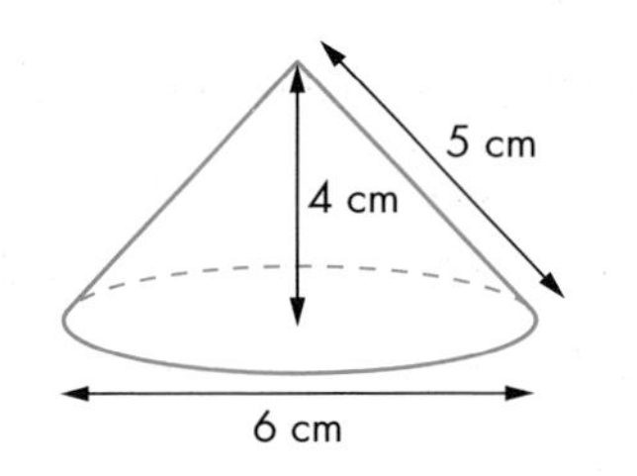

Calculate the total surface area, leaving your answer in terms of π.

PS 7 If the slant height of a cone is equal to the base diameter, show that the area of the curved surface is twice the area of the base.

A*

8 The model shown on the right is made from aluminium.

What is the mass of the model, given that the density of aluminium is 2.7 g/cm^3?

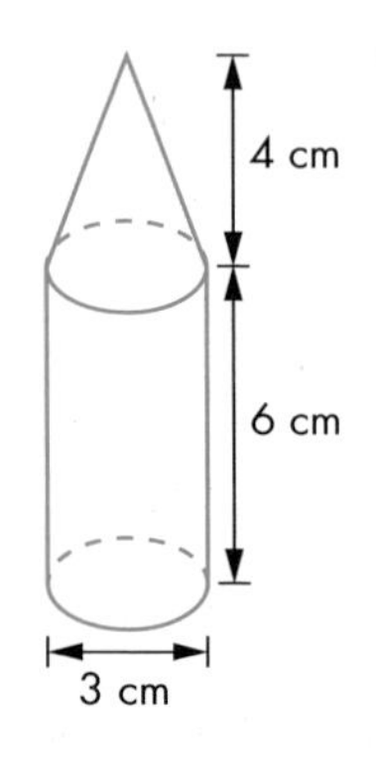

PS 9 A container in the shape of a cone, base radius 10 cm and vertical height 19 cm, is full of water. The water is poured into an empty cylinder of radius 15 cm. How high is the water in the cylinder?

2.3 Spheres

This section will show you how to:
- calculate the volume and surface area of a sphere

Key words
sphere
surface area
volume

The **volume** of a **sphere**, radius r, is given by:

$$V = \tfrac{4}{3}\pi r^3$$

Its **surface area** is given by:

$$A = 4\pi r^2$$

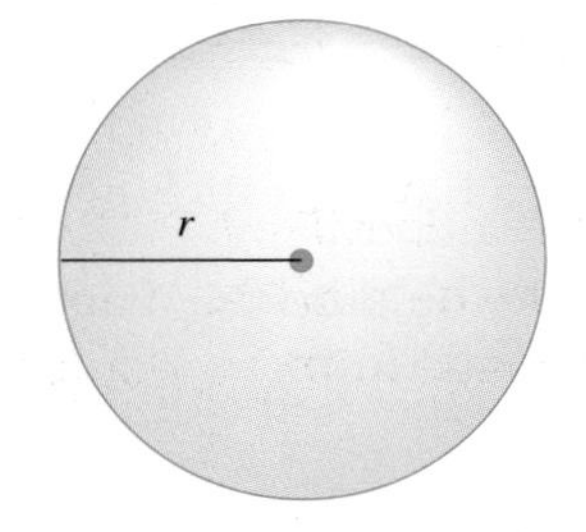

EXAMPLE 4

For a sphere of radius of 8 cm, calculate **i** its volume and **ii** its surface area.

i The volume is given by:

$$V = \frac{4}{3}\pi r^3$$

$$= \frac{4}{3} \times \pi \times 8^3 = \frac{2048}{3} \times \pi = 2140 \text{ cm}^3 \quad \text{(3 significant figures)}$$

ii The surface area is given by:

$$A = 4\pi r^2$$

$$= 4 \times \pi \times 8^2 = 256 \times \pi = 804 \text{ cm}^2 \quad \text{(3 significant figures)}$$

EXERCISE 2C

1 Calculate the volume of each of these spheres. Give your answers in terms of π.

a Radius 3 cm

b Radius 6 cm

c Diameter 20 cm

2 Calculate the surface area of each of these spheres. Give your answers in terms of π.

a Radius 3 cm

b Radius 5 cm

c Diameter 14 cm

3 Calculate the volume and the surface area of a sphere with a diameter of 50 cm.

4 A sphere fits exactly into an open cubical box of side 25 cm. Calculate the following.

a The surface area of the sphere

b The volume of the sphere

AU 5 A metal sphere of radius 15 cm is melted down and recast into a solid cylinder of radius 6 cm. Calculate the height of the cylinder.

PS 6 Lead has a density of 11.35 g/cm^3. (This means that 1 cm^3 of lead has a mass of 11.35 g.) Calculate the maximum number of shot (spherical lead pellets) of radius 1.5 mm which can be made from 1 kg of lead.

A

A

FM 7 The standard (size 5) football must be between 68 cm and 70 cm in circumference and weigh between 410 g and 450 g. They are usually made from 32 panels: 12 regular pentagons and 20 regular hexagons.

a Will a maker of footballs be more interested in the surface area or the volume of the ball? Why?

b What variation in the surface area of a football is allowed?

AU 8 A sphere has a radius of 5.0 cm.

A cone has a base radius of 8.0 cm.

The sphere and the cone have the same volume.

Calculate the height of the cone.

PS 9 A sphere of diameter 10 cm is carved out of a wooden block in the shape of a cube of side 10 cm.

What percentage of the wood is wasted?

A*

AU 10 A manufacturer is making cylindrical packaging for a sphere as shown.

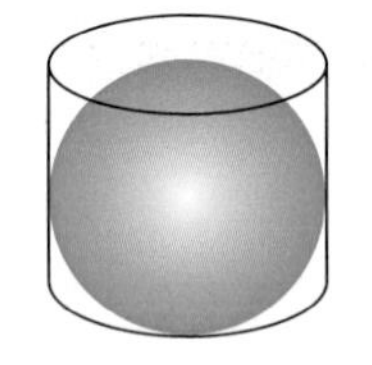

The curved surface of the cylinder is made from card.

Show that the area of the card is the same as the surface area of the sphere.

GRADE BOOSTER

A You can calculate the surface area of cones and spheres

A You can calculate the volume of pyramids, cones and spheres

What you should know now

- The volume of a pyramid is given by $V = \frac{1}{3}Ah$, where A is the area of the base and h is the vertical height of the pyramid
- The volume of a cone is given by $V = \frac{1}{3}\pi r^2 h$, where r is the base radius and h is the vertical height of the cone
- The curved surface area of a cone is given by $S = \pi rl$, where r is the base radius and l is the slant height of the cone
- The volume of a sphere is given by $V = \frac{4}{3}\pi r^3$, where r is its radius
- The surface area of a sphere is given by $A = 4\pi r^2$, where r is its radius

1

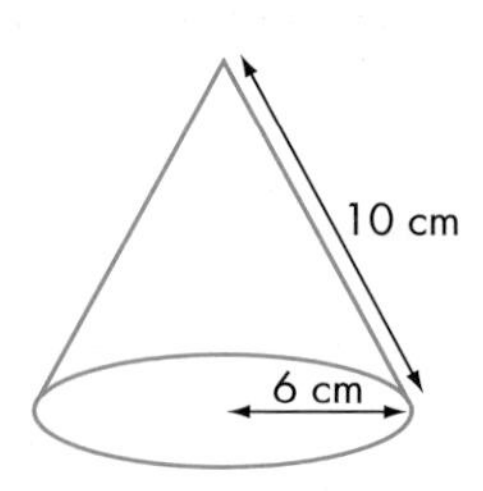

A solid cone has base radius 6 cm and slant height 10 cm. Calculate the total surface area of the cone. Give your answer in terms of π.

2

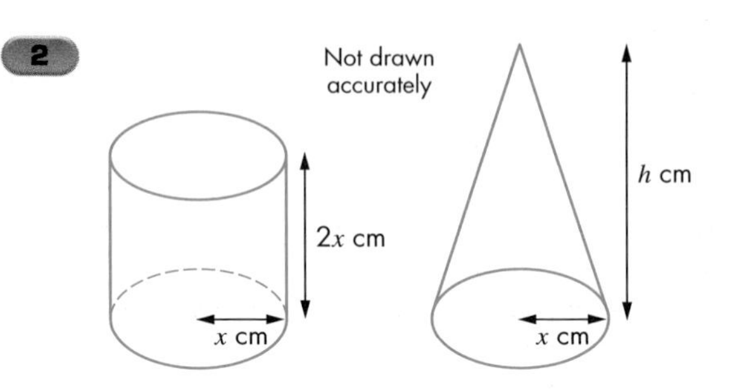

A cylinder has base radius x cm and height $2x$ cm.

A cone has base radius x cm and height h cm.

The volume of the cylinder and the volume of the cone are equal.

Find h in terms of x.

Give your answer in its simplest form. *(3 marks)*

Edexcel, May 2008, Paper 3, Question 26

3 The diagram shows a storage tank.

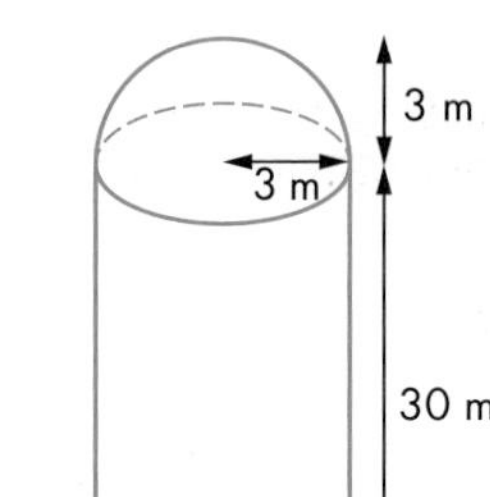

The storage tank consists of a hemisphere on top of a cylinder.

The height of the cylinder is 30 metres.

The radius of the cylinder is 3 metres.

The radius of the hemisphere is 3 metres.

a Calculate the total volume of the storage tank.

Give your answer correct to 3 significant figures. *(3 marks)*

A sphere has a volume of 500 m^3.

b Calculate the radius of the sphere.

Give your answer correct to 3 significant figures. *(3 marks)*

Edexcel, November 2008, Paper 4, Question 24

Worked Examination Questions

AU **1** The diagram shows a pepper pot. The pot consists of a cylinder and a hemisphere. The cylinder has a diameter of 5 cm and a height of 7 cm.

7 cm

x

5 cm

The pepper takes up half the total volume of the pot.
Find the depth of pepper in the pot marked x in the diagram.

1 *Volume of pepper pot*

$= \pi r^2 h + \frac{2}{3}\pi r^3$ — This gets 1 method mark for setting up equation.

$= \pi \times 2.5^2 \times 7 + \frac{2}{3} \times \pi \times 2.5^3 \text{ cm}^3$ — This gets 1 method mark for correct substitutions.

$= 170.2 \text{ cm}^3$ — This gets 1 accuracy mark for correct answers.

So volume of pepper $= 85.1 \text{ cm}^3$

Therefore,

$\pi r^2 x = 85.1 \text{ cm}^3$ — This gets 1 method mark for setting up equation.

and $x = \dfrac{85.1}{\pi \times 2.5^2} = 4.3$ *cm (1 decimal place)* — This gets 1 method mark for correct rearrangement and 1 accuracy mark for correct answer.

Total: 6 marks

2 Functional Maths
Organising a harvest

Farmers have to do mathematical calculations almost every day. For example, an arable farmer may need to know how much seed to buy, how much water is required to irrigate the field each day, how much wheat they expect to grow and how much storage space they need to store wheat once it is harvested.

Farming can be filled with uncertainties, including changes in weather, crop disease and changes in consumption. It is therefore important that farmers correctly calculate variables that are within their control, to minimise the impact of changes that are outside their control.

Grain storage

Wheat is stored in large containers called silos. These are usually big cylinders but can also be various other shapes.

Important information about wheat crops (yield data)

- A 1 kg bag of seeds holds 26 500 seeds
- A 1 kg bag of seeds costs 50p
- I want to plant 60 bags of seed
- I need to plant 100 seeds in each square metre (m^2) of field
- I need to irrigate each square metre of the field with 5 litres of water each day
- I expect to harvest 0.7 kg of wheat from each square metre of the field
- Every cubic metre (m^3) of storage will hold 800 kg of wheat

Your task

Rufus, a crop farmer, is going to grow his first field of wheat next summer. Using all the information that he has gathered, help him to plan for his wheat crop. You should consider:

- the size of the field that he will require
- how much seed he will need
- how much water he will need to irrigate the crops, per day, and how it will be stored
- how he will store the seeds and wheat
- how much profit he could make if grain is sold at £92.25 per tonne.

Getting started

Think about these points to help you create your plan.

- What different shapes and sizes of field could the crops be grown in?
- If Rufus needs a container to hold one day's irrigation water, what size cylinder would he need? How would this change if he chose a cuboid? What other shapes and sizes of container could he use?
- What shapes and sizes could the silos be?

Handy hints

Remember: 1000 litres = $1\ m^3$
1000 kg = 1 tonne

Why this chapter matters

How can you find the height of a mountain?

How do you draw an accurate map?

How can computers take an image and make it rotate so that you can view it from different directions?

How were sailors able to navigate before GPS? And how does GPS work? How can music be produced electronically?

The answer is by looking at the angles and sides of triangles and the connections between them. This important branch of mathematics is called trigonometry and has a huge range of applications in science, engineering, electronics and everyday life. This chapter gives a brief introduction to the subject and introduces some important new mathematical tools.

The first major work of trigonometry that still survives is called the *Almagest*. It was a written by an astronomer called Ptolemy who lived in Alexandria, Egypt, over 1800 years ago. In it, there are tables of numbers, called 'trigonometric ratios', used in making calculations about the positions of stars and planets. Trigonometry also helped Ptolemy to make a map of the known world at that time. Today we no longer need to look up tables of values of trigonometric ratios because they are programmed into calculators and computers.

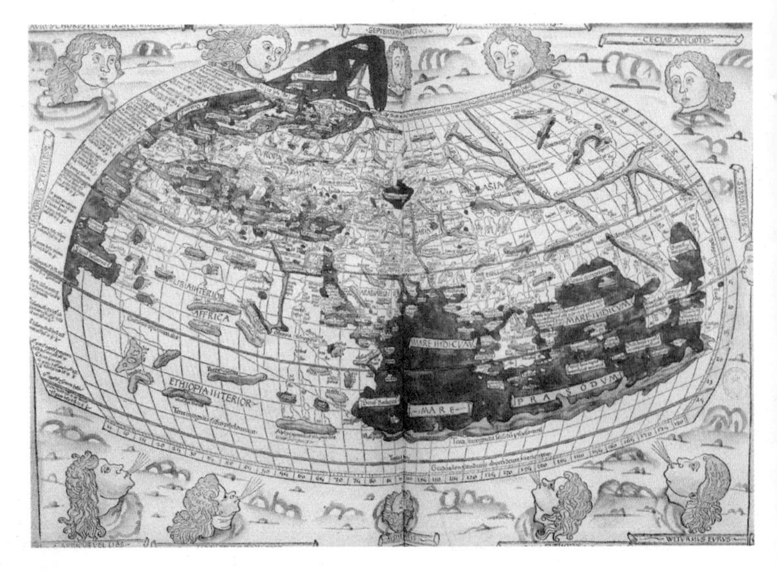

Nearer our own time, the 19th century French mathematician Jean Fourier showed how all musical sounds can be broken down into a combination of pure tones that can be described by trigonometry. His work is what makes it possible to imitate the sound of any instrument electronically – a clear example of how mathematics has a profound influence on other areas of life, including art.

Chapter

Geometry: Trigonometry 1

The grades given in this chapter are target grades.

This chapter will show you ...

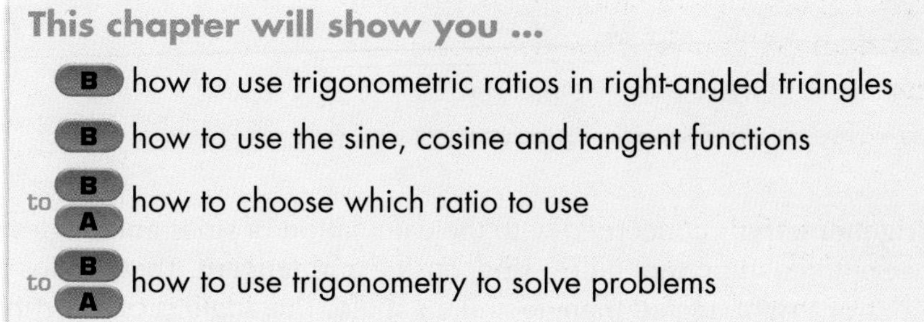

- **B** how to use trigonometric ratios in right-angled triangles
- **B** how to use the sine, cosine and tangent functions
- **B** to **A** how to choose which ratio to use
- **B** to **A** how to use trigonometry to solve problems

Visual overview

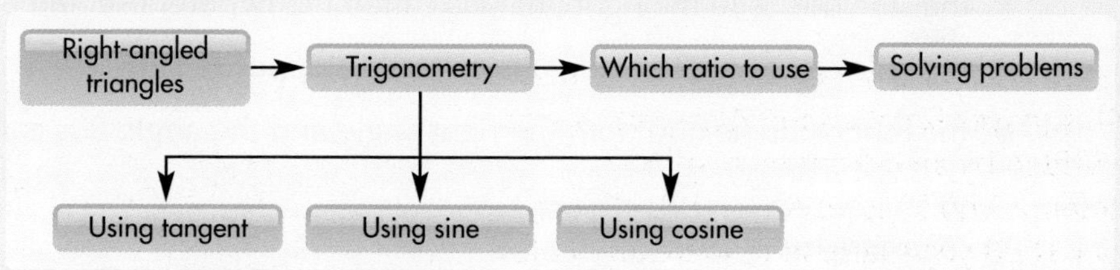

What you should already know

- how to find the square and square root of a number **(KS3 level 5, GCSE grade F)**
- how to round numbers to a suitable degree of accuracy **(KS3 level 6, GCSE grade E)**

Quick check

Use your calculator to evaluate the following, giving your answers to one decimal place.

1 2.3^2

2 15.7^2

3 0.78^2

4 $\sqrt{8}$

5 $\sqrt{260}$

6 $\sqrt{0.5}$

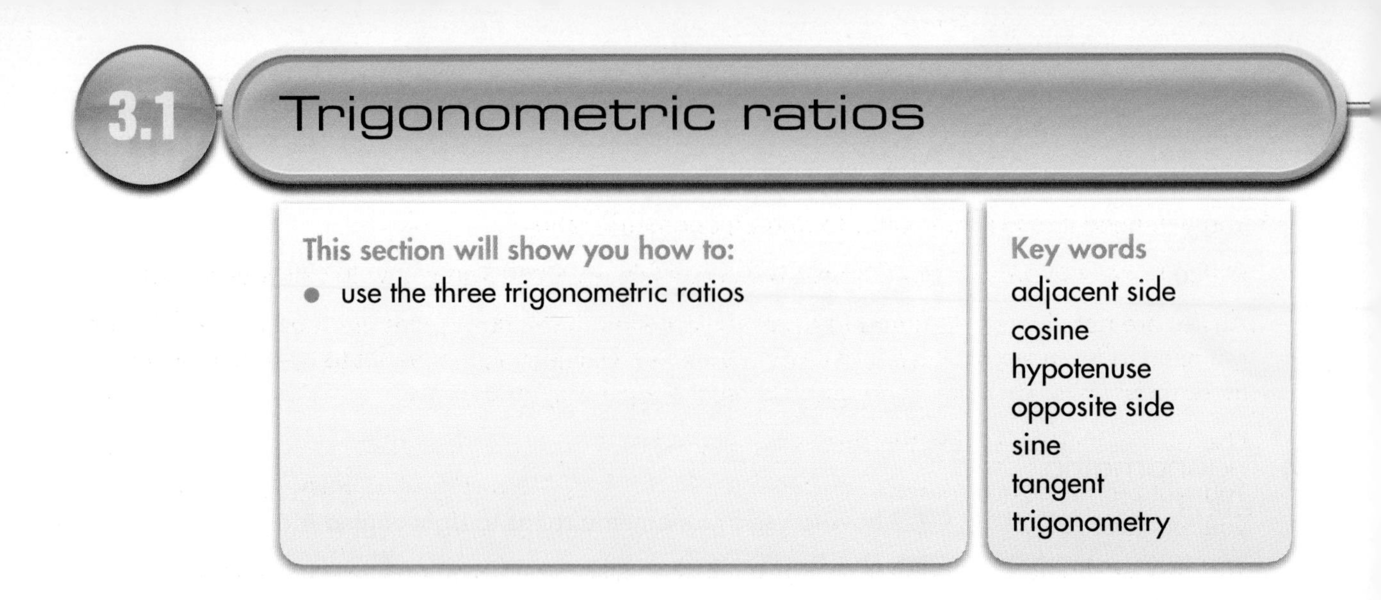

Trigonometry is concerned with the calculation of sides and angles in triangles, and involves the use of three important ratios: **sine**, **cosine** and **tangent**. These ratios are defined in terms of the sides of a right-angled triangle and an angle. The angle is often written as θ.

In a right-angled triangle:

- the side opposite the right angle is called the **hypotenuse** and is the longest side
- the side opposite the angle θ is called the **opposite side**
- the other side next to both the right angle and the angle θ is called the **adjacent side**.

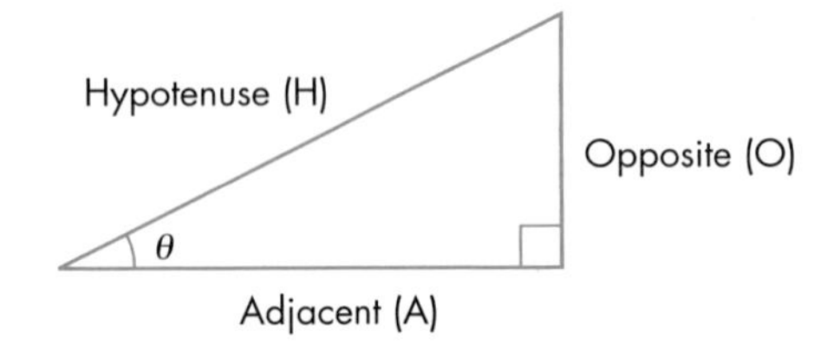

The sine, cosine and tangent ratios for θ are defined as:

$$\text{sine } \theta = \frac{\text{Opposite}}{\text{Hypotenuse}} \qquad \text{cosine } \theta = \frac{\text{Adjacent}}{\text{Hypotenuse}} \qquad \text{tangent } \theta = \frac{\text{Opposite}}{\text{Adjacent}}$$

These ratios are usually abbreviated as:

$$\sin \theta = \frac{\text{O}}{\text{H}} \qquad \cos \theta = \frac{\text{A}}{\text{H}} \qquad \tan \theta = \frac{\text{O}}{\text{A}}$$

These abbreviated forms are also used on calculator keys.

Memorising these formulae may be helped by a mnemonic such as,

Silly **O**ld **H**itler **C**ouldn't **A**dvance **H**is **T**roops **O**ver **A**frica

in which the first letter of each word is taken in order to give

$$\text{S} = \frac{\text{O}}{\text{H}} \qquad \text{C} = \frac{\text{A}}{\text{H}} \qquad \text{T} = \frac{\text{O}}{\text{A}}$$

Using your calculator

You will need to use a calculator to find trigonometric ratios.

Different calculators work in different ways, so make sure you know how to use your model.

Angles are not always measured in degrees. Sometimes radians or grads are used instead. You do not need to learn about those in your GCSE course. Calculators can be set to operate in any of these three units, so make sure your calculator is operating in degrees.

Use your calculator to find the sine of 60 degrees.

You will probably press the keys **sin** **6** **0** **=** in that order, but it might be different on your calculator.

The answer should be 0.8660… or $\frac{\sqrt{3}}{2}$. If it is the latter, make sure you can convert that to the decimal form.

3 cos 57° is a shorthand way of writing 3 × cos 57°.

On most calculators you do not need to use the × button and you can just press the keys in the way it is written: **3** **cos** **5** **7** **=**

Check to see whether your calculator works this way.

The answer should be 1.63.

EXAMPLE 1

Find 5.6 sin 30°.

This means 5.6 × sine of 30 degrees.

Remember that you may not need to press the × button.

5.6 sin 30° = 2.8

EXERCISE 3A

1 Find these values, rounding off your answers to 3 significant figures.

a sin 43° **b** sin 56° **c** sin 67.2° **d** sin 90°
e sin 45° **f** sin 20° **g** sin 22° **h** sin 0°

2 Find these values, rounding off your answers to 3 significant figures.

a cos 43° **b** cos 56° **c** cos 67.2° **d** cos 90°
e cos 45° **f** cos 20° **g** cos 22° **h** cos 0°

3 From your answers to questions **1** and **2**, what angle has the same value for sine and cosine?

B

B

4 **a** **i** What is sin 35°? **ii** What is cos 55°?

b **i** What is sin 12°? **ii** What is cos 78°?

c **i** What is cos 67°? **ii** What is sin 23°?

d What connects the values in parts **a**, **b** and **c**?

e Copy and complete these sentences.

i sin 15° is the same as cos …

ii cos 82° is the same as sin …

iii sin x is the same as cos …

5 Use your calculator to work out the values of the following.

a tan 43° **b** tan 56° **c** tan 67.2° **d** tan 90°

e tan 45° **f** tan 20° **g** tan 22° **h** tan 0°

6 Use your calculator to work out the values of the following.

a sin 73° **b** cos 26° **c** tan 65.2° **d** sin 88°

e cos 35° **f** tan 30° **g** sin 28° **h** cos 5°

7 What is so different about tan compared with both sin and cos?

8 Use your calculator to work out the values of the following.

a 5 sin 65° **b** 6 cos 42° **c** 6 sin 90° **d** 5 sin 0°

9 Use your calculator to work out the values of the following.

a 5 tan 65° **b** 6 tan 42° **c** 6 tan 90° **d** 5 tan 0°

10 Use your calculator to work out the values of the following.

a 4 sin 63° **b** 7 tan 52° **c** 5 tan 80° **d** 9 cos 8°

11 Use your calculator to work out the values of the following.

a $\frac{5}{\sin 63°}$ **b** $\frac{6}{\sin 32°}$ **c** $\frac{6}{\sin 90°}$ **d** $\frac{5}{\sin 30°}$

12 Use your calculator to work out the values of the following.

a $\frac{3}{\tan 64°}$ **b** $\frac{7}{\tan 42°}$ **c** $\frac{5}{\tan 89°}$ **d** $\frac{6}{\tan 40°}$

13 Use your calculator to work out the values of the following.

a 8 sin 75° **b** $\frac{19}{\sin 23°}$ **c** 7 cos 71° **d** $\frac{15}{\sin 81°}$

14 Use your calculator to work out the values of the following.

a 8 tan 75° **b** $\frac{19}{\tan 23°}$ **c** 7 tan 71° **d** $\frac{15}{\tan 81°}$

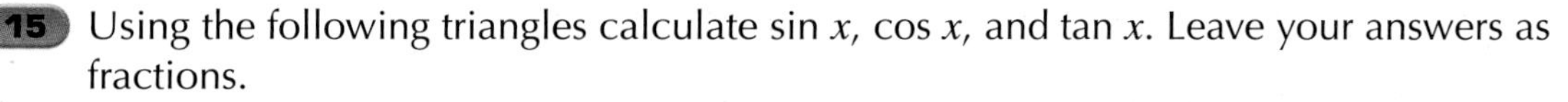

15 Using the following triangles calculate $\sin x$, $\cos x$, and $\tan x$. Leave your answers as fractions.

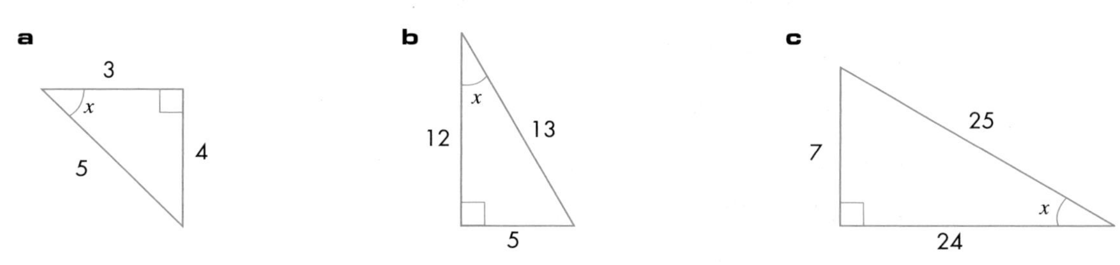

B

3.2 Calculating angles

This section will show you how to:
- use the trigonometric ratios to calculate an angle

Key words
inverse

What angle has a cosine of 0.6? We can use a calculator to find out.

'The angle with a cosine of 0.6' is written as $\cos^{-1} 0.6$ and is called the 'inverse cosine of 0.6'.

Find out where $\cos^{-1}$ is on your calculator.

You will probably find it on the same key as cos, but you will need to press SHIFT or INV or 2ndF first.

Look to see if $\cos^{-1}$ is written above the cos key.

Check that $\cos^{-1} 0.6 = 53.1301... = 53.1°$ (1 decimal place)

Check that $\cos 53.1° = 0.600$ (3 decimal places)

Check that you can find the inverse sine and the inverse tangent in the same way.

EXAMPLE 2

What angle has a sine of $\frac{3}{8}$?

You need to find $\sin^{-1} \frac{3}{8}$.

You could use the fraction button on your calculator or you could calculate $\sin^{-1} (3 \div 8)$.

If you use the fraction key you may not need a bracket, or your calculator may put one in automatically.

Try to do it in both of these ways and then use whichever you prefer.

The answer should be 22.0°.

EXAMPLE 3

Find the angle with a tangent of 0.75.

$\tan^{-1} 0.75 = 36.869\,897\,65 = 36.9°$ (1 decimal place)

EXERCISE 3B

Use your calculator to find the answers to the following. Give your answers to 1 decimal place.

B

1 What angles have the following sines?

a 0.5 **b** 0.785 **c** 0.64 **d** 0.877 **e** 0.999 **f** 0.707

2 What angles have the following cosines?

a 0.5 **b** 0.64 **c** 0.999 **d** 0.707 **e** 0.2 **f** 0.7

3 What angles have the following tangents?

a 0.6 **b** 0.38 **c** 0.895 **d** 1.05 **e** 2.67 **f** 4.38

4 What angles have the following sines?

a $4 \div 5$ **b** $2 \div 3$ **c** $7 \div 10$ **d** $5 \div 6$ **e** $1 \div 24$ **f** $5 \div 13$

5 What angles have the following cosines?

a $4 \div 5$ **b** $2 \div 3$ **c** $7 \div 10$ **d** $5 \div 6$ **e** $1 \div 24$ **f** $5 \div 13$

6 What angles have the following tangents?

a $3 \div 5$ **b** $7 \div 9$ **c** $2 \div 7$ **d** $9 \div 5$ **e** $11 \div 7$ **f** $6 \div 5$

7 What happens when you try to find the angle with a sine of 1.2? What is the largest value of sine you can put into your calculator without getting an error when you ask for the inverse sine? What is the smallest?

PS **8** **a** **i** What angle has a sine of 0.3? (Keep the answer in your calculator memory.)

ii What angle has a cosine of 0.3?

iii Add the two accurate answers of parts **i** and **ii** together.

b Will you always get the same answer to the above no matter what number you start with?

3.3 Using the sine and cosine functions

This section will show you how to:
- find lengths of sides and angles in right-angled triangles using the sine and cosine functions

Key words
cosine
sine

Sine function

Remember sine $\theta = \dfrac{\text{Opposite}}{\text{Hypotenuse}}$

We can use the **sine** ratio to calculate the lengths of sides and angles in right-angled triangles.

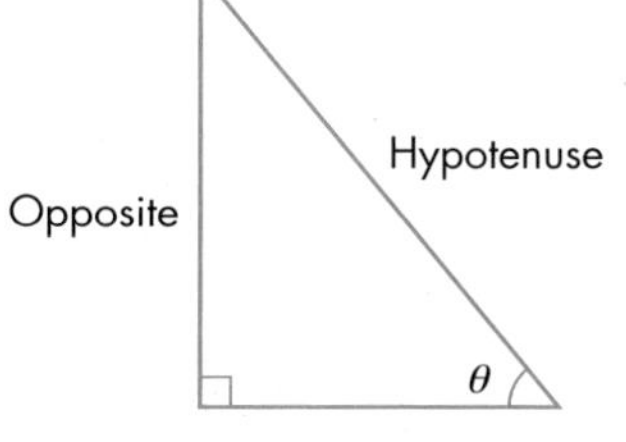

EXAMPLE 4

Find the angle θ, given that the opposite side is 7 cm and the hypotenuse is 10 cm.

Draw a diagram. (This is an essential step.)

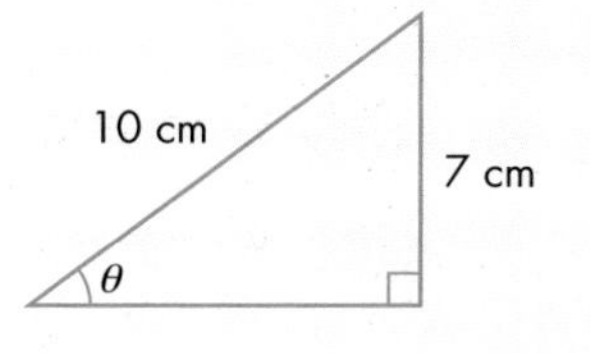

From the information given, use sine.

$$\sin\theta = \frac{O}{H} = \frac{7}{10} = 0.7$$

What angle has a sine of 0.7? To find out, use the inverse sine function on your calculator.

$$\sin^{-1} 0.7 = 44.4° \text{ (1 decimal place)}$$

EXAMPLE 5

Find the length of the side marked a in this triangle.

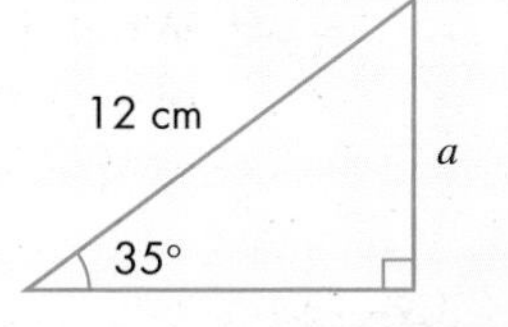

Side a is the opposite side, with 12 cm as the hypotenuse, so use sine.

$$\sin\theta = \frac{O}{H}$$

$$\sin 35° = \frac{a}{12}$$

So $a = 12 \sin 35° = 6.88$ cm (3 significant figures)

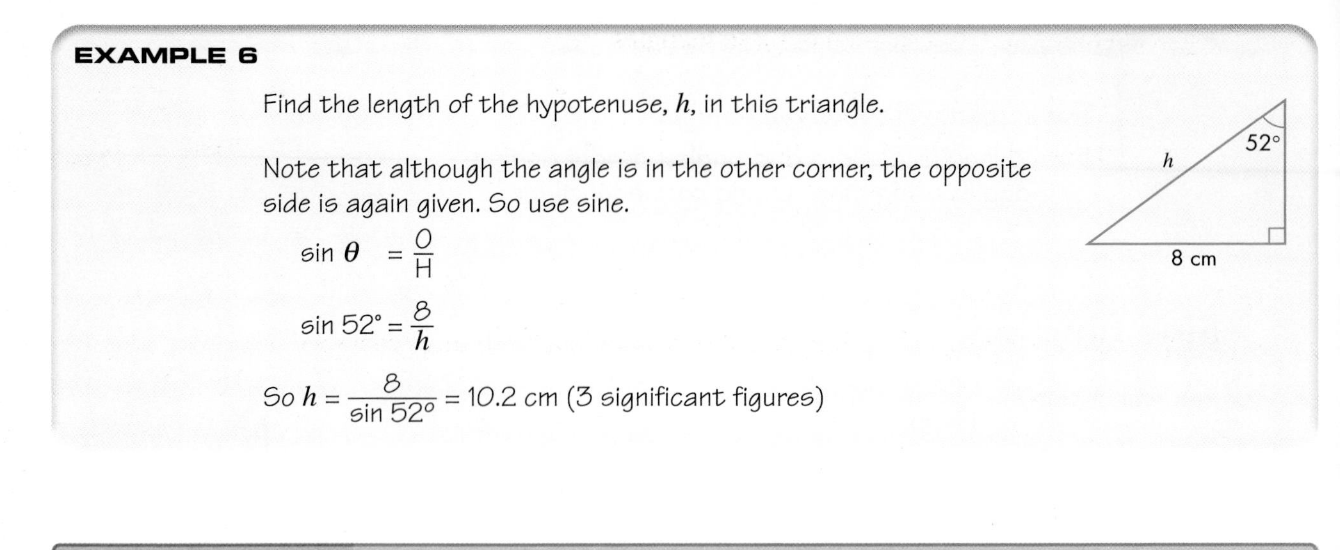

EXAMPLE 6

Find the length of the hypotenuse, h, in this triangle.

Note that although the angle is in the other corner, the opposite side is again given. So use sine.

$$\sin \theta = \frac{O}{H}$$

$$\sin 52° = \frac{8}{h}$$

So $h = \frac{8}{\sin 52°} = 10.2$ cm (3 significant figures)

EXERCISE 3C

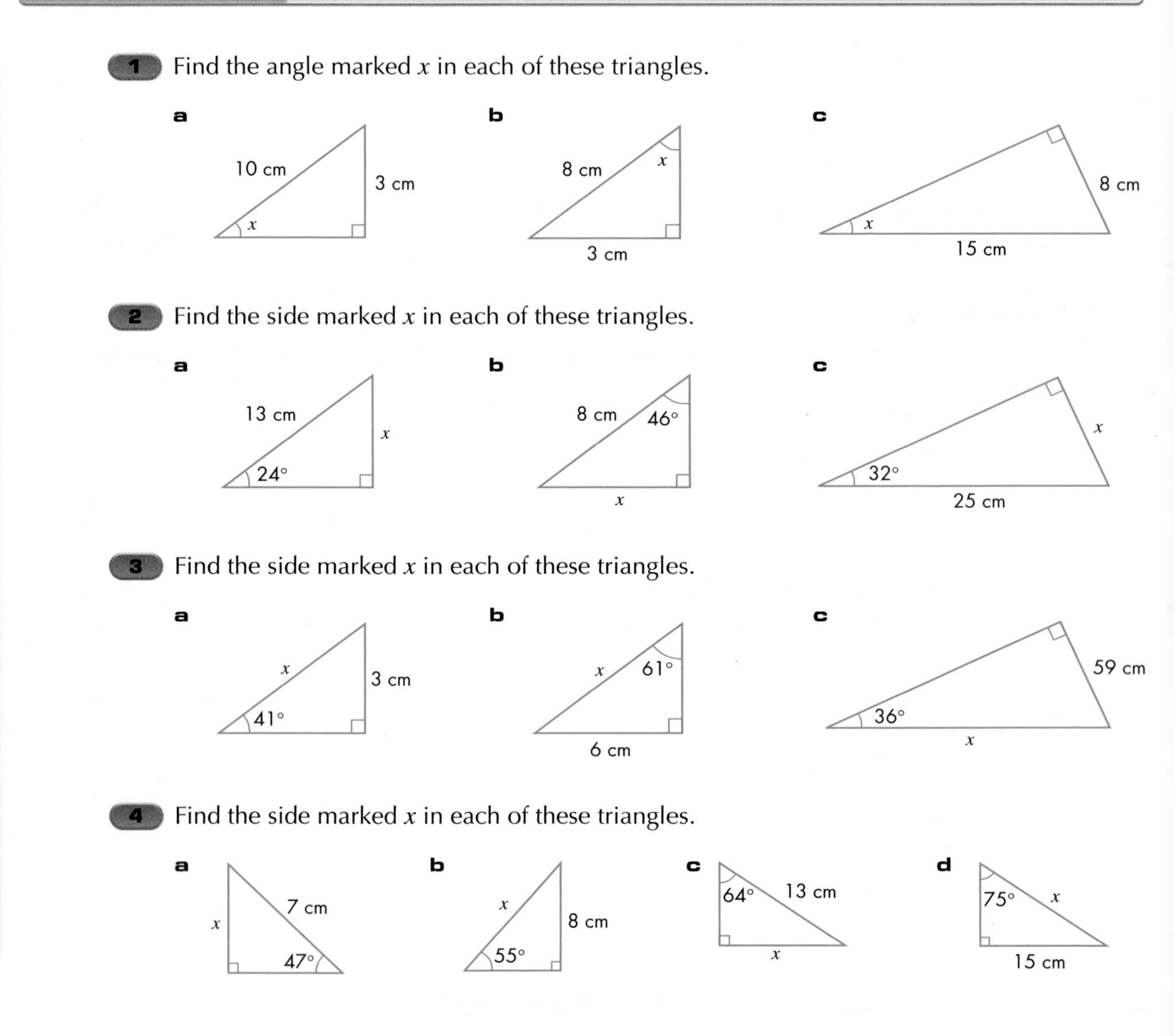

1 Find the angle marked x in each of these triangles.

2 Find the side marked x in each of these triangles.

3 Find the side marked x in each of these triangles.

4 Find the side marked x in each of these triangles.

5 Find the value of x in each of these triangles.

a 11 cm, 15 cm, x

b 9 cm, $37°$, x

c $17°$, x, 4 cm

d 8 cm, 13 cm, x

6 Angle θ has a sine of $\frac{3}{5}$. Calculate the missing lengths in these triangles.

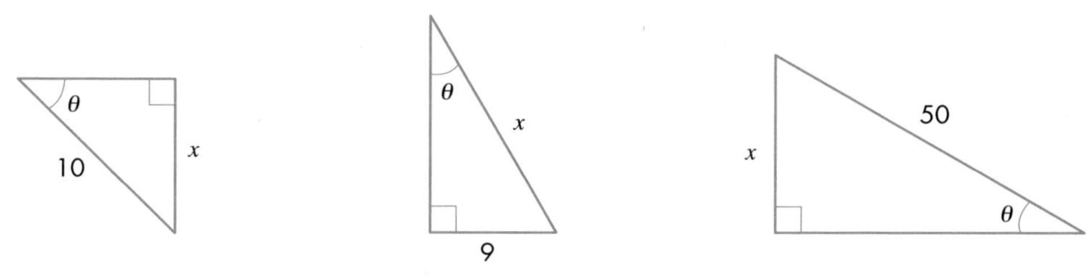

Cosine function

Remember cosine $\theta = \dfrac{\text{Adjacent}}{\text{Hypotenuse}}$

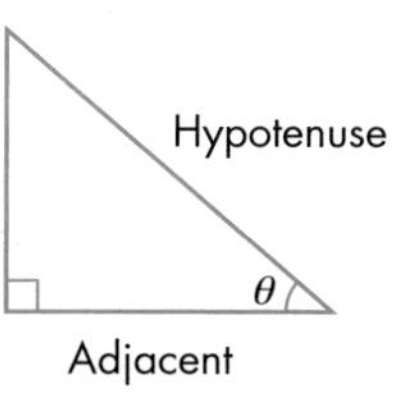

We can use the **cosine** ratio to calculate the lengths of sides and angles in right-angled triangles.

EXAMPLE 7

Find the angle θ, given that the adjacent side is 5 cm and the hypotenuse is 12 cm.

Draw a diagram. (This is an essential step.)

12 cm, 5 cm, θ

From the information given, use cosine.

$$\cos\theta = \frac{A}{H} = \frac{5}{12}$$

What angle has a cosine of $\frac{5}{12}$? To find out, use the inverse cosine function on your calculator.

$$\cos^{-1}\frac{5}{12} = 65.4° \text{ (1 decimal place)}$$

EXAMPLE 8

Find the length of the side marked a in this triangle.

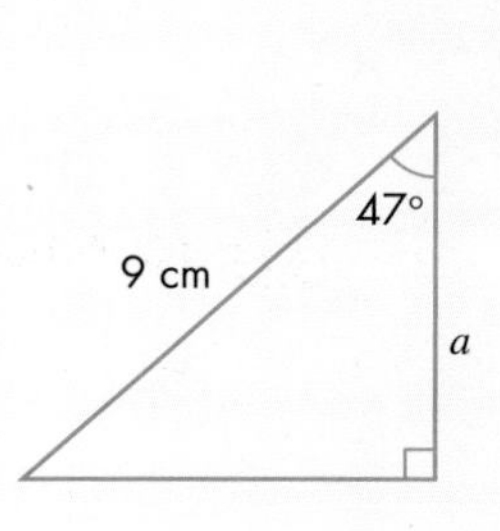

Side a is the adjacent side, with 9 cm as the hypotenuse, so use cosine.

$$\cos\theta = \frac{A}{H}$$

$$\cos 47° = \frac{a}{9}$$

So $a = 9\cos 47° = 6.14$ cm (3 significant figures)

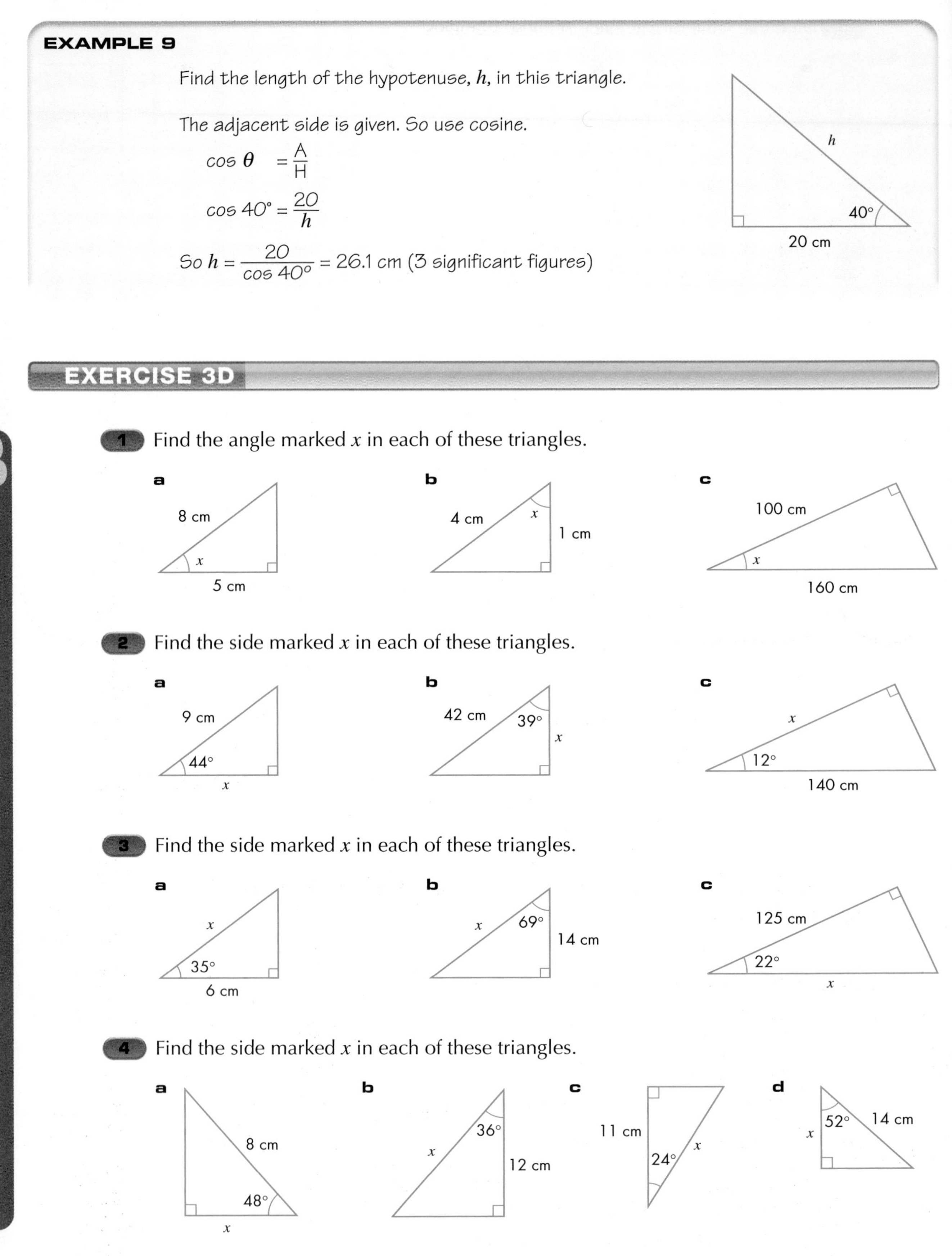

EXAMPLE 9

Find the length of the hypotenuse, h, in this triangle.

The adjacent side is given. So use cosine.

$$\cos \theta = \frac{A}{H}$$

$$\cos 40° = \frac{20}{h}$$

So $h = \dfrac{20}{\cos 40°} = 26.1$ cm (3 significant figures)

EXERCISE 3D

B

1 Find the angle marked x in each of these triangles.

a **b** **c**

2 Find the side marked x in each of these triangles.

a **b** **c**

3 Find the side marked x in each of these triangles.

a **b** **c**

4 Find the side marked x in each of these triangles.

a **b** **c** **d**

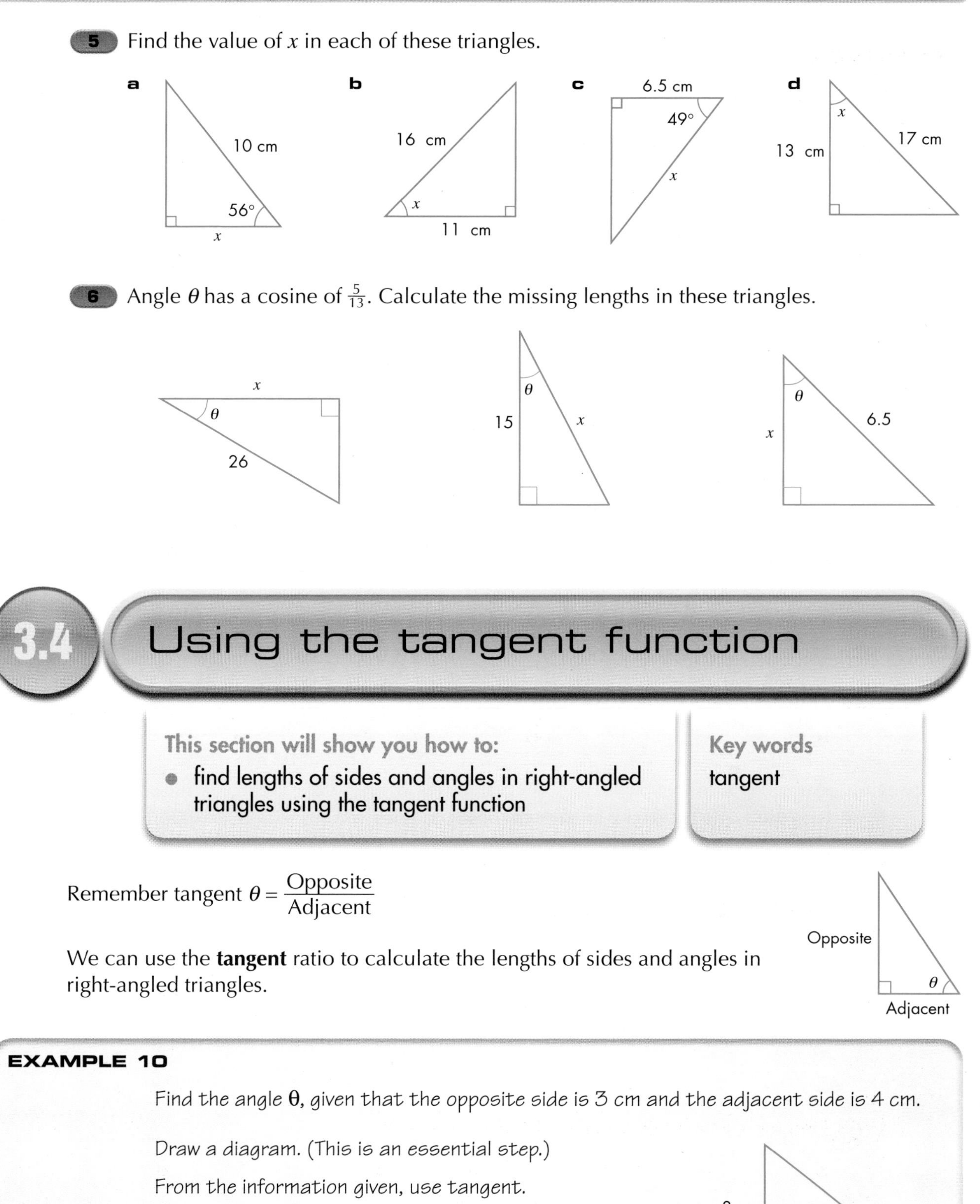

5 Find the value of x in each of these triangles.

6 Angle θ has a cosine of $\frac{5}{13}$. Calculate the missing lengths in these triangles.

B

3.4 Using the tangent function

This section will show you how to:
- find lengths of sides and angles in right-angled triangles using the tangent function

Key words
tangent

Remember tangent $\theta = \dfrac{\text{Opposite}}{\text{Adjacent}}$

We can use the **tangent** ratio to calculate the lengths of sides and angles in right-angled triangles.

EXAMPLE 10

Find the angle θ, given that the opposite side is 3 cm and the adjacent side is 4 cm.

Draw a diagram. (This is an essential step.)

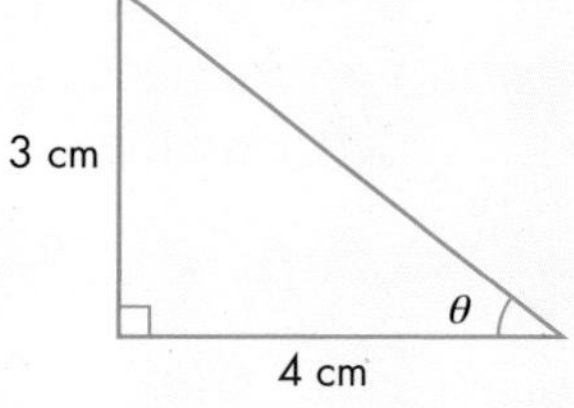

From the information given, use tangent.

$$\tan \theta = \frac{O}{A} = \frac{3}{4} = 0.75$$

What angle has a tangent of 0.75? To find out, use the inverse tangent function on your calculator.

$\tan^{-1} 0.75 = 36.9°$ (1 decimal place)

EXAMPLE 11

Find the length of the side marked x in this triangle.

Side x is the opposite side, with 9 cm as the adjacent side, so use tangent.

$$\tan \theta = \frac{O}{A}$$

$$\tan 62° = \frac{x}{9}$$

So $x = 9 \tan 62° = 16.9$ cm (3 significant figures)

EXAMPLE 12

Find the length of the side marked a in this triangle.

Side a is the adjacent side and the opposite side is given. So use tangent.

$$\tan \theta = \frac{O}{A}$$

$$\tan 35° = \frac{6}{a}$$

So $a = \dfrac{6}{\tan 35°} = 8.57$ cm (3 significant figures)

1 Find the angle marked x in each of these triangles.

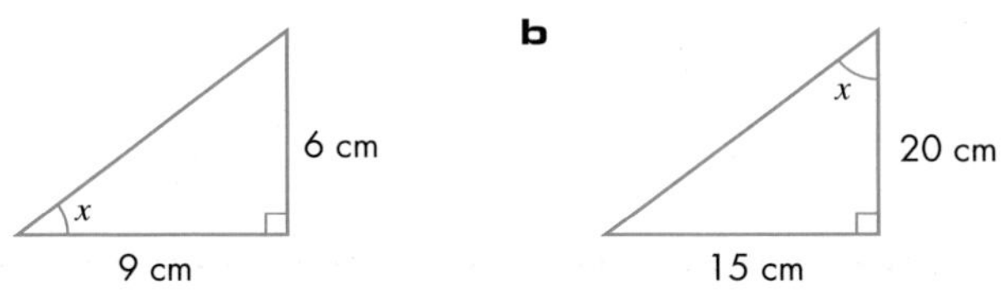

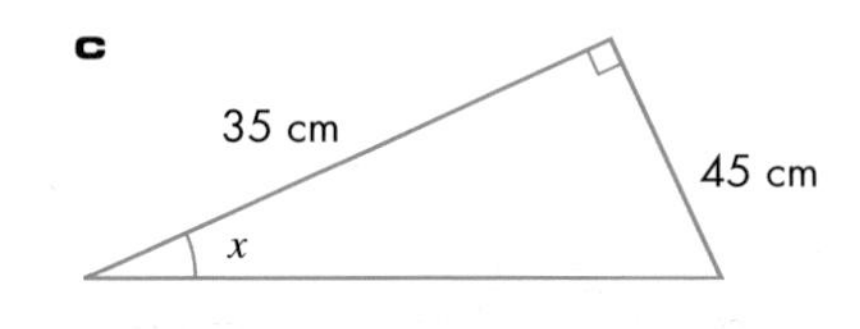

2 Find the side marked x in each of these triangles.

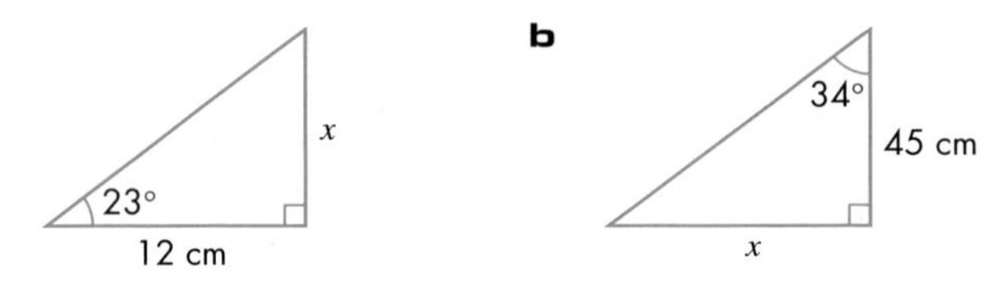

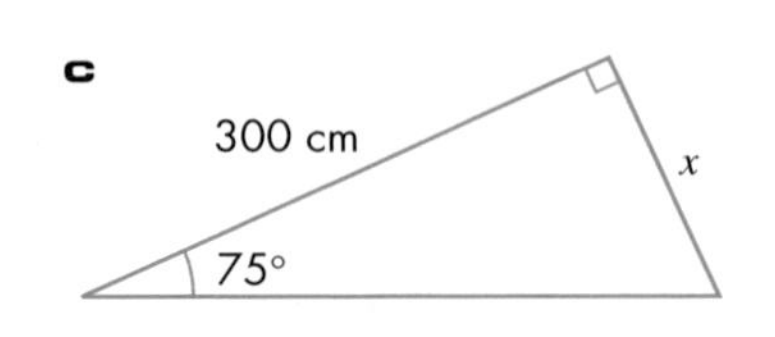

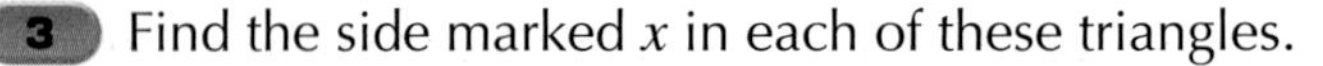

3 Find the side marked x in each of these triangles.

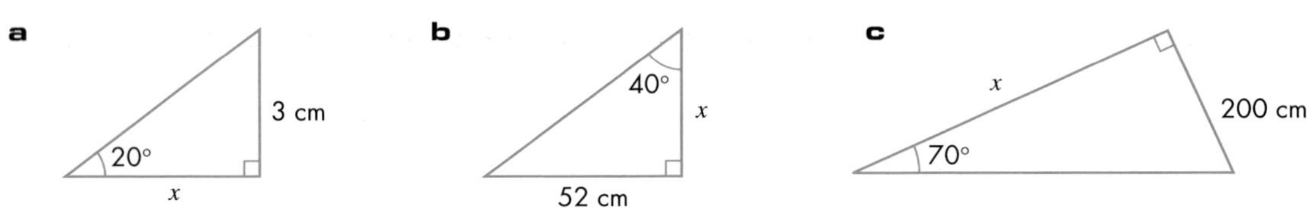

B

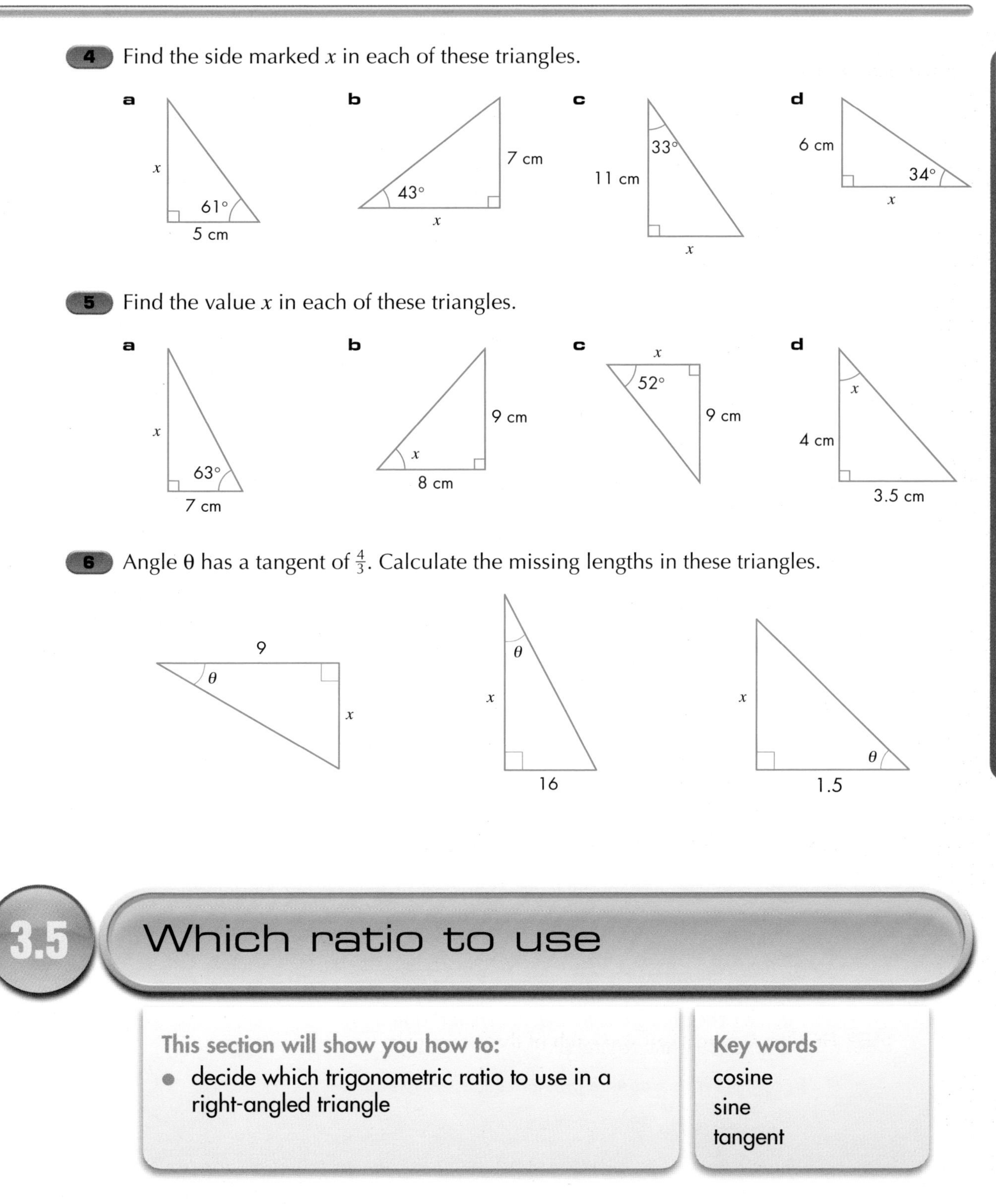

4 Find the side marked x in each of these triangles.

5 Find the value x in each of these triangles.

6 Angle θ has a tangent of $\frac{4}{3}$. Calculate the missing lengths in these triangles.

3.5 Which ratio to use

This section will show you how to:
- decide which trigonometric ratio to use in a right-angled triangle

Key words
cosine
sine
tangent

The difficulty with any trigonometric problem is knowing which ratio to use to solve it.

The following examples show you how to determine which ratio you need in any given situation.

EXAMPLE 13

Find the length of the side marked x in this triangle.

16 cm
x
37°

Step 1 Identify what information is given and what needs to be found. Namely, x is opposite the angle and 16 cm is the hypotenuse.

Step 2 Decide which ratio to use. Only one ratio uses opposite and hypotenuse: **sine.**

Step 3 Remember $\sin \theta = \frac{O}{H}$

Step 4 Put in the numbers and letters: $\sin 37° = \frac{x}{16}$

Step 5 Rearrange the equation and work out the answer:
$x = 16 \sin 37° = 9.629\,040\,371$ cm

Step 6 Give the answer to an appropriate degree of accuracy: $x = 9.63$ cm (3 significant figures)

In reality, you do not write down every step as in Example 13. Step 1 can be done by marking the triangle. Steps 2 and 3 can be done in your head. Steps 4 to 6 are what you write down.

Remember that examiners will want to see evidence of working. Any reasonable attempt at identifying the sides and using a ratio will probably get you some method marks, but only if the fraction is the right way round.

The next examples are set out in a way that requires the *minimum* amount of working but gets *maximum* marks.

EXAMPLE 14

Find the length of the side marked x in this triangle.

7 cm
50°
x

Mark on the triangle the side you know (H) and the side you want to find (A).

H
7 cm
50°
A
x

Recognise it is a **cosine** problem because you have A and H.

So $\cos 50° = \frac{x}{7}$

$x = 7 \cos 50° = 4.50$ cm (3 significant figures)

EXAMPLE 15

Find the angle marked x in this triangle.

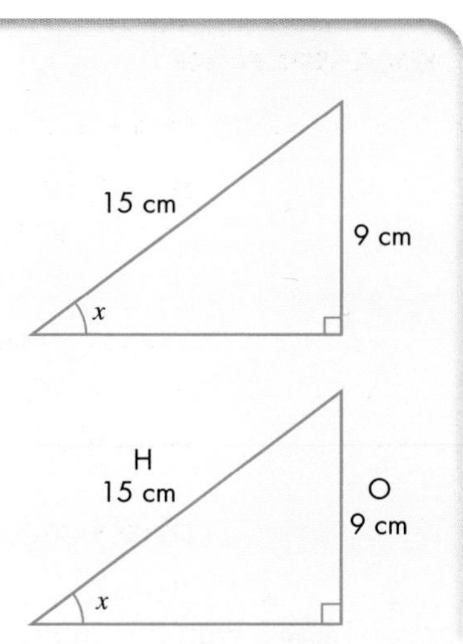

Mark on the triangle the sides you know.

Recognise it is a sine problem because you have O and H.

So $\sin x = \frac{9}{15} = 0.6$

$x = \sin^{-1} 0.6 = 36.9°$ (1 decimal place)

EXAMPLE 16

Find the angle marked x in this triangle.

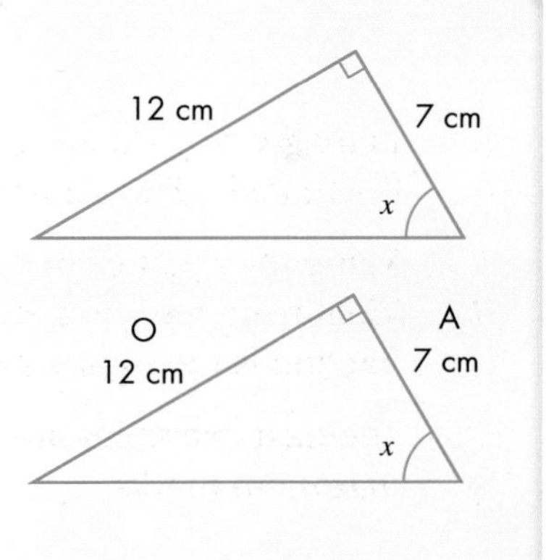

Mark on the triangle the sides you know.

Recognise it is a **tangent** problem because you have O and A.

So $\tan x = \frac{12}{7}$

$x = \tan^{-1} \frac{12}{7} = 59.7°$ (1 decimal place)

EXERCISE 3F

1 Find the length marked x in each of these triangles.

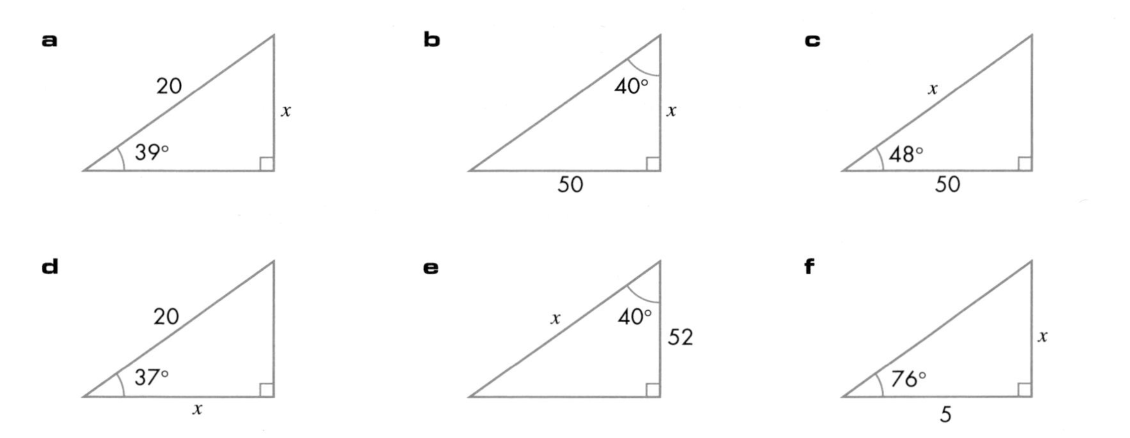

B

B

2 Find the angle marked x in each of these triangles.

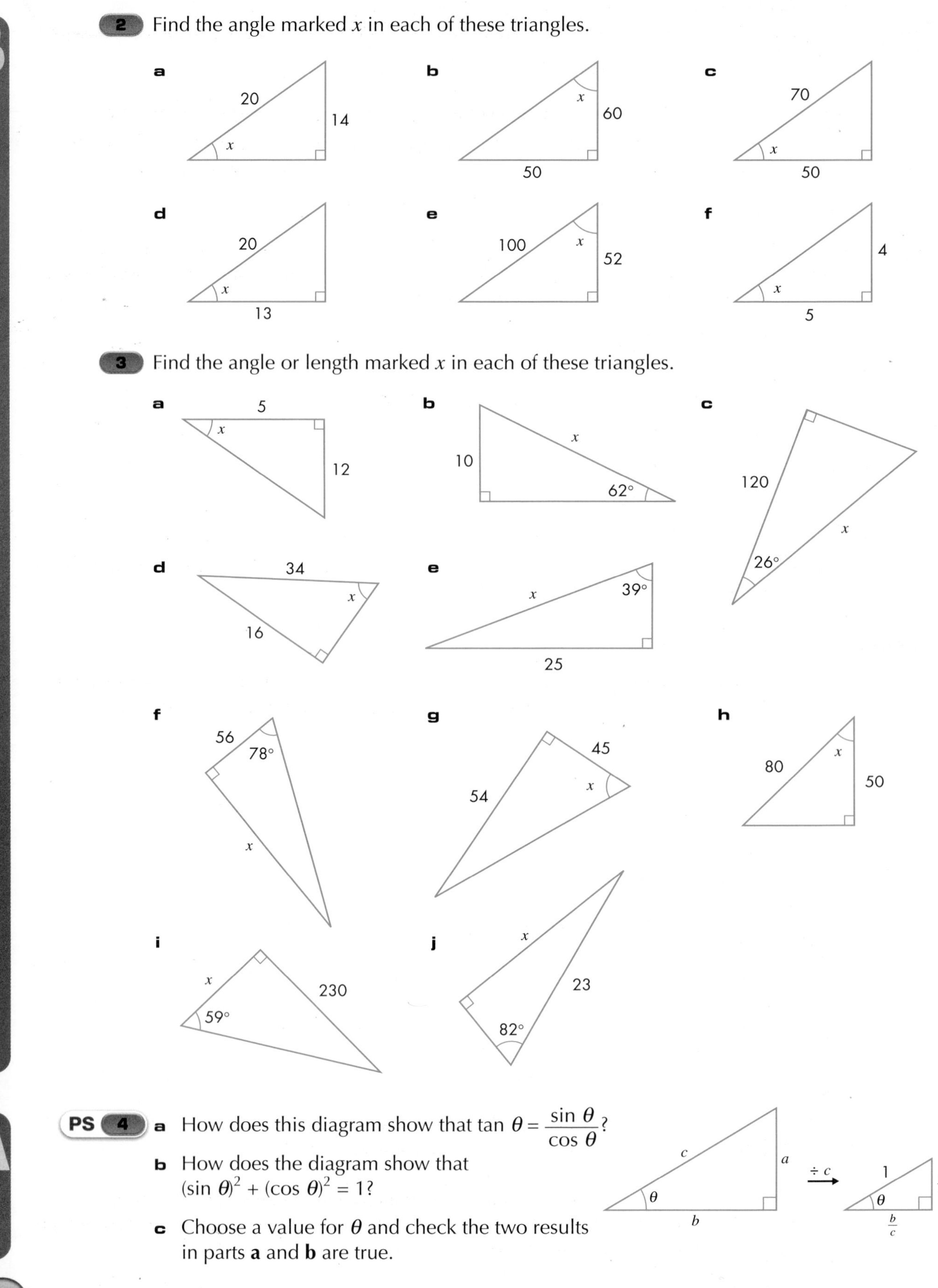

3 Find the angle or length marked x in each of these triangles.

A

PS 4 **a** How does this diagram show that $\tan\theta = \dfrac{\sin\theta}{\cos\theta}$?

b How does the diagram show that $(\sin\theta)^2 + (\cos\theta)^2 = 1$?

c Choose a value for θ and check the two results in parts **a** and **b** are true.

3.6 Solving problems using trigonometry 1

This section will show you how to:
- solve practical problems using trigonometry
- solve problems using an angle of elevation or an angle of depression

Key words
angle of depression
angle of elevation
trigonometry

Many **trigonometry** problems in GCSE examination papers do not come as straightforward triangles. Sometimes, solving a triangle is part of solving a practical problem. You should follow these steps when solving a practical problem using trigonometry.

- Draw the triangle required.
- Put on the information given (angles and sides).
- Put on x for the unknown angle or side.
- Mark on two of O, A or H as appropriate.
- Choose which ratio to use.
- Write out the equation with the numbers in.
- Rearrange the equation if necessary, then work out the answer.
- Give your answer to a sensible degree of accuracy. Answers given to 3 significant figures or to the nearest degree are acceptable in exams.

EXAMPLE 17

A window cleaner has a ladder which is 7 m long. The window cleaner leans it against a wall so that the foot of the ladder is 3 m from the wall. What angle does the ladder make with the wall?

Draw the situation as a right-angled triangle.

Then mark the sides and angle.

Recognise it is a sine problem because you have O and H.

So $\sin x = \frac{3}{7}$

$x = \sin^{-1}\frac{3}{7} = 25°$ (to the nearest degree)

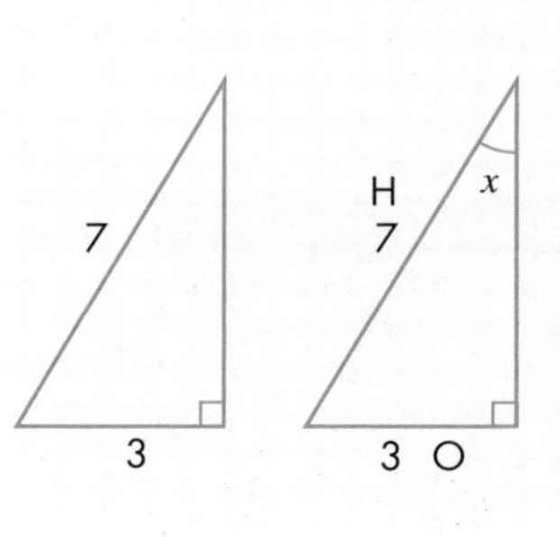

EXERCISE 3G

In these questions, give answers involving angles to the nearest degree.

1 A ladder, 6 m long, rests against a wall. The foot of the ladder is 2.5 m from the base of the wall. What angle does the ladder make with the ground?

2 The ladder in question **1** has a 'safe angle' with the ground of between 70° and 80°. What are the safe limits for the distance of the foot of this ladder from the wall? How high up the wall does the ladder reach?

B

FM 3 A ladder, of length 10 m, is placed so that it reaches 7 m up the wall. What angle does it make with the ground?

FM 4 A ladder is placed so that it makes an angle of 76° with the ground. The foot of the ladder is 1.7 m from the foot of the wall. How high up the wall does the ladder reach?

PS 5 Calculate the angle that the diagonal makes with the long side of a rectangle which measures 10 cm by 6 cm.

FM 6 This diagram shows a frame for a bookcase.

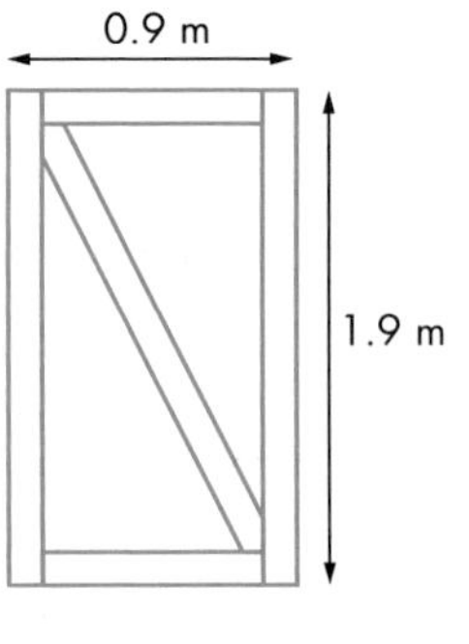

a What angle does the diagonal strut make with the long side?

b Use Pythagoras' theorem to calculate the length of the strut.

c Why might your answers be inaccurate in this case?

FM 7 This diagram shows a roof truss.

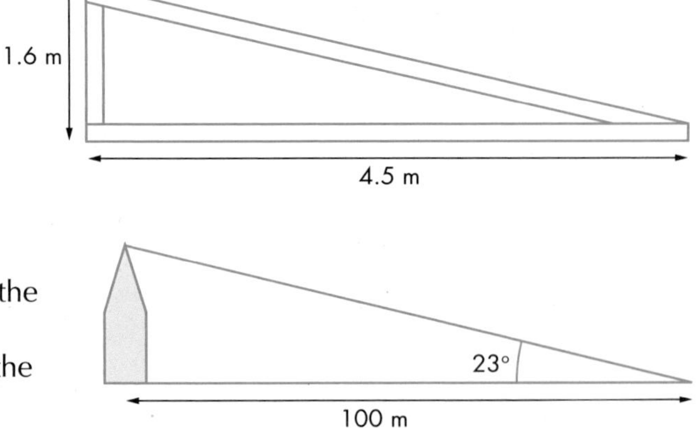

a What angle will the roof make with the horizontal?

b Calculate the length of the sloping strut.

FM 8 Alicia paces out 100 m from the base of a church. She then measures the angle to the top of the spire as 23°. How would Alicia find the height of the church spire?

AU 9 A girl is flying a kite on a string 32 m long. The string, which is being held at 1 m above the ground, makes an angle of 39° with the horizontal. How high is the kite above the ground?

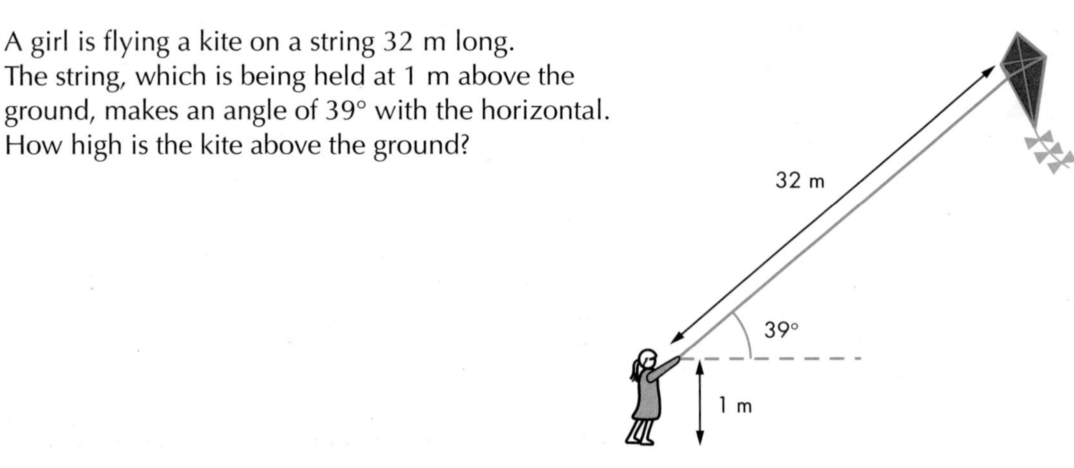

FM 10 Helena is standing on one bank of a wide river. She wants to find the width of the river. She cannot get to the other side.
She asks if you can use trigonometry to find the width of the river.

What can you suggest?

Angles of elevation and depression

When you look *up* at an aircraft in the sky, the angle through which your line of sight turns from looking straight ahead (the horizontal) is called the **angle of elevation**.

When you are standing on a high point and look *down* at a boat, the angle through which your line of sight turns from looking straight ahead (the horizontal) is called the **angle of depression**.

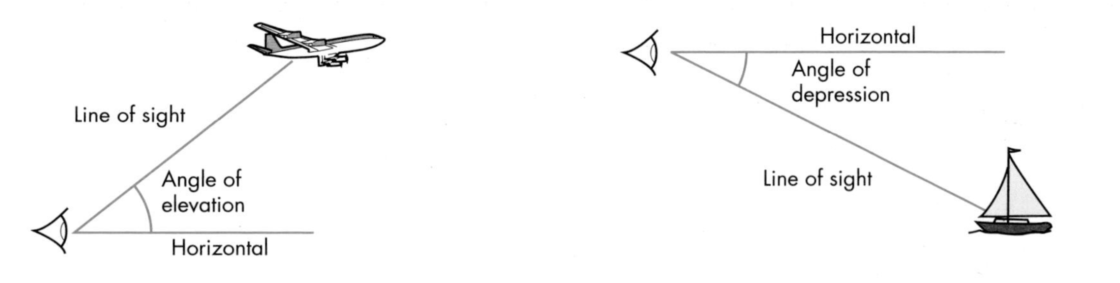

EXAMPLE 18

From the top of a vertical cliff, 100 m high, Andrew sees a boat out at sea. The angle of depression from Andrew to the boat is 42°. How far from the base of the cliff is the boat?

The diagram of the situation is shown in figure **i**.

From this, you get the triangle shown in figure **ii**.

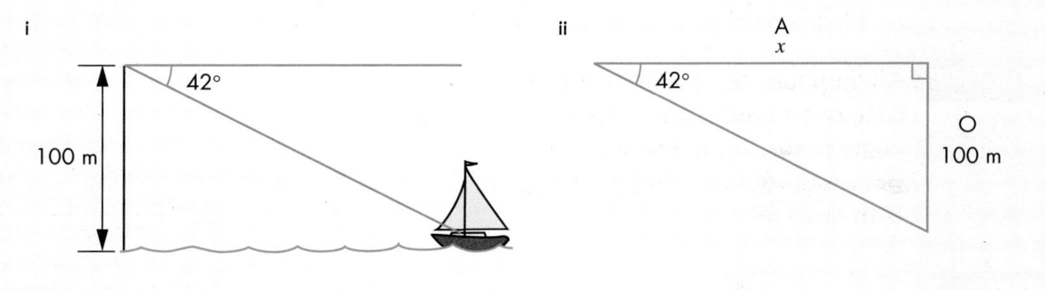

From figure **ii**, you see that this is a tangent problem.

So $\tan 42° = \dfrac{100}{x}$

$$x = \frac{100}{\tan 42°} = 111 \text{ m (3 significant figures)}$$

EXERCISE 3H

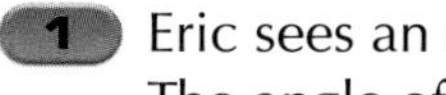

In these questions, give any answers involving angles to the nearest degree.

1 Eric sees an aircraft in the sky. The aircraft is at a horizontal distance of 25 km from Eric. The angle of elevation is 22°. How high is the aircraft?

2 An aircraft is flying at an altitude of 4000 m and is 10 km from the airport. If a passenger can see the airport, what is the angle of depression?

B

B

3 A man standing 200 m from the base of a television transmitter looks at the top of it and notices that the angle of elevation of the top is 65°. How high is the tower?

AU 4 **a** From the top of a vertical cliff, 200 m high, a boat has an angle of depression of 52°. How far from the base of the cliff is the boat?

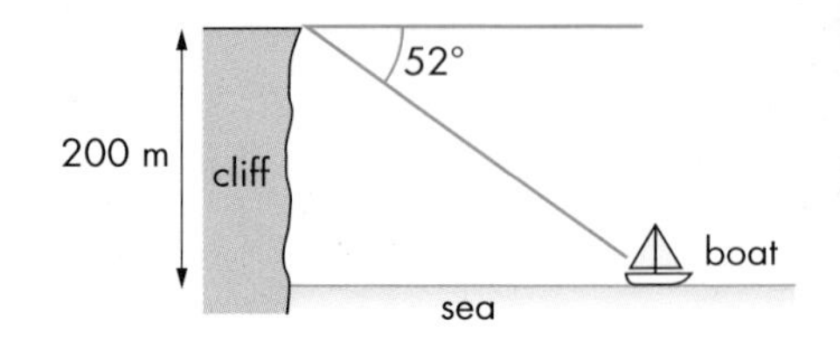

b The boat now sails away from the cliff so that the distance is doubled. Does that mean that the angle of depression is halved? Give a reason for your answer.

FM 5 From a boat, the angle of elevation of the foot of a lighthouse on the edge of a cliff is 34°.

a If the cliff is 150 m high, how far from the base of the cliff is the boat?

b If the lighthouse is 50 m high, what would be the angle of elevation of the top of the lighthouse from the boat?

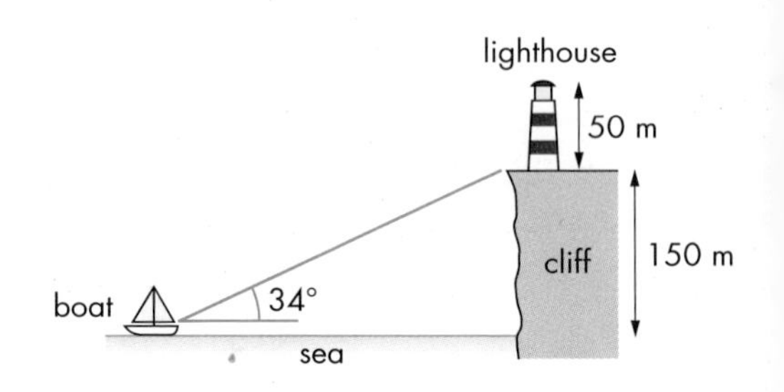

AU 6 A bird flies from the top of a 12 m tall tree, at an angle of depression of 34°, to catch a worm on the ground.

a How far does the bird actually fly?

b How far was the worm from the base of the tree?

FM 7 Sunil wants to work out the height of a building. He stands about 50 m away from a building. The angle of elevation from Sunil to the top of the building is about 15°. How tall is the building?

8 The top of a ski run is 100 m above the finishing line. The run is 300 m long. What is the angle of depression of the ski run?

PS 9 Nessie and Cara are standing on opposite sides of a tree.

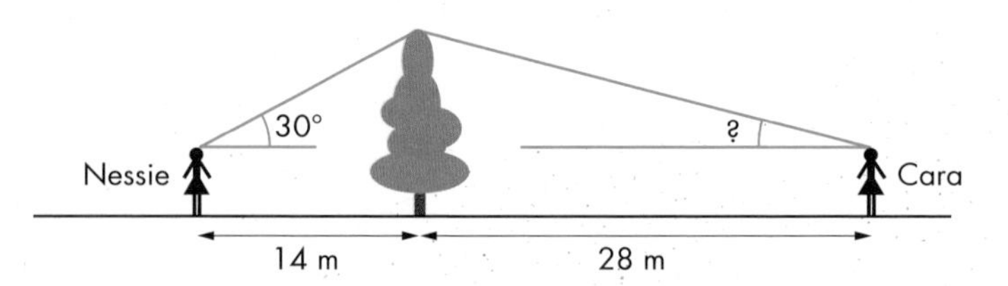

Nessie is 14 m away and the angle of elevation of the top of the tree is 30°.

Cara is 28 m away. She says the angle of elevation for her must be 15° because she is twice as far away.

Is she correct?

What do you think the angle of elevation is?

3.7 Solving problems using trigonometry 2

This section will show you how to:

- solve bearing problems using trigonometry
- use trigonometry to solve problems involving isosceles triangles

Key words

bearing
isosceles triangle
three-figure bearing
trigonometry

Trigonometry and bearings

A **bearing** is the direction to one place from another. The usual way of giving a bearing is as an angle measured from north in a clockwise direction. This is how a navigational compass and a surveyor's compass measure bearings.

A bearing is always written as a three-digit number, known as a **three-figure bearing**.

The diagram shows how this works, using the main compass points as examples.

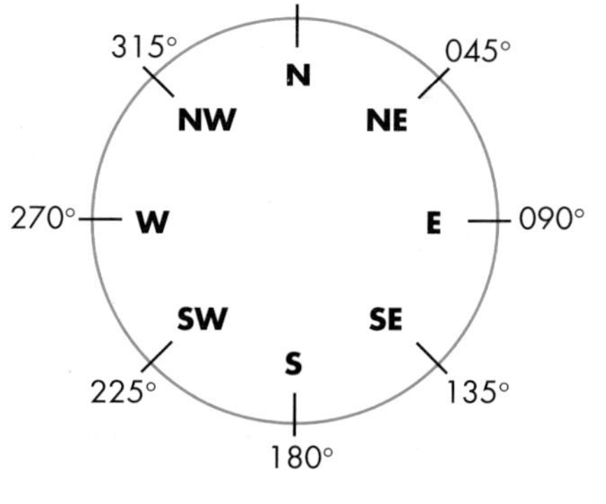

When working with bearings, follow these three rules.

- Always start from *north*.
- Always measure *clockwise*.
- Always give a bearing in degrees and as a *three-figure bearing*.

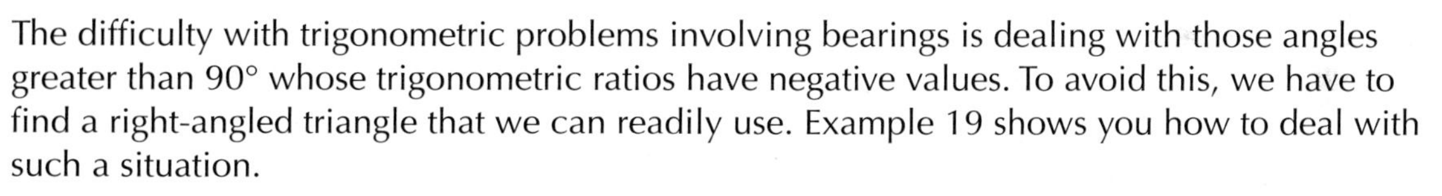

The difficulty with trigonometric problems involving bearings is dealing with those angles greater than 90° whose trigonometric ratios have negative values. To avoid this, we have to find a right-angled triangle that we can readily use. Example 19 shows you how to deal with such a situation.

EXAMPLE 19

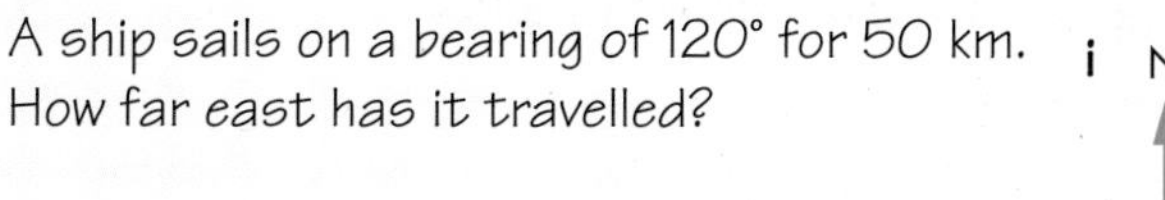

A ship sails on a bearing of 120° for 50 km. How far east has it travelled?

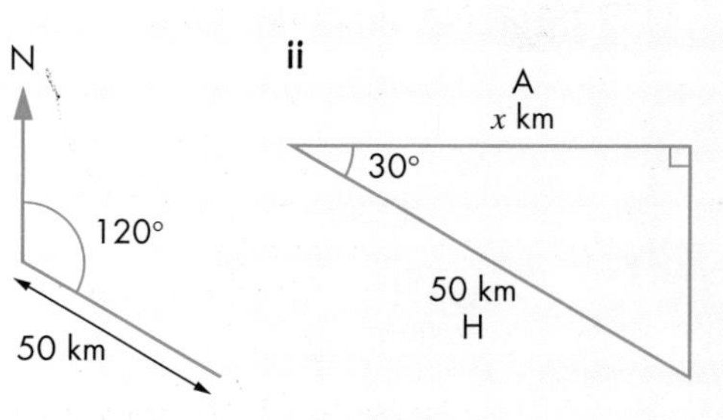

The diagram of the situation is shown in figure **i**. From this, you can get the acute-angled triangle shown in figure **ii**.

From figure **ii**, you see that this is a cosine problem.

So $\cos 30° = \frac{x}{50}$

$x = 50 \cos 30° = 43.301$

Distance east = 43.3 km (to 3 significant figures)

EXERCISE 3I

B

1 A ship sails for 75 km on a bearing of 078°.

a How far east has it travelled? **b** How far north has it travelled?

2 Lopham is 17 miles from Wath on a bearing of 210°.

a How far south of Wath is Lopham? **b** How far east of Lopham is Wath?

FM 3 A plane sets off from an airport and flies due east for 120 km, then turns to fly due south for 70 km before landing at Seddeth. Another pilot decides to fly the direct route from the airport to Seddeth. On what bearing should he fly?

PS 4 A helicopter leaves an army base and flies 60 km on a bearing of 278°.

a How far west has the helicopter flown? **b** How far north has the helicopter flown?

5 A ship sails from a port on a bearing of 117° for 35 km before heading due north for 40 km and docking at Angle Bay.

a How far south had the ship sailed before turning?

b How far north had the ship sailed from the port to Angle Bay?

c How far east is Angle Bay from the port?

d What is the bearing from the port to Angle Bay?

AU 6 Mountain A is due west of a walker. Mountain B is due north of the walker. The guidebook says that mountain B is 4.3 km from mountain A, on a bearing of 058°. How far is the walker from mountain B?

PS 7 The shopping mall is 5.5 km east of my house and the supermarket is 3.8 km south. What is the bearing of the supermarket from the shopping mall?

8 The diagram shows the relative distances and bearings of three ships A, B and C.

a How far north of A is B? (Distance x on diagram.)

b How far north of B is C? (Distance y on diagram.)

c How far west of A is C? (Distance z on diagram.)

d What is the bearing of A from C? (Angle $w°$ on diagram.)

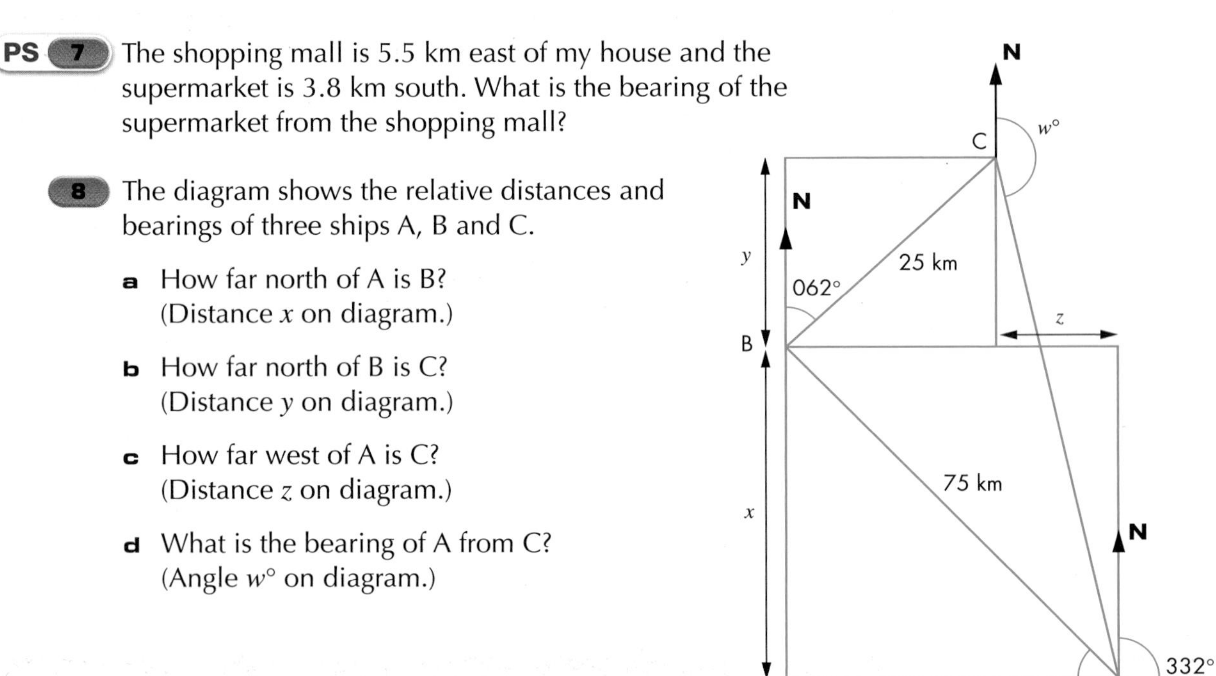

A

AU **9** A ship sails from port A for 42 km on a bearing of 130° to point B. It then changes course and sails for 24 km on a bearing of 040° to point C, where it breaks down and anchors. What distance and on what bearing will a helicopter have to fly from port A to go directly to the ship at C?

Trigonometry and isosceles triangles

Isosceles triangles often feature in **trigonometry** problems because such a triangle can be split into two right-angled triangles that are congruent.

EXAMPLE 20

a Find the length x in this isosceles triangle.

b Calculate the area of the triangle.

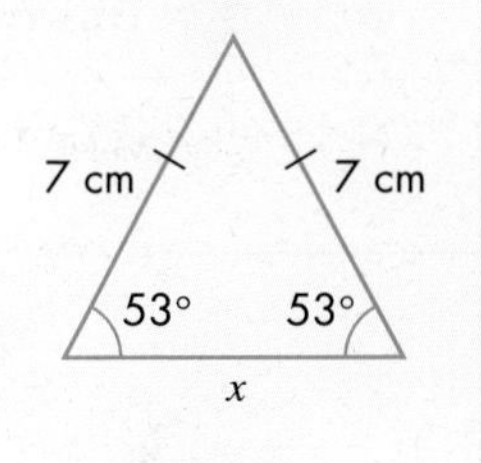

Draw a perpendicular from the apex of the triangle to its base, splitting the triangle into two congruent, right-angled triangles.

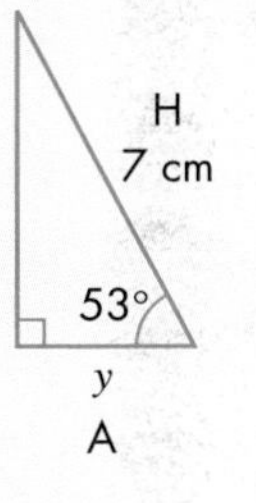

a To find the length y, which is $\frac{1}{2}$ of x, use cosine.

So, $\cos 53° = \frac{y}{7}$

$y = 7 \cos 53° = 4.2127051$ cm

So the length $x = 2y = 8.43$ cm (3 significant figures)

b To calculate the area of the original triangle, you first need to find its vertical height, h.

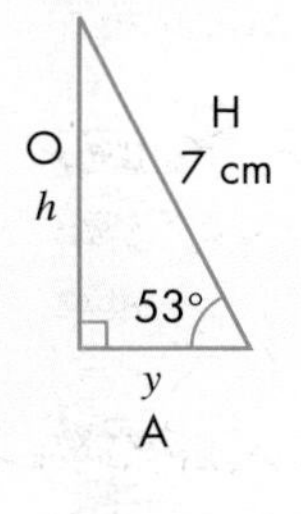

You have two choices, both of which involve the right-angled triangle of part **a**. We can use either Pythagoras' theorem ($h^2 + y^2 = 7^2$) or trigonometry. It is safer to use trigonometry again, since we are then still using known information.

This is a sine problem.

So, $\sin 53° = \frac{h}{7}$

$h = 7 \sin 53° = 5.5904486$ cm

(Keep the accurate figure in the calculator.)

The area of the triangle = $\frac{1}{2}$ × base × height. (We should use the most accurate figures we have for this calculation.)

$A = \frac{1}{2} \times 8.4254103 \times 5.5904486 = 23.6 \text{ cm}^2$ (3 significant figures)

You are not expected to write down these eight-figure numbers, just to use them.

EXERCISE 3J

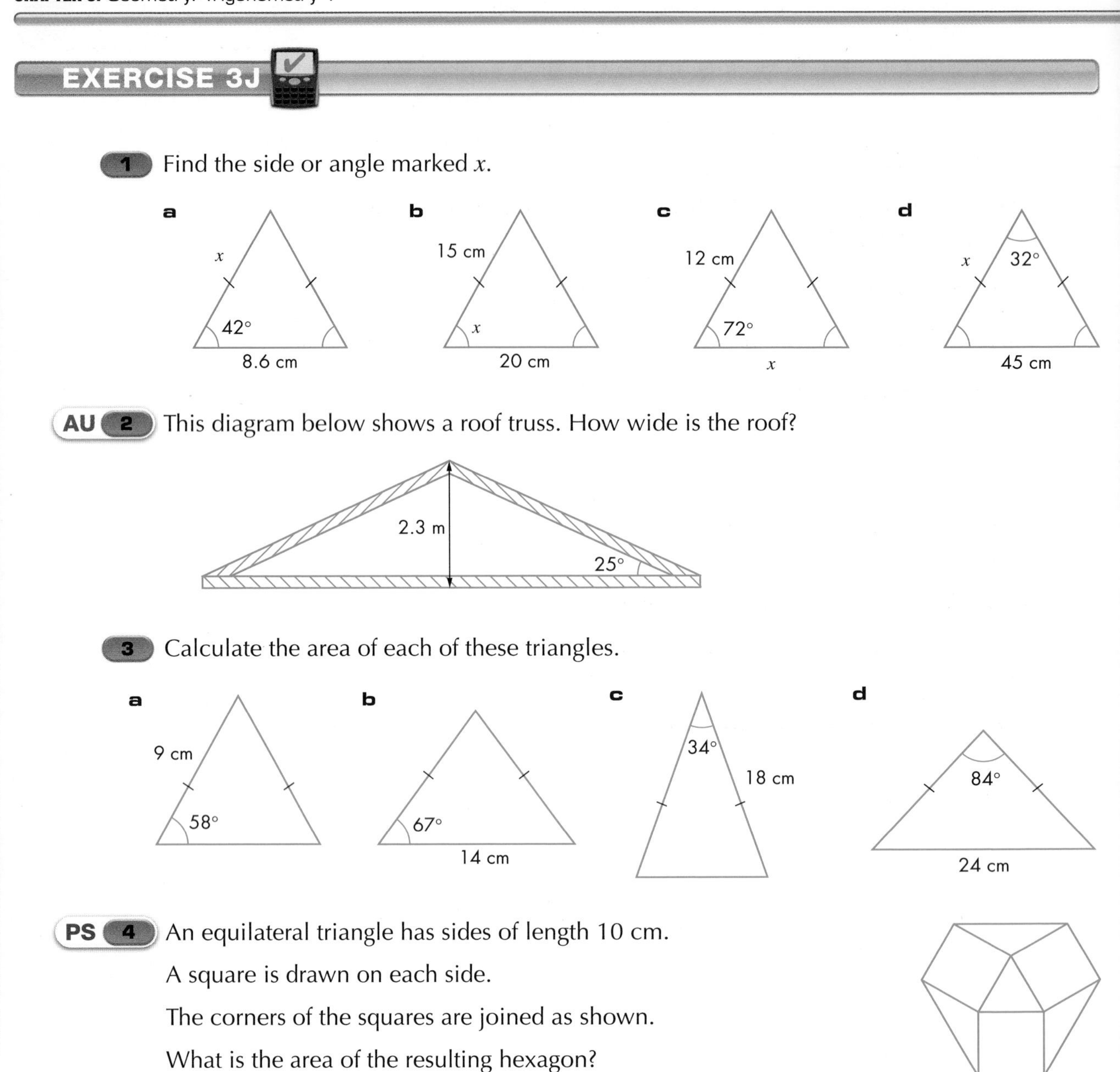

1 Find the side or angle marked x.

AU 2 This diagram below shows a roof truss. How wide is the roof?

3 Calculate the area of each of these triangles.

PS 4 An equilateral triangle has sides of length 10 cm.

A square is drawn on each side.

The corners of the squares are joined as shown.

What is the area of the resulting hexagon?

GRADE BOOSTER

- **B** You can use trigonometry to find lengths of sides and angles in right-angled triangles
- **B** You can use the sine, cosine and tangent functions
- **B** You can choose the correct ratio to use
- **B** You can use trigonometry to solve problems

What you should know now

- How to use the trigonometric ratios for sine, cosine and tangent in right-angled triangles
- How to solve problems using trigonometry
- How to solve problems using angles of elevation, angles of depression and bearings

EXAMINATION QUESTIONS

1 **a** ABC is a right angled triangle. AB = 12 cm, BC = 8 cm. Find the size of angle CAB (marked x in the diagram). Give your answer to 1 decimal place.

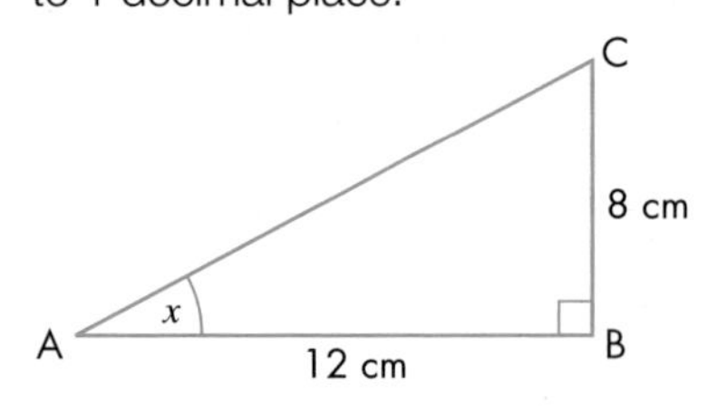

b PQR is a right-angled triangle. PQ = 15 cm, angle QPR = 32°. Find the length of PR (marked y in the diagram). Give your answer to 1 decimal place.

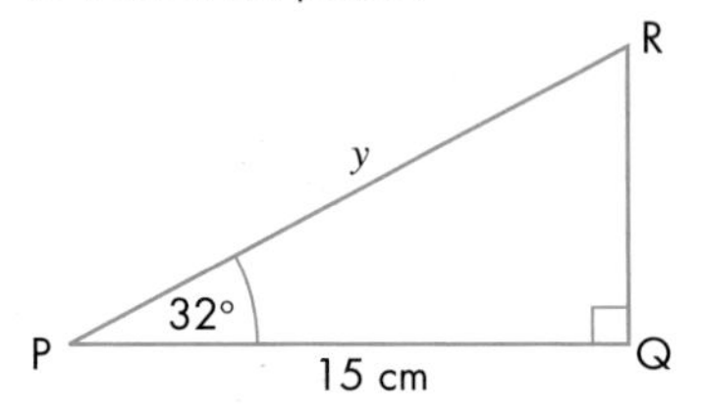

2 DEF is another right-angled triangle.

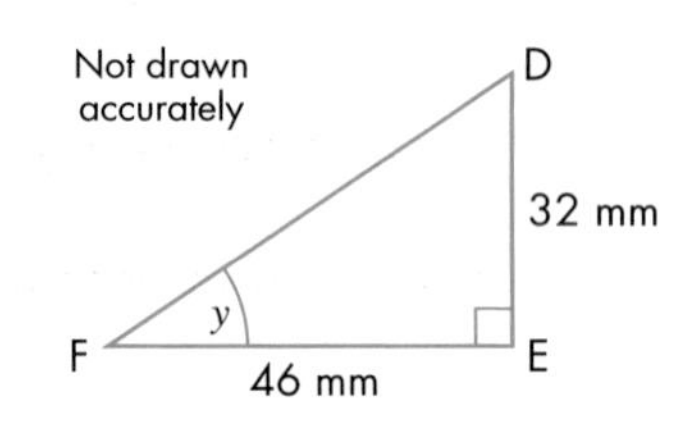

DE = 32 mm

FE = 46 mm

Calculate the size of angle y.

Give your answer correct to 1 decimal place. *(3 marks)*

Edexcel, June 2008, Paper 4, Question 14

3 PQR is a right-angled triangle.

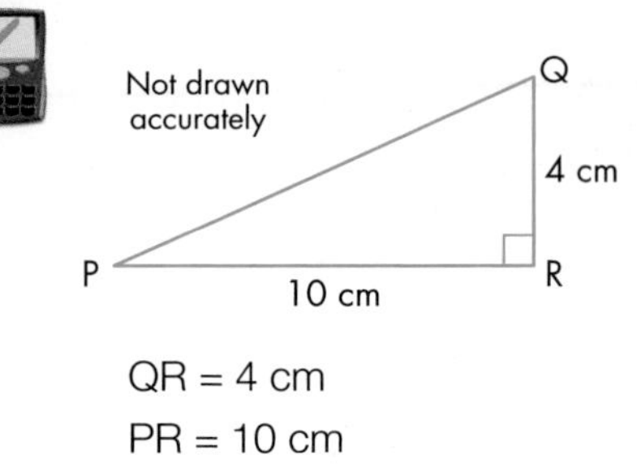

QR = 4 cm

PR = 10 cm

Work out the size of angle RPQ. Give your answer correct to 3 significant figures. *(3 marks)*

Edexcel, November 2008, Paper 15, Question 16

4 A lighthouse, L, is 3.2 km due West of a port, P. A ship, S, is 1.9 km due North of the lighthouse, L.

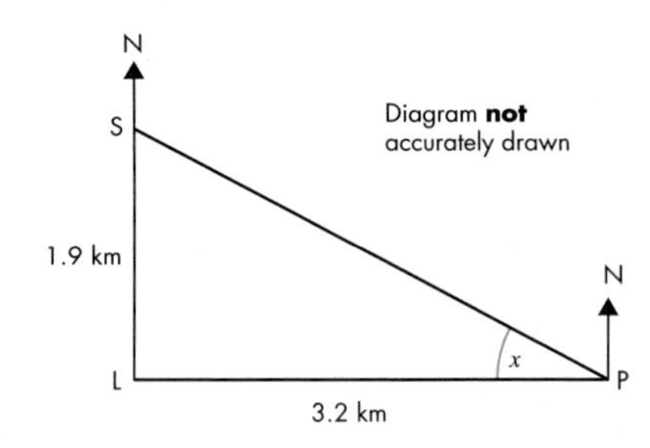

a Calculate the size of the angle marked x. Give your answer correct to 3 significant figures.

b Find the bearing of the port, P, from the ship, S. Give your answer correct to 3 significant figures.

Edexcel, June 2005, Paper 6 Higher, Question 10

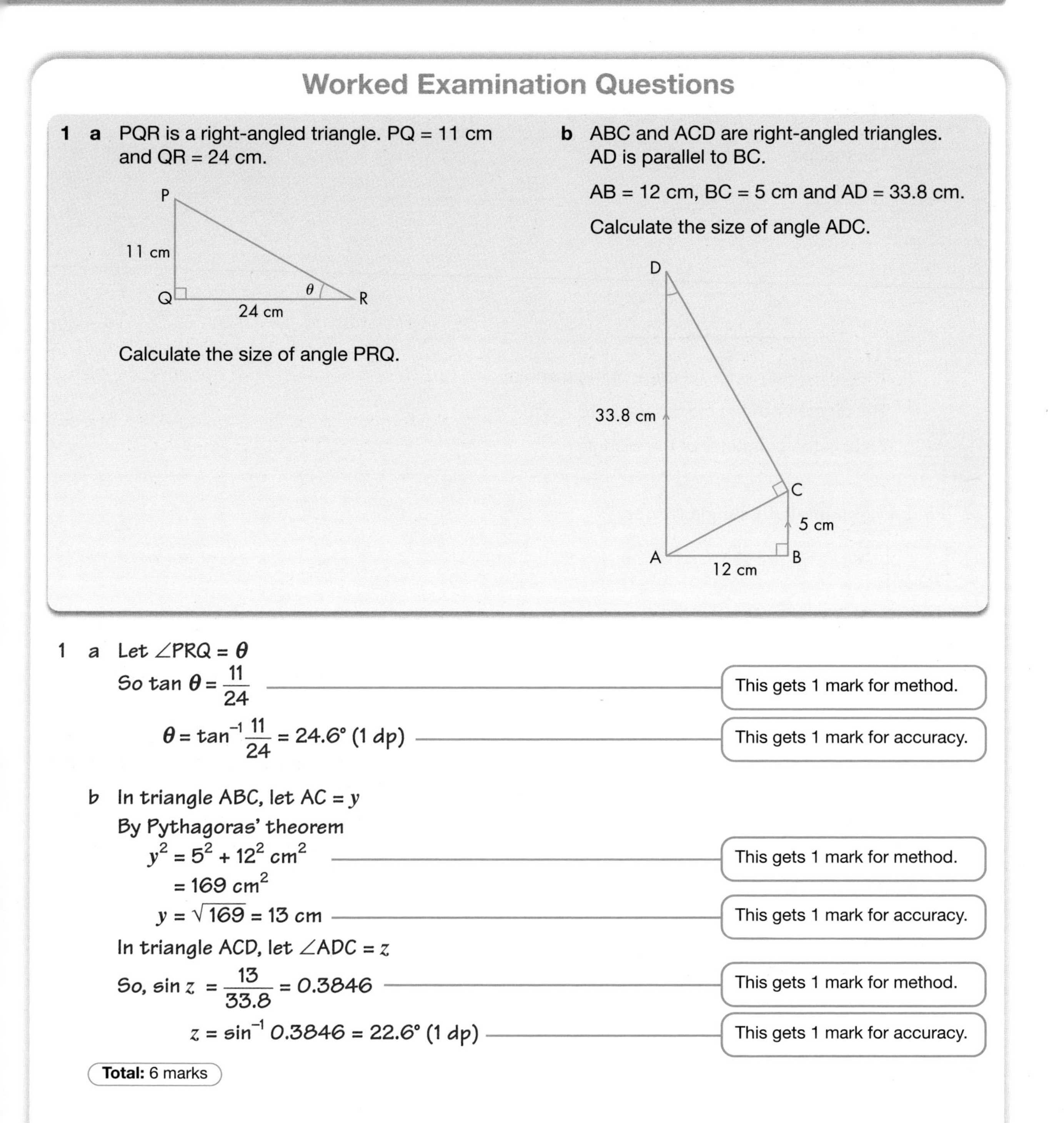

Worked Examination Questions

1 a PQR is a right-angled triangle. PQ = 11 cm and QR = 24 cm.

Calculate the size of angle PRQ.

b ABC and ACD are right-angled triangles. AD is parallel to BC.

AB = 12 cm, BC = 5 cm and AD = 33.8 cm.

Calculate the size of angle ADC.

1 a Let $\angle PRQ = \theta$

So $\tan \theta = \frac{11}{24}$ — This gets 1 mark for method.

$\theta = \tan^{-1} \frac{11}{24} = 24.6°$ (1 dp) — This gets 1 mark for accuracy.

b In triangle ABC, let AC = y

By Pythagoras' theorem

$y^2 = 5^2 + 12^2$ cm^2 — This gets 1 mark for method.

$= 169$ cm^2

$y = \sqrt{169} = 13$ cm — This gets 1 mark for accuracy.

In triangle ACD, let $\angle ADC = z$

So, $\sin z = \frac{13}{33.8} = 0.3846$ — This gets 1 mark for method.

$z = \sin^{-1} 0.3846 = 22.6°$ (1 dp) — This gets 1 mark for accuracy.

Total: 6 marks

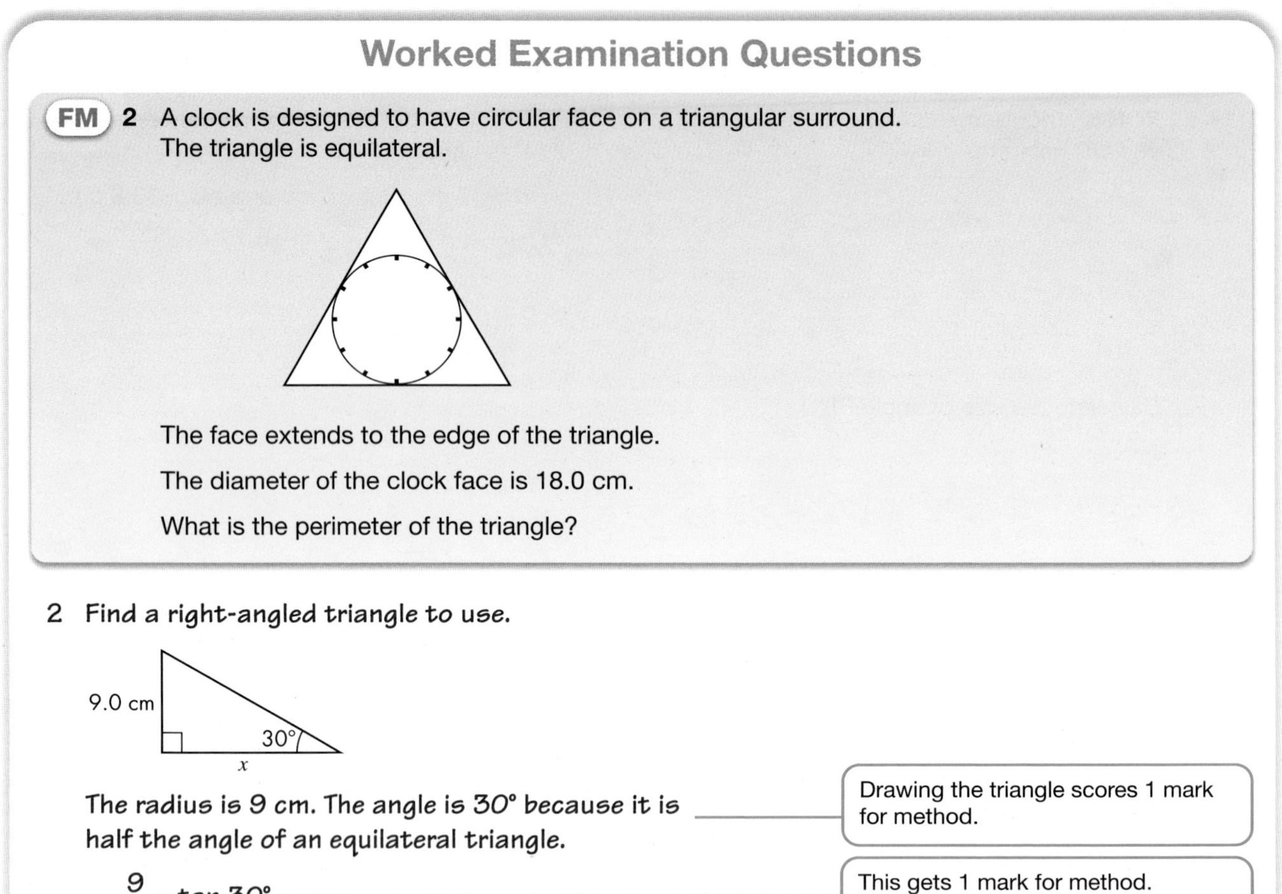

Worked Examination Questions

FM **2** A clock is designed to have circular face on a triangular surround.
The triangle is equilateral.

The face extends to the edge of the triangle.

The diameter of the clock face is 18.0 cm.

What is the perimeter of the triangle?

2 Find a right-angled triangle to use.

The radius is 9 cm. The angle is 30° because it is half the angle of an equilateral triangle. — Drawing the triangle scores 1 mark for method.

$\frac{9}{x} = \tan 30°$ — This gets 1 mark for method.

$x = \frac{9}{\tan 30°} = 15.58\ldots$ — This gets 1 mark for method and 1 mark for accuracy.

Perimeter of triangle = 15.58... x 6 = 93.5 cm or 94 cm — This gets 1 mark for accuracy.

Total: 5 marks

Worked Examination Questions

PS **3** Find the area of a regular hexagon of side 6 cm.

3 Divide the hexagon into smaller parts and add the separate areas together.
One way is to divide it into six equilateral triangles.

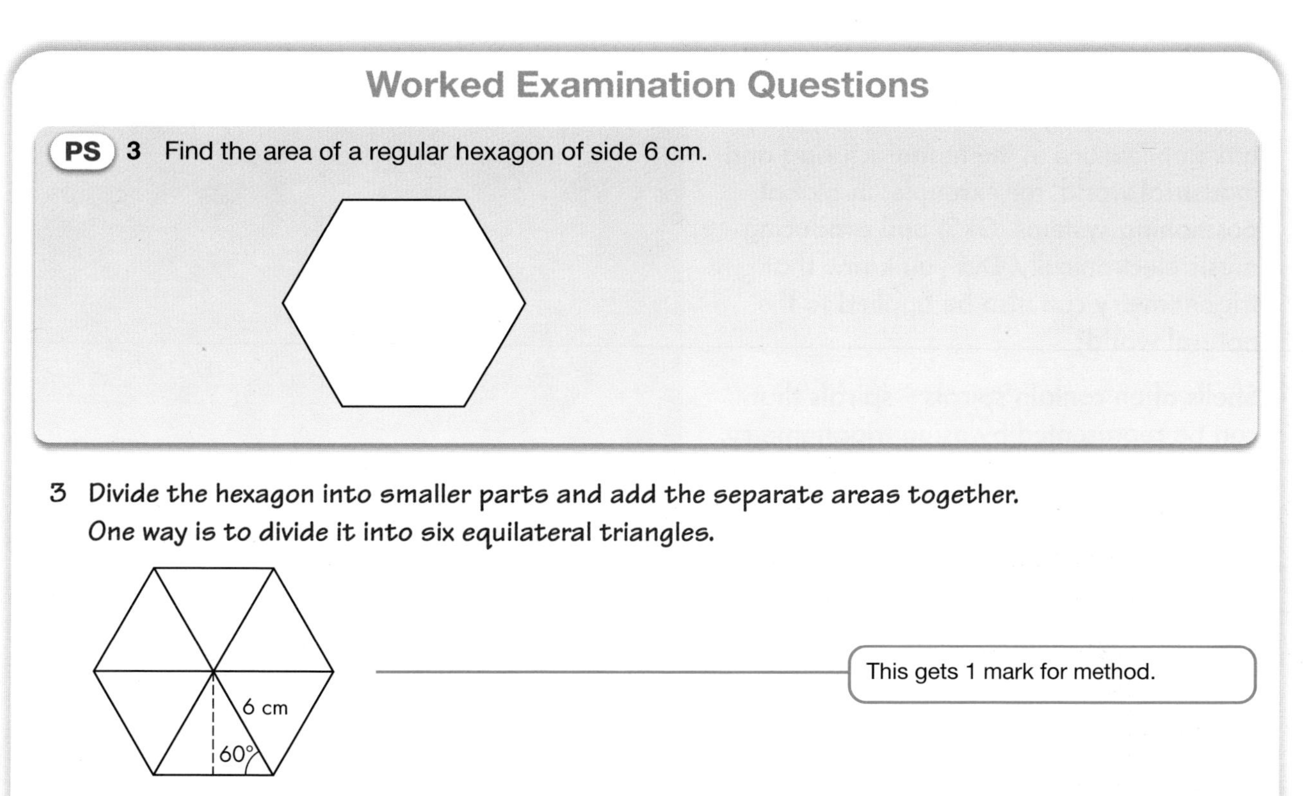

This gets 1 mark for method.

Use Pythagoras' theorem (or trigonometry) to find the height of the triangle.

For example, height = $\sqrt{(6^2 - 3^2)} = 5.19\ldots$

This gets 1 mark for method.

Area of one triangle = $3 \times 5.19\ldots = 15.58\ldots$

Area of hexagon = $6 \times$ area of triangle = 94 cm^2

This gets 1 mark for method and 1 mark for accuracy. Remember not to round your answer until the end.

Total: 4 marks

Problem Solving
Investigating spirals

You have already seen that trigonometry has applications in the manufacturing and industrial world, for example, in global positioning systems (GPS) and producing music electronically. Did you know that trigonometry can also be applied to the natural world?

Shells often contain spirals – spirals that can be represented by using trigonometry. In this task, you will investigate how to draw the spirals that you see in nature.

Getting started

Can you calculate the missing angles and length of the third side in this triangle?

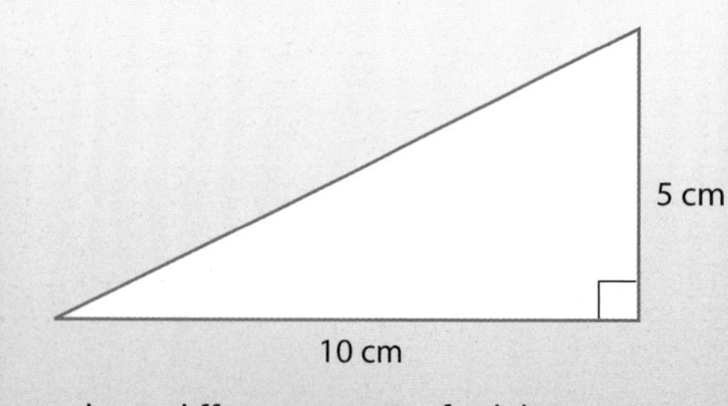

Are there different ways to find the answers?

Your task

1 Here is one way to draw a spiral, by using right-angled triangles.

On the diagram below, the blue lines are always the same length, what we shall call 1 unit. The hypotenuse of the first triangle is used as the base of the next, and so on.

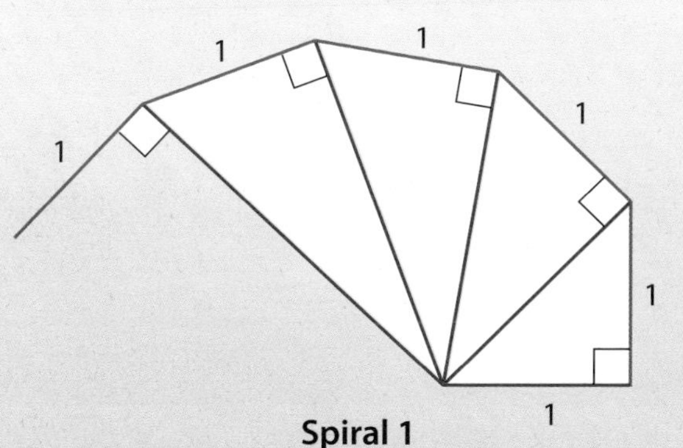

Spiral 1

Investigate how the lengths of the red lines, and the angles between them, change as the spiral grows.

How many triangles can you draw before the spiral overlaps itself?

Your task (continued)

2 An alternative way to draw a spiral is to keep the angle constant.

Start by drawing a spiral in which the angle is always 45°.

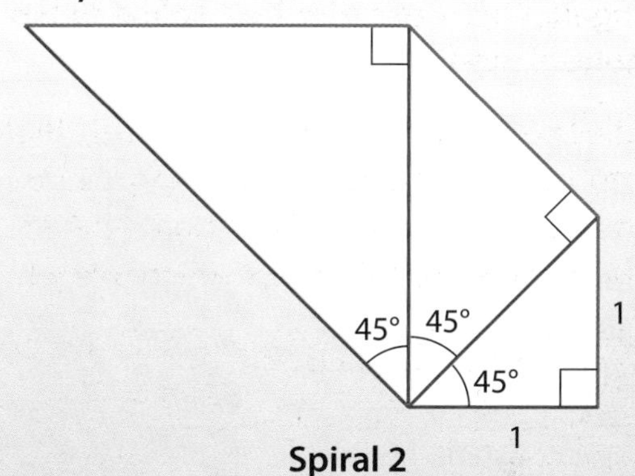

Spiral 2

In this case, how does the length of the blue line change as the spiral grows?

How does this differ from the spiral in part 1 of this task?

Investigate what happens if you use a different constant angle to draw your spiral.

Write a mathematical report of your investigation, describing your method and stating your conclusions. Remember to justify your explanation.

Why this chapter matters

A circle is a simple shape made up of all the points that are all the same distance from a given point. The given point is called the centre and the common distance of the points from the centre is called the radius.

Circles

Humankind has known about circles since before the earliest recorded history. The origins of the discovery that dividing the circumference of a circle by its diameter always gives a constant, known as π (pi), reach so far back in time that nobody can say when the rule was first stated, though the Babylonians understood circles and derived a number very close to π. However, it has been known and applied generally all over the world.

Circles appear everywhere in nature and in architecture, from the giant *Victoria regina* lily, to Stonehenge, the stone circle that was erected in around 2400–2200 BC.

Circles feature in all sorts of human activity, from making pottery and clocks to wheels. The most primitive wheels were probably just logs which simply rolled under the object being moved; one of the largest wheels currently is the London Eye. Engineering owes a lot to the study of circles, since gears and pulleys rely on them. More recently, circles are used in what some see as an artform, others as vandalism, the formation of crop circles! And in philosophy, circles are seen as representing a whole as well as the completion of a cycle.

The Earth itself is a sphere, divided into zones by lines of latitude and longitude, which are imaginary lines along the surface of the planet.

There is a famous story about the thirteenth-century Italian artist, Giotto. Pope Boniface wanted to find the best painter in Italy to work on some paintings for St Peter's in Rome. When Giotto was asked for a sample of his work, he took a brush and, in one even sweep, drew a perfect circle on a sheet of paper. The Pope's messenger was angry, but took the drawing and showed the Pope, explaining how Giotto had drawn it, freehand. Giotto got the commission.

Euclid, who lived around 300 BC, is famous for his study of geometry, including many propositions or theorems based on circles. In this chapter, you will study some of those theorems.

Chapter 4

Geometry: Properties of circles

Circle theorems

Cyclic quadrilaterals

Tangents and chords

Alternate segment theorem

The grades given in this chapter are target grades.

This chapter will show you ...

B to A* how to find angles using circle theorems

B to A* how to find angles in cyclic quadrilaterals

B to A* how to use tangents, chords and alternate segment theorem to find angles in circles

Visual overview

2D shapes → Circles → Circle theorems

What you should already know

- The three interior angles of a triangle add up to 180°. So, $a + b + c = 180°$
  (KS3 level 5, GCSE grade E)

- The four interior angles of a quadrilateral quadrilateral add up to 360°.
  So, $a + b + c + d = 360°$
  (KS3 level 5, GCSE grade E)

- Angles in parallel lines

a and b are equal	a and b are equal	$a + b = 180°$
a and b are alternate angles	a and b are corresponding angles	a and b are allied angles

(KS3 level 6, GCSE grade D)

continued

- Circle terms

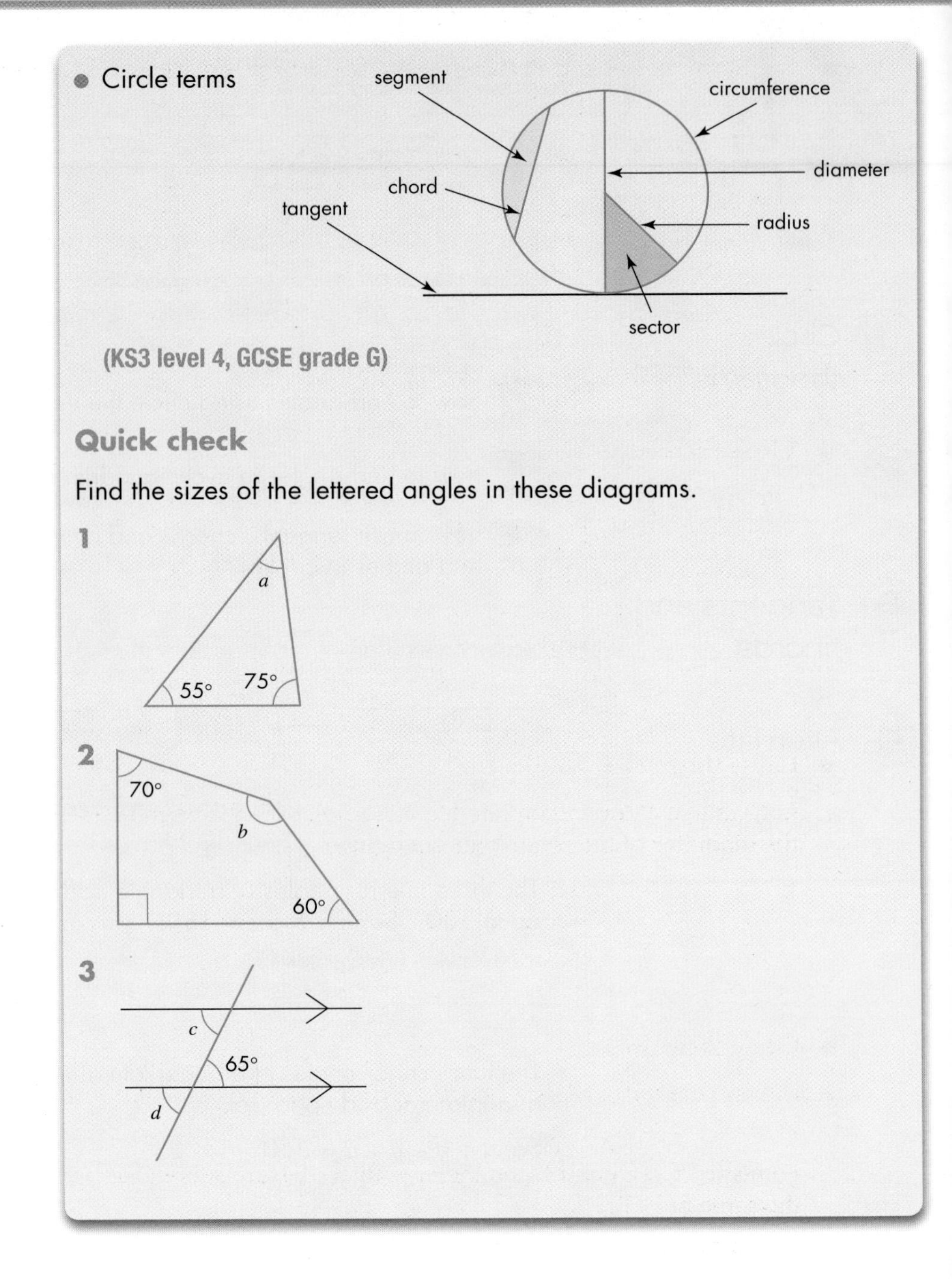

(KS3 level 4, GCSE grade G)

Quick check

Find the sizes of the lettered angles in these diagrams.

1

2

3

4.1 Circle theorems

This section will show you how to:

- work out the sizes of angles in circles

Key words

arc, circle, circumference, diameter, segment, semicircle, subtended

Here are three **circle** theorems you need to know.

- **Circle theorem 1**

  The angle at the centre of a circle is twice the angle at the **circumference** that is **subtended** by the same **arc**.

  $\angle AOB = 2 \times \angle ACB$

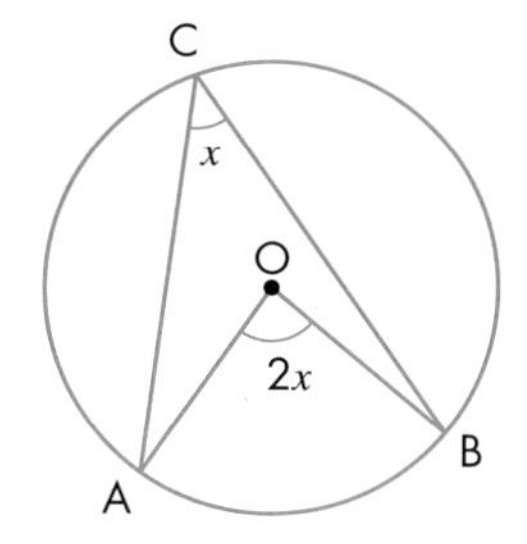

- **Circle theorem 2**

  Every angle at the circumference of a **semicircle** that is subtended by the **diameter** of the semicircle is a right angle.

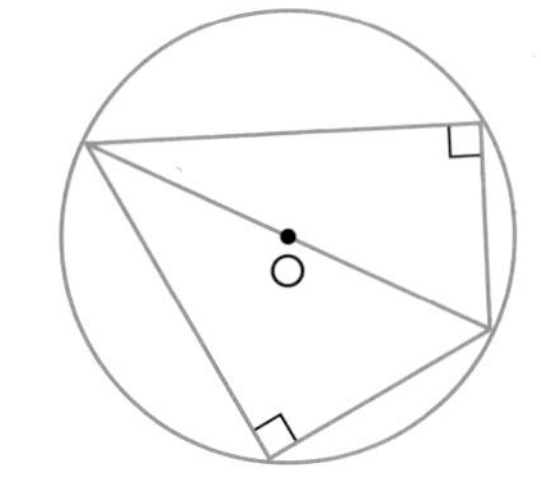

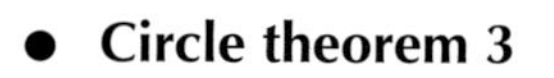

- **Circle theorem 3**

  Angles subtended at the circumference in the same **segment** of a circle are equal.

  Points C_1, C_2, C_3 and C_4 on the circumference are subtended by the same arc AB.

  So $\angle AC_1B = \angle AC_2B = \angle AC_3B = \angle AC_4B$

  Follow through Examples 1–3 to see how these theorems are applied.

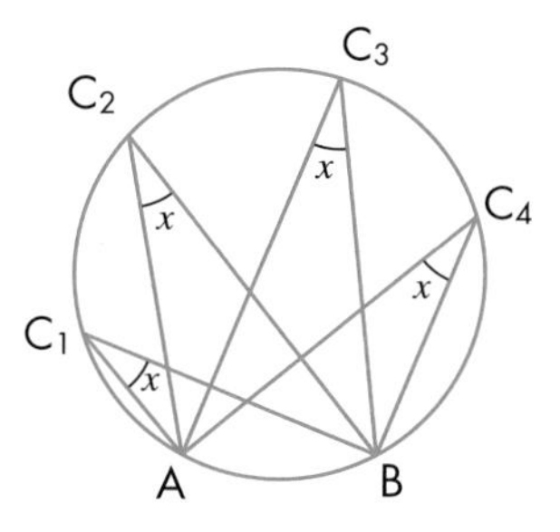

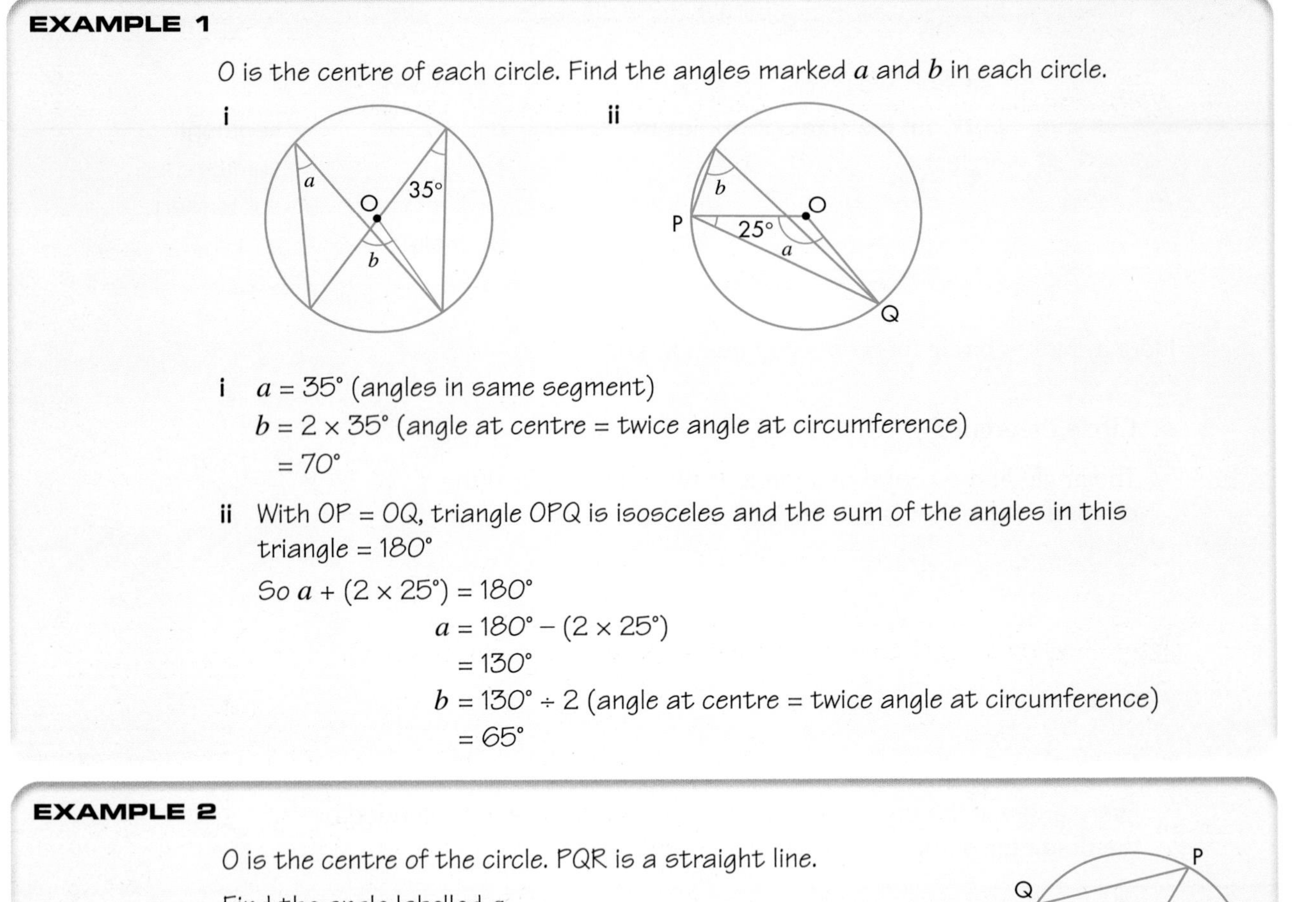

EXAMPLE 1

O is the centre of each circle. Find the angles marked a and b in each circle.

i

ii

i $a = 35°$ (angles in same segment)

$b = 2 \times 35°$ (angle at centre = twice angle at circumference)

$= 70°$

ii With OP = OQ, triangle OPQ is isosceles and the sum of the angles in this triangle = 180°

So $a + (2 \times 25°) = 180°$

$a = 180° - (2 \times 25°)$

$= 130°$

$b = 130° \div 2$ (angle at centre = twice angle at circumference)

$= 65°$

EXAMPLE 2

O is the centre of the circle. PQR is a straight line.

Find the angle labelled a.

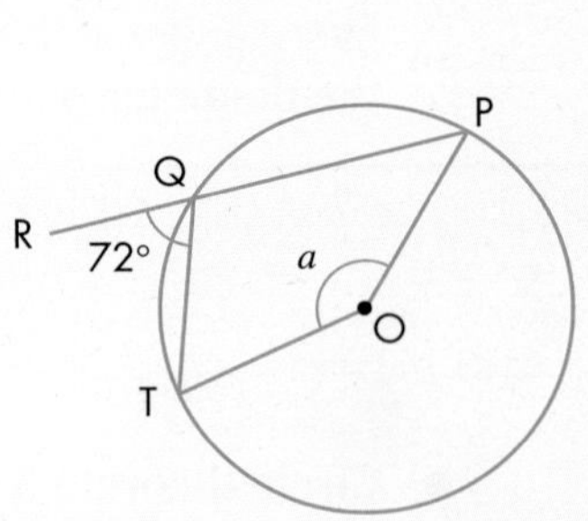

$\angle PQT = 180° - 72° = 108°$ (angles on straight line)

The reflex angle $\angle POT = 2 \times 108°$
(angle at centre = twice angle at circumference)

$= 216°$

$a + 216° = 360°$ (sum of angles around a point)

$a = 360° - 216°$

$a = 144°$

EXAMPLE 3

O is the centre of the circle. POQ is parallel to TR.

Find the angles labelled a and b.

$a = 64° \div 2$ (angle at centre = twice angle at circumference)

$a = 32°$

$\angle TQP = a$ (alternate angles)

$= 32°$

$\angle PTQ = 90°$ (angle in a semicircle)

$b + 90° + 32° = 180°$ (sum of angles in $\triangle PQT$)

$b = 180° - 122°$

$b = 58°$

EXERCISE 4A

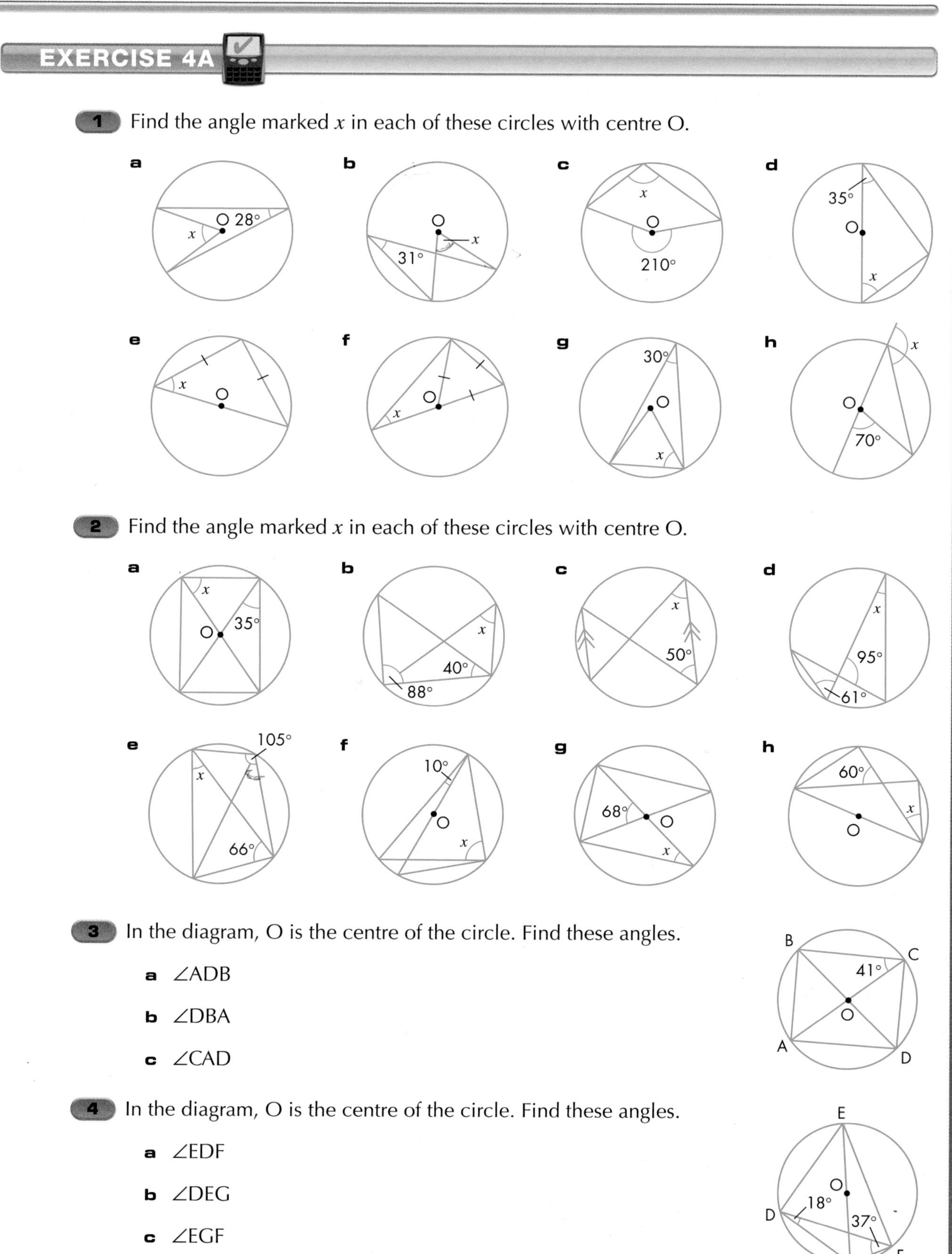

1 Find the angle marked x in each of these circles with centre O.

2 Find the angle marked x in each of these circles with centre O.

3 In the diagram, O is the centre of the circle. Find these angles.

a ∠ADB

b ∠DBA

c ∠CAD

4 In the diagram, O is the centre of the circle. Find these angles.

a ∠EDF

b ∠DEG

c ∠EGF

AU 5 In the diagram XY is a diameter of the circle and ∠AZX is a.

Ben says that the value of a is 55°.

Give reasons to explain why he is wrong.

6 Find the angles marked x and y in each of these circles. O is the centre where shown.

a **b** **c**

d **e** **f**

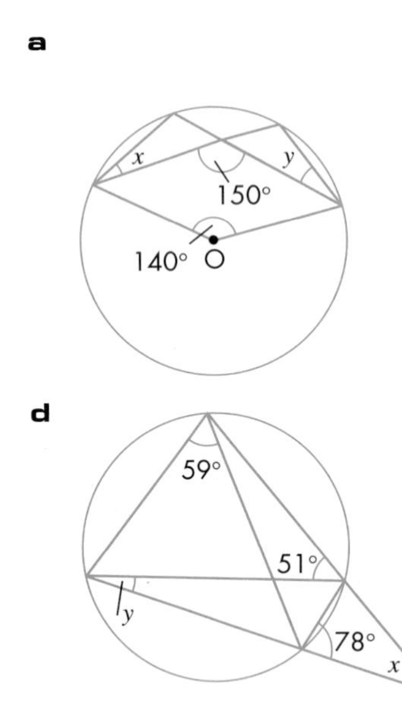

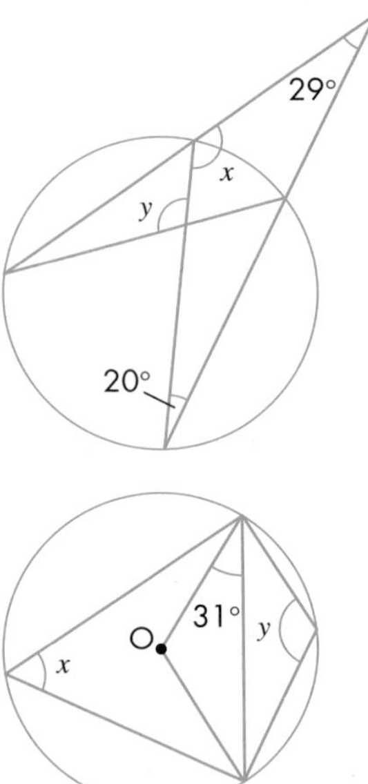

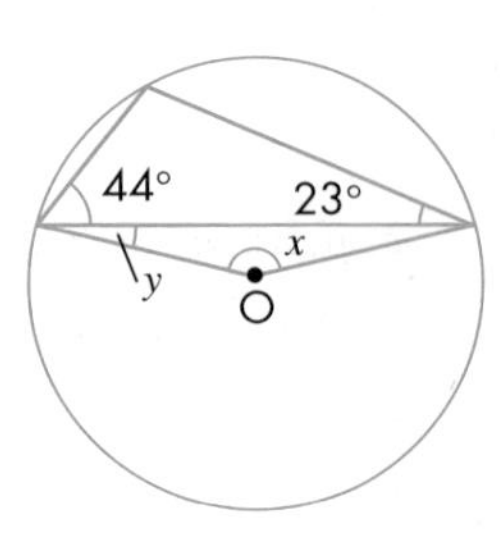

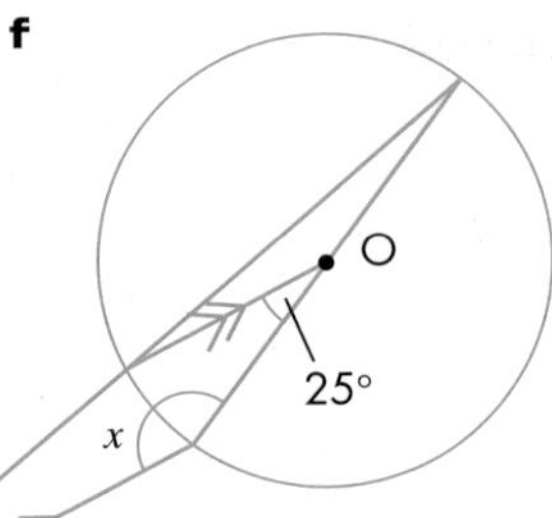

7 In the diagram, O is the centre and AD a diameter of the circle. Find x.

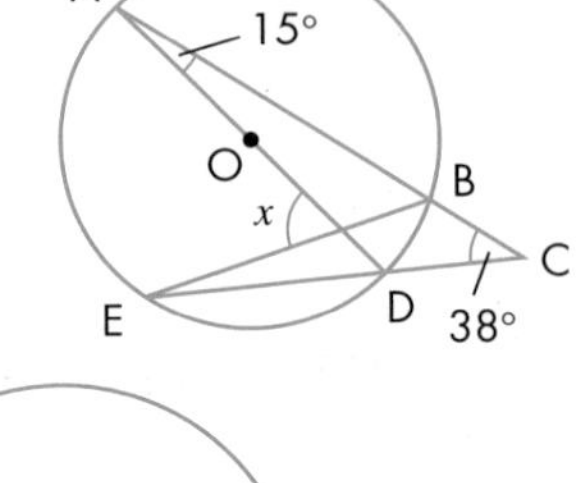

PS 8 In the diagram, O is the centre of the circle and ∠CBD is x.

Show that the reflex ∠AOC is $2x$, giving reasons to explain your answer.

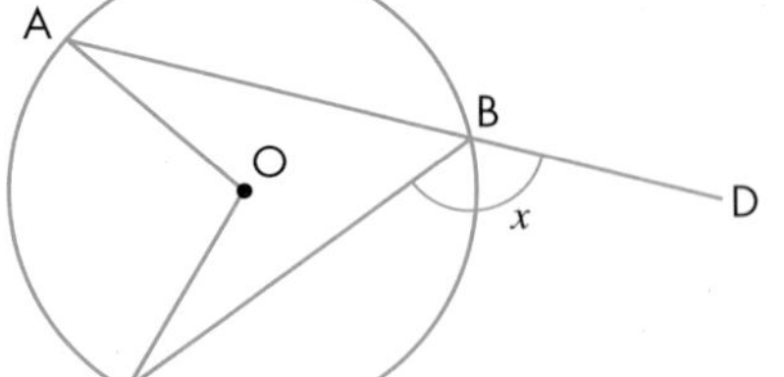

9 A, B, C and D are points on the circumference of a circle with centre O.
Angle ABO is x° and angle CBO is y°.

a State the value of angle BAO.

b State the value of angle AOD.

c Prove that the angle subtended by the chord AC at the centre of a circle is twice the angle subtended at the circumference.

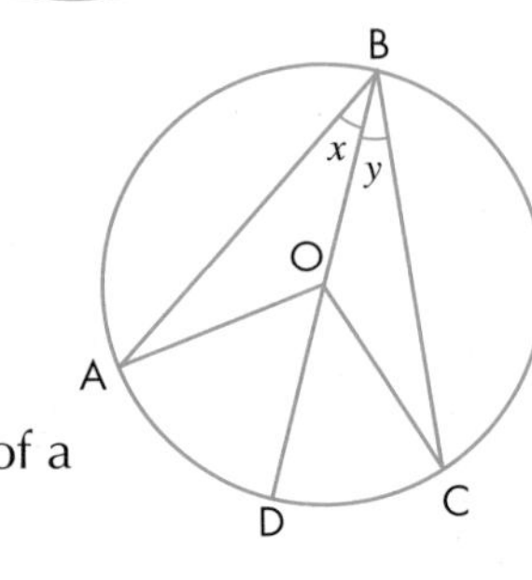

4.2 Cyclic quadrilaterals

This section will show you how to:
- find the sizes of angles in cyclic quadrilaterals

Key words
cyclic quadrilateral

A quadrilateral whose four vertices lie on the circumference of a circle is called a **cyclic quadrilateral**.

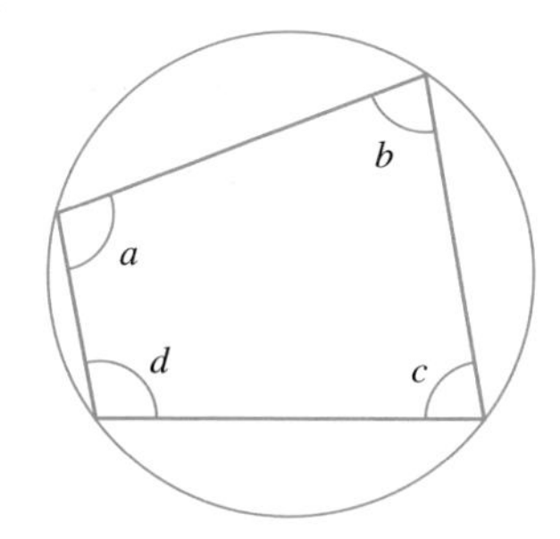

- **Circle theorem 4**

  The sum of the opposite angles of a cyclic quadrilateral is 180°.

  $a + c = 180°$ and $b + d = 180°$

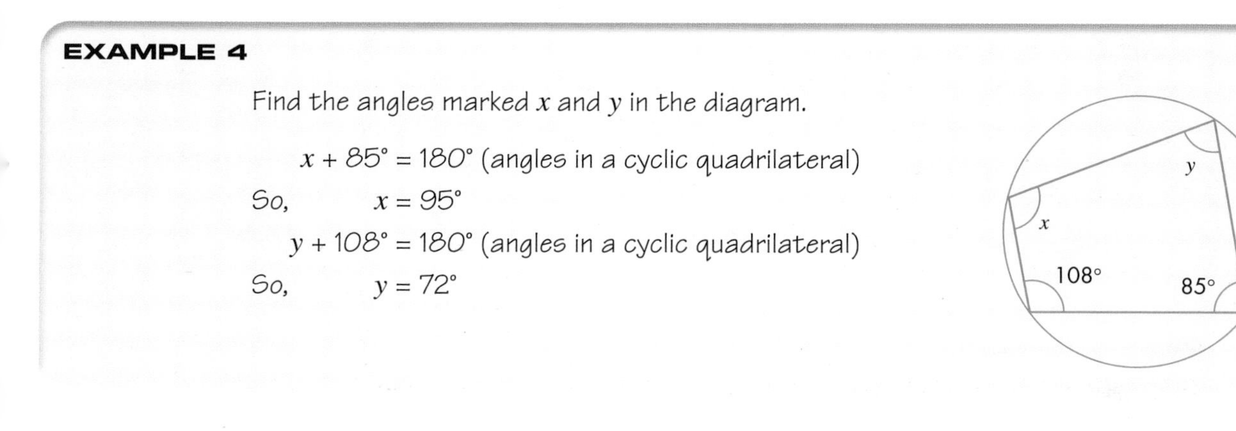

EXAMPLE 4

Find the angles marked x and y in the diagram.

$x + 85° = 180°$ (angles in a cyclic quadrilateral)

So, $x = 95°$

$y + 108° = 180°$ (angles in a cyclic quadrilateral)

So, $y = 72°$

EXERCISE 4B

1 Find the sizes of the lettered angles in each of these circles.

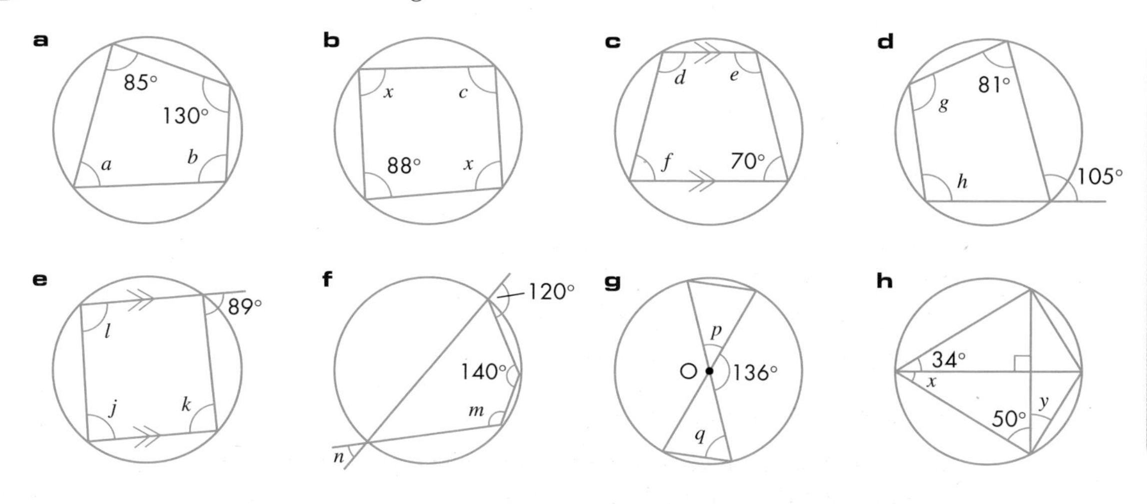

2 Find the values of x and y in each of these circles. Where shown, O marks the centre of the circle.

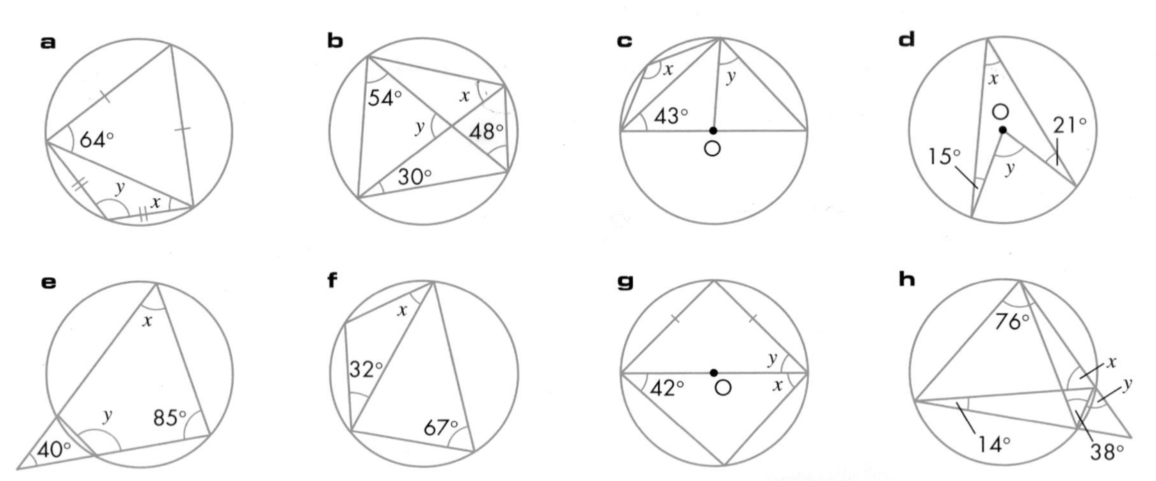

3 Find the values of x and y in each of these circles. Where shown, O marks the centre of the circle.

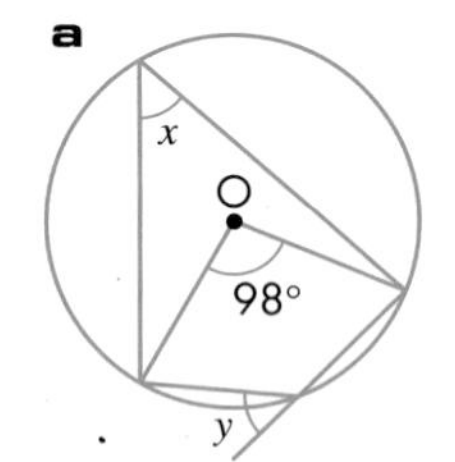

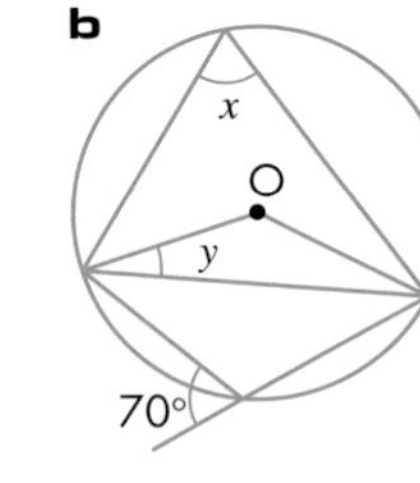

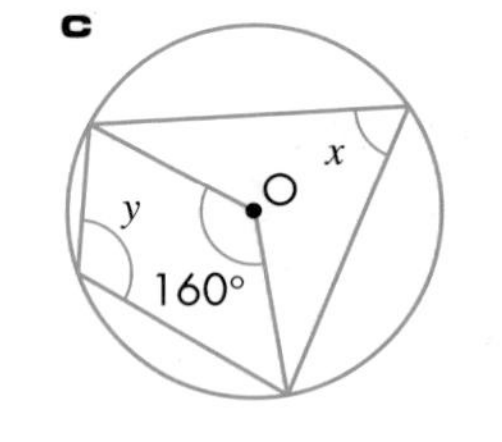

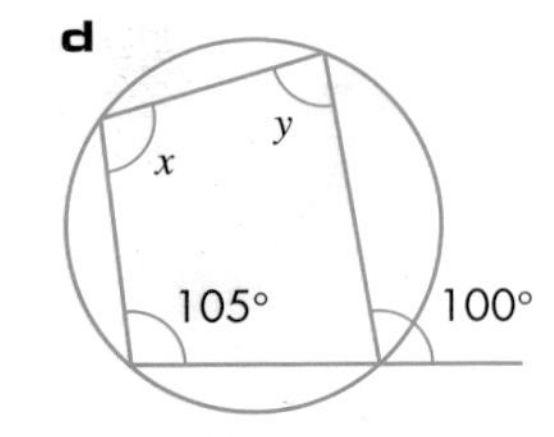

4 Find the values of x and y in each of these circles.

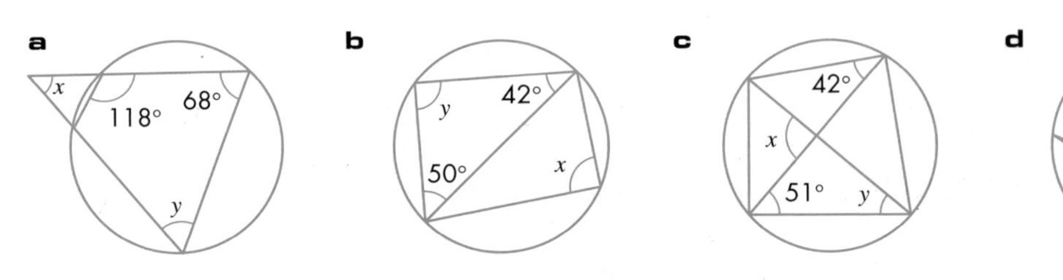

5 Find the values of x and y in each of these circles with centre O.

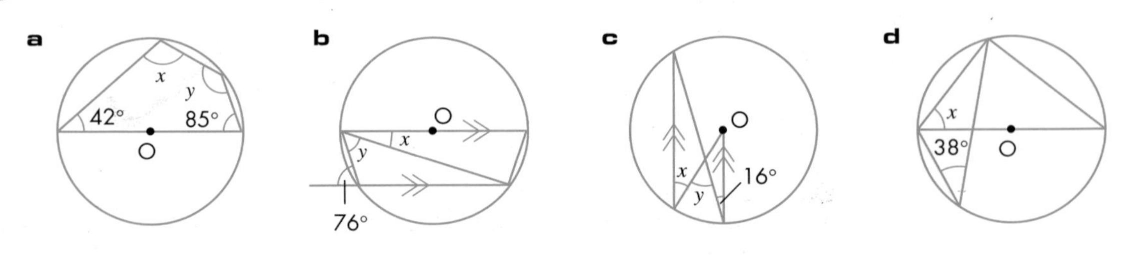

PS 6 The cyclic quadrilateral PQRT has ∠ROQ equal to 38° where O is the centre of the circle. POT is a diameter and parallel to QR. Calculate these angles.

a ∠ROT **b** ∠QRT **c** ∠QPT

AU **7** In the diagram, O is the centre of the circle.

a Explain why $3x - 30° = 180°$.

b Work out the size of $\angle CDO$, marked y on the diagram.

Give reasons in your working.

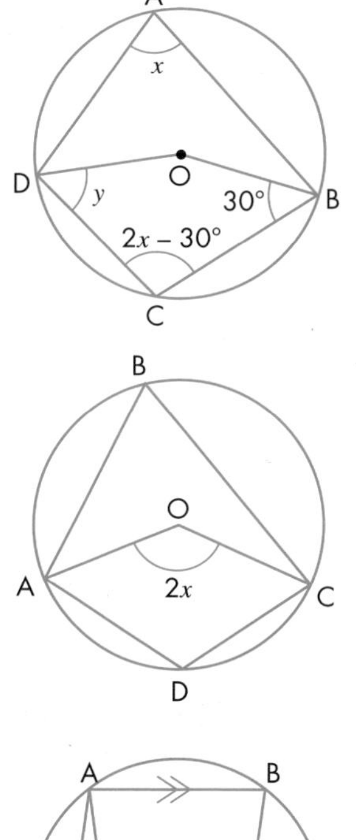

8 ABCD is a cyclic quadrilateral within a circle centre O and $\angle AOC$ is $2x°$.

a Write down the value of $\angle ABC$.

b Write down the value of the reflex angle AOC.

c Prove that the sum of a pair of opposite angles of a cyclic quadrilateral is 180°.

PS **9** In the diagram, ABCE is a parallelogram.

Prove $\angle AED = \angle ADE$.

Give reasons in your working.

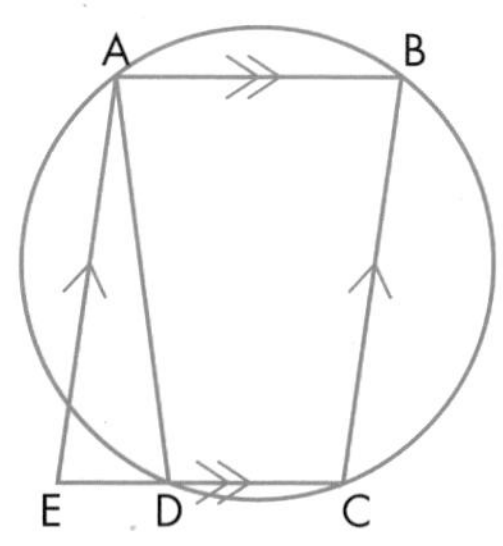

4.3 Tangents and chords

This section will show you how to:

- use tangents and chords to find the sizes of angles in circles

Key words
chord
point of contact
radius
tangent

A **tangent** is a straight line that touches a circle at one point only. This point is called the **point of contact**. A **chord** is a line that joins two points on the circumference.

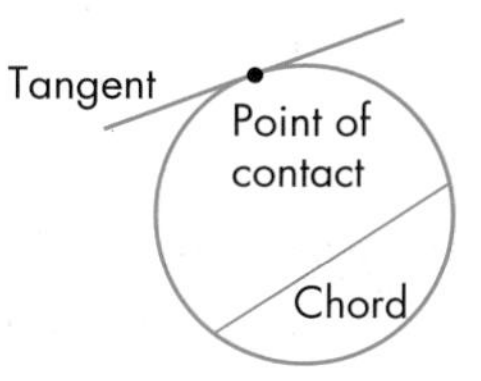

- **Circle theorem 5**

  A tangent to a circle is perpendicular to the **radius** drawn to the point of contact.

  The radius OX is perpendicular to the tangent AB.

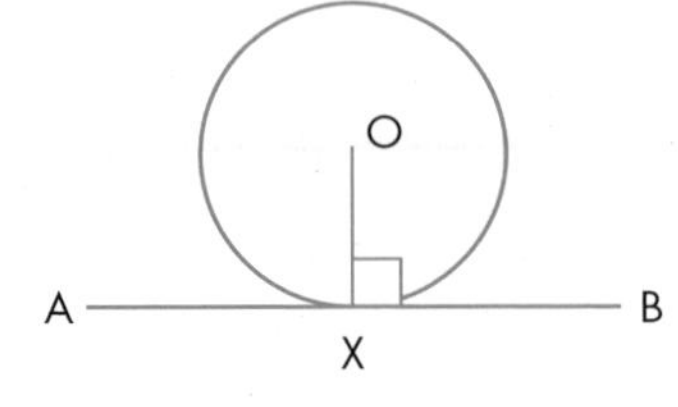

- **Circle theorem 6**

  Tangents to a circle from an external point to the points of contact are equal in length.

  AX = AY

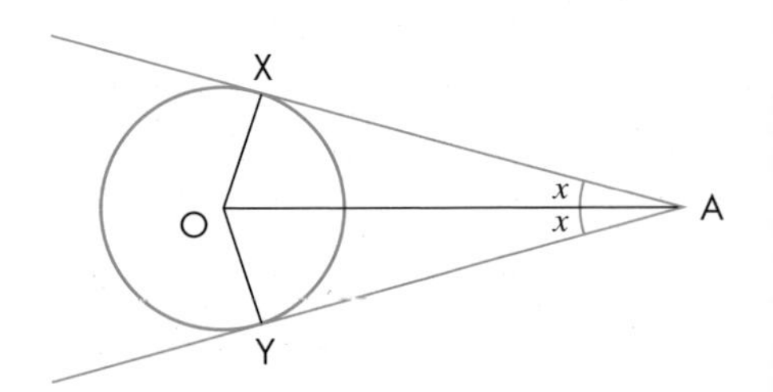

- **Circle theorem 7**

  The line joining an external point to the centre of the circle bisects the angle between the tangents.

  $\angle OAX = \angle OAY$

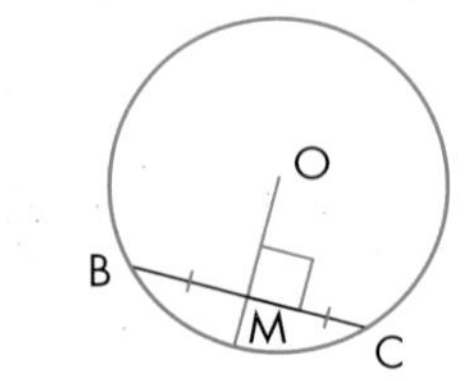

- **Circle theorem 8**

  A radius bisects a chord at 90°.

  If O is the centre of the circle,

  $\angle BMO = 90°$ and BM = CM

EXAMPLE 5

OA is the radius of the circle and AB is a tangent.

OA = 5 cm and AB = 12 cm

Calculate the length OB.

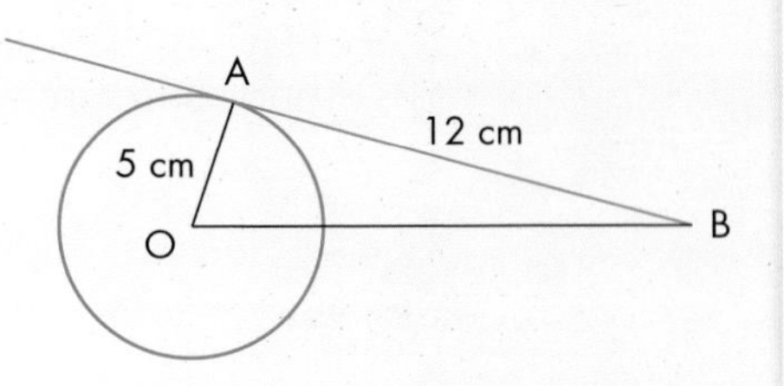

$\angle OAB = 90°$ (radius is perpendicular to a tangent)

Let OB = x

By Pythagoras' theorem,

$x^2 = 5^2 + 12^2$ cm^2

$x^2 = 169$ cm^2

So $x = \sqrt{169} = 13$ cm

EXERCISE 4C

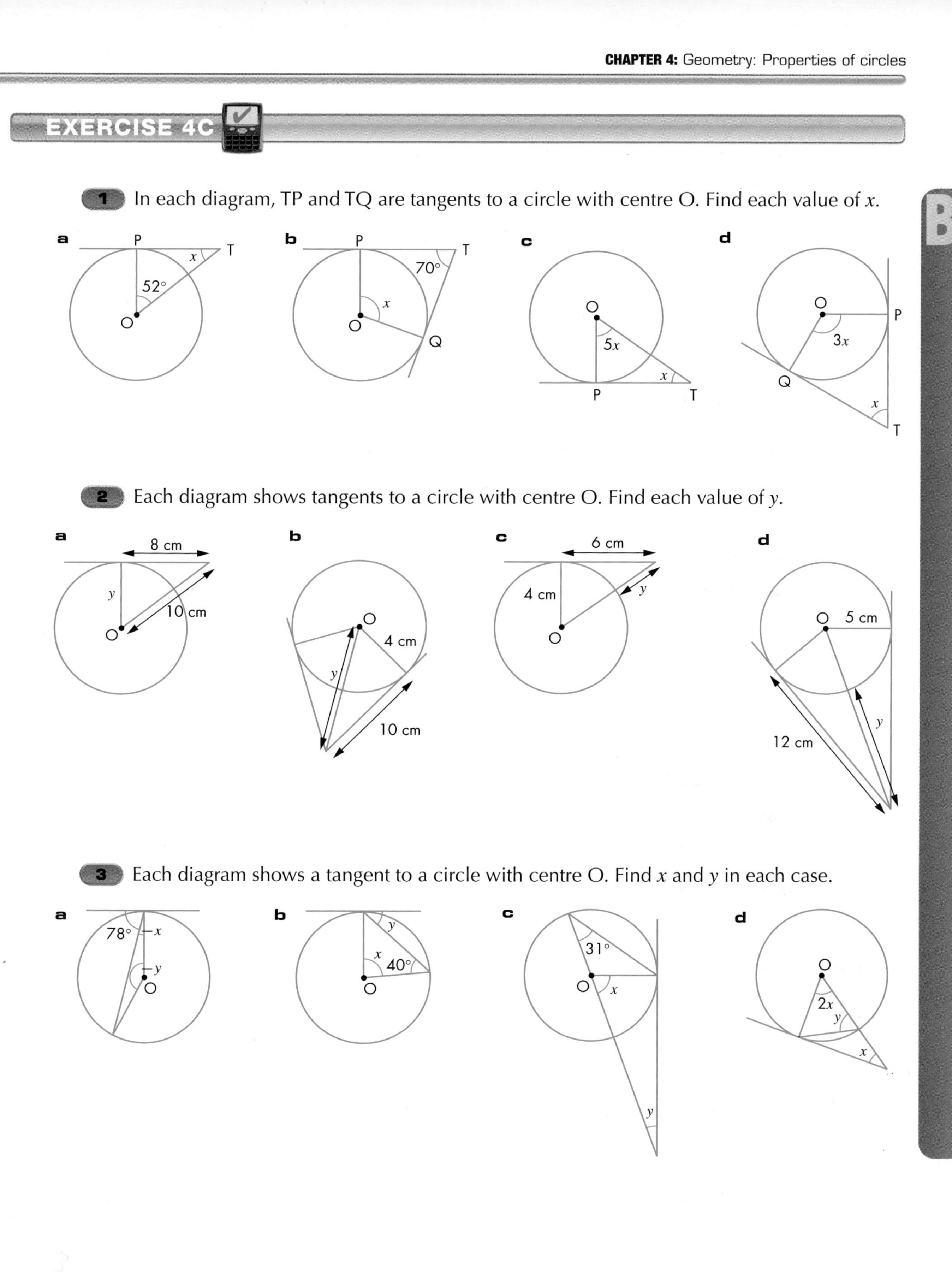

1 In each diagram, TP and TQ are tangents to a circle with centre O. Find each value of x.

2 Each diagram shows tangents to a circle with centre O. Find each value of y.

3 Each diagram shows a tangent to a circle with centre O. Find x and y in each case.

B

4 In each of the diagrams, TP and TQ are tangents to the circle with centre O. Find each value of x.

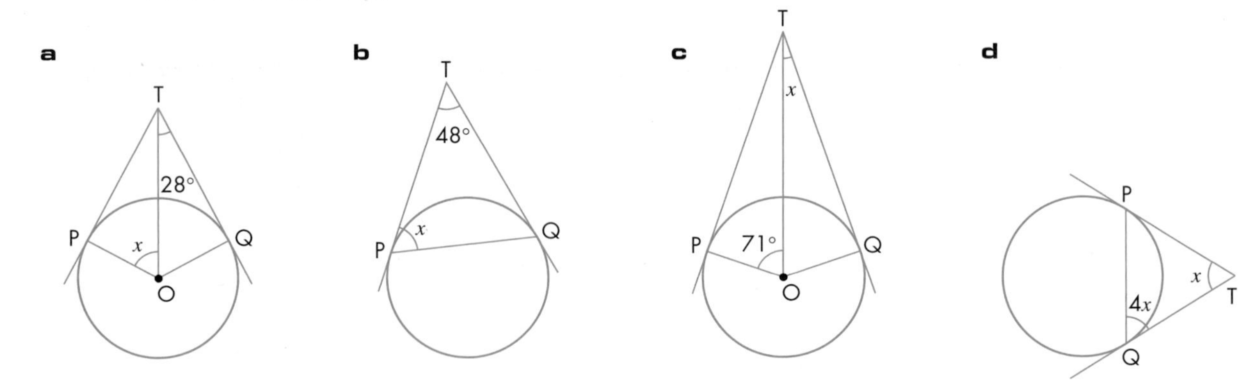

A

PS 5 Two circles with the same centre have radii of 7 cm and 12 cm respectively. A tangent to the inner circle cuts the outer circle at A and B. Find the length of AB.

PS 6 The diagram shows a circle with centre O.
The circle fits exactly inside an equilateral triangle XYZ.
The lengths of the sides of the triangle are 20 cm.

Work out the radius of the circle.

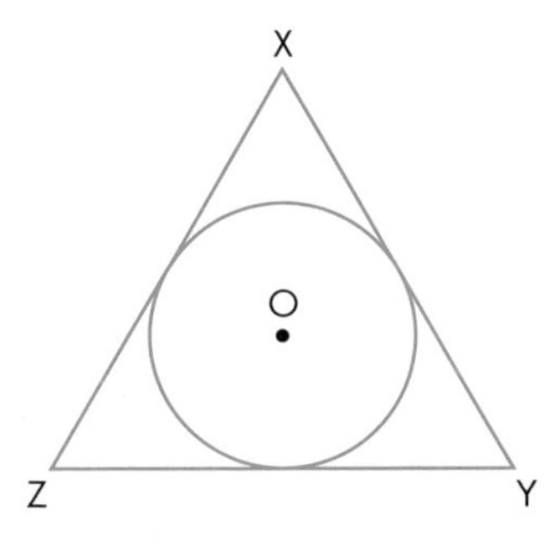

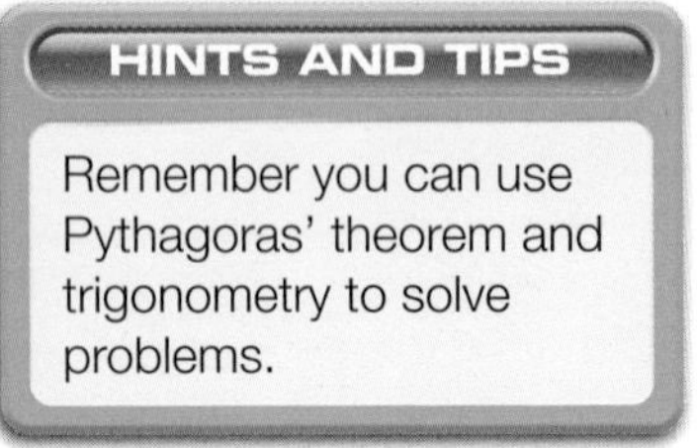

AU 7 In the diagram, O is the centre of the circle and AB is a tangent to the circle at C.

Explain why triangle BCD is isosceles.

Give reasons to justify your answer.

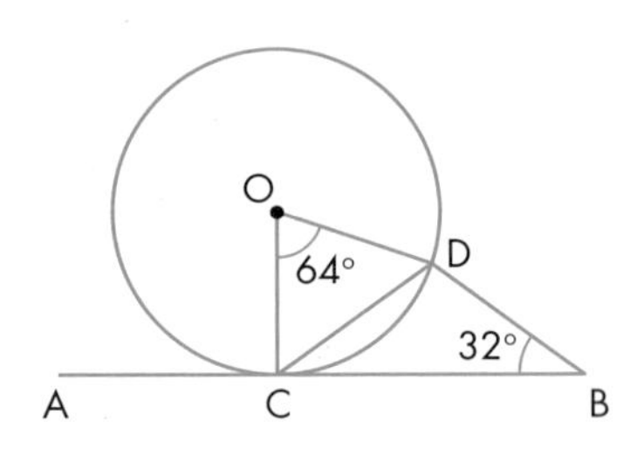

A*

8 AB and CB are tangents from B to the circle with centre O. OA and OC are radii.

a Prove that angles AOB and COB are equal.

b Prove that OB bisects the angle ABC.

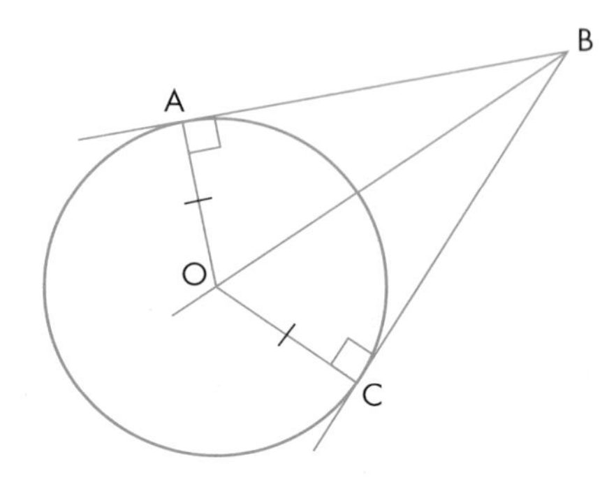

4.4 Alternate segment theorem

This section will show you how to:
- use the alternate segment theorem to find the sizes of angles in circles

Key words
alternate segment
chord
tangent

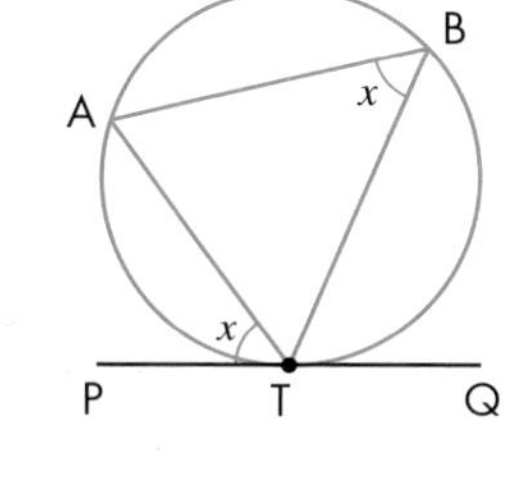

PTQ is the **tangent** to a circle at T. The segment containing ∠TBA is known as the **alternate segment** of ∠PTA, because it is on the other side of the **chord** AT from ∠PTA.

- **Circle theorem 9**

The angle between a tangent and a chord through the point of contact is equal to the angle in the alternate segment.

∠PTA = ∠TBA

EXAMPLE 6

In the diagram, find **a** ∠ATS and **b** ∠TSR.

a ∠ATS = 80° (angle in alternate segment)

b ∠TSR = 70° (angle in alternate segment)

EXERCISE 4D

1 Find the size of each lettered angle.

a **b** **c** **d**

2 In each diagram, find the size of each lettered angle.

a **b** **c** **d**

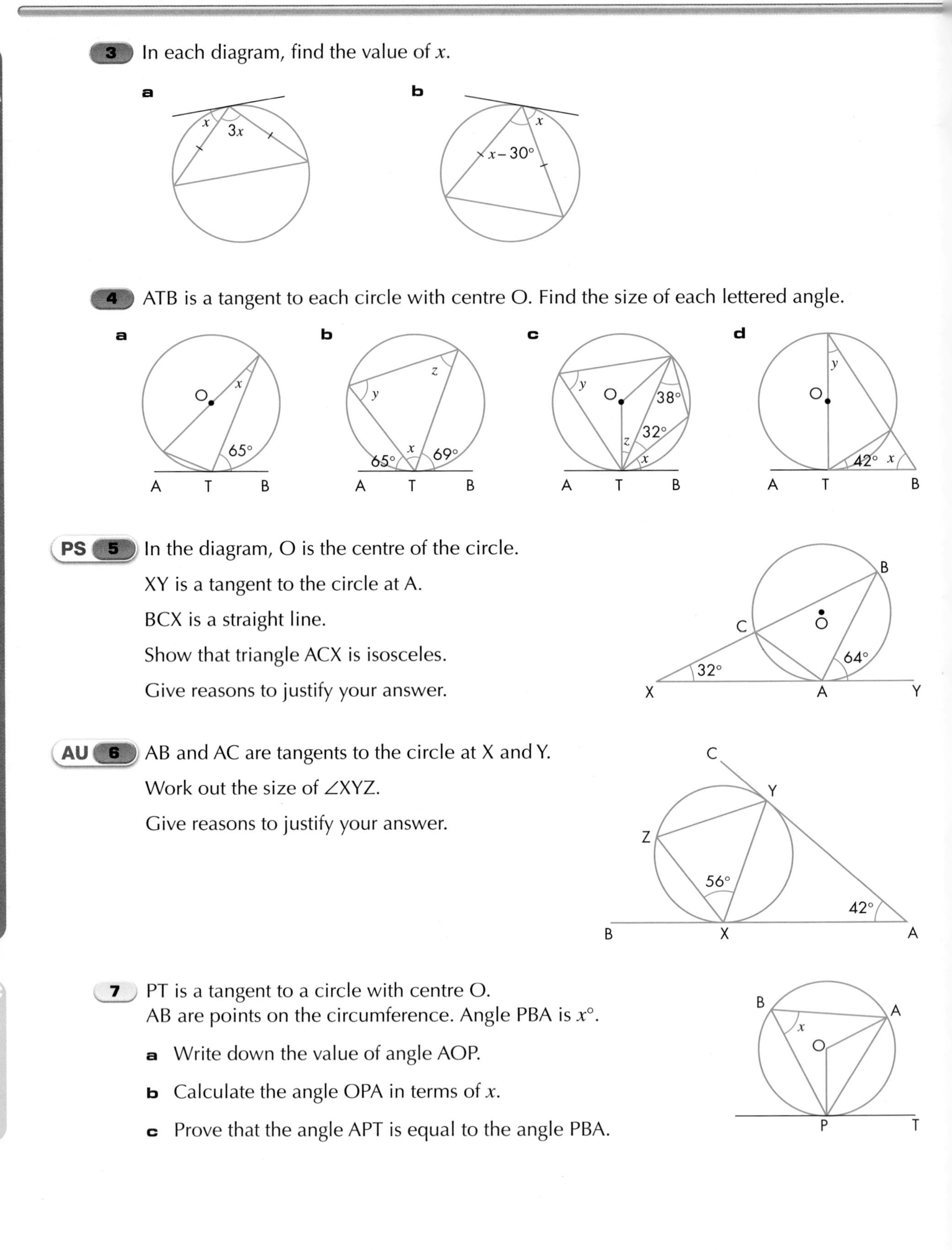

3 In each diagram, find the value of x.

4 ATB is a tangent to each circle with centre O. Find the size of each lettered angle.

PS 5 In the diagram, O is the centre of the circle.

XY is a tangent to the circle at A.

BCX is a straight line.

Show that triangle ACX is isosceles.

Give reasons to justify your answer.

AU 6 AB and AC are tangents to the circle at X and Y.

Work out the size of ∠XYZ.

Give reasons to justify your answer.

7 PT is a tangent to a circle with centre O.
AB are points on the circumference. Angle PBA is $x°$.

a Write down the value of angle AOP.

b Calculate the angle OPA in terms of x.

c Prove that the angle APT is equal to the angle PBA.

GRADE BOOSTER

- **B** You can find angles in circles using tangents and chords
- **A** You can find angles in cyclic quadrilaterals
- **A** You can find angles in circles, using the alternate segment theorem
- **A*** You can use circle theorems to prove geometrical results

What you should know now

- How to use circle theorems to find angles
- How to find angles in cyclic quadrilaterals
- How to use tangents, chords and alternate segment theorem to find angles in circles

1

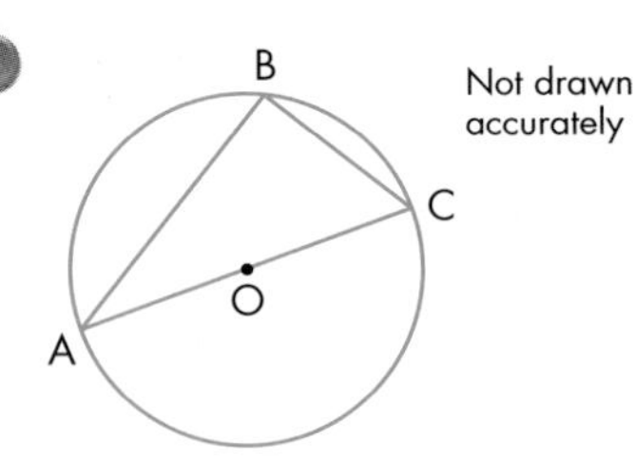

A, B and C are points on the circumference of a circle, centre O.

AC is a diameter of the circle.

a **i** Write down the size of angle ABC.

ii Give a reason for your answer. *(2 marks)*

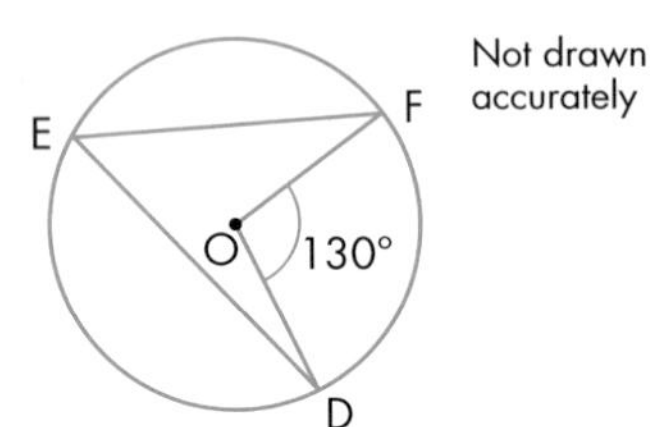

D, E and F are points on the points on the circumference of a circle, centre O.

Angle DOF = 130°.

b **i** Work out the size of angle DEF.

ii Give a reason for your answer. *(2 marks)*

Edexcel, November 2008, Paper 14 Higher, Question 13(a, b)

2

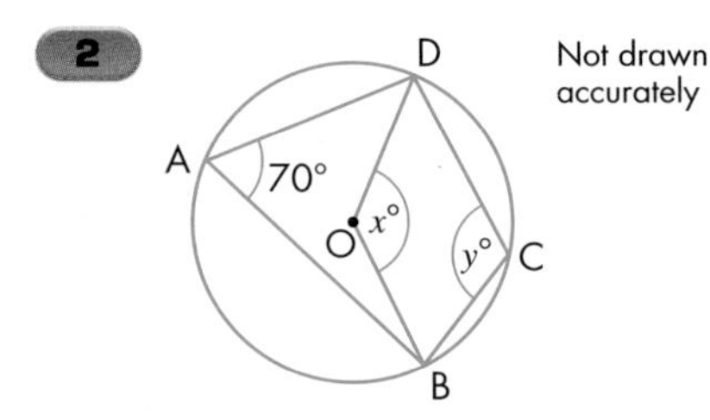

In the diagram, A, B, C and D are points on the circumference of a circle, centre O.

Angle BAD = 70°.
Angle BOD = x.
Angle BCD = y.

a **i** Work out the value of x.

ii Give a reason for your answer. *(2 marks)*

b **i** Work out the value of y.

ii Give a reason for your answer. *(2 marks)*

Edexcel, May 2008, Paper 14 Higher, Question 10(a, b)

3

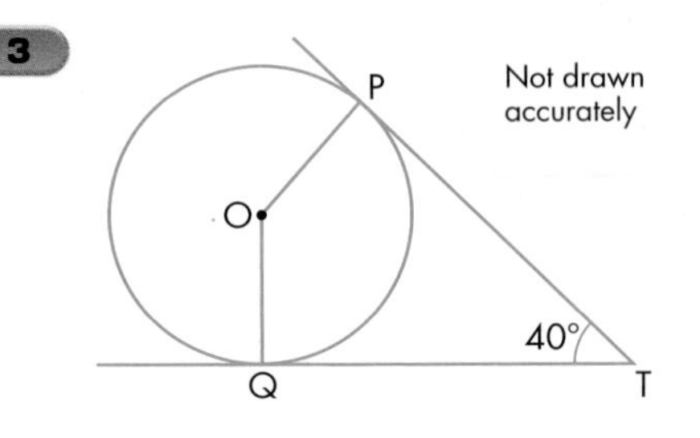

P and Q are two points on a circle, centre O.

The tangents to the circle at P and Q intersect at the point T.

a Write down the size of angle OQT. *(1 mark)*

b Calculate the size of the obtuse angle POQ. *(2 marks)*

c Give reasons why angle PQT is 70°. *(2 marks)*

Edexcel, June 2007, Paper 11 Higher, Question 8

4

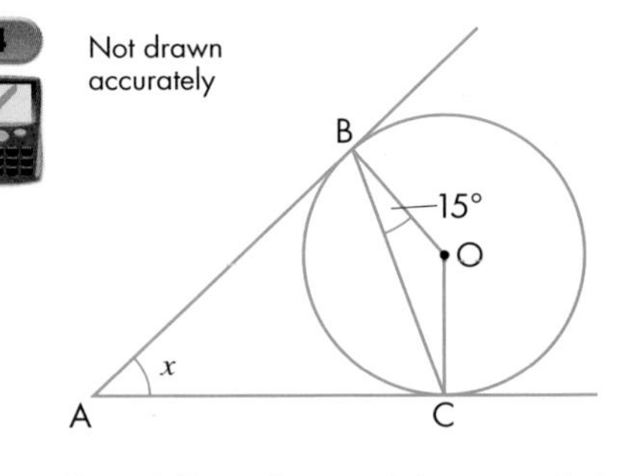

B and C are two points on a circle, centre O.

Angle OBC = 15°.

AB and AC are tangents to the circle.

a Calculate the size of the angle marked x. *(2 marks)*

b Give reasons for your answer. *(2 marks)*

Edexcel, June 2008, Paper 11 Higher, Question 8(a, b)

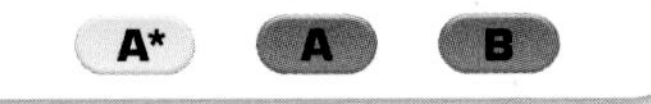

5

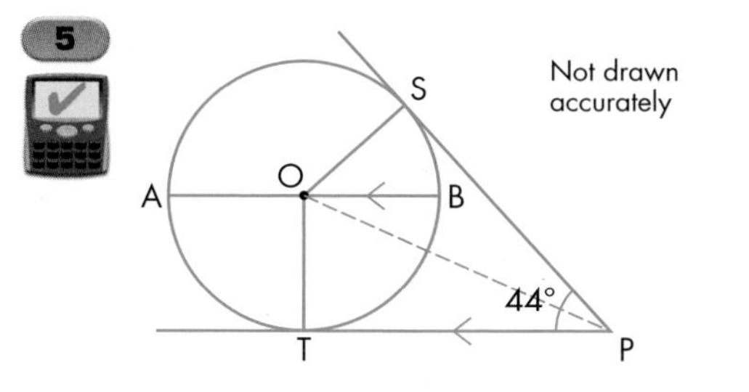

The diagram shows a circle, centre O.

A, S, B and T are points on the circumference of the circle.

PT and PS are tangents to the circle.

AB is parallel to TP.

Angle SPT = 44°.

Work out the size of angle SOB. *(4 marks)*

Edexcel, March 2009, Paper 10 Higher, Question 7

6

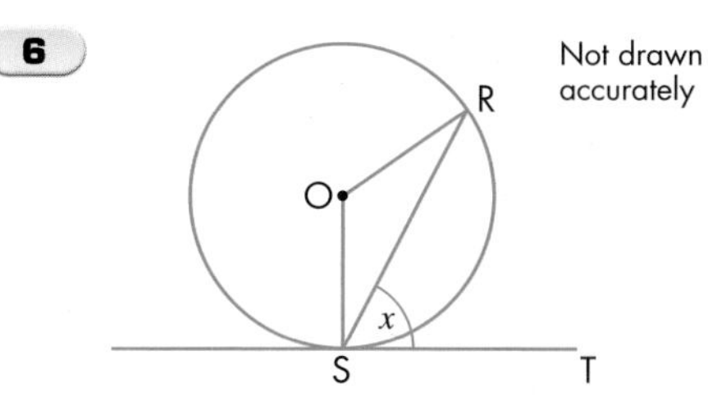

R and S are two points on a circle, centre O.

TS is a tangent to the circle.

Angle RST = x.

Prove that angle ROS = $2x$.

You must give reasons for each stage of your working. *(4 marks)*

Edexcel, March 2008, Paper 10 Higher, Question 9

Worked Examination Questions

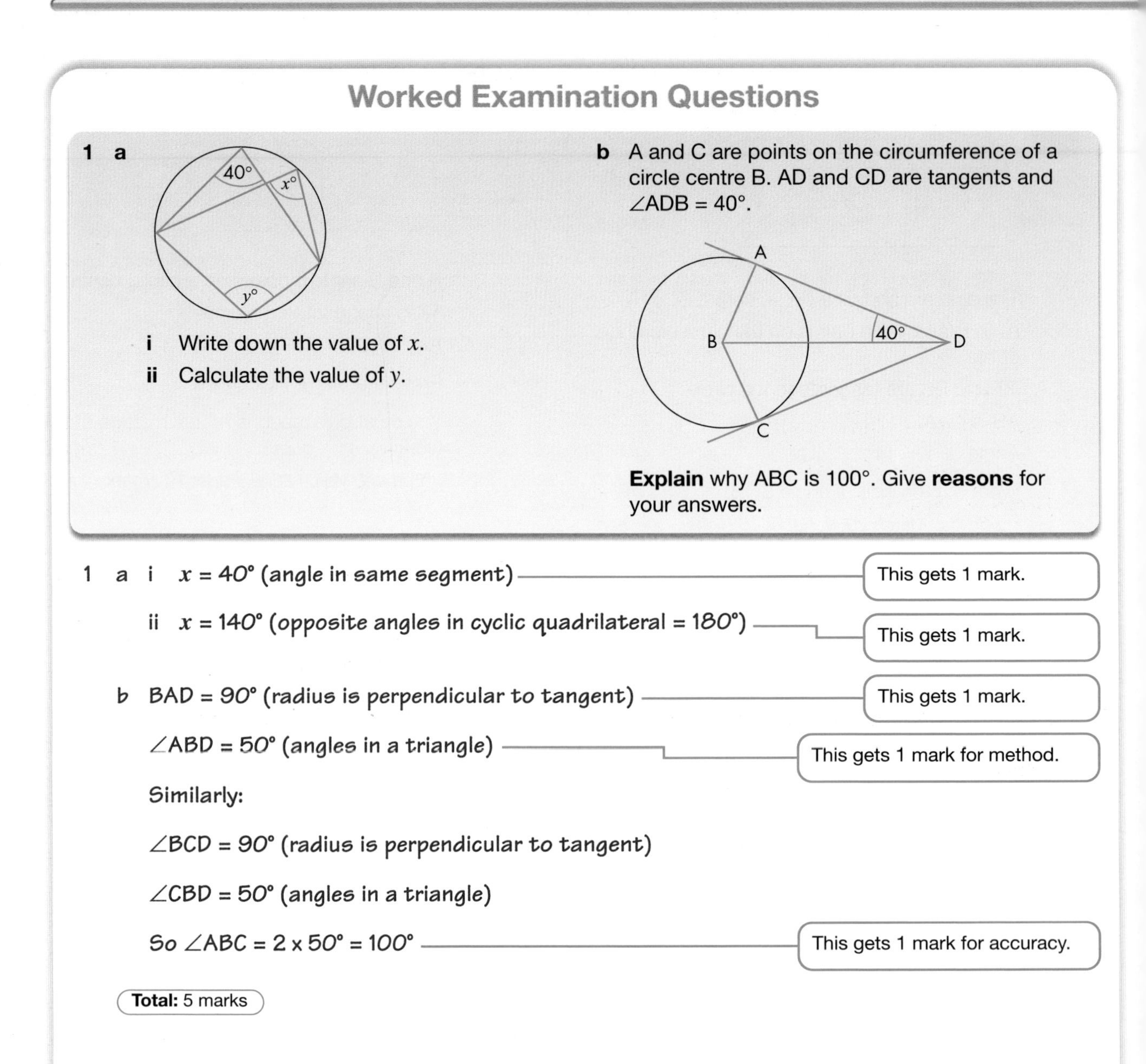

1 a

i Write down the value of x.

ii Calculate the value of y.

b A and C are points on the circumference of a circle centre B. AD and CD are tangents and ∠ADB = 40°.

Explain why ABC is 100°. Give **reasons** for your answers.

1 a i $x = 40°$ (angle in same segment) — This gets 1 mark.

ii $x = 140°$ (opposite angles in cyclic quadrilateral = 180°) — This gets 1 mark.

b BAD = 90° (radius is perpendicular to tangent) — This gets 1 mark.

∠ABD = 50° (angles in a triangle) — This gets 1 mark for method.

Similarly:

∠BCD = 90° (radius is perpendicular to tangent)

∠CBD = 50° (angles in a triangle)

So ∠ABC = 2 x 50° = 100° — This gets 1 mark for accuracy.

Total: 5 marks

Worked Examination Questions

2

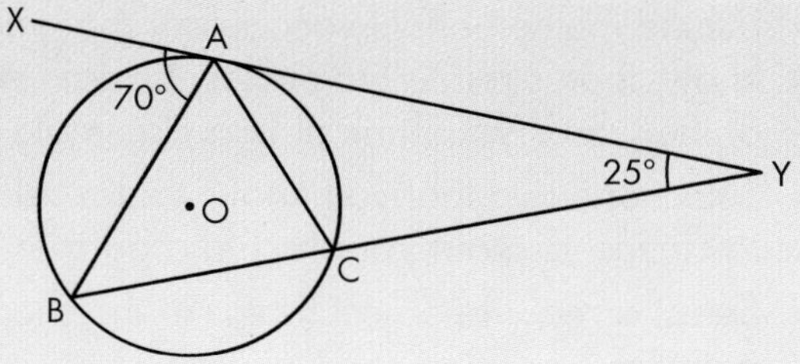

In the diagram, XY is a tangent to the circle at A.

BCY is a straight line.

Work out the size of ∠ABC.

Give reasons to justify your answer.

2 ∠ACB = 70° (angle in alternate segment)

This gets 1 mark.

∠ACY = 110° (angles on a line = 180°)

∠CAY = 45° (angles in a triangle = 180°)

This gets 1 method mark for using angles in a triangle.

so ∠ABC = 45° (angle in alternate segment)

This gets 1 accuracy mark.

Total: 3 marks

4 Problem Solving
Proving properties of circles

Circular shapes form the basis of many of the objects that we see and use every day. For example, we see circular shapes in DVDs, wheels, coins and jewellery. Where else do you see circles?

Given how frequently circles appear in our lives, it is important that we understand them mathematically.

In this task you will investigate the properties of circles, using mathematical theory and proof to help you understand this shape more fully.

Getting started

- List the mathematical vocabulary that you know, that is related to circles. Explain each of the words you think of to a classmate.
- Select one fact that you know, that is related to circles. Explain your fact to a partner.
- Select a real-life object that is in the shape of a circle.

  What mathematical questions could you ask about this object?

  How would you use mathematics to find the answers to your questions?

Your task

1 With a partner, develop a statement about the properties of circles. For example, "A radius that is perpendicular to a chord bisects that chord."

Think carefully about your statement: it will be the hypothesis that forms the basis of a mathematical investigation.

Alternatively, create a question to form the basis of your investigation. For example, "What is the relationship between an angle subtended at the centre of a circle and the angle subtended by the same two points at the circumference, opposite and on the same arc?"

2 Swap your hypothesis or question with that of another pair.

Now, use your problem-solving skills to investigate the hypothesis or question. You should create a presentation, using slides or an interactive whiteboard, to explain the mathematical process that you go through in your investigation, the mathematics that you used and the conclusion that you reach.

In your presentation you must:

- explain the overall approach you took in your investigation
- summarise each step taken during your investigation
- advance a solution to the hypothesis or problem with which you were presented
- find the most effective way of representing your solution
- give examples to support your conclusions
- show what you have now learnt about circles
- or any other aspect of geometry.

Why this chapter matters

Very large and very small numbers can often be difficult to read. Scientists use standard form as a shorthand way of representing numbers.

The planets

Mercury is the closest planet to the Sun (and is very hot). It orbits 60 million km (6×10^7 km) away from the Sun.

Venus rotates the opposite way to the other planets and has a diameter of 12 100 km (12.1×10^4 km).

Earth takes 365 days to orbit the Sun and 24 hours to complete a rotation.

Mars has the largest volcano in the solar system. It is almost 600 km across and rises 24 km above the surface. This is five times bigger than the biggest volcano on Earth.

Jupiter is made of gas. It has no solid land so visiting it is not recommended! It has a huge storm which rages across its surface. This is about 8 km high, 40 000 km long and 14 000 km wide. It looks like a red spot and is called 'the Great Red Spot'.

Saturn is the largest planet in the solar system. It is about 120 000 km across (1.2×10^5 km) and 1400 million km from the Sun (1.4×10^9 km).

Uranus takes 84 days to orbit the Sun.

Neptune is similar to Jupiter in that it is a gas planet and has violent storms. Winds can blow at up to 2000 km per hour, so a cloud can circle Neptune in about 16 hours.

Pluto is the furthest planet from the Sun. Some astronomers dispute whether it can be classed as a planet. The average surface temperature on Pluto is about –230 °C.

The mass of an electron is about 0.000 000 000 000 000 000 000 000 000 000 91 kg.
This is written 9.1×10^{-31} kg.

The mass of the Earth is about 5 970 000 000 000 000 000 000 000 kg.
This is written 5.97×10^{24} kg.

Chapter

5 Number: Powers, standard form and surds

The grades given in this chapter are target grades.

This chapter will show you ...

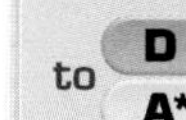
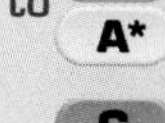
D to A* how to calculate using powers (indices)

C how to work out a reciprocal

D to A how to write numbers in standard form and how to calculate with standard form

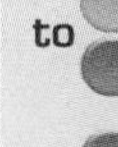
A to A* how to convert fractions to terminating decimals and recurring decimals, and vice versa

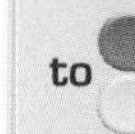
A to A* how to calculate with surds

Visual overview

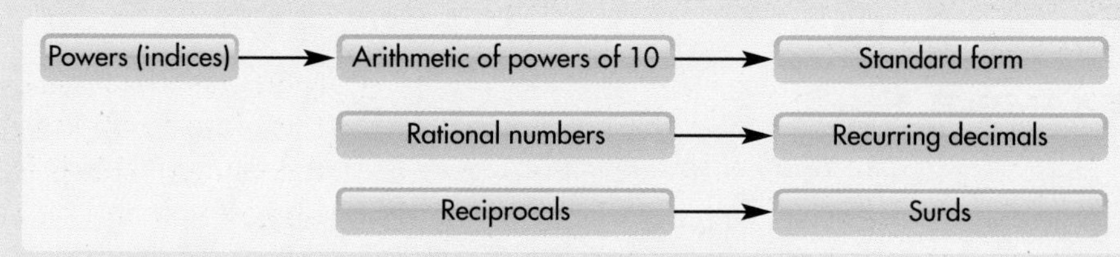

What you should already know

- How to convert a fraction to a decimal **(KS3 level 4, GCSE grade E)**
- How to convert a decimal to a fraction **(KS3 level 4, GCSE grade E)**
- How to find the lowest common denominator of two fractions **(KS3 level 5, GCSE grade D)**
- The meaning of square root and cube root **(KS3 level 5, GCSE grade E)**

Quick check

1 Convert the following fractions to decimals.

a $\frac{6}{10}$ **b** $\frac{11}{25}$ **c** $\frac{3}{8}$

2 Convert the following decimals to fractions.

a 0.17 **b** 0.64 **c** 0.858

3 Work these out. **a** $\frac{2}{3} + \frac{1}{5}$ **b** $\frac{5}{8} - \frac{2}{5}$

4 Write down the values. **a** $\sqrt{25}$ **b** $\sqrt[3]{64}$

5.1 Powers (indices)

This section will show you how to:
- use powers (also known as indices)

Key words
cube, power, index, reciprocal, indices, square

Powers are a convenient way of writing repetitive multiplications. (Powers are also called **indices**, singular **index**.)

The power tells you the number of times a number is multiplied by itself. For example:

$4^6 = 4 \times 4 \times 4 \times 4 \times 4 \times 4$ six lots of 4 multiplied together

$6^4 = 6 \times 6 \times 6 \times 6$ four lots of 6 multiplied together

$7^3 = 7 \times 7 \times 7$

$12^2 = 12 \times 12$

You are expected to know **square** numbers (power 2) up to $15^2 = 225$

You should also know the **cubes** of numbers (power 3):

$1^3 = 1$, $2^3 = 8$, $3^3 = 27$, $4^3 = 64$, $5^3 = 125$ and $10^3 = 1000$

EXAMPLE 1

a What is the value of:

i 7 squared **ii** 5 cubed?

b Write each of these numbers out in full.

i 4^6 **ii** 6^4 **iii** 7^3 **iv** 12^2

c Write the following multiplications using powers.

i $3 \times 3 \times 3 \times 3 \times 3 \times 3 \times 3 \times 3$

ii $13 \times 13 \times 13 \times 13 \times 13$

iii $7 \times 7 \times 7 \times 7$

iv $5 \times 5 \times 5 \times 5 \times 5 \times 5 \times 5$

a The value of 7 squared is $7^2 = 7 \times 7 = 49$

The value of 5 cubed is $5^3 = 5 \times 5 \times 5 = 125$

b **i** $4^6 = 4 \times 4 \times 4 \times 4 \times 4 \times 4$ **ii** $6^4 = 6 \times 6 \times 6 \times 6$

iii $7^3 = 7 \times 7 \times 7$ **iv** $12^2 = 12 \times 12$

c **i** $3 \times 3 \times 3 \times 3 \times 3 \times 3 \times 3 \times 3 = 3^8$

ii $13 \times 13 \times 13 \times 13 \times 13 = 13^5$

iii $7 \times 7 \times 7 \times 7 = 7^4$

iv $5 \times 5 \times 5 \times 5 \times 5 \times 5 \times 5 = 5^7$

Working out powers on your calculator

The power button on your calculator will probably look like this $x^{\square}$.

To work out 5^7 on your calculator use the power key.

5^7 = 5 $x^{\square}$ 7 = 78 125

Two special powers

Power 1

Any number to the power 1 is the same as the number itself. This is always true so normally you do not write the power 1.

For example: $5^1 = 5$ $32^1 = 32$ $(-8)^1 = -8$

Power zero

Any number to the power 0 is equal to 1.

For example: $5^0 = 1$ $32^0 = 1$ $(-8)^0 = 1$

You can check these results on your calculator.

EXERCISE 5A

D

1 Write these expressions using index notation. Do not work them out yet.

a $2 \times 2 \times 2 \times 2$
b $3 \times 3 \times 3 \times 3 \times 3$
c 7×7
d $5 \times 5 \times 5$
e $10 \times 10 \times 10 \times 10 \times 10 \times 10 \times 10$
f $6 \times 6 \times 6 \times 6$
g 4
h $1 \times 1 \times 1 \times 1 \times 1 \times 1 \times 1$
i $0.5 \times 0.5 \times 0.5 \times 0.5$
j $100 \times 100 \times 100$

2 Write these power terms out in full. Do not work them out yet.

a 3^4 **b** 9^3 **c** 6^2 **d** 10^5 **e** 2^{10}
f 8^1 **g** 0.1^3 **h** 2.5^2 **i** 0.7^3 **j** 1000^2

3 Using the power key on your calculator (or another method), work out the values of the power terms in question **1**.

4 Using the power key on your calculator (or another method), work out the values of the power terms in question **2**.

FM **5** A storage container is in the shape of a cube. The length of the container is 5 m.

To work out the volume of a cube, use the formula:

volume = $(\text{length of edge})^3$

Work out the total storage space in the container.

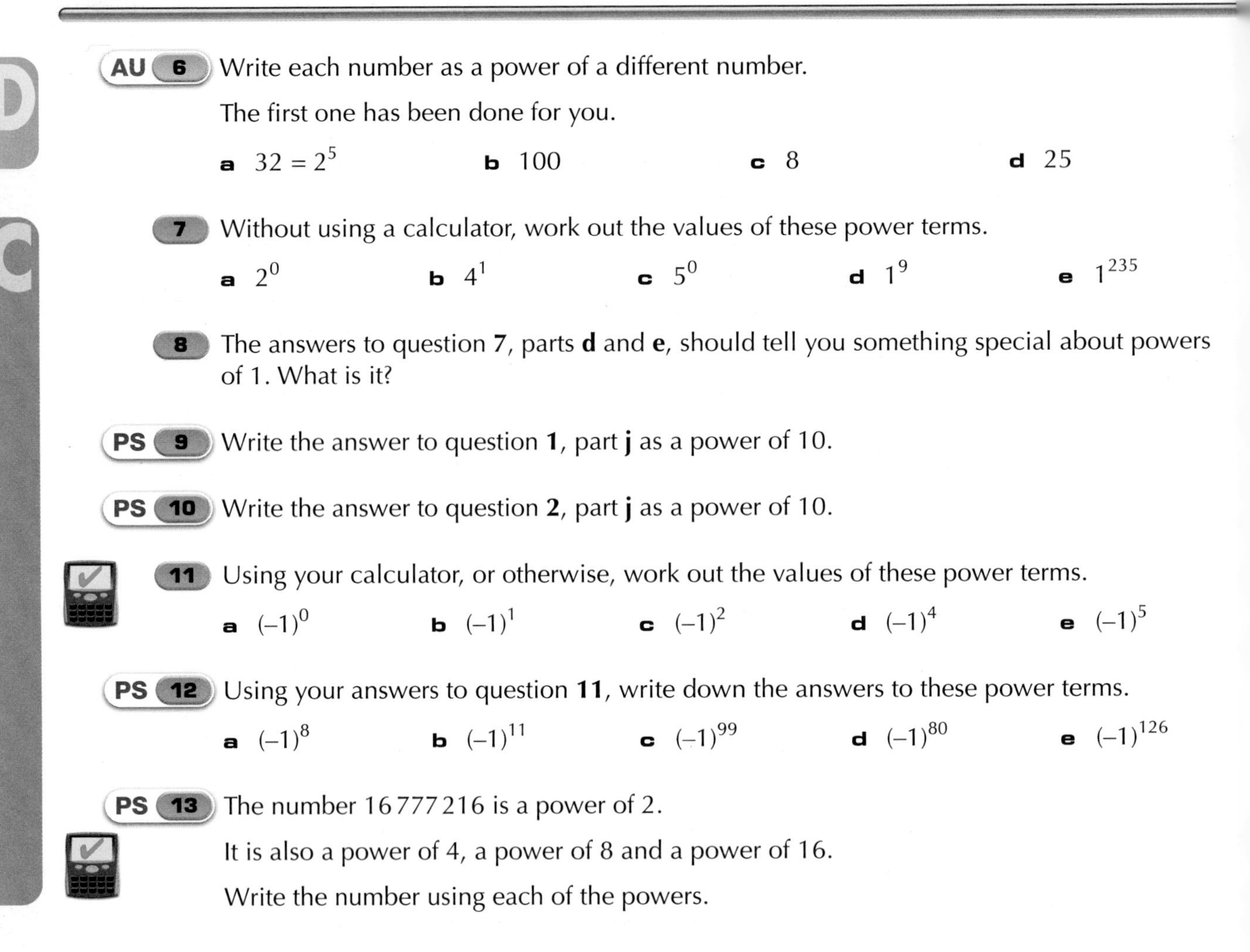

D

AU 6 Write each number as a power of a different number.

The first one has been done for you.

a $32 = 2^5$ **b** 100 **c** 8 **d** 25

C

7 Without using a calculator, work out the values of these power terms.

a 2^0 **b** 4^1 **c** 5^0 **d** 1^9 **e** 1^{235}

8 The answers to question **7**, parts **d** and **e**, should tell you something special about powers of 1. What is it?

PS 9 Write the answer to question **1**, part **j** as a power of 10.

PS 10 Write the answer to question **2**, part **j** as a power of 10.

11 Using your calculator, or otherwise, work out the values of these power terms.

a $(-1)^0$ **b** $(-1)^1$ **c** $(-1)^2$ **d** $(-1)^4$ **e** $(-1)^5$

PS 12 Using your answers to question **11**, write down the answers to these power terms.

a $(-1)^8$ **b** $(-1)^{11}$ **c** $(-1)^{99}$ **d** $(-1)^{80}$ **e** $(-1)^{126}$

PS 13 The number 16 777 216 is a power of 2.

It is also a power of 4, a power of 8 and a power of 16.

Write the number using each of the powers.

Negative powers (or negative indices)

A negative index is a convenient way of writing the **reciprocal** of a number or term. (That is, one divided by that number or term.) For example,

$$x^{-a} = \frac{1}{x^a}$$

Here are some other examples.

$$5^{-2} = \frac{1}{5^2} \qquad 3^{-1} = \frac{1}{3} \qquad 5x^{-2} = \frac{5}{x^2}$$

EXAMPLE 2

Rewrite the following in the form 2^n.

a 8 **b** $\frac{1}{4}$ **c** -32 **d** $-\frac{1}{64}$

a $8 = 2 \times 2 \times 2 = 2^3$ **b** $\frac{1}{4} = \frac{1}{2^2} = 2^{-2}$

c $-32 = -2^5$ **d** $-\frac{1}{64} = -\frac{1}{2^6} = -2^{-6}$

EXERCISE 5B

1 Write down each of these in fraction form.

a 5^{-3} **b** 6^{-1} **c** 10^{-5} **d** 3^{-2} **e** 8^{-2}

f 9^{-1} **g** w^{-2} **h** t^{-1} **i** x^{-m} **j** $4m^{-3}$

2 Write down each of these in negative index form.

a $\frac{1}{3^2}$ **b** $\frac{1}{5}$ **c** $\frac{1}{10^3}$ **d** $\frac{1}{m}$ **e** $\frac{1}{t^n}$

HINTS AND TIPS

If you move a power from top to bottom, or vice versa, the sign changes. Negative power means the reciprocal: it does not mean the answer is negative.

3 Change each of the following expressions into an index form of the type shown.

a All of the form 2^n

i 16 **ii** $\frac{1}{2}$ **iii** $\frac{1}{16}$ **iv** –8

b All of the form 10^n

i 1000 **ii** $\frac{1}{10}$ **iii** $\frac{1}{100}$ **iv** 1 million

c All of the form 5^n

i 125 **ii** $\frac{1}{5}$ **iii** $\frac{1}{25}$ **iv** $\frac{1}{625}$

d All of the form 3^n

i 9 **ii** $\frac{1}{27}$ **iii** $\frac{1}{81}$ **iv** –243

4 Rewrite each of the following expressions in fraction form.

a $5x^{-3}$ **b** $6t^{-1}$ **c** $7m^{-2}$ **d** $4q^{-4}$ **e** $10y^{-5}$

f $\frac{1}{2}x^{-3}$ **g** $\frac{1}{2}m^{-1}$ **h** $\frac{3}{4}t^{-4}$ **i** $\frac{4}{5}y^{-3}$ **j** $\frac{7}{8}x^{-5}$

5 Write each fraction in index form.

a $\frac{7}{x^3}$ **b** $\frac{10}{p}$ **c** $\frac{5}{t^2}$ **d** $\frac{8}{m^5}$ **e** $\frac{3}{y}$

6 Find the value of each of the following.

a $x = 5$

i x^2 **ii** x^{-3} **iii** $4x^{-1}$

b $t = 4$

i t^3 **ii** t^{-2} **iii** $5t^{-4}$

c $m = 2$

i m^3 **ii** m^{-5} **iii** $9m^{-1}$

d $w = 10$

i w^6 **ii** w^{-3} **iii** $25w^{-2}$

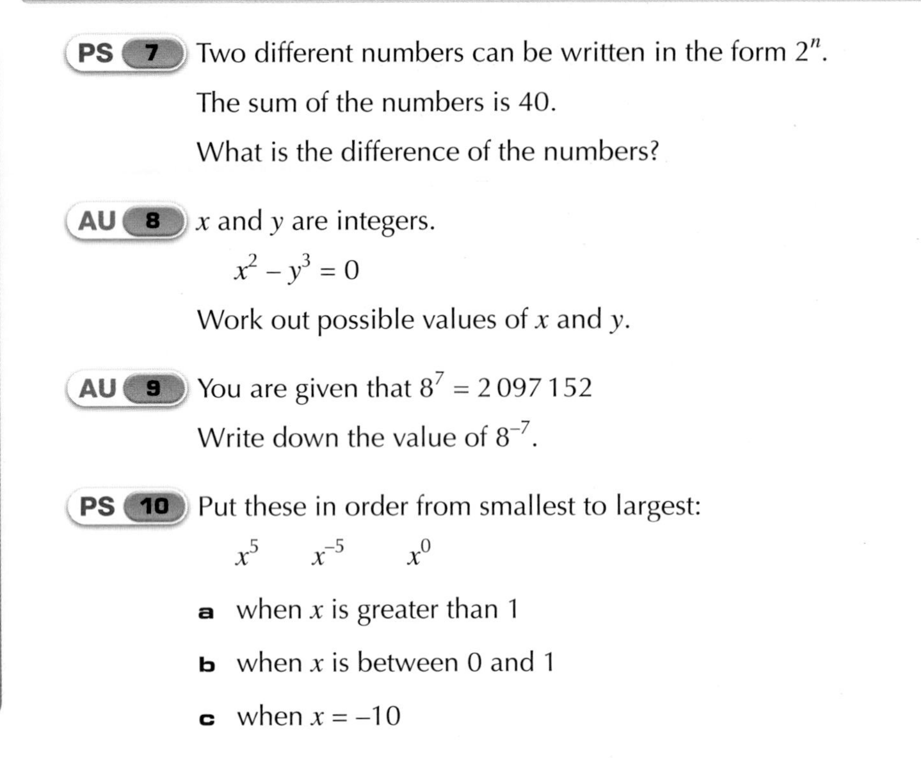

A

PS 7 Two different numbers can be written in the form 2^n.

The sum of the numbers is 40.

What is the difference of the numbers?

AU 8 x and y are integers.

$$x^2 - y^3 = 0$$

Work out possible values of x and y.

AU 9 You are given that $8^7 = 2\,097\,152$

Write down the value of 8^{-7}.

PS 10 Put these in order from smallest to largest:

$$x^5 \qquad x^{-5} \qquad x^0$$

a when x is greater than 1

b when x is between 0 and 1

c when $x = -10$

Rules for multiplying and dividing numbers in index form

When you *multiply* powers of the same number or variable, you *add* the indices. For example,

$3^4 \times 3^5 = 3^{(4 + 5)} = 3^9$

$2^3 \times 2^4 \times 2^5 = 2^{12}$

$10^4 \times 10^{-2} = 10^2$

$10^{-3} \times 10^{-1} = 10^{-4}$

$a^x \times a^y = a^{(x + y)}$

When you *divide* powers of the same number or variable, you *subtract* the indices. For example,

$a^4 \div a^3 = a^{(4 - 3)} = a^1 = a$

$b^4 \div b^7 = b^{-3}$

$10^4 \div 10^{-2} = 10^6$

$10^{-2} \div 10^{-4} = 10^2$

$a^x \div a^y = a^{(x - y)}$

When you *raise* a power to a further power, you *multiply* the indices. For example,

$(a^2)^3 = a^{2 \times 3} = a^6$

$(a^{-2})^4 = a^{-8}$

$(a^2)^6 = a^{12}$

$(a^x)^y = a^{xy}$

Here are some examples of different kinds of expressions using powers.

$2a^2 \times 3a^4 = (2 \times 3) \times (a^2 \times a^4) = 6 \times a^6 = 6a^6$

$4a^2b^3 \times 2ab^2 = (4 \times 2) \times (a^2 \times a) \times (b^3 \times b^2) = 8a^3b^5$

$12a^5 \div 3a^2 = (12 \div 3) \times (a^5 \div a^2) = 4a^3$

$(2a^2)^3 = (2)^3 \times (a^2)^3 = 8 \times a^6 = 8a^6$

EXERCISE 5C

1 Write these as single powers of 5.

a $5^2 \times 5^2$ **b** 5×5^2 **c** $5^{-2} \times 5^4$ **d** $5^6 \times 5^{-3}$ **e** $5^{-2} \times 5^{-3}$

2 Write these as single powers of 6.

a $6^5 \div 6^2$ **b** $6^4 \div 6^4$ **c** $6^4 \div 6^{-2}$ **d** $6^{-3} \div 6^4$ **e** $6^{-3} \div 6^{-5}$

3 Simplify these and write them as single powers of a.

a $a^2 \times a$ **b** $a^3 \times a^2$ **c** $a^4 \times a^3$

d $a^6 \div a^2$ **e** $a^3 \div a$ **f** $a^5 \div a^4$

AU **4** **a** $a^x \times a^y = a^{10}$

Write down a possible pair of values of x and y.

b $a^x \div a^y = a^{10}$

Write down a possible pair of values of x and y.

5 Write these as single powers of 4.

a $(4^2)^3$ **b** $(4^3)^5$ **c** $(4^1)^6$

d $(4^3)^{-2}$ **e** $(4^{-2})^{-3}$ **f** $(4^7)^0$

6 Simplify these expressions.

a $2a^2 \times 3a^3$ **b** $3a^4 \times 3a^{-2}$ **c** $(2a^2)^3$

d $-2a^2 \times 3a^2$ **e** $-4a^3 \times -2a^5$ **f** $-2a^4 \times 5a^{-7}$

7 Simplify these expressions.

a $6a^3 \div 2a^2$ **b** $12a^5 \div 3a^2$

c $15a^5 \div 5a$ **d** $18a^{-2} \div 3a^{-1}$

e $24a^5 \div 6a^{-2}$ **f** $30a \div 6a^5$

HINTS AND TIPS

Deal with numbers and indices separately and do not confuse the rules.
For example: $12a^5 \div 4a^2 = (12 \div 4) \times (a^5 \div a^2)$

8 Simplify these expressions.

a $2a^2b^3 \times 4a^3b$ **b** $5a^2b^4 \times 2ab^{-3}$ **c** $6a^2b^3 \times 5a^{-4}b^{-5}$

d $12a^2b^4 \div 6ab$ **e** $24a^{-3}b^4 \div 3a^2b^{-3}$

C

B

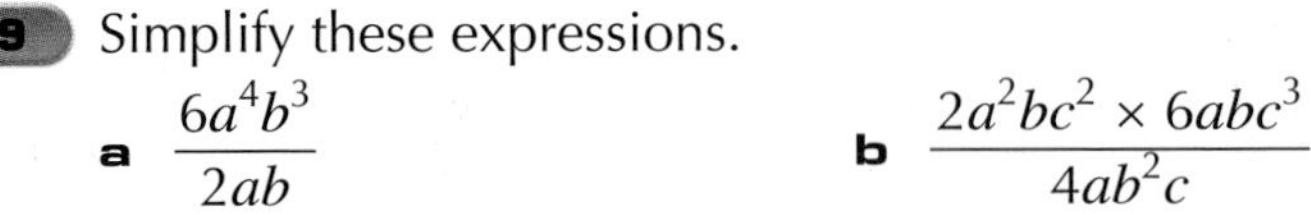

B

9 Simplify these expressions.

a $\dfrac{6a^4b^3}{2ab}$ **b** $\dfrac{2a^2bc^2 \times 6abc^3}{4ab^2c}$ **c** $\dfrac{3abc \times 4a^3b^2c \times 6c^2}{9a^2bc}$

FM 10 Write down **two** possible:

a multiplication questions with an answer of $12x^2y^5$

b division questions with an answer of $12x^2y^5$.

A

PS 11 a, b and c are three different positive integers.

What is the smallest possible value of a^2b^3c?

FM PS 12 Use the general rule for dividing powers of the same number, $\dfrac{a^x}{a^y} = a^{x-y}$, to prove that any number raised to the power zero is 1.

Indices of the form $\frac{1}{n}$

Consider the problem $7^x \times 7^x = 7$. This can be written as:

$$7^{(x+x)} = 7$$

$$7^{2x} = 7^1 \Rightarrow 2x = 1 \Rightarrow x = \tfrac{1}{2}$$

If you now substitute $x = \frac{1}{2}$ back into the original equation, you see that:

$$7^{\frac{1}{2}} \times 7^{\frac{1}{2}} = 7$$

This makes $7^{\frac{1}{2}}$ the same as $\sqrt{7}$.

You can similarly show that $7^{\frac{1}{3}}$ is the same as $\sqrt[3]{7}$. And that, generally,

$$x^{\frac{1}{n}} = \sqrt[n]{x} \text{ (}n\text{th root of } x\text{)}$$

So in summary:

Power $\frac{1}{2}$ is the same as positive square root.

Power $\frac{1}{3}$ is the same as cube root.

Power $\frac{1}{n}$ is the same as nth root.

For example,

$49^{\frac{1}{2}} = \sqrt{49} = 7$ $8^{\frac{1}{3}} = \sqrt[3]{8} = 2$ $10\,000^{\frac{1}{4}} = \sqrt[4]{10\,000} = 10$ $36^{-\frac{1}{2}} = \dfrac{1}{\sqrt{36}} = \dfrac{1}{6}$

EXERCISE 5D

A

1 Evaluate the following.

a $25^{\frac{1}{2}}$ **b** $100^{\frac{1}{2}}$ **c** $64^{\frac{1}{2}}$ **d** $81^{\frac{1}{2}}$ **e** $625^{\frac{1}{2}}$

f $27^{\frac{1}{3}}$ **g** $64^{\frac{1}{3}}$ **h** $1000^{\frac{1}{3}}$ **i** $125^{\frac{1}{3}}$ **j** $512^{\frac{1}{3}}$

k $144^{\frac{1}{2}}$ **l** $400^{\frac{1}{2}}$ **m** $625^{\frac{1}{4}}$ **n** $81^{\frac{1}{4}}$ **o** $100\,000^{\frac{1}{5}}$

p $729^{\frac{1}{6}}$ **q** $32^{\frac{1}{5}}$ **r** $1024^{\frac{1}{10}}$ **s** $1296^{\frac{1}{4}}$ **t** $216^{\frac{1}{3}}$

u $16^{-\frac{1}{2}}$ **v** $8^{-\frac{1}{3}}$ **w** $81^{-\frac{1}{4}}$ **x** $3125^{-\frac{1}{5}}$ **y** $1\,000\,000^{-\frac{1}{6}}$

A*

2 Evaluate the following.

a $\left(\frac{25}{36}\right)^{\frac{1}{2}}$ **b** $\left(\frac{100}{36}\right)^{\frac{1}{2}}$ **c** $\left(\frac{64}{81}\right)^{\frac{1}{2}}$ **d** $\left(\frac{81}{25}\right)^{\frac{1}{2}}$ **e** $\left(\frac{25}{64}\right)^{\frac{1}{2}}$

f $\left(\frac{27}{125}\right)^{\frac{1}{3}}$ **g** $\left(\frac{8}{512}\right)^{\frac{1}{3}}$ **h** $\left(\frac{1000}{64}\right)^{\frac{1}{3}}$ **i** $\left(\frac{64}{125}\right)^{\frac{1}{3}}$ **j** $\left(\frac{512}{343}\right)^{\frac{1}{3}}$

3 Use the general rule for raising a power to another power to prove that $x^{\frac{1}{n}}$ is equivalent to $\sqrt[n]{x}$.

PS 4 Which of these is the odd one out?

$16^{-\frac{1}{4}}$ $64^{-\frac{1}{2}}$ $8^{-\frac{1}{3}}$

Show how you decided.

AU 5 Imagine that you are the teacher.

Write down how you would teach the class that $27^{-\frac{1}{3}}$ is equal to $\frac{1}{3}$.

PS 6 $x^{-\frac{2}{3}} = y^{\frac{1}{3}}$

Find values for x and y that make this equation work.

Indices of the form $\frac{a}{b}$

Here are two examples of this form.

$t^{\frac{2}{3}} = t^{\frac{1}{3}} \times t^{\frac{1}{3}} = (\sqrt[3]{t})^2$ $81^{\frac{3}{4}} = (\sqrt[4]{81})^3 = 3^3 = 27$

EXAMPLE 3

Evaluate the following. **a** $16^{-\frac{1}{4}}$ **b** $32^{-\frac{4}{5}}$

When dealing with the negative index remember that it means reciprocal.

Do problems like these one step at a time.

Step 1: Rewrite the calculation as a fraction by dealing with the negative power.

Step 2: Take the root of the base number given by the denominator of the fraction.

Step 3: Raise the result to the power given by the numerator of the fraction.

Step 4: Write out the answer as a fraction.

a **Step 1:** $16^{-\frac{1}{4}} = \left(\frac{1}{16}\right)^{\frac{1}{4}}$ **Step 2:** $16^{\frac{1}{4}} = \sqrt[4]{16} = 2$ **Step 3:** $2^1 = 2$ **Step 4:** $16^{-\frac{1}{4}} = \frac{1}{2}$

b **Step 1:** $32^{-\frac{4}{5}} = \left(\frac{1}{32}\right)^{\frac{4}{5}}$ **Step 2:** $32^{\frac{1}{5}} = \sqrt[5]{32} = 2$ **Step 3:** $2^4 = 16$ **Step 4:** $32^{-\frac{4}{5}} = \frac{1}{16}$

EXERCISE 5E

A

1 Evaluate the following.

a $32^{\frac{4}{5}}$ **b** $125^{\frac{2}{3}}$ **c** $1296^{\frac{3}{4}}$ **d** $243^{\frac{4}{5}}$

2 Rewrite the following in index form.

a $\sqrt[3]{t^2}$ **b** $\sqrt[4]{m^3}$ **c** $\sqrt[5]{k^2}$ **d** $\sqrt{x^3}$

A
A*

3 Evaluate the following.

a $8^{\frac{2}{3}}$ **b** $27^{\frac{2}{3}}$ **c** $16^{\frac{3}{2}}$ **d** $625^{\frac{5}{4}}$

4 Evaluate the following.

a $25^{-\frac{1}{2}}$ **b** $36^{-\frac{1}{2}}$ **c** $16^{-\frac{1}{4}}$ **d** $81^{-\frac{1}{4}}$

e $16^{-\frac{1}{2}}$ **f** $8^{-\frac{1}{3}}$ **g** $32^{-\frac{1}{5}}$ **h** $27^{-\frac{1}{3}}$

5 Evaluate the following.

a $25^{-\frac{3}{2}}$ **b** $36^{-\frac{3}{2}}$ **c** $16^{-\frac{3}{4}}$ **d** $81^{-\frac{3}{4}}$

e $64^{-\frac{4}{3}}$ **f** $8^{-\frac{2}{3}}$ **g** $32^{-\frac{2}{5}}$ **h** $27^{-\frac{2}{3}}$

6 Evaluate the following.

a $100^{-\frac{5}{2}}$ **b** $144^{-\frac{1}{2}}$ **c** $125^{-\frac{2}{3}}$ **d** $9^{-\frac{3}{2}}$

e $4^{-\frac{5}{2}}$ **f** $64^{-\frac{5}{6}}$ **g** $27^{-\frac{4}{3}}$ **h** $169^{-\frac{1}{2}}$

PS 7 Which of these is the odd one out?

$16^{-\frac{3}{4}}$ $64^{-\frac{1}{2}}$ $8^{-\frac{2}{3}}$

Show how you decided.

AU 8 Imagine that you are the teacher.

Write down how you would teach the class that $27^{-\frac{2}{3}}$ is equal to $\frac{1}{9}$.

5.2 Standard form

This section will show you how to:

- change a number into standard form
- calculate using numbers in standard form

Key words
powers
standard form
standard index form

Arithmetic of powers of 10

Multiplying

You have already done some arithmetic with multiples of 10 in Book 1. You will now look at **powers** of 10.

How many zeros does a million have? What is a million as a power of 10? This table shows some of the pattern of the powers of 10.

Number	0.001	0.01	0.1	1	10	100	1000	10 000	100 000
Powers	10^{-3}	10^{-2}	10^{-1}	10^{0}	10^{1}	10^{2}	10^{3}	10^{4}	10^{5}

What pattern is there in the top row? What pattern is there to the powers in the bottom row?

To multiply by any power of 10, you simply move the digits according to these two rules.

- When the index is *positive*, move the digits to the *left* by the same number of places as the value of the index.
- When the index is *negative*, move the digits to the *right* by the same number of places as the value of the index.

EXAMPLE 4

Write these as ordinary numbers.

a 12.356×10^2 **b** 3.45×10^1

c 753.4×10^{-2} **d** 6789×10^{-1}

a $12.356 \times 10^2 = 1235.6$ **b** $3.45 \times 10^1 = 34.5$

c $753.4 \times 10^{-2} = 7.534$ **d** $6789 \times 10^{-1} = 678.9$

In certain cases, you have to insert the 'hidden' zeros.

EXAMPLE 5

Write these as ordinary numbers.

a 75×10^4 **b** 2.04×10^5

c 6.78×10^{-3} **d** 0.897×10^{-4}

a $75 \times 10^4 = 750\,000$ **b** $2.04 \times 10^5 = 204\,000$

c $6.78 \times 10^{-3} = 0.006\,78$ **d** $0.897 \times 10^{-4} = 0.000\,0897$

Dividing

To divide by any power of 10, you simply move the digits according to these two rules.

- When the index is *positive*, move the digits to the *right* by the same number of places as the value of the index.
- When the index is *negative*, move the digits to the *left* by the same number of places as the value of the index.

EXAMPLE 6

Write these as ordinary numbers.

a $712.35 \div 10^2$ **b** $38.45 \div 10^1$

c $3.463 \div 10^{-2}$ **d** $6.789 \div 10^{-1}$

a $712.35 \div 10^2 = 7.1235$ **b** $38.45 \div 10^1 = 3.845$

c $3.463 \div 10^{-2} = 346.3$ **d** $6.789 \div 10^{-1} = 67.89$

In certain cases, you have to insert the 'hidden' zeros.

EXAMPLE 7

Write these as ordinary numbers.

a $75 \div 10^4$ **b** $2.04 \div 10^5$

c $6.78 \div 10^{-3}$ **d** $0.08 \div 10^{-4}$

a $75 \div 10^4 = 0.0075$ **b** $2.04 \div 10^5 = 0.000\,0204$

c $6.78 \div 10^{-3} = 6780$ **d** $0.08 \div 10^{-4} = 800$

When doing the next exercise, remember:

$10\,000 = 10 \times 10 \times 10 \times 10 = 10^4$ $1 = 10^0$

$1000 = 10 \times 10 \times 10 = 10^3$ $0.1 = 1 \div 10 = 10^{-1}$

$100 = 10 \times 10 = 10^2$ $0.01 = 1 \div 100 = 10^{-2}$

$10 = 10 = 10^1$ $0.001 = 1 \div 1000 = 10^{-3}$

EXERCISE 5F

D

1 Write down the value of each of the following.

a 3.1×10 **b** 3.1×100 **c** 3.1×1000 **d** $3.1 \times 10\,000$

2 Write down the value of each of the following.

a 6.5×10 **b** 6.5×10^2 **c** 6.5×10^3 **d** 6.5×10^4

3 Write down the value of each of the following.

a $3.1 \div 10$ **b** $3.1 \div 100$ **c** $3.1 \div 1000$ **d** $3.1 \div 10\,000$

4 Write down the value of each of the following.

a $6.5 \div 10$ **b** $6.5 \div 10^2$ **c** $6.5 \div 10^3$ **d** $6.5 \div 10^4$

5 Evaluate the following.

a 2.5×100 **b** 3.45×10 **c** 4.67×1000 **d** 34.6×10

e 20.789×10 **f** 56.78×1000 **g** 2.46×10^2 **h** 0.076×10

i 0.999×10^6 **j** 234.56×10^2 **k** 98.7654×10^3 **l** 43.23×10^6

m $0.003\,457\,8 \times 10^5$ **n** 0.0006×10^7 **o** $0.005\,67 \times 10^4$ **p** 56.0045×10^4

D

6 Evaluate the following.

a $2.5 \div 100$ b $3.45 \div 10$ c $4.67 \div 1000$ d $34.6 \div 10$

e $20.789 \div 100$ f $56.78 \div 1000$ g $2.46 \div 10^2$ h $0.076 \div 10$

i $0.999 \div 10^6$ j $234.56 \div 10^2$ k $98.7654 \div 10^3$ l $43.23 \div 10^6$

m $0.003\,4578 \div 10^5$ n $0.0006 \div 10^7$ o $0.005\,67 \div 10^4$ p $56.0045 \div 10^4$

7 Without using a calculator, work out the following.

a 2.3×10^2 b 5.789×10^5 c 4.79×10^3 d 5.7×10^7

e 2.16×10^2 f 1.05×10^4 g 3.2×10^{-4} h 9.87×10^3

8 Which of these statements is true about the numbers in question **7**?

a The first part is always a number between 1 and 10.

b There is always a multiplication sign in the middle of the expression.

c There is always a power of 10 at the end.

d Calculator displays sometimes show numbers in this form.

C

AU PS **9** The mass of Mars is 6.4×10^{23} kg.

The mass of Venus is 4.9×10^{24} kg.

Without working out the answers, explain how you can tell which planet is the heavier.

B

PS **10** A number is between one million and 10 million. It is written in the form 4.7×10^n.

What is the value of n?

Standard form

Standard form is also known as **standard index form**.

Standard form is a way of writing very large and very small numbers using powers of 10. In this form, a number is given a value between 1 and 10 multiplied by a power of 10. That is,

$a \times 10^n$ where $1 \leqslant a < 10$, and n is a whole number.

Follow through these examples to see how numbers are written in standard form.

$52 = 5.2 \times 10 = \mathbf{5.2 \times 10^1}$

$73 = 7.3 \times 10 = \mathbf{7.3 \times 10^1}$

$625 = 6.25 \times 100 = \mathbf{6.25 \times 10^2}$ The numbers in bold are in standard form.

$389 = 3.89 \times 100 = \mathbf{3.89 \times 10^2}$

$3147 = 3.147 \times 1000 = \mathbf{3.147 \times 10^3}$

When writing a number in this way, you must always follow two rules.

- The first part must be a number between 1 and 10 (1 is allowed but 10 isn't).
- The second part must be a whole-number (negative or positive) power of 10. Note that you would *not normally* write the power 1.

Standard form on a calculator

A number such as 123 000 000 000 is obviously difficult to key into a calculator. Instead, you enter it in standard form (assuming you are using a scientific calculator):

$123\,000\,000\,000 = 1.23 \times 10^{11}$

The key strokes to enter this into your calculator will be:

1 • 2 3 ×10ˣ 1 1

Your calculator display will display the number either as an ordinary number, if there is enough space, or in standard form.

Standard form of numbers less than 1

These numbers are written in standard form. Make sure that you understand how they are formed.

a $0.4 = 4 \times 10^{-1}$ **b** $0.05 = 5 \times 10^{-2}$ **c** $0.007 = 7 \times 10^{-3}$

d $0.123 = 1.23 \times 10^{-1}$ **e** $0.007\,65 = 7.65 \times 10^{-3}$ **f** $0.9804 = 9.804 \times 10^{-1}$

g $0.0098 = 9.8 \times 10^{-3}$ **h** $0.000\,0078 = 7.8 \times 10^{-6}$

On a calculator you would enter 1.23×10^{-6}, for example, as:

1 • 2 3 ×10ˣ (–) 6

Try entering some of the numbers **a** to **h** (above) into your calculator for practice.

EXERCISE 5G

D

1 Write down the value of each of the following.

a 3.1×0.1 **b** 3.1×0.01 **c** 3.1×0.001 **d** 3.1×0.0001

2 Write down the value of each of the following.

a 6.5×10^{-1} **b** 6.5×10^{-2} **c** 6.5×10^{-3} **d** 6.5×10^{-4}

PS **3** **a** What is the largest number you can enter into your calculator?

b What is the smallest number you can enter into your calculator?

4 Work out the value of each of the following.

a $3.1 \div 0.1$ **b** $3.1 \div 0.01$ **c** $3.1 \div 0.001$ **d** $3.1 \div 0.0001$

5 Work out the value of each of the following.

a $6.5 \div 10^{-1}$ **b** $6.5 \div 10^{-2}$ **c** $6.5 \div 10^{-3}$ **d** $6.5 \div 10^{-4}$

C

6 Write these numbers out in full.

a 2.5×10^{2} **b** 3.45×10 **c** 4.67×10^{-3} **d** 3.46×10

e 2.0789×10^{-2} **f** 5.678×10^{3} **g** 2.46×10^{2} **h** 7.6×10^{3}

i 8.97×10^{5} **j** 8.65×10^{-3} **k** 6×10^{7} **l** 5.67×10^{-4}

7 Write these numbers in standard form.

a 250 **b** 0.345 **c** 46 700

d 3 400 000 000 **e** 20 780 000 000 **f** 0.000 567 8

g 2460 **h** 0.076 **i** 0.000 76

j 0.999 **k** 234.56 **l** 98.7654

m 0.0006 **n** 0.005 67 **o** 56.0045

In questions **8** to **10**, write the numbers given in each question in standard form.

8 One year, 27 797 runners completed the New York marathon.

9 The largest number of dominoes ever toppled by one person is 281 581, although 30 people set up and toppled 1 382 101.

10 The asteroid *Phaethon* comes within 12 980 000 miles of the Sun, whilst the asteroid *Pholus*, at its furthest point, is a distance of 2997 million miles from the Earth. The closest an asteroid ever came to Earth was 93 000 miles from the planet.

AU **11** How many times bigger is 3.2×10^6 than 3.2×10^4?

PS FM **12** The speed of sound (Mach 1) is 1236 kilometres per hour or about 1 mile in 5 seconds.

A plane travelling at Mach 2 would be travelling at twice the speed of sound.

How many miles would a plane travelling at Mach 3 cover in 1 minute?

Calculating with standard form

Calculations involving very large or very small numbers can be done more easily using standard form.

In these examples, you will see how to work out the area of a pixel on a computer screen, and how long it takes light to reach the Earth from a distant star.

EXAMPLE 8

A pixel on a computer screen is 2×10^{-2} cm long by 7×10^{-3} cm wide.

What is the area of the pixel?

The area is given by length times width.

$$\text{Area} = 2 \times 10^{-2} \text{ cm} \times 7 \times 10^{-3} \text{ cm}$$
$$= (2 \times 7) \times (10^{-2} \times 10^{-3}) \text{ cm}^2 = 14 \times 10^{-5} \text{ cm}^2$$

Note that you multiply the numbers and add the powers of 10. (You should not need to use a calculator to do this calculation.) The answer is not in standard form as the first part is not between 1 and 10, so you have to change it to standard form.

$$14 = 1.4 \times 10^1$$

So area $= 14 \times 10^{-5} \text{ cm}^2 = 1.4 \times 10^1 \times 10^{-5} \text{ cm}^2 = 1.4 \times 10^{-4} \text{ cm}^2$

EXAMPLE 9

The star *Betelgeuse* is 1.8×10^{15} miles from Earth. Light travels at 1.86×10^5 miles per second.

a How many seconds does it take light to travel from *Betelgeuse* to Earth? Give your answer in standard form.

b How many years does it take light to travel from *Betelgeuse* to Earth?

a Time = distance ÷ speed = 1.8×10^{15} miles ÷ 1.86×10^5 miles per second

$= (1.8 \div 1.86) \times (10^{15} \div 10^5)$ seconds

$= 0.967\,741\,935 \times 10^{10}$ seconds

Note that you divide the numbers and subtract the powers of 10. To change the answer to standard form, first round it, which gives:

$0.97 \times 10^{10} = 9.7 \times 10^9$ seconds

b To convert from seconds to years, you have to divide first by 3600 to get to hours, then by 24 to get to days, and finally by 365 to get to years.

$9.7 \times 10^9 \div (3600 \times 24 \times 365) = 307.6$ years

EXERCISE 5H

B

1 These numbers are not in standard form. Write them in standard form.

a 56.7×10^2 **b** 0.06×10^4 **c** 34.6×10^{-2}

d 0.07×10^{-2} **e** 56×10 **f** $2 \times 3 \times 10^5$

g $2 \times 10^2 \times 35$ **h** 160×10^{-2} **i** 23 million

j 0.0003×10^{-2} **k** 25.6×10^5 **l** $16 \times 10^2 \times 3 \times 10^{-1}$

m $2 \times 10^4 \times 56 \times 10^{-4}$ **n** $(18 \times 10^2) \div (3 \times 10^3)$ **o** $(56 \times 10^3) \div (2 \times 10^{-2})$

2 Work out the following. Give your answers in standard form.

a $2 \times 10^4 \times 5.4 \times 10^3$ **b** $1.6 \times 10^2 \times 3 \times 10^4$ **c** $2 \times 10^4 \times 6 \times 10^4$

d $2 \times 10^{-4} \times 5.4 \times 10^3$ **e** $1.6 \times 10^{-2} \times 4 \times 10^4$ **f** $2 \times 10^4 \times 6 \times 10^{-4}$

g $7.2 \times 10^{-3} \times 4 \times 10^2$ **h** $(5 \times 10^3)^2$ **i** $(2 \times 10^{-2})^3$

3 Work out the following. Give your answers in standard form, rounding to an appropriate degree of accuracy where necessary.

a $2.1 \times 10^4 \times 5.4 \times 10^3$ **b** $1.6 \times 10^3 \times 3.8 \times 10^3$ **c** $2.4 \times 10^4 \times 6.6 \times 10^4$

d $7.3 \times 10^{-6} \times 5.4 \times 10^3$ **e** $(3.1 \times 10^4)^2$ **f** $(6.8 \times 10^{-4})^2$

g $5.7 \times 10 \times 3.7 \times 10$ **h** $1.9 \times 10^{-2} \times 1.9 \times 10^9$ **i** $5.9 \times 10^3 \times 2.5 \times 10^{-2}$

j $5.2 \times 10^3 \times 2.2 \times 10^2 \times 3.1 \times 10^3$ **k** $1.8 \times 10^2 \times 3.6 \times 10^3 \times 2.4 \times 10^{-2}$

B

4 Work out the following. Give your answers in standard form.

a $(5.4 \times 10^4) \div (2 \times 10^3)$ **b** $(4.8 \times 10^2) \div (3 \times 10^4)$ **c** $(1.2 \times 10^4) \div (6 \times 10^4)$

d $(2 \times 10^{-4}) \div (5 \times 10^3)$ **e** $(1.8 \times 10^4) \div (9 \times 10^{-2})$ **f** $\sqrt{(36 \times 10^{-4})}$

g $(5.4 \times 10^{-3}) \div (2.7 \times 10^2)$ **h** $(1.8 \times 10^6) \div (3.6 \times 10^3)$ **i** $(5.6 \times 10^3) \div (2.8 \times 10^2)$

5 Work out the following. Give your answers in standard form, rounding to an appropriate degree of accuracy where necessary.

a $(2.7 \times 10^4) \div (5 \times 10^2)$ **b** $(2.3 \times 10^4) \div (8 \times 10^6)$ **c** $(3.2 \times 10^{-1}) \div (2.8 \times 10^{-1})$

d $(2.6 \times 10^{-6}) \div (4.1 \times 10^3)$ **e** $\sqrt{(8 \times 10^4)}$ **f** $\sqrt{(30 \times 10^{-4})}$

g $5.3 \times 10^3 \times 2.3 \times 10^2 \div (2.5 \times 10^3)$ **h** $1.8 \times 10^2 \times 3.1 \times 10^3 \div (6.5 \times 10^{-2})$

6 A typical adult has about 20 000 000 000 000 red corpuscles. Each red corpuscle has a mass of about 0.000 000 000 1 g. Write both of these numbers in standard form and work out the total mass of red corpuscles in a typical adult.

PS 7 A man puts one grain of rice on the first square of a chess board, two on the second square, four on the third, eight on the fourth and so on.

a How many grains of rice will he put on the 64th square of the board?

b How many grains of rice will there be altogether?

Give your answers in standard form.

HINTS AND TIPS

Compare powers of 2 with the running totals. By the fourth square you have 15 grains altogether, and $2^4 = 16$.

8 The surface area of the Earth is approximately 2×10^8 square miles. The area of the Earth's surface that is covered by water is approximately 1.4×10^8 square miles.

a Calculate the area of the Earth's surface *not* covered by water. Give your answer in standard form.

b What percentage of the Earth's surface is not covered by water?

9 The Moon is a sphere with a radius of 1.080×10^3 miles. The formula for working out the surface area of a sphere is:

surface area = $4\pi r^2$

Calculate the surface area of the Moon.

10 Evaluate $\frac{E}{M}$ when $E = 1.5 \times 10^3$ and $M = 3 \times 10^{-2}$, giving your answer in standard form.

11 Work out the value of $\frac{3.2 \times 10^7}{1.4 \times 10^2}$ giving your answer in standard form, correct to 2 significant figures.

12 In 2009, British Airways carried 33 million passengers. Of these, 70% passed through Heathrow Airport. On average, each passenger carried 19.7 kg of luggage. Calculate the total mass of the luggage carried by these passengers.

B

A

FM 13 In 2009 the world population was approximately 6.77×10^9. In 2010 the world population is approximately 6.85×10^9.

By how much did the population rise? Give your answer as an ordinary number.

PS 14 Here are four numbers written in standard form.

1.6×10^4 4.8×10^6 3.2×10^2 6.4×10^3

a Work out the smallest answer when two of these numbers are multiplied together.

b Work out the largest answer when two of these numbers are added together.

Give your answers in standard form.

FM 15 Many people withdraw money from their banks by using hole-in-the-wall machines. Each day there are eight million withdrawals from 32 000 machines. What is the average number of withdrawals per machine?

PS 16 The mass of Saturn is 5.686×10^{26} tonnes. The mass of the Earth is 6.04×10^{21} tonnes. How many times heavier is Saturn than the Earth? Give your answer in standard form to a suitable degree of accuracy.

AU 17 A number, when written in standard form, is greater than 100 million and less than 1000 million.

Write down a possible value of the number, in standard form.

Rational numbers and reciprocals

This section will show you how to:

- recognise rational numbers, reciprocals, terminating decimals and recurring decimals
- convert terminal decimals to fractions
- convert fractions to recurring decimals
- find reciprocals of numbers or fractions

Key words
rational number
reciprocal
recurring decimal
terminating decimal

Rational numbers

A **rational number** is a number that can be written as a fraction, for example, $\frac{1}{4}$ or $\frac{10}{3}$.

When a fraction is converted to a decimal it will either be:

- a **terminating decimal** or
- a **recurring decimal**.

A terminating decimal has a finite number of digits. For example, $\frac{1}{4} = 0.25$, $\frac{1}{8} = 0.125$.

A recurring decimal has a digit, or block of digits, that repeat. For example, $\frac{1}{3} = 0.3333 \ldots$, $\frac{2}{11} = 0.181818 \ldots$

Recurring digits can be shown by putting a dot over the first and last digit of the group that repeats.

0.3333 … becomes $0.\dot{3}$

0.181818 … becomes $0.\dot{1}\dot{8}$

0.123123123 … becomes $0.\dot{1}2\dot{3}$

0.58333 … becomes $0.58\dot{3}$

0.6181818 … becomes $0.6\dot{1}\dot{8}$

0.4123123123 … become $0.4\dot{1}2\dot{3}$

Converting fractions into recurring decimals

A fraction that does not convert to a terminating decimal will give a recurring decimal. You may already know that $\frac{1}{3} = 0.333\ldots = 0.\dot{3}$

This means that the 3s go on forever and the decimal never ends.

To convert the fraction, you can usually use a calculator to divide the numerator by the denominator. Note that calculators round the last digit so it may not always be a true recurring decimal in the display. Use a calculator to check the following recurring decimals.

$$\frac{2}{11} = 0.181\,818\ldots = 0.\dot{1}\dot{8}$$

$$\frac{4}{15} = 0.2666\ldots = 0.2\dot{6}$$

$$\frac{8}{13} = 0.615\,384\,615\,384\,6\ldots = 0.\dot{6}15\,38\dot{4}$$

Converting terminal decimals into fractions

To convert a terminating decimal to a fraction, take the decimal number as the numerator. Then the denominator is 10, 100, 1000 …, depending on the number of decimal places. Because a terminating decimal has a specific number of decimal places, you can use place value to work out exactly where the numerator and the denominator end. For example:

- $0.7 = \frac{7}{10}$
- $0.23 = \frac{23}{100}$
- $0.045 = \frac{45}{1000} = \frac{9}{200}$
- $2.34 = \frac{234}{100} = \frac{117}{50} = 2\frac{17}{50}$
- $0.625 = \frac{625}{1000} = \frac{5}{8}$

Converting recurring decimals into fractions

To convert a recurring decimal to a fraction you have to use the algebraic method shown in the examples below.

EXAMPLE 10

Convert $0.\dot{7}$ to a fraction.

Let x be the fraction. Then:

	$x = 0.777\,777\,777\ldots$	(1)
Multiply (1) by 10	$10x = 7.777\,777\,777\ldots$	(2)
Subtract (2) − (1)	$9x = 7$	
	$\Rightarrow x = \frac{7}{9}$	

EXAMPLE 11

Convert $0.\dot{5}6\dot{4}$ to a fraction.

Let x be the fraction. Then:

	$x = 0.564\,564\,564\ldots$	(1)
Multiply (1) by 1000	$1000x = 564.564\,564\,564\ldots$	(2)
Subtract (2) − (1)	$999x = 564$	
	$\Rightarrow x = \frac{564}{999} = \frac{188}{333}$	

As a general rule, multiply by 10 if one digit recurs, multiply by 100 if two digits recur, multiply by 1000 if three digits recur, and so on.

Finding reciprocals of numbers or fractions

The **reciprocal** of any number is 1 divided by the number.

For example, the reciprocal of 2 is $1 \div 2 = \frac{1}{2} = 0.5$

The reciprocal of 0.25 is $1 \div 0.25 = 4$

You can find the reciprocal of a fraction by inverting it.

For example, the reciprocal of $\frac{2}{3}$ is $\frac{3}{2}$.

The reciprocal of $\frac{7}{4}$ is $\frac{4}{7}$.

EXERCISE 5I

1 Work out each of these fractions as a decimal. Give them as terminating decimals or recurring decimals as appropriate.

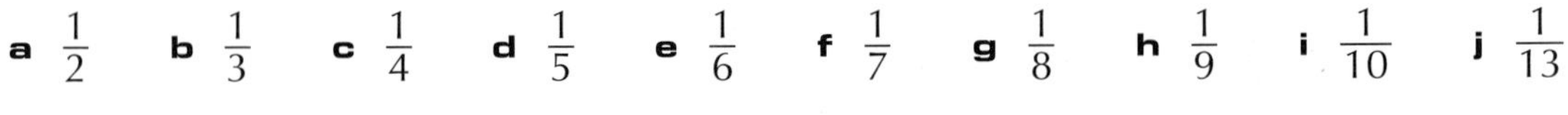

a $\frac{1}{2}$ **b** $\frac{1}{3}$ **c** $\frac{1}{4}$ **d** $\frac{1}{5}$ **e** $\frac{1}{6}$ **f** $\frac{1}{7}$ **g** $\frac{1}{8}$ **h** $\frac{1}{9}$ **i** $\frac{1}{10}$ **j** $\frac{1}{13}$

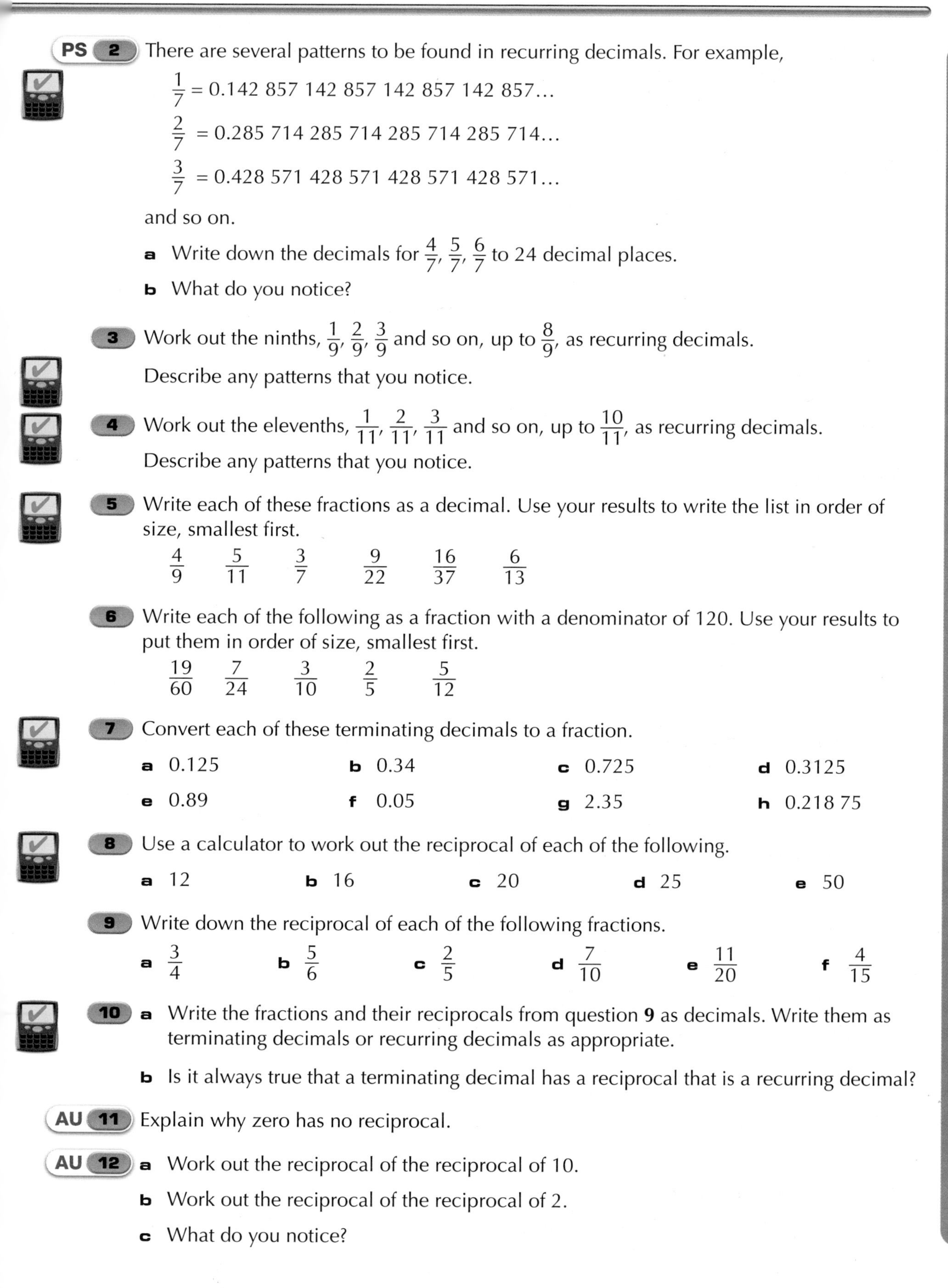

PS 2 There are several patterns to be found in recurring decimals. For example,

$\frac{1}{7} = 0.142\,857\,142\,857\,142\,857\,142\,857...$

$\frac{2}{7} = 0.285\,714\,285\,714\,285\,714\,285\,714...$

$\frac{3}{7} = 0.428\,571\,428\,571\,428\,571\,428\,571...$

and so on.

a Write down the decimals for $\frac{4}{7}, \frac{5}{7}, \frac{6}{7}$ to 24 decimal places.

b What do you notice?

3 Work out the ninths, $\frac{1}{9}, \frac{2}{9}, \frac{3}{9}$ and so on, up to $\frac{8}{9}$, as recurring decimals.

Describe any patterns that you notice.

4 Work out the elevenths, $\frac{1}{11}, \frac{2}{11}, \frac{3}{11}$ and so on, up to $\frac{10}{11}$, as recurring decimals.

Describe any patterns that you notice.

5 Write each of these fractions as a decimal. Use your results to write the list in order of size, smallest first.

$\frac{4}{9}$ $\quad \frac{5}{11}$ $\quad \frac{3}{7}$ $\quad \frac{9}{22}$ $\quad \frac{16}{37}$ $\quad \frac{6}{13}$

6 Write each of the following as a fraction with a denominator of 120. Use your results to put them in order of size, smallest first.

$\frac{19}{60}$ $\quad \frac{7}{24}$ $\quad \frac{3}{10}$ $\quad \frac{2}{5}$ $\quad \frac{5}{12}$

7 Convert each of these terminating decimals to a fraction.

a 0.125 **b** 0.34 **c** 0.725 **d** 0.3125

e 0.89 **f** 0.05 **g** 2.35 **h** 0.218 75

8 Use a calculator to work out the reciprocal of each of the following.

a 12 **b** 16 **c** 20 **d** 25 **e** 50

9 Write down the reciprocal of each of the following fractions.

a $\frac{3}{4}$ **b** $\frac{5}{6}$ **c** $\frac{2}{5}$ **d** $\frac{7}{10}$ **e** $\frac{11}{20}$ **f** $\frac{4}{15}$

10 **a** Write the fractions and their reciprocals from question **9** as decimals. Write them as terminating decimals or recurring decimals as appropriate.

b Is it always true that a terminating decimal has a reciprocal that is a recurring decimal?

AU 11 Explain why zero has no reciprocal.

AU 12 **a** Work out the reciprocal of the reciprocal of 10.

b Work out the reciprocal of the reciprocal of 2.

c What do you notice?

C

AU 13 x and y are two positive numbers.

If x is less than y, which statement is true?

The reciprocal of x is less than the reciprocal of y.

The reciprocal of x is greater than the reciprocal of y.

It is impossible to tell.

Give an example to support your answer.

AU 14 Explain why a number multiplied by its reciprocal is equal to 1. Use examples to show that this is true for negative numbers.

A

15 $x = 0.242\,424\,\ldots$

a What is $100x$?

b By subtracting the original value from your answer to part **a**, work out the value of $99x$.

c What is x as a fraction?

16 Convert each of these recurring decimals to a fraction.

a $0.\dot{8}$ **b** $0.\dot{3}\dot{4}$ **c** $0.\dot{4}\dot{5}$

d $0.\dot{5}6\dot{7}$ **e** $0.\dot{4}$ **f** $0.0\dot{4}$

g $0.1\dot{4}$ **h** $0.0\dot{4}\dot{5}$ **i** $2.\dot{7}$

j $7.\dot{6}\dot{3}$ **k** $3.\dot{3}$ **l** $2.\dot{0}\dot{6}$

A*

PS 17 **a** $\frac{1}{7}$ is a recurring decimal. $(\frac{1}{7})^2 = \frac{1}{49}$ is also a recurring decimal.

Is it true that when you square any fraction that is a recurring decimal, you get another fraction that is also a recurring decimal? Try this with at least four numerical examples before you make a decision.

b $\frac{1}{4}$ is a terminating decimal. $(\frac{1}{4})^2 = \frac{1}{16}$ is also a terminating decimal.

Is it true that when you square any fraction that is a terminating decimal, you get another fraction that is also a terminating decimal? Try this with at least four numerical examples before you make a decision.

c What type of fraction do you get when you multiply a fraction that gives a recurring decimal by another fraction that gives a terminating decimal? Try this with at least four numerical examples before you make a decision.

PS 18 **a** Convert the recurring decimal $0.\dot{9}$ to a fraction.

b Prove that $0.4\dot{9}$ is equal to 0.5

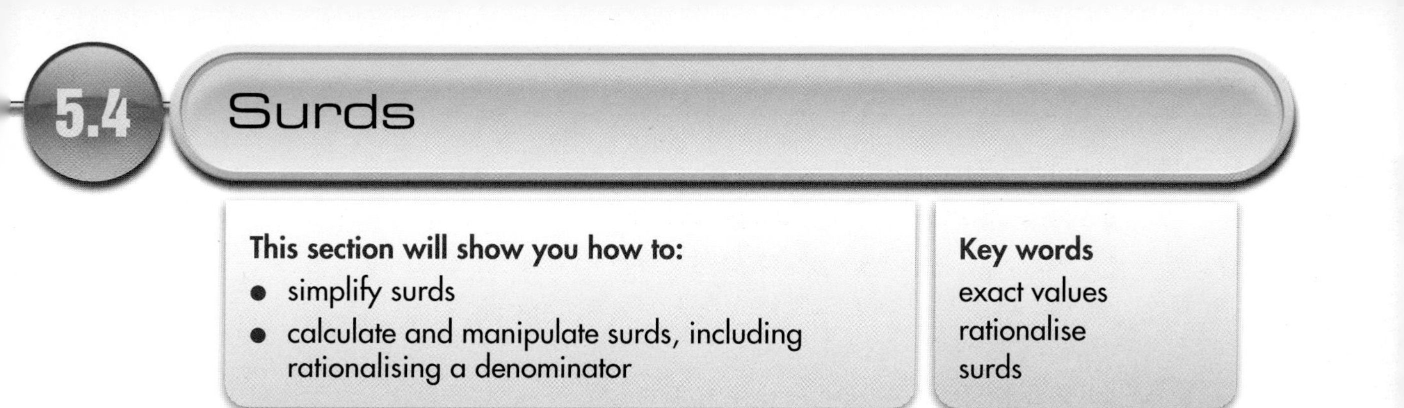

5.4 Surds

This section will show you how to:
- simplify surds
- calculate and manipulate surds, including rationalising a denominator

Key words
exact values
rationalise
surds

It is useful to be able to work with **surds**, which are roots of rational numbers written as, for example,

$\sqrt{2}$ $\sqrt{5}$ $\sqrt{15}$ $\sqrt{9}$ $\sqrt{3}$ $\sqrt{10}$

These are also referred to as **exact values**.

Here are four general rules for simplifying surds.

You can check that these rules work by taking numerical examples.

$\sqrt{a} \times \sqrt{b} = \sqrt{ab}$ $\quad$ $C\sqrt{a} \times D\sqrt{b} = CD\sqrt{ab}$

$\sqrt{a} \div \sqrt{b} = \sqrt{\dfrac{a}{b}}$ $\quad$ $C\sqrt{a} \div D\sqrt{b} = \dfrac{C}{D}\sqrt{\dfrac{a}{b}}$

For example,

$\sqrt{2} \times \sqrt{2} = \sqrt{4} = 2$ $\quad$ $\sqrt{2} \times \sqrt{10} = \sqrt{20} = \sqrt{(4 \times 5)} = \sqrt{4} \times \sqrt{5} = 2\sqrt{5}$

$\sqrt{2} \times \sqrt{3} = \sqrt{6}$ $\quad$ $\sqrt{6} \times \sqrt{15} = \sqrt{90} = \sqrt{9} \times \sqrt{10} = 3\sqrt{10}$

$\sqrt{2} \times \sqrt{8} = \sqrt{16} = 4$ $\quad$ $3\sqrt{5} \times 4\sqrt{3} = 12\sqrt{15}$

EXERCISE 5J

1 Simplify each of the following. Leave your answers in surd form if necessary.

a $\sqrt{2} \times \sqrt{3}$ **b** $\sqrt{5} \times \sqrt{3}$ **c** $\sqrt{2} \times \sqrt{2}$ **d** $\sqrt{2} \times \sqrt{8}$

e $\sqrt{5} \times \sqrt{8}$ **f** $\sqrt{3} \times \sqrt{3}$ **g** $\sqrt{6} \times \sqrt{2}$ **h** $\sqrt{7} \times \sqrt{3}$

i $\sqrt{2} \times \sqrt{7}$ **j** $\sqrt{2} \times \sqrt{18}$ **k** $\sqrt{6} \times \sqrt{6}$ **l** $\sqrt{5} \times \sqrt{6}$

2 Simplify each of the following. Leave your answers in surd form if necessary.

a $\sqrt{12} \div \sqrt{3}$ **b** $\sqrt{15} \div \sqrt{3}$ **c** $\sqrt{12} \div \sqrt{2}$ **d** $\sqrt{24} \div \sqrt{8}$

e $\sqrt{40} \div \sqrt{8}$ **f** $\sqrt{3} \div \sqrt{3}$ **g** $\sqrt{6} \div \sqrt{2}$ **h** $\sqrt{21} \div \sqrt{3}$

i $\sqrt{28} \div \sqrt{7}$ **j** $\sqrt{48} \div \sqrt{8}$ **k** $\sqrt{6} \div \sqrt{6}$ **l** $\sqrt{54} \div \sqrt{6}$

3 Simplify each of the following. Leave your answers in surd form if necessary.

a $\sqrt{2} \times \sqrt{3} \times \sqrt{2}$ **b** $\sqrt{5} \times \sqrt{3} \times \sqrt{15}$ **c** $\sqrt{2} \times \sqrt{2} \times \sqrt{8}$ **d** $\sqrt{2} \times \sqrt{8} \times \sqrt{3}$

e $\sqrt{5} \times \sqrt{8} \times \sqrt{8}$ **f** $\sqrt{3} \times \sqrt{3} \times \sqrt{3}$ **g** $\sqrt{6} \times \sqrt{2} \times \sqrt{48}$ **h** $\sqrt{7} \times \sqrt{3} \times \sqrt{3}$

i $\sqrt{2} \times \sqrt{7} \times \sqrt{2}$ **j** $\sqrt{2} \times \sqrt{18} \times \sqrt{5}$ **k** $\sqrt{6} \times \sqrt{6} \times \sqrt{3}$ **l** $\sqrt{5} \times \sqrt{6} \times \sqrt{30}$

A

A

4 Simplify each of the following. Leave your answers in surd form.

a $\sqrt{2} \times \sqrt{3} \div \sqrt{2}$ **b** $\sqrt{5} \times \sqrt{3} \div \sqrt{15}$ **c** $\sqrt{32} \times \sqrt{2} \div \sqrt{8}$ **d** $\sqrt{2} \times \sqrt{8} \div \sqrt{8}$

e $\sqrt{5} \times \sqrt{8} \div \sqrt{8}$ **f** $\sqrt{3} \times \sqrt{3} \div \sqrt{3}$ **g** $\sqrt{8} \times \sqrt{12} \div \sqrt{48}$ **h** $\sqrt{7} \times \sqrt{3} \div \sqrt{3}$

i $\sqrt{2} \times \sqrt{7} \div \sqrt{2}$ **j** $\sqrt{2} \times \sqrt{18} \div \sqrt{3}$ **k** $\sqrt{6} \times \sqrt{6} \div \sqrt{3}$ **l** $\sqrt{5} \times \sqrt{6} \div \sqrt{30}$

5 Simplify each of these expressions.

a $\sqrt{a} \times \sqrt{a}$ **b** $\sqrt{a} \div \sqrt{a}$ **c** $\sqrt{a} \times \sqrt{a} \div \sqrt{a}$

6 Simplify each of the following surds into the form $a\sqrt{b}$.

a $\sqrt{18}$ **b** $\sqrt{24}$ **c** $\sqrt{12}$ **d** $\sqrt{50}$ **e** $\sqrt{8}$ **f** $\sqrt{27}$

g $\sqrt{48}$ **h** $\sqrt{75}$ **i** $\sqrt{45}$ **j** $\sqrt{63}$ **k** $\sqrt{32}$ **l** $\sqrt{200}$

m $\sqrt{1000}$ **n** $\sqrt{250}$ **o** $\sqrt{98}$ **p** $\sqrt{243}$

7 Simplify each of these.

a $2\sqrt{18} \times 3\sqrt{2}$ **b** $4\sqrt{24} \times 2\sqrt{5}$ **c** $3\sqrt{12} \times 3\sqrt{3}$ **d** $2\sqrt{8} \times 2\sqrt{8}$

e $2\sqrt{27} \times 4\sqrt{8}$ **f** $2\sqrt{48} \times 3\sqrt{8}$ **g** $2\sqrt{45} \times 3\sqrt{3}$ **h** $2\sqrt{63} \times 2\sqrt{7}$

i $2\sqrt{32} \times 4\sqrt{2}$ **j** $\sqrt{1000} \times \sqrt{10}$ **k** $\sqrt{250} \times \sqrt{10}$ **l** $2\sqrt{98} \times 2\sqrt{2}$

8 Simplify each of these.

a $4\sqrt{2} \times 5\sqrt{3}$ **b** $2\sqrt{5} \times 3\sqrt{3}$ **c** $4\sqrt{2} \times 3\sqrt{2}$ **d** $2\sqrt{2} \times 2\sqrt{8}$

e $2\sqrt{5} \times 3\sqrt{8}$ **f** $3\sqrt{3} \times 2\sqrt{3}$ **g** $2\sqrt{6} \times 5\sqrt{2}$ **h** $5\sqrt{7} \times 2\sqrt{3}$

i $2\sqrt{2} \times 3\sqrt{7}$ **j** $2\sqrt{2} \times 3\sqrt{18}$ **k** $2\sqrt{6} \times 2\sqrt{6}$ **l** $4\sqrt{5} \times 3\sqrt{6}$

9 Simplify each of these.

a $6\sqrt{12} \div 2\sqrt{3}$ **b** $3\sqrt{15} \div \sqrt{3}$ **c** $6\sqrt{12} \div \sqrt{2}$ **d** $4\sqrt{24} \div 2\sqrt{8}$

e $12\sqrt{40} \div 3\sqrt{8}$ **f** $5\sqrt{3} \div \sqrt{3}$ **g** $14\sqrt{6} \div 2\sqrt{2}$ **h** $4\sqrt{21} \div 2\sqrt{3}$

i $9\sqrt{28} \div 3\sqrt{7}$ **j** $12\sqrt{56} \div 6\sqrt{8}$ **k** $25\sqrt{6} \div 5\sqrt{6}$ **l** $32\sqrt{54} \div 4\sqrt{6}$

10 Simplify each of these.

a $4\sqrt{2} \times \sqrt{3} \div 2\sqrt{2}$ **b** $4\sqrt{5} \times \sqrt{3} \div \sqrt{15}$ **c** $2\sqrt{32} \times 3\sqrt{2} \div 2\sqrt{8}$

d $6\sqrt{2} \times 2\sqrt{8} \div 3\sqrt{8}$ **e** $3\sqrt{5} \times 4\sqrt{8} \div 2\sqrt{8}$ **f** $12\sqrt{3} \times 4\sqrt{3} \div 2\sqrt{3}$

g $3\sqrt{8} \times 3\sqrt{12} \div 3\sqrt{48}$ **h** $4\sqrt{7} \times 2\sqrt{3} \div 8\sqrt{3}$ **i** $15\sqrt{2} \times 2\sqrt{7} \div 3\sqrt{2}$

j $8\sqrt{2} \times 2\sqrt{18} \div 4\sqrt{3}$ **k** $5\sqrt{6} \times 5\sqrt{6} \div 5\sqrt{3}$ **l** $2\sqrt{5} \times 3\sqrt{6} \div \sqrt{30}$

11 Simplify each of these expressions.

a $a\sqrt{b} \times c\sqrt{b}$ **b** $a\sqrt{b} \div c\sqrt{b}$ **c** $a\sqrt{b} \times c\sqrt{b} \div a\sqrt{b}$

PS 12 Find the value of a that makes each of these surds true.

a $\sqrt{5} \times \sqrt{a} = 10$ **b** $\sqrt{6} \times \sqrt{a} = 12$ **c** $\sqrt{10} \times 2\sqrt{a} = 20$

d $2\sqrt{6} \times 3\sqrt{a} = 72$ **e** $2\sqrt{a} \times \sqrt{a} = 6$ **f** $3\sqrt{a} \times 3\sqrt{a} = 54$

A*

13 Simplify the following.

a $\left(\frac{\sqrt{3}}{2}\right)^2$ **b** $\left(\frac{5}{\sqrt{3}}\right)^2$ **c** $\left(\frac{\sqrt{5}}{4}\right)^2$ **d** $\left(\frac{6}{\sqrt{3}}\right)^2$ **e** $\left(\frac{\sqrt{8}}{2}\right)^2$

AU 14 Decide whether each statement is true or false.

Show your working.

a $\sqrt{(a+b)} = \sqrt{a} + \sqrt{b}$ **b** $\sqrt{(a-b)} = \sqrt{a} - \sqrt{b}$

PS 15 Write down a product of two different surds which has an integer answer.

Calculating with surds

The following two examples show how surds can be used in solving problems.

EXAMPLE 12

In the right-angled triangle ABC, the side BC is $\sqrt{6}$ cm and the side AC is $\sqrt{18}$ cm.

Calculate the length of AB.
Leave your answer in surd form.

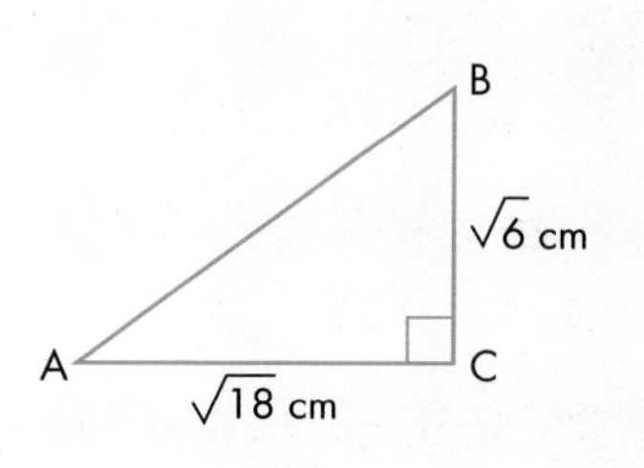

Note: A rule connecting the three sides of a right-angled triangle is

$a^2 + b^2 = c^2$

or $a^2 = c^2 - b^2$

This is known as Pythagoras' theorem.

Using Pythagoras' theorem:

$$AC^2 + BC^2 = AB^2$$

$$(\sqrt{18})^2 + (\sqrt{6})^2 = 18 + 6 = 24$$

$$\Rightarrow AB = \sqrt{24} \text{ cm}$$

$$= 2\sqrt{6} \text{ cm}$$

EXAMPLE 13

Calculate the area of a square with a side of $2 + \sqrt{3}$ cm.

Give your answer in the form $a + b\sqrt{3}$.

$$\text{Area} = (2 + \sqrt{3})^2 \text{ cm}^2$$

$$= (2 + \sqrt{3})(2 + \sqrt{3}) \text{ cm}^2$$

$$= 4 + 2\sqrt{3} + 2\sqrt{3} + 3 \text{ cm}^2$$

$$= 7 + 4\sqrt{3} \text{ cm}^2$$

Rationalising a denominator

When surds are written as fractions in answers they are usually given with a rational denominator.

Multiplying the numerator and denominator by an appropriate square root will make the denominator into a whole number.

EXAMPLE 14

Rationalise the denominator of: **a** $\frac{1}{\sqrt{3}}$ and **b** $\frac{2\sqrt{3}}{\sqrt{8}}$

a Multiply the numerator and denominator by $\sqrt{3}$.

$$\frac{1 \times \sqrt{3}}{\sqrt{3} \times \sqrt{3}} = \frac{\sqrt{3}}{3}$$

b Multiply the numerator and denominator by $\sqrt{8}$.

$$\frac{2\sqrt{3} \times \sqrt{8}}{\sqrt{8} \times \sqrt{8}} = \frac{2\sqrt{24}}{8} = \frac{4\sqrt{6}}{8} = \frac{\sqrt{6}}{2}$$

or

$$\sqrt{8} = 2\sqrt{2}$$

So $\frac{2\sqrt{3}}{\sqrt{8}} = \frac{2\sqrt{3}}{2\sqrt{2}} = \frac{\sqrt{3}}{\sqrt{2}}$

Multiplying the numerator and denominator by $\sqrt{2}$:

$$\frac{\sqrt{3} \times \sqrt{2}}{\sqrt{2} \times \sqrt{2}} = \frac{\sqrt{6}}{2}$$

EXERCISE 5K

A*

1 Show that:

a $(2 + \sqrt{3})(1 + \sqrt{3}) = 5 + 3\sqrt{3}$

b $(1 + \sqrt{2})(2 + \sqrt{3}) = 2 + 2\sqrt{2} + \sqrt{3} + \sqrt{6}$

c $(4 - \sqrt{3})(4 + \sqrt{3}) = 13$

2 Expand and simplify where possible.

a $\sqrt{3}(2 - \sqrt{3})$ **b** $\sqrt{2}(3 - 4\sqrt{2})$ **c** $\sqrt{5}(2\sqrt{5} + 4)$

d $3\sqrt{7}(4 - 2\sqrt{7})$ **e** $3\sqrt{2}(5 - 2\sqrt{8})$ **f** $\sqrt{3}(\sqrt{27} - 1)$

3 Expand and simplify where possible.

a $(1 + \sqrt{3})(3 - \sqrt{3})$ **b** $(2 + \sqrt{5})(3 - \sqrt{5})$ **c** $(1 - \sqrt{2})(3 + 2\sqrt{2})$

d $(3 - 2\sqrt{7})(4 + 3\sqrt{7})$ **e** $(2 - 3\sqrt{5})(2 + 3\sqrt{5})$ **f** $(\sqrt{3} + \sqrt{2})(\sqrt{3} + \sqrt{8})$

g $(2 + \sqrt{5})^2$ **h** $(1 - \sqrt{2})^2$ **i** $(3 + \sqrt{2})^2$

4 Work out the missing lengths in each of these triangles, giving the answer in as simple a form as possible. (Refer to Pythagoras' theorem in Example 12.)

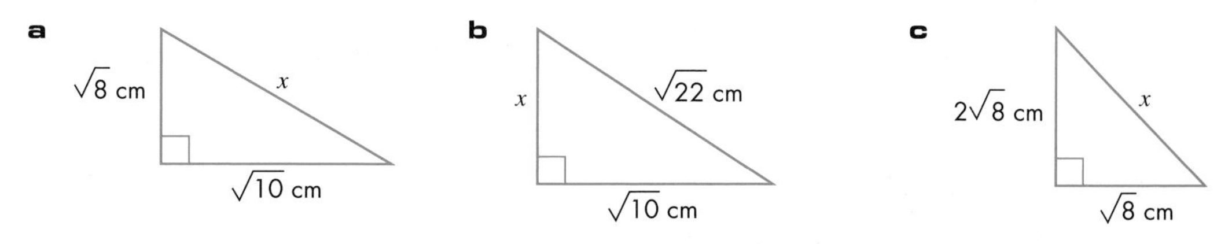

5 Calculate the area of each of these rectangles, simplifying your answers where possible. (The area of a rectangle with length l and width w is $A = l \times w$.)

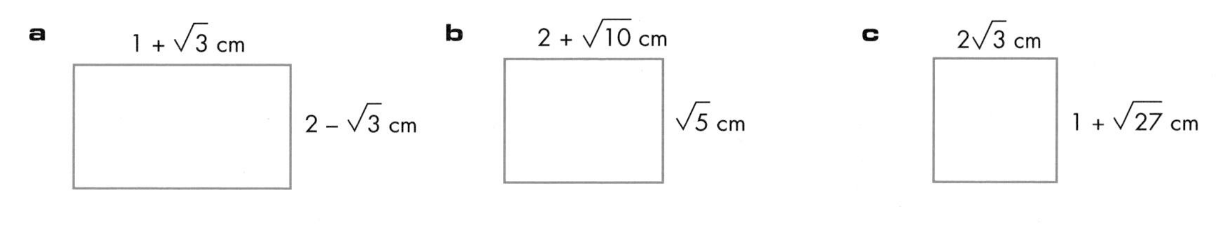

6 Rationalise the denominators of these expressions.

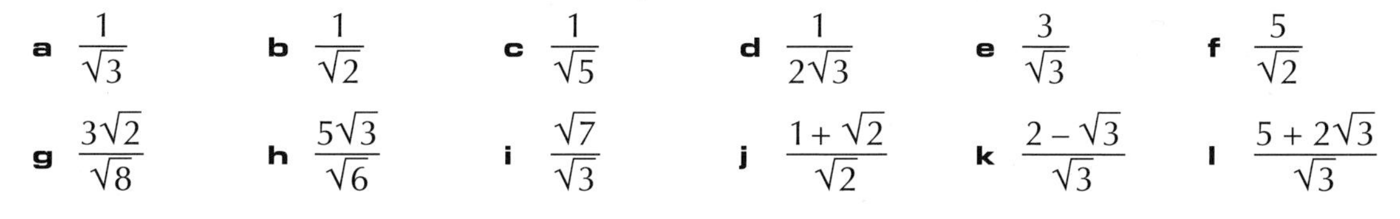

a $\frac{1}{\sqrt{3}}$ **b** $\frac{1}{\sqrt{2}}$ **c** $\frac{1}{\sqrt{5}}$ **d** $\frac{1}{2\sqrt{3}}$ **e** $\frac{3}{\sqrt{3}}$ **f** $\frac{5}{\sqrt{2}}$

g $\frac{3\sqrt{2}}{\sqrt{8}}$ **h** $\frac{5\sqrt{3}}{\sqrt{6}}$ **i** $\frac{\sqrt{7}}{\sqrt{3}}$ **j** $\frac{1+\sqrt{2}}{\sqrt{2}}$ **k** $\frac{2-\sqrt{3}}{\sqrt{3}}$ **l** $\frac{5+2\sqrt{3}}{\sqrt{3}}$

7 **a** Expand and simplify the following.

i $(2 + \sqrt{3})(2 - \sqrt{3})$ **ii** $(1 - \sqrt{5})(1 + \sqrt{5})$ **iii** $(\sqrt{3} - 1)(\sqrt{3} + 1)$

iv $(3\sqrt{2} + 1)(3\sqrt{2} - 1)$ **v** $(2 - 4\sqrt{3})(2 + 4\sqrt{3})$

b What happens in the answers to part **a**? Why?

AU PS **8** **a** Write down two surds that, when multiplied, give a rational number.

b Write down two surds that, when multiplied, do not give a rational number.

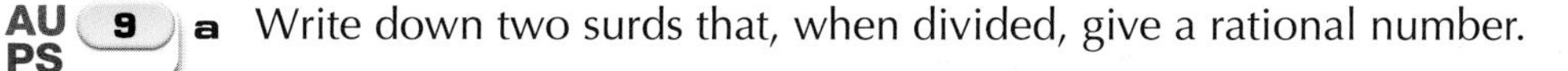

AU PS **9** **a** Write down two surds that, when divided, give a rational number.

b Write down two surds that, when divided, do not give a rational number.

FM **10** An engineer uses a formula to work out the number of metres of cable he needs to complete a job. His calculator displays the answer as $10\sqrt{70}$. The button for converting this to a decimal is not working.

He has 80 metres of cable. Without using a calculator, decide whether he has enough cable. Show clearly how you decide.

11 Write $(3 + \sqrt{2})^2 - (1 - \sqrt{8})^2$ in the form $a + b\sqrt{c}$ where a, b and c are integers.

12 $x^2 - y^2 \equiv (x + y)(x - y)$ is an identity which means it is true for any values of x and y whether they are numeric or algebraic.

Show that it is true for $x = 1 + \sqrt{2}$ and $y = 1 - \sqrt{8}$

SUMMARY

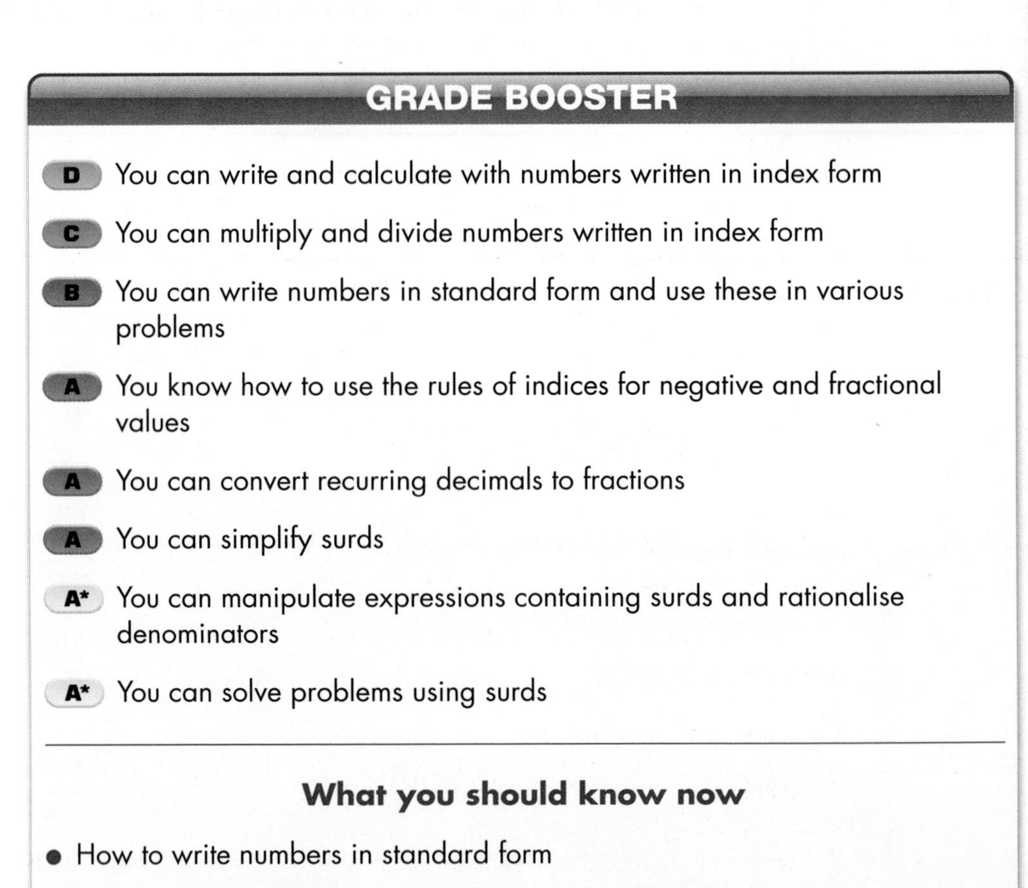

GRADE BOOSTER

- **D** You can write and calculate with numbers written in index form
- **C** You can multiply and divide numbers written in index form
- **B** You can write numbers in standard form and use these in various problems
- **A** You know how to use the rules of indices for negative and fractional values
- **A** You can convert recurring decimals to fractions
- **A** You can simplify surds
- **A*** You can manipulate expressions containing surds and rationalise denominators
- **A*** You can solve problems using surds

What you should know now

- How to write numbers in standard form
- How to solve problems using numbers in standard form
- How to manipulate indices, both integer (positive and negative) and fractional
- How to compare fractions by converting them to decimals
- How to convert decimals into fractions
- What surds are and how to manipulate them

1 $x = \sqrt{\dfrac{p+q}{pq}}$

$p = 4 \times 10^8 \quad q = 3 \times 10^6$

Find the value of x. Give your answer in standard form correct to 2 significant figures.

Edexcel, March 2005, Paper 13 Higher, Question 3

2 **a** Work out $3^6 \div 3^2$ *(1 mark)*

b Write down the value of $36^{\frac{1}{2}}$ *(1 mark)*

c $3^n = \frac{1}{9}$

Find the value of n. *(1 mark)*

Edexcel, June 2007, Paper 11, Question 7

3 **a** Write down the exact value of 3^{-2} *(1 mark)*

b Simplify $\dfrac{7^2 \times 7^4}{7^3}$ *(2 marks)*

Edexcel, May 2008, Paper 14, Question 14

4 **a** Simplify $t^6 \times t^2$ *(1 mark)*

b Simplify $\dfrac{m^8}{m^3}$ *(1 mark)*

c Simplify $(2x)^3$ *(2 marks)*

d Simplify $3a^2h \times 4a^5h^4$ *(2 marks)*

Edexcel, June 2009, Paper 4, Question 16

5 **a** Write the number 45 000 in standard form. *(1 mark)*

b Write 6×10^{-2} as an ordinary number. *(1 mark)*

Edexcel, November 2008, Paper 11, Question 7

6 The number of atoms in one kilogram of helium is 1.51×10^{26}

Calculate the number of atoms in 20 kilograms of helium.

Give your answer in standard form. *(2 marks)*

Edexcel, June 2008, Paper 4, Question 17

7 Simplify

a $t^4 \times t^3$ **b** $\dfrac{m}{m^6}$ **c** $(3k^2m^2) \times (4k^3m)$

8 Simplify **a** $\dfrac{8x^2y \times 3xy^3}{6x^2y^2}$ **b** $(2m^4p^2)^3$

9 Express the recurring decimal 0.5333333… as a fraction. Give your answer in its simplest form.

10 Find values of a and b such that

$(3 + \sqrt{5})(2 - \sqrt{5}) = a + b\sqrt{5}$

11 The area of this rectangle is 40 cm²

Find the value of x. Give your answer in the form $a\sqrt{b}$ where a and b are integers.

$4\sqrt{2}$ cm

x cm

12 **a** Prove that $0.\dot{7}\dot{2} = \dfrac{8}{11}$

b Hence, or otherwise, express $0.3\dot{7}\dot{2}$ as a fraction.

13 Express the recurring decimal $0.2\dot{1}\dot{3}$ as a fraction. *(3 marks)*

Edexcel, November 2008, Paper 3, Question 24

14 **a** Write down the value of $49^{\frac{1}{2}}$ *(1 mark)*

b Write $\sqrt{45}$ in the form $k\sqrt{5}$, where k is an integer. *(1 mark)*

Edexcel, November 2008, Paper 3, Question 25

15 **a** Rationalise the denominator of $\dfrac{1}{\sqrt{3}}$ *(1 mark)*

b Expand $(2 + \sqrt{3})(1 + \sqrt{3})$

Give your answer in the form $a + b\sqrt{3}$, where a and b are integers. *(2 marks)*

Edexcel, May 2008, Paper 3, Question 23

16 Change the recurring decimal $0.\dot{2}\dot{3}$ to a fraction. *(2 marks)*

Edexcel, November 2008, Paper 11, Question 9

17 **a** **i** Show that $\sqrt{32} = 4\sqrt{2}$

ii Expand and simplify $(\sqrt{2} + \sqrt{12})^2$

b Show clearly that this triangle is right-angled

2

$\sqrt{2} + \sqrt{12}$

$2 + \sqrt{6}$

Worked Examination Questions

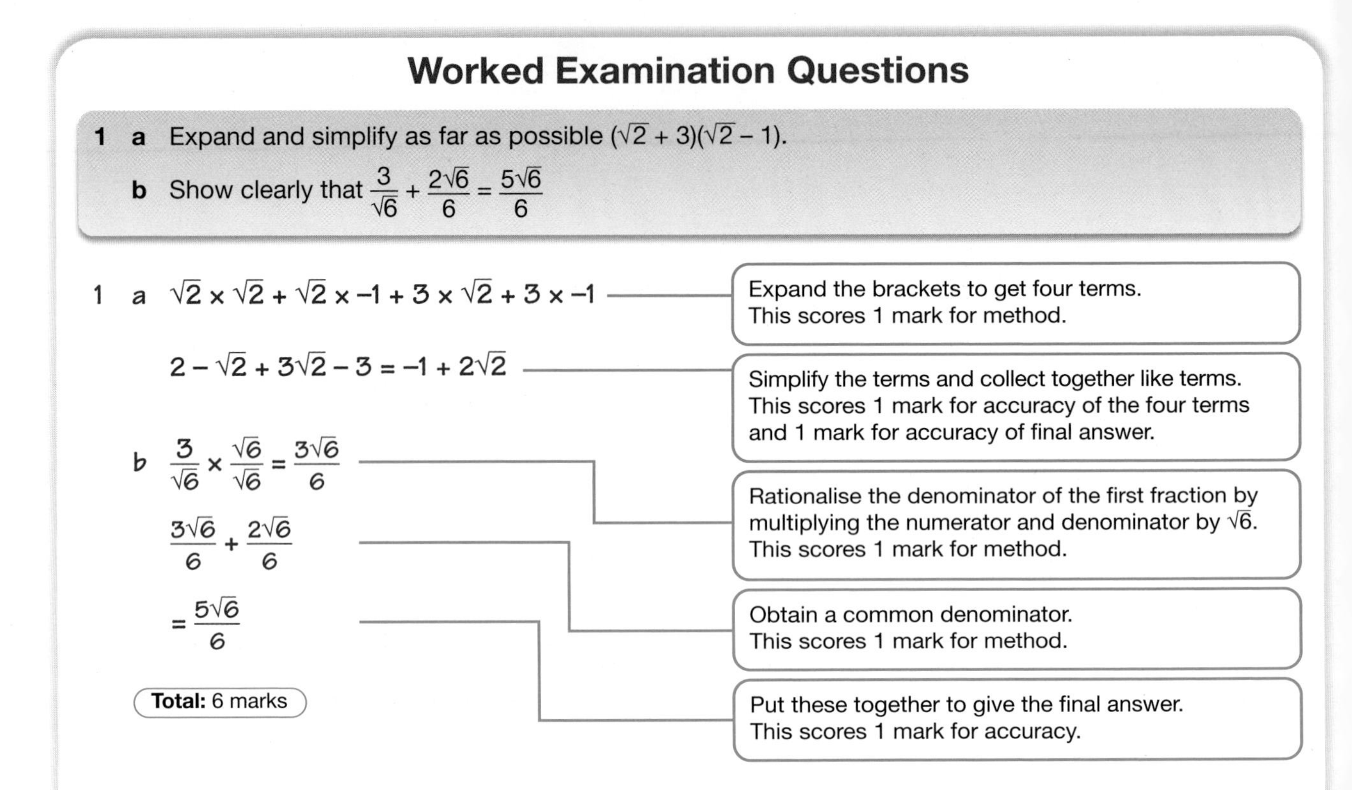

1 a Expand and simplify as far as possible $(\sqrt{2} + 3)(\sqrt{2} - 1)$.

b Show clearly that $\frac{3}{\sqrt{6}} + \frac{2\sqrt{6}}{6} = \frac{5\sqrt{6}}{6}$

1 a $\sqrt{2} \times \sqrt{2} + \sqrt{2} \times -1 + 3 \times \sqrt{2} + 3 \times -1$ — Expand the brackets to get four terms. This scores 1 mark for method.

$2 - \sqrt{2} + 3\sqrt{2} - 3 = -1 + 2\sqrt{2}$ — Simplify the terms and collect together like terms. This scores 1 mark for accuracy of the four terms and 1 mark for accuracy of final answer.

b $\frac{3}{\sqrt{6}} \times \frac{\sqrt{6}}{\sqrt{6}} = \frac{3\sqrt{6}}{6}$ — Rationalise the denominator of the first fraction by multiplying the numerator and denominator by $\sqrt{6}$. This scores 1 mark for method.

$\frac{3\sqrt{6}}{6} + \frac{2\sqrt{6}}{6}$ — Obtain a common denominator. This scores 1 mark for method.

$= \frac{5\sqrt{6}}{6}$ — Put these together to give the final answer. This scores 1 mark for accuracy.

Total: 6 marks

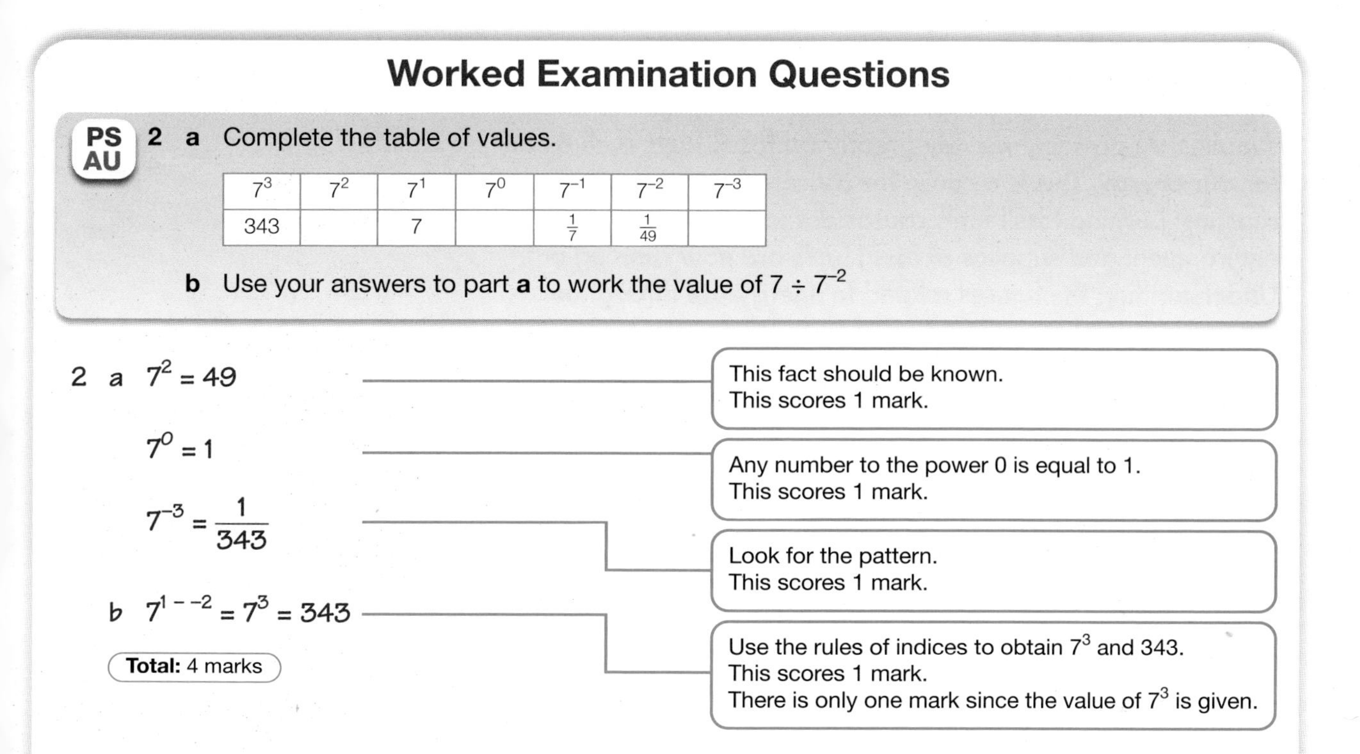

Worked Examination Questions

PS AU **2** **a** Complete the table of values.

7^3	7^2	7^1	7^0	7^{-1}	7^{-2}	7^{-3}
343		7		$\frac{1}{7}$	$\frac{1}{49}$	

b Use your answers to part **a** to work the value of $7 \div 7^{-2}$

2 a $7^2 = 49$ — This fact should be known. This scores 1 mark.

$7^0 = 1$ — Any number to the power 0 is equal to 1. This scores 1 mark.

$7^{-3} = \frac{1}{343}$ — Look for the pattern. This scores 1 mark.

b $7^{1--2} = 7^3 = 343$ — Use the rules of indices to obtain 7^3 and 343. This scores 1 mark. There is only one mark since the value of 7^3 is given.

Total: 4 marks

5 Functional Maths
Oil crisis

Throughout the world, people are concerned about our energy supplies. At present, we rely greatly on fossil fuels such as oil for our energy. This is a cause for concern for two main reasons: burning fossil fuels causes damage to the environment and supplies of fossil fuels are now running out. Understanding the figures related to energy use throughout the world is vital in finding a solution to the energy crisis.

Getting started

Use your knowledge of science and current affairs to bring together what you know about energy resources. The questions below may help you.

- Which sources of energy do we use in the UK?
- Where does energy come from?
- Which forms of energy are considered to be 'green'?

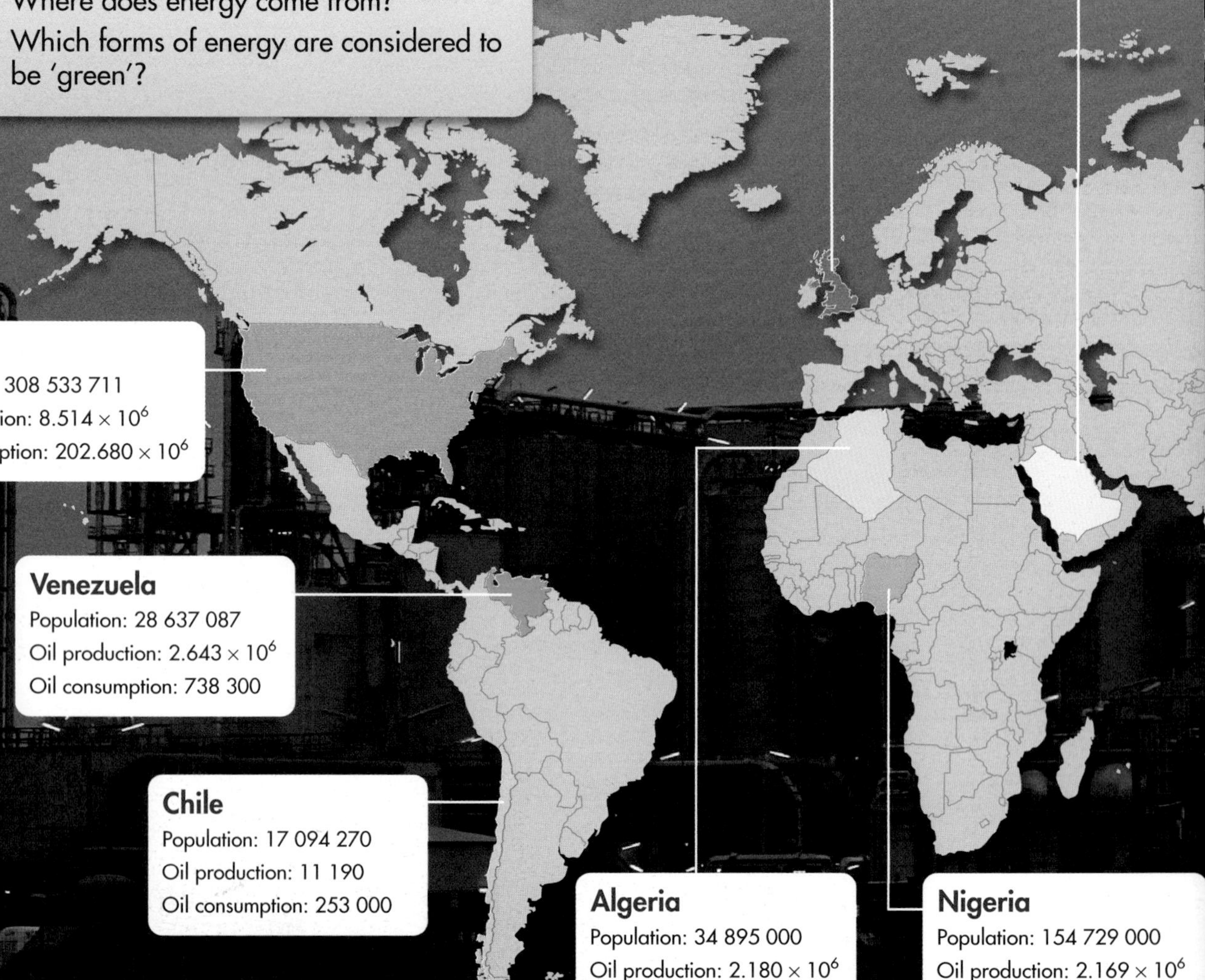

UK
Population: 62 041 708
Oil production: 1.584×10^6
Oil consumption: 1.765×10^6

Saudi Arabia
Population: 25 721 00
Oil production: 10.7×10^6
Oil consumption: 2.34×10^6

USA
Population: 308 533 711
Oil production: 8.514×10^6
Oil consumption: 202.680×10^6

Venezuela
Population: 28 637 087
Oil production: 2.643×10^6
Oil consumption: 738 300

Chile
Population: 17 094 270
Oil production: 11 190
Oil consumption: 253 000

Algeria
Population: 34 895 000
Oil production: 2.180×10^6
Oil consumption: 279 800

Nigeria
Population: 154 729 000
Oil production: 2.169×10^6
Oil consumption: 312 000

The energy that we get from oil can also be generated from more environmentally-friendly, sustainable sources.

Your task

A scientist is researching the production and consumption of oil in the world, in order to inform people about the global energy crisis.

She looks at 10 oil-producing countries and, for each country, finds the most recent figures on the country's population, and its oil production and consumption, measured in barrels per day.

The scientist asks you to write a report about the production and consumption of oil throughout the world for a national newspaper.

Your report should use evidence to explain:

- how many barrels of oil are produced per person, per year
- in which year the world is 'greener'
- which country is the 'best' consumer of oil
- which country is the 'worst' consumer of oil.

Year	World population	World oil production, barrels per day
1984	4.77×10^9	5.45×10^7
1989	5.19×10^9	5.99×10^7
1994	5.61×10^9	6.10×10^7
1999	6.01×10^9	6.58×10^7
2004	6.38×10^9	7.25×10^7
2009	6.79×10^9	8.49×10^7

Oil production and oil consumption figures on this map are measured in barrels.

Japan
Population: 127 530 000
Oil production: 133 100
Oil consumption: 5.007×10^6

Indonesia
Population: 234 181 400
Oil production: 1.051×10^6
Oil consumption: 1.219×10^6

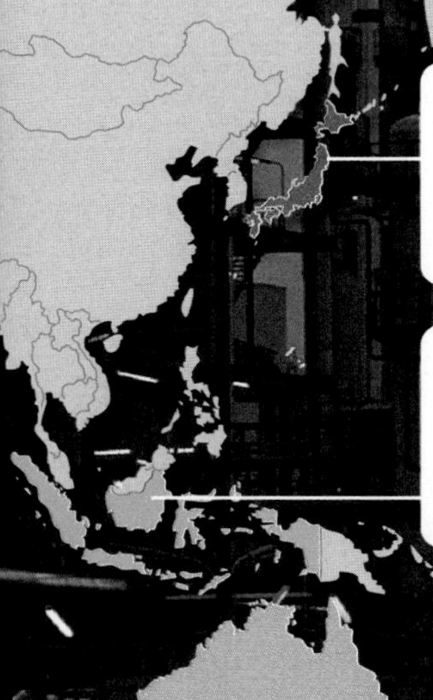

Australia
Population: 22 125 030
Oil production: 86 400
Oil consumption: 966 200

Extension

Develop your report by researching the 'green', sustainable energy resources that many countries produce and consume. How do the figures related to 'green' energy compare to those related to oil?

Why this chapter matters

Like most mathematics, quadratic equations have their origins in ancient Egypt.

The Egyptians did not have a formal system of algebra but could solve problems that involved quadratics.This problem was written in hieroglyphics on the Berlin Papyrus which was written some time around 2160–1700BC:

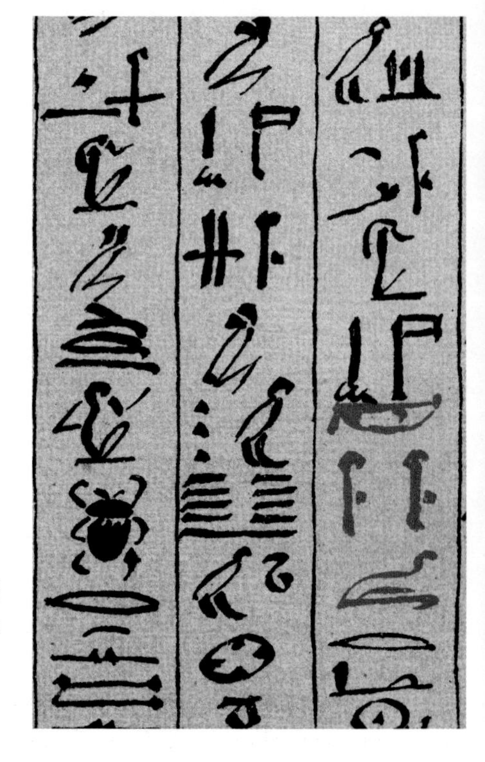

The area of a square of 100 is equal to that of two smaller squares. The side of one is $\frac{1}{2} + \frac{1}{4}$ the side of the other.

Today we would express this as:

$$x^2 + y^2 = 100$$
$$y = \tfrac{3}{4}x$$

Euclid

In about 300BC, Euclid developed a geometrical method for solving quadratics. This work was developed by Hindu mathematicians, but it was not until much later, in 1145AD that the Arabic mathematician Abraham bar Hiyya Ha-Nasi, published the book *Liber embadorum*, which gave a complete solution of the quadratic equation.

On 26 June 2003, the quadratic equation was the subject of a debate in Parliament. The National Union of Teachers had suggested that students should be allowed to give up mathematics at the age of 14, and making them do 'abstract and irrelevant' things such as learning about quadratic equations did not serve any purpose.

Mr Tony McWalter, the MP for Hemel Hempstead, defending the teaching of quadratic equations, said:

> "Someone who thinks that the quadratic equation is an empty manipulation, devoid of any other significance, is someone who is content with leaving the many in ignorance. I believe that he or she is also pleading for the lowering of standards. A quadratic equation is not like a bleak room, devoid of furniture, in which one is asked to squat. It is a door to a room full of the unparalleled riches of human intellectual achievement. If you do not go through that door, or if it is said that it is an uninteresting thing to do, much that passes for human wisdom will be forever denied you."

Chapter 6

Algebra: Quadratic equations

1 Expanding brackets

2 Quadratic factorisation

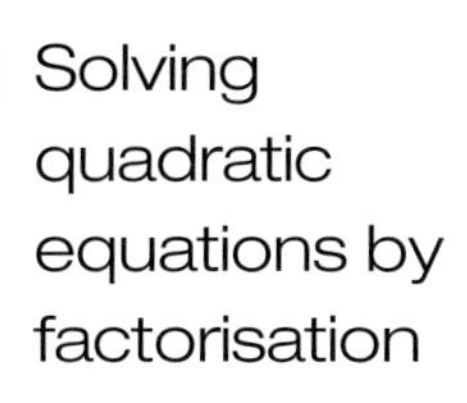
3 Solving quadratic equations by factorisation

4 Solving a quadratic equation by the quadratic formula

5 Solving a quadratic equation by completing the square

6 Problems involving quadratic equations

The grades given in this chapter are target grades.

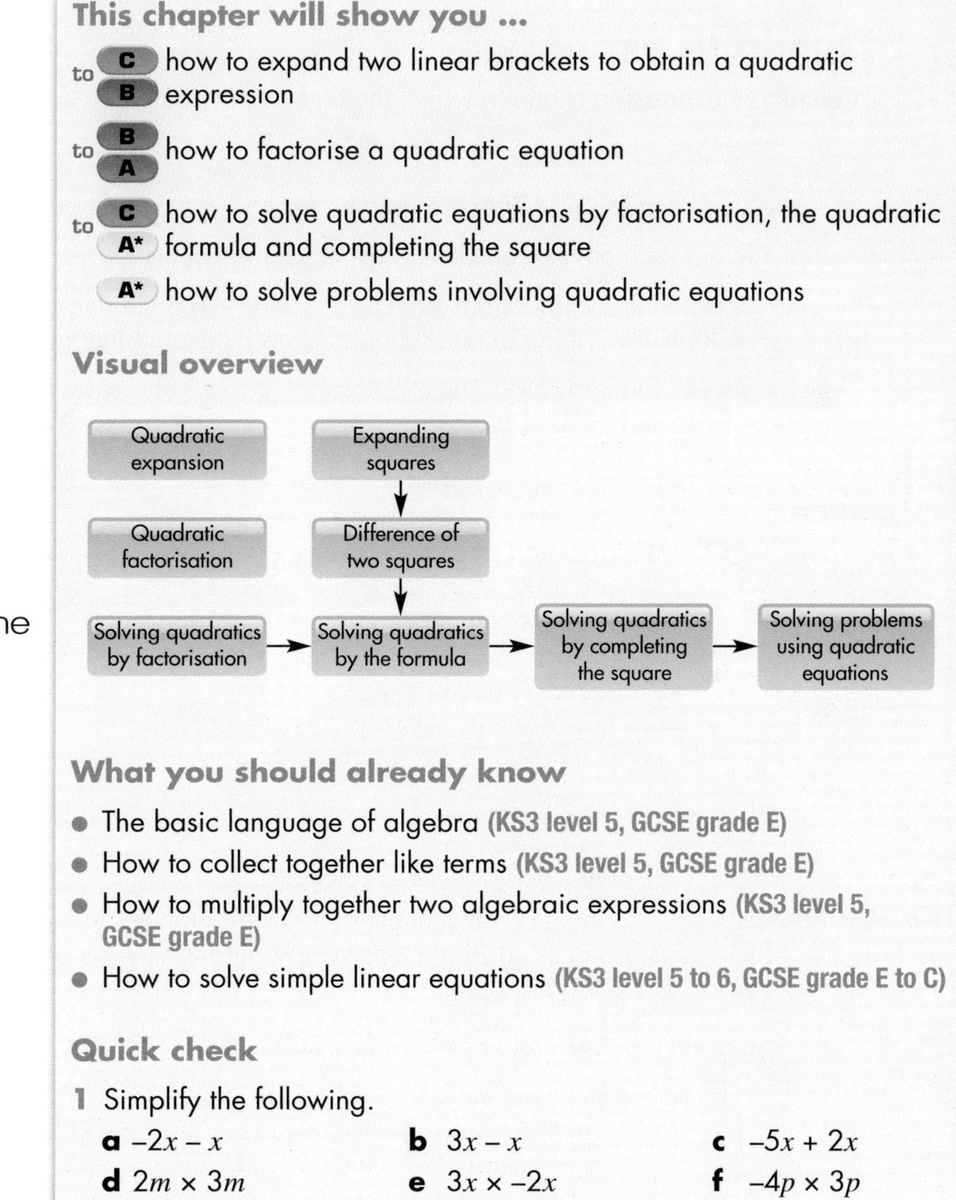

This chapter will show you ...

- C to B: how to expand two linear brackets to obtain a quadratic expression
- B to A: how to factorise a quadratic equation
- C to A*: how to solve quadratic equations by factorisation, the quadratic formula and completing the square
- A*: how to solve problems involving quadratic equations

Visual overview

What you should already know

- The basic language of algebra **(KS3 level 5, GCSE grade E)**
- How to collect together like terms **(KS3 level 5, GCSE grade E)**
- How to multiply together two algebraic expressions **(KS3 level 5, GCSE grade E)**
- How to solve simple linear equations **(KS3 level 5 to 6, GCSE grade E to C)**

Quick check

1 Simplify the following.

a $-2x - x$ **b** $3x - x$ **c** $-5x + 2x$

d $2m \times 3m$ **e** $3x \times -2x$ **f** $-4p \times 3p$

2 Solve these equations.

a $x + 6 = 0$ **b** $2x + 1 = 0$ **c** $3x - 2 = 0$

6.1 Expanding brackets

This section will show you how to:
- expand two linear brackets to obtain a quadratic expression

Key words
coefficient
linear
quadratic expression

Quadratic expansion

A **quadratic expression** is one in which the highest power of the variables is 2. For example,

y^2 $\quad$ $3t^2 + 5t$ $\quad$ $5m^2 + 3m + 8$

An expression such as $(3y + 2)(4y - 5)$ can be expanded to give a quadratic expression.

Multiplying out such pairs of brackets is usually called *quadratic expansion*.

The rule for expanding expressions such as $(t + 5)(3t - 4)$ is similar to that for expanding single brackets: multiply everything in one set of brackets by everything in the other set of brackets.

There are several methods for doing this. Examples 1 to 3 show the three main methods: expansion, FOIL and the box method.

EXAMPLE 1

In the expansion method, split the terms in the first set of brackets, make each of them multiply both terms in the second set of brackets, then simplify the outcome.

Expand $(x + 3)(x + 4)$

$$(x + 3)(x + 4) = x(x + 4) + 3(x + 4)$$
$$= x^2 + 4x + 3x + 12$$
$$= x^2 + 7x + 12$$

EXAMPLE 2

FOIL stands for First, Outer, Inner and Last. This is the order of multiplying the terms from each set of brackets.

Expand $(t + 5)(t - 2)$

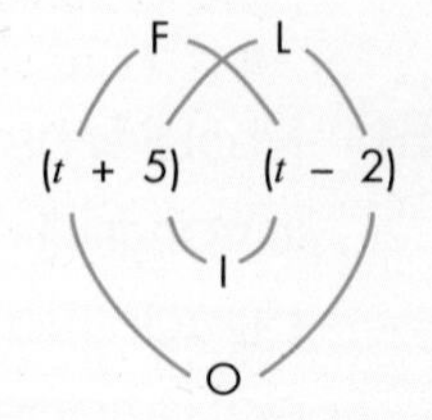

First terms give: $t \times t = t^2$

Outer terms give: $t \times -2 = -2t$

Inner terms give: $5 \times t = 5t$

Last terms give: $+5 \times -2 = -10$

$$(t + 5)(t - 2) = t^2 - 2t + 5t - 10$$
$$= t^2 + 3t - 10$$

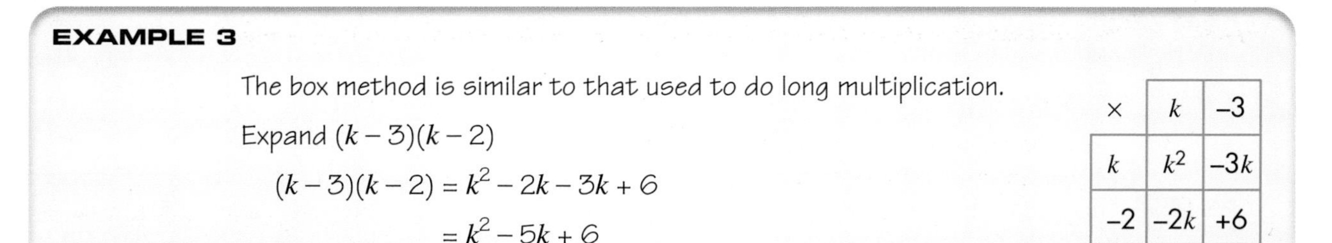

EXAMPLE 3

The box method is similar to that used to do long multiplication.

Expand $(k - 3)(k - 2)$

$$(k - 3)(k - 2) = k^2 - 2k - 3k + 6$$
$$= k^2 - 5k + 6$$

×	k	-3
k	k^2	$-3k$
-2	$-2k$	$+6$

Warning: Be careful with the signs. This is the main place where mistakes are made in questions involving the expansion of brackets.

EXERCISE 6A

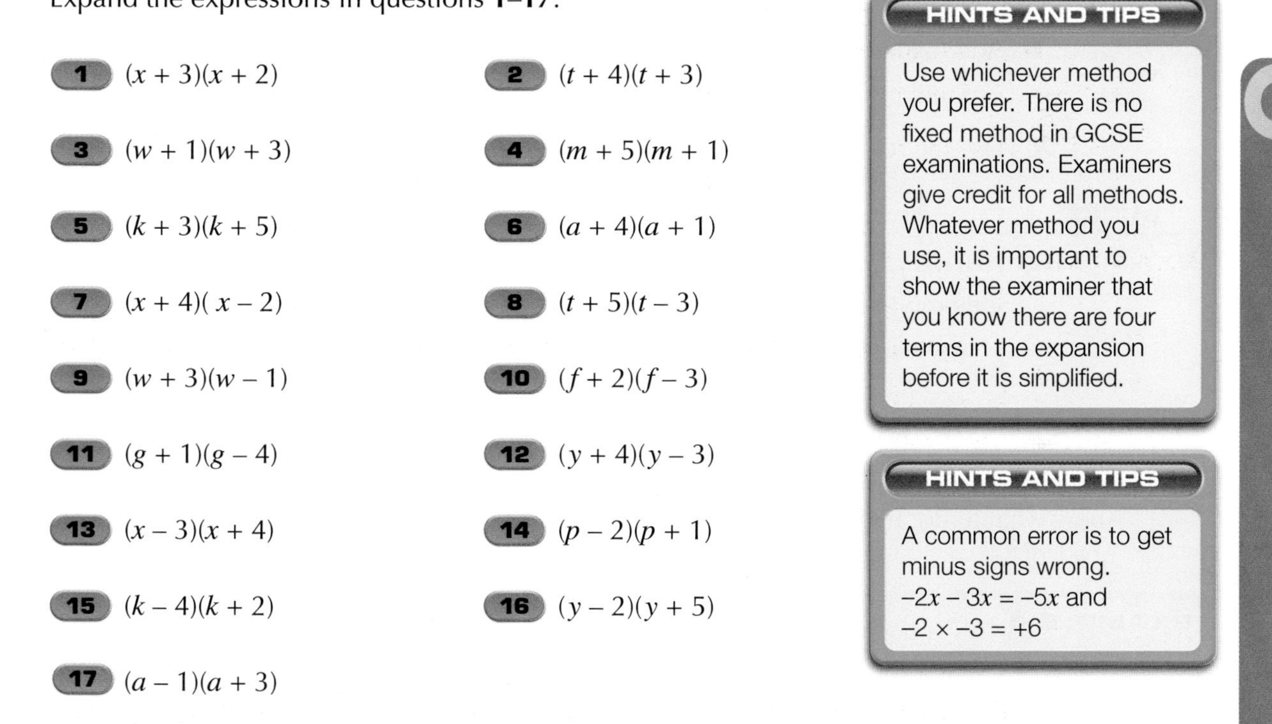

Expand the expressions in questions **1–17**.

1 $(x + 3)(x + 2)$

2 $(t + 4)(t + 3)$

3 $(w + 1)(w + 3)$

4 $(m + 5)(m + 1)$

5 $(k + 3)(k + 5)$

6 $(a + 4)(a + 1)$

7 $(x + 4)(x - 2)$

8 $(t + 5)(t - 3)$

9 $(w + 3)(w - 1)$

10 $(f + 2)(f - 3)$

11 $(g + 1)(g - 4)$

12 $(y + 4)(y - 3)$

13 $(x - 3)(x + 4)$

14 $(p - 2)(p + 1)$

15 $(k - 4)(k + 2)$

16 $(y - 2)(y + 5)$

17 $(a - 1)(a + 3)$

HINTS AND TIPS

Use whichever method you prefer. There is no fixed method in GCSE examinations. Examiners give credit for all methods. Whatever method you use, it is important to show the examiner that you know there are four terms in the expansion before it is simplified.

HINTS AND TIPS

A common error is to get minus signs wrong.
$-2x - 3x = -5x$ and
$-2 \times -3 = +6$

C

The expansions of the expressions in questions **18–26** follow a pattern. Work out the first few and try to spot the pattern that will allow you immediately to write down the answers to the rest.

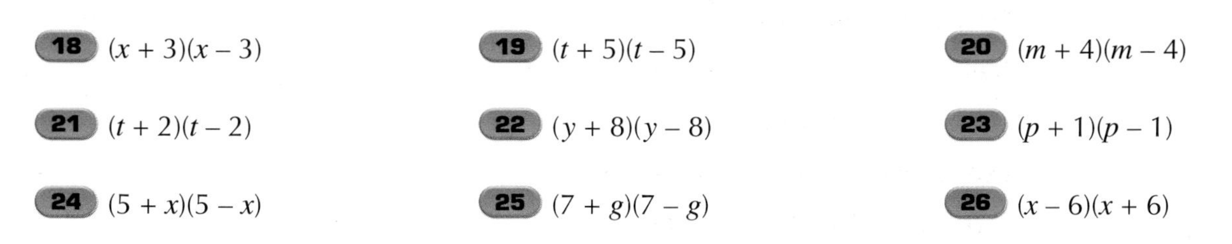

18 $(x + 3)(x - 3)$

19 $(t + 5)(t - 5)$

20 $(m + 4)(m - 4)$

21 $(t + 2)(t - 2)$

22 $(y + 8)(y - 8)$

23 $(p + 1)(p - 1)$

24 $(5 + x)(5 - x)$

25 $(7 + g)(7 - g)$

26 $(x - 6)(x + 6)$

PS 27 This rectangle is made up of four parts with areas of x^2, $2x$, $3x$ and 6 square units.

Work out expressions for the sides of the rectangle, in terms of x.

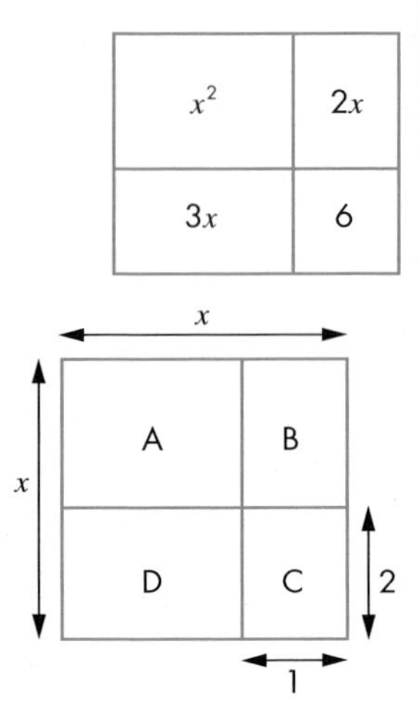

PS 28 This square has an area of x^2 square units.
It is split into four rectangles.

a Fill in the table below to show the dimensions and area of each rectangle.

Rectangle	Length	Width	Area
A	$x - 1$	$x - 2$	$(x - 1)(x - 2)$
B			
C			
D			

b Add together the areas of rectangles B, C and D.

Expand any brackets and collect terms together.

c Use the results to explain why $(x - 1)(x - 2) = x^2 - 3x + 2$.

AU 29 **a** Expand $(x - 3)(x + 3)$

b Use the result in **a** to write down the answers to these. (Do not use a calculator or do a long multiplication.)

i 97×103 **ii** 197×203

Quadratic expansion with non-unit coefficients

All the algebraic terms in x^2 in Exercise 6A have a **coefficient** of 1 or –1. The next two examples show what to do if you have to expand brackets containing terms in x^2 with coefficients that are not 1 or –1.

EXAMPLE 4

Expand $(2t + 3)(3t + 1)$

$$(2t + 3)(3t + 1) = 6t^2 + 2t + 9t + 3$$
$$= 6t^2 + 11t + 3$$

×	$2t$	$+3$
$3t$	$6t^2$	$+9t$
$+1$	$+2t$	$+3$

EXAMPLE 5

Expand $(4x - 1)(3x - 5)$

$$(4x - 1)(3x - 5) = 4x(3x - 5) - (3x - 5)$$ [**Note:** $(3x - 5)$ is the same as $1(3x - 5)$.]

$$= 12x^2 - 20x - 3x + 5$$
$$= 12x^2 - 23x + 5$$

EXERCISE 6B

Expand the expressions in questions **1–21**.

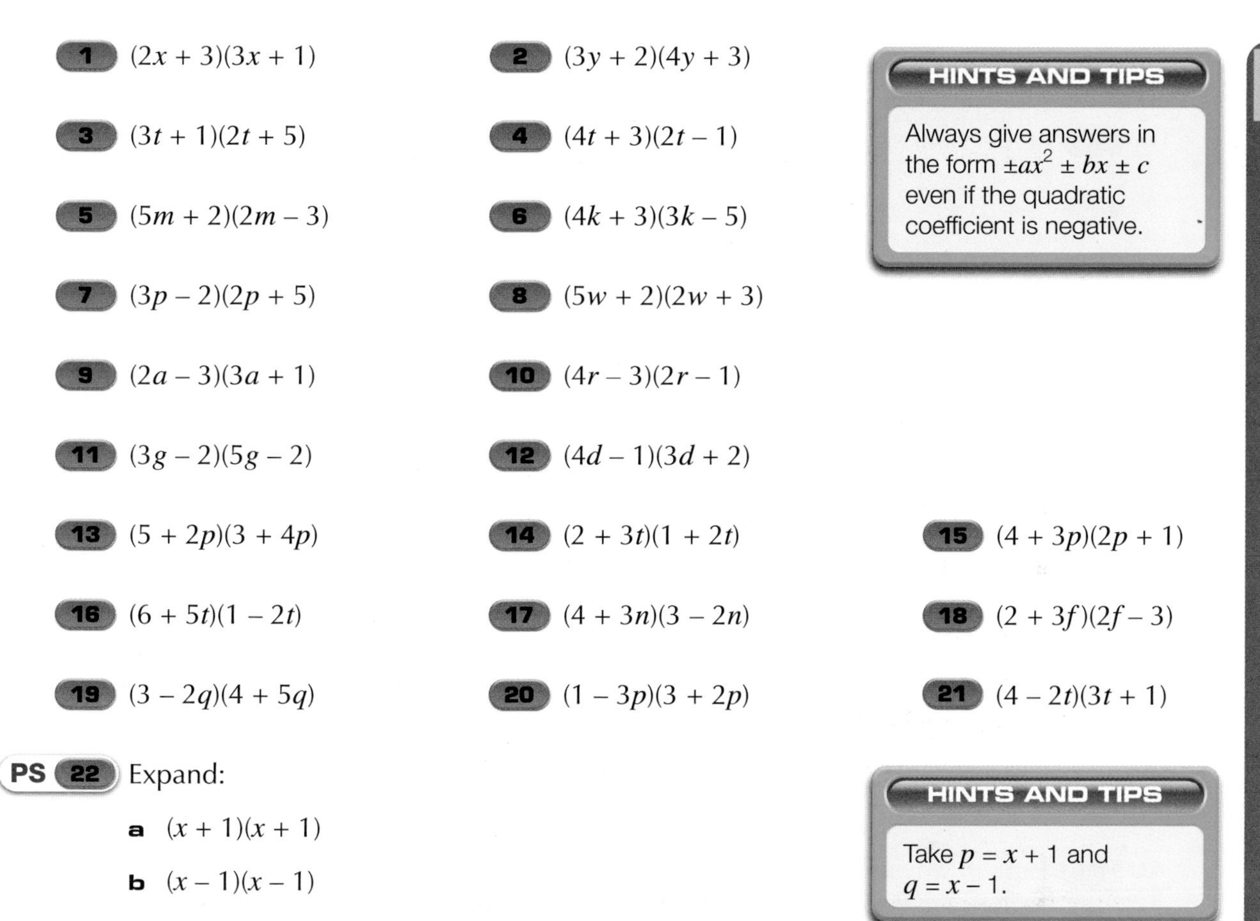

1 $(2x + 3)(3x + 1)$

2 $(3y + 2)(4y + 3)$

3 $(3t + 1)(2t + 5)$

4 $(4t + 3)(2t - 1)$

5 $(5m + 2)(2m - 3)$

6 $(4k + 3)(3k - 5)$

7 $(3p - 2)(2p + 5)$

8 $(5w + 2)(2w + 3)$

9 $(2a - 3)(3a + 1)$

10 $(4r - 3)(2r - 1)$

11 $(3g - 2)(5g - 2)$

12 $(4d - 1)(3d + 2)$

13 $(5 + 2p)(3 + 4p)$

14 $(2 + 3t)(1 + 2t)$

15 $(4 + 3p)(2p + 1)$

16 $(6 + 5t)(1 - 2t)$

17 $(4 + 3n)(3 - 2n)$

18 $(2 + 3f)(2f - 3)$

19 $(3 - 2q)(4 + 5q)$

20 $(1 - 3p)(3 + 2p)$

21 $(4 - 2t)(3t + 1)$

HINTS AND TIPS

Always give answers in the form $\pm ax^2 \pm bx \pm c$ even if the quadratic coefficient is negative.

PS 22 Expand:

a $(x + 1)(x + 1)$

b $(x - 1)(x - 1)$

c $(x + 1)(x - 1)$

d Use the results in parts **a**, **b** and **c** to show that $(p + q)^2 \equiv p^2 + 2pq + q^2$ is an identity.

HINTS AND TIPS

Take $p = x + 1$ and $q = x - 1$.

AU 23 **a** Without expanding the brackets, match each expression on the left with an expression on the right. One is done for you.

$(3x - 2)(2x + 1)$	$4x^2 - 4x + 1$
$(2x - 1)(2x - 1)$	$6x^2 - x - 2$
$(6x - 3)(x + 1)$	$6x^2 + 7x + 2$
$(4x + 1)(x - 1)$	$6x^2 + 3x - 3$
$(3x + 2)(2x + 1)$	$4x^2 - 3x - 1$

b Taking any expression on the left, explain how you can match it with an expression on the right without expanding the brackets.

EXERCISE 6C

Try to spot the pattern in each of the expressions in questions **1–15** so that you can immediately write down the expansion.

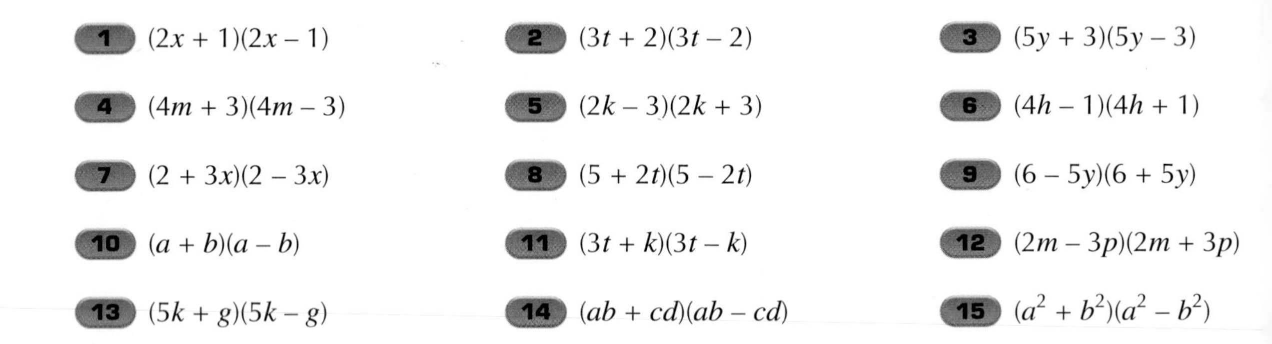

1 $(2x + 1)(2x - 1)$ **2** $(3t + 2)(3t - 2)$ **3** $(5y + 3)(5y - 3)$

4 $(4m + 3)(4m - 3)$ **5** $(2k - 3)(2k + 3)$ **6** $(4h - 1)(4h + 1)$

7 $(2 + 3x)(2 - 3x)$ **8** $(5 + 2t)(5 - 2t)$ **9** $(6 - 5y)(6 + 5y)$

10 $(a + b)(a - b)$ **11** $(3t + k)(3t - k)$ **12** $(2m - 3p)(2m + 3p)$

13 $(5k + g)(5k - g)$ **14** $(ab + cd)(ab - cd)$ **15** $(a^2 + b^2)(a^2 - b^2)$

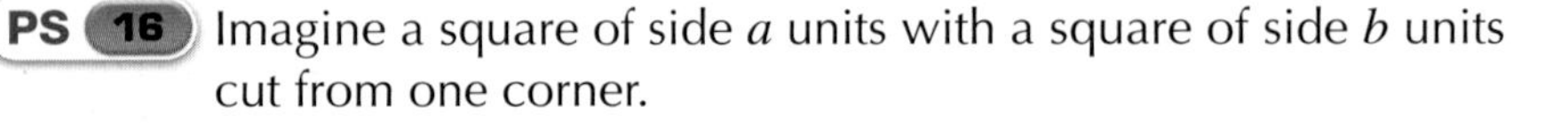

PS 16 Imagine a square of side a units with a square of side b units cut from one corner.

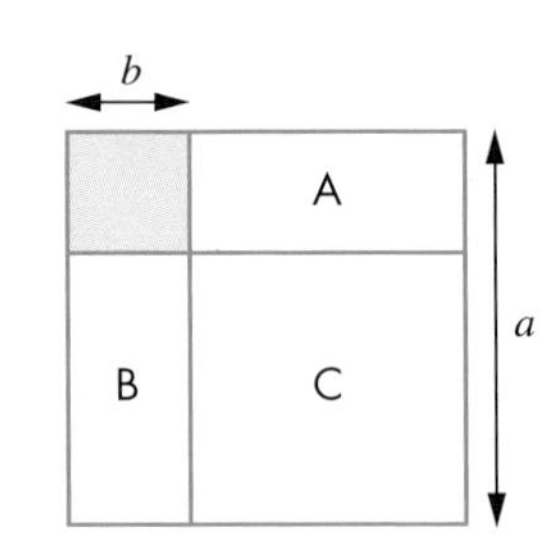

a What is the area remaining after the small square is cut away?

b The remaining area is cut into rectangles, A, B and C, and rearranged as shown.

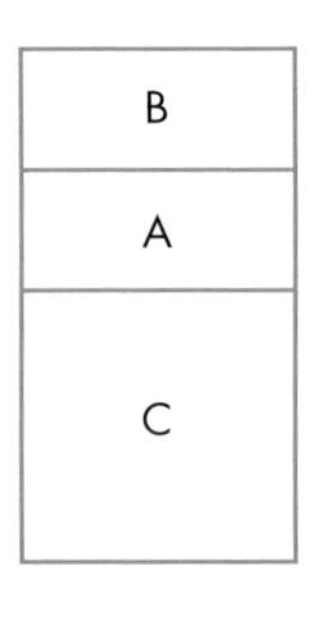

Write down the dimensions and area of the rectangle formed by A, B and C.

c Explain why $a^2 - b^2 = (a + b)(a - b)$.

AU 17 Explain why the areas of the shaded regions are the same.

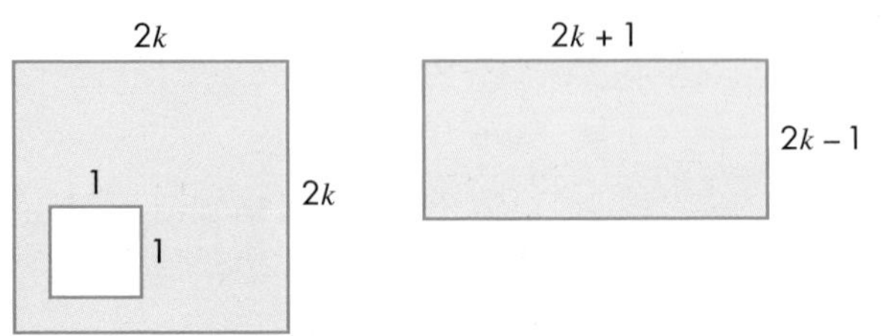

Expanding squares

Whenever you see a **linear** bracketed term squared you must write the brackets down twice and then use whichever method you prefer to expand.

EXAMPLE 6

Expand $(x + 3)^2$

$$(x + 3)^2 = (x + 3)(x + 3)$$
$$= x(x + 3) + 3(x + 3)$$
$$= x^2 + 3x + 3x + 9$$
$$= x^2 + 6x + 9$$

EXAMPLE 7

Expand $(3x-2)^2$

$$(3x-2)^2 = (3x-2)(3x-2)$$
$$= 9x^2 - 6x - 6x + 4$$
$$= 9x^2 - 12x + 4$$

EXERCISE 6D

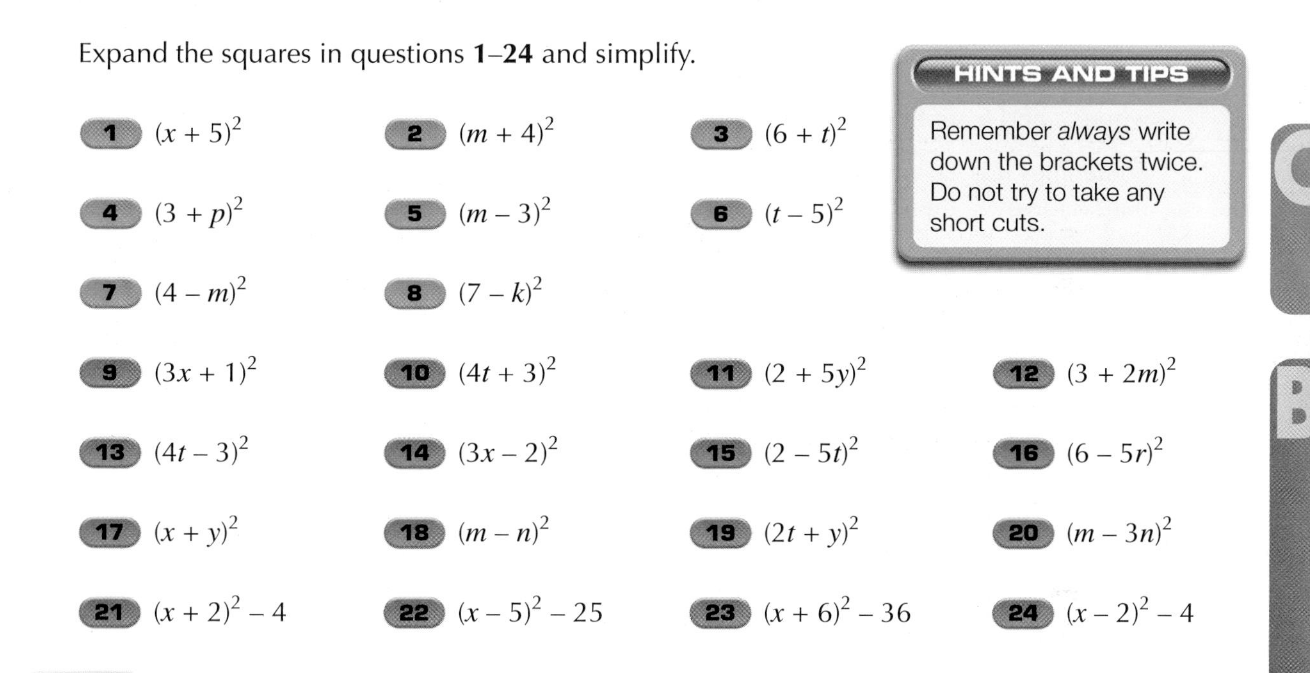

Expand the squares in questions **1–24** and simplify.

HINTS AND TIPS

Remember *always* write down the brackets twice. Do not try to take any short cuts.

1 $(x+5)^2$ **2** $(m+4)^2$ **3** $(6+t)^2$

4 $(3+p)^2$ **5** $(m-3)^2$ **6** $(t-5)^2$

7 $(4-m)^2$ **8** $(7-k)^2$

9 $(3x+1)^2$ **10** $(4t+3)^2$ **11** $(2+5y)^2$ **12** $(3+2m)^2$

13 $(4t-3)^2$ **14** $(3x-2)^2$ **15** $(2-5t)^2$ **16** $(6-5r)^2$

17 $(x+y)^2$ **18** $(m-n)^2$ **19** $(2t+y)^2$ **20** $(m-3n)^2$

21 $(x+2)^2-4$ **22** $(x-5)^2-25$ **23** $(x+6)^2-36$ **24** $(x-2)^2-4$

PS 25 A teacher asks her class to expand $(3x+1)^2$.

Bernice's answer is $9x^2+1$.

Pete's answer is $3x^2+6x+1$.

a Explain the mistakes that Bernice has made.

b Explain the mistakes that Pete has made.

c Work out the correct answer.

AU 26 Use the diagram to show algebraically and diagrammatically that:

$$(2x-1)^2 = 4x^2 - 4x + 1$$

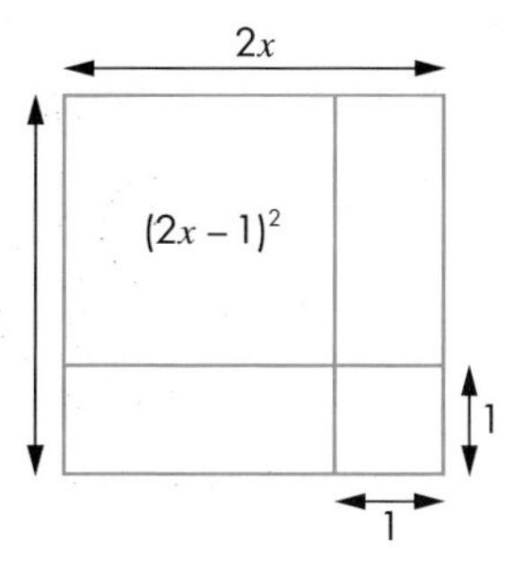

6.2 Quadratic factorisation

This section will show you how to:

- factorise a quadratic expression into two linear brackets

Key words
brackets
coefficient
difference of two squares
factorisation
quadratic expression

Factorisation involves putting a **quadratic expression** back into its **brackets** (if possible). We start with the factorisation of quadratic expressions of the type:

$x^2 + ax + b$

where a and b are integers.

Sometimes it is easy to put a quadratic expression back into its brackets, other times it seems hard. However, there are some simple rules that will help you to factorise.

- The expression inside each set of brackets will start with an x, and the signs in the quadratic expression show which signs to put after the xs.
- When the second sign in the expression is a plus, the signs in both sets of brackets are the same as the first sign.

  $x^2 + ax + b = (x + ?)(x + ?)$ Since everything is positive.

  $x^2 - ax + b = (x - ?)(x - ?)$ Since –ve × –ve = +ve

- When the *second* sign is a *minus*, the signs in the brackets are *different*.

  $x^2 + ax - b = (x + ?)(x - ?)$ Since +ve × –ve = –ve

  $x^2 - ax - b = (x + ?)(x - ?)$

- Next, look at the *last* number, b, in the expression. When multiplied together, the two numbers in the brackets must give b.
- Finally, look at the **coefficient** *of* x, a. The *sum* of the two *numbers* in the brackets will give a.

EXAMPLE 8

Factorise $x^2 - x - 6$

Because of the signs we know the brackets must be $(x + ?)(x - ?)$.

Two numbers that have a product of – 6 and a sum of – 1 are – 3 and + 2.

So, $x^2 - x - 6 = (x + 2)(x - 3)$

EXAMPLE 9

Factorise $x^2 - 9x + 20$

Because of the signs we know the brackets must be $(x - ?)(x - ?)$.

Two numbers that have a product of + 20 and a sum of –9 are –4 and –5.

So, $x^2 - 9x + 20 = (x - 4)(x - 5)$

EXERCISE 6E

Factorise the expressions in questions **1–40**.

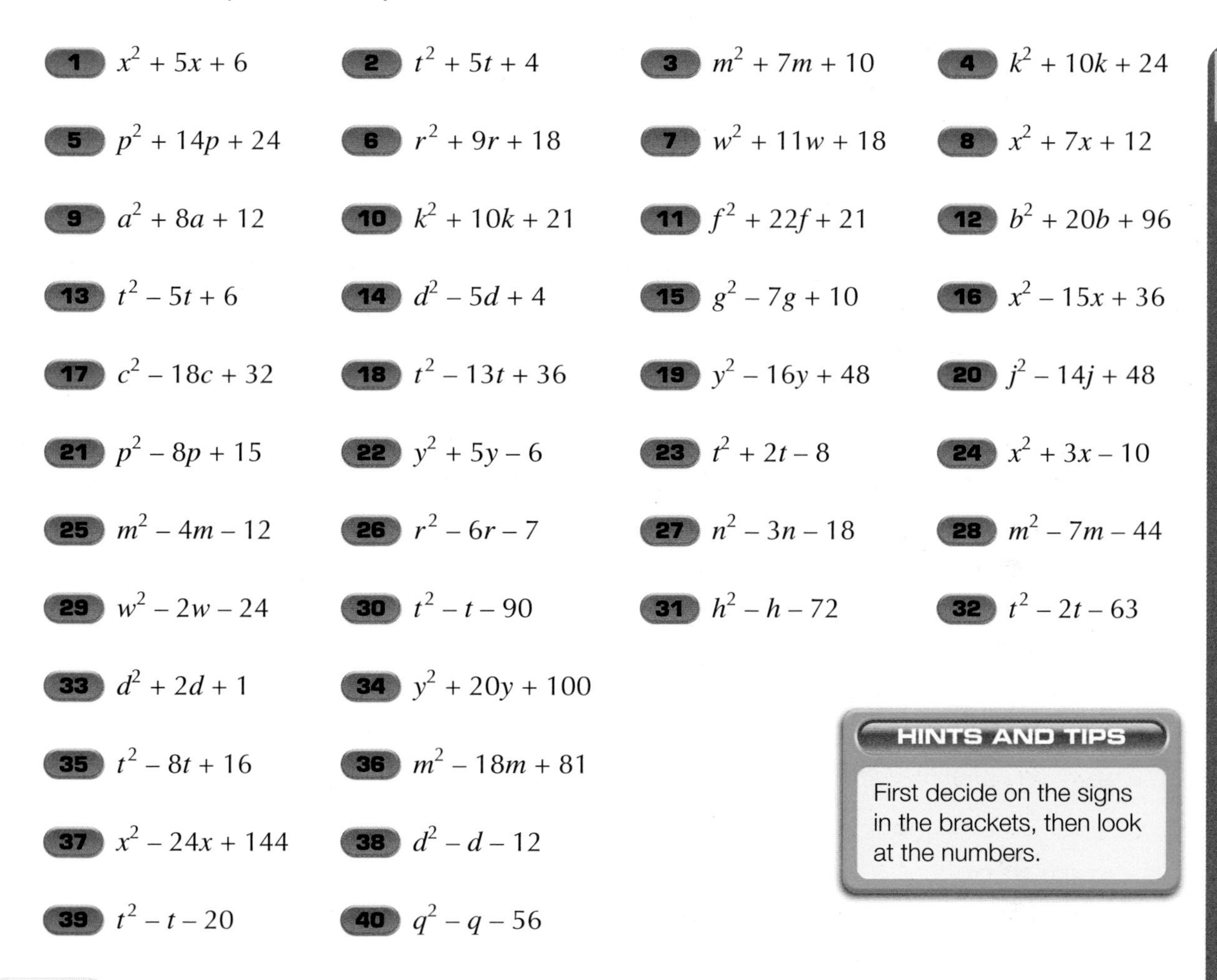

1 $x^2 + 5x + 6$
2 $t^2 + 5t + 4$
3 $m^2 + 7m + 10$
4 $k^2 + 10k + 24$
5 $p^2 + 14p + 24$
6 $r^2 + 9r + 18$
7 $w^2 + 11w + 18$
8 $x^2 + 7x + 12$
9 $a^2 + 8a + 12$
10 $k^2 + 10k + 21$
11 $f^2 + 22f + 21$
12 $b^2 + 20b + 96$
13 $t^2 - 5t + 6$
14 $d^2 - 5d + 4$
15 $g^2 - 7g + 10$
16 $x^2 - 15x + 36$
17 $c^2 - 18c + 32$
18 $t^2 - 13t + 36$
19 $y^2 - 16y + 48$
20 $j^2 - 14j + 48$
21 $p^2 - 8p + 15$
22 $y^2 + 5y - 6$
23 $t^2 + 2t - 8$
24 $x^2 + 3x - 10$
25 $m^2 - 4m - 12$
26 $r^2 - 6r - 7$
27 $n^2 - 3n - 18$
28 $m^2 - 7m - 44$
29 $w^2 - 2w - 24$
30 $t^2 - t - 90$
31 $h^2 - h - 72$
32 $t^2 - 2t - 63$
33 $d^2 + 2d + 1$
34 $y^2 + 20y + 100$
35 $t^2 - 8t + 16$
36 $m^2 - 18m + 81$
37 $x^2 - 24x + 144$
38 $d^2 - d - 12$
39 $t^2 - t - 20$
40 $q^2 - q - 56$

HINTS AND TIPS

First decide on the signs in the brackets, then look at the numbers.

PS 41 This rectangle is made up of four parts. Two of the parts have areas of x^2 and 6 square units.

The sides of the rectangle are of the form $x + a$ and $x + b$.

There are two possible answers for a and b.

Work out both answers and copy and complete the areas in the other parts of the rectangle.

x^2	
	6

AU 42 **a** Expand $(x + a)(x + b)$

b If $x^2 + 7x + 12 = (x + p)(x + q)$, use your answer to part **a** to write down the values of:

i $p + q$ **ii** pq

c Explain how you can tell that $x^2 + 12x + 7$ will not factorise.

Difference of two squares

In Exercise 6C, you multiplied out, for example, $(a + b)(a - b)$ and obtained $a^2 - b^2$. This type of quadratic expression, with only two terms, both of which are perfect squares separated by a minus sign, is called the **difference of two squares**. You should have found that all the expansions in Exercise 6C involved the differences of two squares.

The exercise illustrates a system of factorisation that will *always* work for the difference of two squares such as these.

$x^2 - 9 \qquad x^2 - 25 \qquad x^2 - 4 \qquad x^2 - 100$

There are three conditions that must be met if the difference of two squares works.

- There must be two terms.
- They must separated by a negative sign.
- Each term must be a perfect square, say x^2 and n^2.

When these three conditions are met, the factorisation is:

$$x^2 - n^2 = (x + n)(x - n)$$

EXAMPLE 10

Factorise $x^2 - 36$

- Recognise the difference of two squares x^2 and 6^2.
- So it factorises to $(x + 6)(x - 6)$.

Expanding the brackets shows that they do come from the original expression.

EXAMPLE 11

Factorise $9x^2 - 169$

- Recognise the difference of two squares $(3x)^2$ and 13^2.
- So it factorises to $(3x + 13)(3x - 13)$.

EXERCISE 6F

Each of the expressions in questions **1–9** is the difference of two squares. Factorise them.

HINTS AND TIPS

Learn how to spot the difference of two squares as it occurs a lot in GCSE examinations.

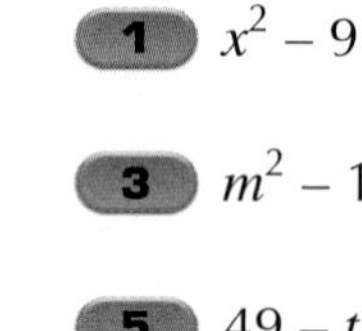

1 $x^2 - 9$ **2** $t^2 - 25$

3 $m^2 - 16$ **4** $9 - x^2$

5 $49 - t^2$ **6** $k^2 - 100$

7 $4 - y^2$ **8** $x^2 - 64$

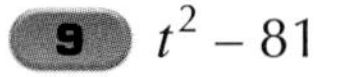

9 $t^2 - 81$

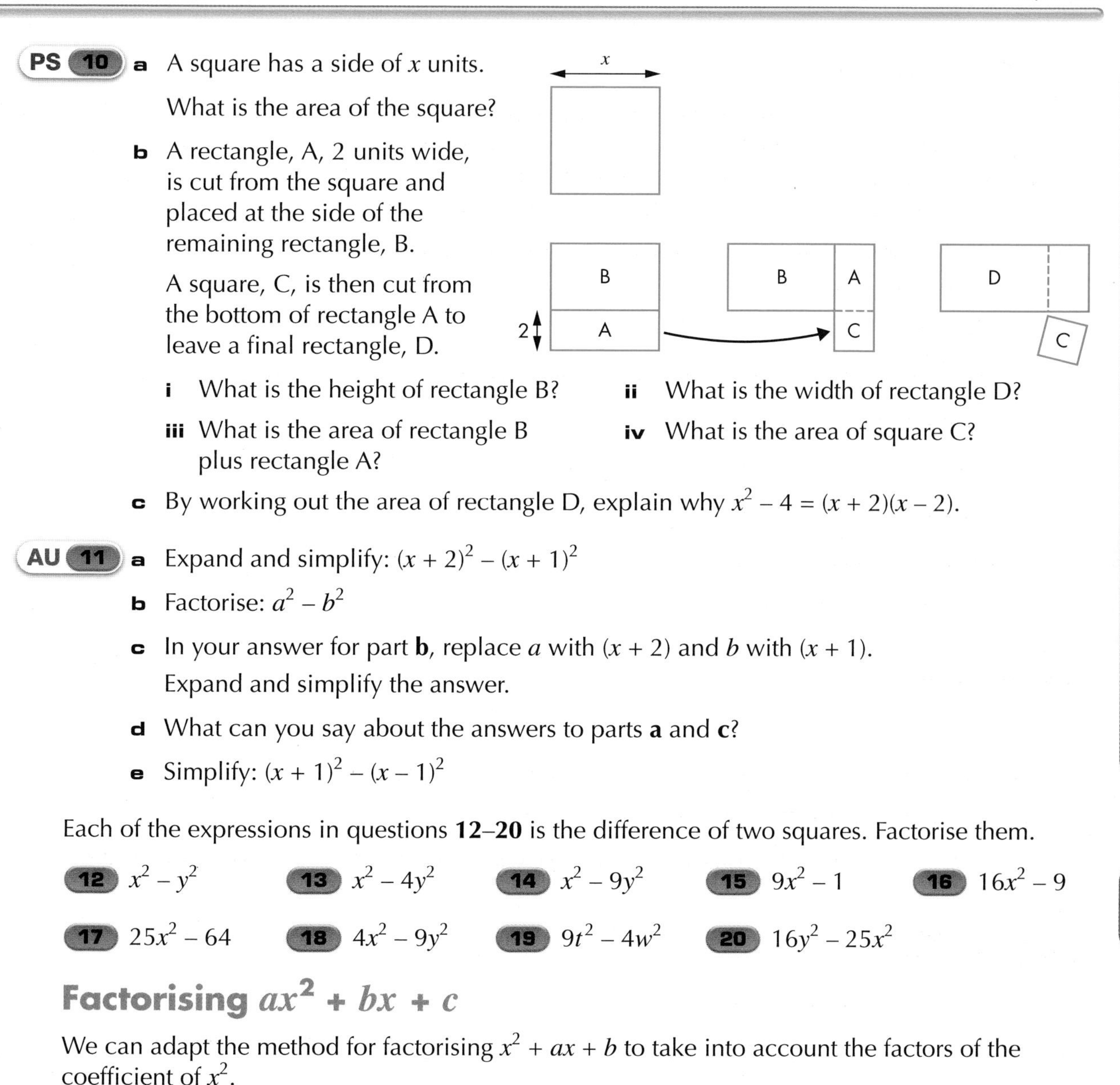

PS 10 **a** A square has a side of x units.

What is the area of the square?

b A rectangle, A, 2 units wide, is cut from the square and placed at the side of the remaining rectangle, B.

A square, C, is then cut from the bottom of rectangle A to leave a final rectangle, D.

i What is the height of rectangle B?

ii What is the width of rectangle D?

iii What is the area of rectangle B plus rectangle A?

iv What is the area of square C?

c By working out the area of rectangle D, explain why $x^2 - 4 = (x + 2)(x - 2)$.

AU 11 **a** Expand and simplify: $(x + 2)^2 - (x + 1)^2$

b Factorise: $a^2 - b^2$

c In your answer for part **b**, replace a with $(x + 2)$ and b with $(x + 1)$.
Expand and simplify the answer.

d What can you say about the answers to parts **a** and **c**?

e Simplify: $(x + 1)^2 - (x - 1)^2$

Each of the expressions in questions **12–20** is the difference of two squares. Factorise them.

12 $x^2 - y^2$ **13** $x^2 - 4y^2$ **14** $x^2 - 9y^2$ **15** $9x^2 - 1$ **16** $16x^2 - 9$

17 $25x^2 - 64$ **18** $4x^2 - 9y^2$ **19** $9t^2 - 4w^2$ **20** $16y^2 - 25x^2$

Factorising $ax^2 + bx + c$

We can adapt the method for factorising $x^2 + ax + b$ to take into account the factors of the coefficient of x^2.

EXAMPLE 12

Factorise $3x^2 + 8x + 4$

- First, note that both signs are positive. So the signs in the brackets must be $(?x + ?)(?x + ?)$.
- As 3 has only 3×1 as factors, the brackets must be $(3x + ?)(x + ?)$.
- Next, note that the factors of 4 are 4×1 and 2×2.
- Now find which pair of factors of 4 combine with 3 and 1 to give 8.

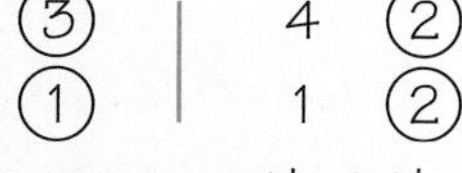

You can see that the combination 3×2 and 1×2 adds up to 8.

- So, the complete factorisation becomes $(3x + 2)(x + 2)$.

EXAMPLE 13

Factorise $6x^2 - 7x - 10$

- First, note that both signs are negative. So the signs in the brackets must be $(?x + ?)(?x - ?)$.
- As 6 has 6×1 and 3×2 as factors, the brackets could be $(6x \pm ?)(x \pm ?)$ or $(3x \pm ?)(2x \pm ?)$.
- Next, note that the factors of 10 are 5×2 and 1×10.
- Now find which pair of factors of 10 combine with the factors of 6 to give -7.

3	(6)	|	±1	(±2)
2	(1)	|	±10	(±5)

You can see that the combination 6×-2 and 1×5 adds up to -7.

- So, the complete factorisation becomes $(6x + 5)(x - 2)$.

Although this seems to be very complicated, it becomes quite easy with practice and experience.

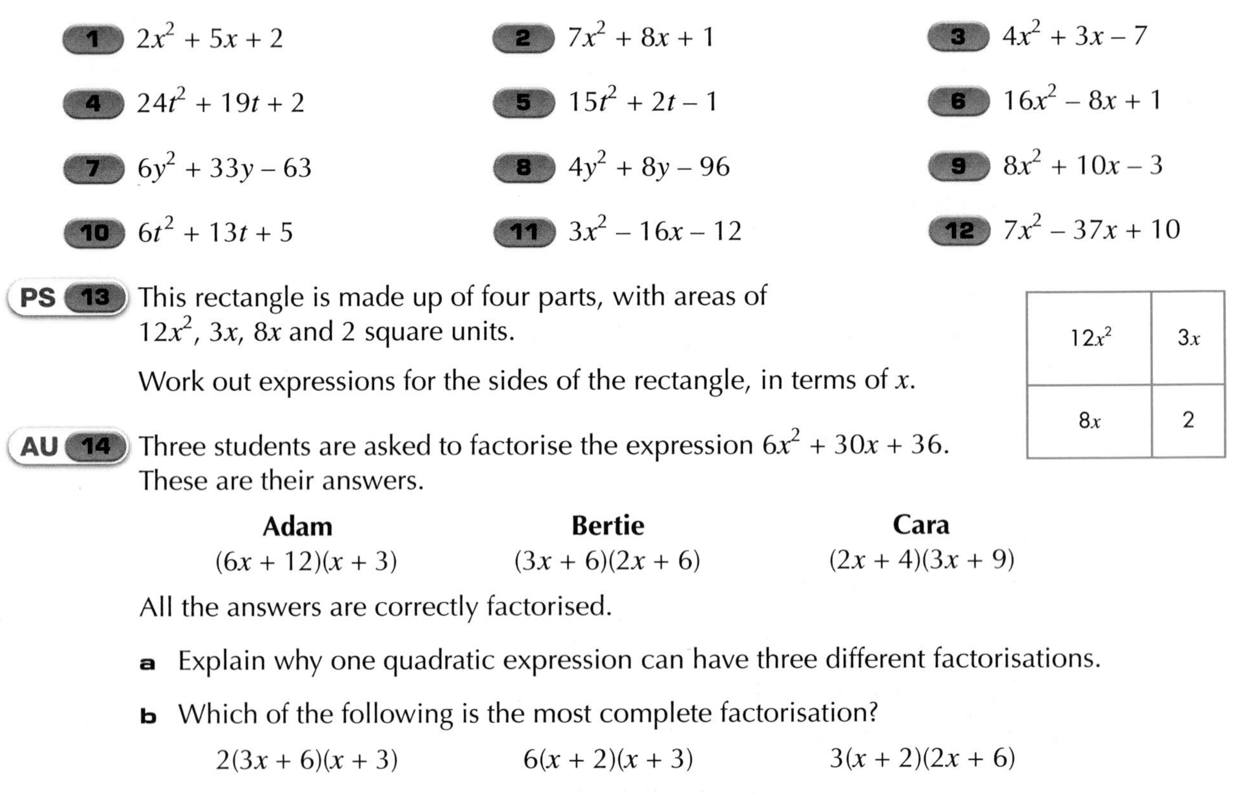

EXERCISE 6G

Factorise the expressions in questions **1–12**.

1 $2x^2 + 5x + 2$

2 $7x^2 + 8x + 1$

3 $4x^2 + 3x - 7$

4 $24t^2 + 19t + 2$

5 $15t^2 + 2t - 1$

6 $16x^2 - 8x + 1$

7 $6y^2 + 33y - 63$

8 $4y^2 + 8y - 96$

9 $8x^2 + 10x - 3$

10 $6t^2 + 13t + 5$

11 $3x^2 - 16x - 12$

12 $7x^2 - 37x + 10$

PS 13 This rectangle is made up of four parts, with areas of $12x^2$, $3x$, $8x$ and 2 square units.

Work out expressions for the sides of the rectangle, in terms of x.

AU 14 Three students are asked to factorise the expression $6x^2 + 30x + 36$. These are their answers.

Adam	**Bertie**	**Cara**
$(6x + 12)(x + 3)$	$(3x + 6)(2x + 6)$	$(2x + 4)(3x + 9)$

All the answers are correctly factorised.

a Explain why one quadratic expression can have three different factorisations.

b Which of the following is the most complete factorisation?

$2(3x + 6)(x + 3)$ $\quad$ $6(x + 2)(x + 3)$ $\quad$ $3(x + 2)(2x + 6)$

Explain your choice.

6.3 Solving quadratic equations by factorisation

This section will show you how to:

- solve a quadratic equation by factorisation

Key words
factors
solve

Solving the quadratic equation $x^2 + ax + b = 0$

To **solve** a quadratic equation such as $x^2 - 2x - 3 = 0$, you first have to be able to factorise it. Work through Examples 14 to 16 below to see how this is done.

EXAMPLE 14

Solve $x^2 + 6x + 5 = 0$

This factorises into $(x + 5)(x + 1) = 0$.

The only way this expression can ever equal 0 is if the value of one of the brackets is 0.
Hence either $(x + 5) = 0$ or $(x + 1) = 0$

$\Rightarrow x + 5 = 0$ or $x + 1 = 0$

$\Rightarrow x = -5$ or $x = -1$

So the solution is $x = -5$ or $x = -1$.

EXAMPLE 15

Solve $x^2 + 3x - 10 = 0$

This factorises into $(x + 5)(x - 2) = 0$.

Hence either $(x + 5) = 0$ or $(x - 2) = 0$

$\Rightarrow x + 5 = 0$ or $x - 2 = 0$

$\Rightarrow x = -5$ or $x = 2$.

So the solution is $x = -5$ or $x = 2$.

EXAMPLE 16

Solve $x^2 - 6x + 9 = 0$

This factorises into $(x - 3)(x - 3) = 0$.

The equation has repeated roots.

That is: $(x - 3)^2 = 0$

Hence, there is only one solution, $x = 3$.

EXERCISE 6H

Solve the equations in questions **1–12**.

1 $(x + 2)(x + 5) = 0$

2 $(t + 3)(t + 1) = 0$

3 $(a + 6)(a + 4) = 0$

4 $(x + 3)(x - 2) = 0$

5 $(x + 1)(x - 3) = 0$

6 $(t + 4)(t - 5) = 0$

7 $(x - 1)(x + 2) = 0$

8 $(x - 2)(x + 5) = 0$

9 $(a - 7)(a + 4) = 0$

10 $(x - 3)(x - 2) = 0$

11 $(x - 1)(x - 5) = 0$

12 $(a - 4)(a - 3) = 0$

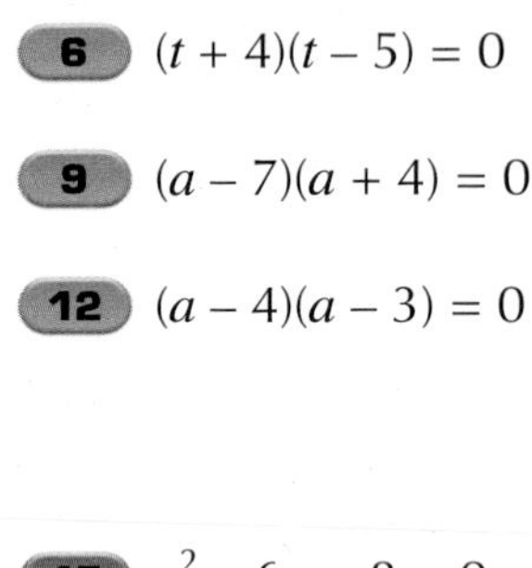

First factorise, then solve the equations in questions **13–26**.

13 $x^2 + 5x + 4 = 0$

14 $x^2 + 11x + 18 = 0$

15 $x^2 - 6x + 8 = 0$

16 $x^2 - 8x + 15 = 0$

17 $x^2 - 3x - 10 = 0$

18 $x^2 - 2x - 15 = 0$

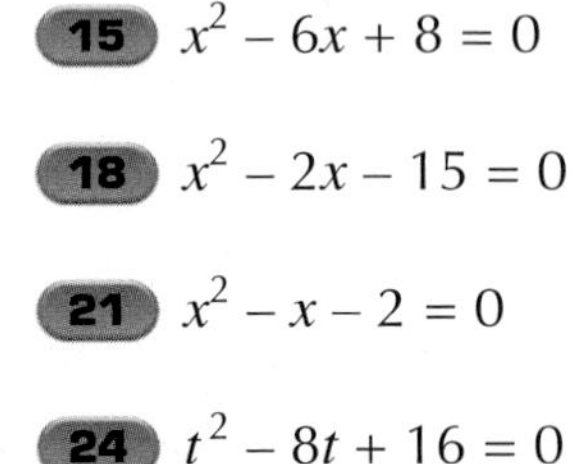

19 $t^2 + 4t - 12 = 0$

20 $t^2 + 3t - 18 = 0$

21 $x^2 - x - 2 = 0$

22 $x^2 + 4x + 4 = 0$

23 $m^2 + 10m + 25 = 0$

24 $t^2 - 8t + 16 = 0$

25 $t^2 + 8t + 12 = 0$

26 $a^2 - 14a + 49 = 0$

PS 27 A woman is x years old. Her husband is three years younger.

The product of their ages is 550.

a Set up a quadratic equation to represent this situation.

b How old is the woman?

HINTS AND TIPS

If one solution to a real-life problem is negative, reject it and only give the positive answer.

PS 28 A rectangular field is 40 m longer than it is wide.
The area is 48 000 square metres.

The farmer wants to place a fence all around the field.

How long will the fence be?

HINTS AND TIPS

Let the width be x, set up a quadratic equation and solve it to get x.

First rearrange the equations in questions **29–37**, then solve them.

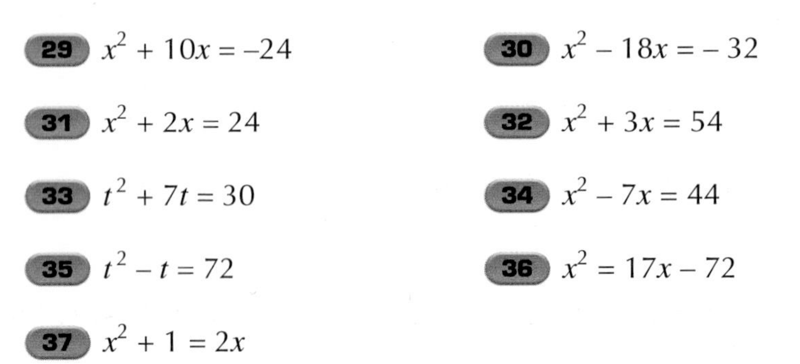

29 $x^2 + 10x = -24$

30 $x^2 - 18x = -32$

31 $x^2 + 2x = 24$

32 $x^2 + 3x = 54$

33 $t^2 + 7t = 30$

34 $x^2 - 7x = 44$

35 $t^2 - t = 72$

36 $x^2 = 17x - 72$

37 $x^2 + 1 = 2x$

HINTS AND TIPS

You cannot solve a quadratic equation by factorisation unless it is in the form
$x^2 + ax + b = 0$

A

AU **38** A teacher asks her class to solve $x^2 - 3x = 4$.

This is Mario's answer.

$x^2 - 3x - 4 = 0$

$(x - 4)(x + 1) = 0$

Hence $x - 4 = 0$ or $x + 1 = 0$

$x = 4$ or -1

This is Sylvan's answer.

$x(x - 3) = 4$

Hence $x = 4$ or $x - 3 = 4 \Rightarrow x = -3 + 4 = -1$

When the teacher reads out the answer of $x = 4$ or -1, both students mark their work as correct.

Who used the correct method and what mistakes did the other student make?

Solving the general quadratic equation by factorisation

The general quadratic equation is of the form $ax^2 + bx + c = 0$ where a, b and c are positive or negative whole numbers. (It is easier to make sure that a is always positive.) Before any quadratic equation can be solved by factorisation, it must be rearranged to this form.

The method is similar to that used to solve equations of the form $x^2 + ax + b = 0$. That is, you have to find two **factors** of $ax^2 + bx + c$ with a product of 0.

EXAMPLE 17

Solve these quadratic equations. **a** $12x^2 - 28x = -15$ **b** $30x^2 - 5x - 5 = 0$

a First, rearrange the equation to the general form.

$12x^2 - 28x + 15 = 0$

This factorises into $(2x - 3)(6x - 5) = 0$.

The only way this product can equal 0 is if the value of one of the brackets is 0. Hence:

either $2x - 3 = 0$ or $6x - 5 = 0$

$\Rightarrow 2x = 3$ or $6x = 5$

$\Rightarrow x = \frac{3}{2}$ or $x = \frac{5}{6}$

So the solution is $x = 1\frac{1}{2}$ or $x = \frac{5}{6}$

Note: It is almost always the case that if a solution is a fraction which is then changed into a rounded-off decimal number, the original equation cannot be evaluated exactly, using that decimal number. So it is preferable to leave the solution in its fraction form. This is called the *rational form.*

b This equation is already in the general form and it will factorise to $(15x + 5)(2x - 1) = 0$ or $(3x + 1)(10x - 5) = 0$.

Look again at the equation. There is a common factor of 5 which can be taken out to give:

$5(6x^2 - x - 1 = 0)$

This is much easier to factorise to $5(3x + 1)(2x - 1) = 0$, which can be solved to give $x = -\frac{1}{3}$ or $x = \frac{1}{2}$

Special cases

Sometimes the values of b and c are zero. (Note that if a is zero the equation is no longer a quadratic equation but a linear equation.)

EXAMPLE 18

Solve these quadratic equations. **a** $3x^2 - 4 = 0$ **b** $4x^2 - 25 = 0$ **c** $6x^2 - x = 0$

a Rearrange to get $3x^2 = 4$.

Divide both sides by 3: $x^2 = \frac{4}{3}$

Take the square root on both sides: $x = \pm\sqrt{\frac{4}{3}} = \pm \frac{2}{\sqrt{3}} = \pm \frac{2\sqrt{3}}{3}$

Note: A square root can be positive or negative. The answer is in *surd form* (see Chapter 9).

b You can use the method of part **a** or you should recognise this as the difference of two squares (page 310). This can be factorised to $(2x - 5)(2x + 5) = 0$.

Each set of brackets can be put equal to zero.

$2x - 5 = 0 \Rightarrow x = +\frac{5}{2}$

$2x + 5 = 0 \Rightarrow x = -\frac{5}{2}$ So the solution is $x = \pm\frac{5}{2}$

c There is a common factor of x, so factorise as $x(6x - 1) = 0$.

There is only one set of brackets this time but each factor can be equal to zero, so $x = 0$ or $6x - 1 = 0$.

Hence, $x = 0$ or $\frac{1}{6}$

EXERCISE 6I

Give your answers either in rational form or as mixed numbers.

1 Solve these equations.

HINTS AND TIPS

Look out for the special cases where b or c is zero.

a $3x^2 + 8x - 3 = 0$ **b** $6x^2 - 5x - 4 = 0$

c $5x^2 - 9x - 2 = 0$ **d** $4t^2 - 4t - 35 = 0$

e $18t^2 + 9t + 1 = 0$ **f** $3t^2 - 14t + 8 = 0$

g $6x^2 + 15x - 9 = 0$ **h** $12x^2 - 16x - 35 = 0$ **i** $15t^2 + 4t - 35 = 0$

j $28x^2 - 85x + 63 = 0$ **k** $24x^2 - 19x + 2 = 0$ **l** $16t^2 - 1 = 0$

m $4x^2 + 9x = 0$ **n** $25t^2 - 49 = 0$ **o** $9m^2 - 24m - 9 = 0$

2 Rearrange these equations into the general form and then solve them.

a $x^2 - x = 42$ **b** $8x(x + 1) = 30$

c $(x + 1)(x - 2) = 40$ **d** $13x^2 = 11 - 2x$

A

A

e $(x + 1)(x - 2) = 4$

f $10x^2 - x = 2$

g $8x^2 + 6x + 3 = 2x^2 + x + 2$

h $25x^2 = 10 - 45x$

i $8x - 16 - x^2 = 0$

j $(2x + 1)(5x + 2) = (2x - 2)(x - 2)$

k $5x + 5 = 30x^2 + 15x + 5$

l $2m^2 = 50$

m $6x^2 + 30 = 5 - 3x^2 - 30x$

n $4x^2 + 4x - 49 = 4x$

o $2t^2 - t = 15$

AU 3 Here are three equations.

A: $(x - 1)^2 = 0$ B: $3x + 2 = 5$ C: $x^2 - 4x = 5$

a Give some mathematical fact that equations A and B have in common.

b Give a mathematical reason why equation B is different from equations A and C.

PS 4 Pythagoras' theorem states that the sum of the squares of the two short sides of a right-angled triangle equals the square of the long side (hypotenuse).

A right-angled triangle has sides $5x - 1$, $2x + 3$ and $x + 1$ cm.

a Show that $20x^2 - 24x - 9 = 0$

b Find the area of the triangle.

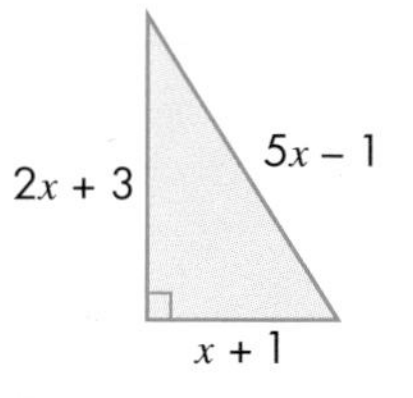

6.4 Solving a quadratic equation by the quadratic formula

This section will show you how to:

- solve a quadratic equation by using the quadratic formula

Key words

coefficient
constant term
quadratic formula
soluble
solve

Many quadratic equations cannot be solved by factorisation because they do not have simple factors. Try to factorise, for example, $x^2 - 4x - 3 = 0$ or $3x^2 - 6x + 2 = 0$. You will find it is impossible.

One way to **solve** this type of equation is to use the **quadratic formula**. This formula can be used to solve *any* quadratic equation that is **soluble**. (Some are not, which the quadratic formula would immediately show.)

The solution of the equation $ax^2 + bx + c = 0$ is given by:

$$x = \frac{-b \pm \sqrt{b^2 - 4ac}}{2a}$$

where a and b are the **coefficients** of x^2 and x respectively and c is the **constant** term.

This is the quadratic formula. It is given on the formula sheet of GCSE examinations but it is best to learn it.

The symbol ± states that the square root has a positive and a negative value, *both* of which must be used in solving for x.

EXAMPLE 19

Solve $5x^2 - 11x - 4 = 0$, giving solutions correct to 2 decimal places.

Take the quadratic formula:

$$x = \frac{-b \pm \sqrt{b^2 - 4ac}}{2a}$$

and put $a = 5$, $b = -11$ and $c = -4$, which gives:

$$x = \frac{-(-11) \pm \sqrt{(-11)^2 - 4(5)(-4)}}{2(5)}$$

Note that the values for a, b and c have been put into the formula in brackets. This is to avoid mistakes in calculation. It is a very common mistake to get the sign of b wrong or to think that -11^2 is -121. Using brackets will help you do the calculation correctly.

$$x = \frac{11 \pm \sqrt{121 + 80}}{10} = \frac{11 \pm \sqrt{201}}{10}$$

$$\Rightarrow x = 2.52 \text{ or } -0.32$$

Note: The calculation has been done in stages. With a calculator it is possible just to work out the answer, but make sure you can use your calculator properly. If not, break the calculation down. Remember the rule 'if you try to do two things at once, you will probably get one of them wrong'.

Examination tip: If you are asked to solve a quadratic equation to one or two decimal places, you can be sure that it can be solved *only* by the quadratic formula.

EXERCISE 6J

Use the quadratic formula to solve the equations in questions **1** to **15**. Give your answers to 2 decimal places.

HINTS AND TIPS

Use brackets when substituting and do not try to work two things out at the same time.

A

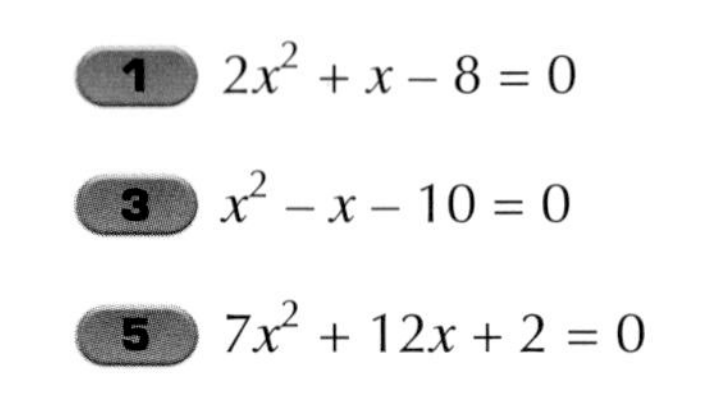

1 $2x^2 + x - 8 = 0$

2 $3x^2 + 5x + 1 = 0$

3 $x^2 - x - 10 = 0$

4 $5x^2 + 2x - 1 = 0$

5 $7x^2 + 12x + 2 = 0$

6 $3x^2 + 11x + 9 = 0$

A

7 $4x^2 + 9x + 3 = 0$

8 $6x^2 + 22x + 19 = 0$

9 $x^2 + 3x - 6 = 0$

10 $3x^2 - 7x + 1 = 0$

11 $2x^2 + 11x + 4 = 0$

12 $4x^2 + 5x - 3 = 0$

13 $4x^2 - 9x + 4 = 0$

14 $7x^2 + 3x - 2 = 0$

15 $5x^2 - 10x + 1 = 0$

FM 16 A rectangular lawn is 2 m longer than it is wide.

The area of the lawn is 21 m^2. The gardener wants to edge the lawn with edging strips, which are sold in lengths of $1\frac{1}{2}$ m. How many will she need to buy?

AU 17 Shaun is solving a quadratic equation, using the formula.

He correctly substitutes values for a, b and c to get:

$$x = \frac{3 \pm \sqrt{37}}{2}$$

What is the equation Shaun is trying to solve?

PS 18 Terry uses the quadratic formula to solve $4x^2 - 4x + 1 = 0$.

June uses factorisation to solve $4x^2 - 4x + 1 = 0$.

They both find something unusual in their solutions.

Explain what this is, and why.

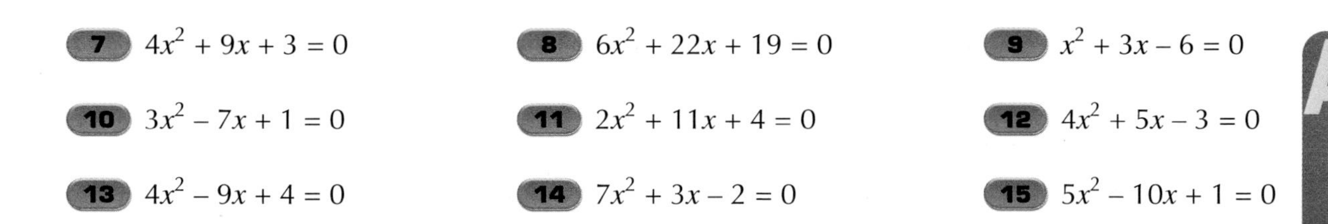

6.5 Solving a quadratic equation by completing the square

This section will show you how to:

- solve a quadratic equation by completing the square

Key words

completing the square
square root
surd

Another method for solving quadratic equations is **completing the square**. This method can be used to give answers to a specified number of decimal places or to leave answers in **surd** form.

You will remember that:

$$(x + a)^2 = x^2 + 2ax + a^2$$

which can be rearranged to give:

$$x^2 + 2ax = (x + a)^2 - a^2$$

This is the basic principle behind completing the square.

There are three basic steps in rewriting $x^2 + px + q$ in the form $(x + a)^2 + b$.

Step 1: Ignore q and just look at the first two terms, $x^2 + px$.

Step 2: Rewrite $x^2 + px$ as $\left(x + \frac{p}{2}\right)^2 - \left(\frac{p}{2}\right)^2$.

Step 3: Bring q back to get $x^2 + px + q = \left(x + \frac{p}{2}\right)^2 - \left(\frac{p}{2}\right)^2 + q$.

Note: p is always even so the numbers involved are whole numbers.

EXAMPLE 20

Rewrite the following in the form $(x \pm a) \pm b$.

a $x^2 + 6x - 7$

b $x^2 - 8x + 3$

a Ignore −7 for the moment.

Rewrite $x^2 + 6x$ as $(x + 3)^2 - 9$

(Expand $(x + 3)^2 - 9 = x^2 + 6x + 9 - 9 = x^2 + 6x$. The 9 is subtracted to get rid of the constant term when the brackets are expanded.)

Now bring the −7 back, so $x^2 + 6x - 7 = (x + 3)^2 - 9 - 7$

Combine the constant terms to get the final answer: $x^2 + 6x - 7 = (x + 3)^2 - 16$

b Ignore +3 for the moment.

Rewrite $x^2 - 8x$ as $(x - 4)^2 - 16$

(Note that you still subtract $(-4)^2$, as $(-4)^2 = +16$)

Now bring the +3 back, so $x^2 - 8x + 3 = (x - 4)^2 - 16 + 3$

Combine the constant terms to get the final answer: $x^2 - 8x + 3 = (x - 4)^2 - 13$

EXAMPLE 21

Rewrite $x^2 + 4x - 7$ in the form $(x + a)^2 - b$. Hence solve the equation $x^2 + 4x - 7 = 0$, giving your answers to 2 decimal places.

Note that:

$$x^2 + 4x = (x + 2)^2 - 4$$

So:

$$x^2 + 4x - 7 = (x + 2)^2 - 4 - 7 = (x + 2)^2 - 11$$

When $x^2 + 4x - 7 = 0$, you can rewrite the equations completing the square as: $(x + 2)^2 - 11 = 0$

Rearranging gives $(x + 2)^2 = 11$

Taking the **square root** of both sides gives:

$x + 2 = \pm\sqrt{11}$ This answer is in surd form and could be left like this, but you

$\Rightarrow x = -2 \pm \sqrt{11}$ are asked to evaluate it to 2 decimal places.

$\Rightarrow x = 1.32$ or -5.32 (to 2 decimal places)

EXAMPLE 22

Solve $x^2 - 6x - 1 = 0$ by completing the square. Leave your answer in the form $a \pm \sqrt{b}$.

$x^2 - 6x = (x - 3)^2 - 9$

So $x^2 - 6x - 1 = (x - 3)^2 - 9 - 1 = (x - 3)^2 - 10$

When $x^2 - 6x - 1 = 0$, then $(x - 3)^2 - 10 = 0$

$\Rightarrow (x - 3)^2 = 10$

Taking the square root of both sides gives:

$x - 3 = \pm\sqrt{10}$

$\Rightarrow x = 3 \pm\sqrt{10}$

EXERCISE 6K

1 Write an equivalent expression in the form $(x \pm a)^2 - b$.

a $x^2 + 4x$ **b** $x^2 + 14x$ **c** $x^2 - 6x$ **d** $x^2 + 6x$

e $x^2 - 4x$ **f** $x^2 - 10x$ **g** $x^2 + 20x$ **h** $x^2 + 10x$

i $x^2 + 8x$ **j** $x^2 - 2x$ **k** $x^2 + 2x$

2 Write an equivalent expression in the form $(x \pm a)^2 - b$.

Question **1** will help with **a** to **e**.

a $x^2 + 4x - 1$ **b** $x^2 + 14x - 5$ **c** $x^2 - 6x + 3$ **d** $x^2 + 6x + 7$

e $x^2 - 4x - 1$ **f** $x^2 + 6x + 3$ **g** $x^2 - 10x - 5$ **h** $x^2 + 20x - 1$

i $x^2 + 8x - 6$ **j** $x^2 + 2x - 1$ **k** $x^2 - 2x - 7$ **l** $x^2 + 2x - 9$

3 Solve the following equations by completing the square. Leave your answers in surd form where appropriate. The answers to question **2** will help.

a $x^2 + 4x - 1 = 0$ **b** $x^2 + 14x - 5 = 0$ **c** $x^2 - 6x + 3 = 0$

d $x^2 + 6x + 7 = 0$ **e** $x^2 - 4x - 1 = 0$ **f** $x^2 + 6x + 3 = 0$

g $x^2 - 10x - 5 = 0$ **h** $x^2 + 20x - 1 = 0$ **i** $x^2 + 8x - 6 = 0$

j $x^2 + 2x - 1 = 0$ **k** $x^2 - 2x - 7 = 0$ **l** $x^2 + 2x - 9 = 0$

4 Solve by completing the square. Give your answers to 2 decimal places.

a $x^2 + 2x - 5 = 0$ **b** $x^2 - 4x - 7 = 0$ **c** $x^2 + 2x - 9 = 0$

5 Prove that the solutions to the equation $x^2 + bx + c = 0$ are:

$$-\frac{b}{2} \pm \sqrt{\left(\frac{b^2}{4} - c\right)}$$

A

A*

A*

AU 6 Dave rewrites the expression $x^2 + px + q$ by completing the square.
He correctly does this and gets $(x - 7)^2 - 52$.

What are the values of p and q?

PS 7 **a** Frankie writes the steps to solve $x^2 + 6x + 7 = 0$ by completing the square on sticky notes. Unfortunately he drops them and they get out of order. Can you put the notes in the correct order?

Take –2 over the equals sign	Take +3 over the equals sign	Write $x^2 + 6x + 7 = 0$ as $(x + 3)^2 - 2 = 0$	Take the square root of both sides

b Write down the stages as in part **a** needed to solve the equation $x^2 - 4x - 3 = 0$

c Solve the equations below, giving the answers in surd form.

i $x^2 + 6x + 7 = 0$ **ii** $x^2 - 4x - 3 = 0$

PS 8 Rearrange the following statements to give the complete solution, using the method of completing the square to the equation: $ax^2 + bx + c = 0$

a $x = -\dfrac{b}{2a} \pm \sqrt{\dfrac{b^2}{4a^2} - \dfrac{c}{a}}$

b $\left(\left(x + \dfrac{b}{2a}\right)^2 - \dfrac{b^2}{4a^2}\right) + \dfrac{c}{a} = 0$

c $a\left(\left(x + \dfrac{b}{2a}\right)^2 - \dfrac{b^2}{4a^2}\right) + c = 0$

d $\left(x + \dfrac{b}{2a}\right)^2 = \dfrac{b^2}{4a^2} - \dfrac{c}{a}$

e $\left(x + \dfrac{b}{2a}\right)^2 - \dfrac{b^2}{4a^2} + \dfrac{c}{a} = 0$

f $x = -\dfrac{b}{2a} \pm \sqrt{\dfrac{b^2}{4a^2} - \dfrac{4ac}{4a^2}}$

g $x = -\dfrac{b}{2a} \pm \dfrac{1}{2a}\sqrt{b^2 - 4ac}$

h $a\left(x^2 + \dfrac{b}{a}x\right) + c = 0$

i $x = \dfrac{-b \pm \sqrt{b^2 - 4ac}}{2a}$

j $x + \dfrac{b}{2a} = \pm\sqrt{\dfrac{b^2}{4a^2} - \dfrac{c}{a}}$

6.6 Problems involving quadratic equations

This section will show you how to:

- recognise why some quadratic equations cannot be factorised
- solve practical problems, using quadratic equations

Key words

discriminant

Quadratic equations with no solution

The quantity $(b^2 - 4ac)$ in the quadratic formula is known as the **discriminant**.

When $b^2 > 4ac$, $(b^2 - 4ac)$ is positive. This has been the case in almost all of the quadratics you have solved so far and it means there are two solutions.

When $b^2 = 4ac$, $(b^2 - 4ac)$ is zero. This has been the case in some of the quadratics you have solved so far. It means there is only one solution (the repeated root).

When $b^2 < 4ac$, $(b^2 - 4ac)$ is negative. So you need to find the square root of a negative number.

Such a square root cannot be found (at GCSE level) and therefore there are no solutions. You will not be asked about this in examinations but if it happens then you will have made a mistake and should check your working.

EXAMPLE 23

Find the discriminant $b^2 - 4ac$ of the equation $x^2 + 3x + 5 = 0$ and explain what the result tells you.

$$b^2 - 4ac = (3)^2 - 4(1)(5) = 9 - 20 = -11.$$

This means there are no solutions for x.

All quadratic equations can be shown as graphs that have a characteristic shape known as a parabola.

Here are the graphs of the three types of quadratic equations: one with two solutions $(b^2 - 4ac > 0)$, one with one solution $(b^2 - 4ac = 0)$ and one with no solutions $(b^2 - 4ac < 0)$.

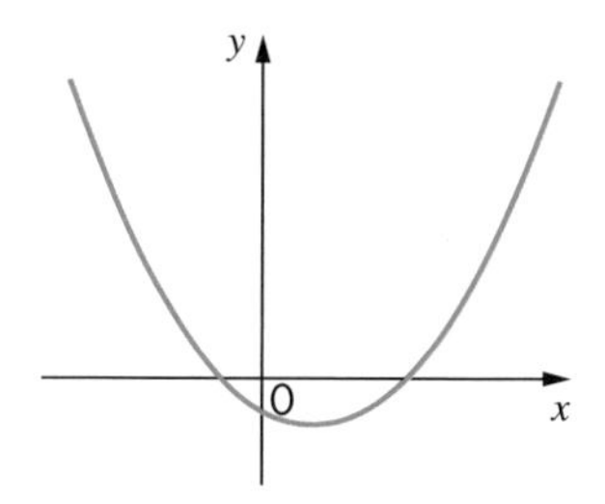

$b^2 - 4ac > 0$ two solutions, crosses the x-axis twice

$b^2 - 4ac = 0$ one solution, just touches the x-axis

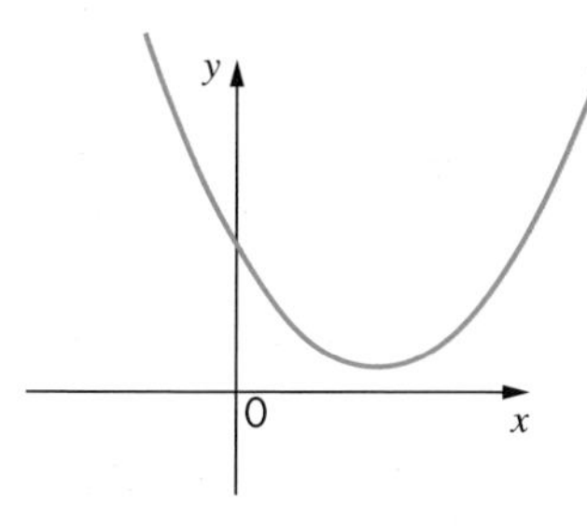

$b^2 - 4ac < 0$ no solution, does not cross the x-axis

EXERCISE 6L

Work out the discriminant $b^2 - 4ac$ of the equations in questions **1** to **12**. In each case say how many solutions the equation has.

1 $3x^2 + 2x - 4 = 0$

2 $2x^2 - 7x - 2 = 0$

3 $5x^2 - 8x + 2 = 0$

4 $3x^2 + x - 7 = 0$

5 $16x^2 - 23x + 6 = 0$

6 $x^2 - 2x - 16 = 0$

7 $5x^2 + 5x + 3 = 0$

8 $4x^2 + 3x + 2 = 0$

9 $5x^2 - x - 2 = 0$

10 $x^2 + 6x - 1 = 0$

11 $17x^2 - x + 2 = 0$

12 $x^2 + 5x - 3 = 0$

PS 13 Bill works out the discriminant of the quadratic equation $x^2 + bx - c = 0$ as $b^2 - 4ac = 13$. There are four possible equations that could lead to this discriminant. What are they?

Problems solved by quadratic equations

You are likely to have to solve a problem which involves generating a quadratic equation and finding its solution.

EXAMPLE 24

Find the sides of the right-angled triangle shown in the diagram.

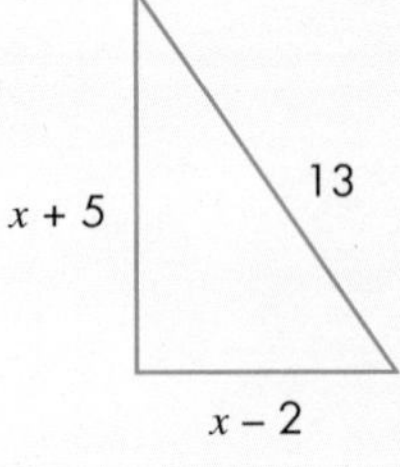

The sides of a right-angled triangle are connected by Pythagoras' theorem, which says that $c^2 = a^2 + b^2$.

$(x + 5)^2 + (x - 2)^2 = 13^2$

$(x^2 + 10x + 25) + (x^2 - 4x + 4) = 169$

$2x^2 + 6x + 29 = 169$

$2x^2 + 6x - 140 = 0$

Divide by a factor of 2: $x^2 + 3x - 70 = 0$

This factorises to: $(x + 10)(x - 7) = 0$

This gives $x = -10$ or 7.

Reject the negative value as it would give negative lengths.

Hence the sides of the triangle are 5, 12 and 13.

Note: You may know the Pythagorean triple 5, 12, 13 and guessed the answer but you would be expected to show working. Most 'real-life' problems will end up with a quadratic that factorises, as the questions are complicated enough without expecting you to use the quadratic formula.

EXAMPLE 25

Solve this equation. $2x - \frac{3}{x} = 5$

Multiply through by x to give: $2x^2 - 3 = 5x$

Rearrange into the general form: $2x^2 - 5x - 3 = 0$

This factorises to: $(2x + 1)(x - 3) = 0$

So $x = -\frac{1}{2}$ or $x = 3$.

EXAMPLE 26

A coach driver undertook a journey of 300 km. Her actual average speed turned out to be 10 km/h less than expected. Therefore, she took 1 hour longer over the journey than expected. Find her actual average speed.

Let the driver's actual average speed be x km/h.

So the estimated speed would have been $(x + 10)$ km/h.

$$\text{Time taken} = \frac{\text{distance travelled}}{\text{speed}}$$

At x km/h, she did the journey in $\frac{300}{x}$ hours.

At $(x + 10)$ km/h, she would have done the journey in $\frac{300}{x + 10}$ hours.

Since the journey took 1 hour longer than expected, then:

$$\text{time taken} = \frac{300}{x + 10} + 1 = \frac{300 + x + 10}{x + 10} = \frac{310 + x}{x + 10}$$

$$\text{So} = \frac{300}{x} = \frac{310 + x}{x + 10} \Rightarrow 300(x + 10) = x(310 + x) \Rightarrow 300x + 3000 = 310x + x^2$$

Rearranging into the general form gives: $x^2 + 10x - 3000 = 0$

This factorises into: $(x + 60)(x - 50) = 0 \Rightarrow x = -60$ or 50

The coach's average speed could not be −60 km/h, so it has to be 50 km/h.

EXERCISE 6M

PS 1 The length of a rectangle is 5 m more than its width. Its area is 300 m^2. Find the actual dimensions of the rectangle.

FM 2 The average mass of a group of people is 45.2 kg. A newcomer to the group weighs 51 kg, which increases the average mass by 0.2 kg. How many people are now in the group?

3 Solve the equation $x + \frac{3}{x} = 7$. Give your answers correct to 2 decimal places.

A*

A*

4 Solve the equation $2x + \frac{5}{x} = 11$.

PS 5 A tennis court has an area of 224 m^2. If the length were decreased by 1 m and the width increased by 1 m, the area would be increased by 1 m^2. Find the dimensions of the court.

PS 6 On a journey of 400 km, the driver of a train calculates that if he were to increase his average speed by 2 km/h, he would take 20 minutes less. Find his average speed.

PS 7 The difference of the squares of two positive numbers, whose difference is 2, is 184. Find these two numbers.

PS 8 The length of a carpet is 1 m more than its width. Its area is 9 m^2. Find the dimensions of the carpet to 2 decimal places.

FM 9 Helen worked out that she could save 30 minutes on a 45 km journey if she travelled at an average speed which was 15 km/h faster than that at which she had planned to travel. Find the speed at which Helen had originally planned to travel.

FM 10 Claire intended to spend £3.20 on balloons for her party. But each balloon cost her 2p more than she expected, so she had to buy eight fewer balloons. Find the cost of each balloon.

PS 11 The sum of a number and its reciprocal is 2.05. What are the two numbers?

FM 12 A woman buys goods for £$60x$ and sells them for £$(600 - 6x)$ at a loss of x%. Find x.

FM 13 A train has a scheduled time for its journey. If the train averages 50 km/h, it arrives 12 minutes early. If the train averages 45 km/h, it arrives 20 minutes late. Find how long the train should take for the journey.

14 A rectangular garden measures 15 m by 11 m and is surrounded by a path of uniform width of area 41.25 m^2. Find the width of the path.

15 A rectangular room is 3 m longer than it is wide.

It cost £364 to carpet the room. Carpet costs £16 per square metre.

How wide is the room?

HINTS AND TIPS

Calculate the area and set up a quadratic equation to work out the width.

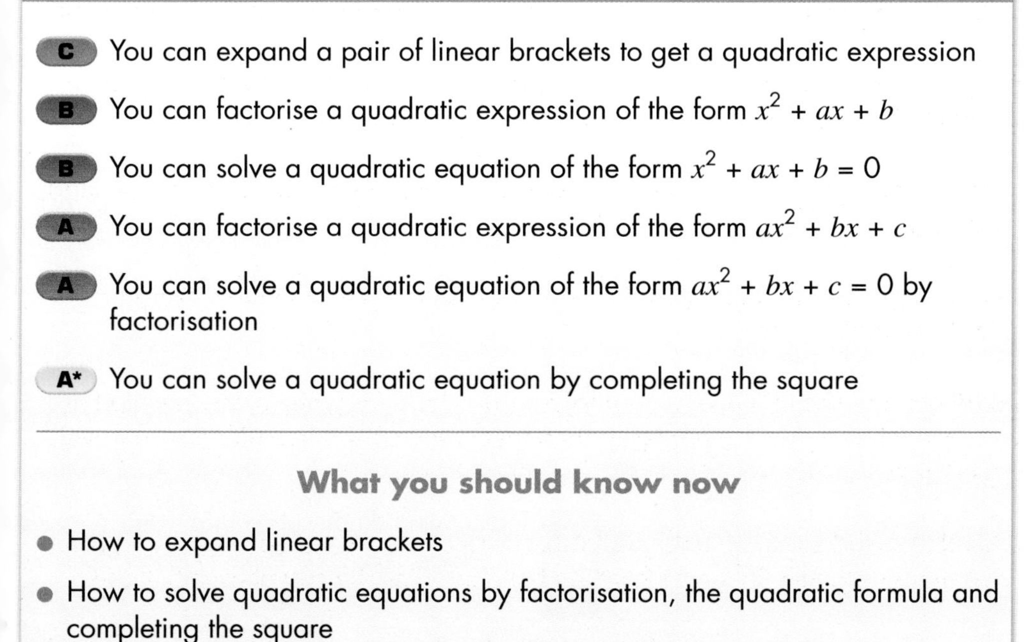

GRADE BOOSTER

- **C** You can expand a pair of linear brackets to get a quadratic expression
- **B** You can factorise a quadratic expression of the form $x^2 + ax + b$
- **B** You can solve a quadratic equation of the form $x^2 + ax + b = 0$
- **A** You can factorise a quadratic expression of the form $ax^2 + bx + c$
- **A** You can solve a quadratic equation of the form $ax^2 + bx + c = 0$ by factorisation
- **A*** You can solve a quadratic equation by completing the square

What you should know now

- How to expand linear brackets
- How to solve quadratic equations by factorisation, the quadratic formula and completing the square
- How to solve problems involving quadratic equations

1 Expand and simplify $(e + 3)(e + 4)$ *(2 marks)*

Edexcel, June 2007, Paper 11 Higher, Question 6(c)

2 Expand and simplify $(x + 3)(x - 5)$ *(2 marks)*

Edexcel, March 2008, Paper 11 Higher, Question 5(b)

3 Expand and simplify $(x - 2)(x + 1)$ *(2 marks)*

Edexcel, March 2008, Paper 10 Higher, Question 7

4 Expand and simplify $(2x + 5)(x - 2)$ *(2 marks)*

Edexcel, November 2008, Paper 3 Higher, Question 22(b)

5 Expand and simplify $(y + 2)(y + 3)$ *(2 marks)*

Edexcel, November 2008, Paper 10 Higher, Question 6(a)

6 Factorise $t^2 - 16$ *(1 mark)*

Edexcel, November 2008, Paper 11 Higher, Question 6(c)

7 Expand and simplify $(x + 4)(x - 3)$ *(2 marks)*

Edexcel, November 2008, Paper 11 Higher, Question 6

8 Expand and simplify $(x + 5)(x + 8)$ *(2 marks)*

Edexcel, March 2009, Paper 10 Higher, Question 4(b)

9 Expand and simplify $(3x - 5)(x + 1)$ *(2 marks)*

Edexcel, March 2007, Paper 11 Higher, Question 7

10 **a** Expand and simplify $(x + 3)(x + 4)$ *(2 marks)*

b Factorise $y^2 + 8y + 15$ *(2 marks)*

Edexcel, June 2008, Paper 4 Higher, Question 11(d & e)

11 Factorise $a^2 - 16a + 64$ *(2 marks)*

Edexcel, June 2007, Paper 11 Higher, Question 6(c)

12 Factorise $x^2 + 2x - 15$ *(2 marks)*

Edexcel, November 2007, Paper 11 Higher, Question 6

13 Expand and simplify $(3x + 4)(5x - 1)$ *(2 marks)*

Edexcel, November 2007, Paper 11 Higher, Question 9

14 Factorise $x^2 - 6x + 5$ *(2 marks)*

Edexcel, March 2008, Paper 11 Higher (5543H/11A), Question 8

15 Factorise $x^2 - 36$ *(1 mark)*

Edexcel, May 2008, Paper 3 Higher, Question 15(b)

16 Expand and simplify $(2x - 1)(5x + 3)$ *(3 marks)*

Edexcel, June 2008, Paper 11 Higher, Question 7

17 Expand and simplify $(2x - 3)(3x + 2)$ *(2 marks)*

Edexcel, March 2008, Paper 11 Higher, Question 9(a)

18 **a** Factorise $x^2 - y^2$ *(1 mark)*

Hence, or otherwise,

b Factorise $(x + 1)^2 - (y + 1)^2$ *(2 marks)*

Edexcel, March 2009, Paper 10 Higher, Question 8

19 Solve $3x^2 + 7x - 13 = 0$

Give your solutions correct to 2 decimal places. *(3 marks)*

Edexcel, June 2008, Paper 15 Higher, Question 17

20 **a** Show that the equation

$$\frac{5}{x + 2} = \frac{4 - 3x}{x - 1}$$

Can be rearranged to give

$3x^2 + 7x - 13 = 0$ *(3 marks)*

b Solve $3x^2 + 7x - 13 = 0$

Give your solutions correct to 2 decimal places. *(3 marks)*

Edexcel, June 2008, Paper 4 Higher, Question 23

21 The diagram below shows a six-sided shape.

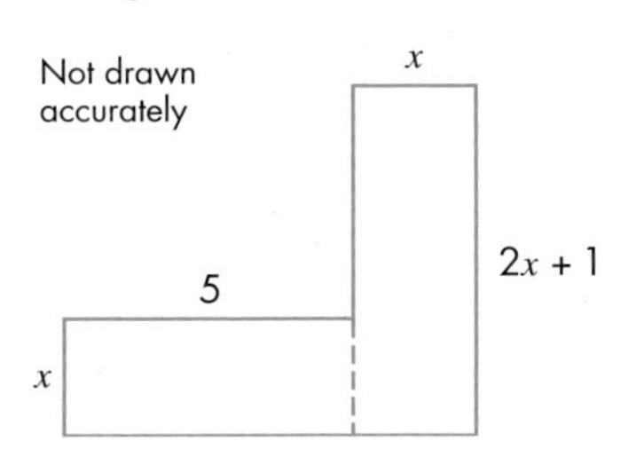

All the corners are right angles.

All the measurements are given in centimetres.

The area of the shape is 95 cm^2.

a Show that $2x^2 + 6x - 95 = 0$ *(3 marks)*

b Solve the equation $2x^2 + 6x - 95 = 0$

Give your solutions correct to 3 significant figures. *(3 marks)*

Edexcel, November 2008, Paper 4 Higher, Question 19

22 Solve the equation $2x^2 + 6x - 95 = 0$

Give your solutions correct to 3 significant figures. *(3 marks)*

Edexcel, November 2008, Paper 15 Higher, Question 15

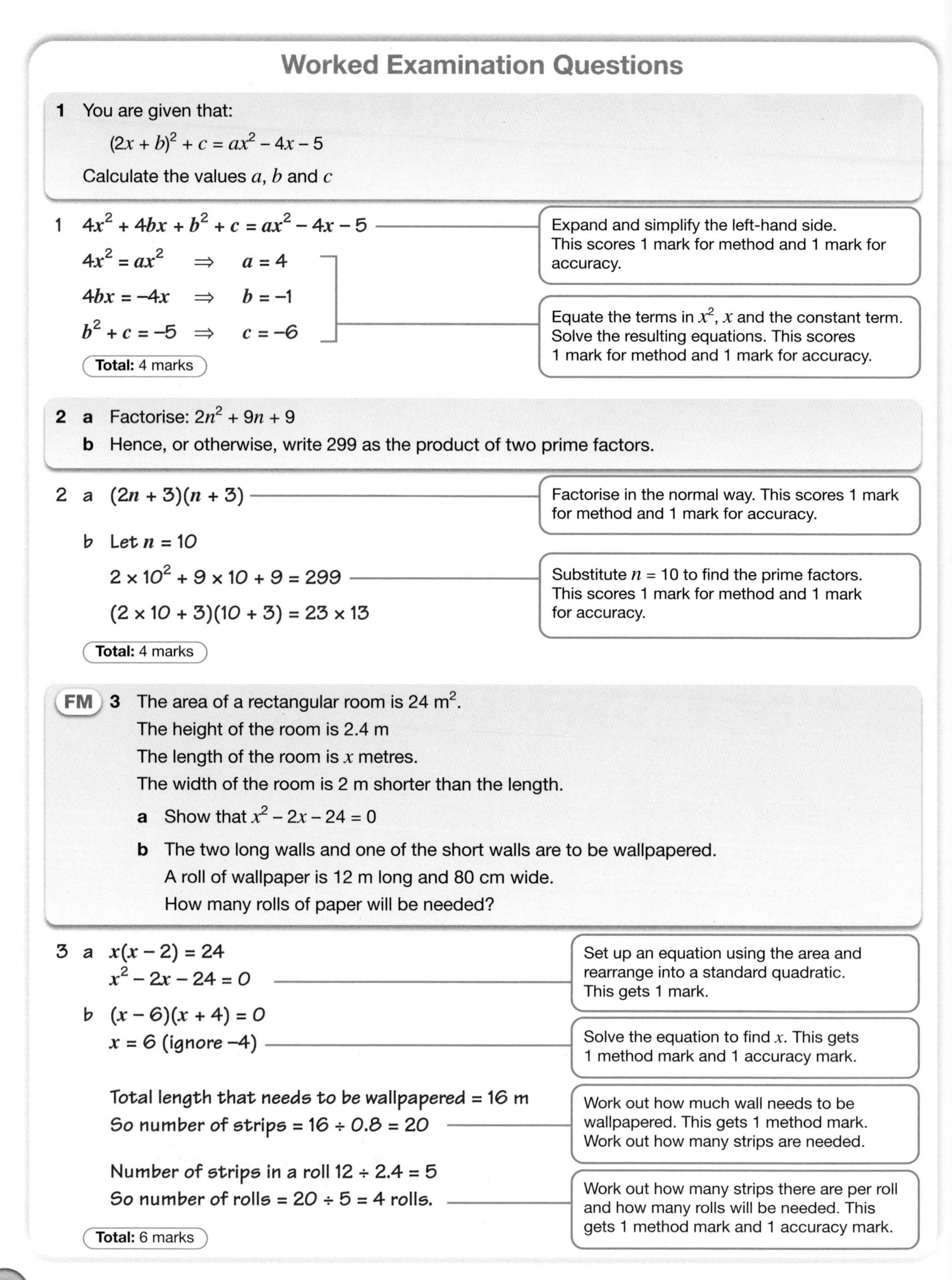

Worked Examination Questions

1 You are given that:

$(2x + b)^2 + c = ax^2 - 4x - 5$

Calculate the values a, b and c

1 $4x^2 + 4bx + b^2 + c = ax^2 - 4x - 5$ — Expand and simplify the left-hand side. This scores 1 mark for method and 1 mark for accuracy.

$4x^2 = ax^2 \Rightarrow a = 4$

$4bx = -4x \Rightarrow b = -1$

$b^2 + c = -5 \Rightarrow c = -6$

Equate the terms in x^2, x and the constant term. Solve the resulting equations. This scores 1 mark for method and 1 mark for accuracy.

Total: 4 marks

2 **a** Factorise: $2n^2 + 9n + 9$

b Hence, or otherwise, write 299 as the product of two prime factors.

2 a $(2n + 3)(n + 3)$ — Factorise in the normal way. This scores 1 mark for method and 1 mark for accuracy.

b Let $n = 10$

$2 \times 10^2 + 9 \times 10 + 9 = 299$

$(2 \times 10 + 3)(10 + 3) = 23 \times 13$

Substitute $n = 10$ to find the prime factors. This scores 1 mark for method and 1 mark for accuracy.

Total: 4 marks

FM **3** The area of a rectangular room is 24 m².

The height of the room is 2.4 m

The length of the room is x metres.

The width of the room is 2 m shorter than the length.

a Show that $x^2 - 2x - 24 = 0$

b The two long walls and one of the short walls are to be wallpapered.

A roll of wallpaper is 12 m long and 80 cm wide.

How many rolls of paper will be needed?

3 a $x(x - 2) = 24$

$x^2 - 2x - 24 = 0$

Set up an equation using the area and rearrange into a standard quadratic. This gets 1 mark.

b $(x - 6)(x + 4) = 0$

$x = 6$ (ignore -4)

Solve the equation to find x. This gets 1 method mark and 1 accuracy mark.

Total length that needs to be wallpapered = 16 m

So number of strips = 16 ÷ 0.8 = 20

Work out how much wall needs to be wallpapered. This gets 1 method mark. Work out how many strips are needed.

Number of strips in a roll 12 ÷ 2.4 = 5

So number of rolls = 20 ÷ 5 = 4 rolls.

Work out how many strips there are per roll and how many rolls will be needed. This gets 1 method mark and 1 accuracy mark.

Total: 6 marks

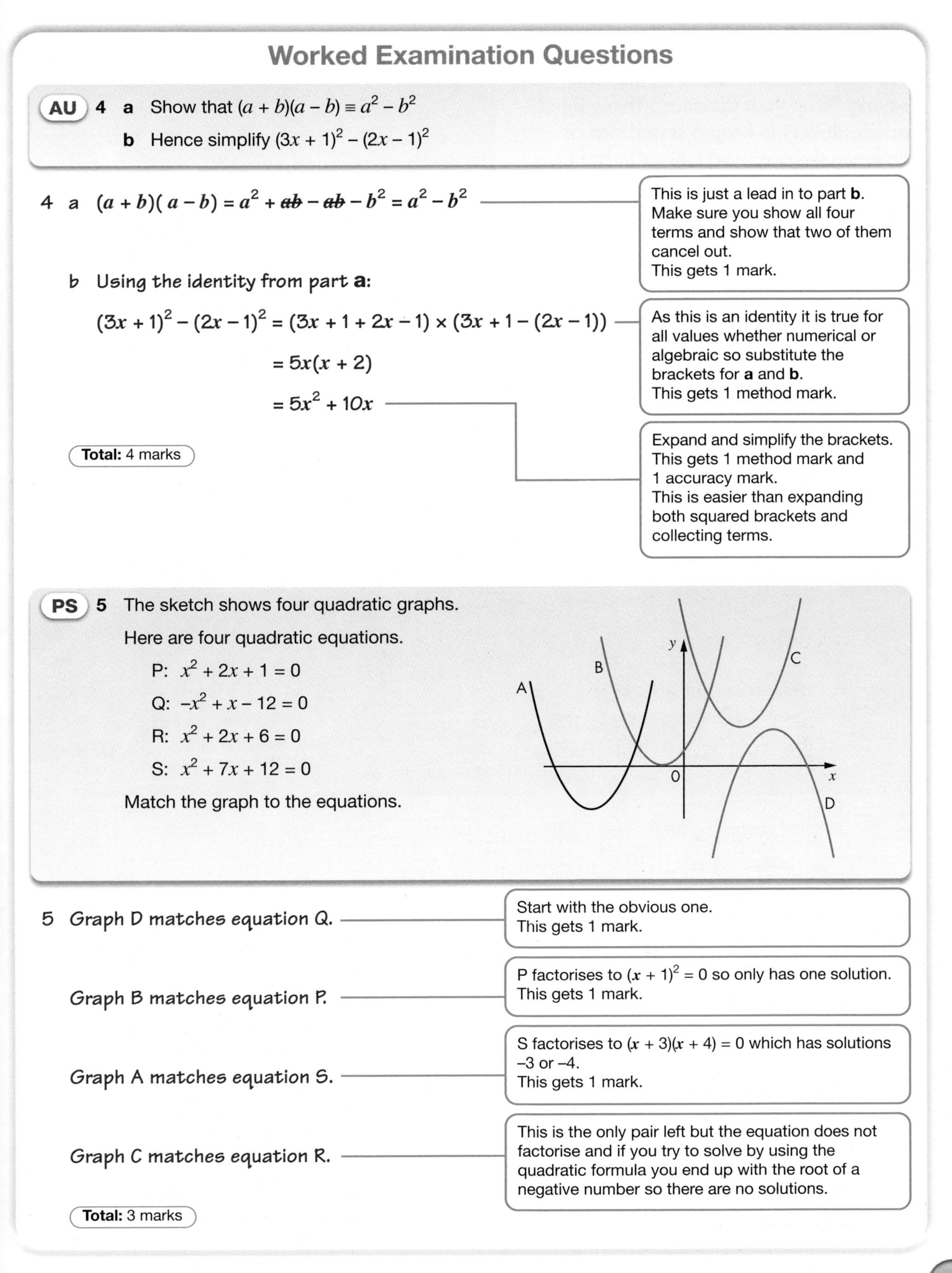

Worked Examination Questions

AU **4** **a** Show that $(a + b)(a - b) \equiv a^2 - b^2$

b Hence simplify $(3x + 1)^2 - (2x - 1)^2$

4 a $(a + b)(a - b) = a^2 + \cancel{ab} - \cancel{ab} - b^2 = a^2 - b^2$

> This is just a lead in to part **b**. Make sure you show all four terms and show that two of them cancel out.
> This gets 1 mark.

b Using the identity from part **a**:

$(3x + 1)^2 - (2x - 1)^2 = (3x + 1 + 2x - 1) \times (3x + 1 - (2x - 1))$

$= 5x(x + 2)$

$= 5x^2 + 10x$

> As this is an identity it is true for all values whether numerical or algebraic so substitute the brackets for **a** and **b**.
> This gets 1 method mark.

> Expand and simplify the brackets. This gets 1 method mark and 1 accuracy mark.
> This is easier than expanding both squared brackets and collecting terms.

Total: 4 marks

PS **5** The sketch shows four quadratic graphs.

Here are four quadratic equations.

P: $x^2 + 2x + 1 = 0$

Q: $-x^2 + x - 12 = 0$

R: $x^2 + 2x + 6 = 0$

S: $x^2 + 7x + 12 = 0$

Match the graph to the equations.

5 Graph D matches equation Q.

> Start with the obvious one.
> This gets 1 mark.

Graph B matches equation P.

> P factorises to $(x + 1)^2 = 0$ so only has one solution.
> This gets 1 mark.

Graph A matches equation S.

> S factorises to $(x + 3)(x + 4) = 0$ which has solutions -3 or -4.
> This gets 1 mark.

Graph C matches equation R.

> This is the only pair left but the equation does not factorise and if you try to solve by using the quadratic formula you end up with the root of a negative number so there are no solutions.

Total: 3 marks

6

Functional Maths
Stopping distances

You may have seen signs on the motorway saying 'Keep your distance'. These signs advise drivers to keep a safe distance between their car and the car in front, so that they have time to stop if, for example, the car in front comes to a rapid stop or if the engine fails. The distance that should be left between cars is commonly known as the 'stopping distance'.

The stopping distance is made up of two parts. The first part is the 'thinking distance', which is the time it takes for the brain to react and for you to apply the brakes. The second part is the braking distance, which is the distance it takes for the car to come to a complete stop, once the brakes have been applied.

KEEP YOUR DISTANCE

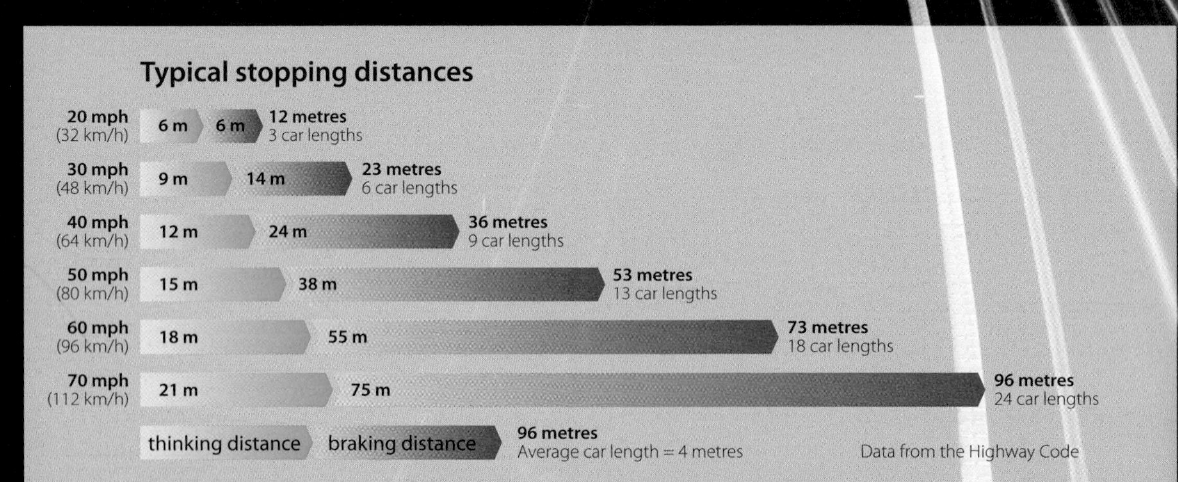

Notice that this diagram is based on 'typical' stopping distances. There are many factors that affect the stopping distance, such as road conditions, how good the brakes are and how heavy the car is, and so the stopping distance will never be the same in every situation.

Your task

This table shows the probability of a crash being fatal at certain speeds.

Speed at which crash happens (mph)	Probability of crash being fatal
70	0.60
60	0.50
50	0.42
40	0.34
30	0.26
20	0.17
10	0.09
5	0.05

James is learning to drive. His driving instructor has been teaching him about typical stopping distances, but James is not convinced that the distances given by his instructor, which are taken from the *Highway Code*, are right.

His instructor sets him a challenge: show whether the distances given in the *Highway Code* are correct.

James finds this formula for calculating stopping distances:

$$d = \frac{s^2}{20} + s$$

where d is the stopping distance (in feet) and s is the speed (in miles per hour).

Remember: 1 foot = 30 cm.

Does he prove his instructor to be right or wrong?

Getting started

Consider the following points to help you complete this task.

- How could you represent thinking, braking and stopping distances as a graph?
- Can you apply James's formula to real-life scenarios?
  - For example:
    If you are driving down a straight road and suddenly, 175 feet in front of you, a pallet falls off a lorry, what is the maximum speed from which you would be able to stop safely?
- How might the weather affect James's formula?
- What are the risks of fatality?

Why this chapter matters

Thales of Miletus (624–547 BC) was a Greek philosopher and one of the Seven Sages of Greece. He is believed to have been the first person to use similar triangles to find the height of tall objects.

Thales discovered that, at a particular time of day, the height of an object and the length of its shadow were the same. He used this observation to calculate the height of the Egyptian pyramids. Later, he took this knowledge back to Greece. His observations are considered to be the forerunner of the technique of using similar triangles to solve such problems.

A clinometer is an instrument used to measure the height of objects from a distance. Using a clinometer, you can apply the geometry of triangles to determine the height visually, rather than by physically measuring it. Clinometers are commonly used to measure the heights of trees, buildings and towers, mountains and other objects for which taking physical measurements might be impractical.

Astronomers use the geometry of triangles to measure the distance to nearby stars. They take advantage of the Earth's journey in its orbit around the Sun to obtain the maximum distance between two measurements. They observe the star twice, from the same point on Earth and at the same time of day, but six months apart.

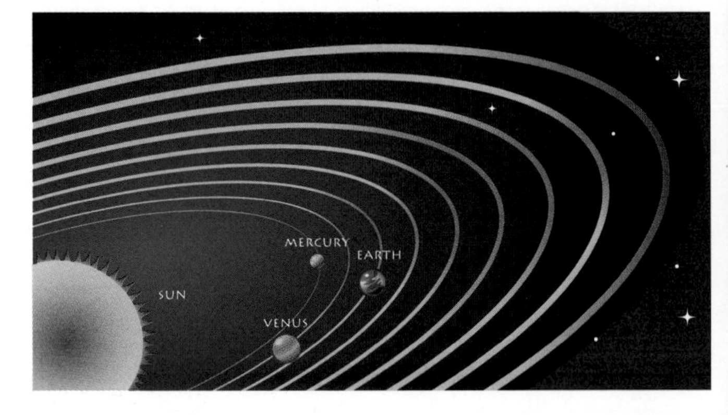

Telescopes and binoculars also use the geometry of triangles. The Hubble Space Telescope took this image of the Eagle Nebula. This star-forming region is located 6500 light years from Earth. It is only about 6 million years old and the dense clouds of interstellar gas are still collapsing to form new stars.

A light year is the distance that a ray of light travels in one year.

It is about 5 878 630 000 000 miles or just under 10^{13} km.

Chapter

7 Measures: Similarity

The grades given in this chapter are target grades.

This chapter will show you ...

- **C** how to work out the scale factor for two similar shapes
- **B** how to work out lengths of sides in similar figures
- **A** to **A*** how to work out areas and volumes of similar shapes

Visual overview

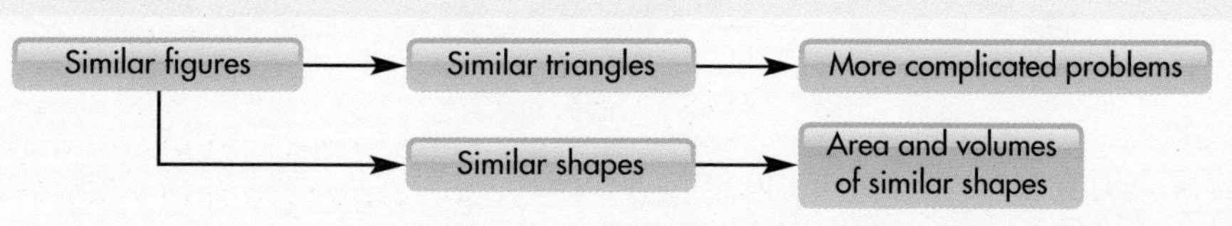

What you should already know

- How to use and simplify ratios **KS3 level 6, GCSE grade D**
- How to enlarge a shape by a given scale factor **KS3 level 6, GCSE grade D–C**
- How to solve equations **KS3 level 6, GCSE grade D–C**

Quick check

1 Simplify the following ratios.

a 15 : 20 **b** 24 : 30

c $6 : 1\frac{1}{2}$ **d** 7.5 : 5

2 Solve the following equations.

a $\frac{x}{4} = \frac{7}{2}$ **b** $\frac{x}{2} = \frac{5}{4}$

c $\frac{x}{4} = \frac{x-2}{3}$ **d** $\frac{x+4}{x} = \frac{5}{3}$

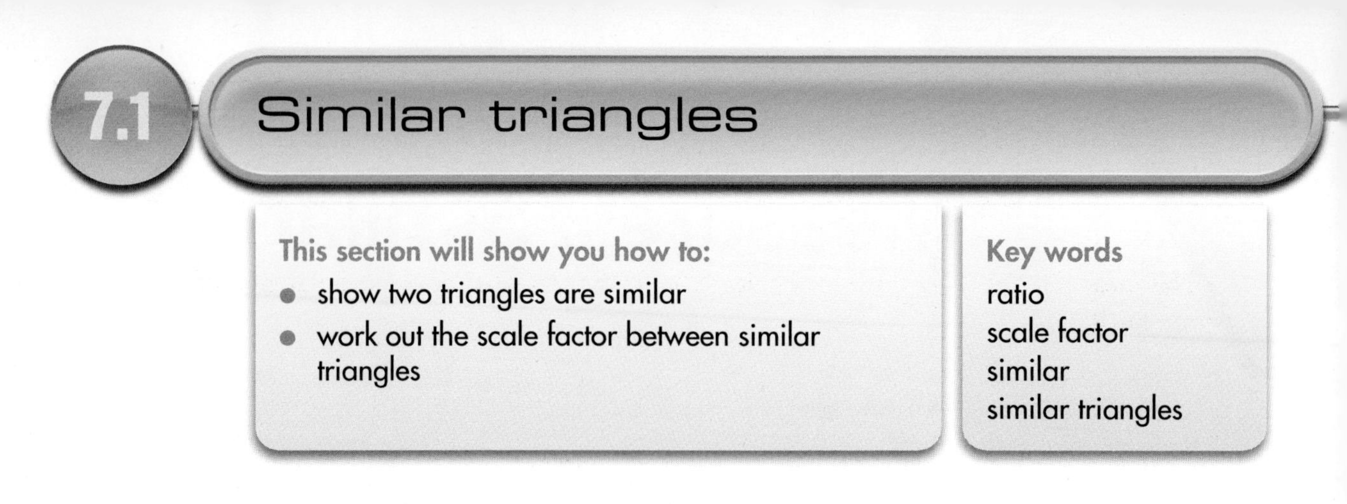

7.1 Similar triangles

This section will show you how to:
- show two triangles are similar
- work out the scale factor between similar triangles

Key words
ratio
scale factor
similar
similar triangles

Triangles are **similar** if their corresponding angles are equal. Their corresponding sides are then in the same **ratio**.

These two right-angled triangles are **similar triangles**.

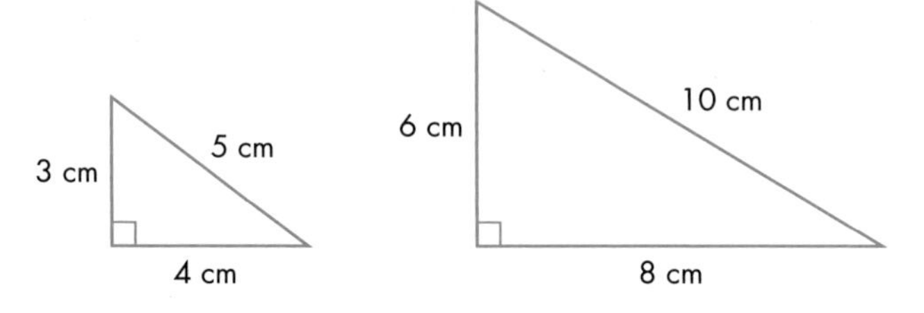

The scale factor of the enlargement = 2

The ratios of the lengths of corresponding sides all cancel to the same ratio.

$3 : 6 = 4 : 8 = 5 : 10 = 1 : 2$

All corresponding angles are equal.

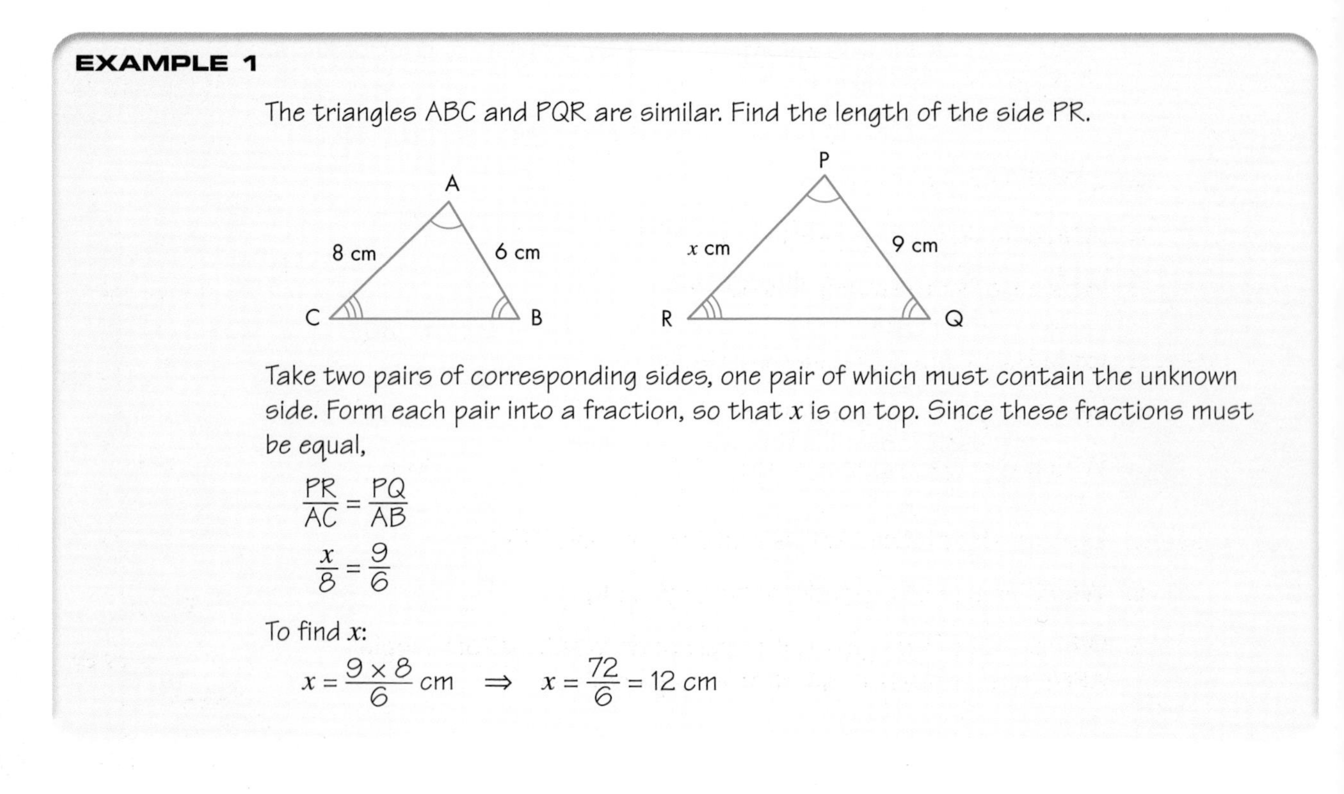

EXAMPLE 1

The triangles ABC and PQR are similar. Find the length of the side PR.

Take two pairs of corresponding sides, one pair of which must contain the unknown side. Form each pair into a fraction, so that x is on top. Since these fractions must be equal,

$$\frac{PR}{AC} = \frac{PQ}{AB}$$

$$\frac{x}{8} = \frac{9}{6}$$

To find x:

$$x = \frac{9 \times 8}{6} \text{ cm} \quad \Rightarrow \quad x = \frac{72}{6} = 12 \text{ cm}$$

EXERCISE 7A

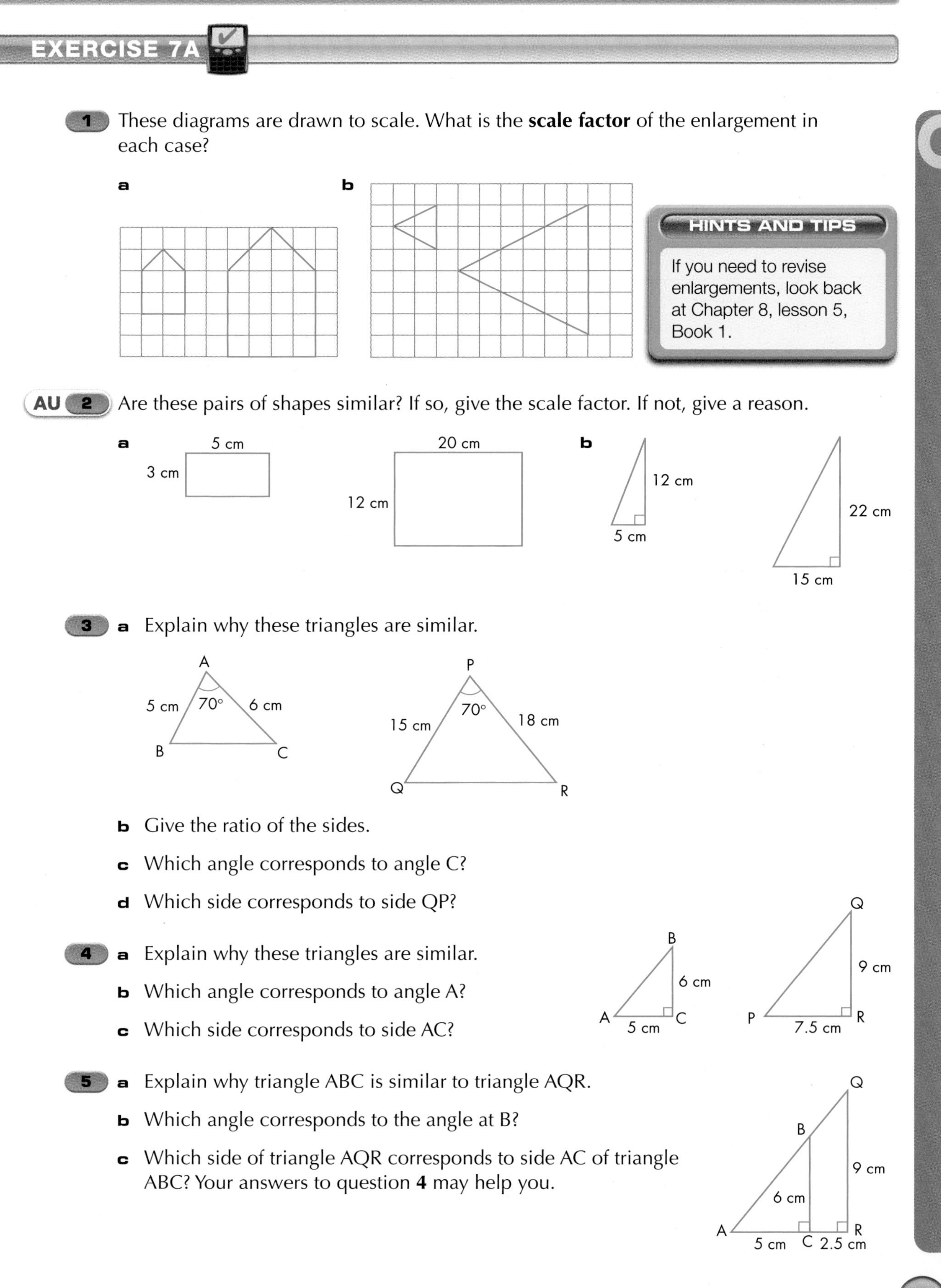

1 These diagrams are drawn to scale. What is the **scale factor** of the enlargement in each case?

a

b

> **HINTS AND TIPS**
>
> If you need to revise enlargements, look back at Chapter 8, lesson 5, Book 1.

AU 2 Are these pairs of shapes similar? If so, give the scale factor. If not, give a reason.

a 5 cm × 3 cm rectangle; 20 cm × 12 cm rectangle

b Right-angled triangles: 12 cm, 5 cm; 22 cm, 15 cm

3 **a** Explain why these triangles are similar.

b Give the ratio of the sides.

c Which angle corresponds to angle C?

d Which side corresponds to side QP?

4 **a** Explain why these triangles are similar.

b Which angle corresponds to angle A?

c Which side corresponds to side AC?

5 **a** Explain why triangle ABC is similar to triangle AQR.

b Which angle corresponds to the angle at B?

c Which side of triangle AQR corresponds to side AC of triangle ABC? Your answers to question **4** may help you.

B

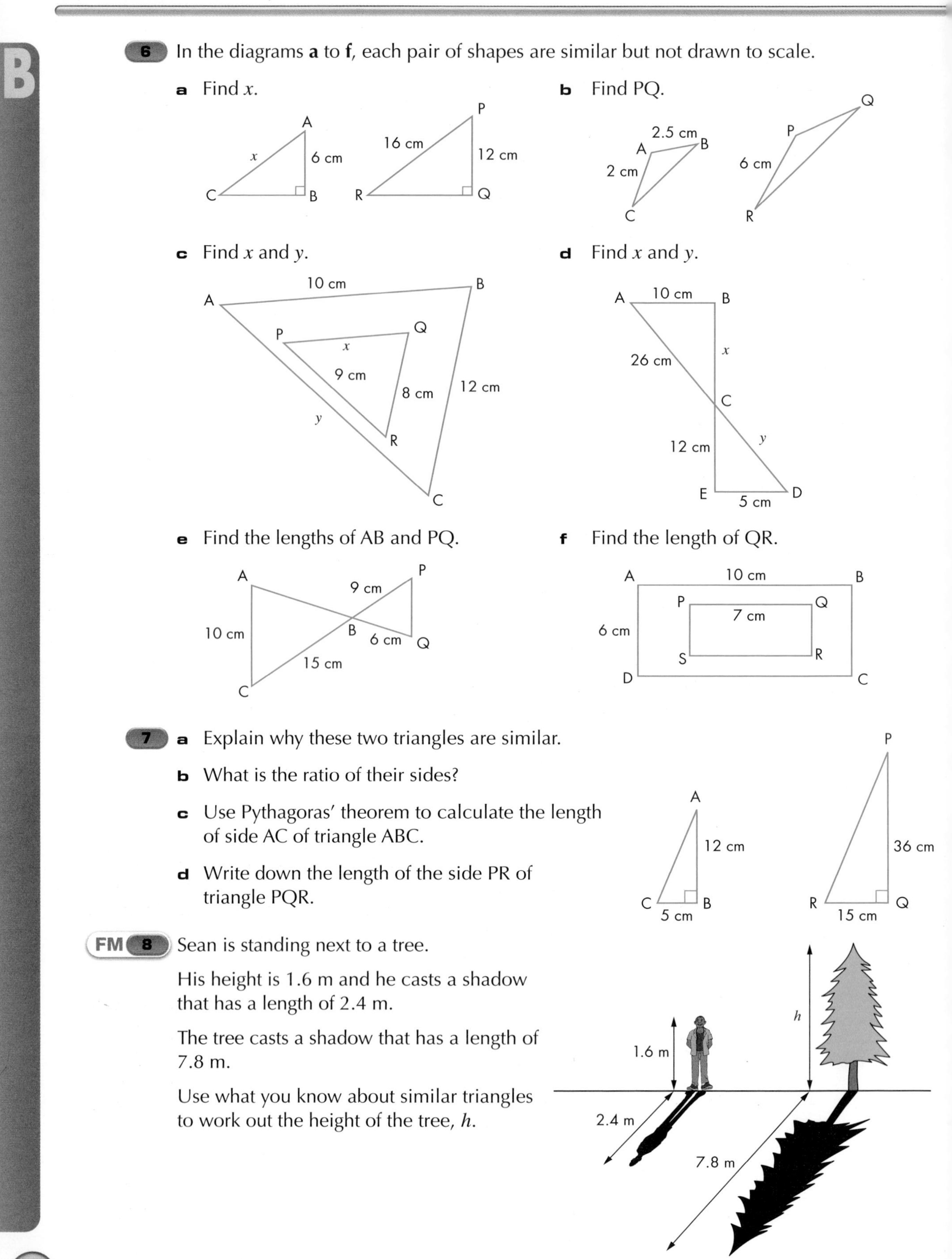

6 In the diagrams **a** to **f**, each pair of shapes are similar but not drawn to scale.

a Find x.

b Find PQ.

c Find x and y.

d Find x and y.

e Find the lengths of AB and PQ.

f Find the length of QR.

7 **a** Explain why these two triangles are similar.

b What is the ratio of their sides?

c Use Pythagoras' theorem to calculate the length of side AC of triangle ABC.

d Write down the length of the side PR of triangle PQR.

FM 8 Sean is standing next to a tree.

His height is 1.6 m and he casts a shadow that has a length of 2.4 m.

The tree casts a shadow that has a length of 7.8 m.

Use what you know about similar triangles to work out the height of the tree, h.

B

PS 9 Here are two rectangles.

Explain why the two rectangles are not similar.

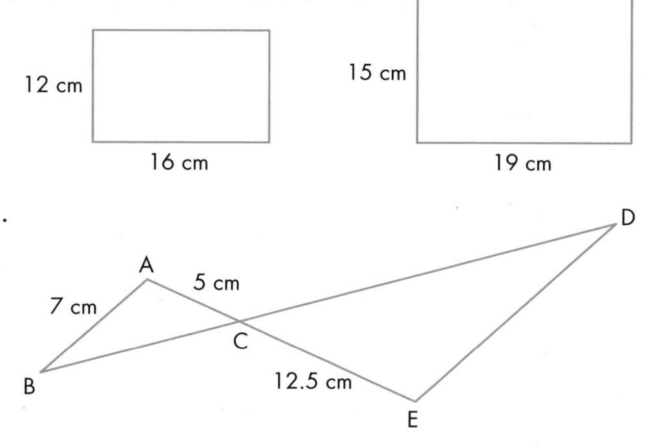

AU 10 Triangle ABC is similar to triangle CDE.

Jay says that the length of DE is 14 cm.

Explain why Jay is wrong.

Further examples of similar triangles

EXAMPLE 2

Find the lengths marked x and y in the diagram (not drawn to scale).

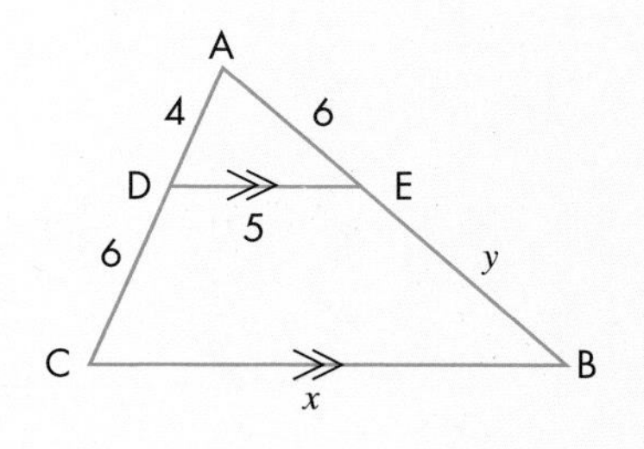

Triangles AED and ABC are similar. So using the corresponding sides CB, DE with AC, AD gives,

$$\frac{x}{5} = \frac{10}{4}$$

$$\Rightarrow x = \frac{10 \times 5}{4} = 12.5$$

Using the corresponding sides AE, AB with AD, AC gives,

$$\frac{y+6}{6} = \frac{10}{4} \Rightarrow y + 6 = \frac{10 \times 6}{4} = 15$$

$$\Rightarrow \quad y = 15 - 6 = 9$$

EXAMPLE 3

Ahmed wants to work out the height of a tall building. He walks 100 paces from the building and sticks a pole, 2 m long, vertically into the ground. He then walks another 10 paces on the same line and notices that when he looks from ground level, the top of the pole and the top of the building are in line. How tall is the building?

First, draw a diagram of the situation and label it.

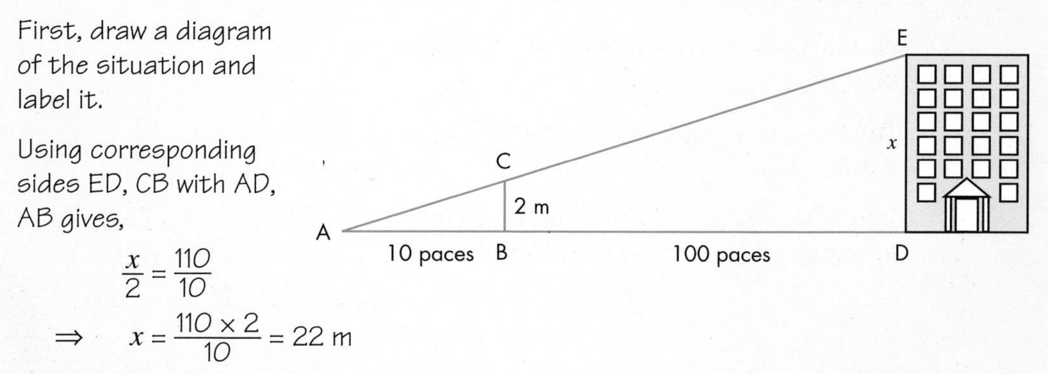

Using corresponding sides ED, CB with AD, AB gives,

$$\frac{x}{2} = \frac{110}{10}$$

$$\Rightarrow \quad x = \frac{110 \times 2}{10} = 22 \text{ m}$$

Hence the building is 22 m high.

EXERCISE 7B

1 In each of the cases below, state a pair of similar triangles and find the length marked x. Separate the similar triangles if it makes it easier for you.

a

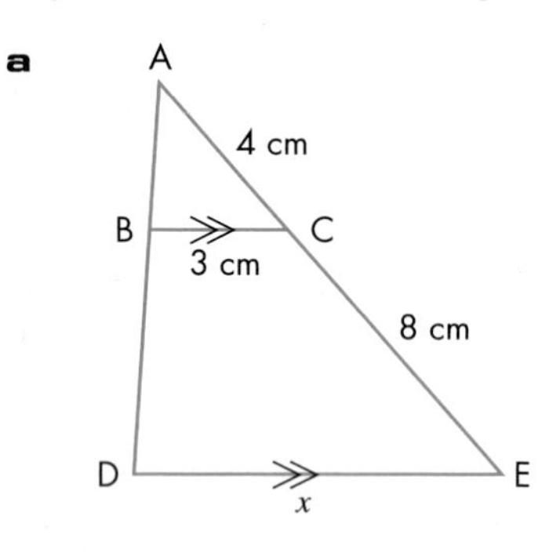

b

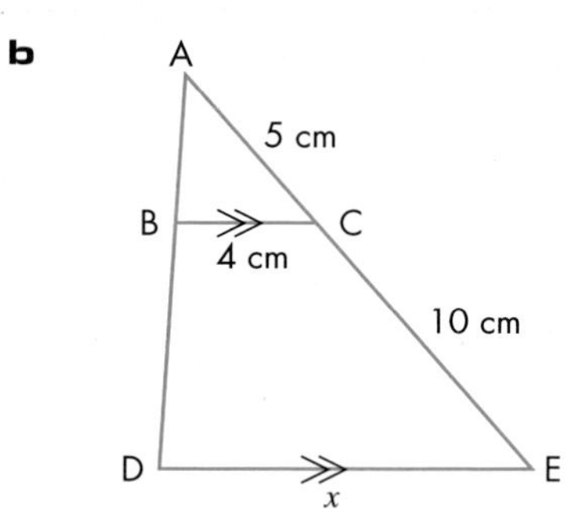

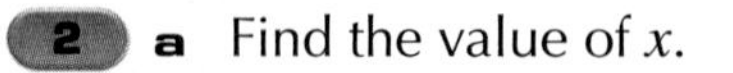

2 **a** Find the value of x.

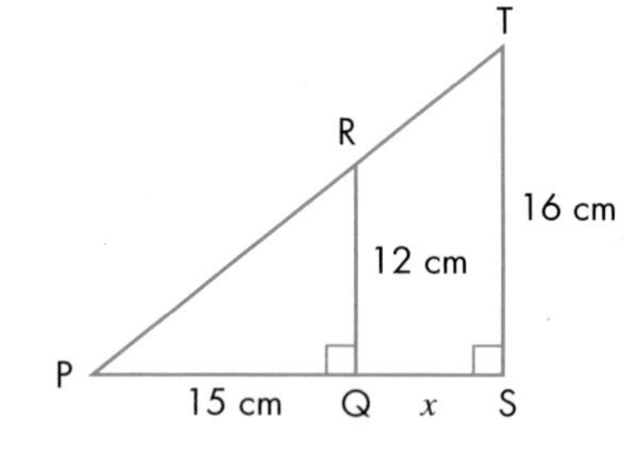

b Find the length of CE.

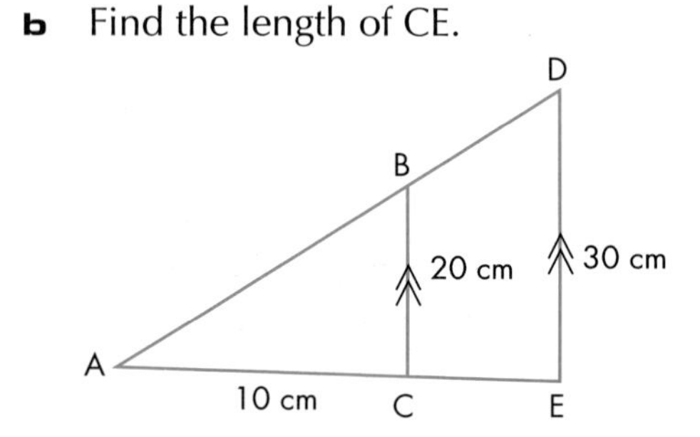

c Find the values of x and y.

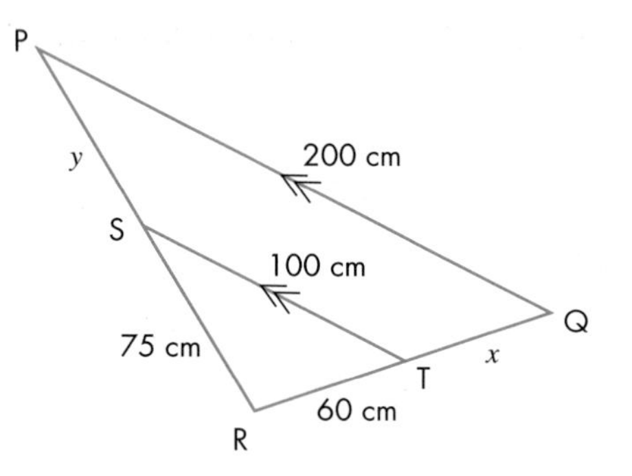

d Find the values of x and y.

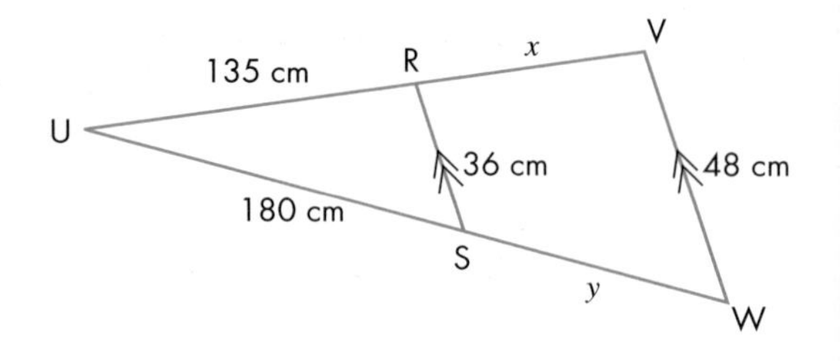

e Find the lengths of DC and EB.

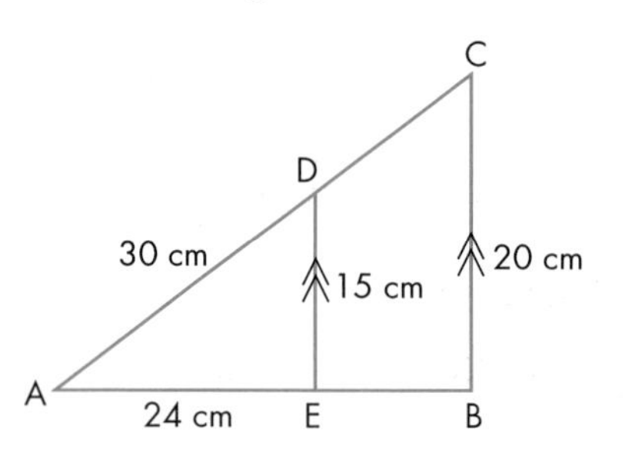

FM 3 This diagram shows a method of working out the height of a tower.

A stick, 2 m long, is placed vertically 120 m from the base of a tower so that the top of the tower and the top of the stick are in line with a point on the ground 3 m from the base of the stick. How high is the tower?

h
2 m
3 m
120 m

FM 4 It is known that a factory chimney is 330 feet high. Patrick paces out distances as shown in the diagram, so that the top of the chimney and the top of the flag pole are in line with each other. How high is the flag pole?

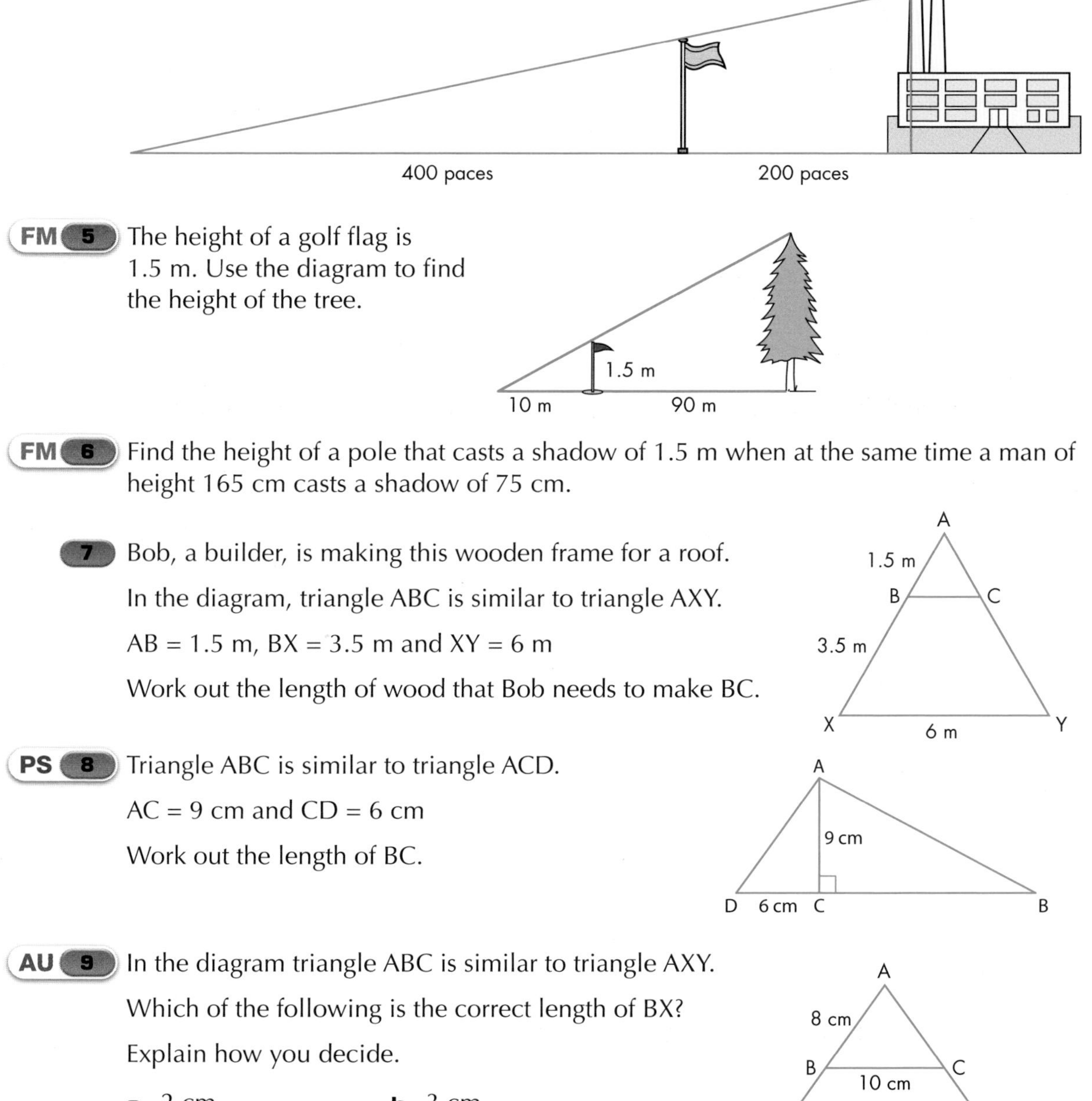

FM 5 The height of a golf flag is 1.5 m. Use the diagram to find the height of the tree.

FM 6 Find the height of a pole that casts a shadow of 1.5 m when at the same time a man of height 165 cm casts a shadow of 75 cm.

7 Bob, a builder, is making this wooden frame for a roof.

In the diagram, triangle ABC is similar to triangle AXY.

AB = 1.5 m, BX = 3.5 m and XY = 6 m

Work out the length of wood that Bob needs to make BC.

PS 8 Triangle ABC is similar to triangle ACD.

AC = 9 cm and CD = 6 cm

Work out the length of BC.

AU 9 In the diagram triangle ABC is similar to triangle AXY.

Which of the following is the correct length of BX?

Explain how you decide.

a 2 cm **b** 3 cm

c 4 cm **d** 5 cm

More complicated problems

The information given in a similar triangle situation can be more complicated than anything you have met so far, and you will need to have good algebraic skills to deal with it. Example 4 is typical of the more complicated problem you may be asked to solve, so follow it through carefully.

EXAMPLE 4

Find the value of x in this triangle.

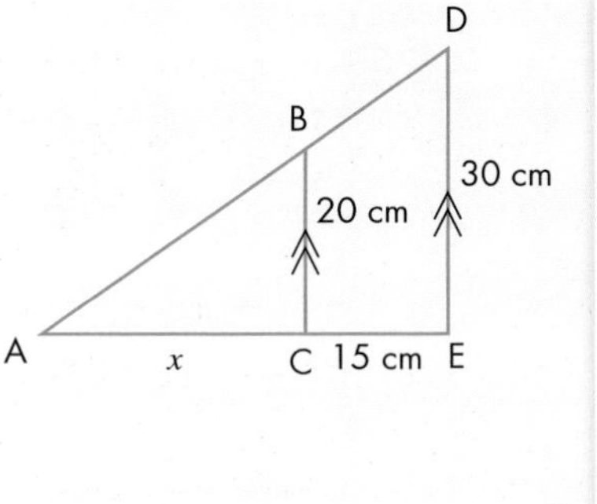

You know that triangle ABC is similar to triangle ADE.

Splitting up the triangles may help you to see what will be needed.

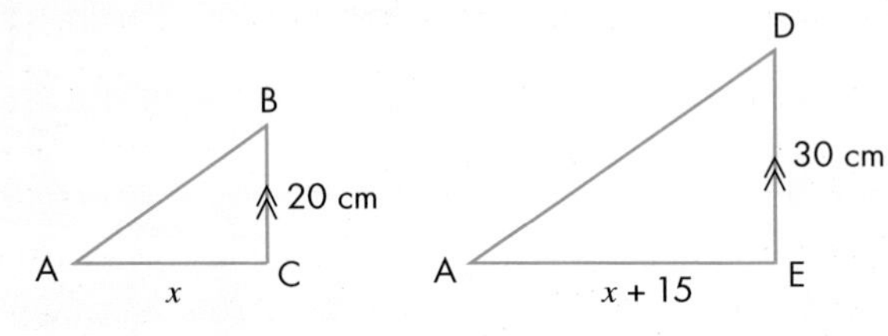

So your equation will be:

$$\frac{x + 15}{x} = \frac{30}{20}$$

Cross multiplying (moving each of the two bottom terms to the opposite side and multiplying) gives:

$$20x + 300 = 30x$$

$$\Rightarrow \quad 300 = 10x \Rightarrow x = 30 \text{ cm}$$

EXERCISE 7C

Find the lengths x or x and y in the diagrams **1** to **6**.

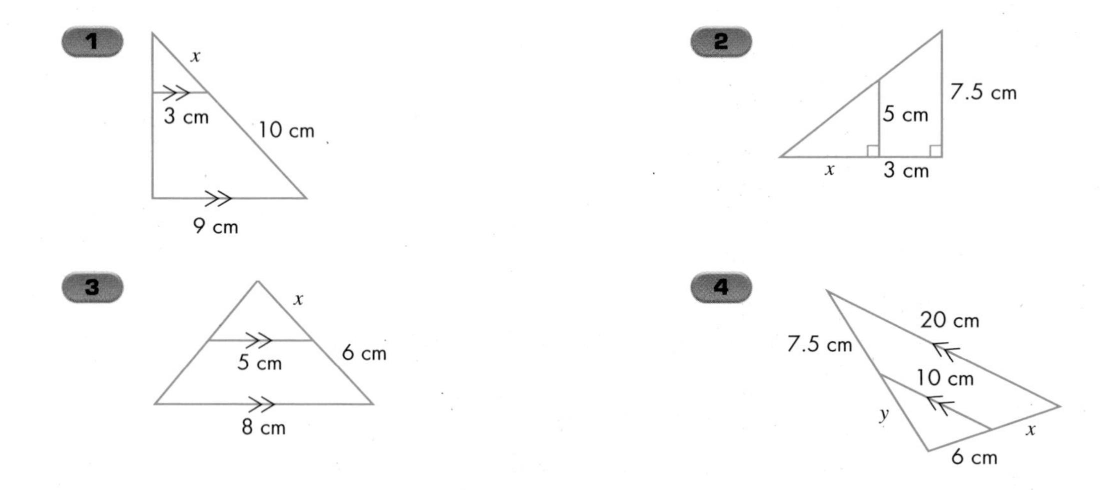

B

5

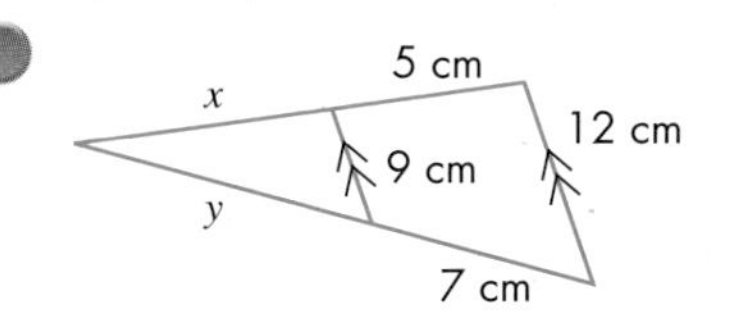

6

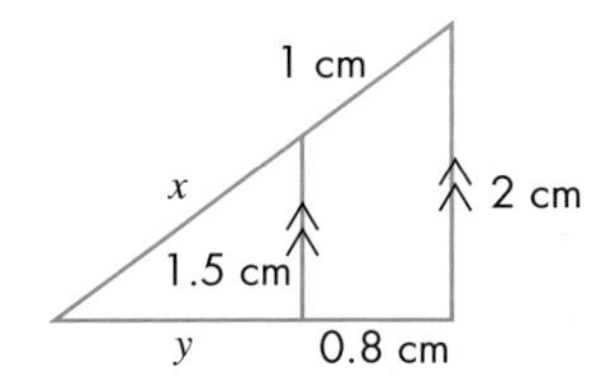

7.2 Areas and volumes of similar shapes

This section will show you how to:

- solve problems involving the area and volume of similar shapes

Key words

area ratio
area scale factor
length ratio
linear scale factor
volume ratio
volume scale factor

There are relationships between the lengths, areas and volumes of similar shapes.

You saw in Chapter 8, lesson 5, Book 1, that when a 2D shape is enlarged by a given scale factor to form a new, similar shape, the corresponding lengths of the original shape and the new shape are all in the same ratio, which is equal to the scale factor. This scale factor of the lengths is called the **length ratio** or **linear scale factor**.

Two similar shapes also have an **area ratio**, which is equal to the ratio of the squares of their corresponding lengths. The area ratio, or **area scale factor**, is the square of the length ratio.

Likewise, two 3D shapes are similar if their corresponding lengths are in the same ratio. Their **volume ratio** is equal to the ratio of the cubes of their corresponding lengths. The volume ratio, or **volume scale factor**, is the cube of the length ratio.

Generally, the relationship between similar shapes can be expressed as:

Length ratio $x : y$ Area ratio $x^2 : y^2$ Volume ratio $x^3 : y^3$

EXAMPLE 5

A model yacht is made to a scale of $\frac{1}{20}$ of the size of the real yacht. The area of the sail of the model is 150 cm^2. What is the area of the sail of the real yacht?

At first sight, it may appear that you do not have enough information to solve this problem, but it can be done as follows.

Linear scale factor = 1 : 20

Area scale factor = 1 : 400 (square of the linear scale factor)

Area of real sail = 400 × area of model sail

= 400 × 150 cm^2

= 60 000 cm^2 = 6 m^2

EXAMPLE 6

A bottle has a base radius of 4 cm, a height of 15 cm and a capacity of 650 cm^3. A similar bottle has a base radius of 3 cm.

a What is the length ratio?

b What is the volume ratio?

c What is the volume of the smaller bottle?

a The length ratio is given by the ratio of the two radii, that is 4 : 3.

b The volume ratio is therefore $4^3 : 3^3 = 64 : 27$.

c Let v be the volume of the smaller bottle. Then the volume ratio is:

$$\frac{\text{volume of smaller bottle}}{\text{volume of larger bottle}} = \frac{v}{650} = \frac{27}{64}$$

$$\Rightarrow v = \frac{27 \times 650}{64} = 274 \text{ cm}^3 \text{ (3 significant figures)}$$

EXAMPLE 7

The cost of a tin of paint , height 12 cm, is £3.20 and its label has an area of 24 cm^2.

a If the cost is based on the amount of paint in the tin, what is the cost of a similar tin, 18 cm high?

b Assuming the labels are similar, what will be the area of the label on the larger tin?

a The cost of the paint is proportional to the volume of the tin.

Length ratio = 12 : 18 = 2 : 3

Volume ratio = $2^3 : 3^3 = 8 : 27$

Let P be the cost of the larger tin. Then the cost ratio is:

$$\frac{\text{cost of larger tin}}{\text{cost of smaller tin}} = \frac{P}{3.2}$$

Therefore,

$$\frac{P}{3.2} = \frac{27}{8}$$

$$\Rightarrow P = \frac{27 \times 3.2}{8} = £10.80$$

b Area ratio = $2^2 : 3^2 = 4 : 9$

Let A be the area of the larger label. Then the area ratio is:

$$\frac{\text{larger label area}}{\text{smaller label area}} = \frac{A}{24}$$

Therefore,

$$\frac{A}{24} = \frac{9}{4}$$

$$\Rightarrow A = \frac{9 \times 24}{4} = 54 \text{ cm}^2$$

EXERCISE 7D

1 The length ratio between two similar solids is 2 : 5.

a What is the area ratio between the solids?

b What is the volume ratio between the solids?

2 The length ratio between two similar solids is 4 : 7.

a What is the area ratio between the solids?

b What is the volume ratio between the solids?

3 Copy and complete this table.

Linear scale factor	Linear ratio	Linear fraction	Area scale factor	Volume scale factor
2	1 : 2	$\frac{2}{1}$		
3				
$\frac{1}{4}$	4 : 1	$\frac{1}{4}$		$\frac{1}{64}$
			25	
				$\frac{1}{1000}$

4 A shape has an area of 15 cm^2. What is the area of a similar shape with lengths that are three times the corresponding lengths of the first shape?

FM **5** A toy brick has a surface area of 14 cm^2. What would be the surface area of a similar toy brick with lengths that are:

a twice the corresponding lengths of the first brick

b three times the corresponding lengths of the first brick?

6 A rug has an area of 12 m^2. What area would be covered by rugs with lengths that are:

a twice the corresponding lengths of the first rug

b half the corresponding lengths of the first rug?

7 A brick has a volume of 300 cm^3. What would be the volume of a similar brick whose lengths are:

a twice the corresponding lengths of the first brick

b three times the corresponding lengths of the first brick?

FM **8** A tin of paint, 6 cm high, holds a half a litre of paint. How much paint would go into a similar tin which is 12 cm high?

FM **9** A model statue is 10 cm high and has a volume of 100 cm^3. The real statue is 2.4 m high. What is the volume of the real statue? Give your answer in m^3.

A

10 A small tin of paint costs 75p. What is the cost of a larger similar tin with height twice that of the smaller tin? Assume that the cost is based only on the volume of paint in the tin.

11 A small trinket box of width 2 cm has a volume of 10 cm^3. What is the width of a similar trinket box with a volume of 80 cm^3?

FM 12 A cinema sells popcorn in two different-sized tubs that are similar in shape.

Show that it is true that the big tub is better value.

PS 13 The diameters of two ball bearings are given below.

Work out:

a the ratio of their radii

b the ratio of their surface areas

c the ratio of their volumes.

6 mm 8 mm

AU 14 Cuboid A is similar to cuboid B.

The length of cuboid A is 10 cm and the length of cuboid B is 5 cm.

The volume of cuboid A is 720 cm^3.

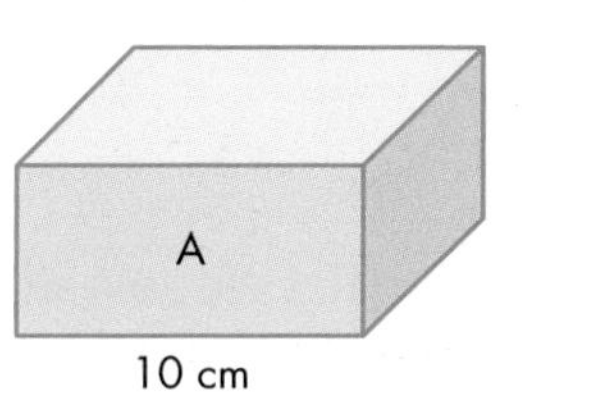

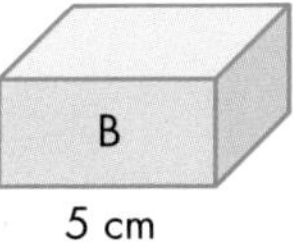

Shona says that the volume of cuboid B must be 360 cm^3.

Explain why she is wrong.

More complex problems using area and volume ratios

In some problems involving similar shapes, the length ratio is not given, so we have to start with the area ratio or the volume ratio. We usually then need to find the length ratio in order to proceed with the solution.

EXAMPLE 8

A manufacturer makes a range of clown hats that are all similar in shape. The smallest hat is 8 cm tall and uses 180 cm^2 of card. What will be the height of a hat made from 300 cm^2 of card?

The area ratio is 180 : 300

Therefore, the length ratio is $\sqrt{180} : \sqrt{300}$ (do not calculate these yet)

Let the height of the larger hat be H, then

$$\frac{H}{8} = \frac{\sqrt{300}}{\sqrt{180}} = \sqrt{\frac{300}{180}}$$

$$\Rightarrow H = 8 \times \sqrt{\frac{300}{180}} = 10.3 \text{ cm (1 decimal place)}$$

EXAMPLE 9

A supermarket stocks similar small and large tins of soup. The areas of their labels are 110 cm² and 190 cm² respectively. The weight of a small tin is 450 g. What is the weight of a large tin?

The area ratio is 110 : 190

Therefore, the length ratio is $\sqrt{110} : \sqrt{190}$ (do not calculate these yet)

So the volume (weight) ratio is $(\sqrt{110})^3 : (\sqrt{190})^3$.

Let the weight of a large tin be W, then

$$\frac{W}{450} = \frac{(\sqrt{190})^3}{(\sqrt{110})^3} = \left(\sqrt{\frac{190}{110}}\right)^3$$

$$\Rightarrow \quad W = 450 \times \left(\sqrt{\frac{190}{110}}\right)^3 = 1020 \text{ g} \qquad \text{(3 significant figures)}$$

EXAMPLE 10

Two similar tins hold respectively 1.5 litres and 2.5 litres of paint. The area of the label on the smaller tin is 85 cm². What is the area of the label on the larger tin?

The volume ratio is 1.5 : 2.5

Therefore, the length ratio is $\sqrt[3]{1.5} : \sqrt[3]{2.5}$ (do not calculate these yet)

So the area ratio is $(\sqrt[3]{1.5})^2 : (\sqrt[3]{2.5})^2$

Let the area of the label on the larger tin be A, then

$$\frac{A}{85} = \frac{(\sqrt[3]{2.5})^2}{(\sqrt[3]{1.5})^2} = \left(\sqrt[3]{\frac{2.5}{1.5}}\right)^2$$

$$\Rightarrow \quad A = 85 \times \left(\sqrt[3]{\frac{2.5}{1.5}}\right)^2 = 119 \text{ cm}^2 \qquad \text{(3 significant figures)}$$

EXERCISE 7E

FM 1 A firm produces three sizes of similar-shaped labels for its products. Their areas are 150 cm², 250 cm² and 400 cm². The 250 cm² label just fits around a can of height 8 cm. Find the heights of similar cans around which the other two labels would just fit.

2 A firm makes similar gift boxes in three different sizes: small, medium and large. The areas of their lids are as follows.

Small: 30 cm²   Medium: 50 cm²   Large: 75 cm²

The medium box is 5.5 cm high. Find the heights of the other two sizes.

A*

A*

3 A cone of height 8 cm can be made from a piece of card with an area of 140 cm^2. What is the height of a similar cone made from a similar piece of card with an area of 200 cm^2?

4 It takes 5.6 litres of paint to paint a chimney which is 3 m high. What is the tallest similar chimney that can be painted with 8 litres of paint?

5 A piece of card, 1200 cm^2 in area, will make a tube 13 cm long. What is the length of a similar tube made from a similar piece of card with an area of 500 cm^2?

6 All television screens (of the same style) are similar. If a screen of area 220 cm^2 has a diagonal length of 21 cm, what will be the diagonal length of a screen of area 350 cm^2?

7 Two similar statues, made from the same bronze, are placed in a school. One weighs 300 g, the other weighs 2 kg. The height of the smaller statue is 9 cm. What is the height of the larger statue?

FM 8 A supermarket sells similar cans of pasta rings in three different sizes: small, medium and large. The sizes of the labels around the cans are as follows.

Small can: 24 cm^2 Medium can: 46 cm^2 Large can: 78 cm^2

The medium size can is 6 cm tall with a weight of 380 g. Calculate these quantities.

a The heights of the other two sizes

b The weights of the other two sizes

9 A statue weighs 840 kg. A similar statue was made out of the same material but two-fifths the height of the first one. What was the weight of the smaller statue?

10 A model stands on a base of area 12 cm^2. A smaller but similar model, made of the same material, stands on a base of area 7.5 cm^2. Calculate the weight of the smaller model if the larger one is 3.5 kg.

FM 11 Steve fills two similar jugs with orange juice.

The first jug holds 1.5 litres of juice and has a base diameter of 8 cm.

The second jug holds 2 litres of juice. Work out the base diameter of the second jug.

PS 12 The total surface areas of two similar cuboids are 500 cm^2 and 800 cm^2.

If the width of one of the cuboids is 10 cm, calculate the two possible widths for the other cuboid.

AU 13 The volumes of two similar cylinders are 256 cm^3 and 864 cm^3.

Which of the following gives the ratio of their surface areas?

a 2 : 3 **b** 4 : 9 **c** 8 : 27

GRADE BOOSTER

C You can work out unknown lengths in 2D shapes, using scale factors

B You can use ratios and equations to find unknown lengths in similar triangles and shapes

A You can solve problems, using area and volume scale factors

A* You can solve more complex problems, using area and volume scale factors

What you should know now

- How to find the ratios between two similar shapes
- How to work out unknown lengths, areas and volumes of similar 3D shapes
- How to solve practical problems, using similar shapes
- How to solve problems, using area and volume ratios

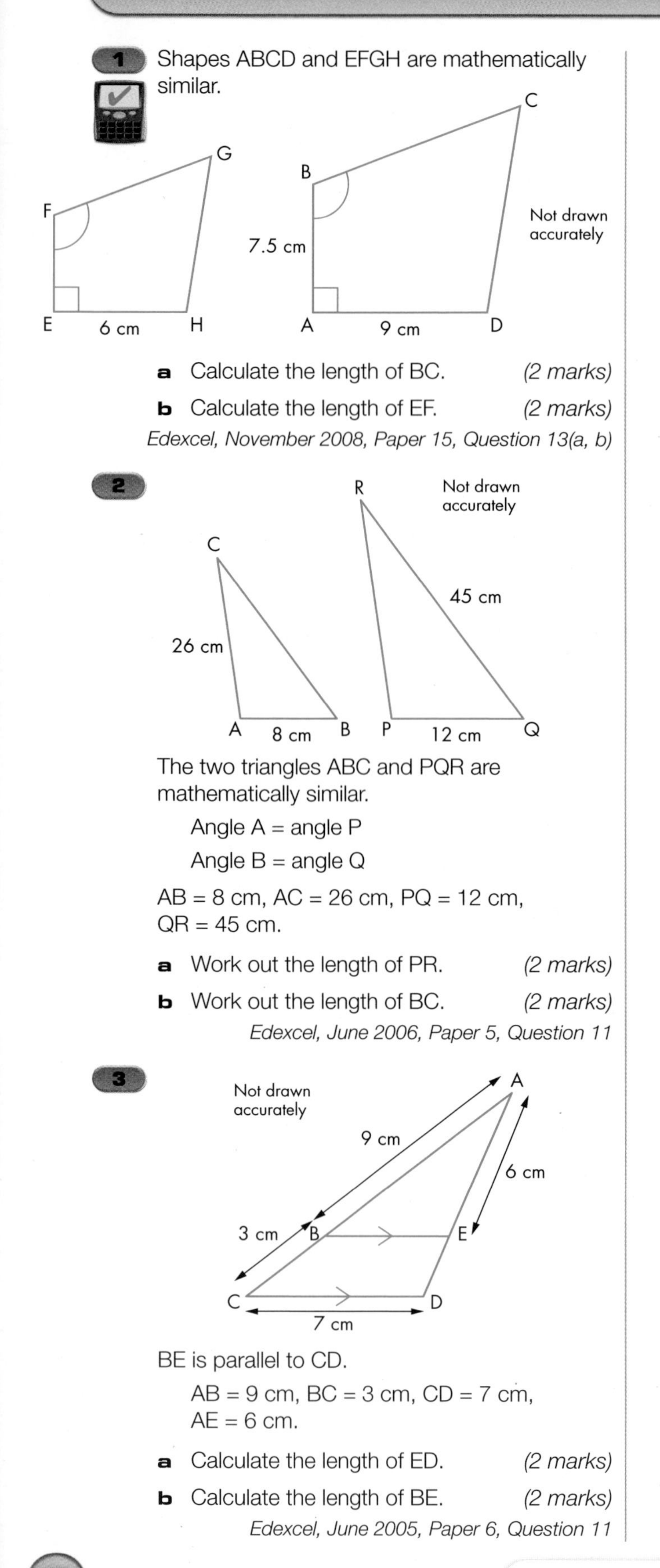

1 Shapes ABCD and EFGH are mathematically similar.

a Calculate the length of BC. *(2 marks)*

b Calculate the length of EF. *(2 marks)*

Edexcel, November 2008, Paper 15, Question 13(a, b)

2 The two triangles ABC and PQR are mathematically similar.

Angle A = angle P

Angle B = angle Q

AB = 8 cm, AC = 26 cm, PQ = 12 cm, QR = 45 cm.

a Work out the length of PR. *(2 marks)*

b Work out the length of BC. *(2 marks)*

Edexcel, June 2006, Paper 5, Question 11

3 BE is parallel to CD.

AB = 9 cm, BC = 3 cm, CD = 7 cm, AE = 6 cm.

a Calculate the length of ED. *(2 marks)*

b Calculate the length of BE. *(2 marks)*

Edexcel, June 2005, Paper 6, Question 11

4

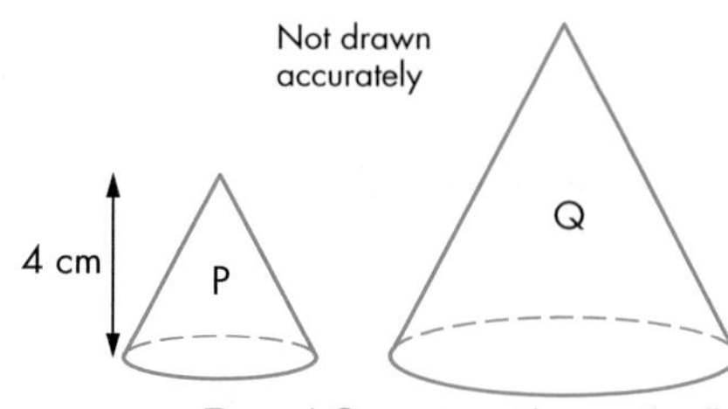

Two cones, P and Q, are mathematically similar.

The total surface area of cone P is 24 cm^2.

The total surface area of cone Q is 96 cm^2.

The height of cone P is 4 cm.

a Work out the height of cone Q. *(2 marks)*

The volume of cone P is 12 cm^3.

b Work out the volume of cone Q. *(2 marks)*

Edexcel, June 2007, Paper 5, Question 20

5

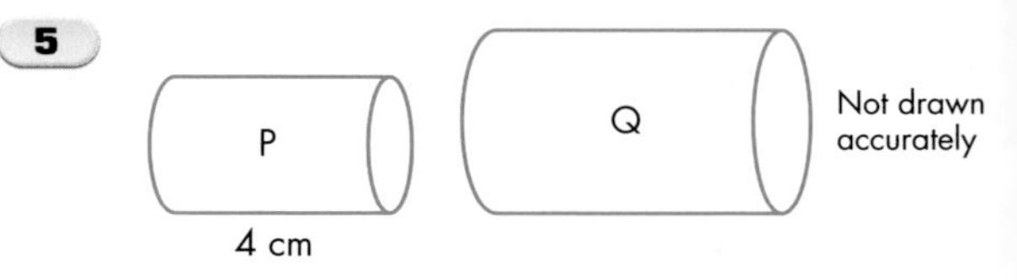

The cylinders P and Q are mathematically similar.

The total surface area of cylinder P is 90π cm^2.

The total surface area of cylinder Q is 810π cm^2.

The length of cylinder P is 4 cm.

a Work out the length of cylinder Q. *(3 marks)*

The volume of cylinder P is 100π cm3.

b Work out the volume of cylinder Q. Give your answer as a multiple of π. *(2 marks)*

Edexcel, June 2005, Paper 5, Question 18

Worked Examination Questions

PS **1** Triangles ABC and CDE are similar.

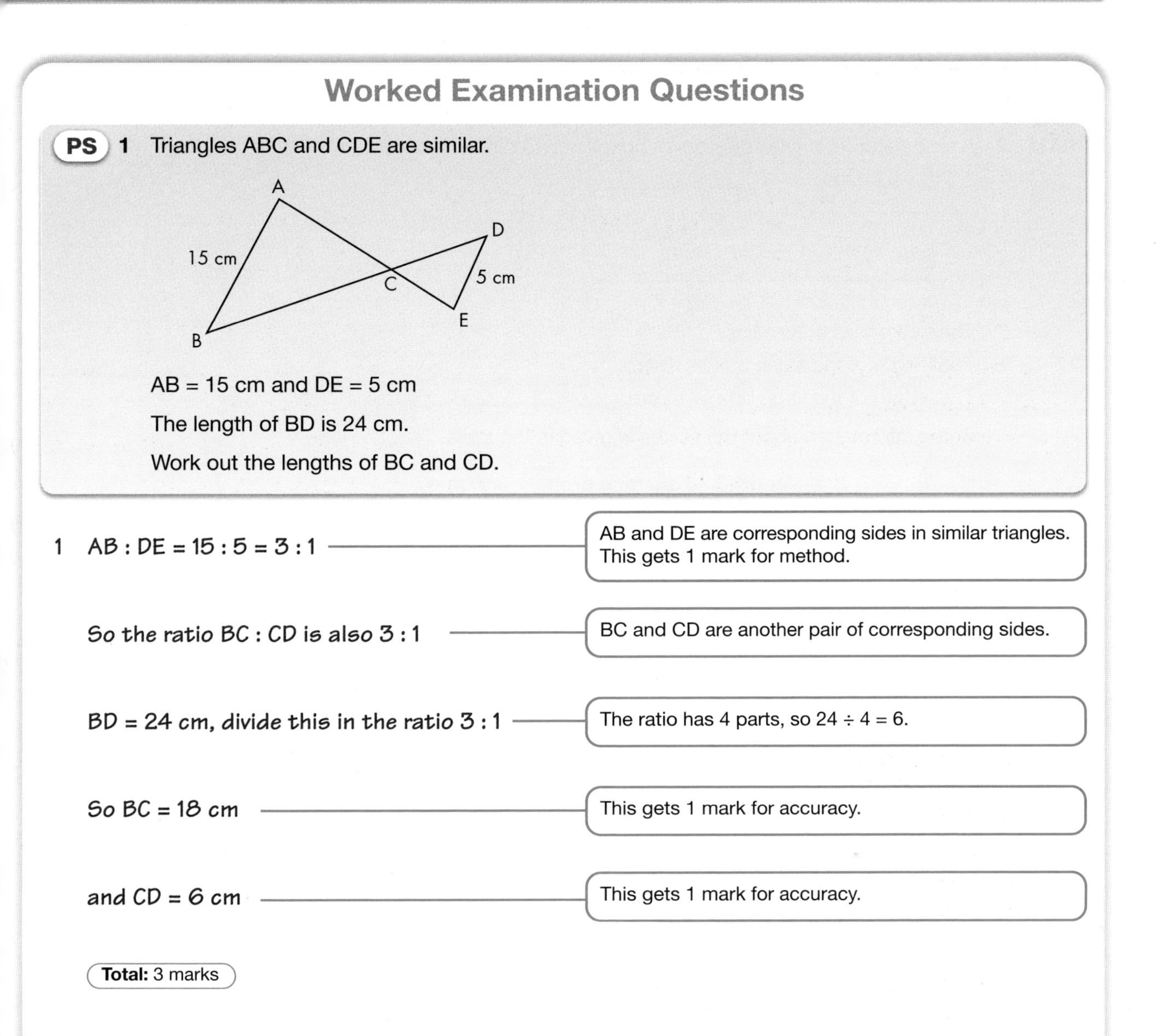

AB = 15 cm and DE = 5 cm

The length of BD is 24 cm.

Work out the lengths of BC and CD.

1 AB : DE = 15 : 5 = 3 : 1 — AB and DE are corresponding sides in similar triangles. This gets 1 mark for method.

So the ratio BC : CD is also 3 : 1 — BC and CD are another pair of corresponding sides.

BD = 24 cm, divide this in the ratio 3 : 1 — The ratio has 4 parts, so 24 ÷ 4 = 6.

So BC = 18 cm — This gets 1 mark for accuracy.

and CD = 6 cm — This gets 1 mark for accuracy.

Total: 3 marks

Worked Examination Questions

AU **2** A manufacturer makes cylindrical boxes in two different sizes.

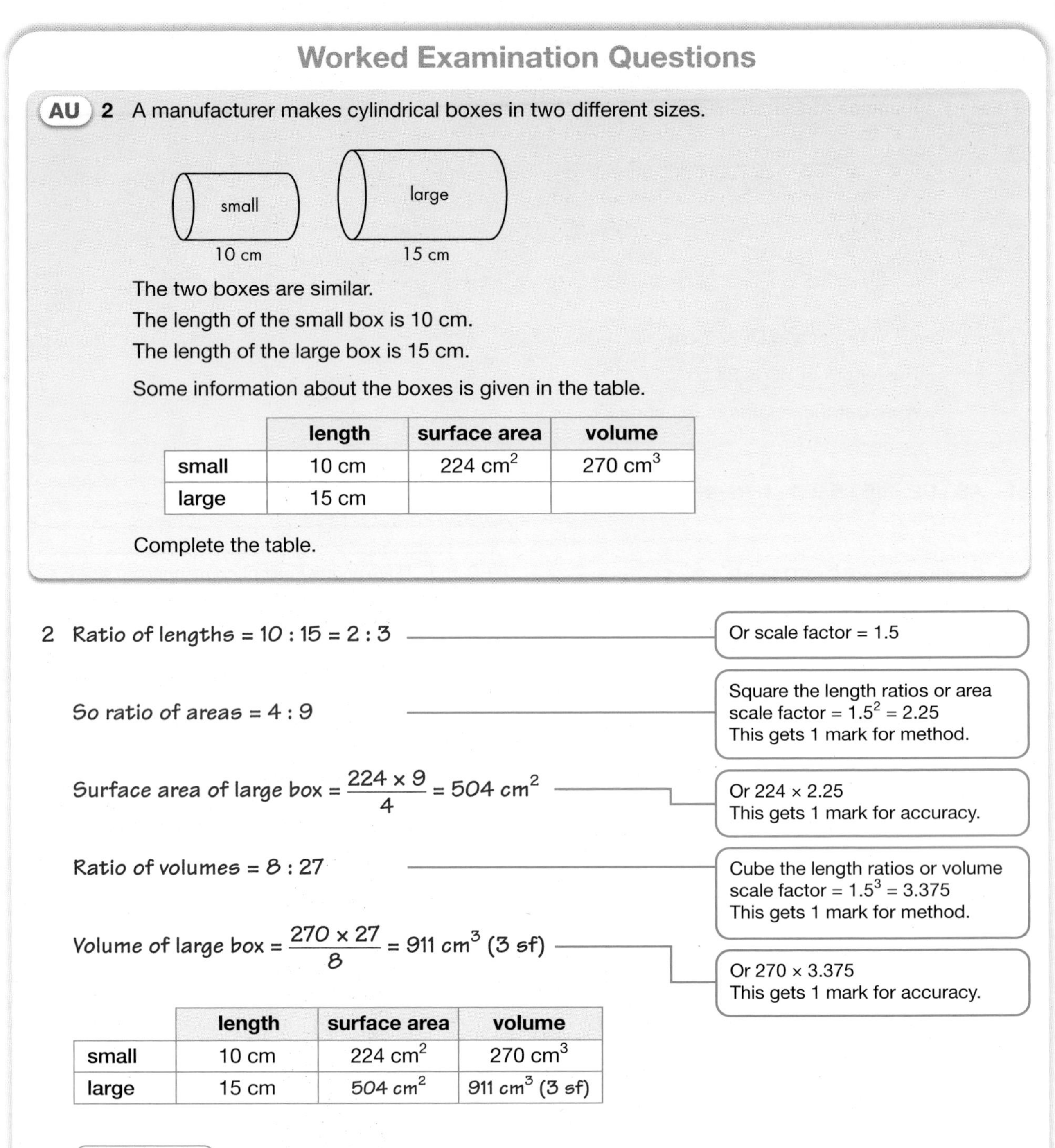

The two boxes are similar.

The length of the small box is 10 cm.

The length of the large box is 15 cm.

Some information about the boxes is given in the table.

	length	surface area	volume
small	10 cm	224 cm^2	270 cm^3
large	15 cm		

Complete the table.

2 Ratio of lengths = 10 : 15 = 2 : 3

Or scale factor = 1.5

So ratio of areas = 4 : 9

Square the length ratios or area scale factor = $1.5^2 = 2.25$
This gets 1 mark for method.

Surface area of large box = $\dfrac{224 \times 9}{4} = 504\text{ cm}^2$

Or 224×2.25
This gets 1 mark for accuracy.

Ratio of volumes = 8 : 27

Cube the length ratios or volume scale factor = $1.5^3 = 3.375$
This gets 1 mark for method.

Volume of large box = $\dfrac{270 \times 27}{8} = 911\text{ cm}^3$ (3 sf)

Or 270×3.375
This gets 1 mark for accuracy.

	length	surface area	volume
small	10 cm	224 cm^2	270 cm^3
large	15 cm	504 cm^2	911 cm^3 (3 sf)

Total: 4 marks

Worked Examination Questions

FM **3** A camping gas container is in the shape of a cylinder with a hemispherical top.
The dimensions of the container are shown in the diagram.

It is decided to increase the volume of the container by **20%**.
The new container is mathematically **similar** to the old one.

Calculate the base **diameter** of the new container.

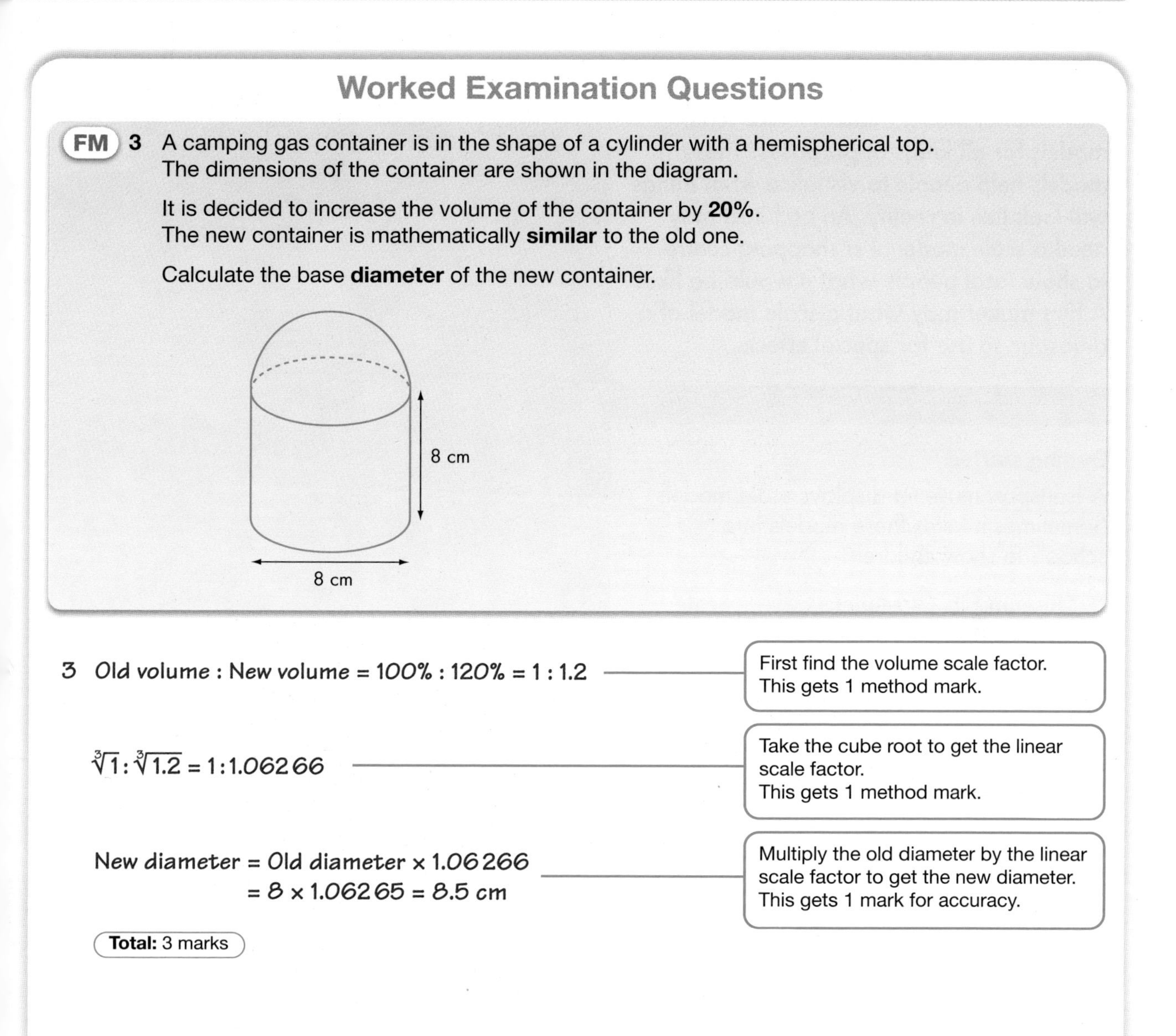

3 Old volume : New volume = 100% : 120% = 1 : 1.2

> First find the volume scale factor.
> This gets 1 method mark.

$\sqrt[3]{1} : \sqrt[3]{1.2} = 1 : 1.06266$

> Take the cube root to get the linear scale factor.
> This gets 1 method mark.

New diameter = Old diameter × 1.06266
= 8 × 1.06265 = 8.5 cm

> Multiply the old diameter by the linear scale factor to get the new diameter.
> This gets 1 mark for accuracy.

Total: 3 marks

7 Functional Maths
Making scale models

Professional model-makers make scale models for all kinds of purposes. These models help people to visualise what things will look like in reality. An architect may need a scale model of a shopping centre, to show local people what it would be like. A film-maker may want a scale model of a dinosaur, to use for special effects.

Getting started

A transport museum displays scale models. Sometimes it takes these models into schools to show children.

Museum department	Model	Scale
Aeroplanes	*Concorde*	1 : 50
Trains	*Orient Express* single cabin	1 : 5
Automobiles	Rolls Royce *Phantom*	1 : 8

- Why didn't the model-maker use the same scale for *Concorde* as for the Rolls Royce *Phantom*?
- The dimensions of the bed in the model of the *Orient Express* cabin are 38 cm × 17 cm. Work out the dimensions of a bed on the real *Orient Express*.
- The wing area of the model *Concorde* is 1432 cm^2. Work out the wing area for the real *Concorde*. Give your answer in square metres (m^2).

Your task

The transport museum decides to have two further models made: one for its spacecraft department and one for its bicycle department.

It is your task to commission the two models. You must decide what type of spacecraft and bicycle you wish to exhibit. You must also choose appropriate scales for the models, so that they may sometimes be taken into schools.

Then write the descriptions that will appear on display boards next to each of the models in the museum. Include some facts about the modes of transport and also compare some of the dimensions (for example, lengths, heights, areas and volumes) of each model and the real thing.

Extension

Try using the internet to search for 'spacecraft dimensions' or 'bicycle dimensions'.

Why this chapter matters

Trigonometry has an enormous variety of applications. It is used in practical fields such as navigation, land surveying, building, engineering and astronomy.

In surveying, trigonometry is used extensively in triangulation. This is a process for establishing the location of a point by measuring angles or bearings to this point, from two other known points at either end of a fixed baseline.

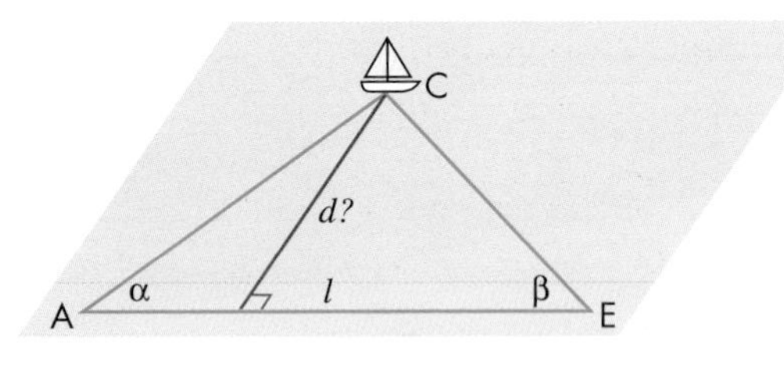

Triangulation can be used to calculate the position and distance from the shore to a ship. The observer at A measures the angle α between the shore and the ship, and the observer at B does likewise for angle β. With the length, l, or the position of A and B known, then the sine rule can be applied to find the position of the ship at C and the distance, d.

Marine sextants like this are used to measure the angle of the Sun or stars with respect to the horizon. Using trigonometry and a marine chronometer, the ship's navigator can determine the position of the ship on the sea.

The Canadarm2 robotic manipulator in the International Space Station is operated by controlling the angles of its joints. Calculating the final position of the astronaut at the end of the arm requires repeated use of trigonometry in three dimensions.

Chapter

Geometry: Trigonometry 2

1 Some 2D problems

2 Some 3D problems

3 Trigonometric ratios of angles between 90° and 360°

4 Solving any triangle

5 Trigonometric ratios in surd form

6 Using sine to find the area of a triangle

The grades given in this chapter are target grades.

This chapter will show you ...

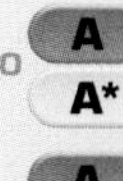
A to A* how to use the sine and cosine rules to solve problems involving non right-angled triangles

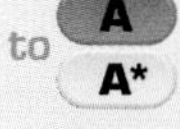
A to A* how to find the area of a triangle using formula $A = \frac{1}{2}ab \sin C$

A* how to use trigonometric ratios to solve more complex 2D problems and 3D problems

A* how to find the sine, cosine and tangent of any angle from 90° to 360°

A* how to work out trigonometric ratios in surd form

Visual overview

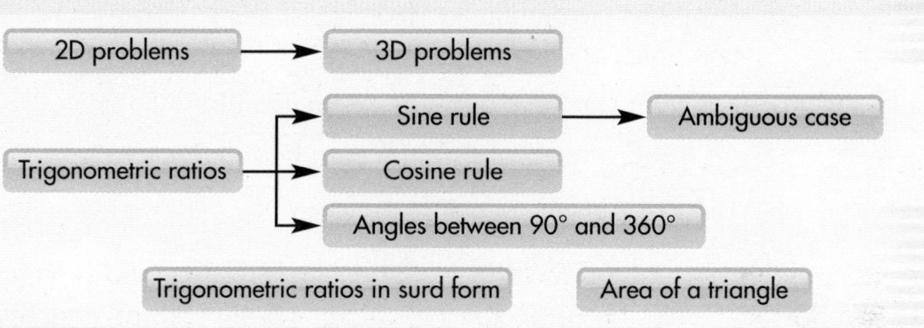

What you should already know

- How to find the sides of right-angled triangles using Pythagoras' theorem **(KS3 level 7, GCSE grade C)**
- How to find angles and sides of right-angled triangles using sine, cosine and tangent **(KS3 level 8, GCSE grade B)**

Quick check

Calculate the value of x in each of these right-angled triangles.
Give your answers to 3 significant figures.

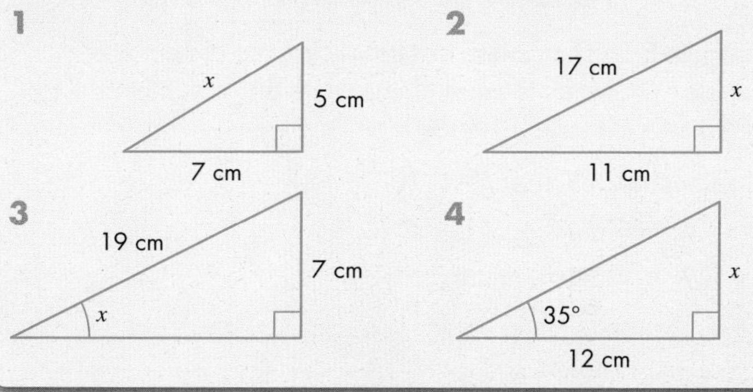

8.1 Some 2D problems

This section will show you how to:
- use trigonometric ratios and Pythagoras' theorem to solve more complex two-dimensional problems

Key words
cosine
Pythagoras' theorem
sine
tangent

This lesson brings together previous work on **Pythagoras' theorem**, circle theorems and trigonometric ratios – **sine** (sin), **cosine** (cos) and **tangent** (tan).

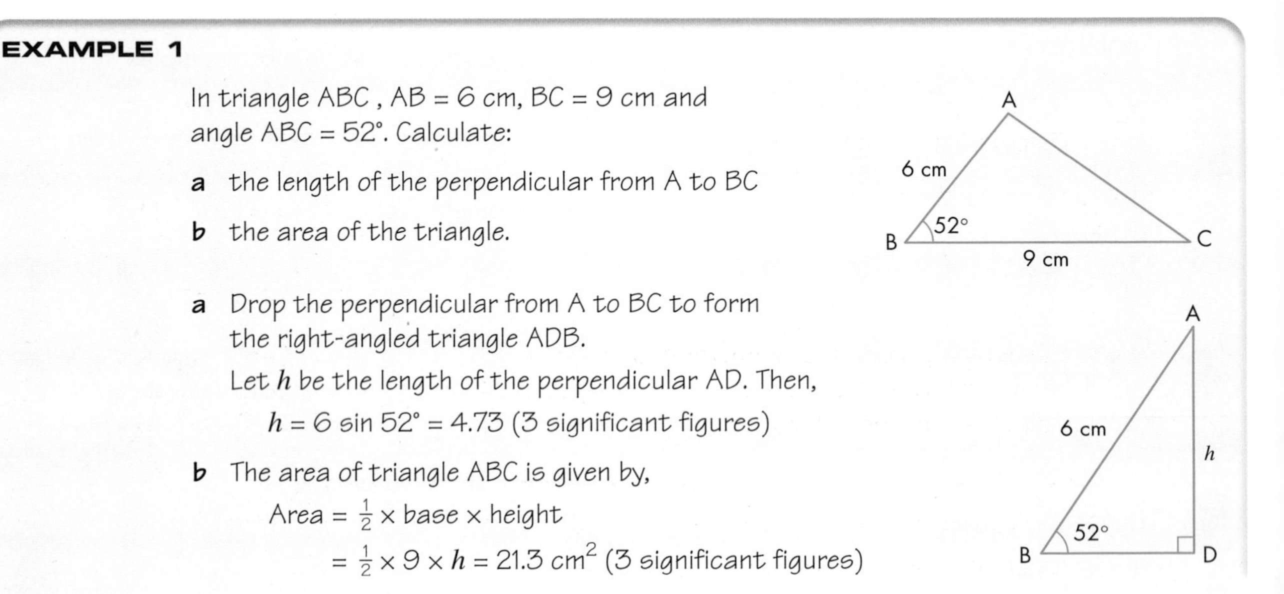

EXAMPLE 1

In triangle ABC , AB = 6 cm, BC = 9 cm and angle ABC = 52°. Calculate:

a the length of the perpendicular from A to BC

b the area of the triangle.

a Drop the perpendicular from A to BC to form the right-angled triangle ADB.

Let h be the length of the perpendicular AD. Then,

$h = 6 \sin 52° = 4.73$ (3 significant figures)

b The area of triangle ABC is given by,

Area = $\frac{1}{2} \times \text{base} \times \text{height}$

$= \frac{1}{2} \times 9 \times h = 21.3 \text{ cm}^2$ (3 significant figures)

EXAMPLE 2

SR is a diameter of a circle of radius 25 cm.
PQ is a chord at right angles to SR. X is the midpoint of PQ. The length of XR is 1 cm. Calculate the length of the arc PQ.

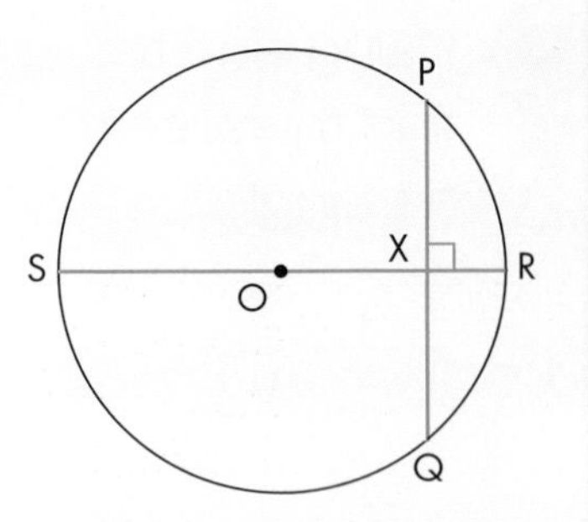

To find the length of the arc PQ, you need first to find the angle it subtends at the centre of the circle.

So join P to the centre of the circle O to obtain the angle POX, which is equal to half the angle subtended by PQ at O.

In right-angled triangle POX,

OX = OR – XR

OX = 25 – 1 = 24 cm

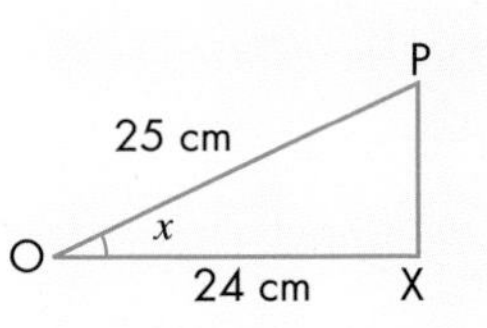

Therefore,

$$\cos x = \frac{24}{25}$$

$$\Rightarrow x = \cos^{-1} 0.96 = 16.26°$$

So, the angle subtended at the centre by the arc PQ is $2 \times 16.26° = 32.52°$, giving the length of the arc PQ as:

$$\frac{32.52}{360} \times 2 \times \pi \times 25 = 14.2 \text{ cm} \qquad \text{(3 significant figures)}$$

EXERCISE 8A

1 AC and BC are tangents to a circle of radius 7 cm. Calculate the length of AB.

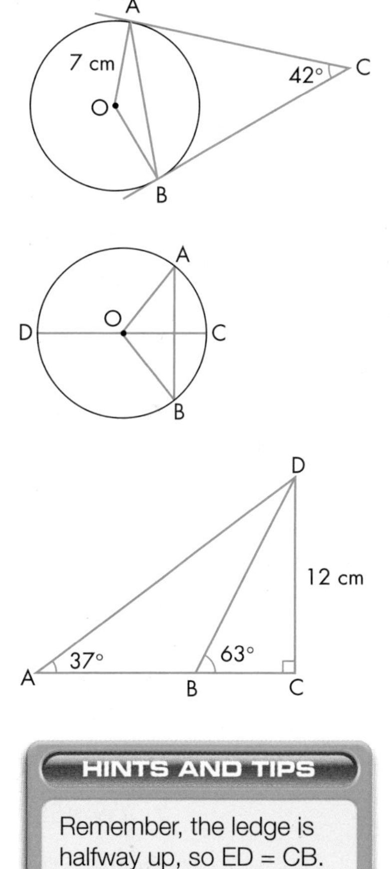

2 CD, length 20 cm, is a diameter of a circle. AB, length 12 cm, is a chord at right angles to DC. Calculate the angle AOB.

3 Calculate the length of AB in the diagram.

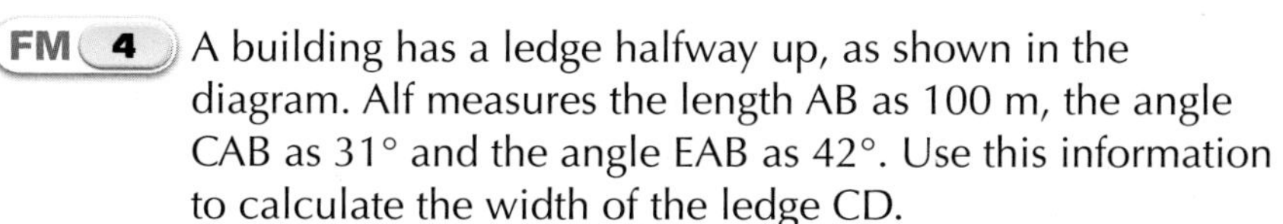

FM 4 A building has a ledge halfway up, as shown in the diagram. Alf measures the length AB as 100 m, the angle CAB as 31° and the angle EAB as 42°. Use this information to calculate the width of the ledge CD.

HINTS AND TIPS

Remember, the ledge is halfway up, so ED = CB.

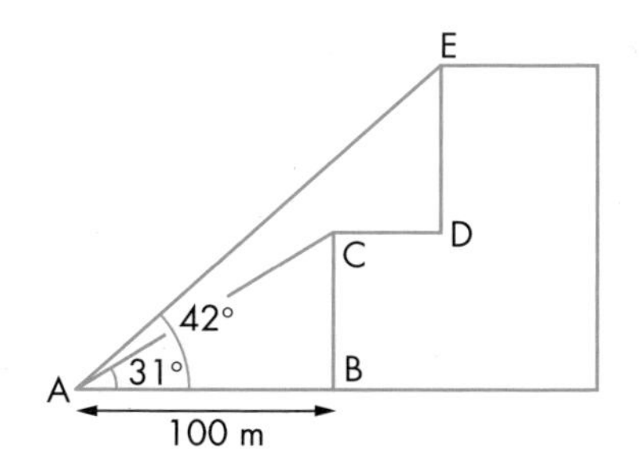

5 AB and CD are two equal, perpendicular chords of a circle that intersect at X. The circle is of radius 6 cm and the angle COA is 113°. Calculate:

a the length AC

b the angle XAO

c the length XB.

HINTS AND TIPS

AX = XC

6 A vertical flagpole PQ is held by a wooden framework, as shown in the diagram. The framework is in the same vertical plane. Angle SRP = 25°, SQ = 6 m and PR = 4 m. Calculate the size of the angle QRP.

FM 7 A mine descends from ground level for 500 m at an angle of 13° to the horizontal and then continues for another 300 m at an angle of 17° to the horizontal, as shown in the diagram. A mining company decides to drill a vertical shaft to join up with the bottom of the mine as shown. How far along the surface from the opening, marked x on the diagram, do they need to drill down?

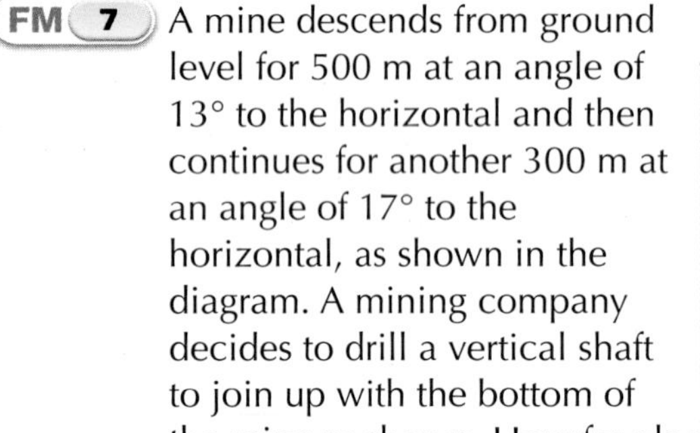

PS 8 **a** Use Pythagoras' theorem to work out the length of BC.

Leave your answer in surd form.

b Write down the values of:

i cos 45° **ii** sin 45° **iii** tan 45°

leaving your answers in surd form.

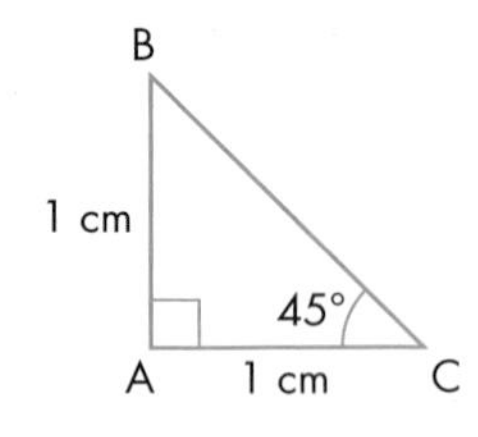

AU 9 In the diagram, AD = 5 cm, AC = 8 cm and AB = 12 cm.

Calculate angle CAB.

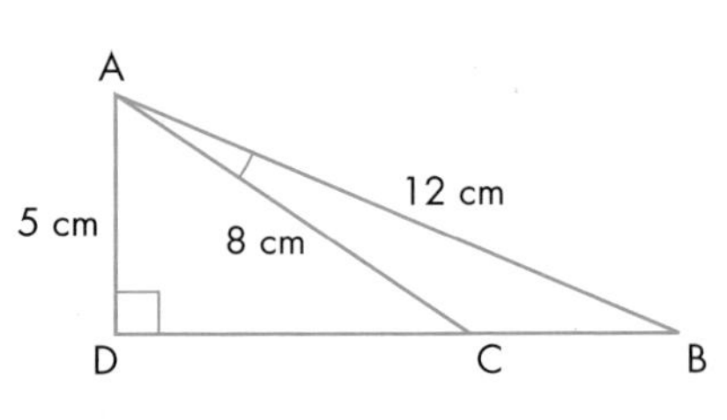

8.2 Some 3D problems

This section will show you how to:

- use trigonometric ratios and Pythagoras' theorem to solve more complex three-dimensional problems

Key words
cosine
Pythagoras' theorem
sine
tangent

Solving a problem set in three dimensions nearly always involves identifying a right-angled triangle that contains the length or angle required. This triangle will have to contain (apart from the right angle) two known measures from which the required calculation can be made.

It is essential to extract the triangle you are going to use from its 3D situation and redraw it as a separate, plain, right-angled triangle. (It is rarely the case that the required triangle appears as a true right-angled triangle in its 3D representation. Even if it does, you should still redraw it as a separate figure.)

Annotate the redrawn triangle with the known quantities and the unknown quantity that is to be found. Then use the trigonometric ratios **sine** (sin), **cosine** (cos) and **tangent** (tan), **Pythagoras' theorem** and the circle theorems to solve the triangle.

EXAMPLE 3

A, B and C are three points at ground level. They are in the same horizontal plane. C is 50 km east of B. B is north of A. C is on a bearing of 050° from A.

An aircraft, flying in an easterly direction, passes over B and over C at the same height. When it passes over B, the angle of elevation from A is 12°. Find the angle of elevation of the aircraft from A when it is over C.

First, draw a diagram containing all the known information.

Next, use the right-angled triangle ABC to calculate AB and AC.

$AB = \dfrac{50}{\tan 50°} = 41.95$ km (4 significant figures)

$AC = \dfrac{50}{\sin 50°} = 65.27$ km (4 significant figures)

Then use the right-angled triangle ABX to calculate BX, and hence CY.

$BX = 41.95 \tan 12° = 8.917$ km (4 significant figures)

Finally, use the right-angled triangle ACY to calculate the required angle of elevation, θ.

$\tan \theta = \dfrac{8.917}{65.27} = 0.1366$

$\Rightarrow \theta = \tan^{-1} 0.1366 = 7.8°$ (1 decimal place)

Always write down intermediate working values to at least 4 significant figures, or use the answer on your calculator display to avoid inaccuracy in the final answer.

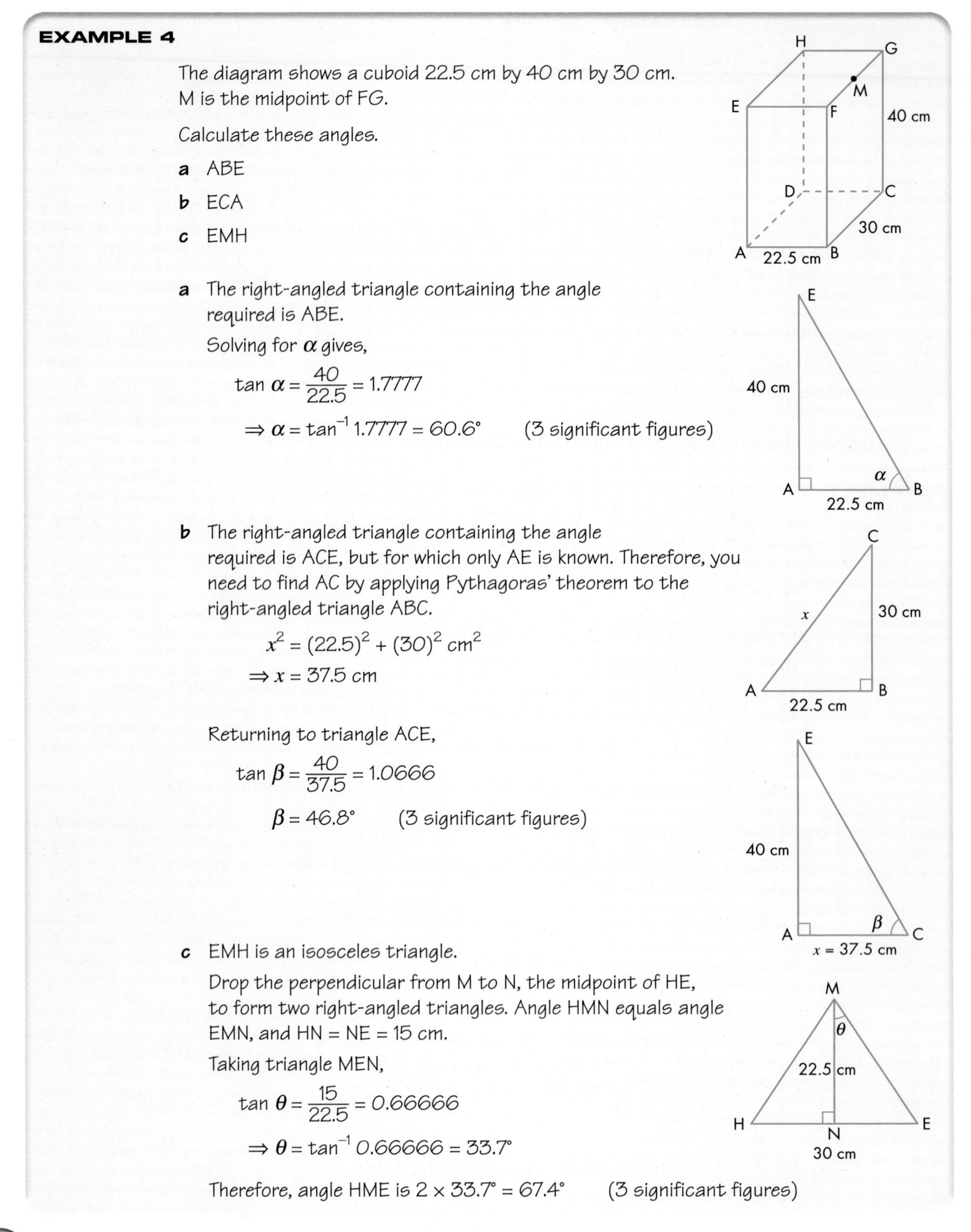

EXAMPLE 4

The diagram shows a cuboid 22.5 cm by 40 cm by 30 cm.
M is the midpoint of FG.

Calculate these angles.

a ABE

b ECA

c EMH

a The right-angled triangle containing the angle required is ABE.
Solving for α gives,

$$\tan \alpha = \frac{40}{22.5} = 1.7777$$

$$\Rightarrow \alpha = \tan^{-1} 1.7777 = 60.6^\circ \quad \text{(3 significant figures)}$$

b The right-angled triangle containing the angle required is ACE, but for which only AE is known. Therefore, you need to find AC by applying Pythagoras' theorem to the right-angled triangle ABC.

$$x^2 = (22.5)^2 + (30)^2 \text{ cm}^2$$

$$\Rightarrow x = 37.5 \text{ cm}$$

Returning to triangle ACE,

$$\tan \beta = \frac{40}{37.5} = 1.0666$$

$$\beta = 46.8^\circ \quad \text{(3 significant figures)}$$

c EMH is an isosceles triangle.
Drop the perpendicular from M to N, the midpoint of HE, to form two right-angled triangles. Angle HMN equals angle EMN, and HN = NE = 15 cm.

Taking triangle MEN,

$$\tan \theta = \frac{15}{22.5} = 0.66666$$

$$\Rightarrow \theta = \tan^{-1} 0.66666 = 33.7^\circ$$

Therefore, angle HME is $2 \times 33.7^\circ = 67.4^\circ$ (3 significant figures)

EXERCISE 8B

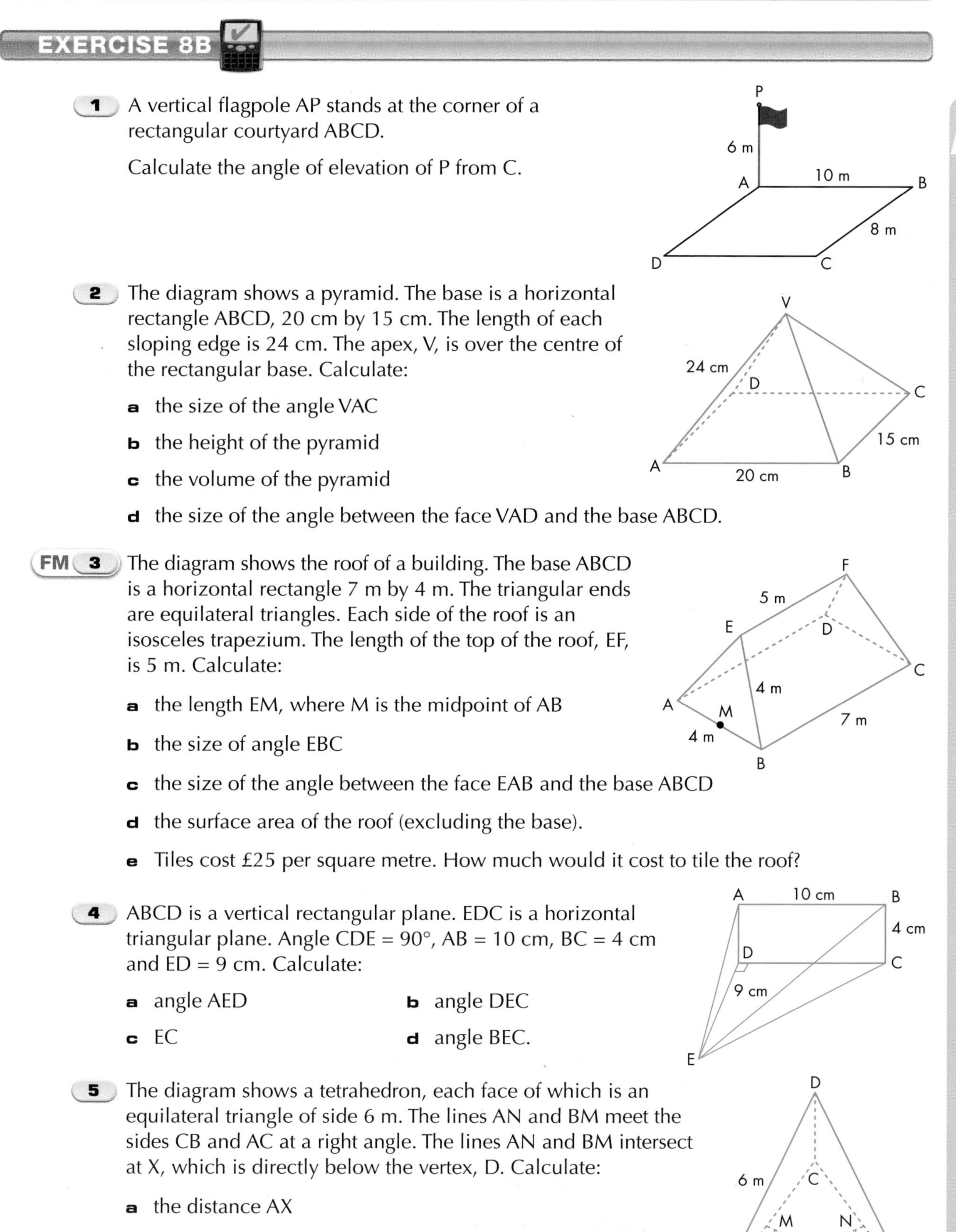

1 A vertical flagpole AP stands at the corner of a rectangular courtyard ABCD.

Calculate the angle of elevation of P from C.

2 The diagram shows a pyramid. The base is a horizontal rectangle ABCD, 20 cm by 15 cm. The length of each sloping edge is 24 cm. The apex, V, is over the centre of the rectangular base. Calculate:

- **a** the size of the angle VAC
- **b** the height of the pyramid
- **c** the volume of the pyramid
- **d** the size of the angle between the face VAD and the base ABCD.

FM **3** The diagram shows the roof of a building. The base ABCD is a horizontal rectangle 7 m by 4 m. The triangular ends are equilateral triangles. Each side of the roof is an isosceles trapezium. The length of the top of the roof, EF, is 5 m. Calculate:

- **a** the length EM, where M is the midpoint of AB
- **b** the size of angle EBC
- **c** the size of the angle between the face EAB and the base ABCD
- **d** the surface area of the roof (excluding the base).
- **e** Tiles cost £25 per square metre. How much would it cost to tile the roof?

4 ABCD is a vertical rectangular plane. EDC is a horizontal triangular plane. Angle CDE = 90°, AB = 10 cm, BC = 4 cm and ED = 9 cm. Calculate:

- **a** angle AED
- **b** angle DEC
- **c** EC
- **d** angle BEC.

5 The diagram shows a tetrahedron, each face of which is an equilateral triangle of side 6 m. The lines AN and BM meet the sides CB and AC at a right angle. The lines AN and BM intersect at X, which is directly below the vertex, D. Calculate:

- **a** the distance AX
- **b** the angle between the side DBC and the base ABC.

A*

A*

PS 6 The lengths of the sides of a cuboid are a, b and c.

Show that the length of the diagonal XY is:

$\sqrt{a^2 + b^2 + c^2}$

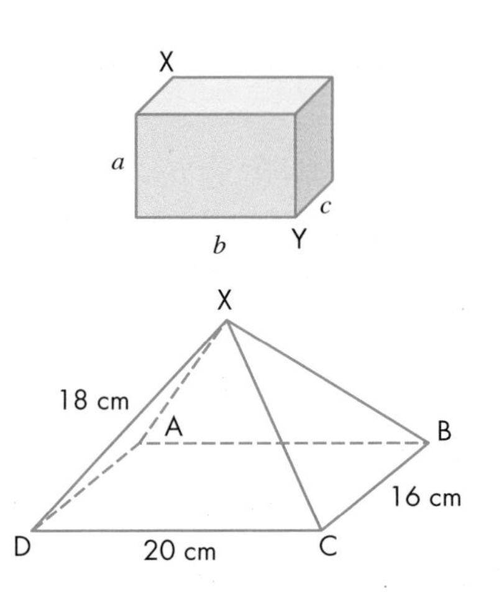

AU 7 In the diagram, XABCD is a right pyramid with a rectangular base.

Ellie says that the angle between the edge XD and the base ABCD is 56.3°.

Work out the correct answer to show that Ellie is wrong.

8.3 Trigonometric ratios of angles between 90° and 360°

This section will show you how to:

- find the sine, cosine and tangent of any angle from 0° to 360°

Key words
circular function
cosine
sine
tangent

ACTIVITY

a Copy and complete this table, using your calculator and rounding to three decimal places.

x	$\sin x$	x	$\sin x$	x	$\sin x$	x	$\sin x$
0°		180°		180°		360°	
15°		165°		195°		335°	
30°		150°		210°		320°	
45°		135°		225°		315°	
60°		120°		240°		300°	
75°		105°		255°		285°	
90°		90°		270°		270°	

b Comment on what you notice about the **sine** of each acute angle, and the sines of its corresponding non-acute angles.

c Draw a graph of $\sin x$ against x. Take x from 0° to 360° and $\sin x$ from –1 to 1.

d Comment on any symmetries your graph has.

You should have discovered these three facts.

- When $90° < x < 180°$, $\sin x = \sin (180° - x)$
  For example, $\sin 153° = \sin (180° - 153°) = \sin 27° = 0.454$
- When $180° < x < 270°$, $\sin x = - \sin (x - 180°)$
  For example, $\sin 214° = - \sin (214° - 180°) = - \sin 34° = - 0.559$
- When $270° < x < 360°$, $\sin x = - \sin (360° - x)$
  For example, $\sin 287° = - \sin (360° - 287°) = - \sin 73° = - 0.956$

Note:

- Each and every value of sine between –1 and 1 gives *two* angles between 0° and 360°.
- When the value of sine is positive, both angles are between 0° and 180°.
- When the value of sine is negative, both angles are between 180° and 360°.
- You can use the sine graph from 0° to 360° to check values approximately.

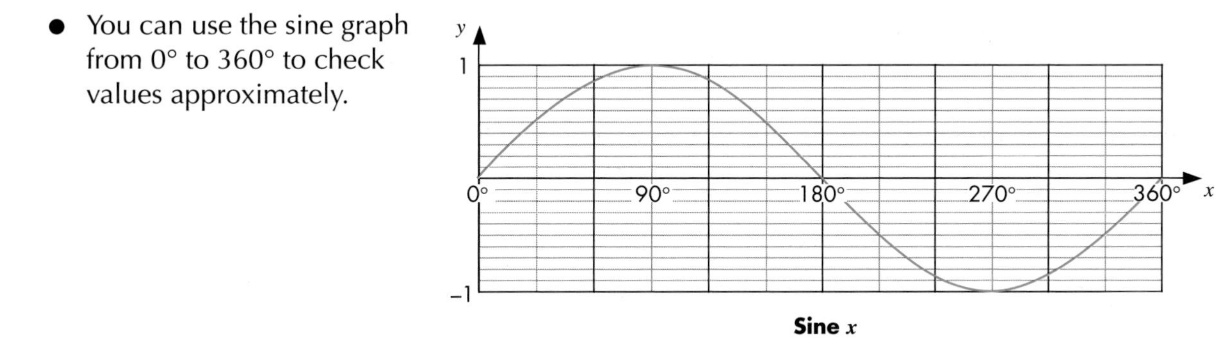

Sine x

EXAMPLE 5

Find the angles with a sine of 0.56.

You know that both angles are between 0° and 180°.

Using your calculator to find $\sin^{-1} 0.56$, you obtain 34.1°.

The other angle is, therefore,

$180° - 34.1° = 145.9°$

So, the angles are 34.1° and 145.9°.

EXAMPLE 6

Find the angles with a sine of –0.197.

You know that both angles are between 180° and 360°.

Using your calculator to find $\sin^{-1} 0.197$, you obtain 11.4°.

So the angles are

$180° + 11.4°$ and $360° - 11.4°$

which give 191.4° and 348.6°.

You can always use your calculator to check your answer to this type of problem by first keying in the angle and the appropriate trigonometric function (which would be sine in the above examples).

EXERCISE 8C

A*

1 State the two angles between 0° and 360° for each of these sine values.

a 0.6 **b** 0.8 **c** 0.75 **d** –0.7

e –0.25 **f** –0.32 **g** –0.175 **h** –0.814

i 0.471 **j** –0.097 **k** 0.553 **l** –0.5

AU 2 Which of these values is the odd one out and why?

sin 36° sin 144° sin 234° sin 324°

PS 3 The graph of sine x is cyclic, which means that it repeats forever in each direction.

a Write down one value of x greater than 360° for which the sine value is 0.978 147 600 73.

b Write down one value of x less than 0° for which the sine value is 0.978 147 600 73.

c Describe any symmetries of the graph of $y = \sin x$.

ACTIVITY

a Copy and complete this table, using your calculator and rounding to 3 decimal places.

x	$\cos x$	x	$\cos x$	x	$\cos x$	x	$\cos x$
0°		180°		180°		360°	
15°		165°		195°		335°	
30°		150°		210°		320°	
45°		135°		225°		315°	
60°		120°		240°		300°	
75°		105°		255°		285°	
90°		90°		270°		270°	

b Comment on what you notice about the cosines of the angles.

c Draw a graph of $\cos x$ against x. Take x from 0° to 360° and $\cos x$ from –1 to 1.

d Comment on the symmetry of the graph.

You should have discovered these three facts.

- When $90° < x < 180°$, $\cos x = -\cos(180 - x)°$
  For example, $\cos 161° = -\cos(180° - 161°) = -\cos 19° = -0.946$ (3 significant figures)
- When $180° < x < 270°$, $\cos x = -\cos(x - 180°)$
  For example, $\cos 245° = -\cos(245° - 180°) = -\cos 65° = -0.423$ (3 significant figures)
- When $270° < x < 360°$, $\cos x = \cos(360° - x)$
  For example, $\cos 310° = \cos(360° - 310°) = \cos 50° = 0.643$ (3 significant figures)

Note:

- Each and every value of cosine between –1 and 1 gives *two* angles between 0° and 360°.
- When the value of cosine is positive, one angle is between 0° and 90°, and the other is between 270° and 360°.
- When the value of cosine is negative, both angles are between 90° and 270°.
- You can use the cosine graph from 0° to 360° to check values approximately.

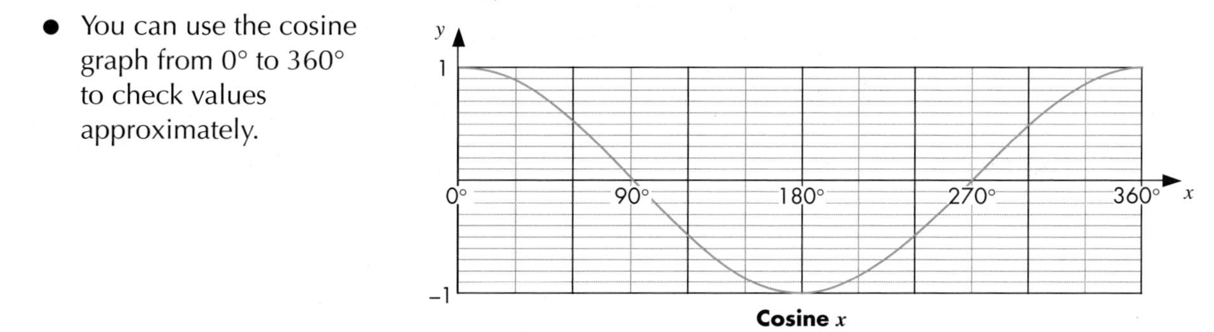

Cosine x

EXAMPLE 7

Find the angles with a cosine of 0.75.

One angle is between 0° and 90°, and the other is between 270° and 360°.

Using your calculator to find $\cos^{-1} 0.75$, you obtain 41.4°.

The other angle is, therefore,

$360° - 41.4° = 318.6°$

So, the angles are 41.4° and 318.6°.

EXAMPLE 8

Find the angles with a cosine of –0.285.

You know that both angles are between 90° and 270°.

Using your calculator to find $\cos^{-1} 0.285$, you obtain 73.4°.

The two angles are, therefore,

$180° - 73.4°$ and $180° + 73.4°$

which give 106.6° and 253.4°.

Here again, you can use your calculator to check your answer, by keying in cosine.

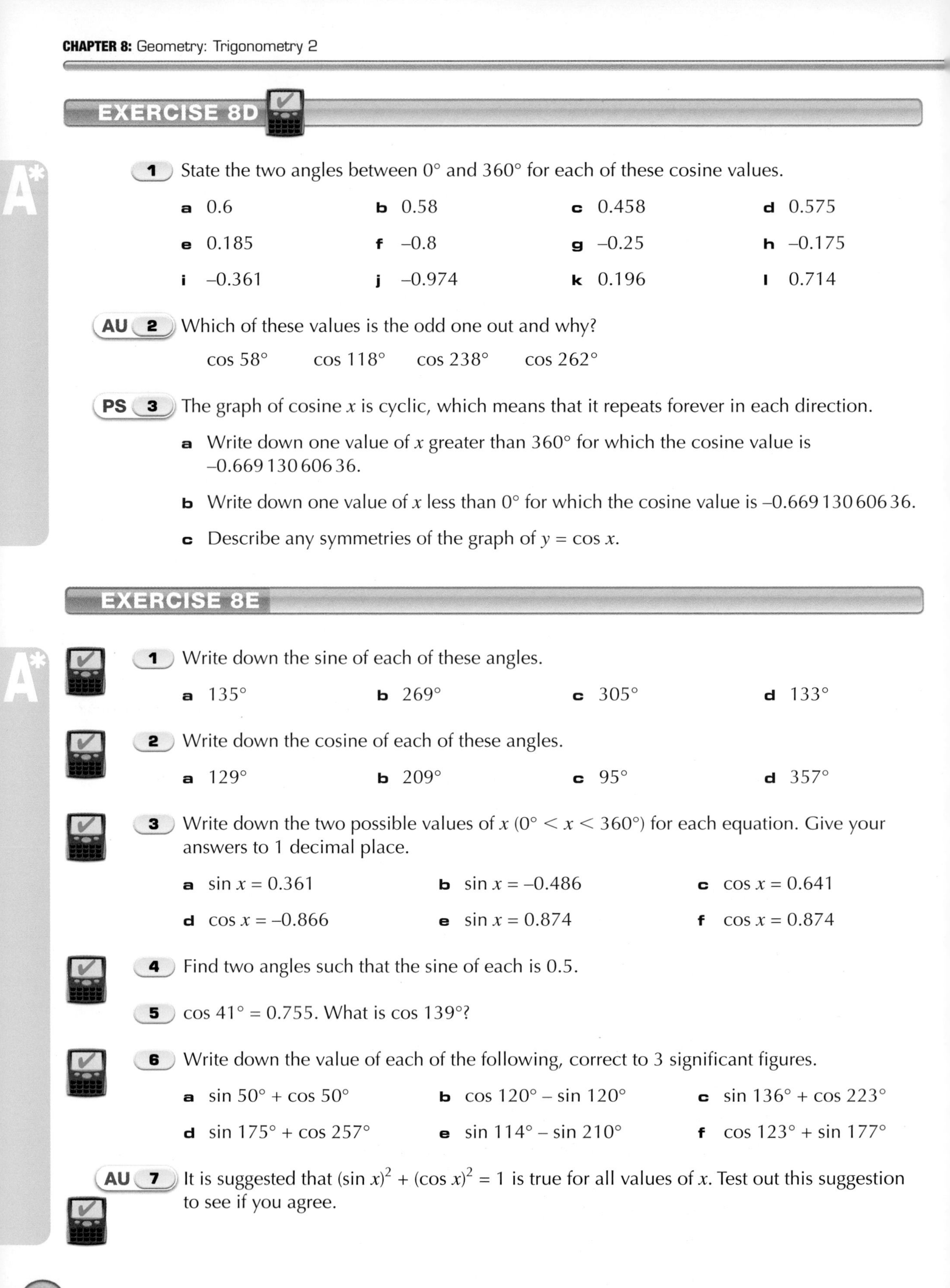

EXERCISE 8D

1 State the two angles between 0° and 360° for each of these cosine values.

a 0.6 **b** 0.58 **c** 0.458 **d** 0.575

e 0.185 **f** –0.8 **g** –0.25 **h** –0.175

i –0.361 **j** –0.974 **k** 0.196 **l** 0.714

AU 2 Which of these values is the odd one out and why?

cos 58° cos 118° cos 238° cos 262°

PS 3 The graph of cosine x is cyclic, which means that it repeats forever in each direction.

a Write down one value of x greater than 360° for which the cosine value is –0.669 130 606 36.

b Write down one value of x less than 0° for which the cosine value is –0.669 130 606 36.

c Describe any symmetries of the graph of $y = \cos x$.

EXERCISE 8E

1 Write down the sine of each of these angles.

a 135° **b** 269° **c** 305° **d** 133°

2 Write down the cosine of each of these angles.

a 129° **b** 209° **c** 95° **d** 357°

3 Write down the two possible values of x ($0° < x < 360°$) for each equation. Give your answers to 1 decimal place.

a $\sin x = 0.361$ **b** $\sin x = -0.486$ **c** $\cos x = 0.641$

d $\cos x = -0.866$ **e** $\sin x = 0.874$ **f** $\cos x = 0.874$

4 Find two angles such that the sine of each is 0.5.

5 cos 41° = 0.755. What is cos 139°?

6 Write down the value of each of the following, correct to 3 significant figures.

a sin 50° + cos 50° **b** cos 120° – sin 120° **c** sin 136° + cos 223°

d sin 175° + cos 257° **e** sin 114° – sin 210° **f** cos 123° + sin 177°

AU 7 It is suggested that $(\sin x)^2 + (\cos x)^2 = 1$ is true for all values of x. Test out this suggestion to see if you agree.

A*

PS 8 Suppose the sine key on your calculator is broken, but not the cosine key. Show how you could calculate these.

a sin 25° **b** sin 130°

PS 9 Find a solution to each of these equations.

a $\sin (x + 20°) = 0.5$ **b** $\cos (5x) = 0.45$

PS 10 Use any suitable method to find the solution to the equation $\sin x = (\cos x)^2$.

ACTIVITY

a Try to find tan 90°. What do you notice?

Which is the closest angle to 90° for which you can find the **tangent** on your calculator?

What is the largest value for tangent that you can get on your calculator?

b Find values of $\tan x$ where $0° < x < 360°$. Draw a graph of your results.

State some rules for finding both angles between 0° and 360° that have any given tangent.

EXAMPLE 9

Find the angles between 0° and 360° with a tangent of 0.875.

One angle is between 0° and 90°, and the other is between 180° and 270°.

Using your calculator to find $\tan^{-1} 0.875$, you obtain 41.2°.

The other angle is, therefore,

$180° + 41.2° = 221.2°$

So, the angles are 41.2° and 221.2°.

EXAMPLE 10

Find the angles between 0° and 360° with a tangent of –1.5.

You know that one angle is between 90° and 180°, and that the other is between 270° and 360°.

Using your calculator to find $\tan^{-1} 1.5$, you obtain 56.3°.

The angles are, therefore,

$180° - 56.3°$ and $360° - 56.3°$

which give 123.7° and 303.7°.

EXERCISE 8F

A*

1 State the angles between 0° and 360° which have each of these tangent values.

a 0.258	**b** 0.785	**c** 1.19	**d** 1.875	**e** 2.55
f −0.358	**g** −0.634	**h** −0.987	**i** −1.67	**j** −3.68
k 1.397	**l** 0.907	**m** −0.355	**n** −1.153	**o** 4.15
p −2.05	**q** −0.098	**r** 0.998	**s** 1.208	**t** −2.5

AU 2 Which of these values is the odd one out and why?

tan 45°   tan 135°   tan 235°   tan 315°

PS 3 The graph of tan x is cyclic, which means that it repeats forever in each direction.

a Write down one value of x greater than 360° for which the tangent value is 2.144 506 920 51.

b Write down one value of x less than 0° for which the tangent value is 2.144 506 920 51.

c Describe any symmetries of the graph of $y = \tan x$.

You may see the trigonometric functions sine, cosine and tangent referred to as **circular functions**. You could use the internet to find out more.

8.4 Solving any triangle

This section will show you how to:

- use the sine rule and the cosine rule to find sides and angles in any triangle

Key words

cosine rule
included angle
sine rule

We have already established that any triangle has six measurements: three sides and three angles. To solve a triangle (that is, to find any unknown angles or sides), we need to know at least three of the measurements. Any combination of three measurements – except that of all three angles – is enough to work out the rest. In a right-angled triangle, one of the known measurements is, of course, the right angle.

When we need to solve a triangle which contains no right angle, we can use one or the other of two rules, depending on what is known about the triangle. These are the **sine rule** and the **cosine rule**.

The sine rule

Take a triangle ABC and draw the perpendicular from A to the opposite side BC.

From right-angled triangle ADB, $h = c \sin B$

From right-angled triangle ADC, $h = b \sin C$

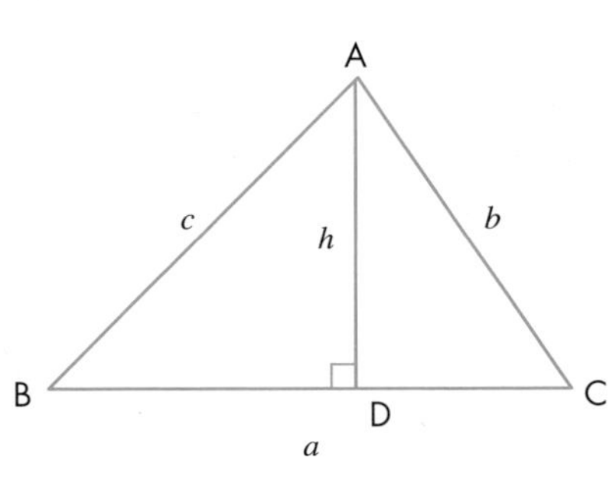

Therefore,

$$c \sin B = b \sin C$$

which can be rearranged to give:

$$\frac{c}{\sin C} = \frac{b}{\sin B}$$

By drawing a perpendicular from each of the other two vertices to the opposite side (or by algebraic symmetry), we see that

$$\frac{a}{\sin A} = \frac{c}{\sin C} \quad \text{and that} \quad \frac{a}{\sin A} = \frac{b}{\sin B}$$

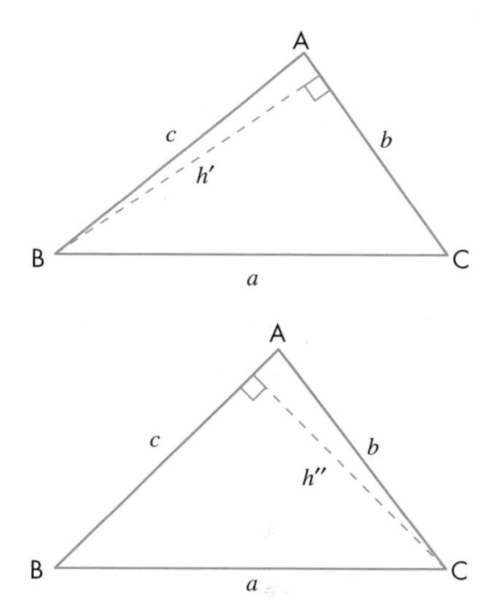

These are usually combined in the form

$$\frac{a}{\sin A} = \frac{b}{\sin B} = \frac{c}{\sin C}$$

which can be inverted to give:

$$\frac{\sin A}{a} = \frac{\sin B}{b} = \frac{\sin C}{c}$$

Usually, a triangle is not conveniently labelled, as in these diagrams. So, when using the sine rule, it is easier to remember to proceed as follows: take each side in turn, divide it by the sine of the angle opposite and then equate the resulting quotients.

Note:

- When you are calculating a *side*, use the rule with the *sides on top*.
- When you are calculating an *angle*, use the rule with the *sines on top*.

EXAMPLE 11

In triangle ABC, find the value of x.

Use the sine rule with sides on top, which gives:

$$\frac{x}{\sin 84°} = \frac{25}{\sin 47°}$$

$$\Rightarrow x = \frac{25 \sin 84°}{\sin 47°} = 34.0 \text{ cm} \quad \text{(3 significant figures)}$$

EXAMPLE 12

In the triangle ABC, find the value of the acute angle x.

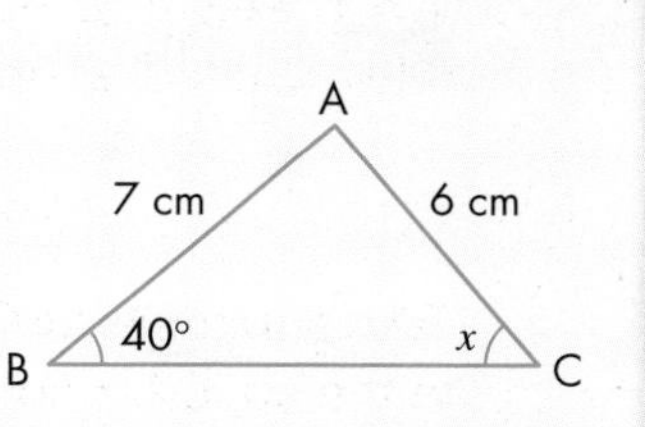

Use the sine rule with sines on top, which gives:

$$\frac{\sin x}{7} = \frac{7 \sin 40°}{6}$$

$$\Rightarrow \sin x = \frac{7 \sin 40°}{6} = 0.7499$$

$$\Rightarrow x = \sin^{-1} 0.7499 = 48.6° \quad \text{(3 significant figures)}$$

The ambiguous case

It is possible to find the sine of an angle that is greater than 90° (see Section 8.3).

For example, sin 30° = sin 150° = 0.5. (Notice that the two angles add up to 180°.)

So sin 25° = sin 155° and sin 100° = sin 80°

EXAMPLE 13

In triangle ABC, AB = 9 cm, AC = 7 cm and angle ABC = 40°. Find the angle ACB.

As you sketch triangle ABC, note that C can have two positions, giving two different configurations.

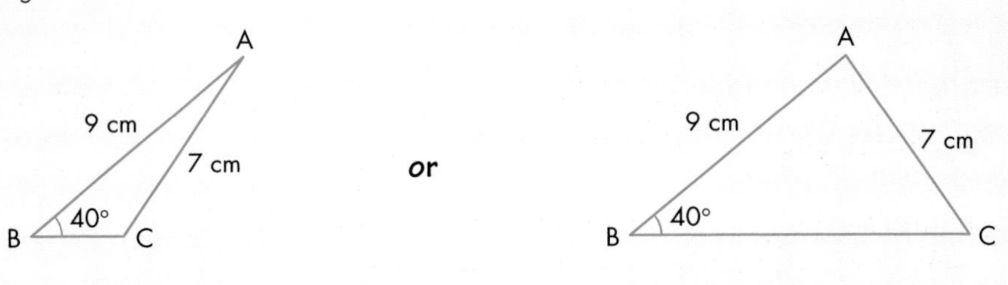

But you still proceed as in the normal sine rule situation, obtaining:

$$\frac{\sin C}{9} = \frac{\sin 40°}{7}$$

$$\Rightarrow \sin C = \frac{9 \sin 40°}{7}$$

$$= 0.8264$$

Keying inverse sine on the calculator gives C = 55.7°. But there is another angle with a sine of 0.8264, given by (180° – 55.7°) = 124.3°.

These two values for C give the two different situations shown above.

When an illustration of the triangle is given, it will be clear whether the required angle is acute or obtuse. When an illustration is not given, the more likely answer is an acute angle.

Examiners will not try to catch you out with the ambiguous case. They will indicate clearly, either with the aid of a diagram or by stating it, what is required.

EXERCISE 8G

1 Find the length x in each of these triangles.

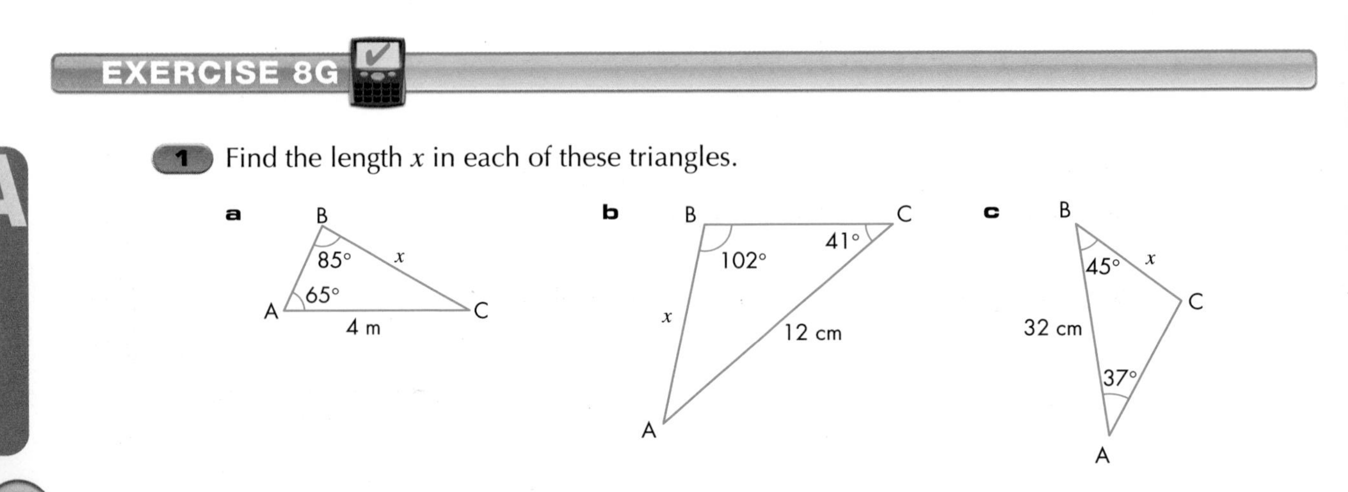

2 Find the angle x in each of these triangles.

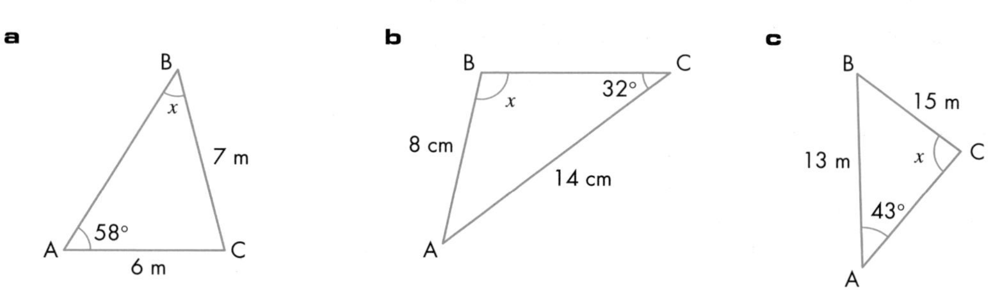

3 In triangle ABC, the angle at A is 38°, the side AB is 10 cm and the side BC is 8 cm. Find the two possible values of the angle at C.

4 In triangle ABC, the angle at A is 42°, the side AB is 16 cm and the side BC is 14 cm. Find the two possible values of the side AC.

FM 5 To find the height of a tower standing on a small hill, Mary made some measurements (see diagram).

From a point B, the angle of elevation of C is 20°, the angle of elevation of A is 50°, and the distance BC is 25 m.

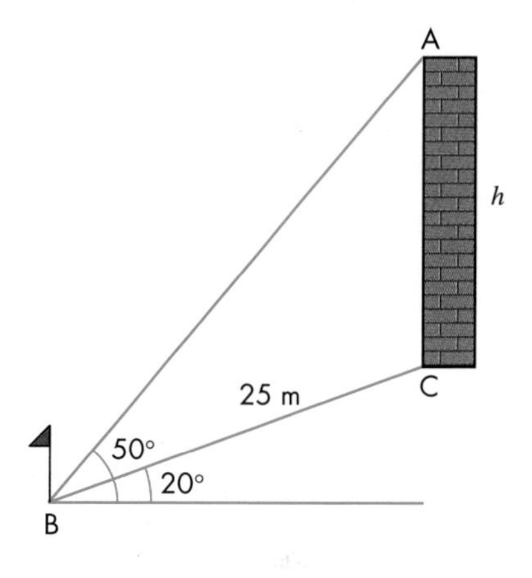

a Calculate these angles.

i ABC

ii BAC

b Using the sine rule and triangle ABC, calculate the height h of the tower.

PS 6 Use the information on this sketch to calculate the width, w, of the river.

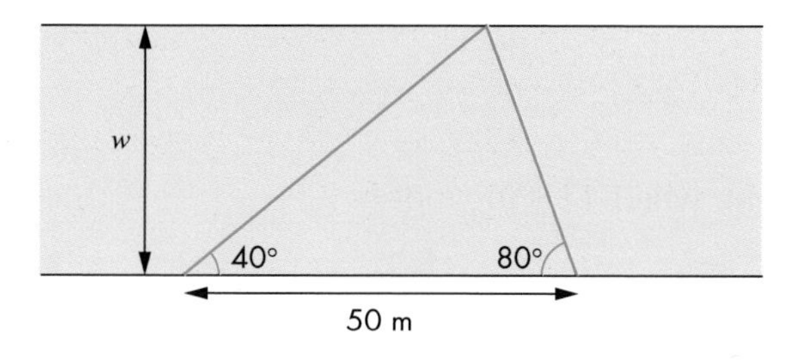

FM 7 An old building is unsafe and is protected by a fence. A demolition company is employed to demolish the building and has to work out the height BD, marked h on the diagram.

Calculate the value of h, using the given information.

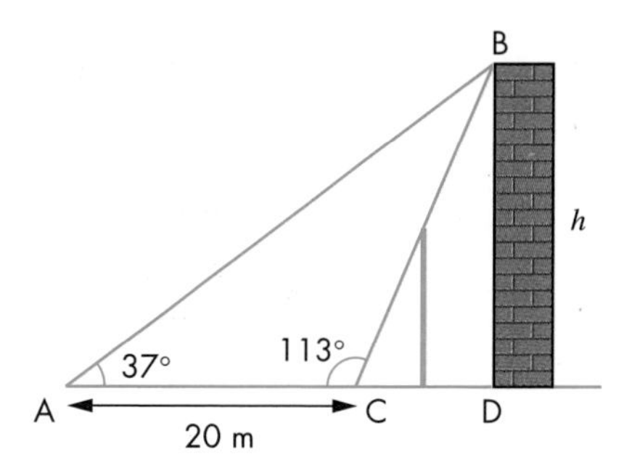

A*

8 A weight is hung from a horizontal beam using two strings. The shorter string is 2.5 m long and makes an angle of 71° with the horizontal. The longer string makes an angle of 43° with the horizontal. What is the length of the longer string?

FM 9 An aircraft is flying over an army base. Suddenly, two searchlights, 3 km apart, are switched on. The two beams of light meet on the aircraft at an angle of 125° vertically above the line joining the searchlights. One of the beams of light makes an angle of 31° with the horizontal. Calculate the height of the aircraft.

FM 10 Two ships leave a port in directions that are 41° from each other. After half an hour, the ships are 11 km apart. If the speed of the slower ship is 7 km/h, what is the speed of the faster ship?

FM 11 A rescue helicopter is based at an airfield at A.

The helicopter is sent out to rescue a man who has had an accident on a mountain at M, due north of A.

The helicopter then flies on a bearing of 145° to a hospital at H as shown on the diagram.

Calculate the direct distance from the mountain to the hospital.

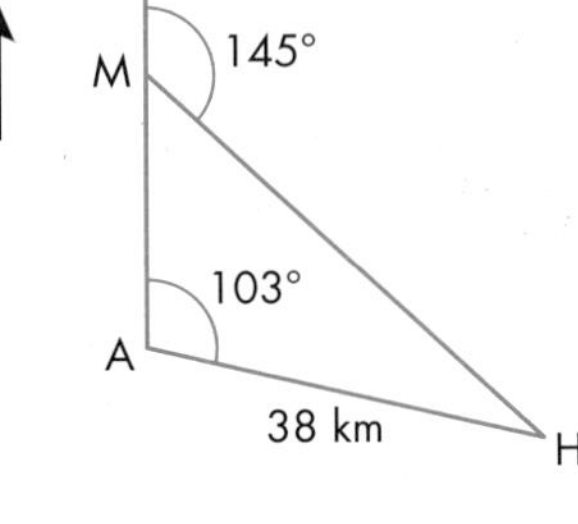

AU 12 Choose four values of θ, $0° < \theta < 90°$, to show that $\sin \theta = \sin (180° - \theta)$.

13 Triangle ABC has an obtuse angle at B.

Calculate the size of angle ABC.

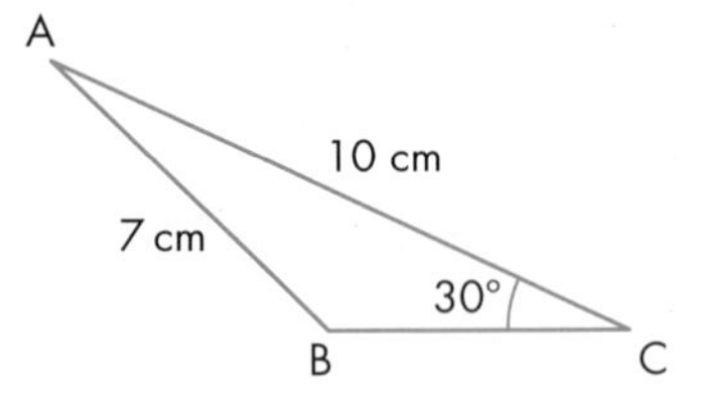

PS 14 For any triangle ABC, prove the sine rule:

$$\frac{a}{\sin A} = \frac{b}{\sin B} = \frac{c}{\sin C}$$

The cosine rule

Take the triangle, shown on the right, where D is the foot of the perpendicular to BC from A.

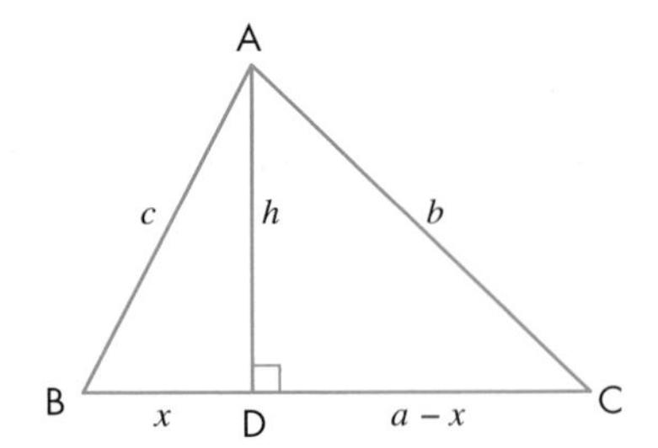

Using Pythagoras' theorem on triangle BDA:

$$h^2 = c^2 - x^2$$

Using Pythagoras' theorem on triangle ADC:

$$h^2 = b^2 - (a - x)^2$$

Therefore,

$$c^2 - x^2 = b^2 - (a - x)^2$$

$$c^2 - x^2 = b^2 - a^2 + 2ax - x^2$$

$$\Rightarrow c^2 = b^2 - a^2 + 2ax$$

From triangle BDA, $x = c \cos B$.

Hence,

$$c^2 = b^2 - a^2 + 2ac \cos B$$

Rearranging gives:

$$b^2 = a^2 + c^2 - 2ac \cos B$$

By algebraic symmetry:

$$a^2 = b^2 + c^2 - 2bc \cos A \quad \text{and} \quad c^2 = a^2 + b^2 - 2ab \cos C$$

This is the cosine rule, which can be best remembered by the diagram on the right, where:

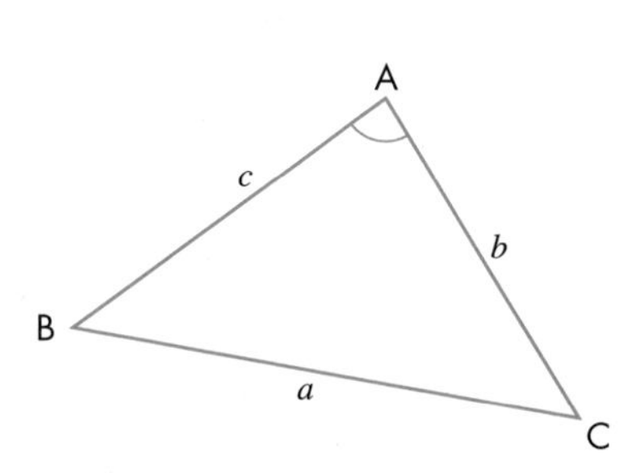

$$a^2 = b^2 + c^2 - 2bc \cos A$$

Note the symmetry of the rule and how the rule works using two adjacent sides and the angle between them (the **included angle**).

The formula can be rearranged to find any of the three angles.

$$\cos A = \frac{b^2 + c^2 - a^2}{2bc}$$

$$\cos B = \frac{a^2 + c^2 - b^2}{2ac}$$

$$\cos C = \frac{a^2 + b^2 - c^2}{2ab}$$

Note that the cosine rule $a^2 = b^2 + c^2 - 2bc \cos A$ is given in the formula sheets in the GCSE examination but the rearranged formula for the angle is not given. You are advised to learn this as trying to rearrange usually ends up with an incorrect formula.

EXAMPLE 14

Find x in this triangle.

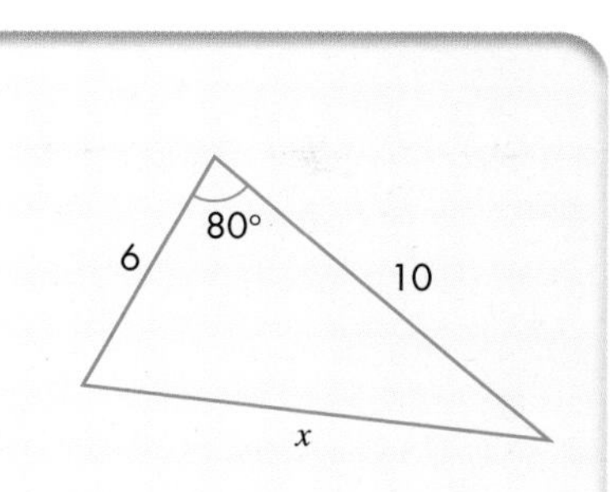

By the cosine rule:

$$x^2 = 6^2 + 10^2 - 2 \times 6 \times 10 \times \cos 80°$$

$$x^2 = 115.16$$

$$\Rightarrow x = 10.7 \quad \text{(3 significant figures)}$$

EXAMPLE 15

Find x in this triangle.

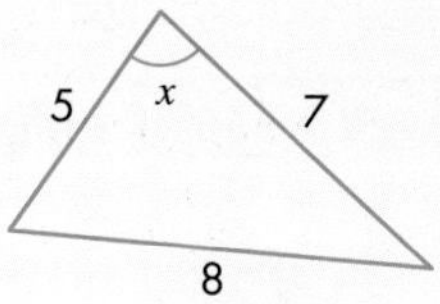

By the cosine rule:

$$\cos x = \frac{5^2 + 7^2 - 8^2}{2 \times 5 \times 7} = 0.1428$$

$$\Rightarrow x = 81.8° \quad \text{(3 significant figures)}$$

It is possible to find the cosine of an angle that is greater than 90° (see Section 8.3). For example, cos 120° = –cos 60° = –0.5. (Notice the minus sign; the two angles add up to 180°.)

So cos 150° = –cos 30° = –0.866

EXAMPLE 16

A ship sails from a port on a bearing of 055° for 40 km. It then changes course to 123° for another 50 km. On what course should the ship be steered to get it straight back to the port?

Previously, you have solved this type of problem using right-angled triangles. This method could be applied here but it would involve at least six separate calculations.

With the aid of the cosine and sine rules, however, you can reduce the solution to two separate calculations, as follows.

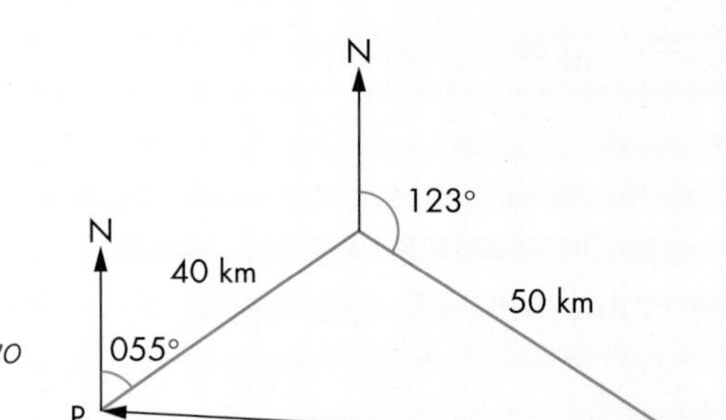

The course diagram gives the triangle PAB (on the right), where angle PAB is found by using alternate angles and angles on a line.
55° + (180° – 123°) = 112°

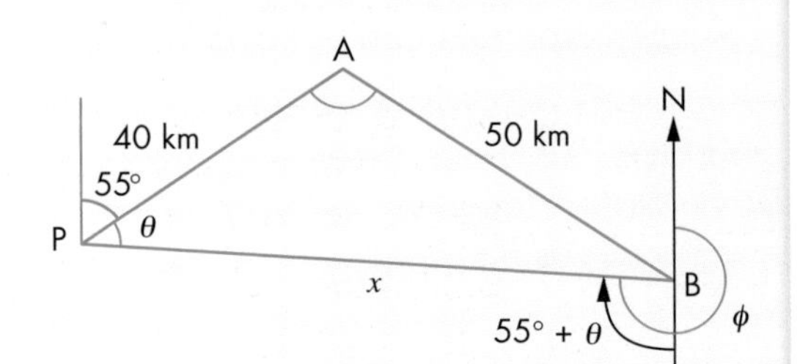

Let ϕ be the bearing to be steered, then

$$\phi = \theta + 55° + 180°$$

To find θ, you first have to obtain PB($= x$), using the cosine rule.

$$x^2 = 40^2 + 50^2 - 2 \times 40 \times 50 \times \cos 112° \text{ km}^2$$

(Remember: the cosine of 112° is negative.)

$$\Rightarrow x^2 = 5598.43 \text{ km}^2$$

$$\Rightarrow x = 74.82 \text{ km}$$

You can now find θ from the sine rule.

$$\frac{\sin \theta}{50} = \frac{\sin 112°}{74.82}$$

$$\Rightarrow \sin \theta = \frac{50 \times \sin 112°}{74.82} = 0.6196$$

$$\Rightarrow \theta = 38.3°$$

So the ship should be steered on a bearing of:

38.3° + 55° + 180° = 273.3°

EXERCISE 8H

1 Find the length x in each of these triangles.

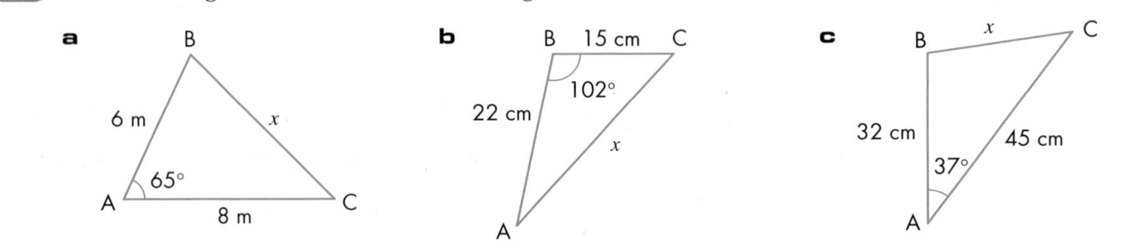

A

2 Find the angle x in each of these triangles.

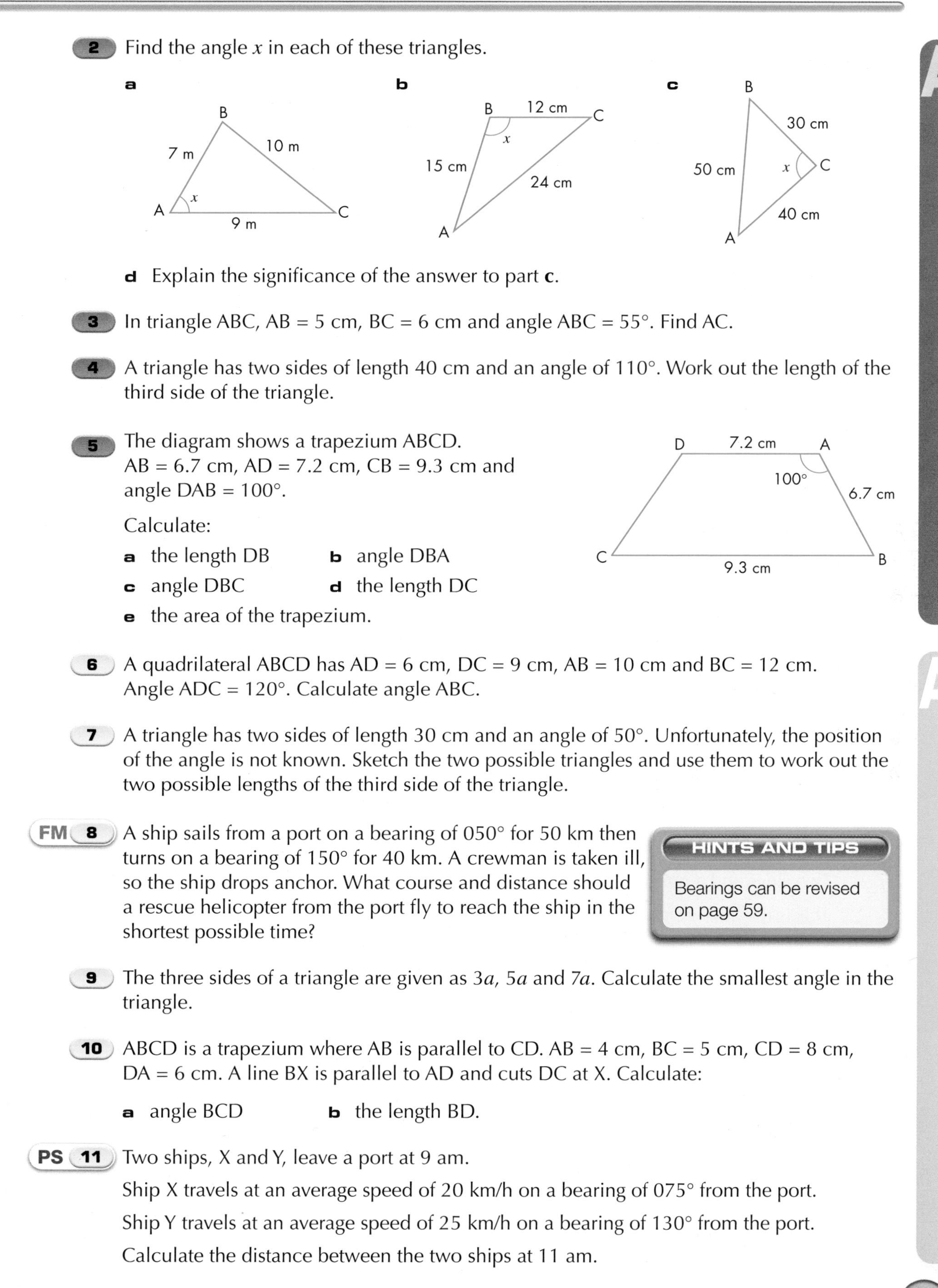

d Explain the significance of the answer to part **c**.

3 In triangle ABC, AB = 5 cm, BC = 6 cm and angle ABC = 55°. Find AC.

4 A triangle has two sides of length 40 cm and an angle of 110°. Work out the length of the third side of the triangle.

5 The diagram shows a trapezium ABCD.
AB = 6.7 cm, AD = 7.2 cm, CB = 9.3 cm and angle DAB = 100°.

Calculate:

a the length DB **b** angle DBA

c angle DBC **d** the length DC

e the area of the trapezium.

6 A quadrilateral ABCD has AD = 6 cm, DC = 9 cm, AB = 10 cm and BC = 12 cm. Angle ADC = 120°. Calculate angle ABC.

7 A triangle has two sides of length 30 cm and an angle of 50°. Unfortunately, the position of the angle is not known. Sketch the two possible triangles and use them to work out the two possible lengths of the third side of the triangle.

FM 8 A ship sails from a port on a bearing of 050° for 50 km then turns on a bearing of 150° for 40 km. A crewman is taken ill, so the ship drops anchor. What course and distance should a rescue helicopter from the port fly to reach the ship in the shortest possible time?

HINTS AND TIPS

Bearings can be revised on page 59.

9 The three sides of a triangle are given as $3a$, $5a$ and $7a$. Calculate the smallest angle in the triangle.

10 ABCD is a trapezium where AB is parallel to CD. AB = 4 cm, BC = 5 cm, CD = 8 cm, DA = 6 cm. A line BX is parallel to AD and cuts DC at X. Calculate:

a angle BCD **b** the length BD.

PS 11 Two ships, X and Y, leave a port at 9 am.

Ship X travels at an average speed of 20 km/h on a bearing of 075° from the port.

Ship Y travels at an average speed of 25 km/h on a bearing of 130° from the port.

Calculate the distance between the two ships at 11 am.

A*

AU 12 Choose four values of θ, $0° < \theta < 90°$, to show that $\cos\theta = -\cos(180° - \theta)$.

AU 13 Calculate the size of the largest angle in the triangle ABC.

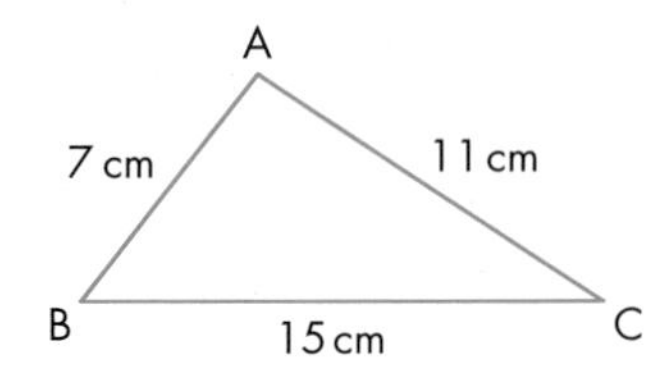

Choosing the correct rule

When solving triangles, there are only four situations that can occur, each of which can be solved completely in three stages.

Two sides and the included angle

1 Use the cosine rule to find the third side.

2 Use the sine rule to find either of the other angles.

3 Use the sum of the angles in a triangle to find the third angle.

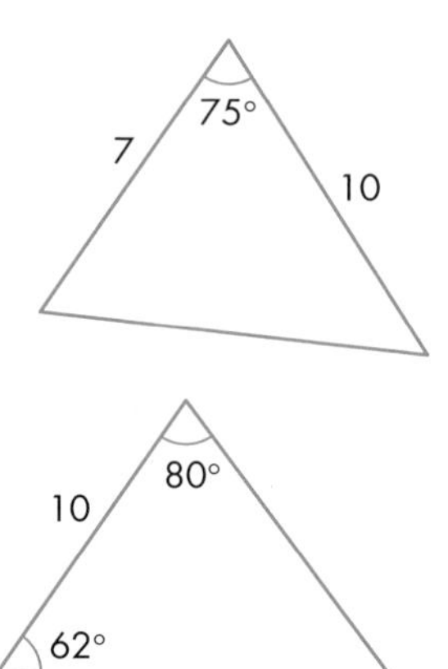

Two angles and a side

1 Use the sum of the angles in a triangle to find the third angle.

2, 3 Use the sine rule to find the other two sides.

Three sides

1 Use the cosine rule to find one angle.

2 Use the sine rule to find another angle.

3 Use the sum of the angles in a triangle to find the third angle.

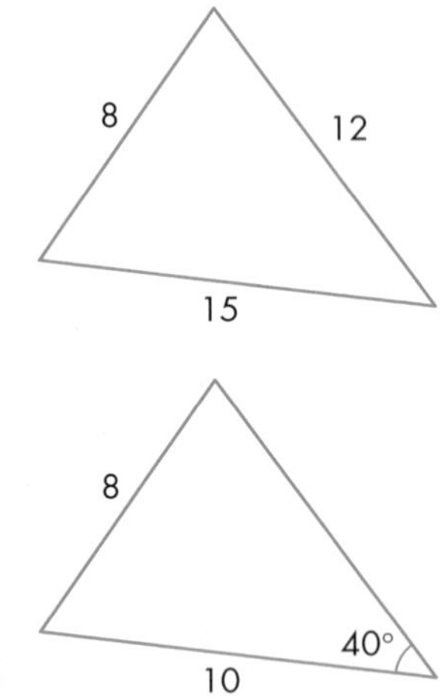

Two sides and a non-included angle

This is the ambiguous case already covered (page 198).

1 Use the sine rule to find the two possible values of the appropriate angle.

2 Use the sum of the angles in a triangle to find the two possible values of the third angle.

3 Use the sine rule to find the two possible values for the length of the third side.

Note: Apply the sine rule wherever you can – it is always easier to use than the cosine rule. You should never need to use the cosine rule more than once.

EXERCISE 8I

1 Find the length or angle x in each of these triangles.

a B, 8 m, x, 65°, A, C, 8 m

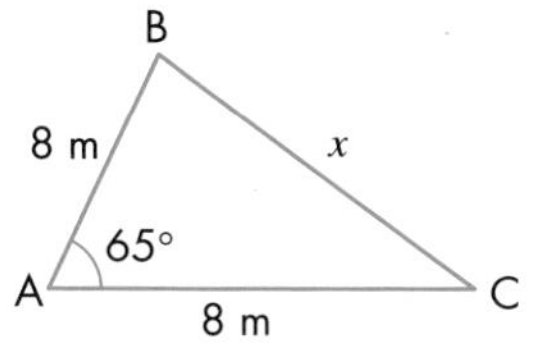

b B, 12 cm, C, x, 16 cm, 20 cm, A

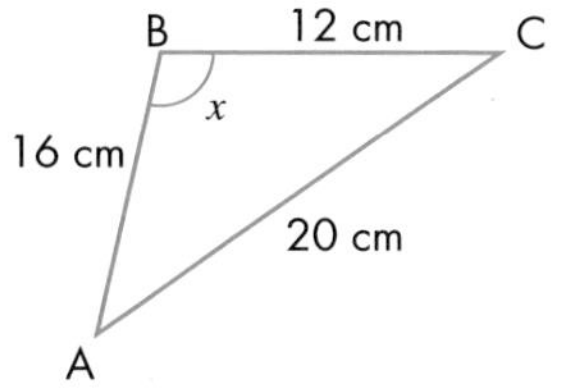

c B, x, C, 84°, 45 cm, 37°, A

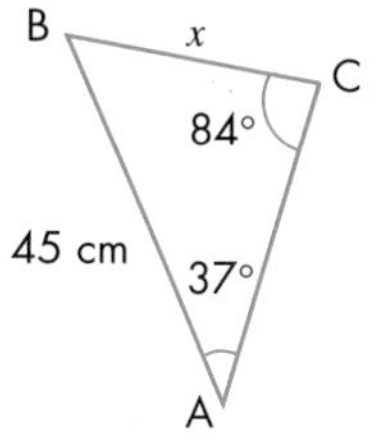

d B, 6 m, 12 m, x, 65°, C, A

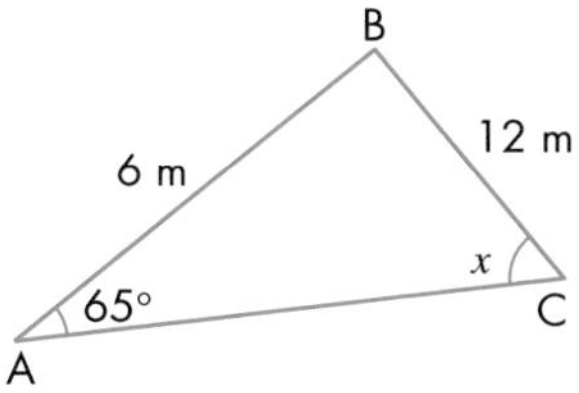

e B, 15 cm, C, 65°, 22 cm, x, A

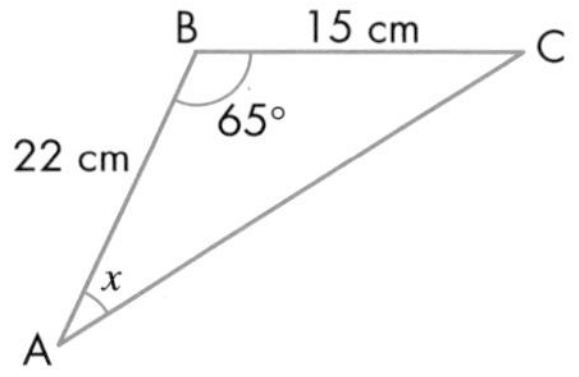

f B, x, C, 91 cm, 43°, 73 cm, A

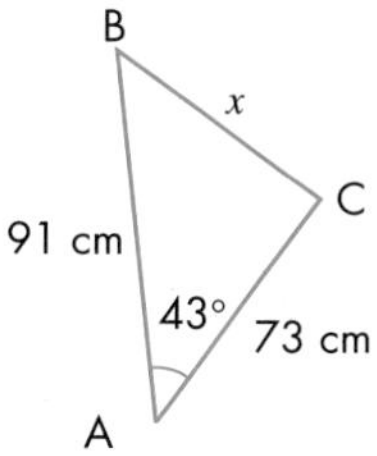

g B, x, 6 m, 53.1°, C, A, 10 m

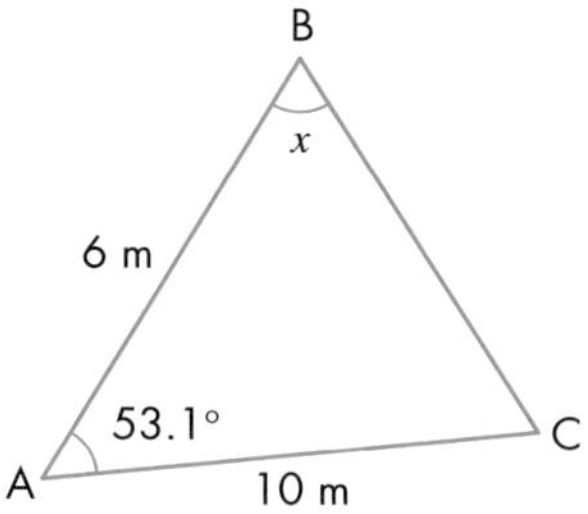

h 530 cm, C, B, 124°, 450 cm, x, A

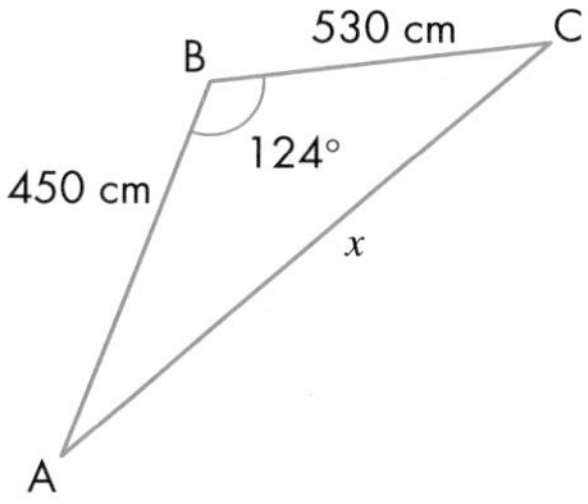

i B, 45 cm, x, 124°, C, 53 cm, A

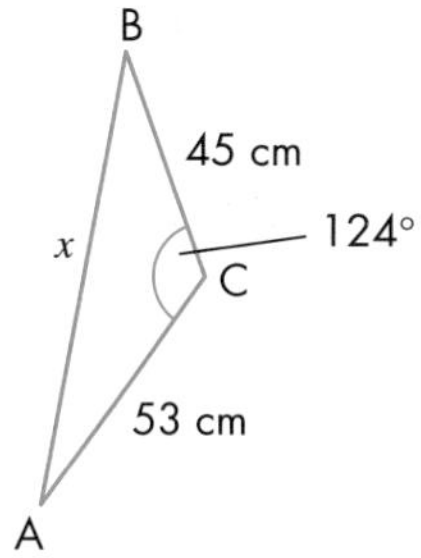

2 The hands of a clock have lengths 3 cm and 5 cm. Find the distance between the tips of the hands at 4 o'clock.

3 A spacecraft is seen hovering at a point which is in the same vertical plane as two towns, X and F, which are on the same level. Its distances from X and F are 8.5 km and 12 km respectively. The angle of elevation of the spacecraft when observed from F is 43°. Calculate the distance between the two towns.

FM 4 Two boats, Mary Jo and Suzie, leave port at the same time.

Mary Jo sails at 10 knots on a bearing of 065°. Suzie sails on a bearing of 120° and after 1 hour Mary Jo is on a bearing of 330° from Suzie. What is Suzie's speed? (A knot is a nautical mile per hour.)

5 Two ships leave port at the same time, Darling Dave sailing at 12 knots on a bearing of 055°, and Merry Mary at 18 knots on a bearing of 280°.

a How far apart are the two ships after 1 hour?

b What is the bearing of Merry Mary from Darling Dave?

A*

PS 6 Triangle ABC has sides with lengths a, b and c, as shown in the diagram.

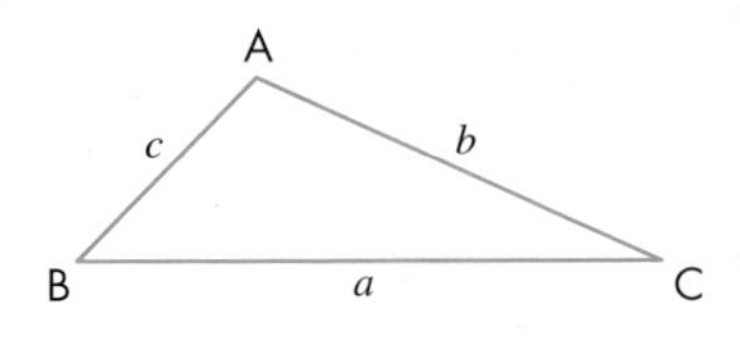

a What can you say about the angle BAC, if $b^2 + c^2 - a^2 = 0$?

b What can you say about the angle BAC, if $b^2 + c^2 - a^2 > 0$?

c What can you say about the angle BAC, if $b^2 + c^2 - a^2 < 0$?

AU 7 The diagram shows a sketch of a field ABCD.
A farmer wants to put a new fence round the perimeter of the field.

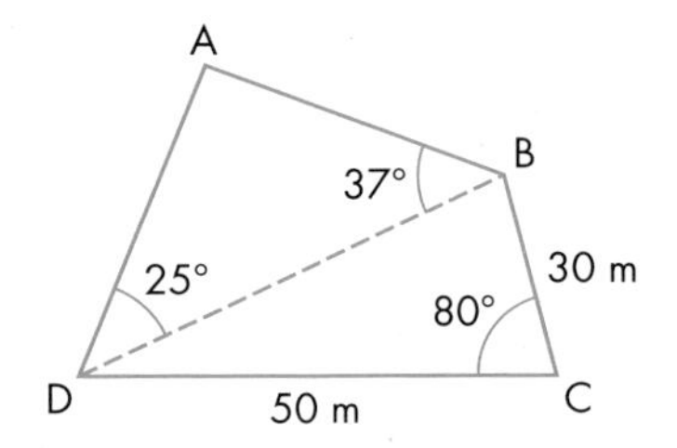

Calculate the perimeter of the field.

Give your answer to an appropriate degree of accuracy.

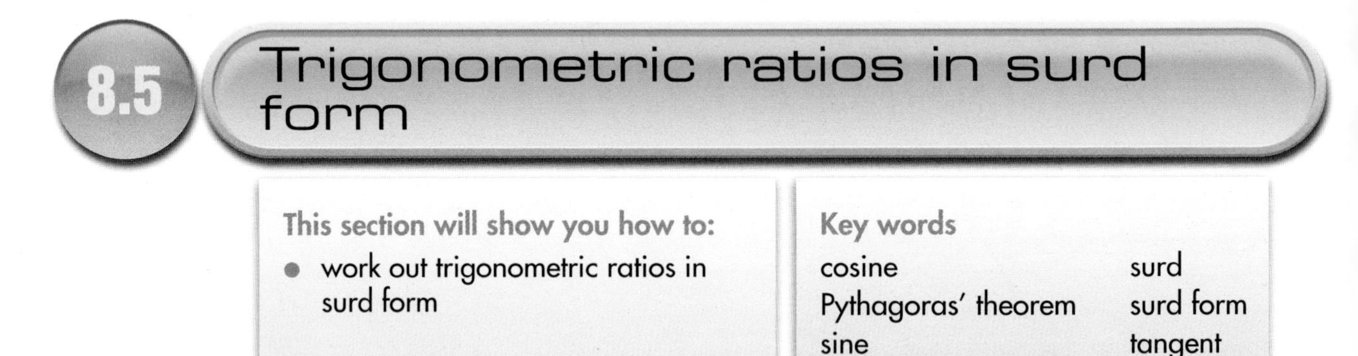

8.5 Trigonometric ratios in surd form

This section will show you how to:
- work out trigonometric ratios in surd form

Key words
cosine
Pythagoras' theorem
sine
surd
surd form
tangent

Solving triangles often involves finding square roots. Unless a number has an exact square root, the value on a calculator can only be an approximation. The exact value of the square root of 2, for example, is often written as $\sqrt{2}$. This expression, using the square root symbol, is a **surd**, and answers given in this way are in **surd form**.

EXAMPLE 17

Using an equilateral triangle with sides of 2 units, write down expressions for the sine, cosine and tangent of 60° and 30°. Give answers in surd form.

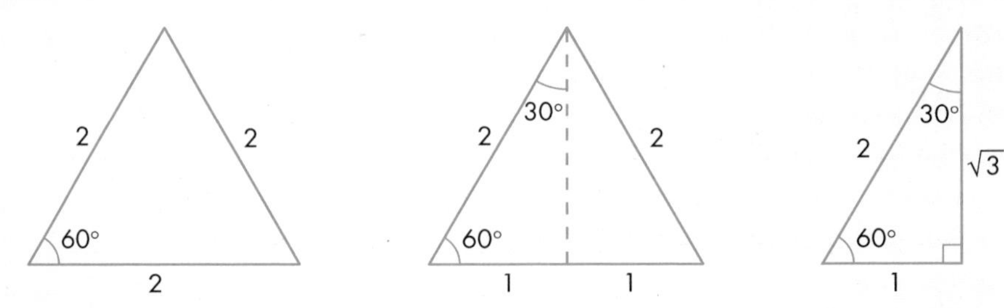

Divide the equilateral triangle into two equal right-angled triangles. Taking one of them, use **Pythagoras' theorem** and the definition of **sine, cosine** and **tangent** to obtain:

$\sin 60° = \frac{\sqrt{3}}{2}$ $\cos 60° = \frac{1}{2}$ $\tan 60° = \sqrt{3}$

and $\sin 30° = \frac{1}{2}$ $\cos 30° = \frac{\sqrt{3}}{2}$ $\tan 30° = \frac{1}{\sqrt{3}} = \frac{\sqrt{3}}{3}$

EXAMPLE 18

Using a right-angled isosceles triangle in which the equal sides are 1 unit, find the sine, cosine and tangent of 45°. Give answers in surd form.

By Pythagoras' theorem, the hypotenuse of the triangle is $\sqrt{2}$ units.

From the definition of sine, cosine and tangent,

$$\sin 45° = \frac{1}{\sqrt{2}} = \frac{\sqrt{2}}{2} \qquad \cos 45° = \frac{1}{\sqrt{2}} = \frac{\sqrt{2}}{2} \qquad \tan 45° = 1$$

These results can be summarised in a table.

You do not need to learn these results, but it is useful to know how to work them out.

θ	$\cos\theta$	$\sin\theta$	$\tan\theta$
30°	$\frac{\sqrt{3}}{2}$	$\frac{1}{2}$	$\frac{\sqrt{3}}{3}$
45°	$\frac{\sqrt{2}}{2}$	$\frac{\sqrt{2}}{2}$	1
60°	$\frac{1}{2}$	$\frac{\sqrt{3}}{2}$	$\sqrt{3}$

When solving problems, you can write trigonometric ratios as numerical values or in surd form, in which case you do not need a calculator.

EXAMPLE 19

ABC is a right-angled triangle.

a Write down the value of $\tan x$.

b Work out the values of $\cos x$ and $\sin x$, giving your answers in simplified surd form.

a $\tan x = \frac{3}{5}$

b Using Pythagoras' theorem, AC = $\sqrt{34}$.

So $\cos x = \frac{5}{\sqrt{34}} = \frac{5\sqrt{34}}{34}$

and $\sin x = \frac{3}{\sqrt{34}} = \frac{3\sqrt{34}}{34}$

EXERCISE 8J

AU 1 The sine of angle x is $\frac{4}{5}$. Work out the cosine of angle x.

AU 2 The cosine of angle x is $\frac{3}{\sqrt{15}}$. Work out the sine of angle x.

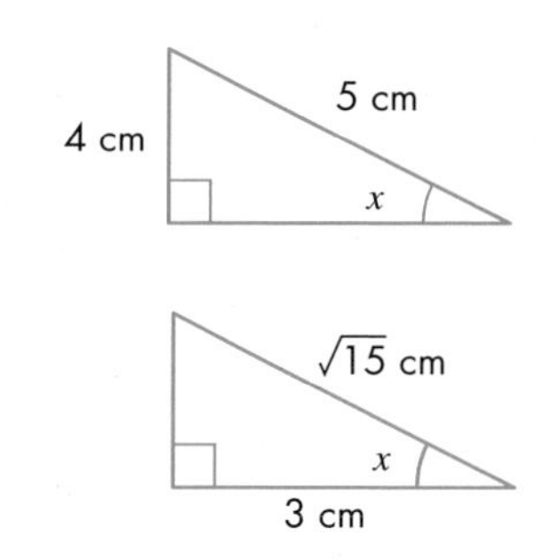

A*

PS 3 The lengths of the two short sides of a right-angled triangle are $\sqrt{6}$ and $\sqrt{13}$. Write down the exact value of the hypotenuse of this triangle, and the exact value of the sine, cosine and tangent of the smallest angle in the triangle.

PS 4 The tangent of angle A is $\frac{6}{11}$. Use this to write down possible lengths of two sides of the triangle.

a Calculate the length of the third side of the triangle.

b Write down the exact values of sin A and cos A.

PS 5 Calculate the exact value of the area of an equilateral triangle of side 6 cm.

6 Work out the exact value of the area of a right-angled isosceles triangle with hypotenuse 40 cm.

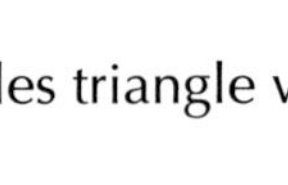

7 Work out the length of AB in the triangle ABC.

Give your answer in simplified surd form.

8.6 Using sine to find the area of a triangle

This section will show you how to:

- work out the area of a triangle if you know two sides and the included angle

Key words

area
area sine rule
cosine rule
included angle
sine rule

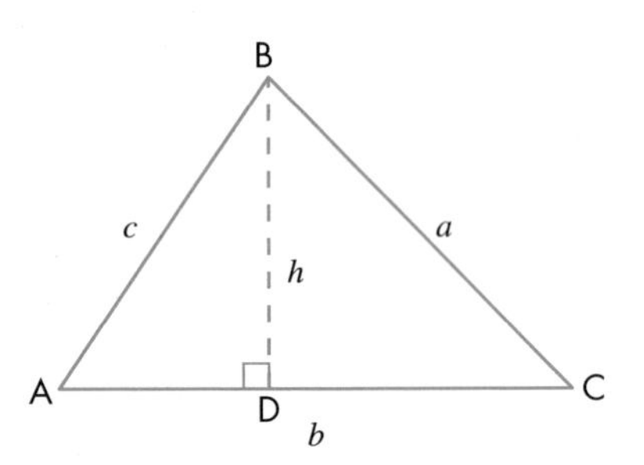

In triangle ABC, the vertical height is BD and the base is AC.

Let BD = h and AC = b, then the **area** of the triangle is given by:

$\frac{1}{2} \times \text{AC} \times \text{BD} = \frac{1}{2}bh$

However, in triangle BCD,

$h = \text{BC} \sin C = a \sin C$

where BC = a.

Substituting into $\frac{1}{2}bh$ gives:

$\frac{1}{2}b \times (a \sin C) = \frac{1}{2}ab \sin C$

as the area of the triangle.

By taking the perpendicular from A to its opposite side BC, and the perpendicular from C to its opposite side AB, we can show that the area of the triangle is also given by:

$\frac{1}{2}ac \sin B$ and $\frac{1}{2}bc \sin A$

Note the pattern: the area is given by the product of two sides multiplied by the sine of the **included angle**. This is the **area sine rule**. Starting from any of the three forms, it is also possible to use the **sine rule** to establish the other two.

EXAMPLE 20

Find the area of triangle ABC.

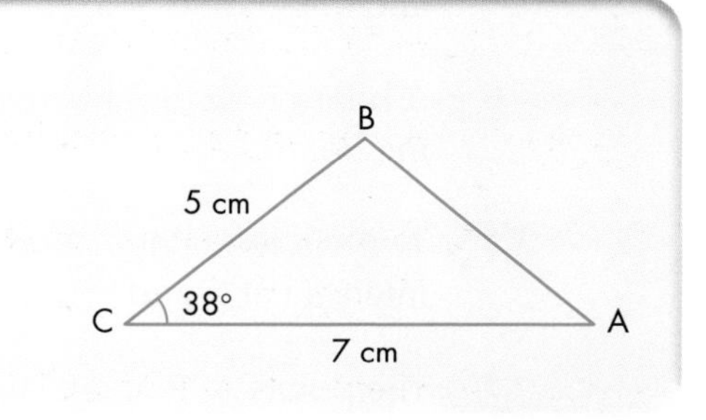

Area = $\frac{1}{2}ab \sin C$

Area = $\frac{1}{2} \times 5 \times 7 \times \sin 38° = 10.8 \text{ cm}^2$

(3 significant figures)

EXAMPLE 21

Find the area of triangle ABC.

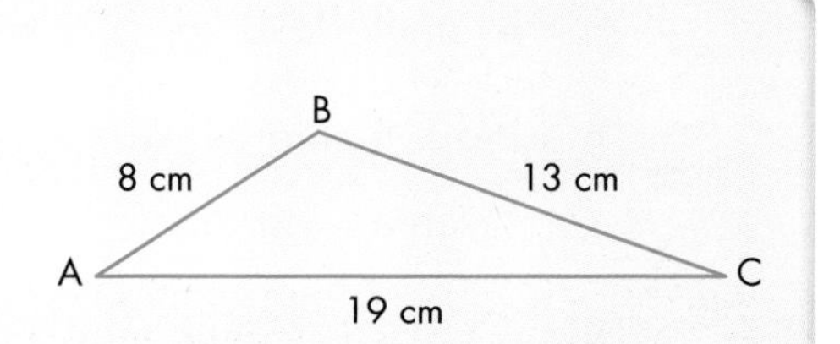

You have all three sides but no angle. So first you must find an angle in order to apply the area sine rule.

Find angle C, using the **cosine rule**.

$$\cos C = \frac{a^2 + b^2 - c^2}{2ab}$$

$$= \frac{13^2 + 19^2 - 8^2}{2 \times 13 \times 19} = 0.9433$$

$$\Rightarrow C = \cos^{-1} 0.9433 = 19.4°$$

(Keep the exact value in your calculator memory.)

Now apply the area sine rule.

$$\frac{1}{2}ab \sin C = \frac{1}{2} \times 13 \times 19 \times \sin 19.4°$$

$$= 41.0 \text{ cm}^2 \quad \text{(3 significant figures)}$$

EXERCISE 8K

1 Find the area of each of the following triangles.

a Triangle ABC where BC = 7 cm, AC = 8 cm and angle ACB = 59°

b Triangle ABC where angle BAC = 86°, AC = 6.7 cm and AB = 8 cm

c Triangle PQR where QR = 27 cm, PR = 19 cm and angle QRP = 109°

d Triangle XYZ where XY = 231 cm, XZ = 191 cm and angle YXZ = 73°

e Triangle LMN where LN = 63 cm, LM = 39 cm and angle NLM = 85°

A

A

2 The area of triangle ABC is 27 cm^2. If BC = 14 cm and angle BCA = 115°, find AC.

3 The area of triangle LMN is 113 cm^2, LM = 16 cm and MN = 21 cm. Angle LMN is acute. Calculate these angles.

a LMN **b** MNL

A*

4 In a quadrilateral ABCD, DC = 4 cm, BD = 11 cm, angle BAD = 32°, angle ABD = 48° and angle BDC = 61°. Calculate the area of the quadrilateral.

5 A board is in the shape of a triangle with sides 60 cm, 70 cm and 80 cm. Find the area of the board.

6 Two circles, centres P and Q, have radii of 6 cm and 7 cm respectively. The circles intersect at X and Y. Given that PQ = 9 cm, find the area of triangle PXQ.

7 The points A, B and C are on the circumference of a circle, centre O and radius 7 cm. AB = 4 cm and BC = 3.5 cm. Calculate:

a angle AOB **b** the area of quadrilateral OABC.

PS 8 Prove that for any triangle ABC,

area = $\frac{1}{2}ab \sin C$

9 **a** ABC is a right-angled isosceles triangle with short sides of 1 cm. Write down the value of sin 45°.

A
1 cm
B
1 cm
C

b Calculate the area of triangle PQR. Give your answer in surd form.

Q
6 cm
P
45°
$7\sqrt{2}$ cm
R

10 Sanjay is making a kite.
The diagram shows a sketch of his kite.

Calculate the area of the material required to make the kite.

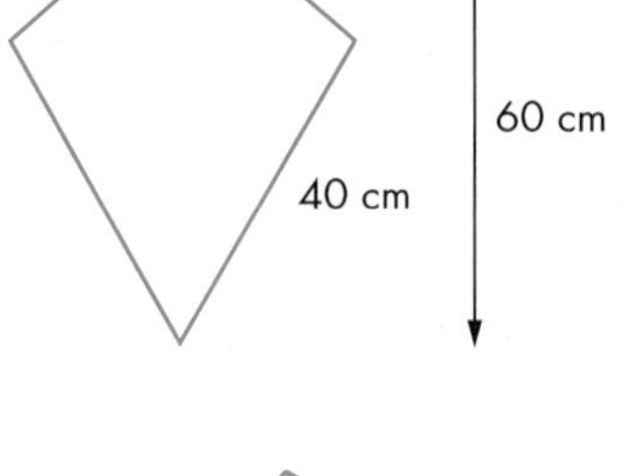

PS 11 An equilateral triangle has sides of length a.

Work out the area of the triangle, giving your answer in surd form.

AU 12 The lengths of all the sides of a square-based pyramid are 10 cm.

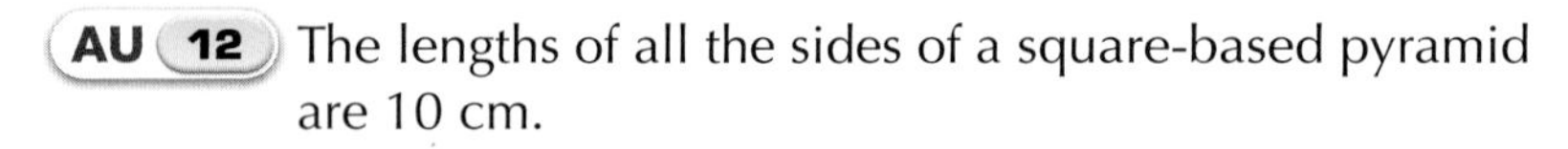

Which of these possible answers correctly gives the total surface area of the pyramid?

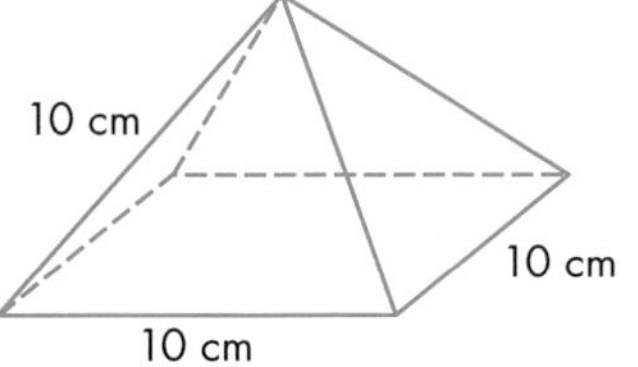

a $100(1 + 2\sqrt{3})$ cm^2 **b** $100(2 + \sqrt{3})$ cm^2

c $100(1 + \sqrt{3})$ cm^2 **d** $100(1 + 2\sqrt{2})$cm^2

GRADE BOOSTER

A You can solve more complex 2D problems, using Pythagoras' theorem and trigonometry

A You can use the sine and cosine rules to calculate missing angles or sides in non-right-angled triangles

A You can find the area of a triangle using the formula area = $\frac{1}{2}ab$ sin C

A* You can use the sine and cosine rules to solve more complex problems involving non-right-angled triangles

A* You can solve 3D problems, using Pythagoras' theorem and trigonometric ratios

A* You can find the sine, cosine and tangent of any angle from 90° to 360°

A* You can work out trigonometric ratios in surd form

What you should know now

- How to find the trigonometric ratios for any angle up to 360°
- How to use the sine and cosine rules
- How to find the area of a triangle, using area = $\frac{1}{2}ab$ sin C

1

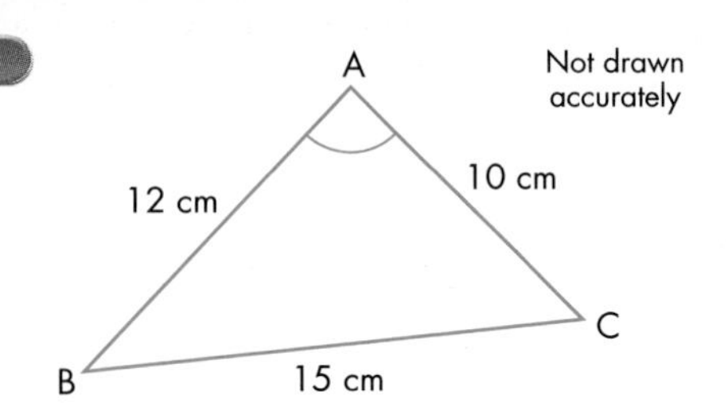

ABC is a triangle.

AB = 12 m, AC = 10 m, BC = 15 m.

Calculate the size of angle BAC.

Give your answer correct to one decimal place. *(3 marks)*

Edexcel, June 2006, Paper 4 Higher, Question 24

2

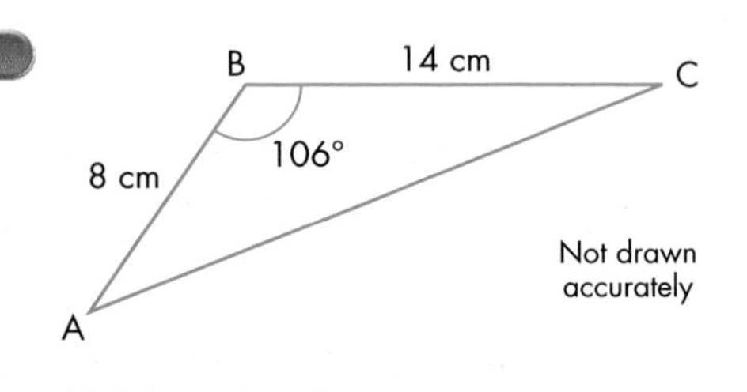

ABC is a triangle.

AB = 8 cm

BC = 14 cm

Angle ABC = 106°

Calculate the area of the triangle.

Give your answer correct to 3 significant figures. *(3 marks)*

Edexcel, June 2005, Paper 6, Question 13

3

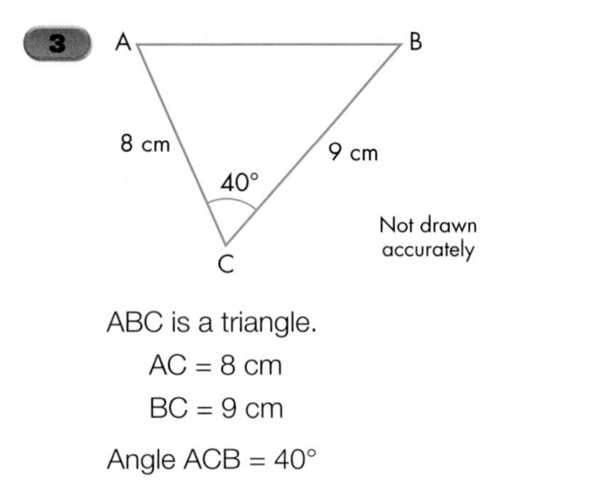

ABC is a triangle.

AC = 8 cm

BC = 9 cm

Angle ACB = 40°

Calculate the length of AB.

Give your answer correct to 3 significant figures. *(3 marks)*

Edexcel, June 2007, Paper 6, Question 21

4 The diagram shows an equilateral triangle of side 2 m.

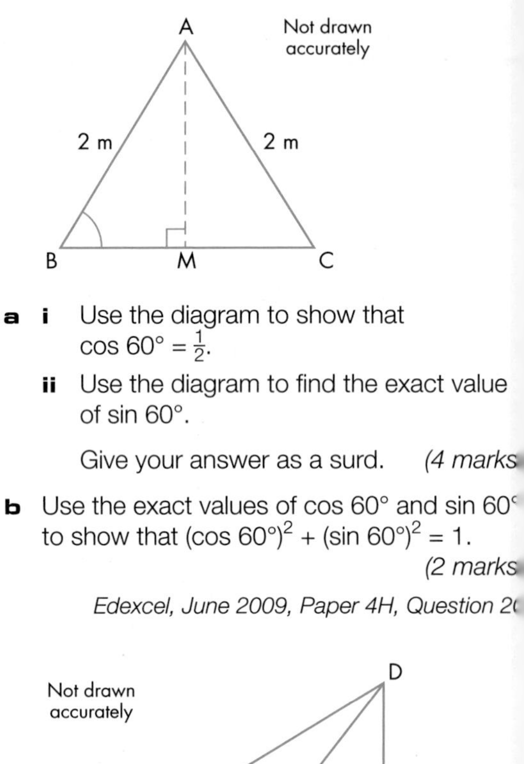

a **i** Use the diagram to show that $\cos 60° = \frac{1}{2}$.

ii Use the diagram to find the exact value of $\sin 60°$.

Give your answer as a surd. *(4 marks*

b Use the exact values of $\cos 60°$ and $\sin 60$ to show that $(\cos 60°)^2 + (\sin 60°)^2 = 1$. *(2 marks*

Edexcel, June 2009, Paper 4H, Question 2

5

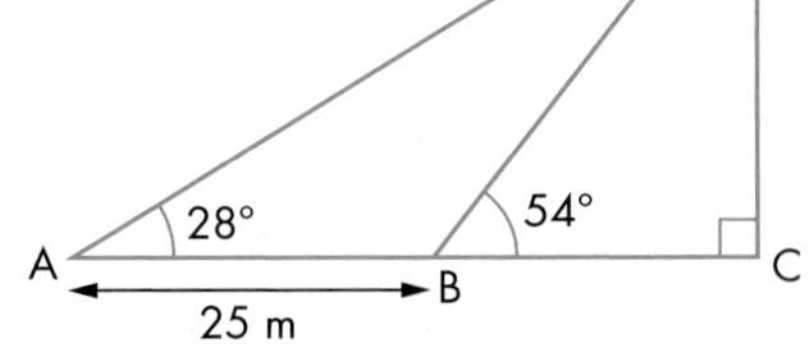

The diagram shows a vertical tower DC on horizontal ground ABC.

ABC is a straight line.

The angle of elevation of D from A is 28°.

The angle of elevation of D from B is 54°.

AB = 25 m.

Calculate the height of the tower.

Give your answer correct to 3 significant figures. *(5 marks*

Edexcel, June 2006, Paper 6, Question 2

Worked Examination Questions

AU **1** The diagram shows a cuboid ABCDEFGH.
Calculate the size of angle AGE.

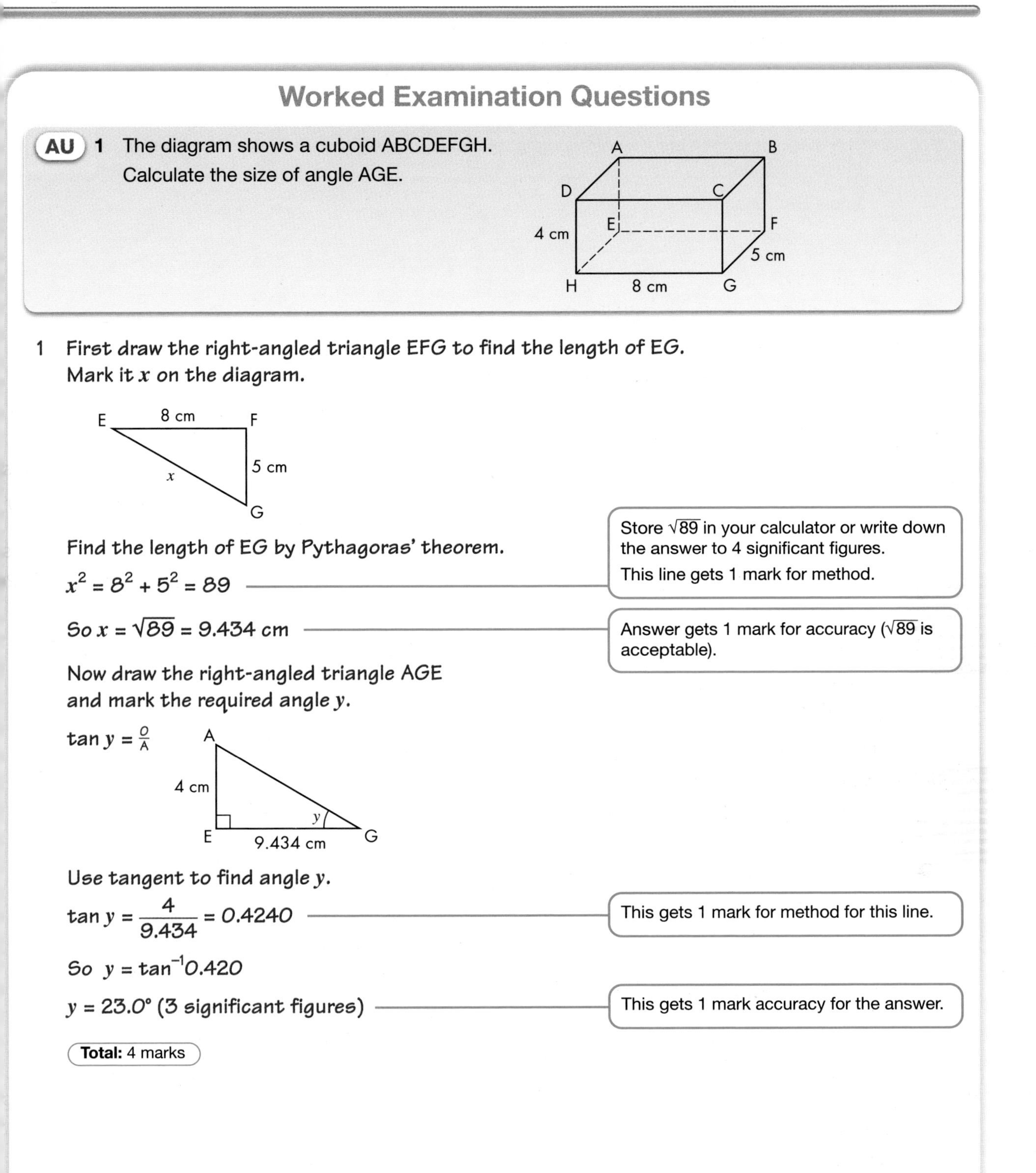

1 First draw the right-angled triangle EFG to find the length of EG.
Mark it x on the diagram.

E 8 cm F
5 cm
x
G

Find the length of EG by Pythagoras' theorem.

$x^2 = 8^2 + 5^2 = 89$ — Store $\sqrt{89}$ in your calculator or write down the answer to 4 significant figures. This line gets 1 mark for method.

So $x = \sqrt{89} = 9.434$ cm — Answer gets 1 mark for accuracy ($\sqrt{89}$ is acceptable).

Now draw the right-angled triangle AGE and mark the required angle y.

$\tan y = \frac{O}{A}$

A
4 cm
y
E 9.434 cm G

Use tangent to find angle y.

$\tan y = \frac{4}{9.434} = 0.4240$ — This gets 1 mark for method for this line.

So $y = \tan^{-1} 0.420$

$y = 23.0°$ (3 significant figures) — This gets 1 mark accuracy for the answer.

Total: 4 marks

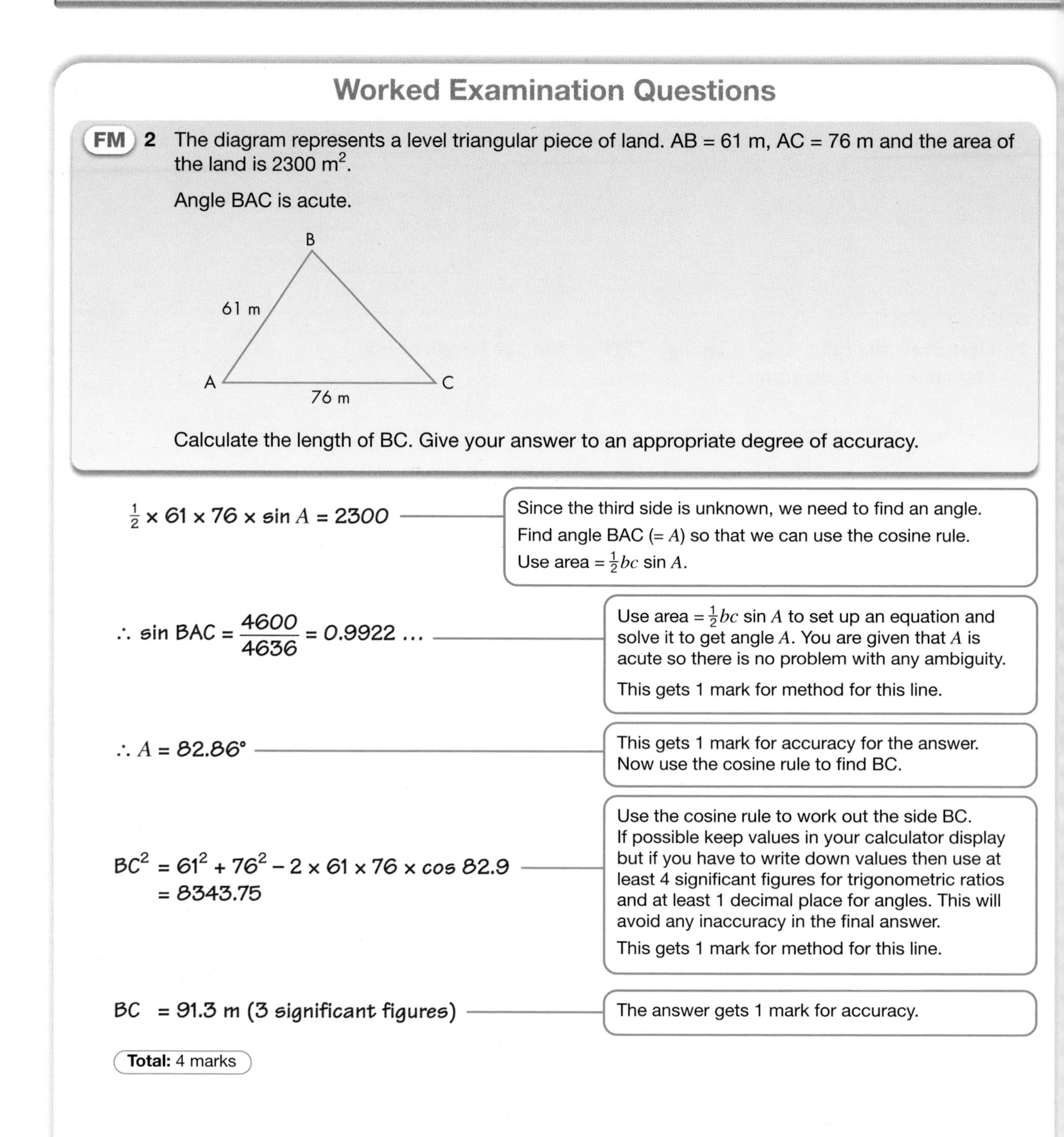

Worked Examination Questions

FM **2** The diagram represents a level triangular piece of land. AB = 61 m, AC = 76 m and the area of the land is 2300 m^2.

Angle BAC is acute.

Calculate the length of BC. Give your answer to an appropriate degree of accuracy.

$\frac{1}{2} \times 61 \times 76 \times \sin A = 2300$

> Since the third side is unknown, we need to find an angle.
> Find angle BAC (= A) so that we can use the cosine rule.
> Use area = $\frac{1}{2}bc \sin A$.

$\therefore \sin \text{BAC} = \frac{4600}{4636} = 0.9922 \ldots$

> Use area = $\frac{1}{2}bc \sin A$ to set up an equation and solve it to get angle A. You are given that A is acute so there is no problem with any ambiguity.
>
> This gets 1 mark for method for this line.

$\therefore A = 82.86°$

> This gets 1 mark for accuracy for the answer.
> Now use the cosine rule to find BC.

$\text{BC}^2 = 61^2 + 76^2 - 2 \times 61 \times 76 \times \cos 82.9$
$= 8343.75$

> Use the cosine rule to work out the side BC.
> If possible keep values in your calculator display but if you have to write down values then use at least 4 significant figures for trigonometric ratios and at least 1 decimal place for angles. This will avoid any inaccuracy in the final answer.
>
> This gets 1 mark for method for this line.

BC = 91.3 m (3 significant figures)

> The answer gets 1 mark for accuracy.

Total: 4 marks

Worked Examination Questions

PS **3** A **tetrahedron** has one face which is an equilateral triangle of side 6 cm and three faces which are isosceles triangles with sides 6 cm, 9 cm and 9 cm.

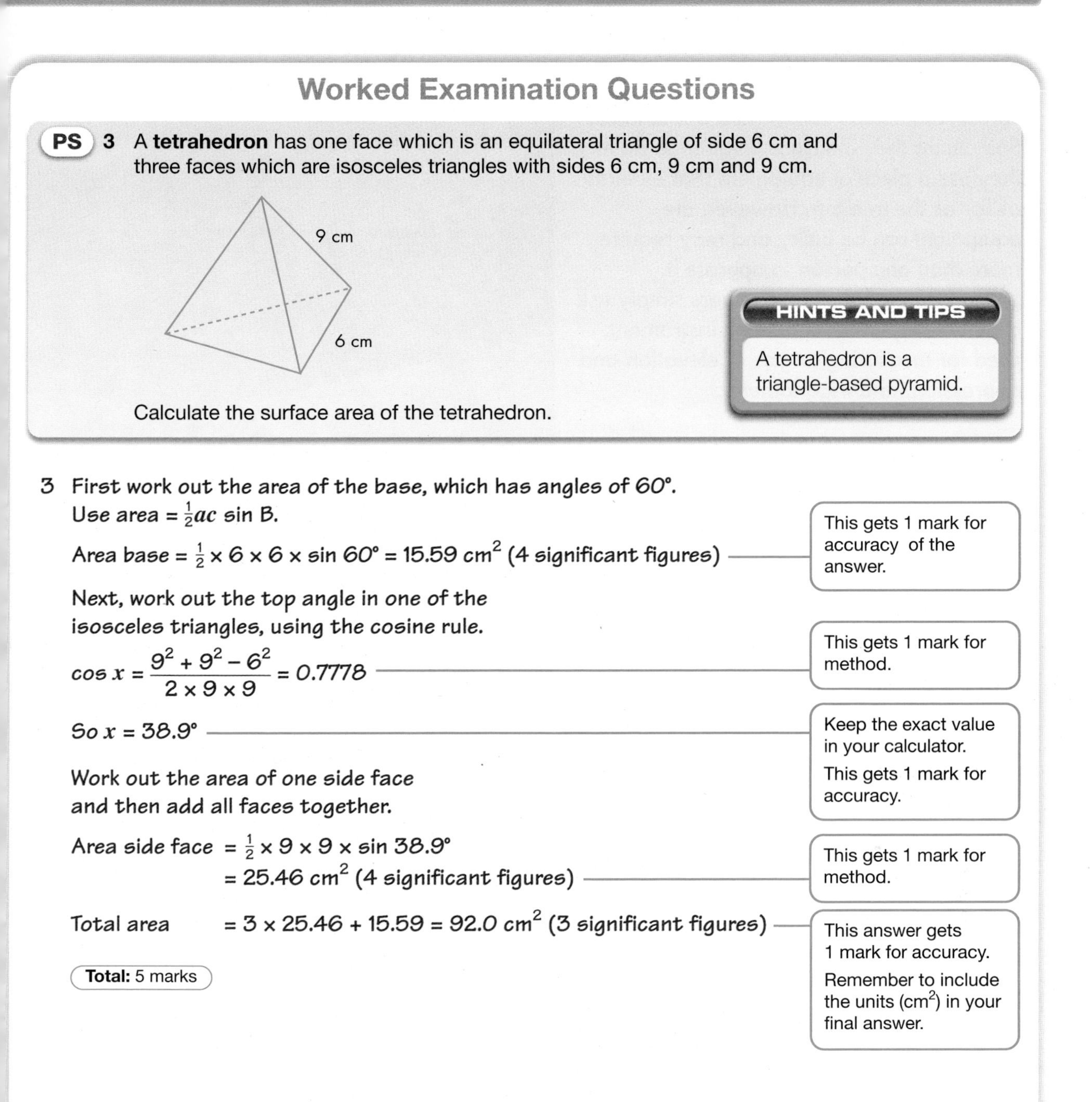

Calculate the surface area of the tetrahedron.

HINTS AND TIPS

A tetrahedron is a triangle-based pyramid.

3 First work out the area of the base, which has angles of 60°.

Use area = $\frac{1}{2}ac \sin B$.

Area base = $\frac{1}{2} \times 6 \times 6 \times \sin 60° = 15.59$ cm^2 (4 significant figures)

> This gets 1 mark for accuracy of the answer.

Next, work out the top angle in one of the isosceles triangles, using the cosine rule.

$$\cos x = \frac{9^2 + 9^2 - 6^2}{2 \times 9 \times 9} = 0.7778$$

> This gets 1 mark for method.

So $x = 38.9°$

> Keep the exact value in your calculator.
>
> This gets 1 mark for accuracy.

Work out the area of one side face and then add all faces together.

Area side face = $\frac{1}{2} \times 9 \times 9 \times \sin 38.9°$

= 25.46 cm^2 (4 significant figures)

> This gets 1 mark for method.

Total area = $3 \times 25.46 + 15.59 = 92.0$ cm^2 (3 significant figures)

> This answer gets 1 mark for accuracy.
>
> Remember to include the units (cm^2) in your final answer.

Total: 5 marks

8 Functional Maths
Building tree-houses

Forest researchers collect data about trees. Sometimes they measure heights. To do this, they use a piece of equipment that extends as far as the tree top. However, the equipment can be bulky and may require more than one person to operate it. Instead, many forest researchers simply use a clinometer, which is a small instrument used for measuring angles of elevation and depression, and trigonometry.

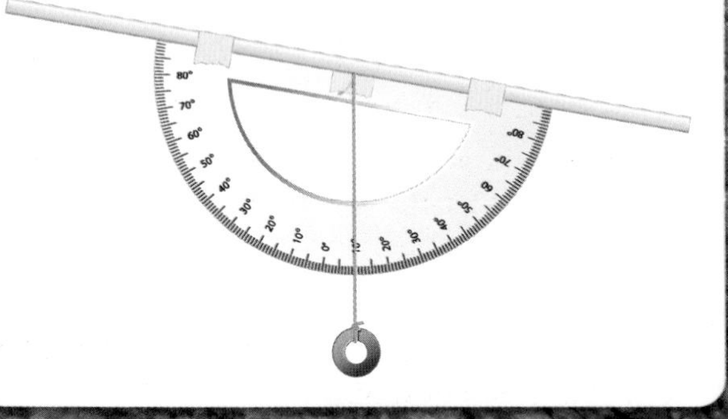

Getting started

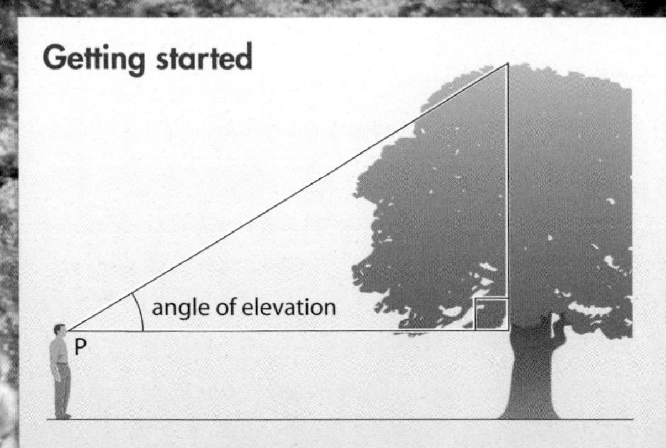

A forest researcher measures the angle of elevation of the top of the tree.

- What other measurement should the researcher find, to help him calculate the height of the tree?
- Which trigonometric ratio would the researcher use?

English oaks			
Tree number	Distance from researcher to base of trunk (m)	Angle of elevation of top of tree	Angle of elevation of lowest branch
001	16.3	45°	15°
002	10	58°	20°
003	24.5	49°	14°
004	15	55°	17°
005	12.4	52°	21°

Your task

The table shows information a researcher collected for the English oaks in one particular forest.

The forest owner would like to install camping tree-houses in three of the English oak trees. She would like each tree-house to be at a different height, with a different incline for each ladder, to appeal to a range of holiday-makers. For safety reasons:

- a tree-house must be no higher than 5.5 m from the ground
- the angle between the tree-house ladder and the ground must be 75° or less.

Look at the table and choose three trees for the tree-houses. Decide on the height above the ground and the length of the ladder required for each tree-house.

Then, write a leaflet advertising your three tree-houses to holiday-makers.

Remember to include information about heights and ladders. What other mathematical information might you include?

Why this chapter matters

A famous saying is 'A picture is worth a thousand words' and graphs in mathematics are worth many lines of algebra as they are a visual way of showing the relationship between two variables. This chapter deals with many differnt types of graph and shows how to solve equations using them.

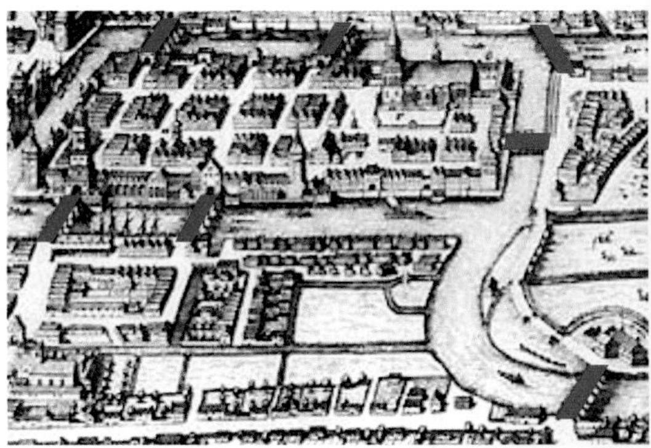

Bridges of Konigsberg

Many years ago the city of Konigsberg (now known as Kalingrad) had seven bridges joining the four separate parts of the city.

The citizens had a challenge to see if anyone could walk across all seven bridges without crossing any of them twice.

The problem is the same as trying to draw this diagram without taking your pencil off the paper.

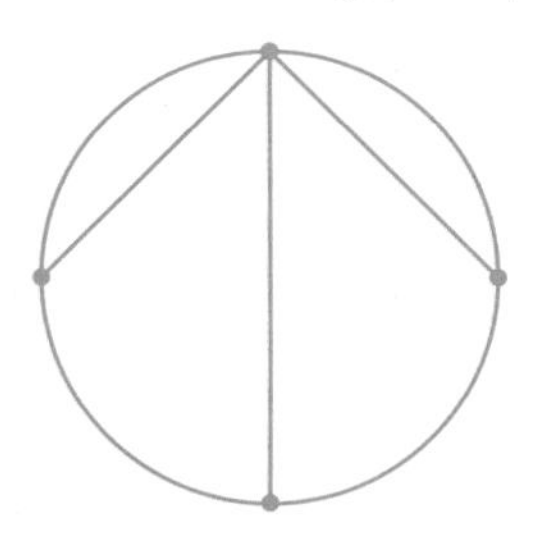

Copy the diagram and see if it can be done.

Like the citizens of Konigsberg and their problem with the bridges, you will find that it is impossible.

This problem was investigated by the Swiss mathematician Leonhard Euler (1707–1783).

Leonard Euler

He proved that it was not possible and started a new branch of mathematics called 'Graph theory'. This is not the same as the study of linear graphs, but is more concerned with where lines meet (vertices) and the lines joining them (arcs).

Copy the following diagrams and see if you can draw any of them without taking your pencil off the paper.

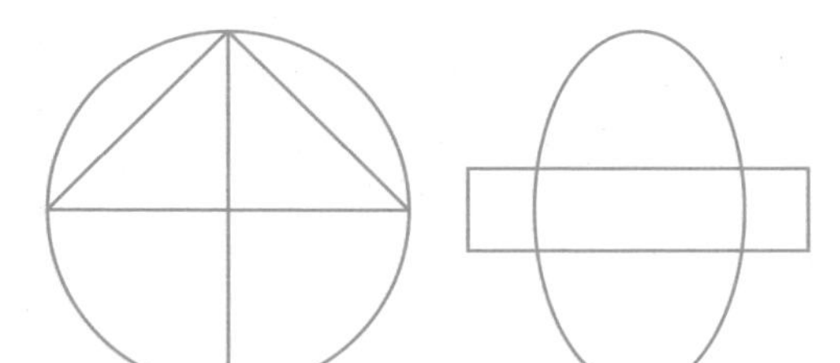

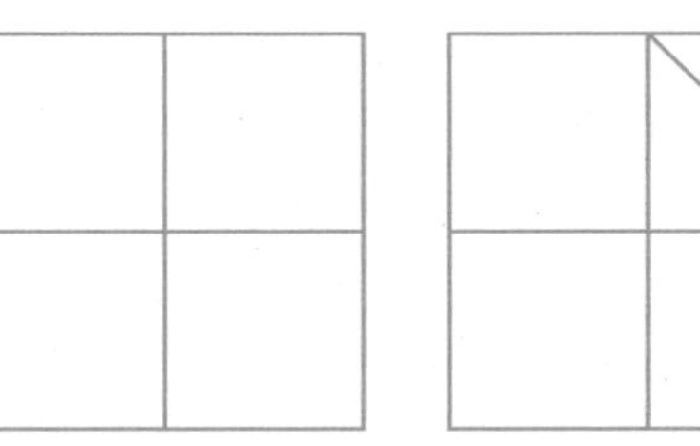

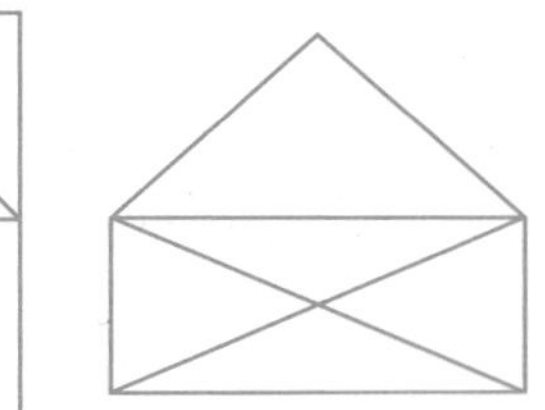

Layout of Kalingrad

In the Second World War, Konigsberg was badly damaged and many of the bridges were destroyed. On the left is a rough layout of Kalingrad (Konigsberg) as it is today.

Can you trace a route, starting from the island, that crosses all the remaining bridges without crossing a bridge twice?

Chapter

Algebra: Graphs and equations

1 Uses of graphs

2 Parallel and perpendicular lines

3 Other graphs

4 The circular function graphs

5 Solving equations, one linear and one non-linear, with graphs

6 Solving equations by the method of intersection

The grades given in this chapter are target grades.

This chapter will show you ...

- **B** how to use graphs to find formulae and solve simultaneous linear equations
- **B to A*** how to draw linear graphs parallel or perpendicular to other lines
- **B to A*** how to recognise and draw cubic, reciprocal and exponential graphs
- **A to A*** how to use graphs to find the solutions to a pair of simultaneous equations, one linear and one non-linear
- **A*** how to use the sine and cosine graphs to find angles with the same sine and cosine between 0° and 360°
- **A*** how to use the method of intersection to solve one quadratic equation, using the graph of another quadratic equation and an appropriate straight line

Visual overview

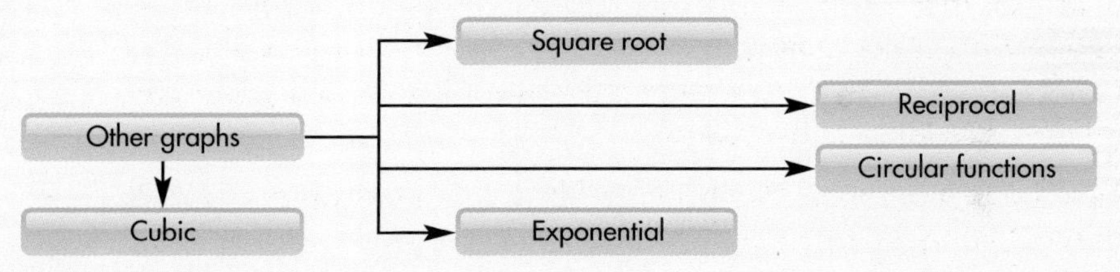

What you should already know

- How to draw linear graphs **(KS3 level 6, GCSE grade D)**
- How to find the equation of a graph, using the gradient-intercept method **(KS3 level 8, GCSE grade B)**

Quick check

1 Draw the graph of $y = 3x - 1$ for values of x from -2 to $+3$.

2 Give the equation of the graph shown.

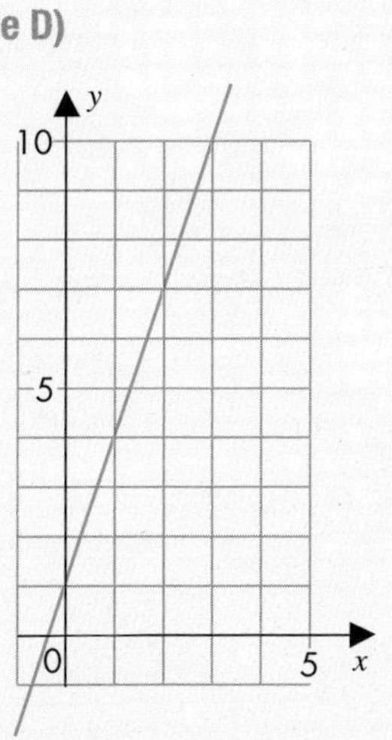

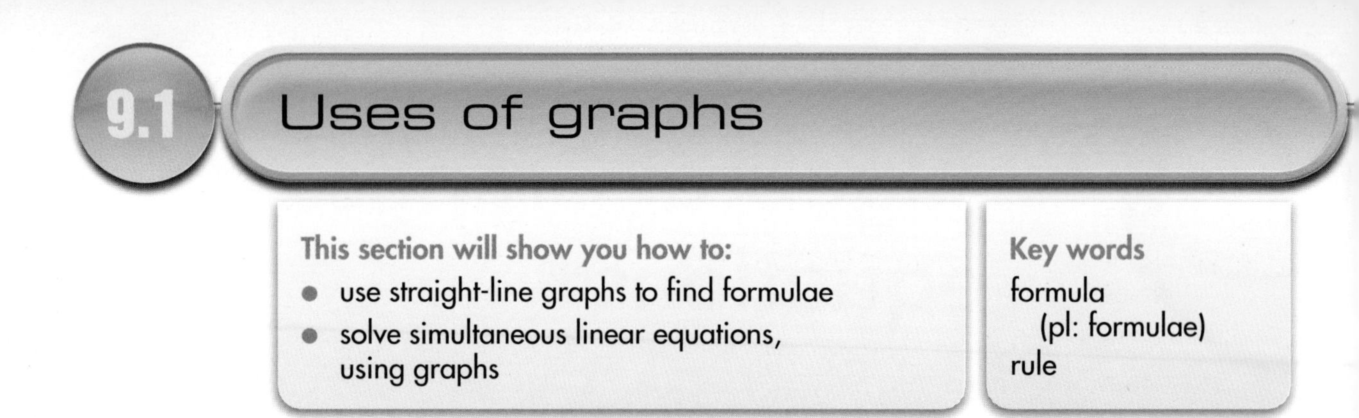

9.1 Uses of graphs

This section will show you how to:
- use straight-line graphs to find formulae
- solve simultaneous linear equations, using graphs

Key words
formula (pl: formulae)
rule

Two uses of graphs that you will now consider are finding **formulae** and solving simultaneous equations. Solving other equations by graphical methods is covered later in this chapter.

Finding formulae or rules

EXAMPLE 1

A taxi fare will cost more, the further you go.
The graph below illustrates the fares in one part of England.

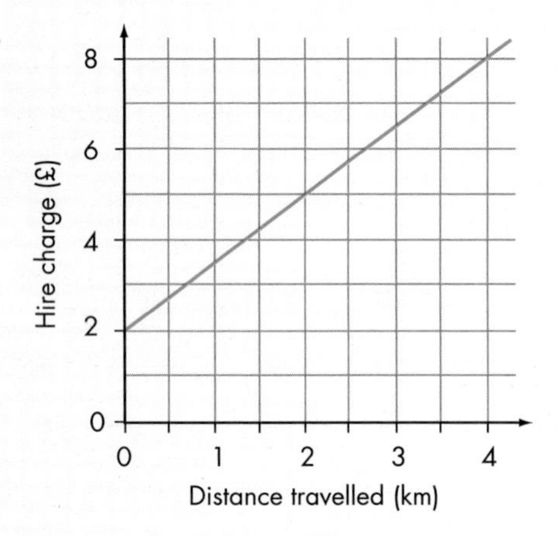

The taxi company charges a basic hire fee to start with of £2.00. This is shown on the graph as the point where the line cuts through the hire-charge axis (when distance travelled is 0).

The gradient of the line is:

$$\frac{8-2}{4} = \frac{6}{4} = 1.5$$

This represents the hire charge per kilometre travelled.

So the total hire charge is made up of two parts: a basic hire charge of £2.00 and an additional charge of £1.50 per kilometre travelled. This can be put in a formula as

hire charge = £2.00 + £1.50 per kilometre

In this example, £2.00 is the constant term in the formula (the equation of the graph).

EXERCISE 9A

FM 1 This graph is a conversion graph between temperatures in °C and °F.

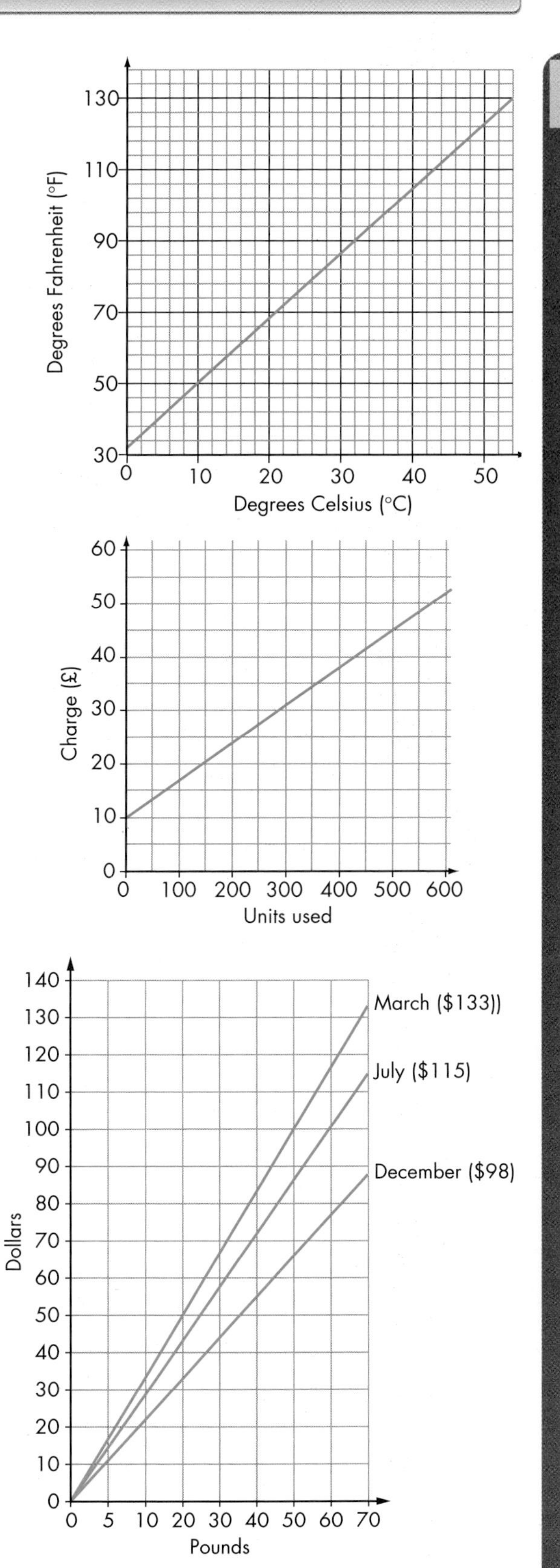

a What temperature in °F is equivalent to a temperature of 0 °C?

b What is the gradient of the line?

c From your answers to parts **a** and **b**, write down a **rule** that can be used to convert °C to °F.

FM 2 This graph illustrates charges for fuel.

a What is the gradient of the line?

b The standing charge is the basic charge before the cost per unit is added. What is the standing charge?

c Write down the rule used to work out the total charge for different amounts of units used.

FM 3 In 2008, the exchange rate between the American dollar and the British pound changed.

The graph shows the exchange rate for three different months of the year.

a If Mr Bush changed £1000 into dollars in March and another £1000 into dollars in December, approximately how much less did he get in December than in March?

b George went to America in March and stayed until July. In March, he changed £5000 into dollars. In July, he still had $2000 dollars left and he changed them back into pounds.

i How much, in dollars, did George spend between March and July?

ii How much, in pounds, did George actually spend between March and July?

B

B

AU 4 This graph is a sketch of the rate charged for taxi journeys by a firm during weekdays from 6 am to 8 pm.

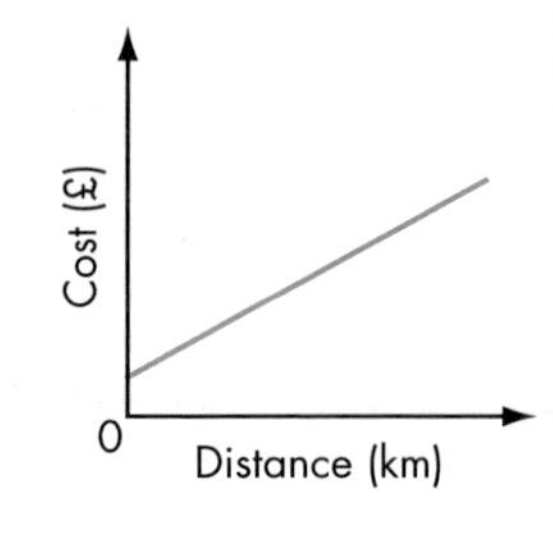

a At weekends from 6 am to 8 am, the company charges the same rate per kilometre but increases the basic charge.

Sketch a graph to show this. Mark it with A.

b During weekdays from 8 pm to 6 am, the company charges the same basic charge but an increased charge per kilometre.

On the same axes as in part **a**, sketch a graph to show this. Mark it with B.

c During weekends from 8 pm to 6 am, the company increases the basic charge and increases the charge per kilometre.

On the same axes as in part **a**, sketch a graph to show this. Mark it with C.

PS 5 A motorcycle courier will deliver packages, up to a weight of 22 pounds, within a city centre.

The courier has three charging bands: packages up to 5 pounds, packages from 5 to 12 pounds and packages over 12 pounds.

This graph shows how much he charges.

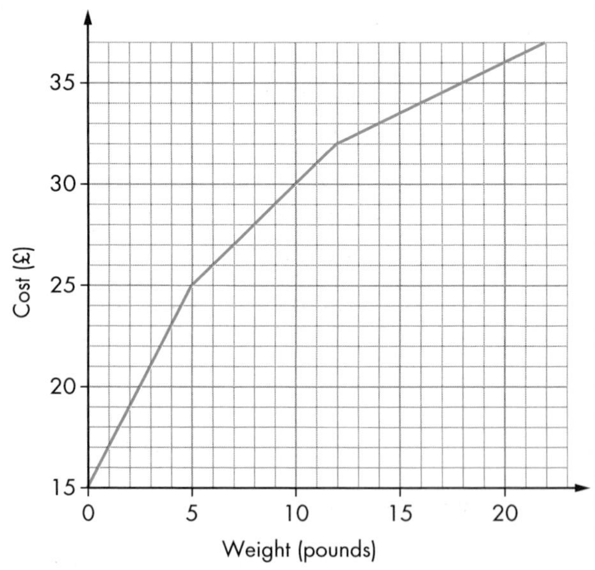

Work out the values of a, b, c, d, e and f to show his charges as equations.

$y = ax + b \qquad 0 < x \leqslant 5$

$y = cx + d \qquad 5 < x \leqslant 12$

$y = ex + f \qquad 12 < x \leqslant 25$

FM 6 This graph shows the hire charge for heaters over a number of days.

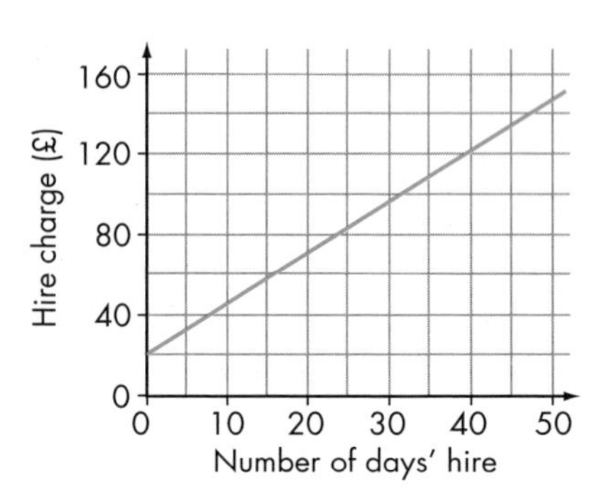

a Calculate the gradient of the line.

b What is the basic charge before the daily hire charge is added on?

c Write down the rule used to work out the total hire charge.

FM 7 This graph shows the hire charge for a conference centre, depending on the number of people at the conference.

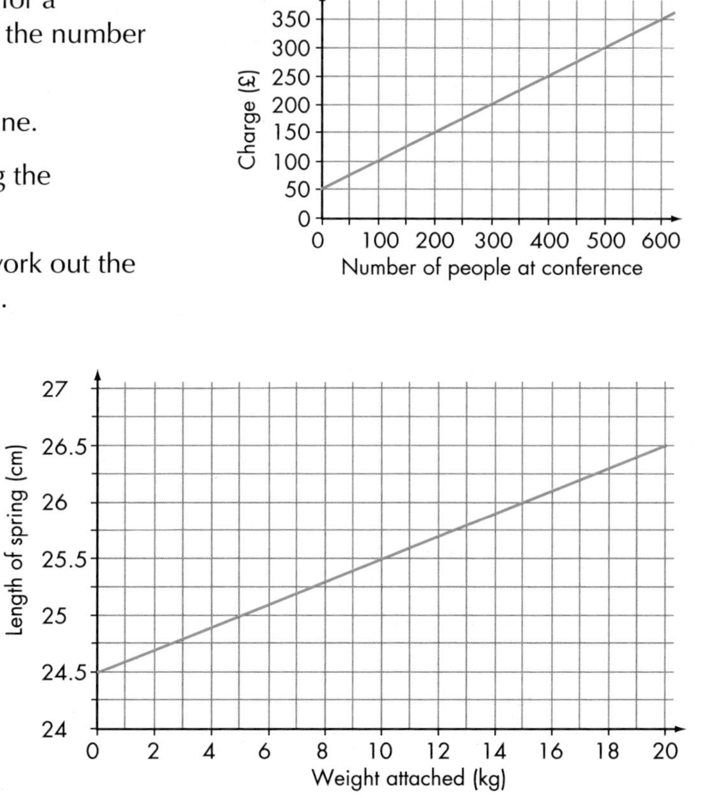

a Calculate the gradient of the line.

b What is the basic fee for hiring the conference centre?

c Write down the rule used to work out the total hire charge for the centre.

FM 8 This graph shows the length of a spring when different weights are attached to it.

a Calculate the gradient of the line.

b How long is the spring when no weight is attached to it?

c By how much does the spring extend per kilogram?

d Write down the rule for finding the length of the spring for different weights.

Solving simultaneous equations

EXAMPLE 2

By drawing their graphs on the same grid, find the solution of these simultaneous equations.

a $3x + y = 6$ **b** $y = 4x - 1$

a The first graph is drawn using the cover-up method. It crosses the x-axis at (2, 0) and the y-axis at (0, 6).

b This graph can be drawn by finding some points or by the gradient-intercept method. If you use the gradient-intercept method, you find the graph crosses the y-axis at –1 and has a gradient of 4.

The point where the graphs intersect is (1, 3). So the solution to the simultaneous equations is $x = 1$, $y = 3$.

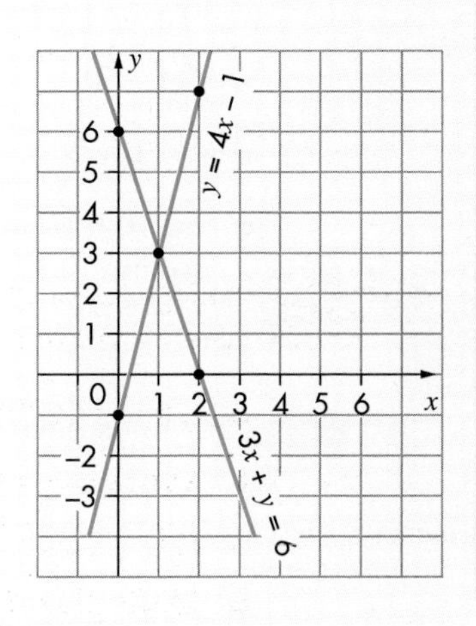

EXERCISE 9B

In questions **1–12**, draw the graphs to find the solution of each pair of simultaneous equations.

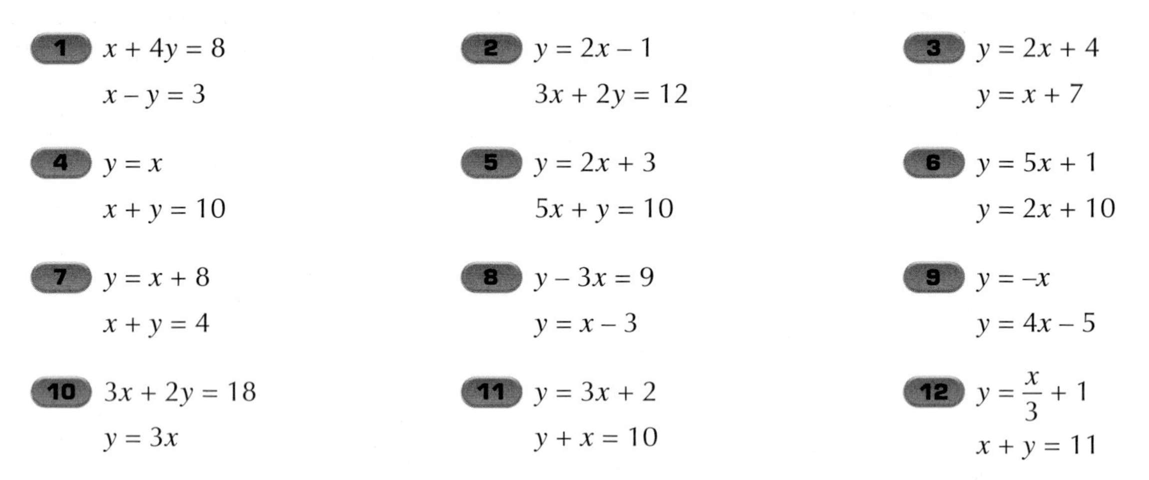

1 $x + 4y = 8$
$x - y = 3$

2 $y = 2x - 1$
$3x + 2y = 12$

3 $y = 2x + 4$
$y = x + 7$

4 $y = x$
$x + y = 10$

5 $y = 2x + 3$
$5x + y = 10$

6 $y = 5x + 1$
$y = 2x + 10$

7 $y = x + 8$
$x + y = 4$

8 $y - 3x = 9$
$y = x - 3$

9 $y = -x$
$y = 4x - 5$

10 $3x + 2y = 18$
$y = 3x$

11 $y = 3x + 2$
$y + x = 10$

12 $y = \frac{x}{3} + 1$
$x + y = 11$

13 One cheesecake and two chocolate gateaux cost £9.50.

Two cheesecakes and one chocolate gateau cost £8.50.

Using x to represent the price of a cheesecake and y to represent the cost of a gateau, set up a pair of simultaneous equations.

On a set of axes with the x-axis numbered from 0 to 20 and the y-axis numbered from 0 to 10, draw the graphs of the two equations.

Use the graphs to write down the cost of a cheesecake and the cost of a gateau.

PS 14 The graph shows four lines.

P: $y = 4x + 1$ Q: $y = 2x + 2$

R: $y = x - 2$ S: $x + y + 1 = 0$

Which pairs of lines intersect at the following points?

a $(-1, -3)$ **b** $(\frac{1}{2}, -1\frac{1}{2})$

c $(\frac{1}{2}, 3)$ **d** $(-1, 0)$

e Solve the simultaneous equations P and S to find the exact solution.

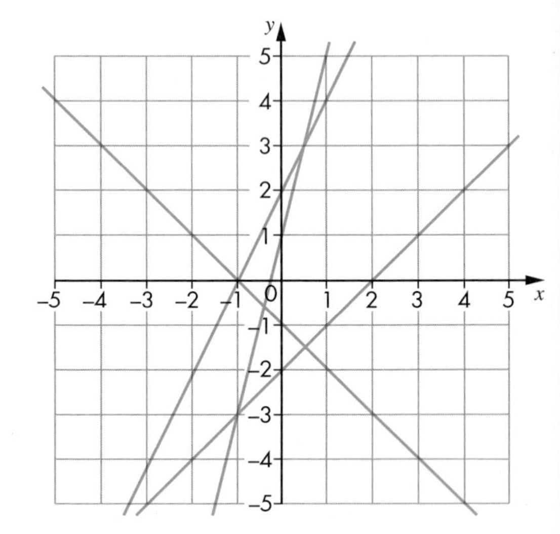

AU 15 Four lines have the following equations.

A: $y = x$ B: $y = 2$

C: $x = -3$ D: $y = -x$

These lines intersect at six different points.

Without drawing the lines accurately, write down the coordinates of the six intersection points.

9.2 Parallel and perpendicular lines

This section will show you how to:

- draw linear graphs parallel or perpendicular to other lines and passing through a specific point

Key words
negative reciprocal
parallel
perpendicular

EXAMPLE 3

In each of these grids, there are two lines.

a

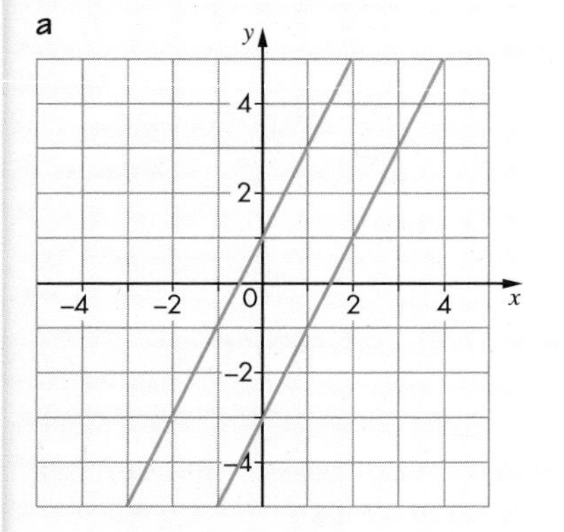

b

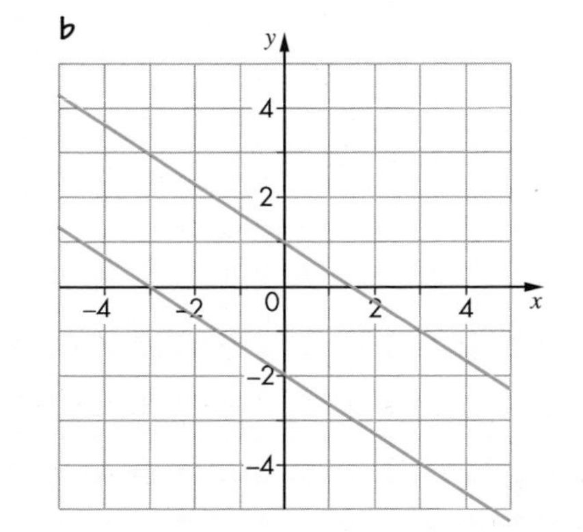

c

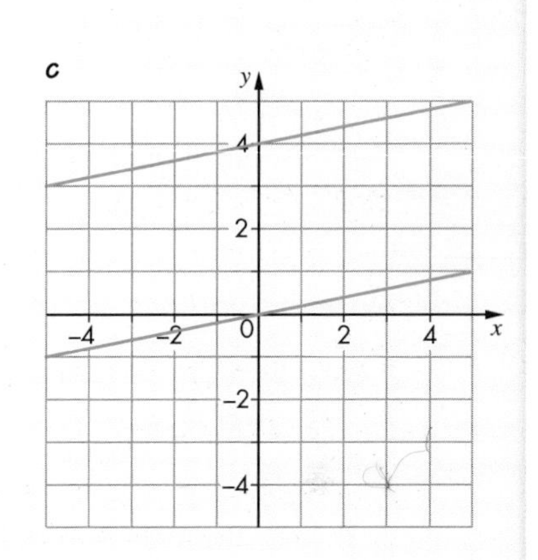

For each grid:

i find the equation of each line

ii describe the geometrical relationship between the lines

iii describe the numerical relationships between their gradients.

i Grid **a**: the lines have equations $y = 2x + 1$, $y = 2x - 3$

Grid **b**: the lines have equations $y = -\frac{2}{3}x + 1$, $y = -\frac{2}{3}x - 2$

Grid **c**: the lines have equations $y = \frac{1}{5}x$, $y = \frac{1}{5}x + 4$

ii In each case, the lines are parallel.

iii In each case, the gradients are equal.

EXAMPLE 4

In each of these grids, there are two lines.

a

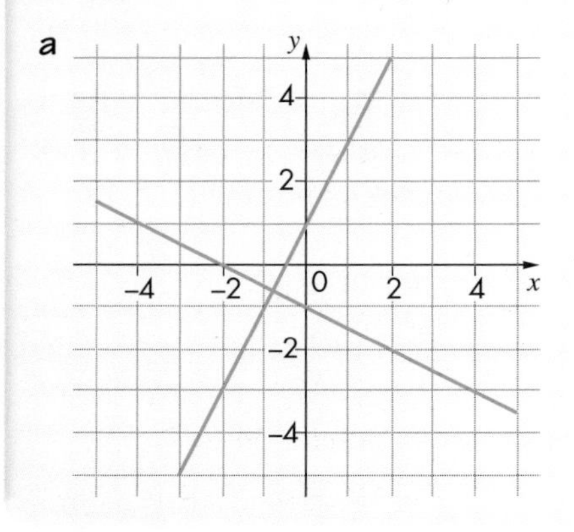

b

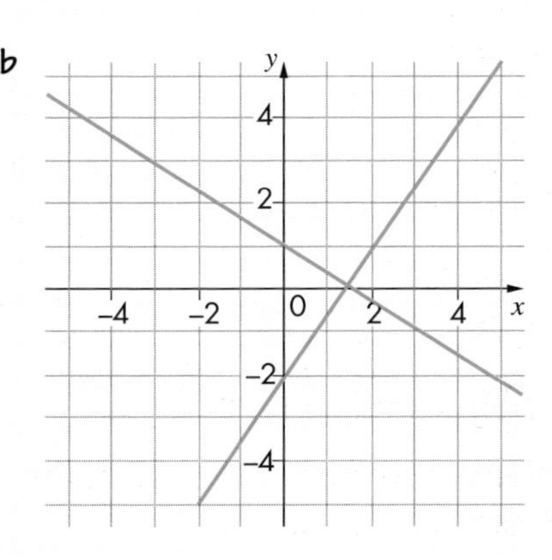

c

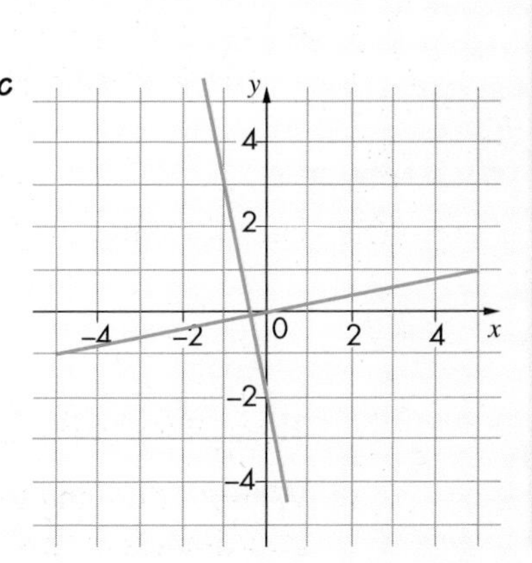

EXAMPLE 4 continued

For each grid:

i find the equation of each line

ii describe the geometrical relationship between the lines

iii describe the numerical relationships between their gradients.

i Grid a: the lines have equations $y = 2x + 1$, $y = -\frac{1}{2}x - 1$

Grid b: the lines have equations $y = \frac{3}{2}x - 2$, $y = -\frac{2}{3}x + 1$

Grid c: the lines have equations $y = \frac{1}{5}x$, $y = -5x - 2$

ii In each case the lines are perpendicular (at right angles).

iii In each case the gradients are reciprocals of each other but with different signs.

Note: If two lines are **parallel**, then their gradients are equal.

If two lines are **perpendicular**, their gradients are **negative reciprocals** of each other.

EXAMPLE 5

Two points A and B are A (0, 1) and B (2, 4).

a Work out the equation of the line AB.

b Write down the equation of the line parallel to AB and passing through the point (0, 5).

c Write down the gradient of a line perpendicular to AB.

d Write down the equation of a line perpendicular to AB and passing through the point (0, 2).

a The gradient of AB is $\frac{3}{2}$ and passes through (0, 1) so the equation is $y = \frac{3}{2}x + 1$.

b The gradient is the same and the intercept is (0, 5) so the equation is $y = \frac{3}{2}x + 5$.

c The perpendicular gradient is the negative reciprocal $-\frac{2}{3}$.

d The gradient is $-\frac{2}{3}$ and the intercept is (0, 2) so the equation is $y = -\frac{2}{3}x + 2$.

EXAMPLE 6

Find the line that is perpendicular to the line $y = \frac{1}{2}x - 3$ and passes through (0, 5).

The gradient of the new line will be the negative reciprocal of $\frac{1}{2}$ which is -2.

The point (0, 5) is the intercept on the y-axis so the equation of the line is:
$y = -2x + 5$

EXAMPLE 7

The point A is (2, 1) and the point B is (4, 4).

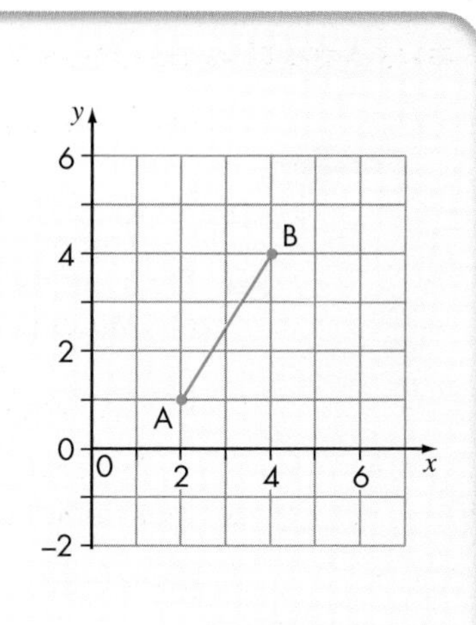

a Find the equation of the line parallel to AB and passing through (6, 11).

b Find the equation of the line parallel to AB and passing through (8, 0).

a The gradient of AB is $\frac{3}{2}$, so the new equation is of the form $y = \frac{3}{2}x + c$

The new line passes through (6, 11), so $11 = \frac{3}{2} \times 6 + c$

$\Rightarrow c = 2$

Hence the new line is $y = \frac{3}{2}x + 2$

b The gradient of AB is $\frac{3}{2}$, so the new equation is of the form $y = \frac{3}{2}x + c$.

The new line passes through (8, 0), so $0 = \frac{3}{2} \times 8 + c$

$\Rightarrow c = -12$

Hence the new line is $y = \frac{3}{2}x - 12$

EXAMPLE 8

The point A is (2, –1) and the point B is (4, 5).

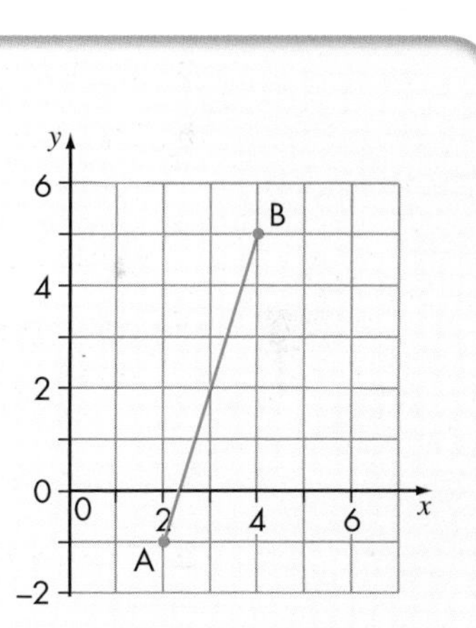

a Find the equation of the line parallel to AB and passing through (2, 8).

b Find the equation of the line perpendicular to the midpoint of AB.

a The gradient of AB is 3, so the new equation is of the form

$y = 3x + c$

The new line passes through (2, 8), so $8 = 3 \times 2 + c$

$\Rightarrow c = 2$

Hence the line is $y = 3x + 2$

b The midpoint of AB is (3, 2).

The gradient of the perpendicular line is the negative reciprocal of 3, which is $-\frac{1}{3}$.

We could find c as in part **a** but we can also do a sketch on the grid. This will show that the perpendicular line passes through (0, 3).

Hence the equation of the line is $y = -\frac{1}{3}x + 3$

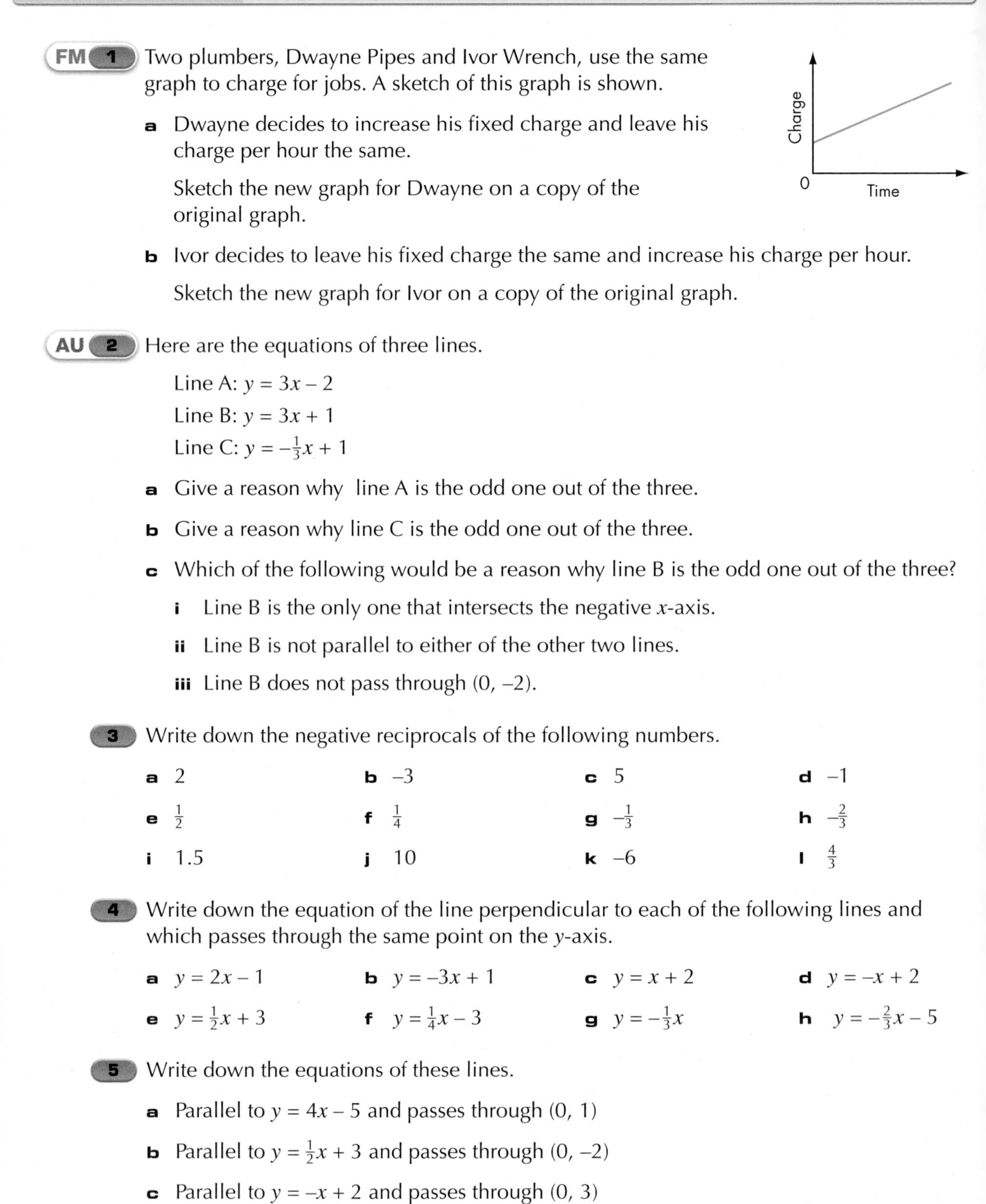

EXERCISE 9C

B

FM 1 Two plumbers, Dwayne Pipes and Ivor Wrench, use the same graph to charge for jobs. A sketch of this graph is shown.

a Dwayne decides to increase his fixed charge and leave his charge per hour the same.

Sketch the new graph for Dwayne on a copy of the original graph.

b Ivor decides to leave his fixed charge the same and increase his charge per hour.

Sketch the new graph for Ivor on a copy of the original graph.

A

AU 2 Here are the equations of three lines.

Line A: $y = 3x - 2$

Line B: $y = 3x + 1$

Line C: $y = -\frac{1}{3}x + 1$

a Give a reason why line A is the odd one out of the three.

b Give a reason why line C is the odd one out of the three.

c Which of the following would be a reason why line B is the odd one out of the three?

i Line B is the only one that intersects the negative x-axis.

ii Line B is not parallel to either of the other two lines.

iii Line B does not pass through $(0, -2)$.

3 Write down the negative reciprocals of the following numbers.

a 2 **b** -3 **c** 5 **d** -1

e $\frac{1}{2}$ **f** $\frac{1}{4}$ **g** $-\frac{1}{3}$ **h** $-\frac{2}{3}$

i 1.5 **j** 10 **k** -6 **l** $\frac{4}{3}$

4 Write down the equation of the line perpendicular to each of the following lines and which passes through the same point on the y-axis.

a $y = 2x - 1$ **b** $y = -3x + 1$ **c** $y = x + 2$ **d** $y = -x + 2$

e $y = \frac{1}{2}x + 3$ **f** $y = \frac{1}{4}x - 3$ **g** $y = -\frac{1}{3}x$ **h** $y = -\frac{2}{3}x - 5$

5 Write down the equations of these lines.

a Parallel to $y = 4x - 5$ and passes through $(0, 1)$

b Parallel to $y = \frac{1}{2}x + 3$ and passes through $(0, -2)$

c Parallel to $y = -x + 2$ and passes through $(0, 3)$

6 Write down the equations of these lines.

a Perpendicular to $y = 3x + 2$ and passes through (0, –1)

b Perpendicular to $y = -\frac{1}{3}x - 2$ and passes through (0, 5)

c Perpendicular to $y = x - 5$ and passes through (0, 1)

7 A is the point (1, 5). B is the point (3, 3).

a Find the equation of the line parallel to AB and passing through (5, 9).

b Find the equation of the line perpendicular to AB and passing through the midpoint of AB.

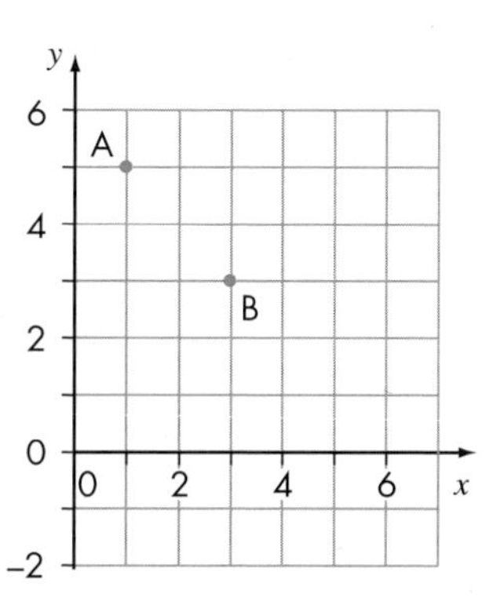

8 Find the equation of the line that passes through the midpoint of AB, where A is (–5, –3) and B is (–1, 3), and has a gradient of 2.

9 Find the equation of the line perpendicular to $y = 4x - 3$, passing though (–4, 3).

10 A is the point (0, 6), B is the point (5, 5) and C is the point (4, 0).

a Write down the point where the line BC intercepts the y-axis.

b Work out the equation of the line AB.

c Write down the equation of the line BC.

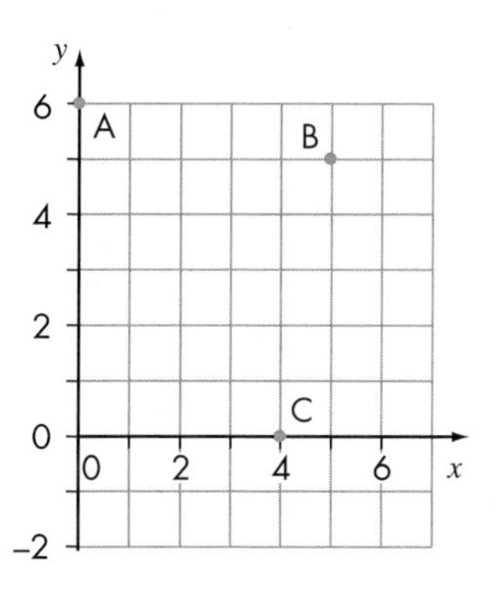

PS 11 Find the equation of the perpendicular bisector of the points A (1, 2) and B (3, 6).

12 A is the point (0, 4), B is the point (4, 6) and C is the point (2, 0).

a Find the equation of the line BC.

b Show that the point of intersection of the perpendicular bisectors of AB and AC is (3, 3).

c Show algebraically that this point lies on the line BC.

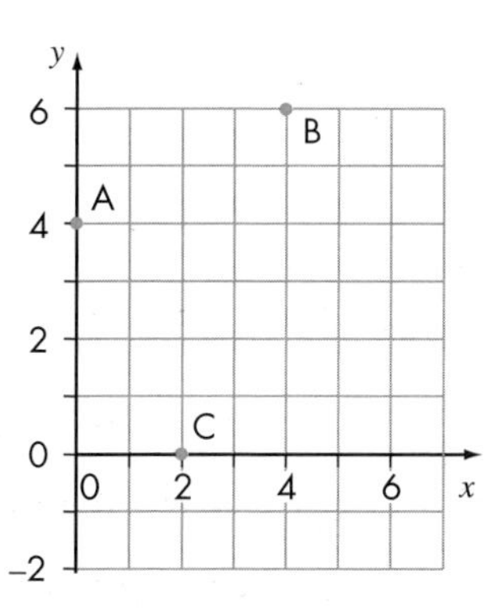

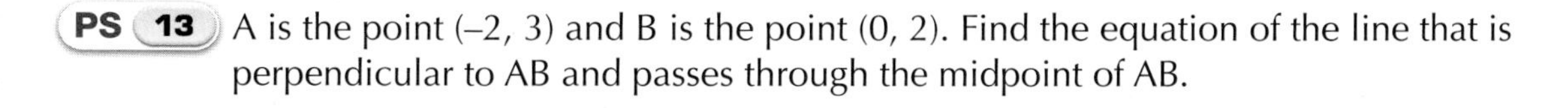
PS 13 A is the point (–2, 3) and B is the point (0, 2). Find the equation of the line that is perpendicular to AB and passes through the midpoint of AB.

9.3 Other graphs

This section will show you how to:

- recognise and plot cubic, exponential and reciprocal graphs

Key words
asymptote
cubic
exponential function
reciprocal

Cubic graphs

A **cubic** function or graph is one that contains a term in x^3. The following are examples of cubic graphs.

$y = x^3$ $\qquad$ $y = x^3 + 3x$ $\qquad$ $y = x^3 + x^2 + x + 1$

The techniques used to draw them are exactly the same as those for quadratic and reciprocal graphs.

Questions requiring an accurate drawing of a cubic graph are not very common in examinations, but questions asking if you can recognise a cubic graph occur quite often.

This is the basic graph $y = x^3$.

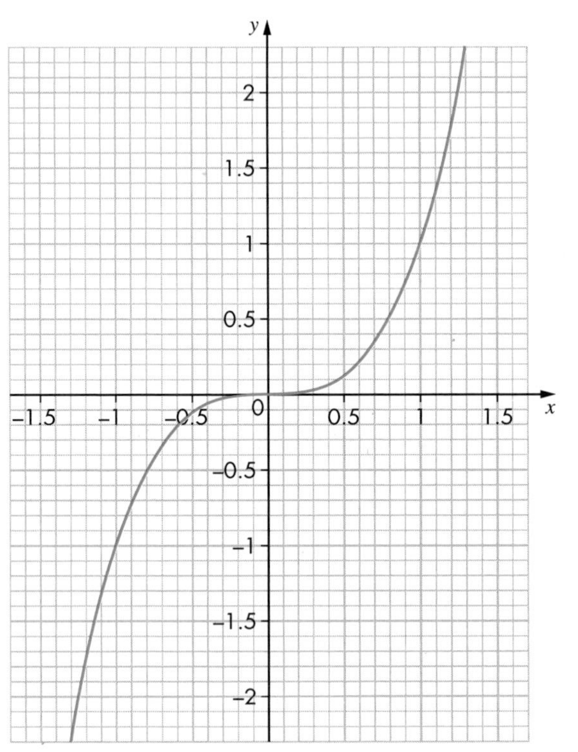

It has a characteristic shape which you should learn to recognise.

Example 9 shows you how to draw a cubic graph accurately.

You should use a calculator to work out the value of y and round to 1 or 2 decimal places.

EXAMPLE 9

a Complete the table to draw the graph of $y = x^3 - x^2 - 4x + 4$ for $-3 \leqslant x \leqslant 3$.

x	−3	−2.5	−2	−1.5	−1	−0.5	0	0.5	1	1.5	2	2.5	3
y	−20.00		0.00		6.00		4.00	1.88				3.38	10.00

b Use your graph to give the roots of the equation $x^3 - x^2 - 4x + 4 = 0$.

c Write down the coordinates of:

i the minimum vertex **ii** the maximum vertex.

d Write down the coordinates of the point where the graph intersects the y-axis.

a The completed table (to 2 decimal places) is given below and the graph is shown below, right.

x	−3	−2.5	−2	−1.5	−1	−0.5	0	0.5	1	1.5	2	2.5	3
y	−20.00	−7.88	0.00	4.38	6.00	5.63	4.00	1.88	0.00	−0.88	0.00	3.38	10.00

b Just as in quadratic graphs, the roots are the points where the graph crosses the x-axis.

So the roots are $x = -2$, 1 and 2.

Note that, in the table, these are the x-values where the y-value is 0.

c **i** The minimum vertex is at the point (1.5, −0.88).

ii The maximum vertex is at the point (−1, 6).

Note that the minimum and maximum values of the function are ± infinity, as the arms of the curve continue forever.

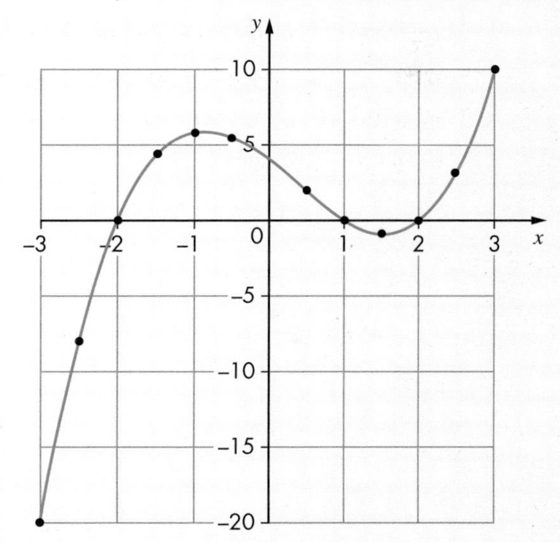

d Just as in the quadratic, this is the constant term in the equation, so the point is (0, 4).

Reciprocal graphs

A **reciprocal** equation has the form $y = \frac{a}{x}$.

Examples of reciprocal equations are:

$$y = \frac{1}{x} \qquad y = \frac{4}{x} \qquad y = -\frac{3}{x}$$

All reciprocal graphs have a similar shape and some symmetry properties.

EXAMPLE 10

Complete the table to draw the graph of $y = \frac{1}{x}$ for $-4 \leqslant x \leqslant 4$.

x	−4	−3	−2	−1	1	2	3	4
y								

Values are rounded to two decimal places, as it is unlikely that you could plot a value more accurately than this. The completed table looks like this.

x	−4	−3	−2	−1	1	2	3	4
y	−0.25	−0.33	−0.5	−1	1	0.5	0.33	0.25

The graph plotted from these values is shown in **A**. This is not much of a graph and does not show the properties of the reciprocal function. If you take x-values from −0.8 to 0.8 in steps of 0.2, you get the next table.

Note that you cannot use $x = 0$ since $\frac{1}{0}$ is infinity.

x	−0.8	−0.6	−0.4	−0.2	0.2	0.4	0.6	0.8
y	−1.25	−1.67	−2.5	−5	5	2.5	1.67	1.25

Plotting these points as well gives the graph in **B**.

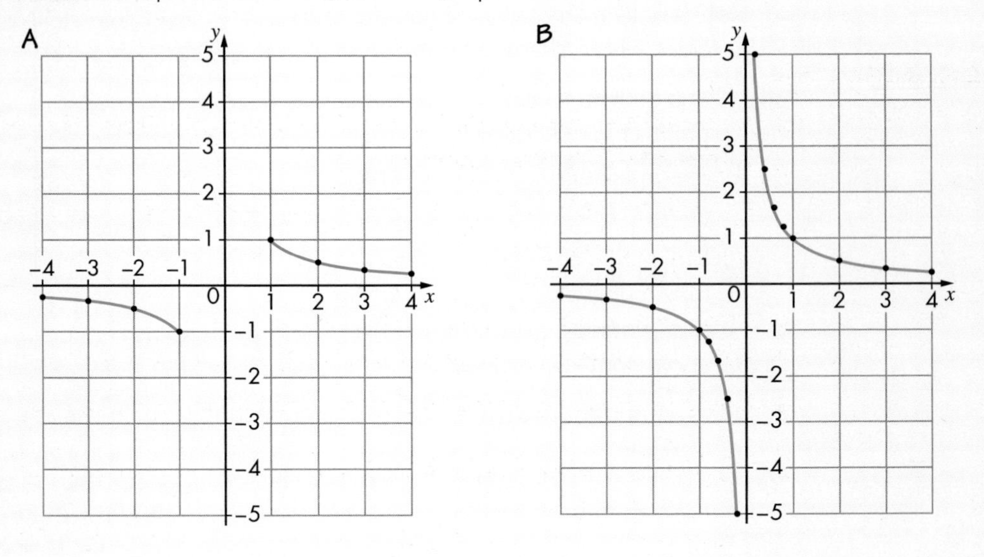

From the graph in **B**, the following properties can be seen.

- The lines $y = x$ and $y = -x$ are lines of symmetry.
- The closer x gets to zero, the nearer the graph gets to the y-axis.
- As x increases, the graph gets closer to the x-axis.

The graph never actually touches the axes, it just gets closer and closer to them. A line to which a graph gets closer but never touches or crosses is an **asymptote**.

These properties are true for *all reciprocal graphs*.

Exponential graphs

Equations that have the form $y = k^x$, where k is a positive number, are called **exponential functions**.

Exponential functions share the following properties.

- When k is greater than 1, the value of y increases steeply as x increases, which you can see from the graph on the right.
- Also when k is greater than 1, as x takes on increasingly large negative values, the closer y gets to zero, and so the graph gets nearer and nearer to the negative x-axis. y never actually becomes zero and so the graph never touches the negative x-axis. That is, the negative x-axis is an asymptote to the graph. (See also previous page.)
- Whatever the value of k, the graph always intercepts the y-axis at 1, because here $y = k^0$.
- The reciprocal graph, $y = k^{-x}$, is the reflection in the y-axis of the graph of $y = k^x$, as you can see from the graph on the right.

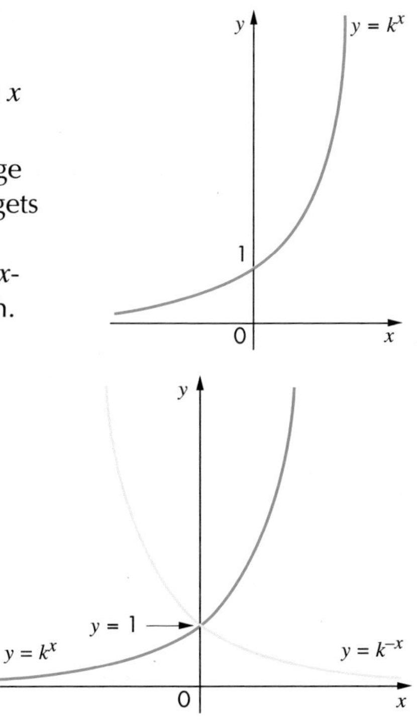

EXAMPLE 11

a Complete the table below for $y = 2^x$ for values of x from –5 to +5. (Values are rounded to 2 decimal places.)

x	–5	–4	–3	–2	–1	0	1	2	3	4	5
$y = 2^x$	0.03	0.06	0.13			1	2	4			32

b Plot the graph of $y = 2^x$ for $-5 \leq x \leq 5$.

c Use your graph to estimate the value of y when $x = 2.5$.

d Use your graph to estimate the value of x when $y = 0.75$.

a The values missing from the table are:

0.25, 0.5, 8 and 16

b Part of the graph (drawn to scale) is shown on the right.

c Draw a line vertically from $x = 2.5$ until it meets the graph and then read across.
The y-value is 5.7.

d Draw a line horizontally from $y = 0.75$, the x-value is –0.4.

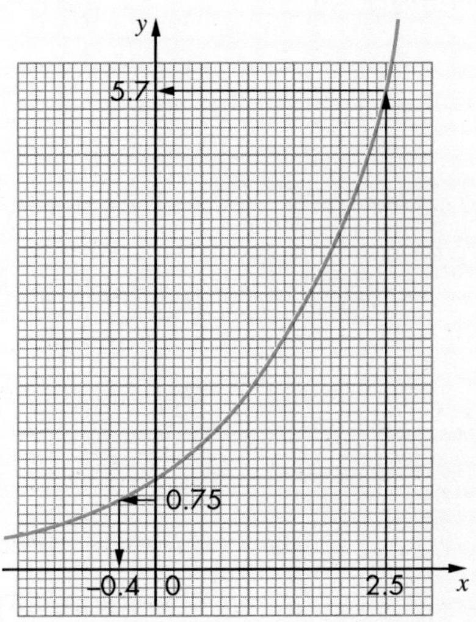

EXERCISE 9D

B

1 Sketch the graph of $y = -x^3$.

2 **a** Copy and complete the table to draw the graph of $y = 2x^3$ for $-3 \leqslant x \leqslant 3$.

x	−3	−2.5	−2	−1.5	−1	−0.5	0	0.5	1	1.5	2	2.5	3
y		−31.25		−6.75			0.00	0.25			16.00		

b Use your graph to find the y-value for an x-value of 2.7.

3 **a** Copy and complete the table to draw the graph of $y = x^3 + 3$ for $-3 \leqslant x \leqslant 3$.

x	−3	−2.5	−2	−1.5	−1	−0.5	0	0.5	1	1.5	2	2.5	3
y	−24.00	−12.63			2.00		3.00	3.13			11.00		30.00

b Use your graph to find the y-value for an x-value of 1.2.

c Find the root of the equation $x^3 + 3 = 0$.

A

4 **a** Copy and complete the table to draw the graph of $y = x^3 - 2x + 5$ for $-3 \leqslant x \leqslant 3$.

x	−3	−2.5	−2	−1.5	−1	−0.5	0	0.5	1	1.5	2	2.5	3
y	−16.00		1.00	4.63			5.00	4.13				15.63	

b Use the graph to find:

i the root of $x^3 - 2x + 5 = 0$

ii the approximate value of the coordinate of the maximum vertex

iii the approximate value of the coordinate of the minimum vertex

iv the coordinates of the point where the graph crosses the y-axis.

5 **a** Copy and complete the table to draw the graph of $y = \frac{2}{x}$ for $-4 \leqslant x \leqslant 4$.

x	0.2	0.4	0.5	0.8	1	1.5	2	3	4
y	10		4	2.5			1		0.5

b Use your graph to find:

i the y-value when $x = 2.5$ **ii** the x-value when $y = -1.25$.

6 **a** Copy and complete the table to draw the graph of $y^2 = 25x$ for $0 \leqslant x \leqslant 5$.

x	0	1	2	3	4	5
$\sqrt{x}$					2 and −2	
$y = 5\sqrt{x}$					10 and −10	

b Use your graph to find:

i the values of y when $x = 3.5$ **ii** the value of x when $y = 8$.

A

7 **a** Copy and complete the table to draw the graph of $y = \frac{5}{x}$ for $-20 \leqslant x \leqslant 20$.

x	0.2	0.4	0.5	1	2	5	10	15	20
y	25		10						0.25

b On the same axes, draw the line $y = x + 10$.

c Use your graph to find the x-values of the points where the graphs cross.

8 **a** Complete the table below for $y = 3^x$ for values of x from –4 to +3. (Values are rounded to 2 decimal places.)

x	–4	–3	–2	–1	0	1	2	3
$y = 3^x$	0.01	0.04			1	3		

b Plot the graph of $y = 3^x$ for $-4 \leqslant x \leqslant 3$. (Take the y-axis from 0 to 30.)

c Use your graph to estimate the value of y when $x = 2.5$.

d Use your graph to estimate the value of x when $y = 0.5$.

9 **a** Complete the table below for $y = (\frac{1}{2})^x$ for values of x from –5 to +5. (Values are rounded to 2 decimal places.)

x	–5	–4	–3	–2	–1	0	1	2	3	4	5
$y = (\frac{1}{2})^x$			8			1				0.06	0.03

b Plot the graph of $y = (\frac{1}{2})^x$ for $-5 \leqslant x \leqslant 5$. (Take the y-axis from 0 to 35.)

c Use your graph to estimate the value of y when $x = 2.5$.

d Use your graph to estimate the value of x when $y = 0.75$.

10 Write down whether each of these graphs is 'linear', 'quadratic', 'reciprocal', 'cubic', 'exponential' or 'none of these'.

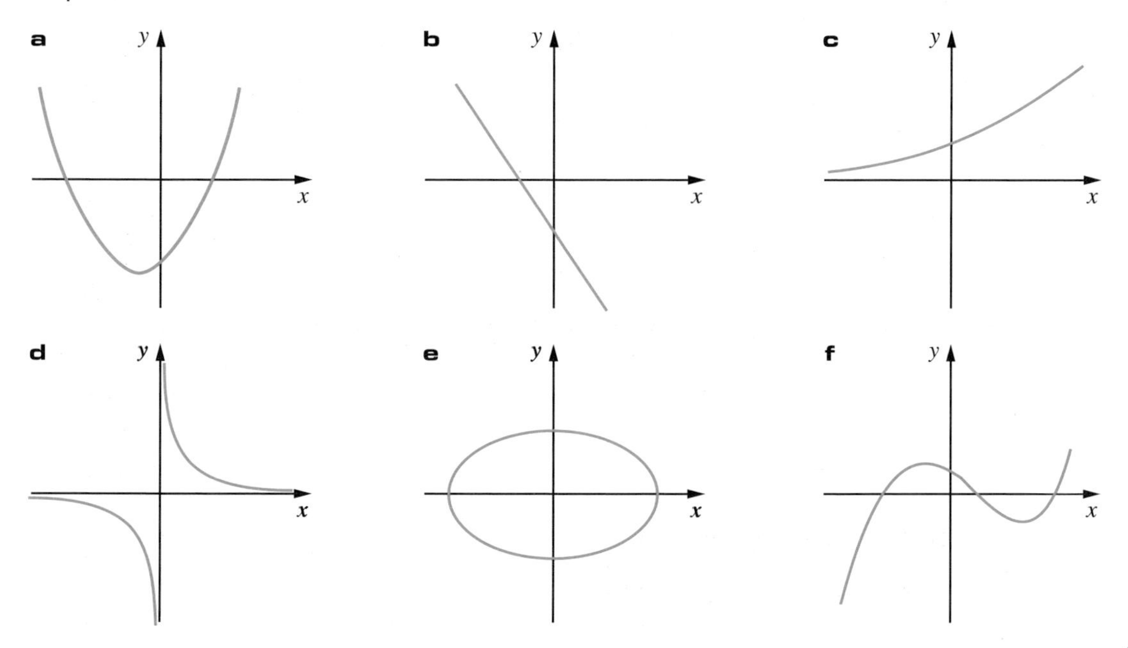

A

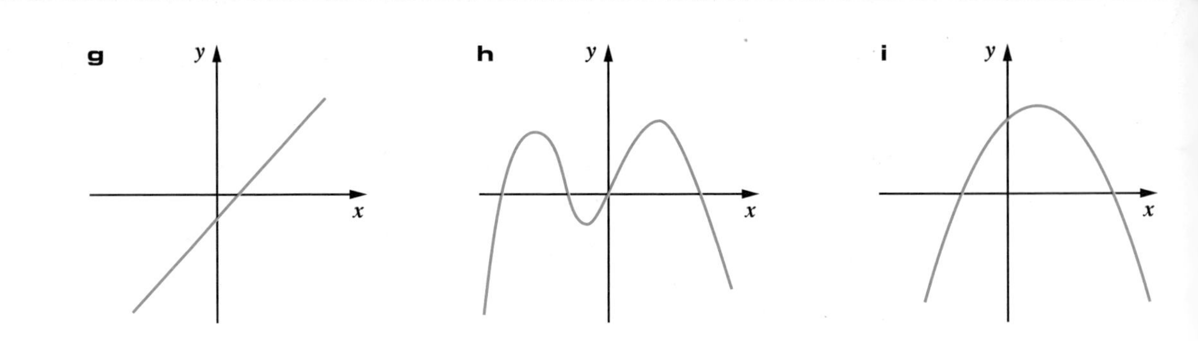

A*

PS 11 One grain of rice is placed on the first square of a chess board. Two grains of rice are placed on the second square, four grains on the third square and so on.

a Explain why $y = 2^{(n-1)}$ gives the number of grains of rice on the nth square.

b Complete the table for the number of grains of rice on the first 10 squares.

Square	1	2	3	4	5	6	7	8	9	10
Grains	1	2	4							

c Use the rule to work out how many grains of rice there are on the 64th square.

d If 1000 grains of rice are worth 5p, how much is the rice on the 64th square worth?

PS 12 An extremely large sheet of paper is 0.01 cm thick. It is torn in half and one piece placed on top of the other. These two pieces are then torn in half and one half is placed on top of the other half to give a pile four sheets thick. This process is repeated 50 times.

a Complete the table to show how many pieces there are in the pile after each tear.

Tears	1	2	3	4	5	6	7	8
Pieces	2	4						

b Write down a rule for the number of pieces after n tears.

c How many pieces will there be piled up after 50 tears?

d How thick is this pile?

AU 13 A curve of the form $y = ab^x$ passes through the points (0, 5) and (2, 45).

Work out the values of a and b.

PS 14 The graph of $y^2 = x$ is shown below.

On a copy of this graph, sketch the following graphs.

a $y^2 + 2 = x$

b $2y^2 = x$

c $(y - 2)^2 = x$

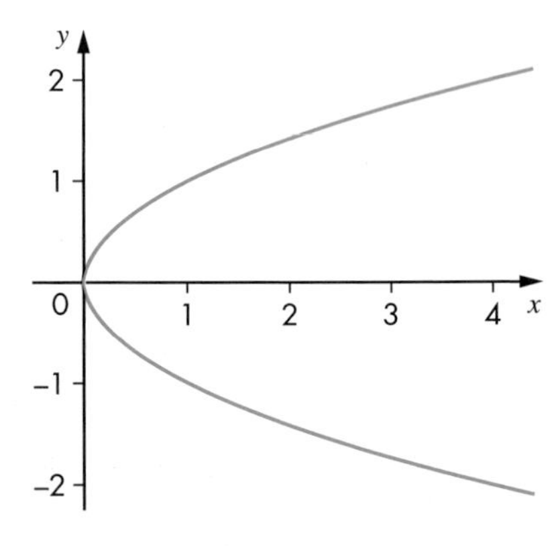

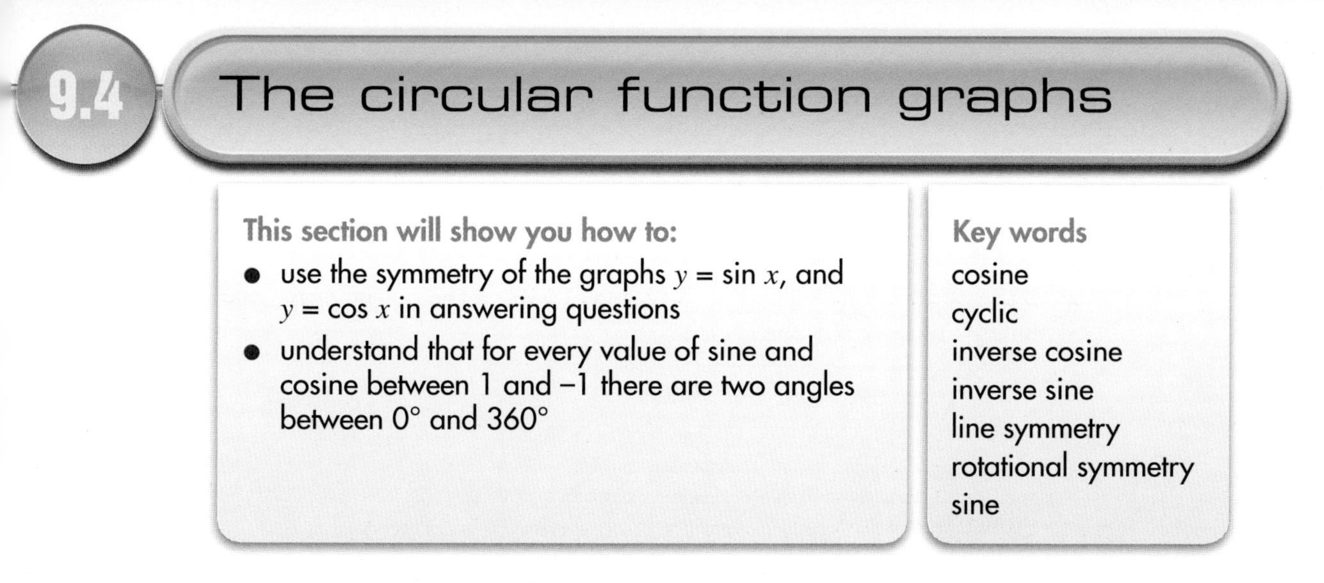

9.4 The circular function graphs

This section will show you how to:

- use the symmetry of the graphs $y = \sin x$, and $y = \cos x$ in answering questions
- understand that for every value of sine and cosine between 1 and –1 there are two angles between 0° and 360°

Key words
cosine
cyclic
inverse cosine
inverse sine
line symmetry
rotational symmetry
sine

You have met **sine** and **cosine** graphs in Chapter 8.

These graphs have some special properties.

- They are **cyclic**. This means that they repeat indefinitely in both directions.
- For every value of sine or cosine between –1 and 1 there are two angles between 0° and 360°, and an infinite number of angles altogether.
- The sine graph has **rotational symmetry** about (180°, 0) and has **line symmetry** between 0° and 180° about $x = 90°$, and between 180° and 360° about $x = 270°$.
- The cosine graph has line symmetry about $x = 180°$, and has rotational symmetry between 0° and 180° about (90°, 0) and between 180° and 360° about (270°, 0).

The graphs can be used to find angles with certain values of sine and cosine.

EXAMPLE 12

Given that sin 42° = 0.669, find another angle between 0° and 360° that also has a sine of 0.669.

Plot the approximate value 0.669 on the sine graph and use the symmetry to work out the other value.

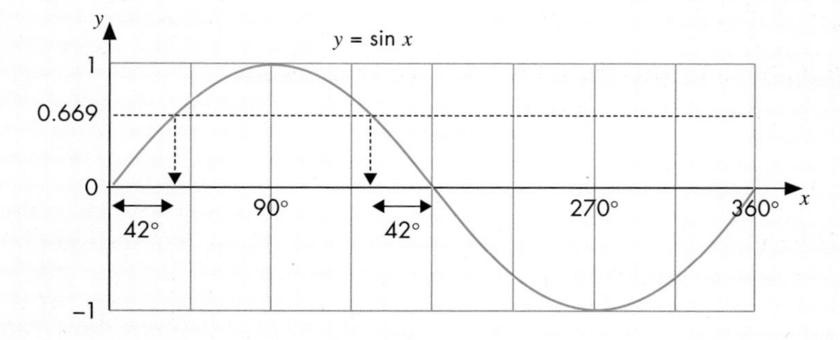

The other value is 180° – 42° = 138°

EXAMPLE 13

Given that cos 110° = –0.342, find two angles between 0° and 360° that have a cosine of +0.342.

Plot the approximate values –0.342 and 0.342 on the cosine graph and use the symmetry to work out the values.

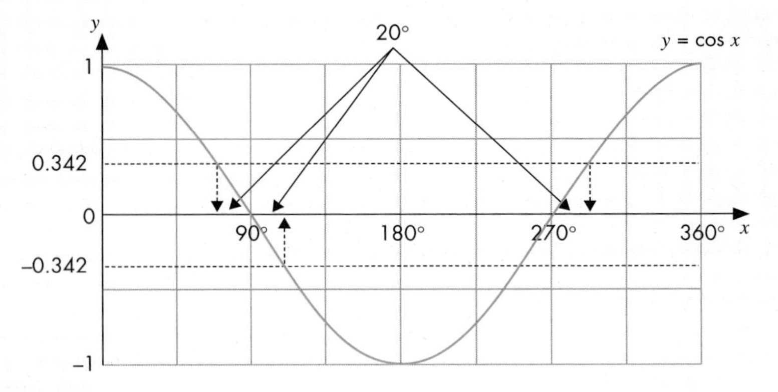

The required values are 90° – 20° = 70° and 270° + 20° = 290°

EXERCISE 9E

A*

1 Given that sin 65° = 0.906, find another angle between 0° and 360° that also has a sine of 0.906.

2 Given that sin 213° = –0.545, find another angle between 0° and 360° that also has a sine of –0.545.

3 Given that cos 36° = 0.809, find another angle between 0° and 360° that also has a cosine of 0.809.

4 Given that cos 165° = –0.966, find another angle between 0° and 360° that also has a cosine of –0.966.

5 Given that sin 30° = 0.5, find two angles between 0° and 360° that have a sine of –0.5.

6 Given that cos 45° = 0.707, find two angles between 0° and 360° that have a cosine of –0.707.

PS 7 **a** Choose an acute angle a. Write down the values of:

i sin a **ii** cos (90° – a).

b Repeat with another acute angle b.

c Write down a rule connecting the sine of an acute angle x and the cosine of the complementary angle (i.e. the difference with 90°).

d Find a similar rule for the cosine of x and the sine of its complementary angle.

8 Given that $\sin 26° = 0.438$:

a write down an angle between 0° and 90° that has a cosine of 0.438

b find two angles between 0° and 360° that have a sine of –0.438

c find two angles between 0° and 360° that have a cosine of –0.438.

AU 9 A formula used to work out the angle of a triangle is

$$\cos A = \frac{b^2 + c^2 - a^2}{2bc}$$

where a, b and c are the sides of the triangle and angle A is the angle opposite side a.

a Work out the value of $\cos A$ for a triangle where $a = 20$, $b = 11$ and $c = 13$.

b What is the size of angle A to the nearest degree? Use the **inverse cosine function** on your calculator.

PS 10 Another formula that can be used to work out the angle of a triangle is

$$\sin A = \frac{a\sin B}{b}$$

where a and b are sides of the triangle and A and B are angles opposite the sides a and b respectively.

a Work out the value of $\sin A$ for this triangle, where $a = 16$, $b = 14$ and $B = 46°$.

b Use your calculator and the **inverse sine** function to find the value of B.

c Does your value from the calculator match the obtuse angle A?

d If not, explain why not.

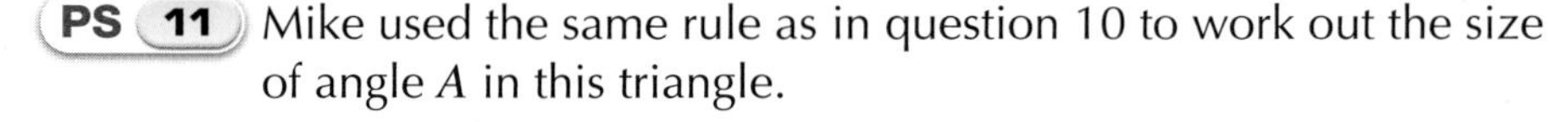

PS 11 Mike used the same rule as in question 10 to work out the size of angle A in this triangle.

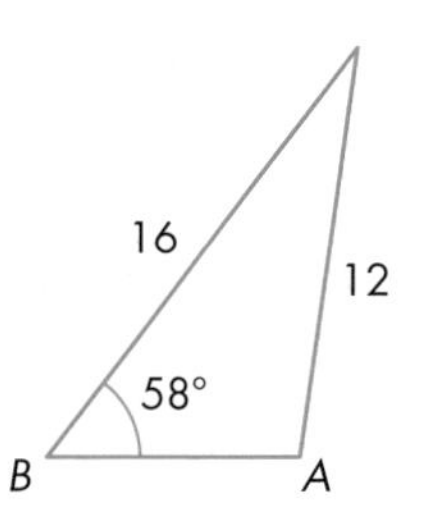

a Work out the value of $\sin A$ for the triangle shown, where $a = 16$, $b = 12$ and $B = 58°$.

b Use your calculator and the inverse sine function to find the value of B.

c What happens? Can you explain why?

AU 12 State if the following rules are true or false.

a $\sin x = \sin (180° - x)$ **b** $\sin x = -\sin (360° - x)$ **c** $\cos x = \cos (360° - x)$

d $\sin x = -\sin (180° + x)$ **e** $\cos(180° - x) = \cos (180° + x)$

9.5 Solving equations, one linear and one non-linear, with graphs

This section will show you how to:

- solve a pair of simultaneous equations where one is linear and one is non-linear, using graphs

Key words
linear
non-linear
simultaneous equations

You will see how to use an algebraic method for solving a pair of **simultaneous equations** where one is **linear** (a straight line) and one is **non-linear** (a curve) in Chapter 12. In this section, you will learn how to do this graphically. The point where the graphs cross gives the solution. However, in most cases, there are two solutions, because the straight line will cross the curve twice.

Most of the non-linear graphs will be quadratic graphs, but there is one other type you can meet. This is an equation of the form $x^2 + y^2 = r^2$, which is a circle, with the centre as the origin and a radius of r.

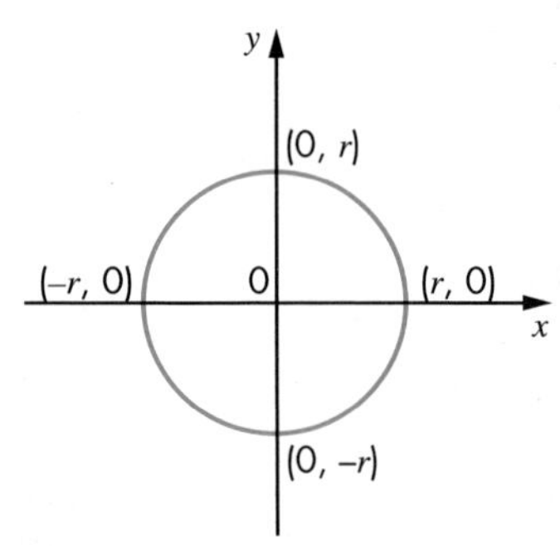

EXAMPLE 14

Find the approximate solutions of the pair of equations $y = x^2 + x - 2$ and $y = 2x + 3$ by graphical means.

Set up a table for the quadratic.

x	–4	–3	–2	–1	0	1	2	3	4
y	10	4	0	–2	–2	0	4	10	18

Draw both graphs on the same set of axes.

From the graph, the approximate solutions can be seen to be (–1.8, –0.6) and (2.8, 8.6).

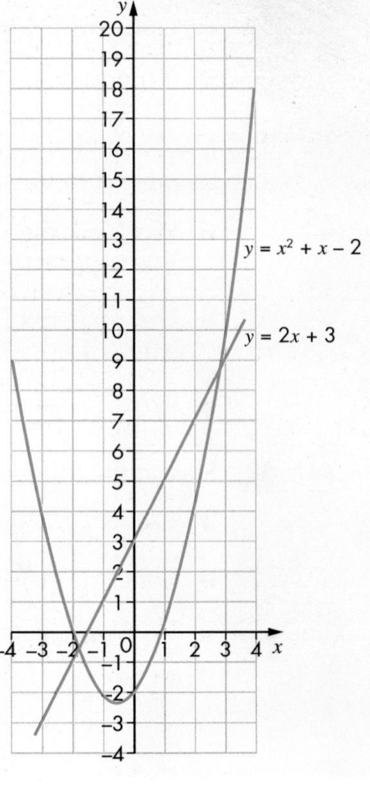

EXAMPLE 15

Find the approximate solutions of the pair of equations $x^2 + y^2 = 25$ and $y = x + 2$ by graphical means.

The curve is a circle of radius 5 centred on the origin.

From the graph, the approximate solutions can be seen to be (–4.4, –2.4), (2.4, 4.4).

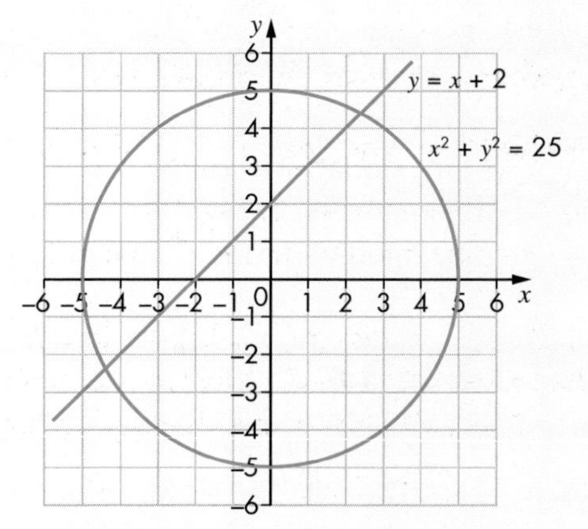

EXERCISE 9F

1 Use graphical methods to find the approximate or exact solutions to the following pairs of simultaneous equations. In this question, suitable ranges for the axes are given. In an examination a grid will be supplied.

- **a** $y = x^2 + 3x - 2$ and $y = x$ $(-5 \leqslant x \leqslant 5, -5 \leqslant y \leqslant 5)$
- **b** $y = x^2 - 3x - 6$ and $y = 2x$ $(-4 \leqslant x \leqslant 8, -10 \leqslant y \leqslant 20)$
- **c** $x^2 + y^2 = 25$ and $x + y = 1$ $(-6 \leqslant x \leqslant 6, -6 \leqslant y \leqslant 6)$
- **d** $x^2 + y^2 = 4$ and $y = x + 1$ $(-5 \leqslant x \leqslant 5, -5 \leqslant y \leqslant 5)$
- **e** $y = x^2 - 3x + 1$ and $y = 2x - 1$ $(0 \leqslant x \leqslant 6, -4 \leqslant y \leqslant 12)$
- **f** $y = x^2 - 3$ and $y = x + 3$ $(-5 \leqslant x \leqslant 5, -4 \leqslant y \leqslant 8)$
- **g** $y = x^2 - 3x - 2$ and $y = 2x - 3$ $(-5 \leqslant x \leqslant 5, -5 \leqslant y \leqslant 10)$
- **h** $x^2 + y^2 = 9$ and $y = x - 1$ $[-5 \leqslant x \leqslant 5, -5 \leqslant y \leqslant 5)$

PS 2
- **a** Solve the simultaneous equations $y = x^2 + 3x - 4$ and $y = 5x - 5$ $(-5 \leqslant x \leqslant 5, -8 \leqslant y \leqslant 8)$.
- **b** What is special about the intersection of these two graphs?
- **c** Show that $5x - 5 = x^2 + 3x - 4$ can be rearranged to $x^2 - 2x + 1 = 0$.
- **d** Factorise and solve $x^2 - 2x + 1 = 0$.
- **e** Explain how the solution in part **d** relates to the intersection of the graphs.

AU 3
- **a** Solve the simultaneous equations $y = x^2 + 2x + 3$ and $y = x - 1$ $(-5 \leqslant x \leqslant 5, -5 \leqslant y \leqslant 8)$.
- **b** What is special about the intersection of these two graphs?
- **c** Rearrange $x - 1 = x^2 + 2x + 3$ into the general quadratic form $ax^2 + bx + c = 0$.
- **d** Work out the discriminant $b^2 - 4ac$ for the quadratic in part **c**.
- **e** Explain how the value of the discriminant relates to the intersection of the graphs.

A

A*

9.6 Solving equations by the method of intersection

This section will show you how to:

- solve equations by the method of intersecting graphs

Many equations can be solved by drawing two intersecting graphs on the same axes and using the x-value(s) of their point(s) of intersection. (In the GCSE examination, you are likely to be presented with one drawn graph and asked to draw a straight line to solve a new equation.)

EXAMPLE 16

Show how each equation given below can be solved using the graph of $y = x^3 - 2x - 2$ and its intersection with another graph. In each case, give the equation of the other graph and the solution(s).

a $x^3 - 2x - 4 = 0$ **b** $x^3 - 3x - 1 = 0$

a This method will give the required graph.

Step 1: Write down the original (given) equation. $y = x^3 - 2x - 2$

Step 2: Write down the (new) equation to be solved in reverse. $0 = x^3 - 2x - 4$

Step 3: Subtract these equations. $y = \quad + 2$

Step 4: Draw this line on the original graph to solve the new equation.

The graphs of $y = x^3 - 2x - 2$ and $y = 2$ are drawn on the same axes.

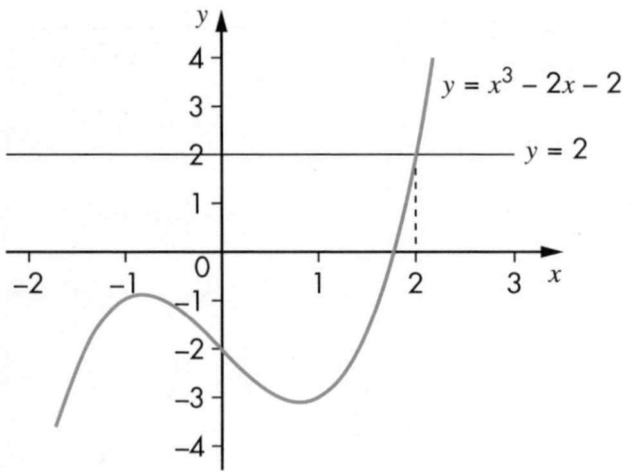

$y = x^3 - 2x - 2$

$y = 2$

The intersection of these two graphs is the solution of

$$x^3 - 2x - 4 = 0.$$

The solution is $x = 2$.

This works because you are drawing a straight line on the same axes as the original graph, and solving for x and y where they intersect.

At the points of intersection the y-values will be the same and so will the x-values.

So you can say: original equation = straight line

Rearranging this gives: (original equation) – (straight line) = 0

You have been asked to solve: (new equation) = 0

So (original equation) – (straight line) = (new equation)

Rearranging this again gives: (original equation) – (new equation) = straight line

Note: In GCSE exams the curve is always drawn already and you will only have to draw the straight line.

b Write down given graph: $y = x^3 - 2x - 2$

Write down new equation: $0 = x^3 - 3x - 1$

Subtract: $y = \quad + x - 1$

The graphs of $y = x^3 - 2x - 2$ and $y = x - 1$ are then drawn on the same axes.

The intersection of the two graphs is the solution of $x^3 - 3x - 1 = 0$.

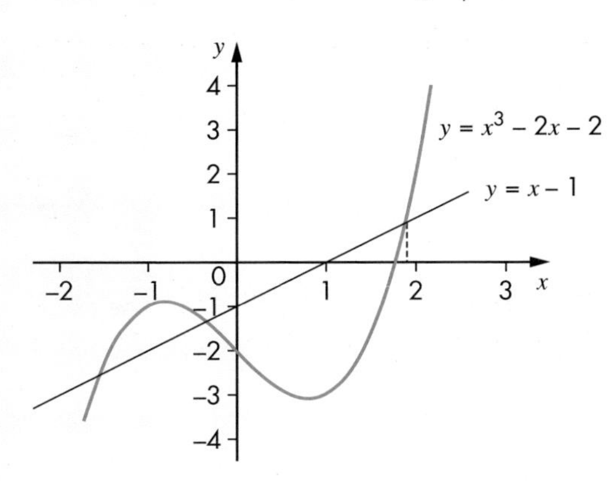

The solutions are $x = -1.5, -0.3$ and 1.9.

EXAMPLE 17

The graph shows the curve $y = x^2 + 3x - 2$.

By drawing a suitable straight line, solve these equations.

a $x^2 + 3x - 1 = 0$ **b** $x^2 + 2x - 3 = 0$

a Given graph: $y = x^2 + 3x - 2$

New equation: $0 = x^2 + 3x - 1$

Subtract: $y = \quad -1$

Draw: $y = -1$

Solutions: $x = 0.3, -3.3$

b Given graph: $y = x^2 + 3x - 2$

New equation: $0 = x^2 + 2x - 3$

Subtract: $y = \quad +x \quad +1$

Draw: $y = x + 1$

Solutions: $x = 1, -3$

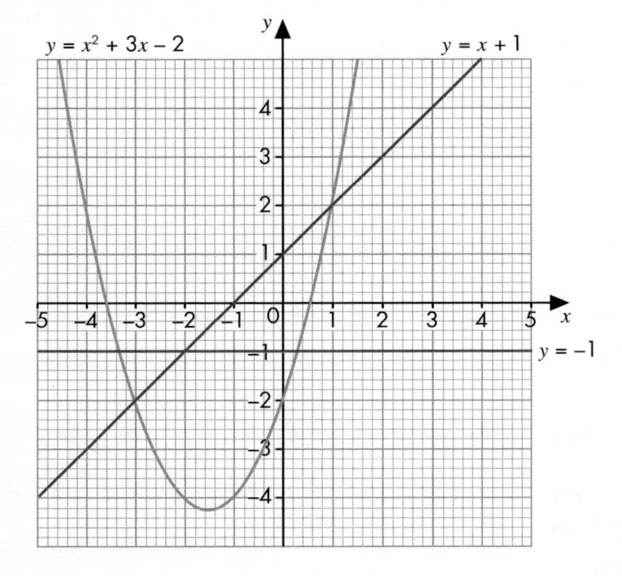

EXERCISE 9G

In questions **1** to **5**, use the graphs given here. Trace the graphs or place a ruler over them in the position of the line. Solution values only need to be given to 1 decimal place. In questions **6** to **10**, either draw the graphs yourself or use a graphics calculator to draw them.

A*

1 Below is the graph of $y = x^2 - 3x - 6$.

a Solve these equations.

i $x^2 - 3x - 6 = 0$ **ii** $x^2 - 3x - 6 = 4$ **iii** $x^2 - 3x - 2 = 0$

b By drawing a suitable straight line solve $2x^2 - 6x + 2 = 0$.

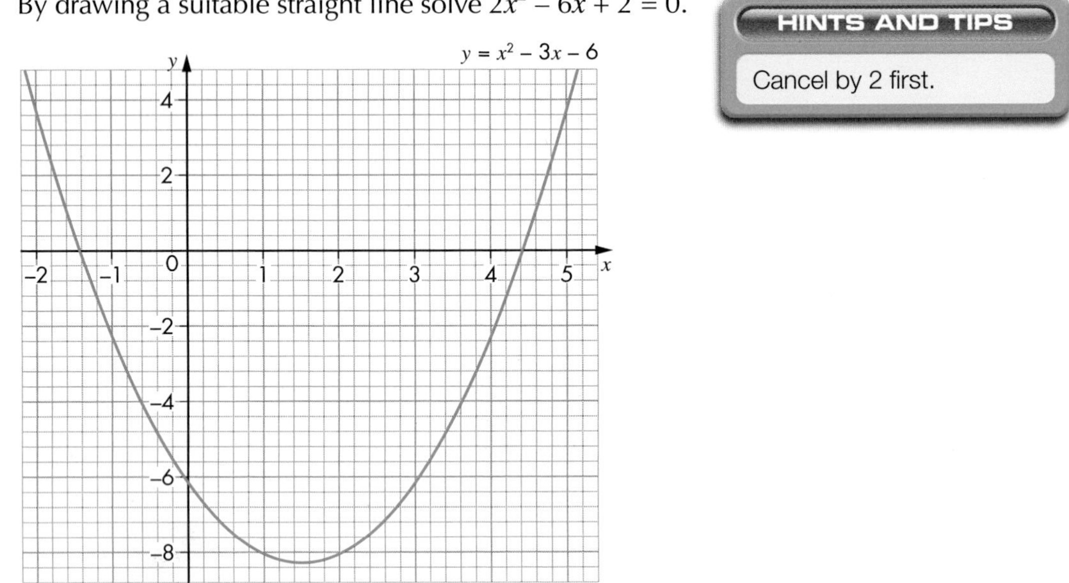

2 Below is the graph of

$y = x^2 + 4x - 5$.

a Solve $x^2 + 4x - 5 = 0$.

b By drawing suitable straight lines solve these equations.

i $x^2 + 4x - 5 = 2$

ii $x^2 + 4x - 4 = 0$

iii $3x^2 + 12x + 6 = 0$

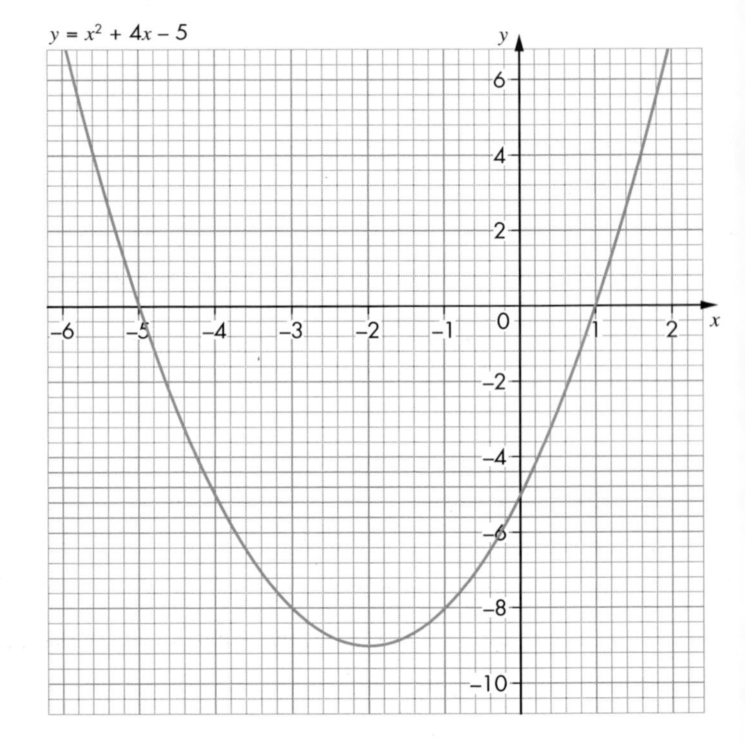

A*

3 Below are the graphs of $y = x^2 - 5x + 3$ and $y = x + 3$.

a Solve these equations. **i** $x^2 - 6x = 0$ **ii** $x^2 - 5x + 3 = 0$

b By drawing suitable straight lines solve these equations.

i $x^2 - 5x + 3 = 2$ **ii** $x^2 - 5x - 2 = 0$

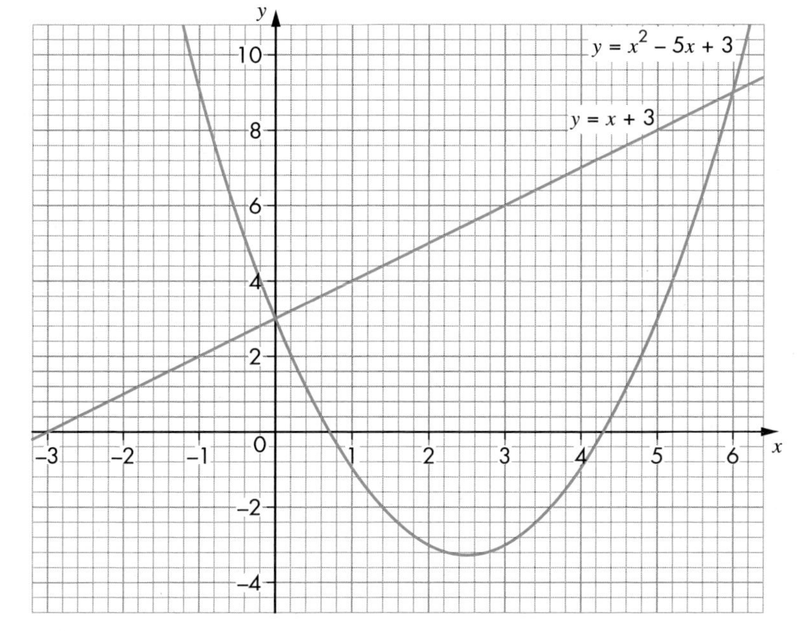

4 Below are the graphs of $y = x^2 - 2$ and $y = x + 2$.

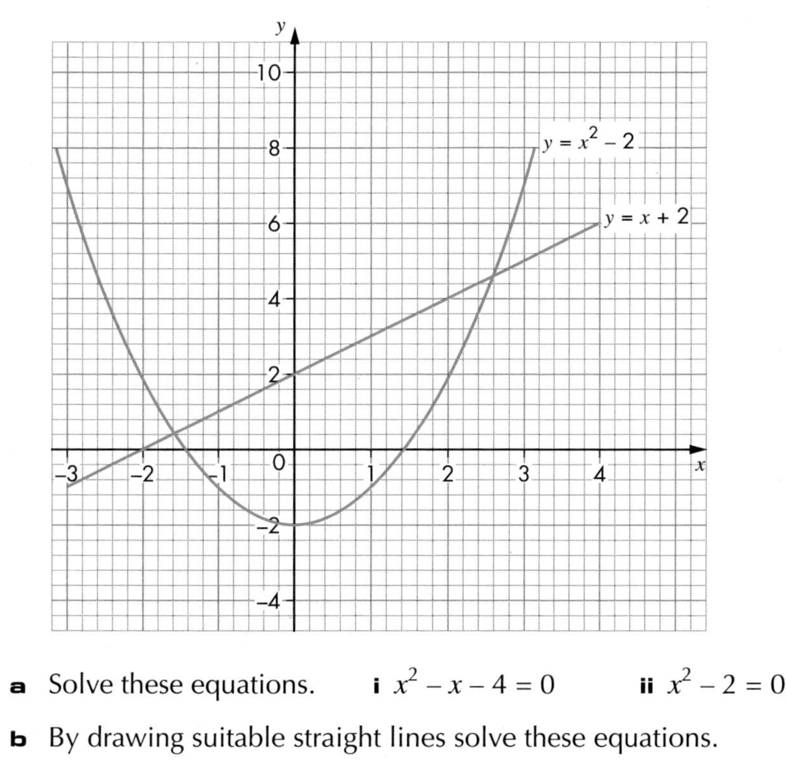

a Solve these equations. **i** $x^2 - x - 4 = 0$ **ii** $x^2 - 2 = 0$

b By drawing suitable straight lines solve these equations.

i $x^2 - 2 = 3$ **ii** $x^2 - 4 = 0$

A*

PS 5 Below are the graphs of $y = x^3 - 2x^2$, $y = 2x + 1$ and $y = x - 1$.

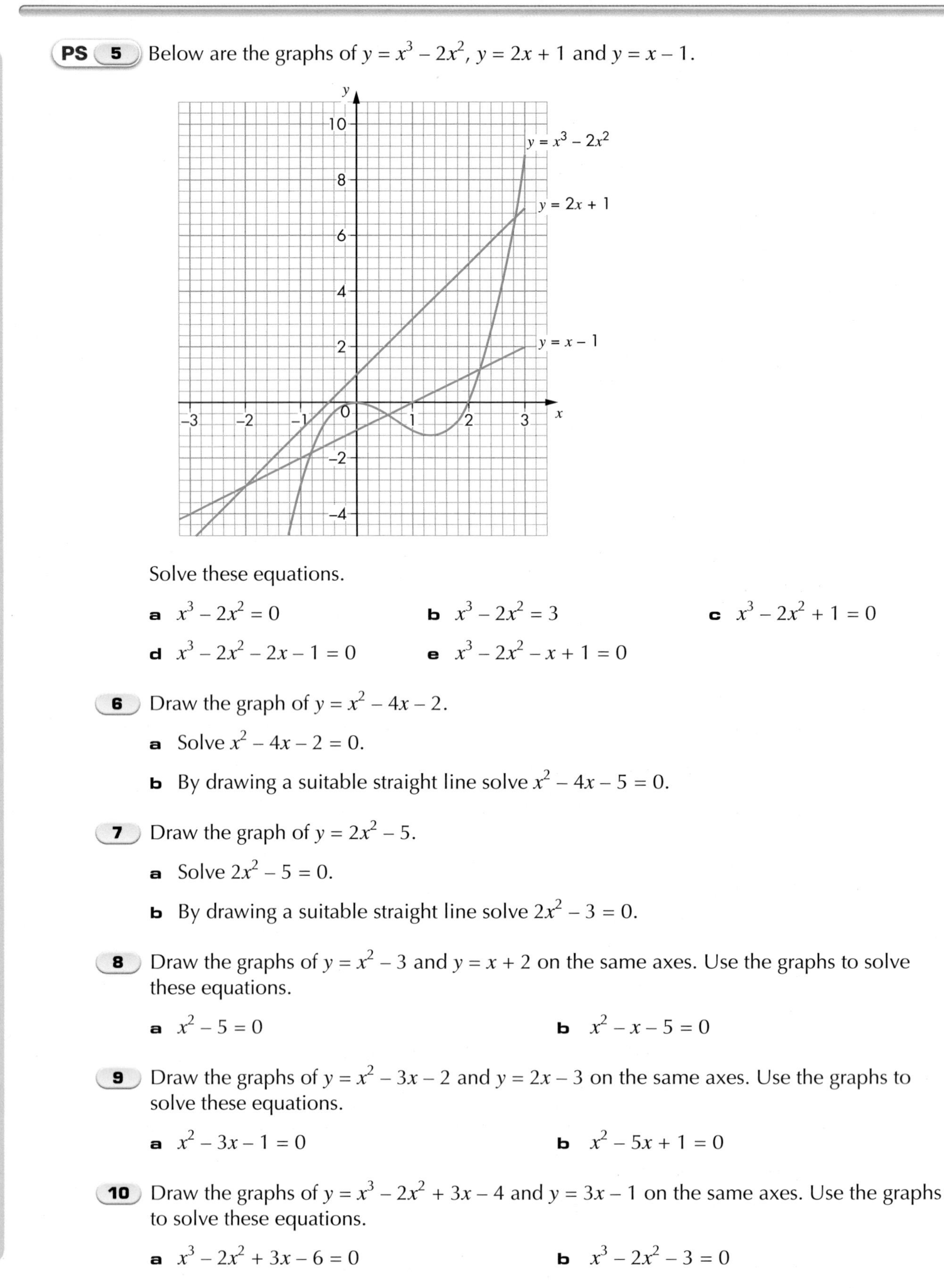

Solve these equations.

a $x^3 - 2x^2 = 0$ **b** $x^3 - 2x^2 = 3$ **c** $x^3 - 2x^2 + 1 = 0$

d $x^3 - 2x^2 - 2x - 1 = 0$ **e** $x^3 - 2x^2 - x + 1 = 0$

6 Draw the graph of $y = x^2 - 4x - 2$.

a Solve $x^2 - 4x - 2 = 0$.

b By drawing a suitable straight line solve $x^2 - 4x - 5 = 0$.

7 Draw the graph of $y = 2x^2 - 5$.

a Solve $2x^2 - 5 = 0$.

b By drawing a suitable straight line solve $2x^2 - 3 = 0$.

8 Draw the graphs of $y = x^2 - 3$ and $y = x + 2$ on the same axes. Use the graphs to solve these equations.

a $x^2 - 5 = 0$ **b** $x^2 - x - 5 = 0$

9 Draw the graphs of $y = x^2 - 3x - 2$ and $y = 2x - 3$ on the same axes. Use the graphs to solve these equations.

a $x^2 - 3x - 1 = 0$ **b** $x^2 - 5x + 1 = 0$

10 Draw the graphs of $y = x^3 - 2x^2 + 3x - 4$ and $y = 3x - 1$ on the same axes. Use the graphs to solve these equations.

a $x^3 - 2x^2 + 3x - 6 = 0$ **b** $x^3 - 2x^2 - 3 = 0$

AU 11 The graph shows the lines A: $y = x^2 + 3x - 2$; B: $y = x$; C: $y = x + 2$; D: $y + x = 3$ and E: $y + x + 1 = 0$.

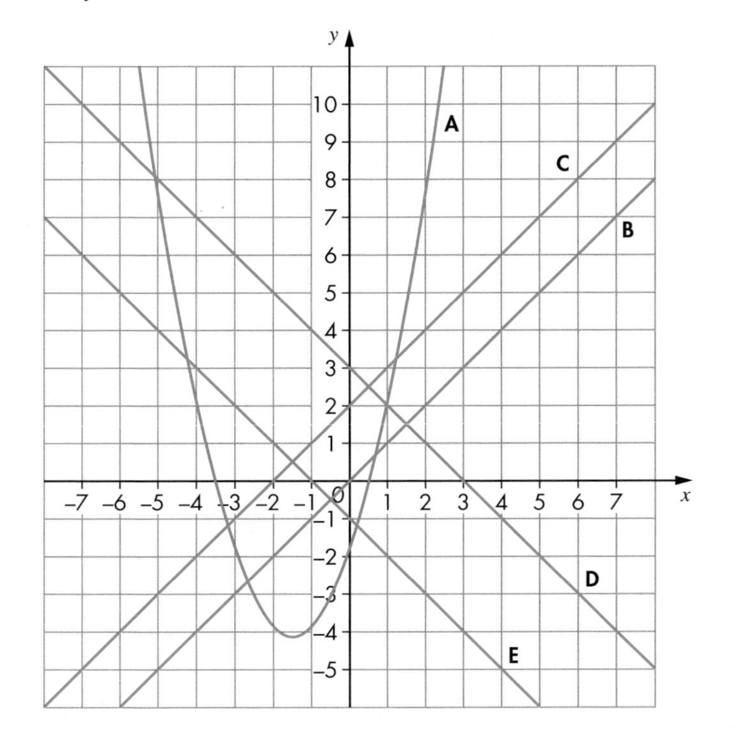

a Which pair of lines has a common solution of (0.5, 2.5)?

b Which pair of lines has the solutions of (1, 2) and (–5, 8)?

c What quadratic equation has an approximate solution of (–4.2, 3.2) and (0.2, –1.2)?

d The minimum point of the graph $y = x^2 + 3x - 2$ is at (–1.5, –4.25).

What is the minimum point of the graph $y = x^2 + 3x - 8$?

PS 12 Jamil was given a sketch of the graph $y = x^2 + 3x + 5$ and asked to draw an appropriate straight line to solve $x^2 + x - 2 = 0$.

This is Jamil's working:

$$\begin{array}{ll} \text{Original} & y = x^2 + 3x + 5 \\ \text{New} & \underline{0 = x^2 + x - 2} \\ & y = 2x - 7 \end{array}$$

When Jamil drew the line $y = 2x - 7$, it did not intersect with the parabola $y = x^2 + 3x + 5$.

He concluded that the equation $x^2 + x - 2 = 0$ did not have any solutions.

a Show by factorisation that the equation $x^2 + x - 2 = 0$ has solutions –2 and 1.

b Explain the error that Jamil made.

c What line should Jamil have drawn?

GRADE BOOSTER

- **B** You can use graphs to find formulae and solve simultaneous linear equations
- **B** You can plot cubic graphs, using a table of values
- **B** You can recognise the shapes of the graphs $y = x^3$ and $y = \frac{1}{x}$
- **A** You can draw linear graphs parallel or perpendicular to other lines
- **A** You can draw a variety of graphs, such as exponential graphs and reciprocal graphs, using a table of values
- **A*** You can solve equations, using the intersection of two graphs
- **A*** You can use trigonometric graphs to solve sine and cosine problems
- **A*** You can find two angles between 0° and 360° for any given value of a trigonometric ratio (positive or negative)

What you should know now

- How to draw non-linear graphs
- How to solve equations by finding the intersection points of the graphs of the equations with the x-axis or other related equations

1

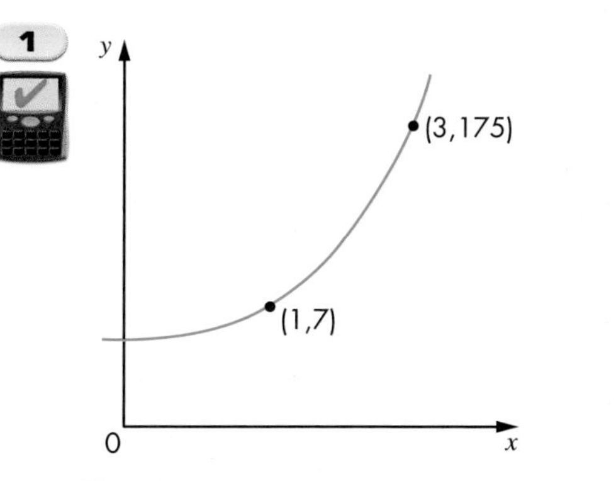

The sketch shows a curve with equation $y = ka^x$

where k and a are constants and $a > 0$.

The curve passes through the points (1, 7) and (3, 175).

Calculate the value of k and the value of a.

(3 marks)

Edexcel, June 2008, Paper 15 Higher, Question 18

2 The diagram shows a sketch of the graph $y = ab^x$.

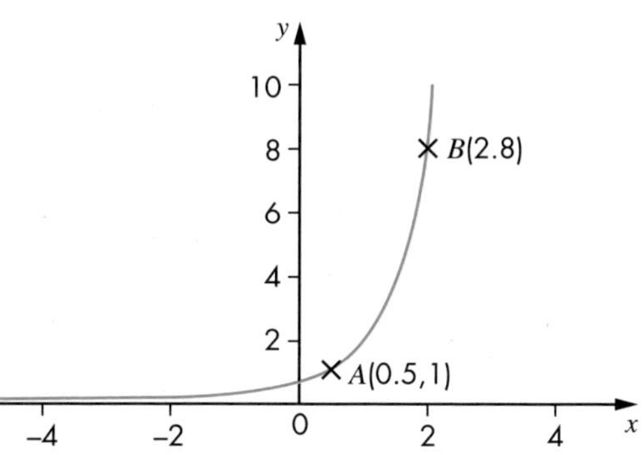

The curve passes through the points A (0.5, 1) and B (2, 8).

The point C (–0.5, k) lies on the curve.

Find the value of k. *(4 marks)*

Edexcel, November 2008, Paper 4 Higher, Question 2

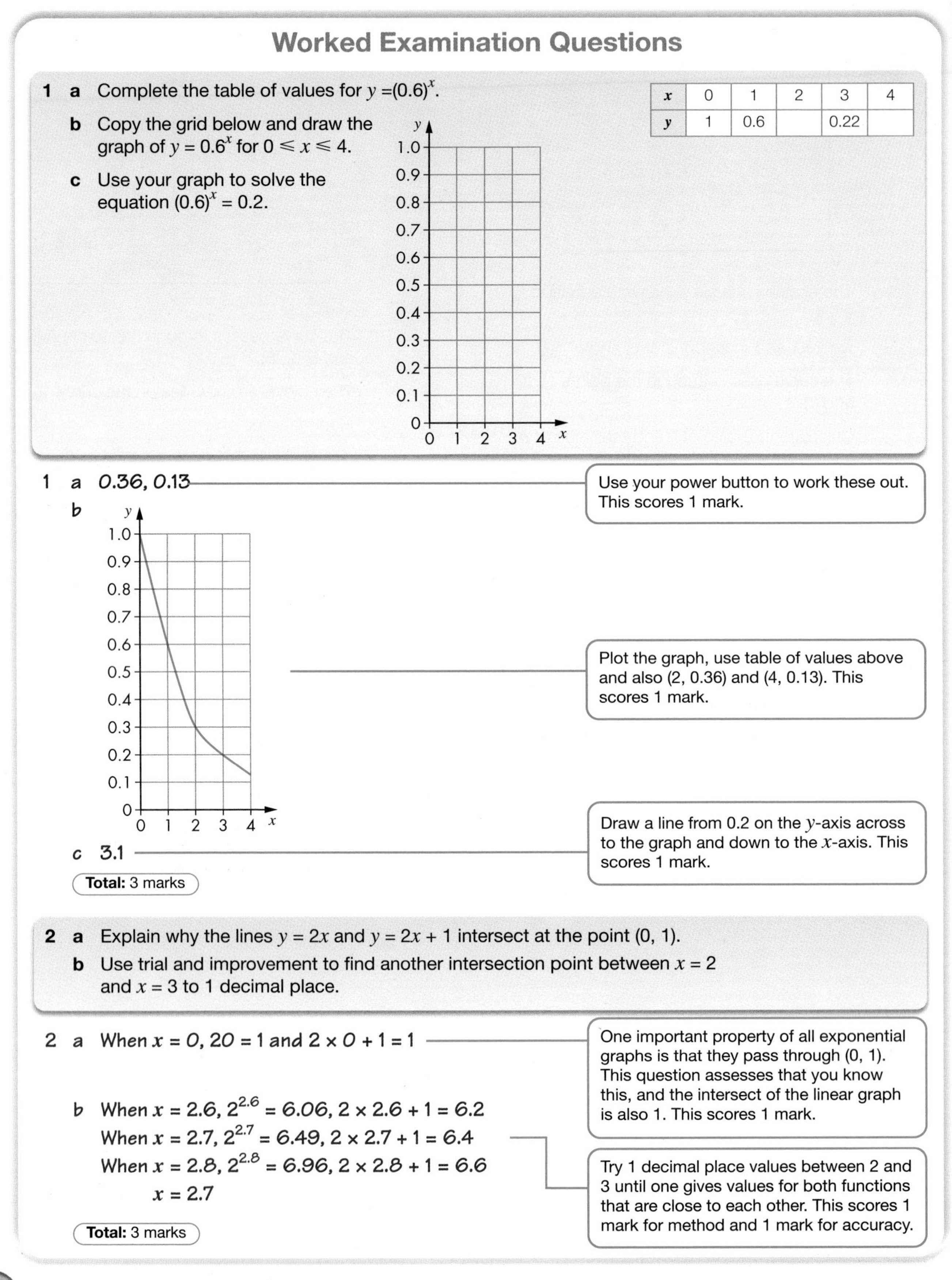

Worked Examination Questions

1 a Complete the table of values for $y = (0.6)^x$.

x	0	1	2	3	4
y	1	0.6		0.22	

b Copy the grid below and draw the graph of $y = 0.6^x$ for $0 \leqslant x \leqslant 4$.

c Use your graph to solve the equation $(0.6)^x = 0.2$.

1 a 0.36, 0.13

Use your power button to work these out. This scores 1 mark.

b

Plot the graph, use table of values above and also (2, 0.36) and (4, 0.13). This scores 1 mark.

c 3.1

Draw a line from 0.2 on the y-axis across to the graph and down to the x-axis. This scores 1 mark.

Total: 3 marks

2 a Explain why the lines $y = 2x$ and $y = 2x + 1$ intersect at the point (0, 1).

b Use trial and improvement to find another intersection point between $x = 2$ and $x = 3$ to 1 decimal place.

2 a When $x = 0$, 20 = 1 and $2 \times 0 + 1 = 1$

One important property of all exponential graphs is that they pass through (0, 1). This question assesses that you know this, and the intersect of the linear graph is also 1. This scores 1 mark.

b When $x = 2.6$, $2^{2.6} = 6.06$, $2 \times 2.6 + 1 = 6.2$

When $x = 2.7$, $2^{2.7} = 6.49$, $2 \times 2.7 + 1 = 6.4$

When $x = 2.8$, $2^{2.8} = 6.96$, $2 \times 2.8 + 1 = 6.6$

$x = 2.7$

Try 1 decimal place values between 2 and 3 until one gives values for both functions that are close to each other. This scores 1 mark for method and 1 mark for accuracy.

Total: 3 marks

Worked Examination Questions

3 The grid below shows the graph of $y = x^2 - x - 6$.

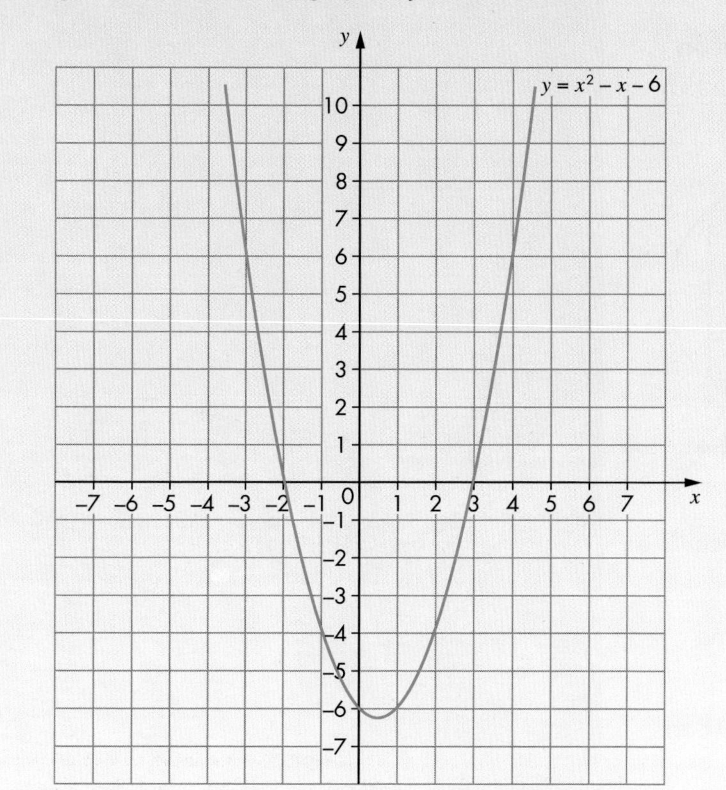

a Deduce the coordinates of the minimum point of the graph $y = x^2 - x - 12$.

b Use the graph to find the approximate solutions to the simultaneous equations
$y = x^2 - x - 6$ and $y = x + 3$.

3 a (0.5, –12.25)

The minimum point of the given graph is (0.5, –6.25) and the required graph is 6 lower than the given graph, as the constant terms have a difference of 6. This scores 1 mark.

b (4.2, 7.2) and (–2.2, 1.2)

Draw the line $y = x + 3$ and read off the points of intersection. This scores 1 mark each.

Total: 3 marks

9 Problem Solving
Rectangles in a circle

If you were given a circle, how would you draw a rectangle inside it?

Each corner of the rectangle must be on the circle.

Here are two possible solutions.

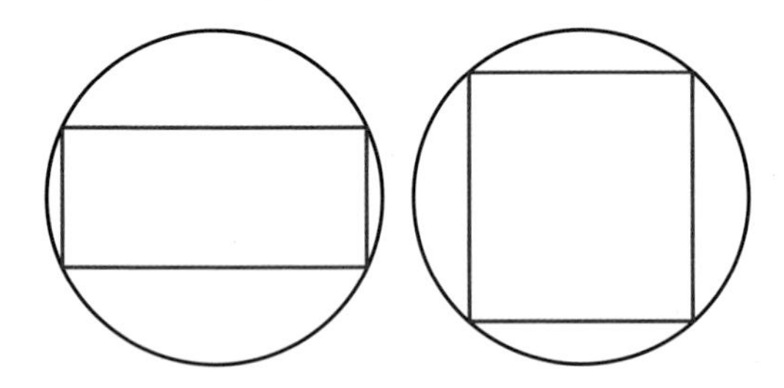

Do you think both rectangles have the same area?

Do you think they both have the same perimeter?

Do you think they both have the same diagonal length?

In this task, you will investigate questions like this in much greater depth.

Getting started

Use trigonometrical ratios to calculate the lengths of the sides of this rectangle.

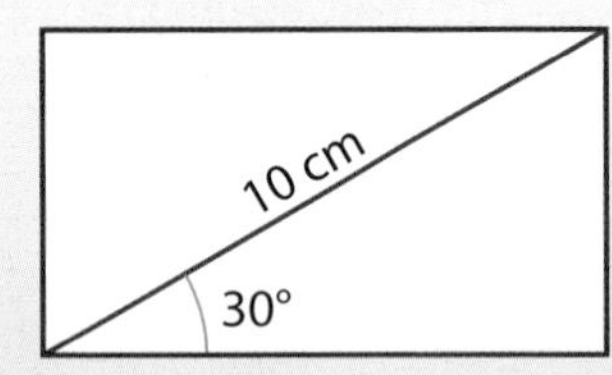

Now do the same for this rectangle.

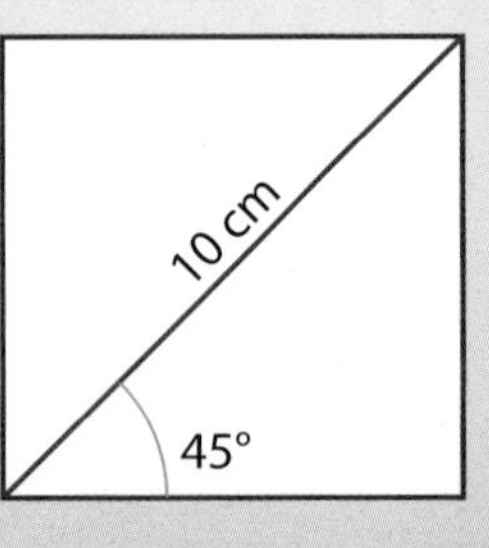

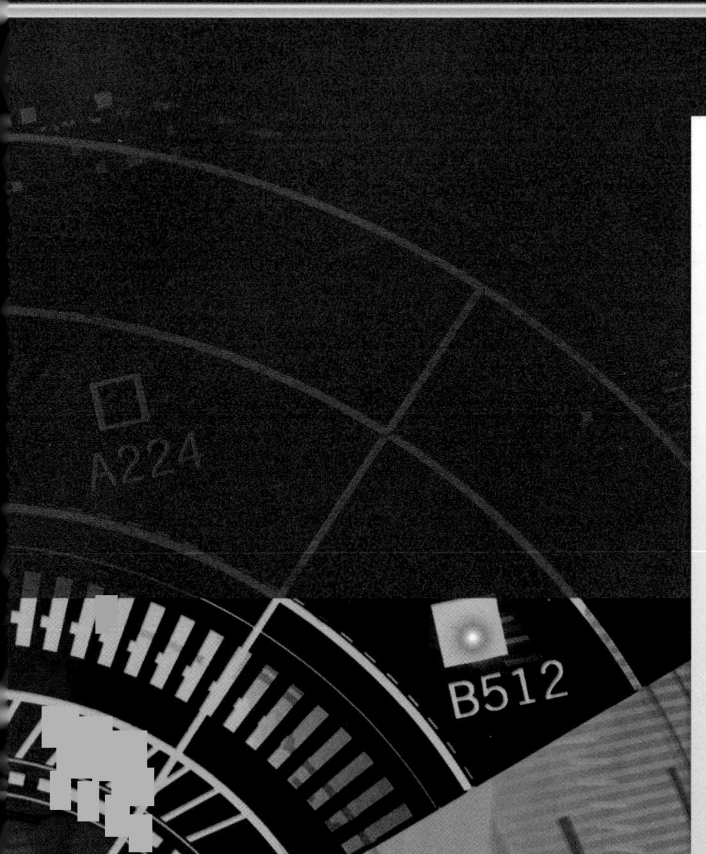

Your task

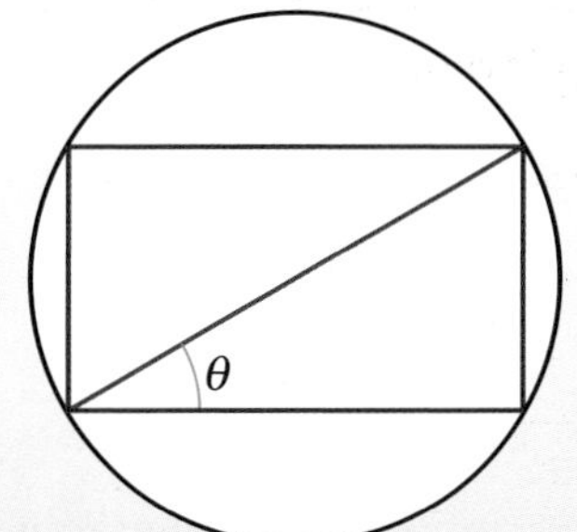

This circle has a diameter of 10 cm.

Investigate how the perimeter and area of the rectangle change as the size of the angle, θ, varies.

Find the most appropriate way to present your findings. You must give details of:

- the method used in your investigation
- an analysis of your method
- the conclusions that you reach
- evidence to support you conclusions

Why this chapter matters

For centuries, statistical graphs have been used in many different areas, from science to politics. They provide a way to represent, analyse and interpret information. But what sort of information is useful? How should this information be displayed? And how should information be interpreted?

Developing statistical analysis

The development of statistical graphs was spurred on by the need, starting as long ago as the 17th century to base policies on demographic and economic data, as well as understanding social conditions, and for this information to be shared with a large number of people.

Since the later part of the 20th century, statistical graphs have become an important way of analysing information, and computer-generated statistical graphs are seen every day on TV, in newspapers and in magazines.

Who invented statistical diagrams?

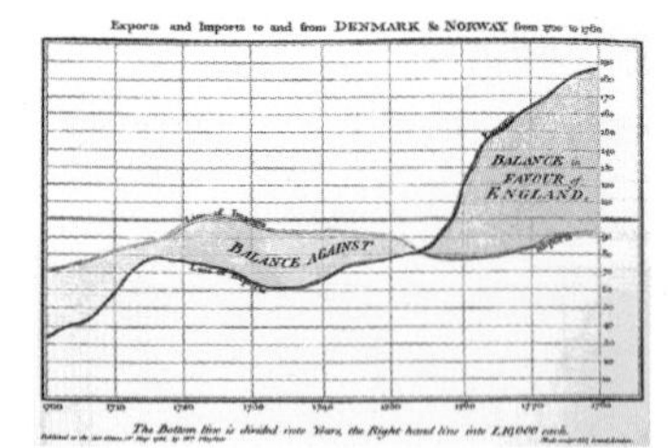

William Playfair was a Scottish engineer, who is thought to be the founder of representing statistics in a graphical way. He invented three types of diagrams: the line graph and bar chart in 1786, then the pie chart in 1801.

Florence Nightingale

Florence Nightingale was born in 1820. She was a famous English nurse and very good at mathematics, becoming a pioneer in presenting information visually. She developed a form of the pie chart now known as the 'polar area diagram' or the 'Nightingale rose diagram', which was like a modern circular bar chart. It illustrated monthly patient deaths in military field hospitals.

John Tukey was born in 1915, in Massachusetts, USA. Tukey was responsible for the introduction of many statistical tools, including the stem-and-leaf diagram, which can be seen in everyday life, such as in this Japanese train timetable.

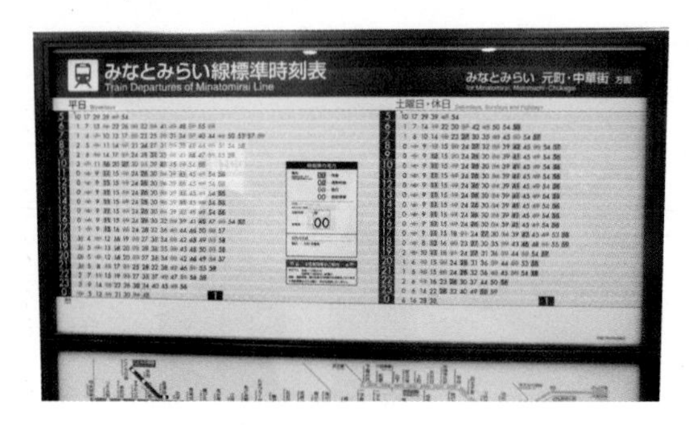

He believed that mathematicians should start from the data and look for a theorem. He worked in a field known as robust analysis, in which researchers can arrive at credible conclusions even when they are working with flawed data. In 1970, Tukey published *Exploratory Data Analysis*, offering new ways to analyse and present data clearly. Between 1960 and 1980, he worked on devising the polls that TV networks use to predict and analyse elections. It was Tukey who first introduced the idea of a box plot.

This chapter introduces you to statistical representation in the form of frequency diagrams, box plots and histograms.

Chapter

Statistics: Data distributions

The grades given in this chapter are target grades.

1 Cumulative frequency diagrams

2 Box plots

3 Histograms with bars of unequal width

This chapter will show you ...

B how to draw cumulative frequency diagrams

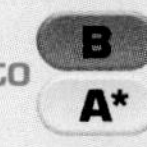

B to A* how to draw and read box plots

B how to draw and read histograms

A* how to find the median, quartiles and interquartile range from a histogram

Visual overview

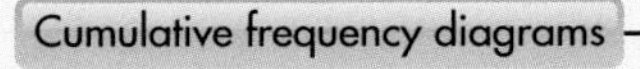

What you should already know

- How to plot coordinate points **(KS3 level 4, GCSE grade F)**
- How to read information from charts and tables **(KS3 level 4, GCSE grade F)**
- How to calculate the mean of a set of data from a frequency table **(KS3 level 5, GCSE grade E)**
- How to recognise a positive or negative gradient **(KS3 level 6, GCSE grade D)**

Quick check

The table shows the numbers of children in 10 classes in a primary school.

Calculate the mean number of children in each class.

Number of children	27	28	29	30	31
Frequency	1	2	4	2	1

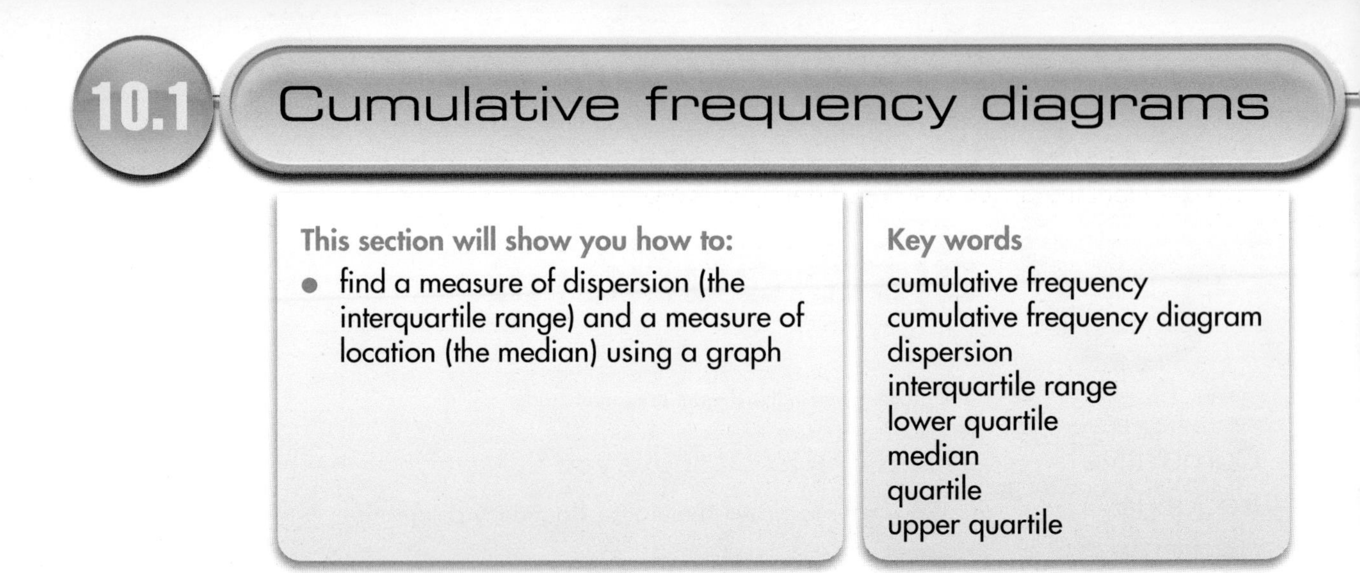

10.1 Cumulative frequency diagrams

This section will show you how to:

- find a measure of dispersion (the interquartile range) and a measure of location (the median) using a graph

Key words

cumulative frequency
cumulative frequency diagram
dispersion
interquartile range
lower quartile
median
quartile
upper quartile

The **interquartile range** is a measure of the **dispersion** of a set of data. The advantage of the interquartile range is that it eliminates extreme values, and bases the measure of spread on the middle 50% of the data. This section will show how to find the interquartile range and the median of a set of data by drawing a **cumulative frequency diagram**.

The grouped table below shows the marks of 50 students in a mathematics test. Note that it includes a column for the **cumulative frequency**, which you can find by adding each frequency to the sum of all preceding frequencies.

Mark	No. of students	Cumulative frequency
21 to 30	1	1
31 to 40	6	7
41 to 50	6	13
51 to 60	8	21
61 to 70	8	29
71 to 80	6	35
81 to 90	7	42
91 to 100	6	48
101 to 110	1	49
111 to 120	1	50

This data can then be used to plot a graph of the top value of each group against its cumulative frequency. The points to be plotted are (30, 1), (40, 7), (50, 13), (60, 21), etc., which will give the graph shown below. Note that the cumulative frequency is *always* the vertical (y) axis.

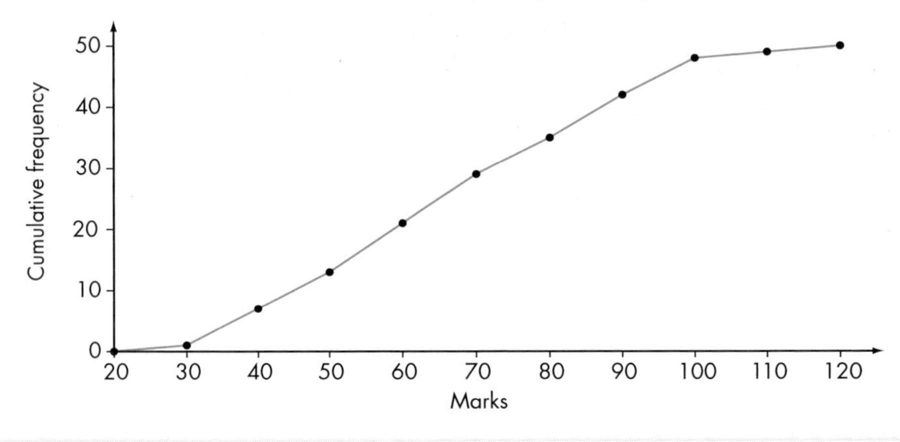

Also note that the scales on both axes are labelled at each graduation mark, in the usual way. **Do not** label the scales as shown here. It is **wrong**.

21–30 31–40 41–50

The plotted points can be joined in two different ways:

- by straight lines, to give a cumulative frequency polygon
- by a freehand curve, to give a cumulative frequency curve or ogive.

They are both called cumulative frequency diagrams.

In an examination you are most likely to be asked to draw a cumulative frequency diagram, and the type (polygon or curve) is up to you. Both will give similar results. The cumulative frequency diagram can be used in several ways, as you will now see.

The median

The **median** is the middle item of data once all the items have been put in order of size, from lowest to highest. So, if you have n items of data plotted as a cumulative frequency diagram, you can find the median from the middle value of the cumulative frequency, that is the $\frac{1}{2}n$th value.

But remember, if you want to find the median from a simple list of discrete data, you *must* use the $\frac{1}{2}(n + 1)$th value. The reason for the difference is that the cumulative frequency diagram treats the data as continuous, even when using data such as examination marks, which are discrete. The reason you can use the $\frac{1}{2}n$th value when working with cumulative frequency diagrams is that you are only looking for an estimate of the median.

There are 50 values in the table on the previous page. The middle value will be the 25th value. Draw a horizontal line from the 25th value to meet the graph, then go down to the horizontal axis. This will give an estimate of the median. In this example, the median is about 65 marks.

The interquartile range

By dividing the cumulative frequency into four parts, you can obtain **quartiles** and the interquartile range.

The **lower quartile** is the item one-quarter of the way up the cumulative frequency axis and is given by the $\frac{1}{4}n$th value.

The **upper quartile** is the item three-quarters of the way up the cumulative frequency axis and is given by the $\frac{3}{4}n$th value.

The interquartile range is the difference between the lower and upper quartiles.

These are illustrated on the graph below.

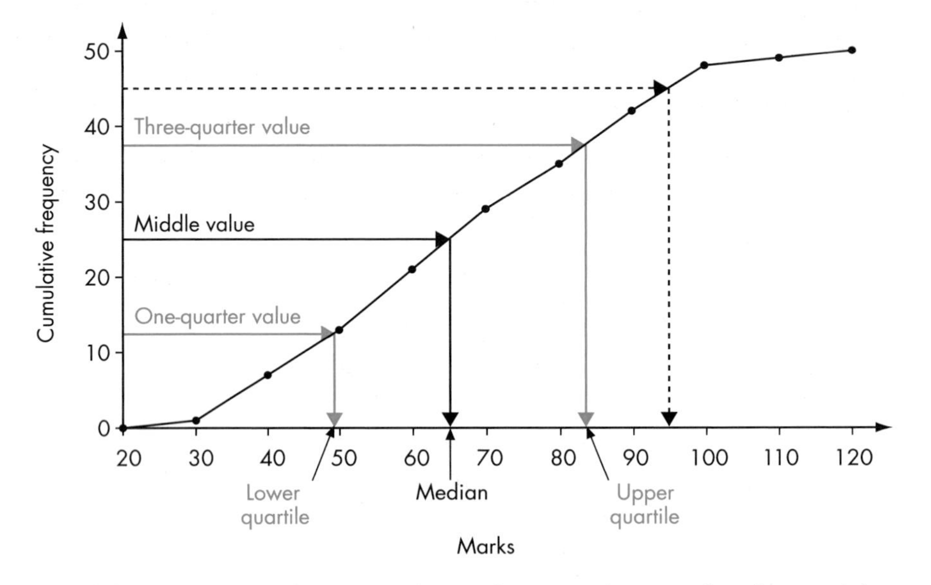

The quarter and three-quarter values out of 50 values are the 12.5th value and the 37.5th value. Draw lines across to the cumulative frequency curve from these values and down to the horizontal axis. These give the lower and upper quartiles. In this example, the lower quartile is 49 marks, the upper quartile is 83 marks and the interquartile range is 83 – 49 = 34 marks.

Note that problems like these are often followed up with an extra question such as: *The head of mathematics decides to give a special award to the top 10% of students. What would the cut-off mark be?*

The top 10% would be the top 5 students (10% of 50 is 5). Draw a line across from the 45th student to the graph and down to the horizontal axis. This gives a cut-off mark of 95.

EXAMPLE 1

This table shows the marks of 100 students in a mathematics test.

a Draw a cumulative frequency curve.

b Use your graph to find the median and the interquartile range.

c Students who score less than 44 have to have extra teaching. How many students will have to have extra teaching?

Mark	No. of students	Cumulative frequency
$21 \leqslant x \leqslant 30$	3	3
$31 \leqslant x \leqslant 40$	9	12
$41 \leqslant x \leqslant 50$	12	24
$51 \leqslant x \leqslant 60$	15	39
$61 \leqslant x \leqslant 70$	22	61
$71 \leqslant x \leqslant 80$	16	77
$81 \leqslant x \leqslant 90$	10	87
$91 \leqslant x \leqslant 100$	8	95
$101 \leqslant x \leqslant 110$	3	98
$111 \leqslant x \leqslant 120$	2	100

The groups are given in a different way to those in the table on page 256. You will meet several ways of giving groups (for example, 21–30, $20 < x \leqslant 30$, $21 < x < 30$) but the important thing to remember is to plot the top point of each group against the corresponding cumulative frequency.

a and **b** Draw the graph and put on the lines for the median (50th value), lower and upper quartiles (25th and 75th values).

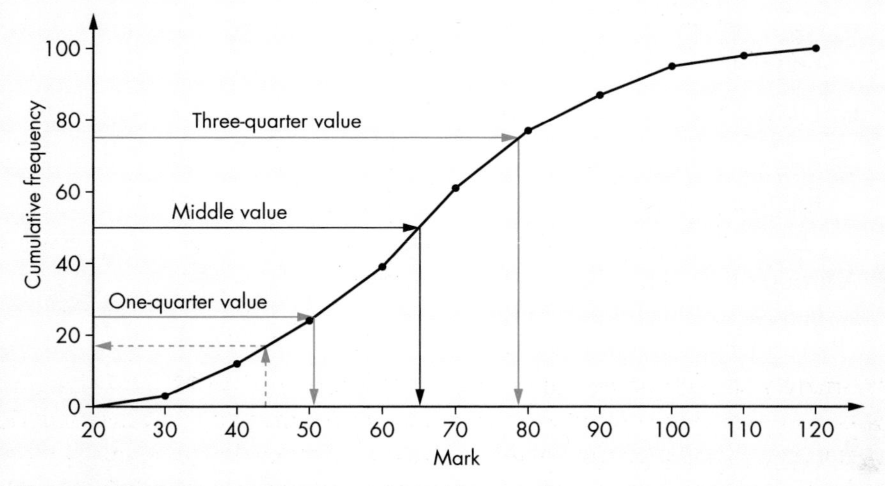

The required answers are read from the graph.

Median = 65 marks
Lower quartile = 51 marks
Upper quartile = 79 marks
Interquartile range = 79 – 51 = 28 marks

c At 44 on the mark axis, draw a perpendicular line to intersect the graph, and at the point of intersection draw a horizontal line across to the cumulative frequency axis, as shown. Number of students needing extra teaching is 18.

Note: An alternative way in which the table in Example 1 could have been set out is shown below. This arrangement has the advantage that the points to be plotted are taken straight from the last two columns. You have to decide which method you prefer. In examination papers, the columns of tables are sometimes given without headings, so you will need to be familiar with all the different ways in which the data can be set out.

Mark	No. of students	Less than	Cumulative frequency
$21 \leqslant x \leqslant 30$	3	30	3
$31 \leqslant x \leqslant 40$	9	40	12
$41 \leqslant x \leqslant 50$	12	50	24
$51 \leqslant x \leqslant 60$	15	60	39
$61 \leqslant x \leqslant 70$	22	70	61
$71 \leqslant x \leqslant 80$	16	80	77
$81 \leqslant x \leqslant 90$	10	90	87
$91 \leqslant x \leqslant 100$	8	100	95
$101 \leqslant x \leqslant 110$	3	110	98
$111 \leqslant x \leqslant 120$	2	120	100

EXERCISE 10A

B

FM 1 A class of 30 students was asked to estimate one minute. The teacher recorded the times the students actually said. The table on the right shows the results.

Time (seconds)	No. of students
$20 < x \leqslant 30$	1
$30 < x \leqslant 40$	3
$40 < x \leqslant 50$	6
$50 < x \leqslant 60$	12
$60 < x \leqslant 70$	3
$70 < x \leqslant 80$	3
$80 < x \leqslant 90$	2

a Copy the table and complete a cumulative frequency column.

b Draw a cumulative frequency diagram.

c Use your diagram to estimate the median time and the interquartile range.

FM 2 A group of 50 pensioners was given the same task as the children in question **1**. The results are shown in the table on the right.

Time (seconds)	No. of pensioners
$10 < x \leqslant 20$	1
$20 < x \leqslant 30$	2
$30 < x \leqslant 40$	2
$40 < x \leqslant 50$	9
$50 < x \leqslant 60$	17
$60 < x \leqslant 70$	13
$70 < x \leqslant 80$	3
$80 < x \leqslant 90$	2
$90 < x \leqslant 100$	1

a Copy the table and complete a cumulative frequency column.

b Draw a cumulative frequency diagram.

c Use your diagram to estimate the median time and the interquartile range.

d Which group, the students or the pensioners, would you say was better at estimating time? Give a reason for your answer.

FM 3 The sizes of 360 secondary schools in South Yorkshire are recorded in the table on the right.

No. of students	No. of schools
100–199	12
200–299	18
300–399	33
400–499	50
500–599	63
600–699	74
700–799	64
800–899	35
900–999	11

a Copy the table and complete a cumulative frequency column.

b Draw a cumulative frequency diagram.

c Use your diagram to estimate the median size of the schools and the interquartile range.

d Schools with fewer than 350 students are threatened with closure. About how many schools are threatened with closure?

B

4 The temperature at a seaside resort was recorded over a period of 50 days. The temperature was recorded to the nearest degree. The table on the right shows the results.

a Copy the table and complete a cumulative frequency column.

b Draw a cumulative frequency diagram. Note that as the temperature is to the nearest degree the top values of the groups are 7.5 °C, 10.5 °C, 13.5 °C, 16.5 °C, etc.

c Use your diagram to estimate the median temperature and the interquartile range.

Temperature (°C)	No. of days
5–7	2
8–10	3
11–13	5
14–16	6
17–19	6
20–22	9
23–25	8
26–28	6
29–31	5

FM **5** At the school charity fête, a game consists of throwing three darts and recording the total score. The results of the first 80 people to throw are recorded in the table on the right.

a Draw a cumulative frequency diagram to show the data.

b Use your diagram to estimate the median score and the interquartile range.

c People who score over 90 get a prize. About what percentage of the people get a prize?

Total score	No. of players
$1 \leqslant x \leqslant 20$	9
$21 \leqslant x \leqslant 40$	13
$41 \leqslant x \leqslant 60$	23
$61 \leqslant x \leqslant 80$	15
$81 \leqslant x \leqslant 100$	11
$101 \leqslant x \leqslant 120$	7
$121 \leqslant x \leqslant 140$	2

6 One hundred children in a primary school were asked to say how much pocket money they each got in a week. The results are in the table on the right.

a Copy the table and complete a cumulative frequency column.

b Draw a cumulative frequency diagram.

c Use your diagram to estimate the median amount of pocket money and the interquartile range.

Amount of pocket money (p)	No. of children
51–100	6
101–150	10
151–200	20
201–250	28
251–300	18
301–350	11
351–400	5
401–450	2

B

FM 7 James set his class an end of course test with two papers, A and B. He produced the cumulative frequency graphs below.

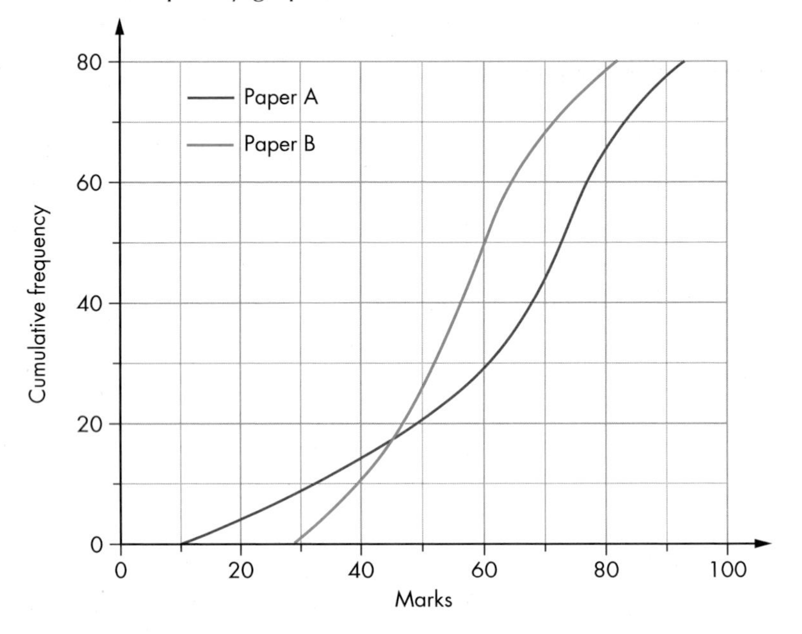

a What is the median score for each paper?

b What is the interquartile range for each paper?

c Which is the harder paper? Explain how you know.

James wanted 80% of the students to pass each paper and 20% of the students to get a top grade in each paper.

d What marks for each paper give:

i a pass **ii** the top grade?

FM PS 8 The lengths of time, in minutes, of 60 helpline telephone calls were recorded. A cumulative frequency diagram of this data is shown on the right.

Calculate the estimated mean length of telephone call.

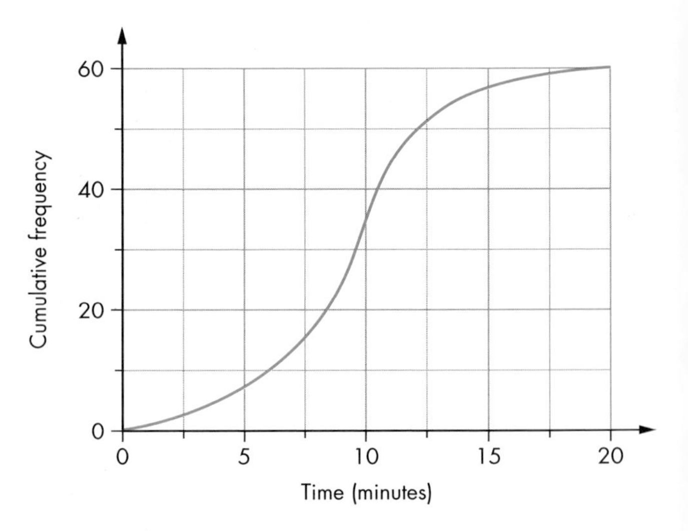

AU 9 Byron was given a cumulative frequency diagram showing the marks obtained by students in a mental maths test.

He was told the top 10% were given the top grade.

How would he find the marks needed to gain this top award?

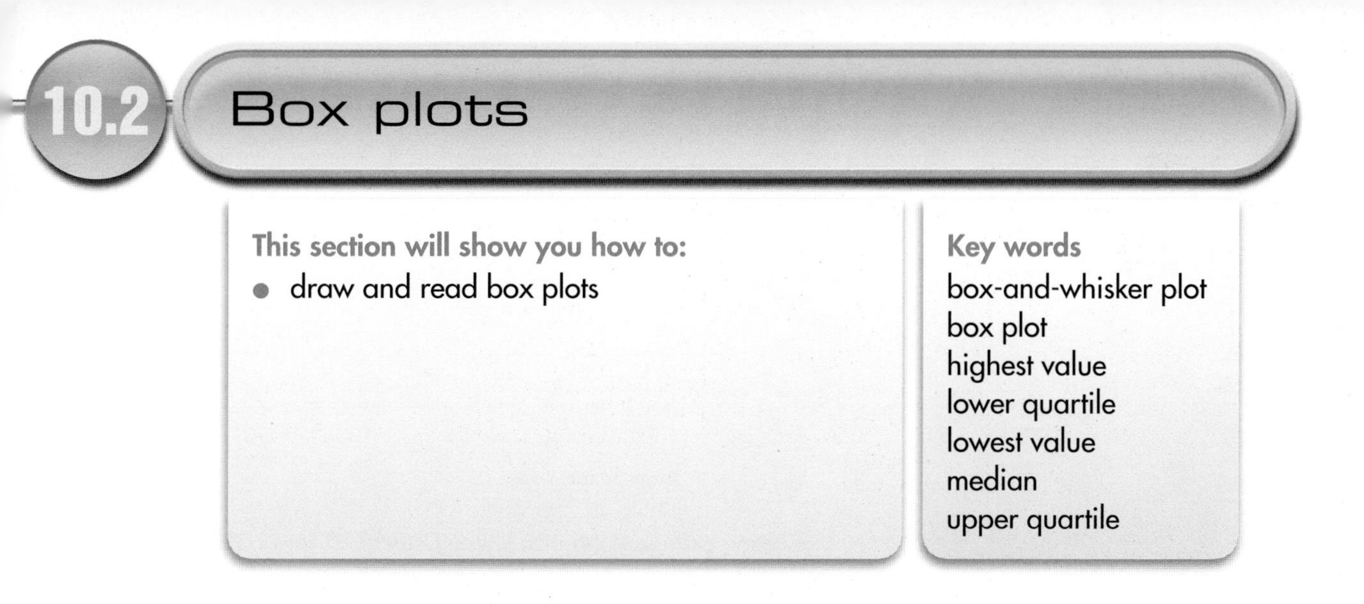

10.2 Box plots

This section will show you how to:

- draw and read box plots

Key words
box-and-whisker plot
box plot
highest value
lower quartile
lowest value
median
upper quartile

Another way of displaying data for comparison is by means of a **box-and-whisker plot** (or just **box plot**). This requires five pieces of data. These are the **lowest value**, the **lower quartile** (Q_1), the **median** (Q_2), the **upper quartile** (Q_3) and the **highest value**. They are drawn in the following way.

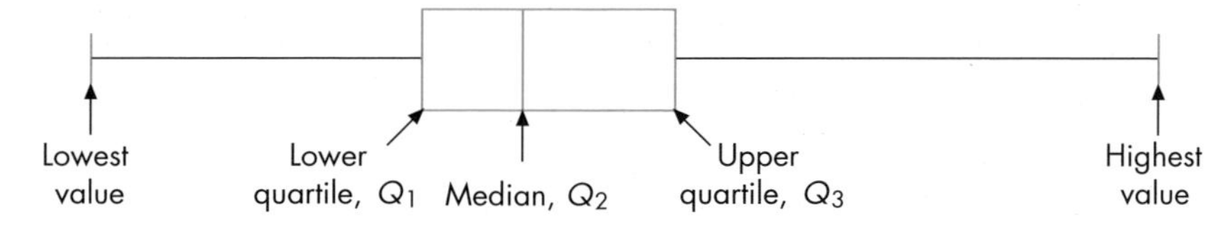

These data values are always placed against a scale so that they are accurately plotted.

The following diagrams show how the cumulative frequency curve, the frequency curve and the box plot are connected for three common types of distribution.

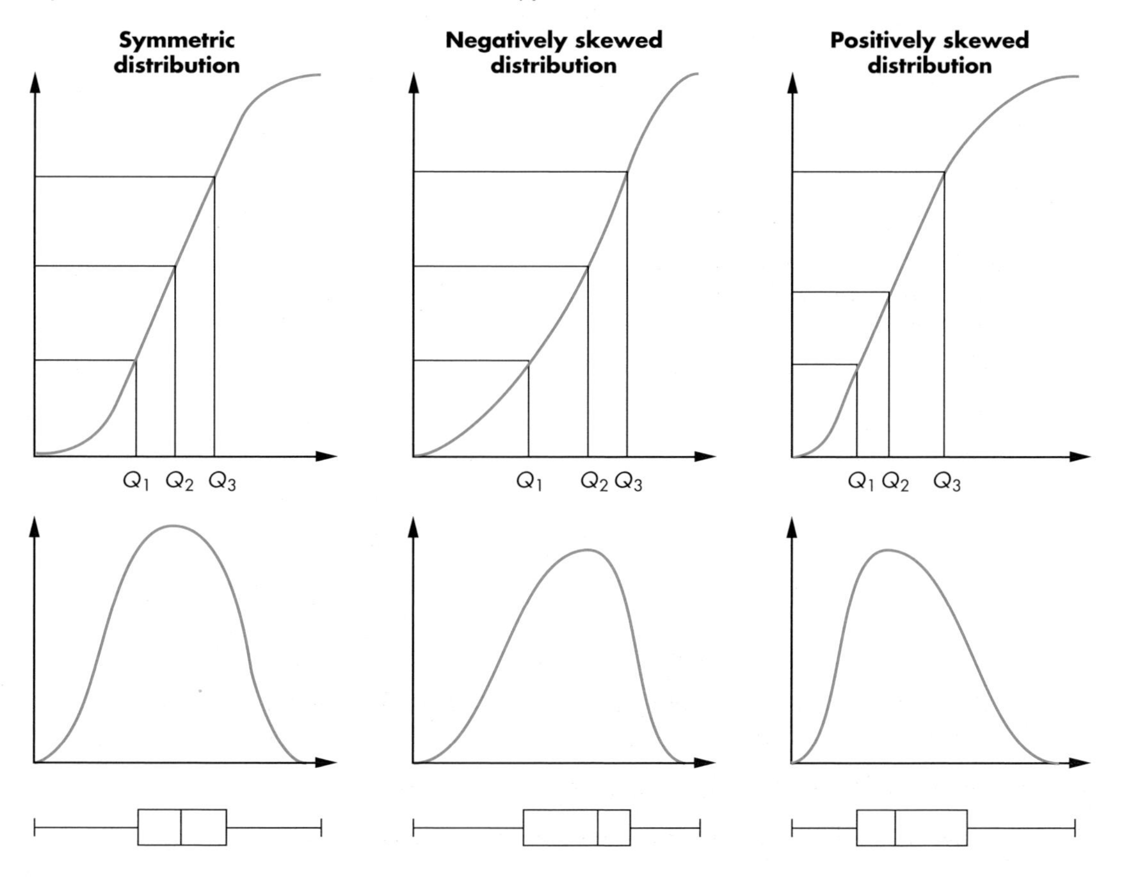

EXAMPLE 2

The box plot for the girls' marks in last year's examination is shown below.

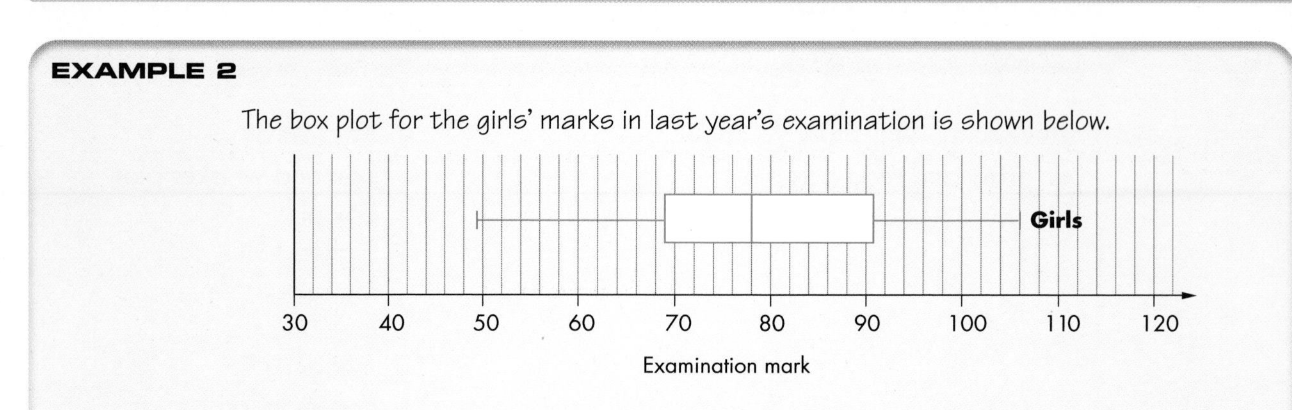

The boys' results for the same examination are: lowest mark 39, lower quartile 65, median 78, upper quartile 87, highest mark 112.

a On the same grid, draw the box plot for the boys' marks.

b Comment on the differences between the two distributions of marks.

a The data for boys and girls is plotted on the grid below.

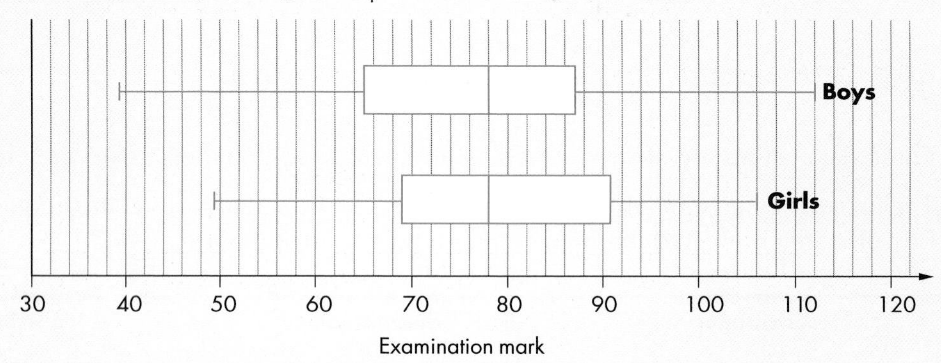

b The girls and boys have the same median mark but both the lower and upper quartiles for the girls are higher than those for the boys, and the girls' range is slightly smaller than the boys'.

This suggests that the girls did better than the boys overall, even though a boy got the highest mark.

EXERCISE 10B

FM 1 The box plot shows the times taken for a group of pensioners to do a set of 10 long-division calculations.

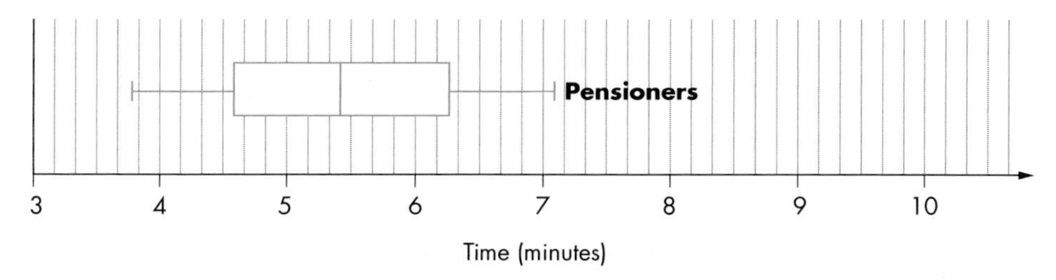

The same set of calculations was given to some students in Year 11. Their results are: shortest time 3 minutes 20 seconds, lower quartile 6 minutes 10 seconds, median 7 minutes, upper quartile 7 minutes 50 seconds and longest time 9 minutes 40 seconds.

a Copy the diagram and draw a box plot for the students' times.

b Comment on the differences between the two distributions.

FM 2 The box plot shows the sizes of secondary schools in Dorset.

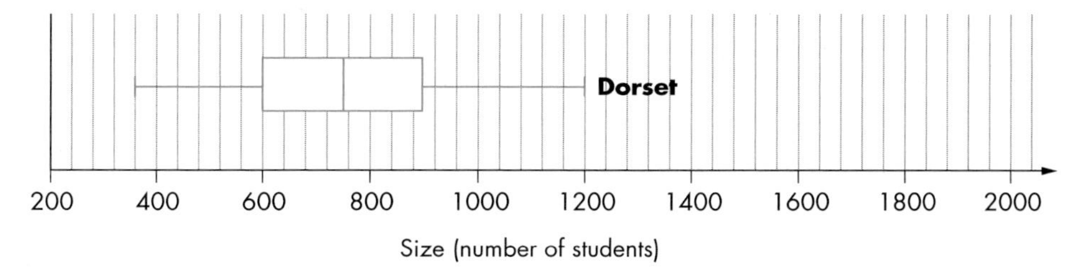

The data for schools in Rotherham is: smallest 280 students, lower quartile 1100 students, median 1400 students, upper quartile 1600 students, largest 1820 students.

a Copy the diagram and draw a box plot for the sizes of schools in Rotherham.

b Comment on the differences between the two distributions.

FM 3 The box plots for the noon temperature at two resorts, recorded over a year, are shown on the grid below.

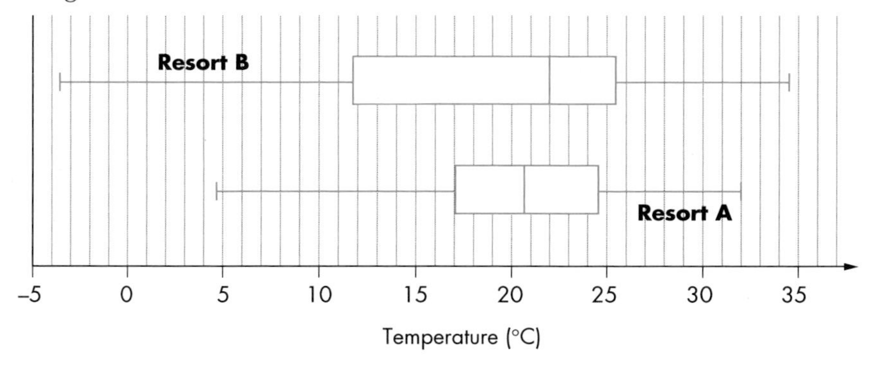

a Comment on the differences in the two distributions.

b Mary wants to go on holiday in July. Which resort would you recommend? Why?

B

FM 4 The following table shows some data on the annual salaries for 100 men and 100 women.

	Lowest salary	Lower quartile	Median salary	Upper quartile	Highest salary
Men	£6500	£16 000	£20 000	£22 000	£44 500
Women	£7000	£14 000	£16 000	£21 500	£33 500

a Draw box plots to compare both sets of data.

b Comment on the differences between the distributions.

FM 5 The table shows the monthly salaries of 100 families.

Monthly salary (£)	No. of families
1451–1500	8
1501–1550	14
1551–1600	25
1601–1650	35
1651–1700	14
1701–1750	4

a Draw a cumulative frequency diagram to show the data.

b Estimate the median monthly salary and the interquartile range.

c The lowest monthly salary was £1480 and the highest was £1740.

i Draw a box plot to show the distribution of salaries.

ii Is the distribution symmetric, negatively skewed or positively skewed?

FM 6 A health practice had two doctors, Dr Excel and Dr Collins.

The following box plots were created to illustrate the waiting times for their patients during October.

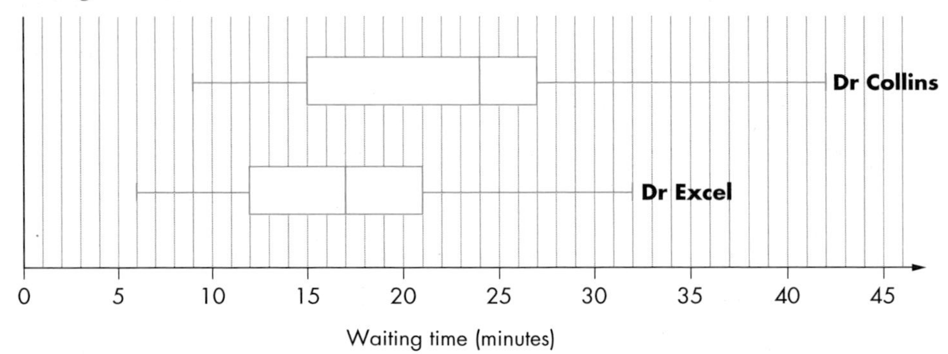

a For Dr Collins, what is:

i the median waiting time

ii the interquartile range for his waiting time

iii the longest time a patient had to wait in October?

b For Dr Excel, what is:

i the shortest waiting time for any patient in October

ii the median waiting time

iii the interquartile range for his waiting time?

c Anwar was deciding which doctor to try to see. Which one would you advise he sees? Why?

PS 7 The box plot for a school's end-of-year mathematics tests are shown below.

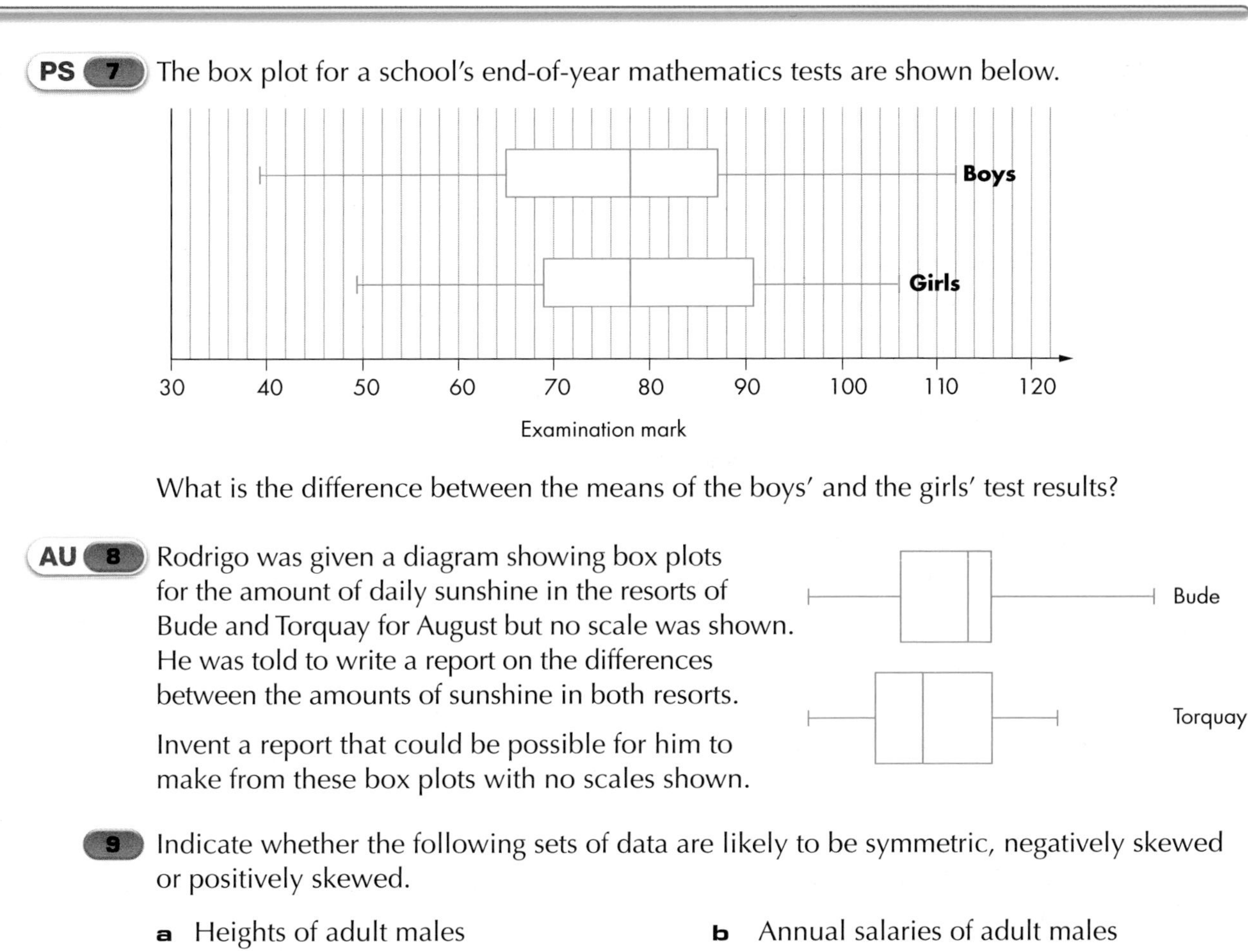

What is the difference between the means of the boys' and the girls' test results?

AU 8 Rodrigo was given a diagram showing box plots for the amount of daily sunshine in the resorts of Bude and Torquay for August but no scale was shown. He was told to write a report on the differences between the amounts of sunshine in both resorts.

Invent a report that could be possible for him to make from these box plots with no scales shown.

9 Indicate whether the following sets of data are likely to be symmetric, negatively skewed or positively skewed.

a Heights of adult males

b Annual salaries of adult males

c Shoe sizes of adult males

d Weights of babies born in Britain

e Speeds of cars on a motorway in the middle of the night

f Speeds of cars on a motorway in the rush hour

g Shopping bills in a supermarket the week before Christmas

h Number of letters in the words in a teenage magazine

i Time taken for students to get to school in the morning

j Time taken for students to run 1 mile

AU 10 Below are four cumulative frequency diagrams and four box plots.

Match each cumulative frequency diagram with a box plot.

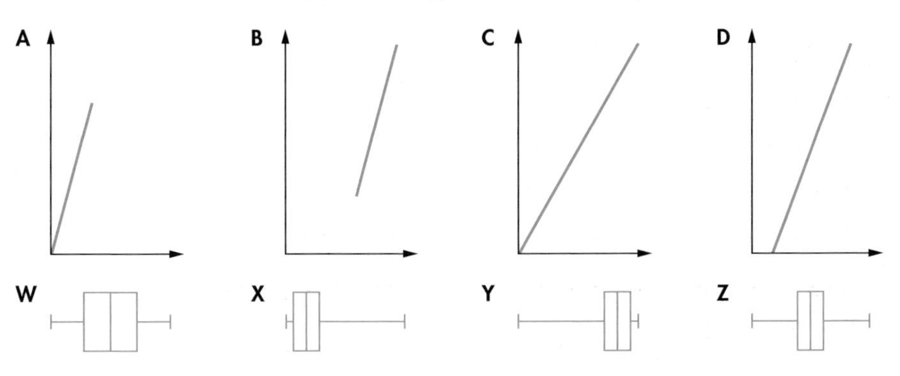

10.3 Histograms with bars of unequal width

This section will show you how to:

- draw and read histograms where the bars are of unequal width
- find the median, quartiles and interquartile range from a histogram

Key words

class interval
interquartile range
lower quartile
median
upper quartile

Sometimes the data in a frequency distribution are grouped into classes with intervals that are different. In this case, the resulting histogram has bars of unequal width.

The key fact that you should always remember is that the area of a bar in a histogram represents the class frequency of the bar. So, in the case of an unequal-width histogram, you find the height to draw each bar by dividing its class frequency by its **class interval** width (bar width), which is the difference between the lower and upper bounds for each interval. Conversely, given a histogram, you can find any of its class frequencies by multiplying the height of the corresponding bar by its width.

It is for this reason that the scale on the vertical axes of histograms is nearly always labelled 'frequency density', where

$$\text{frequency density} = \frac{\text{frequency of class interval}}{\text{width of class interval}}$$

EXAMPLE 3

The heights of a group of girls were measured. The results were classified as shown in the table.

Height, h (cm)	$151 \leqslant h < 153$	$153 \leqslant h < 154$	$154 \leqslant h < 155$	$155 \leqslant h < 159$	$159 \leqslant h < 160$
Frequency	64	43	47	96	12

It is convenient to write the table vertically and add two columns, class width and frequency density.

The class width is found by subtracting the lower class boundary from the upper class boundary. The frequency density is found by dividing the frequency by the class width.

Height, h (cm)	Frequency	Class width	Frequency density
$151 \leqslant h < 153$	64	2	32
$153 \leqslant h < 154$	43	1	43
$154 \leqslant h < 155$	47	1	47
$155 \leqslant h < 159$	96	4	24
$159 \leqslant h < 160$	12	1	12

The histogram can now be drawn. The horizontal scale should be marked off as normal, from a value below the lowest value in the table to a value above the largest value in the table. In this case, mark the scale from 150 cm to 160 cm. The vertical scale is always frequency density and is marked up to at least the largest frequency density in the table. In this case, 50 is a sensible value.

Each bar is drawn between the lower class interval and the upper class interval horizontally, and up to the frequency density vertically.

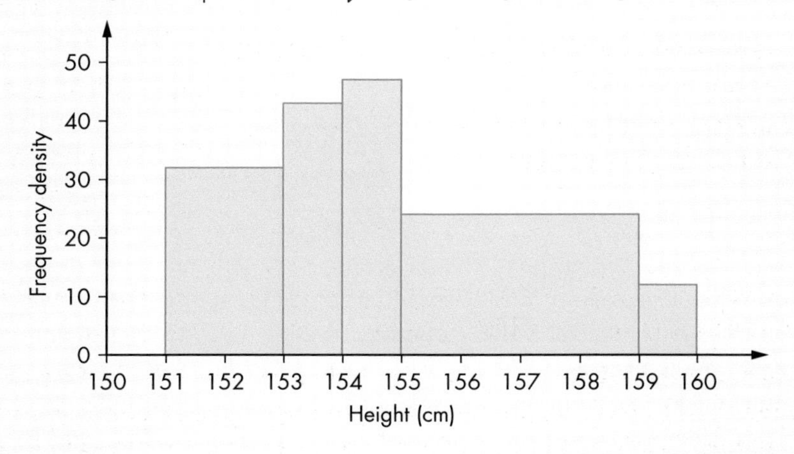

EXAMPLE 4

This histogram shows the distribution of heights of daffodils in a greenhouse.

a Complete a frequency table for the heights of the daffodils, and show the cumulative frequency.

b Find the **median** height.

c Find the **interquartile range** of the heights.

d Estimate the mean of the distribution.

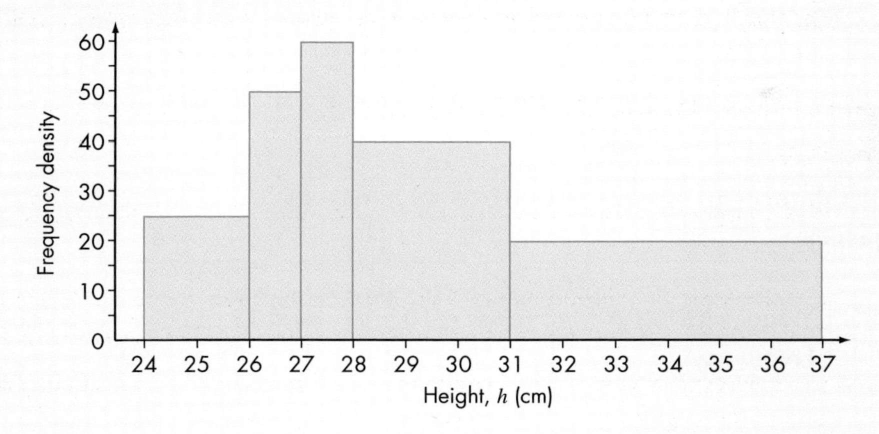

a The frequency table will have groups of $24 \leqslant h < 26$, $26 \leqslant h < 27$, etc. These are read from the height axis. The frequencies will be found by multiplying the width of each bar by the frequency density. Remember that the value on the vertical axis is not the frequency.

Height, h (cm)	$24 \leqslant h < 26$	$26 \leqslant h < 27$	$27 \leqslant h < 28$	$28 \leqslant h < 31$	$31 \leqslant h < 37$
Frequency	50	50	60	120	120
Cumulative frequency	50	100	160	280	400

continued

EXAMPLE 4 (continued)

b There are 400 values so the median will be the 200th value. Counting up the frequencies from the beginning you reach the third column of the table on the previous page.

The median occurs in the $28 \leq h < 31$ group. There are 160 values before this group and 120 in it. To get to the 200th value you need to go 40 more values into this group. 40 out of 120 is one-third. One-third of the way through this group is the value 29 cm. Hence the median is 29 cm.

c The interquartile range is the difference between the **upper quartile** and the **lower quartile**, the quarter and three-quarter values respectively. In this case, the lower quartile is the 100th value (found by dividing 400, the total number of values, by 4) and the upper quartile is the 300th value. So, in the same way that you found the median, you can find the lower (100th value) and upper (300th value) quartiles. The 100th value is at 27 cm and the 300th value is at 32 cm. The interquartile range is 32 cm – 27 cm = 5 cm.

d To estimate the mean, use the table to get the midway values of the groups and multiply these by the frequencies. The sum of these divided by 400 will give the estimated mean.

So, the mean is:

$$(25 \times 50 + 26.5 \times 50 + 27.5 \times 60 + 29.5 \times 120 + 34 \times 120) \div 400$$
$$= 11\,845 \div 400 = 29.6 \text{ cm (3 significant figures)}$$

EXERCISE 10C

A

1 Draw histograms for these grouped frequency distributions.

a

Temperature, t (°C)	$8 \leq t < 10$	$10 \leq t < 12$	$12 \leq t < 15$	$15 \leq t < 17$	$17 \leq t < 20$	$20 \leq t < 24$
Frequency	5	13	18	4	3	6

b

Wage, w (£1000)	$6 \leq w < 10$	$10 \leq w < 12$	$12 \leq w < 16$	$16 \leq w < 24$
Frequency	16	54	60	24

c

Age, a (nearest year)	$11 \leq a < 14$	$14 \leq a < 16$	$16 \leq a < 17$	$17 \leq a < 20$
Frequency	51	36	12	20

d

Pressure, p (mm)	$745 \leq p < 755$	$755 \leq p < 760$	$760 \leq p < 765$	$765 \leq p < 775$
Frequency	4	6	14	10

e

Time, t (min)	$0 \leq t < 8$	$8 \leq t < 12$	$12 \leq t < 16$	$16 \leq t < 20$
Frequency	72	84	54	36

FM **2** The following information was gathered about the weekly pocket money given to 14-year-olds.

Pocket money, p (£)	$0 \leqslant p < 2$	$2 \leqslant p < 4$	$4 \leqslant p < 5$	$5 \leqslant p < 8$	$8 \leqslant p < 10$
Girls	8	15	22	12	4
Boys	6	11	25	15	6

a Represent the information about the boys on a histogram.

b Represent both sets of data with a frequency polygon, using the same pair of axes.

c What is the mean amount of pocket money given to each sex? Comment on your answer.

3 The sales of the *Star* newspaper over 70 years are recorded in this table.

Years	1940–60	1961–80	1981–90	1991–2000	2001–05	2006–2010
Copies	62 000	68 000	71 000	75 000	63 000	52 000

Illustrate this information on a histogram. Take the class boundaries as 1940, 1960, 1980, 1990, 2000, 2005, 2010.

4 The London trains were always late, so one month a survey was undertaken to find how many trains were late, and by how many minutes (to the nearest minute). The results are illustrated by this histogram.

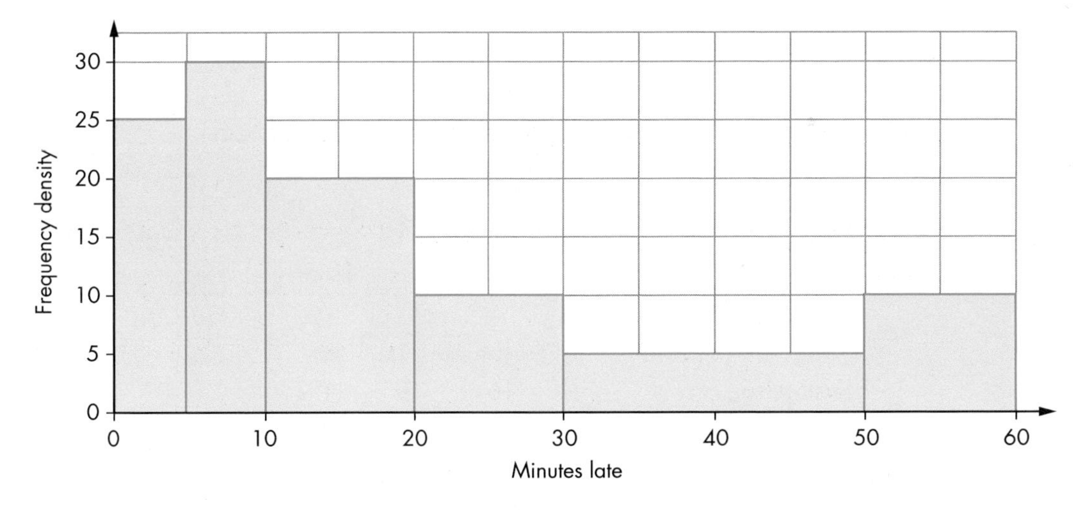

a How many trains were in the survey?

b How many trains were delayed for longer than 15 minutes?

AU **5** Hannah was asked to create a histogram.

Explain how Hannah will find the height of each bar on the frequency density scale.

A

A*

6 For each of the frequency distributions illustrated in the histograms:

i write down the grouped frequency table

ii state the modal group

iii estimate the median

iv find the lower and upper quartiles and the interquartile range

v estimate the mean of the distribution.

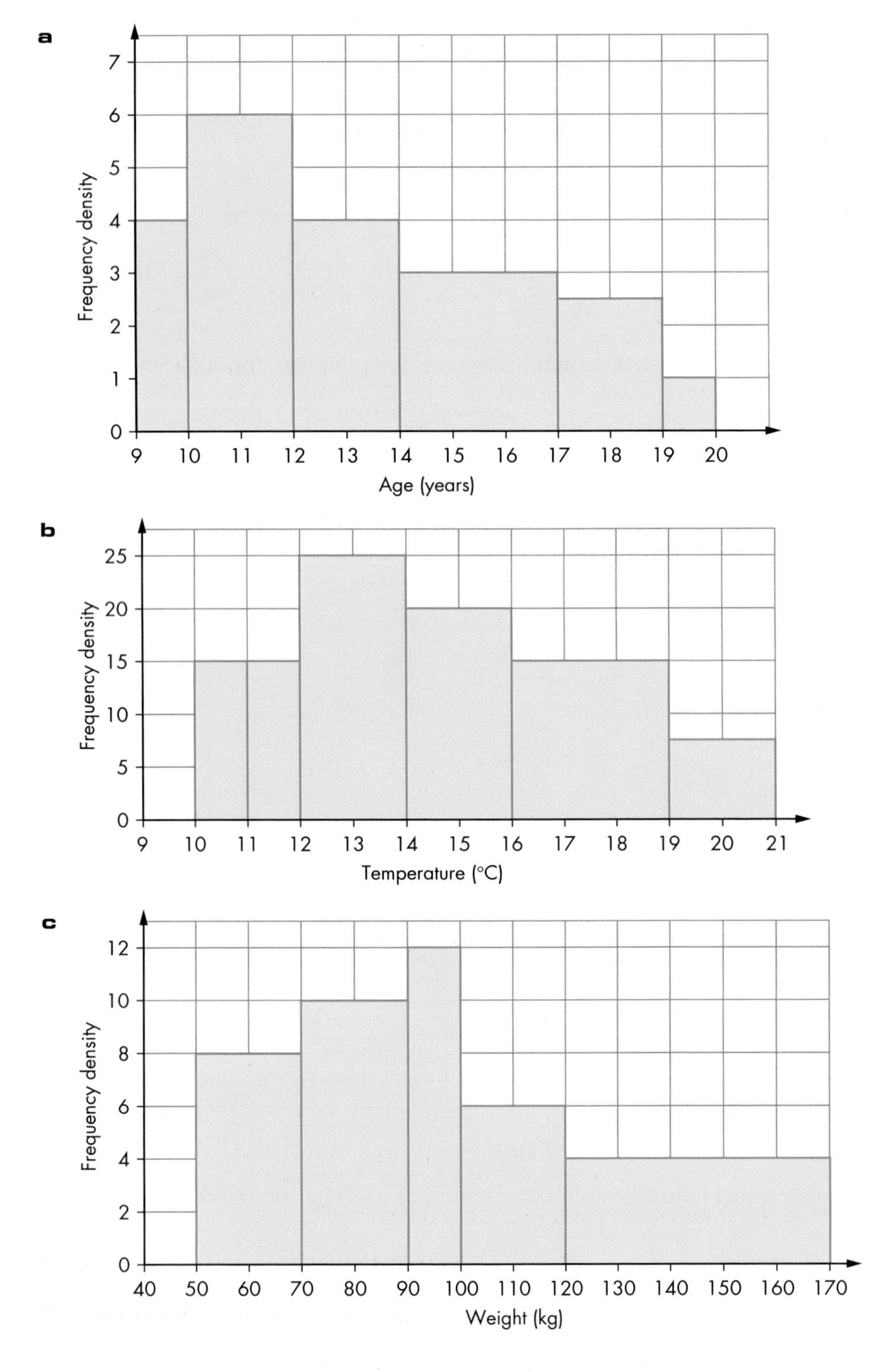

A*

7 All the patients in a hospital were asked how long it was since they last saw a doctor. The results are shown in the table.

Hours, *h*	$0 \leqslant h < 2$	$2 \leqslant h < 4$	$4 \leqslant h < 6$	$6 \leqslant h < 10$	$10 \leqslant h < 16$	$16 \leqslant h < 24$
Frequency	8	12	20	30	20	10

a Find the median time since a patient last saw a doctor.

b Estimate the mean time since a patient last saw a doctor.

c Find the interquartile range of the times.

8 One summer, Albert monitored the weight of the tomatoes grown on each of his plants. His results are summarised in this table.

Weight, *w* (kg)	$6 \leqslant w < 10$	$10 \leqslant w < 12$	$12 \leqslant w < 16$	$16 \leqslant w < 20$	$20 \leqslant w < 25$
Frequency	8	15	28	16	10

a Draw a histogram for this distribution.

b Estimate the median weight of tomatoes the plants produced.

c Estimate the mean weight of tomatoes the plants produced.

d How many plants produced more than 15 kg?

9 A survey was carried out to find the speeds of cars passing a particular point on the M1. The histogram illustrates the results of the survey.

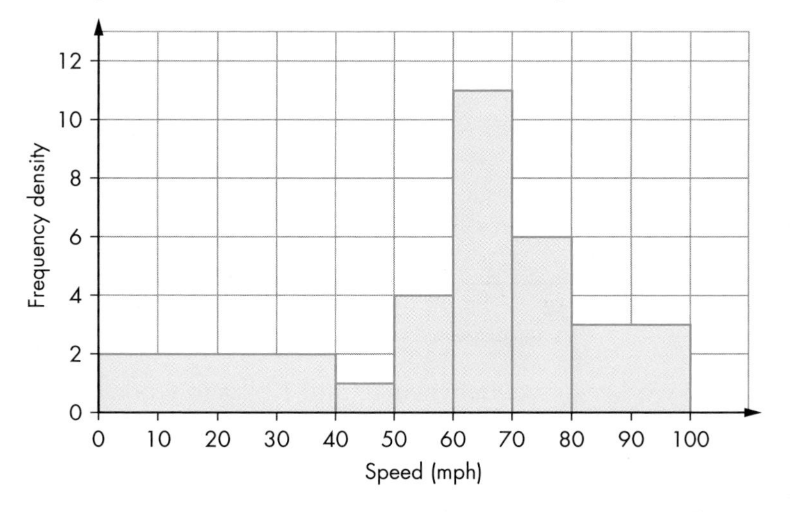

a Copy and complete this table.

Speed, *v* (mph)	$0 < v \leqslant 40$	$40 < v \leqslant 50$	$50 < v \leqslant 60$	$60 < v \leqslant 70$	$70 < v \leqslant 80$	$80 < v \leqslant 100$
Frequency		10	40	110		

b Find the number of cars included in the survey.

c Work out an estimate of the median speed of the cars on this part of the M1.

d Work out an estimate of the mean speed of the cars on this part of the M1.

A*

10 The histogram shows the test scores for 320 students in a school.

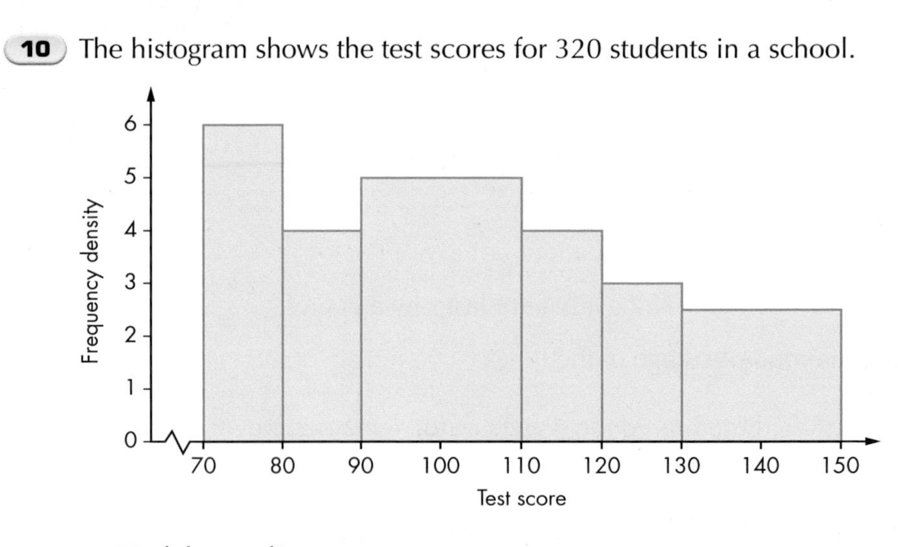

a Find the median score.

b Find the interquartile range of the scores.

c Find an estimate for the mean score.

FM **d** What was the pass score if 90% of the students passed this test?

FM PS **11** The distances employees of a company travel to work are shown in the histogram below.

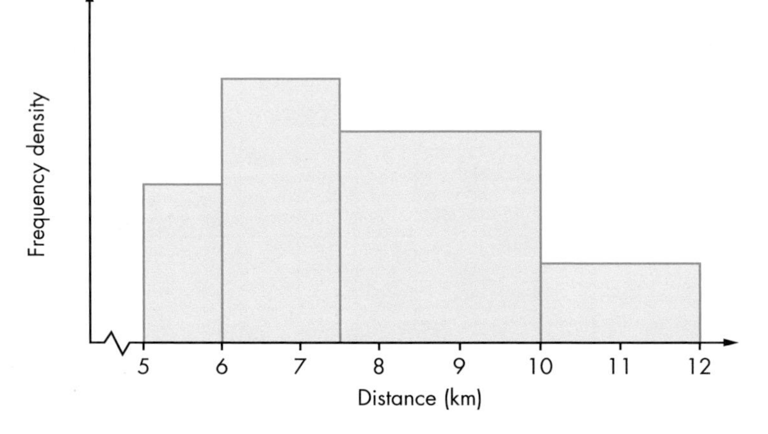

It is known that 18 workers travel between 10 and 12 km to work. What is the probability of choosing a worker at random who travels less than 7.5 km to work?

GRADE BOOSTER

- **B** You can draw a cumulative frequency diagram
- **B** You can find medians and interquartile ranges from cumulative frequency diagrams
- **A** You can draw and interpret box plots
- **A** You can draw and read histograms where the boxes are of unequal width
- **A*** You can find the median, quartile and interquartile range from a histogram

What you should know now

- How to construct a cumulative frequency diagram
- How to draw and interpret box plots
- How to draw and interpret histograms including finding the median, quartile and interquartile range
- How to draw histograms for discrete and continuous data

1 The cumulative frequency graph shows information about the numbers of goals scored by 160 players.

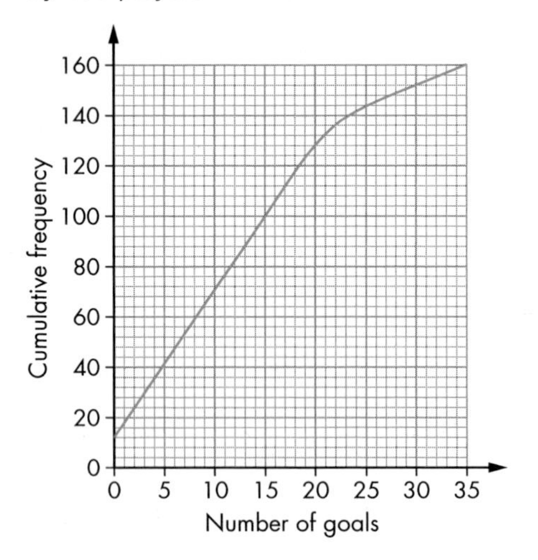

a Use this graph to find an estimate for

i the median

ii the lower quartile. *(2 marks)*

The lowest number of goals scored was 0

The highest number of goals scored was 32

b On a grid, draw a box plot to show information about the numbers of goals scored.

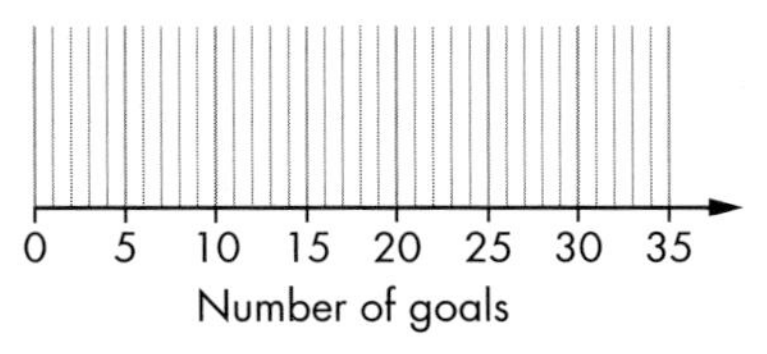

(3 marks)

Edexcel, March 2007, Paper 9 HIgher, Question 3

2 60 people went on a boat.

The grouped frequency table shows information about their ages.

Age (A years)	Frequency
$0 < A \leqslant 10$	4
$10 < A \leqslant 20$	8
$20 < A \leqslant 30$	11
$30 < A \leqslant 40$	16
$40 < A \leqslant 50$	9
$50 < A \leqslant 60$	7
$60 < A \leqslant 70$	5

a Copy and complete the cumulative frequency table.

Age (A years)	Cumulative frequency
$0 < A \leqslant 10$	4
$10 < A \leqslant 20$	
$20 < A \leqslant 30$	
$30 < A \leqslant 40$	
$40 < A \leqslant 50$	
$50 < A \leqslant 60$	
$60 < A \leqslant 70$	

(1 mark)

b On a grid, draw a cumulative frequency graph for your table. *(2 marks)*

c Use your graph to find an estimate for the median age of these 60 people. *(1 mark)*

Edexcel, June 2007, Paper 9 Higher, Question 2

3 Verity records the heights of the girls in her class.

The height of the shortest girl is 1.38 m.

The height of the tallest girl is 1.81 m.

The median height is 1.63 m.

The lower quartile is 1.54 m.

The interquartile range is 0.14 m.

a On a grid, draw a box plot for this information.

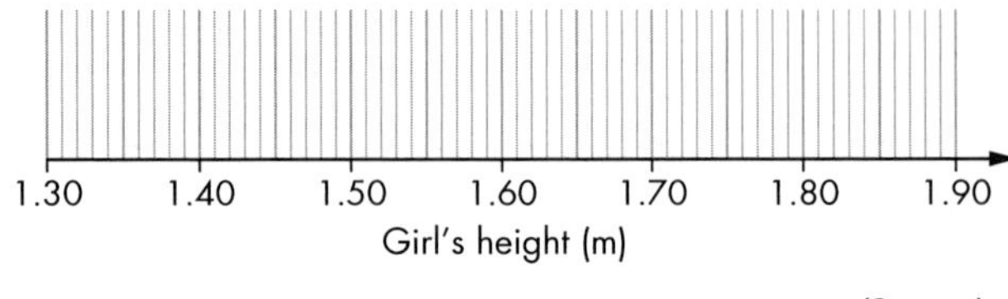

(3 marks)

The box plot below shows information about the heights of the boys in Verity's class.

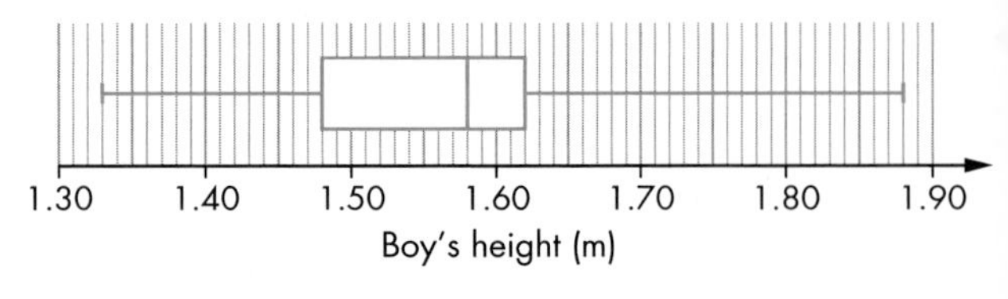

b Compare the distributions of the boys' heights and the girls' heights. *(2 marks)*

Edexcel, March 2008, Paper 9 Higher, Question 4

4 John and Peter each own a garage.

They both sell used cars.

The box plots show some information about the prices of cars at their garages.

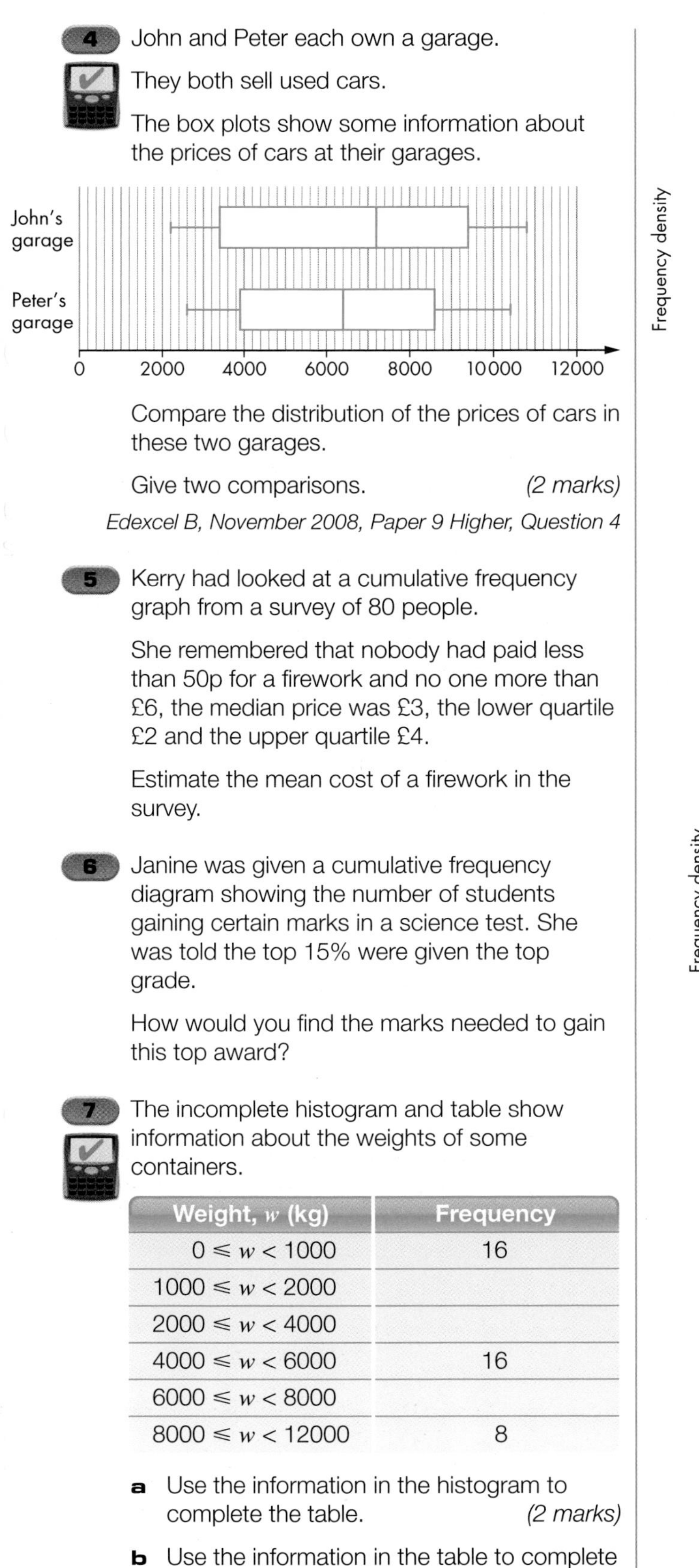

Compare the distribution of the prices of cars in these two garages.

Give two comparisons. *(2 marks)*

Edexcel B, November 2008, Paper 9 Higher, Question 4

5 Kerry had looked at a cumulative frequency graph from a survey of 80 people.

She remembered that nobody had paid less than 50p for a firework and no one more than £6, the median price was £3, the lower quartile £2 and the upper quartile £4.

Estimate the mean cost of a firework in the survey.

6 Janine was given a cumulative frequency diagram showing the number of students gaining certain marks in a science test. She was told the top 15% were given the top grade.

How would you find the marks needed to gain this top award?

7 The incomplete histogram and table show information about the weights of some containers.

Weight, w (kg)	Frequency
$0 \leqslant w < 1000$	16
$1000 \leqslant w < 2000$	
$2000 \leqslant w < 4000$	
$4000 \leqslant w < 6000$	16
$6000 \leqslant w < 8000$	
$8000 \leqslant w < 12000$	8

a Use the information in the histogram to complete the table. *(2 marks)*

b Use the information in the table to complete the histogram. *(2 marks)*

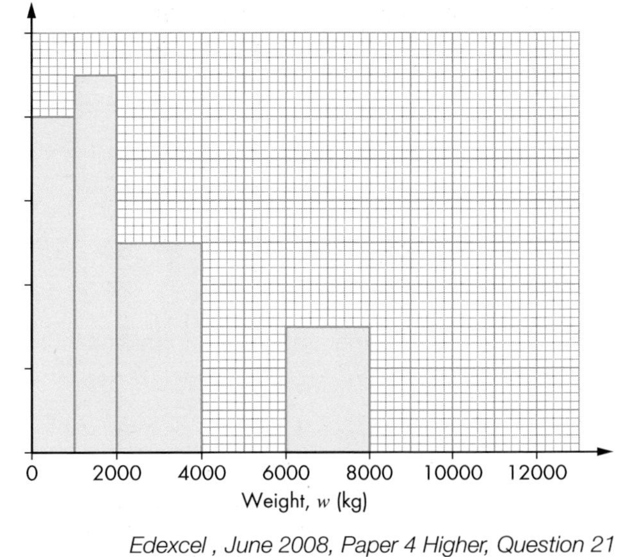

Edexcel , June 2008, Paper 4 Higher, Question 21

8 The incomplete histogram and table give some information about distances some teachers travel to school.

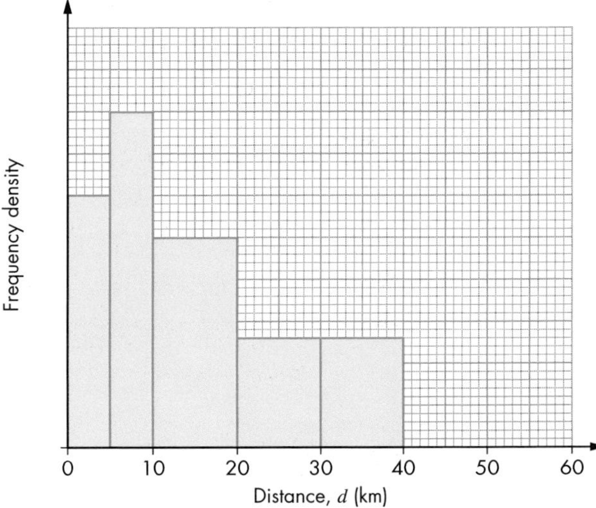

a Use the information in the histogram to complete the table.

Distance, d (km)	Frequency
$0 < d \leqslant 5$	15
$5 < d \leqslant 10$	20
$10 < d \leqslant 20$	
$20 < d \leqslant 40$	
$40 < d \leqslant 60$	10

(2 marks)

b Use the information in the table to complete the histogram. *(1 mark)*

Edexcel, November 2008, Paper 3 Higher, Question 23

9 

The table gives some information about the lengths of time some boys took to run a race.

Time taken, t (min)	Frequency
$40 < t \leqslant 50$	16
$50 < t \leqslant 55$	18
$55 < t \leqslant 65$	32
$65 < t \leqslant 80$	30
$80 < t \leqslant 100$	24

Draw a histogram for the information in the table.

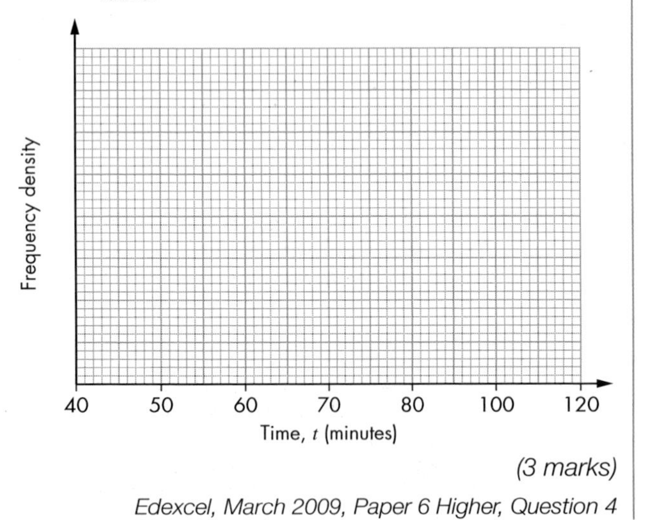

(3 marks)

Edexcel, March 2009, Paper 6 Higher, Question 4

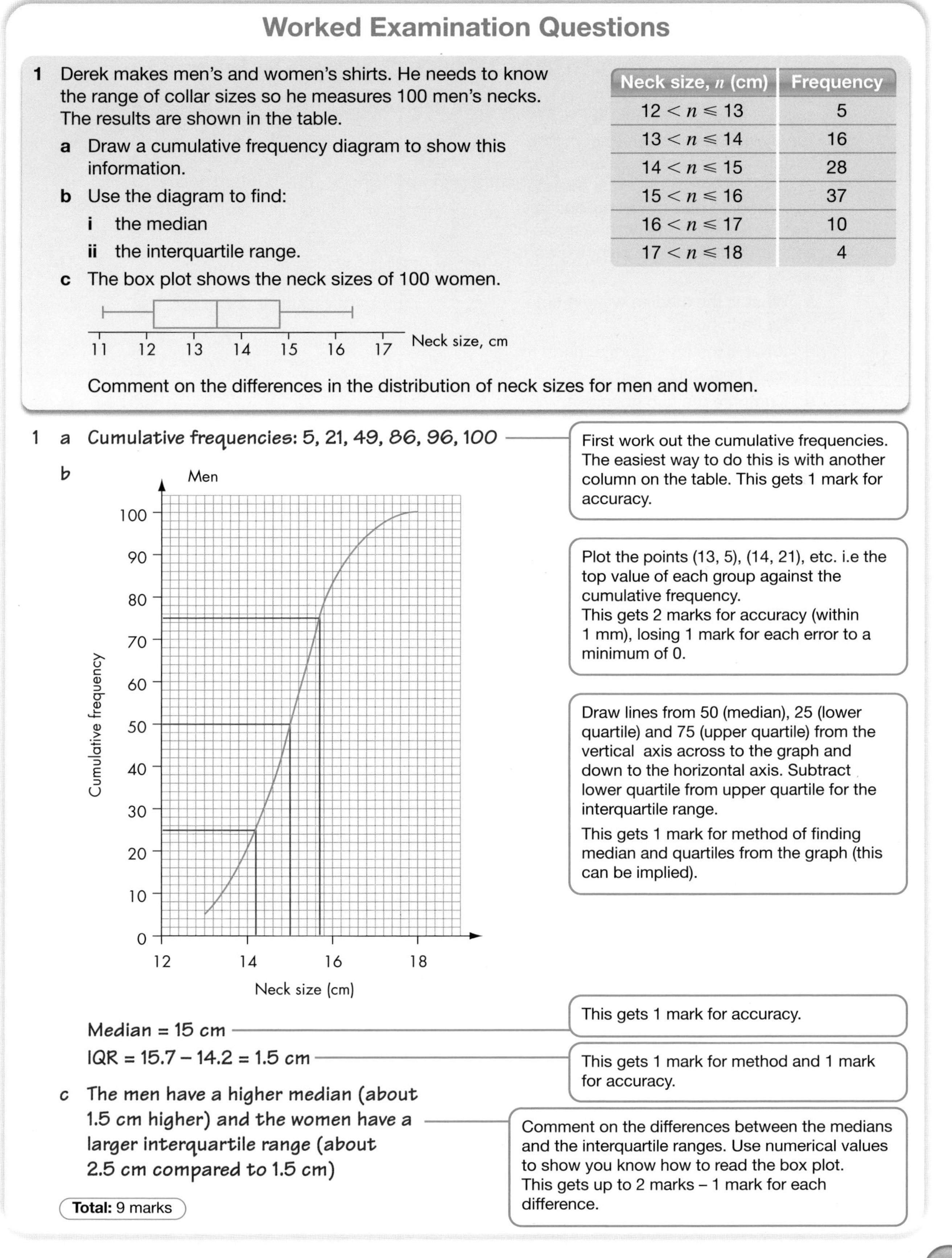

Worked Examination Questions

1 Derek makes men's and women's shirts. He needs to know the range of collar sizes so he measures 100 men's necks. The results are shown in the table.

Neck size, n (cm)	Frequency
$12 < n \leqslant 13$	5
$13 < n \leqslant 14$	16
$14 < n \leqslant 15$	28
$15 < n \leqslant 16$	37
$16 < n \leqslant 17$	10
$17 < n \leqslant 18$	4

a Draw a cumulative frequency diagram to show this information.

b Use the diagram to find:

i the median

ii the interquartile range.

c The box plot shows the neck sizes of 100 women.

Comment on the differences in the distribution of neck sizes for men and women.

1 a Cumulative frequencies: 5, 21, 49, 86, 96, 100

First work out the cumulative frequencies. The easiest way to do this is with another column on the table. This gets 1 mark for accuracy.

b

Plot the points (13, 5), (14, 21), etc. i.e the top value of each group against the cumulative frequency.
This gets 2 marks for accuracy (within 1 mm), losing 1 mark for each error to a minimum of 0.

Draw lines from 50 (median), 25 (lower quartile) and 75 (upper quartile) from the vertical axis across to the graph and down to the horizontal axis. Subtract lower quartile from upper quartile for the interquartile range.

This gets 1 mark for method of finding median and quartiles from the graph (this can be implied).

Median = 15 cm — This gets 1 mark for accuracy.

IQR = 15.7 – 14.2 = 1.5 cm — This gets 1 mark for method and 1 mark for accuracy.

c The men have a higher median (about 1.5 cm higher) and the women have a larger interquartile range (about 2.5 cm compared to 1.5 cm)

Comment on the differences between the medians and the interquartile ranges. Use numerical values to show you know how to read the box plot.
This gets up to 2 marks – 1 mark for each difference.

Total: 9 marks

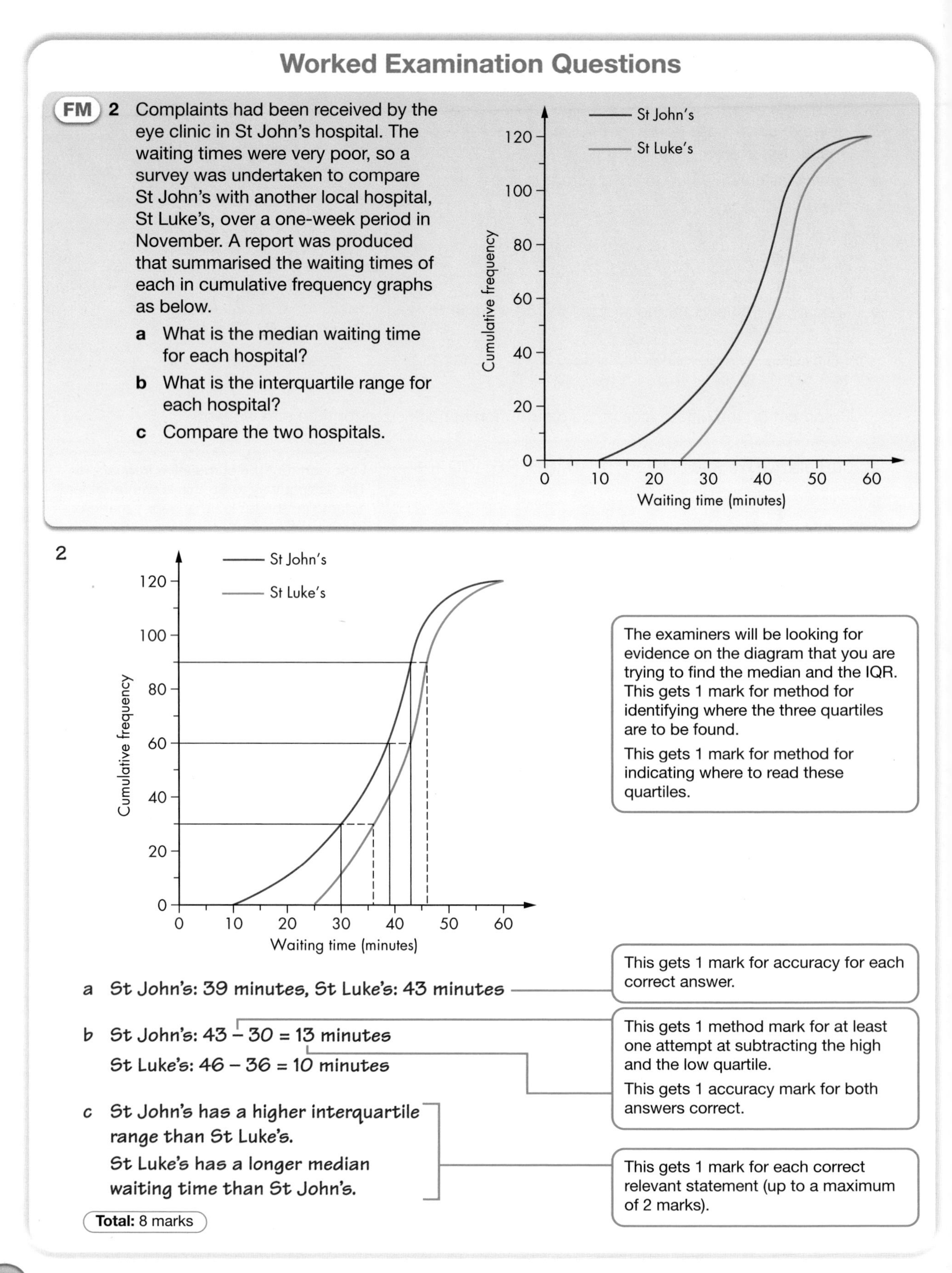

Worked Examination Questions

FM 2 Complaints had been received by the eye clinic in St John's hospital. The waiting times were very poor, so a survey was undertaken to compare St John's with another local hospital, St Luke's, over a one-week period in November. A report was produced that summarised the waiting times of each in cumulative frequency graphs as below.

a What is the median waiting time for each hospital?

b What is the interquartile range for each hospital?

c Compare the two hospitals.

2

The examiners will be looking for evidence on the diagram that you are trying to find the median and the IQR. This gets 1 mark for method for identifying where the three quartiles are to be found.

This gets 1 mark for method for indicating where to read these quartiles.

a St John's: 39 minutes, St Luke's: 43 minutes

This gets 1 mark for accuracy for each correct answer.

b St John's: 43 – 30 = 13 minutes

St Luke's: 46 – 36 = 10 minutes

This gets 1 method mark for at least one attempt at subtracting the high and the low quartile.

This gets 1 accuracy mark for both answers correct.

c St John's has a higher interquartile range than St Luke's.

St Luke's has a longer median waiting time than St John's.

This gets 1 mark for each correct relevant statement (up to a maximum of 2 marks).

Total: 8 marks

Worked Examination Questions

AU **3** Gabriela was given a cumulative frequency diagram showing the number of students gaining marks in an English test. She was told that 85% of the students got passable scores. How would you find the pass mark?

3 Find 15% (100 – 85) of the total frequency on the cumulative frequency scale, read along to the graph and read down to the marks. The mark seen will be the pass mark.

This gets 1 mark for stating the need to find 15% of the frequency.

This gets 1 mark for for indicating how you would read the relevant mark for 15% of the frequency.

Total: 2 marks

10 Functional Maths
Population pyramids

Population pyramids are often used as a way of illustrating the make-up of a country's population. It is possible to use these pyramids to compare the populations of different countries around the world. This makes them a useful tool for social statisticians and world organisations such as the United Nations.

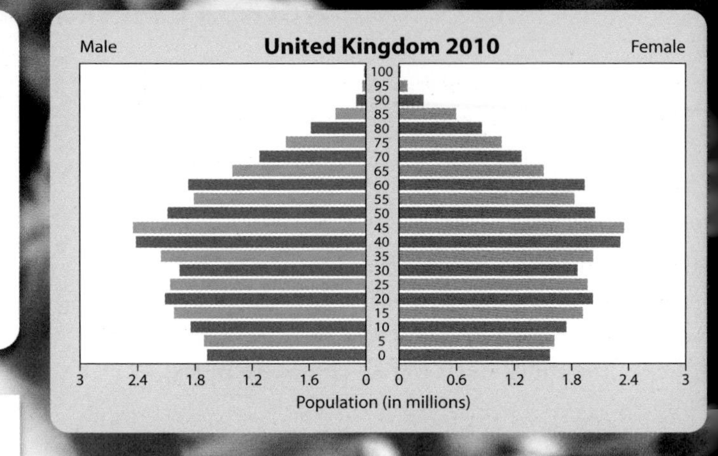

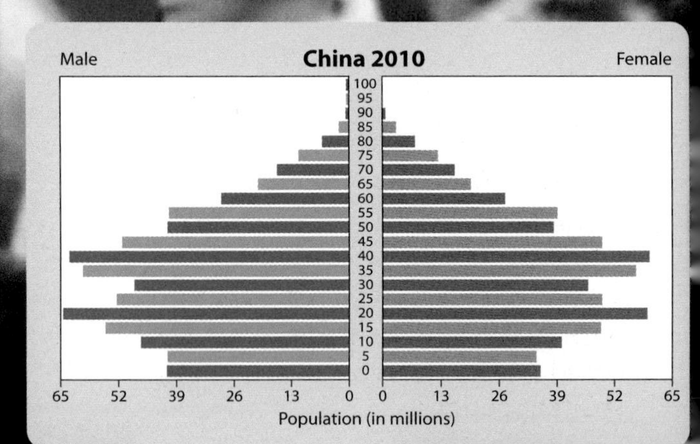

Getting started

Look at the population pyramid for the UK in 2010. It shows the number of people in the country in five-year age bands: 0 to 4, 5 to 9, 10 to 14, and so on.

1. Which age group has the largest number of people? Is the same true for men and women?

   Estimate the number of people in this age group.
2. Do you think the number of people born in the UK each year in the past 20 years has been increasing, decreasing or staying about the same? Give a reason for your answer, based on the chart.
3. Does the chart show that women are likely to live longer than men in the UK?

Now look at the population pyramid for China in 2010.

4. How can you tell from the chart that China has a much bigger population than the UK?
5. Some years ago China introduced a one-child policy, by which every family was encouraged to have no more than one child. When do you think this happened?
6. Which country, China or the UK, has a bigger proportion of people who are over 80 years old? Give a reason for your answer.

Your task

Imagine that you are a member of a United Nations agency, given the task of analysing populations to find the social composition and needs of countries around the globe. Using all the population pyramids on these pages, describe and compare the populations of different countries. If possible, support your conclusions by researching why the population pyramids look as they do.

Write your findings as a presentation, to be made at the next meeting of the agency. In your presentation you must:

- compare the population pyramids of at least two countries, describing their main features, similarities and differences
- represent your comparisons mathematically
- suggest what the social and economic conditions of each country might be, based on the evidence of the population pyramids
- provide any further evidence you can find, based in your own research, to justify your conclusions
- identify any implications your findings have for the countries concerned, now and in the future
- analyse the effectiveness of population pyramids as a statistical tool, compared to other statistical diagrams.

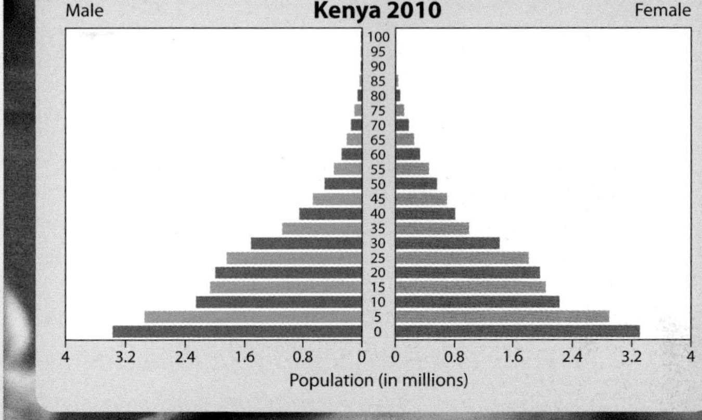

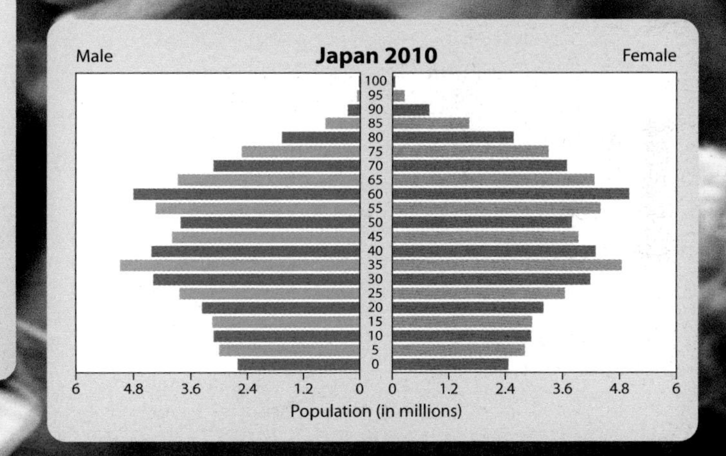

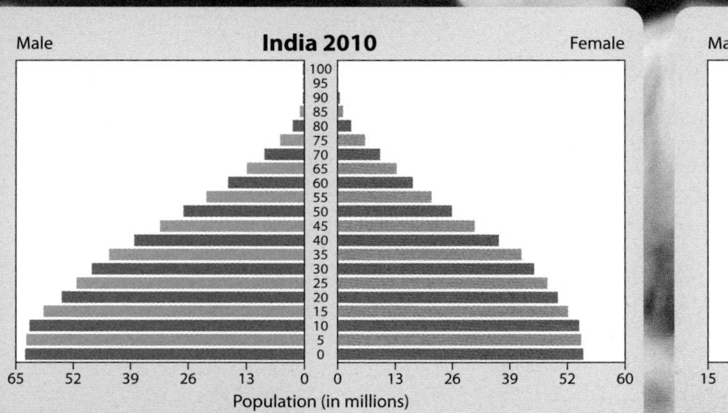

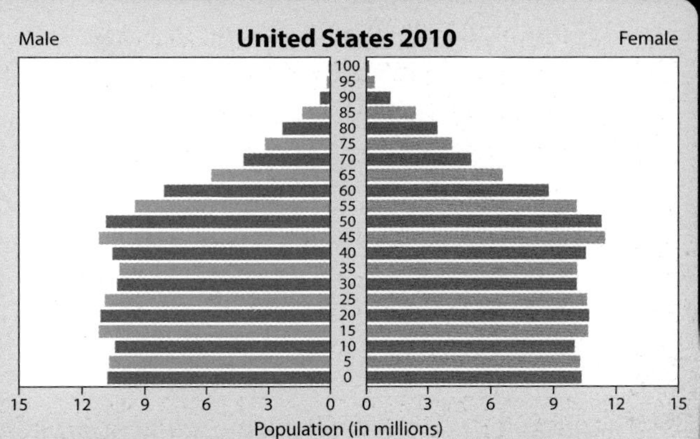

Why this chapter matters

Do you like playing games? In the last 100 years, mathematicians and others have developed game theory that takes probability into account.

Blaise Pascal

In fact, understanding probability was recognized as essential for success in games based on chance and so probability theory developed as gaming became more popular and sophisticated. As long ago as the 17th century, two French mathematicians, Blaise Pascal and Pierre de Fermat, helped develop probability theory at the request of gamblers.

However, it is not only mathematicians who have helped developed the subject. In the 20th century, Jacques-Marie-Émile Lacan (1901–81), a French psychoanalyst, investigated the probability of chance and game theory and devised what is known as Lacan's problem, which introduces logic and intuition into probability.

The governor of a jail has three prisoners. He gives them the following instructions.

I am going to release one of you. To decide which one, I will give you a test.

I have five discs. Three are white and two are black.

I am going to pin one disc on the back of each of you. No one will be able to see the disc on their own back.

You may observe your companions and look at the discs on their backs but you may not communicate with each other. The first person to identify the colour of the disc on his own back will be released.

As soon as any of you is ready to draw a conclusion, he may come to the door, to give his answer.

Then the governor pinned a white disc on the back of each prisoner. He did not use the black ones.

Would you be able to deduce what colour disc was on your back?

Each prisoner could see the colour of the discs on the backs of the other two. In each case, the discs on both the other prisoners were white.

Each might realise that the colour of his own disc was more likely to be black than white since, of the remaining discs, two were black and one was white.

But this is to apply only the rules of probability, where all is assumed to be fair.

Suppose one of the prisoners had a black disc on his back. Then the other two would quickly realise that they must each have a white disc since, if either of them had a black disc, the other would have realised his was white – remember, there were only two black discs.

The *logical* conclusion is therefore that they must all have white discs on their backs.

But did any of the prisoners have the reasoning to see that? Surely the probabilities were the same for each of them – or were they?

This chapter takes you through a study of probability, where all outcomes are taken to be random and equally likely – not influenced by a crafty prison governor.

Chapter 11

Probability: Calculating probabilities

The grades given in this chapter are target grades.

This chapter will show you ...

C to A* how to calculate probabilities for combined events

Visual overview

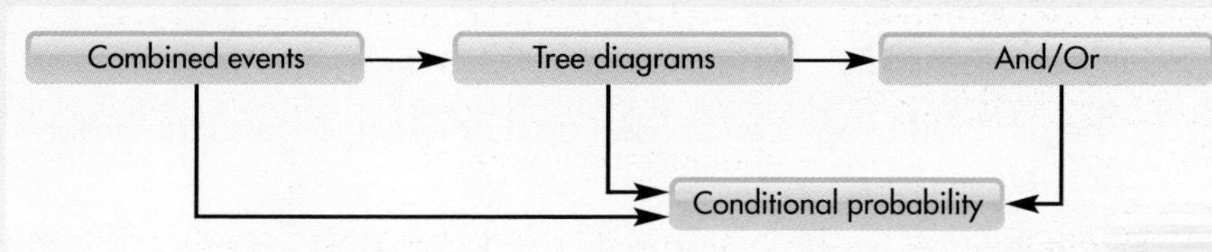

What you should already know

- That the probability scale goes from 0 to 1 **(KS3 level 4, GCSE grade F)**
- How to use the probability scale and to assess the likelihood of events, depending on their position on the scale **(KS3 level 4, GCSE grade F)**
- How to cancel, add and subtract fractions **(KS3 level 5, GCSE grade E)**

Quick check

Draw a probability scale and put an arrow to show the approximate probability of each of the following events happening.

a The next TV programme you watch will have been made in Britain.

b A person in your class will have been born in April.

c It will snow in July in Spain.

d In the next Olympic Games, a man will run the 100 m race in less than 20 seconds.

e During this week, you will drink some water or pop.

Terminology

The topic of probability has its own special terminology, which will be explained as it arises. For example, a **trial** is one go at performing something, such as throwing a dice or tossing a coin. So, if we throw a dice 10 times, we perform 10 trials.

Two other probability terms are **event** and **outcome**. An event is anything of which we want to measure the probability. An outcome is a result of the event.

Another probability term is **at random**. This means 'without looking' or 'not knowing what the outcome is in advance'.

Note: 'Dice' is used in this book in preference to 'die' for the singular form of the noun, as well as for the plural. This is in keeping with growing common usage, including in examination papers.

Probability facts

The probability of a *certain* outcome or event is 1 and the probability of an *impossible* outcome or event is 0.

Probability is *never greater than 1 or less than 0.*

Many probability examples involve coins, dice and packs of cards. Here is a reminder of their outcomes.

- Tossing a coin has two possible outcomes: head or tail.
- Throwing an ordinary six-sided dice has six possible outcomes: 1, 2, 3, 4, 5, 6.
- A pack of cards consists of 52 cards divided into four suits: Hearts (red), Spades (black), Diamonds (red) and Clubs (black). Each suit consists of 13 cards bearing the following values: 2, 3, 4, 5, 6, 7, 8, 9, 10, Jack, Queen, King and Ace. The Jack, Queen and King are called 'picture cards'. (The Ace is sometimes also called a picture card.) So the total number of outcomes is 52.

Probability is defined as:

$$\text{P(event)} = \frac{\text{number of ways the event can happen}}{\text{total number of all possible outcomes}}$$

This definition always leads to a fraction, which should be cancelled to its simplest form. Make sure that you know how to cancel fractions, with or without a calculator. It is acceptable to give a probability as a decimal or a percentage but a fraction is better.

This definition can be used to work out the probability of outcomes or events, as the following example shows.

EXAMPLE 1

A card is drawn from a normal pack of cards. What is the probability that it is one of the following?

a A red card **b** A Spade **c** A seven

d A picture card **e** A number less than 5 **f** A red King

a There are 26 red cards, so P(red card) = $\frac{26}{52} = \frac{1}{2}$

b There are 13 Spades, so P(Spade) = $\frac{13}{52} = \frac{1}{4}$

c There are four sevens, so P(seven) = $\frac{4}{52} = \frac{1}{13}$

d There are 12 picture cards, so P(picture card) = $\frac{12}{52} = \frac{3}{13}$

e If you count the value of an Ace as 1, there are 16 cards with a value less than 5. So, P(number less than 5) = $\frac{16}{52} = \frac{4}{13}$

f There are two red Kings, so P(red King) = $\frac{2}{52} = \frac{1}{26}$

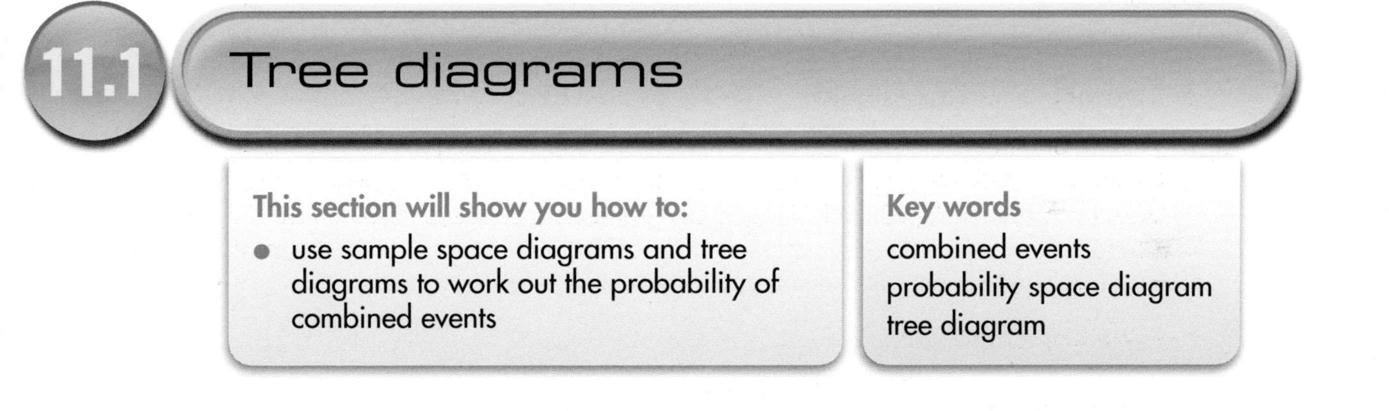

11.1 Tree diagrams

This section will show you how to:

- use sample space diagrams and tree diagrams to work out the probability of combined events

Key words
combined events
probability space diagram
tree diagram

Imagine you have to draw two cards from this pack of six cards, but you must replace the first card before you select the second card.

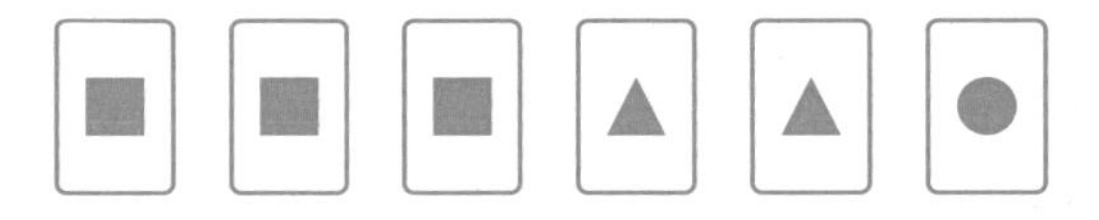

One way you could show all the outcomes of this experiment is to construct a **probability space diagram**. For example, this could be an array set in a pair of axes (see page 288), or a pictogram, or simply a list of all the outcomes.

By showing all the outcomes of the experiment as an array, you obtain the diagram below.

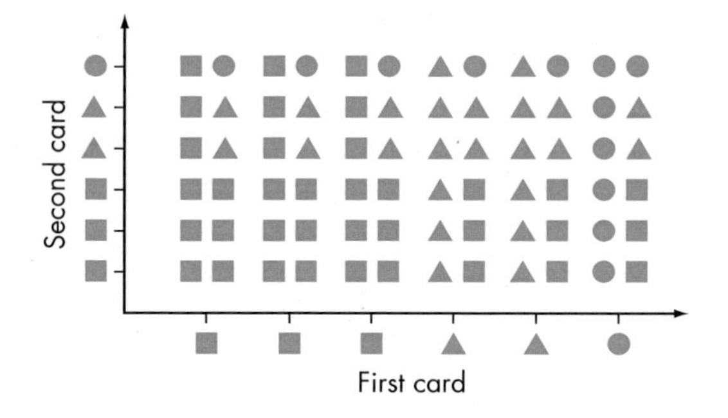

From the diagram, you can see immediately that the probability of picking, say, two squares, is 9 out of 36 pairs of cards. So:

$$P(2 \text{ squares}) = \frac{9}{36} = \frac{1}{4}$$

EXAMPLE 2

Using the probability space diagram above, what is the probability of getting each of these outcomes?

a A square and a triangle (in any order)

b Two circles

c Two shapes the same

a There are 12 combinations that give a square and a triangle together. There are six when a square is chosen first and six when a triangle is chosen first. So:

$$P(\text{square and triangle, in any order}) = \frac{12}{36} = \frac{1}{3}$$

b There is only one combination which gives two circles. So:

$$P(\text{two circles}) = \frac{1}{36}$$

c There are nine combinations of two squares together, four combinations of two triangles together and one combination of two circles together. These give a total of 14 combinations with two shapes the same. So:

$$P(\text{two shapes the same}) = \frac{14}{36} = \frac{7}{18}$$

An alternative method to tackling problems involving **combined events** is to use a **tree diagram**.

When you pick the first card, there are three possible outcomes: a square, a triangle or a circle. For a single event:

$$P(\text{square}) = \frac{3}{6} \qquad P(\text{triangle}) = \frac{2}{6} \qquad P(\text{circle}) = \frac{1}{6}$$

You can show this by depicting each event as a branch and writing its probability on the branch.

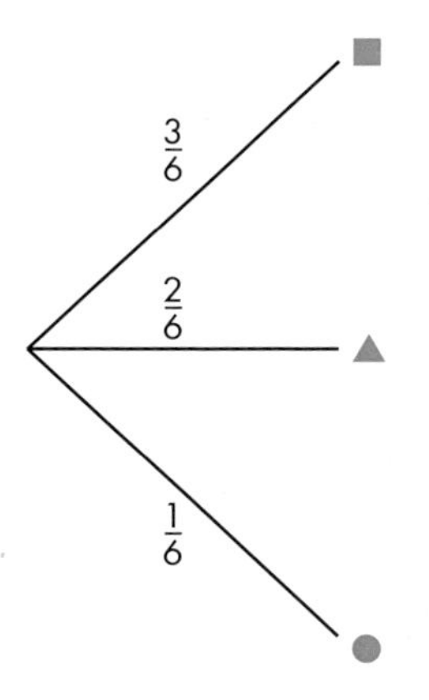

The diagram can then be extended to take into account a second choice. Because the first card has been replaced, you can still pick a square, a triangle or a circle. This is true no matter what is chosen the first time. You can demonstrate this by adding three more branches to the 'squares' branch in the diagram.

Here is the complete tree diagram.

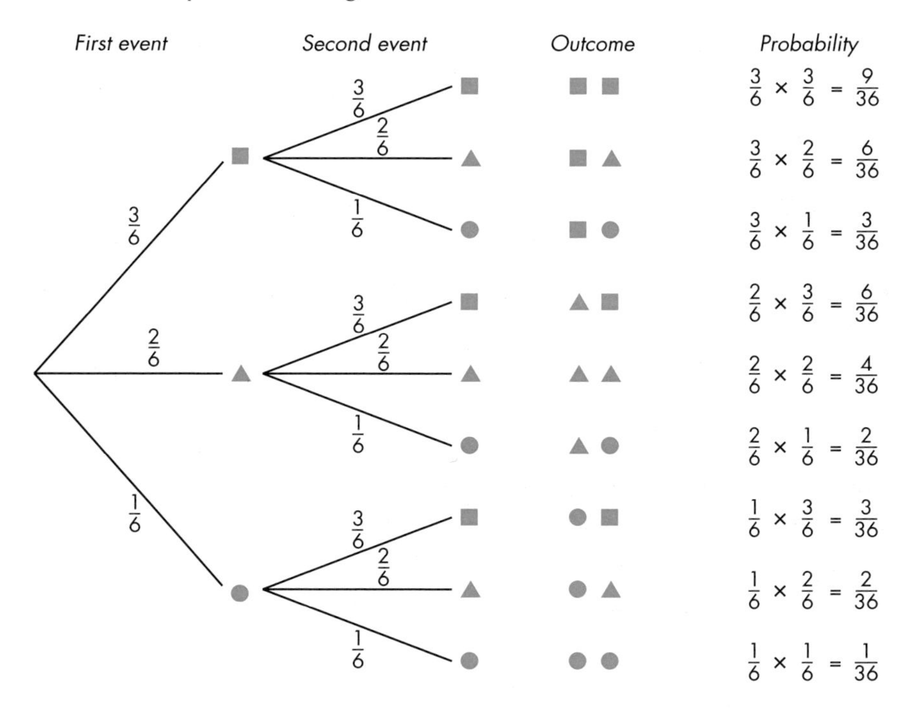

The probability of any outcome is calculated by multiplying all the probabilities on its branches. For instance:

$$\text{P(two squares)} = \frac{3}{6} \times \frac{3}{6} = \frac{9}{36}$$

$$\text{P(triangle followed by circle)} = \frac{2}{6} \times \frac{1}{6} = \frac{2}{36}$$

EXAMPLE 3

Using the tree diagram on the previous page, what is the probability of obtaining:

a two triangles

b a circle followed by a triangle

c a square and a triangle, in any order

d two circles

e two shapes the same?

a P(two triangles) $= \frac{4}{36} = \frac{1}{9}$

b P(circle followed by triangle) $= \frac{2}{36} = \frac{1}{18}$

c There are two places in the outcome column that have a square and a triangle. These are the second and fourth rows. The probability of each is $\frac{1}{6}$. Their combined probability is given by the addition rule.

P(square and triangle, in any order) $= \frac{6}{36} + \frac{6}{36} = \frac{12}{36} = \frac{1}{3}$

d P(two circles) $= \frac{1}{36}$

e There are three places in the outcome column that have two shapes the same. These are the first, fifth and last rows. The probabilities are respectively $\frac{1}{4}$, $\frac{1}{9}$ and $\frac{1}{36}$. Their combined probability is given by the addition rule.

P(two shapes the same) $= \frac{9}{36} + \frac{4}{36} + \frac{1}{36} = \frac{14}{36} = \frac{7}{18}$

Note that the answers to parts **c**, **d** and **e** are the same as the answers obtained in Example 2.

EXERCISE 11A

B

1 A coin is tossed twice. Copy and complete this tree diagram to show all the outcomes.

Use your tree diagram to work out the probability of each of these outcomes.

a Getting two heads

b Getting a head and a tail

c Getting at least one tail

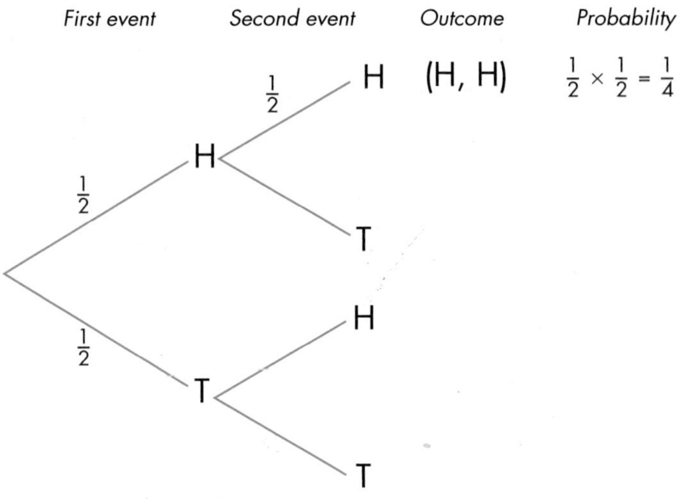

First event *Second event* *Outcome* *Probability*

$\frac{1}{2}$ H — $\frac{1}{2}$ H (H, H) $\frac{1}{2} \times \frac{1}{2} = \frac{1}{4}$

H — T

$\frac{1}{2}$ T — H

T — T

B

2 A card is drawn from a pack of cards. It is replaced, the pack is shuffled and another card is drawn.

a What is the probability that either card was an Ace?

b What is the probability that either card was not an Ace?

c Draw a tree diagram to show the outcomes of two cards being drawn as described. Use the tree diagram to work out the probability of each of these.

i Both cards will be Aces.

ii At least one of the cards will be an Ace.

FM 3 On my way to work, I drive through two sets of road works with traffic lights which only show green or red. I know that the probability of the first set being green is $\frac{1}{3}$ and the probability of the second set being green is $\frac{1}{2}$.

a What is the probability that the first set of lights will be red?

b What is the probability that the second set of lights will be red?

c Copy and complete the tree diagram below, showing the possible outcomes when passing through both sets of lights.

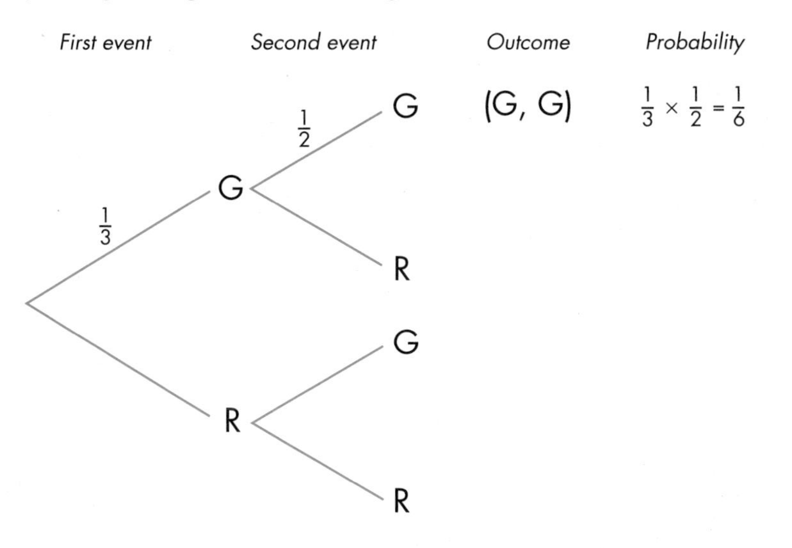

d Using the tree diagram, what is the probability of each of the following outcomes?

i I do not get held up at either set of lights.

ii I get held up at exactly one set of lights.

iii I get held up at least once.

e Over a school term I make 90 journeys to work. On how many days can I expect to get two green lights?

B

FM **4** Six out of every 10 cars in Britain are foreign made.

a What is the probability that any car will be British made?

b Two cars can be seen approaching in the distance. Draw a tree diagram to work out the probability of each of these outcomes.

i Both cars will be British made.

ii One car will be British and the other car will be foreign made.

5 Three coins are tossed. Copy and complete the tree diagram below and use it to answer the questions.

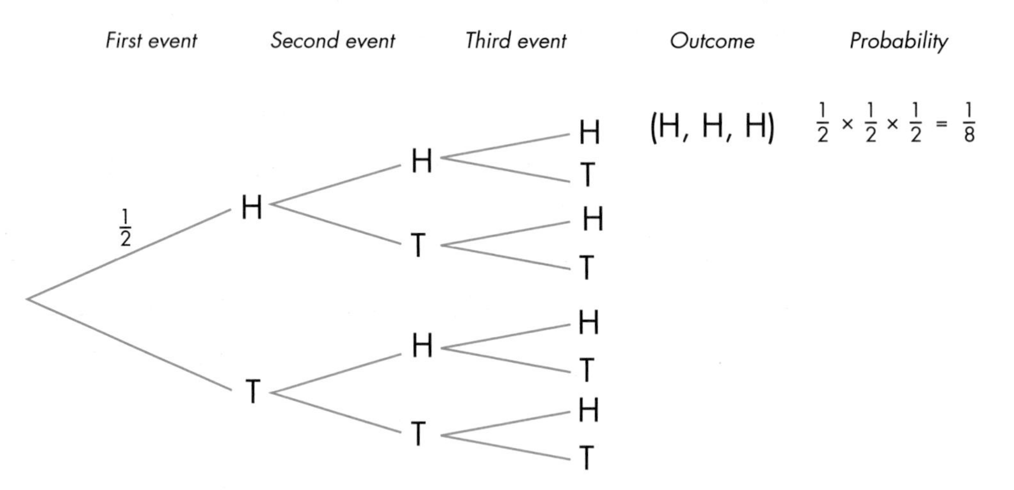

If a coin is tossed three times, what is the probability of each of these outcomes?

a Three heads

b Two heads and a tail

c At least one tail

FM **6** Thomas has to take a three-part language examination paper. The first part is speaking. He has a 0.4 chance of passing this part. The second is listening. He has a 0.5 chance of passing this part. The third part is writing. He has a 0.7 chance of passing this part. Draw a tree diagram covering three events, where the first event is passing or failing the speaking part of the examination, the second event is passing or failing the listening part and the third event is passing or failing the writing part.

a If he passes all three parts, his father will give him £20. What is the probability that he gets the money?

b If he passes two parts only, he can resit the other part. What is the chance he will have to resit?

c If he fails all three parts, he will be thrown off the course. What is the chance he is thrown off the course?

B

FM **7** In a group of 10 girls, six like the pop group Smudge and four like the pop group Mirage. Two girls are to be chosen for a pop quiz.

a What is the probability that the first girl chosen will be a Smudge fan?

b Draw a tree diagram to show the outcomes of choosing two girls and which pop groups they like. (Remember, once a girl has been chosen the first time she cannot be chosen again.)

c Use your tree diagram to work out the probability that:

i both girls chosen will like Smudge

ii both girls chosen will like the same group

iii both girls chosen will like different groups.

8 Look at all the tree diagrams that you have seen so far.

a What do the probabilities across any set of branches (outlined in the diagram below) always add up to?

b What do the final probabilities (outlined in the diagram below) always add up to?

c You should now be able to fill in all of the missing values in the diagram.

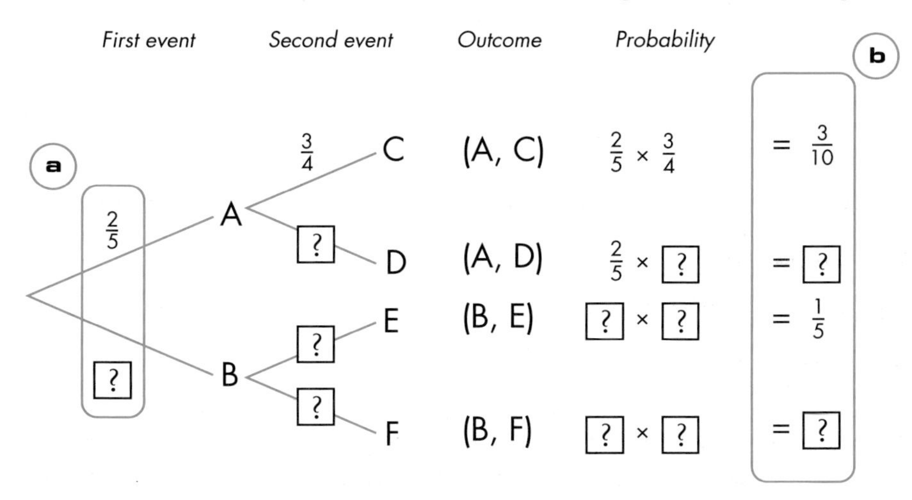

PS **9** When playing the game *Pontoon*, you are dealt two cards. If you get an Ace and a King, Queen or Jack you have been dealt a 'Royal Pontoon'.

What is the probability of being dealt a Royal Pontoon? Give your answer to 3 decimal places.

AU **10** I have a bag containing white, blue and green jelly babies. Explain how a tree diagram can help me find the probability of picking at random three sweets of different colours.

11.2 Independent events 1

This section will show you how to:
- use the connectors 'and' and 'or' to find the probability of combined events

Key words
and
independent events
or

If the outcome of event A does not effect the outcome of event B, then events A and B are called **independent events**. Most of the combined events you have looked at so far have been independent events.

It is possible to work out problems on combined events without using tree diagrams. The method explained in Example 4 is basically the same as that of a tree diagram but uses the words **and** and **or**.

EXAMPLE 4

The chance that Ashley hits a target with an arrow is $\frac{1}{4}$. He has two shots at the target. What is the probability of each of these outcomes?

a He hits the target both times.

b He hits the target once only.

c He hits the target at least once.

a P(hits both times) = P(first shot hits **and** second shot hits) $= \frac{1}{4} \times \frac{1}{4} = \frac{1}{16}$

b P(hits the target once only) = P(first hits **and** second misses **or** first misses **and** second hits) $= \left(\frac{1}{4} \times \frac{3}{4}\right) + \left(\frac{3}{4} \times \frac{1}{4}\right) = \frac{6}{16} = \frac{3}{8}$

c P(hits at least once) = P(both hit **or** one hits) $= \frac{1}{16} + \frac{3}{8} = \frac{7}{16}$

Note the connections between the word 'and' and the operation 'times', and the word 'or' and the operation 'add'.

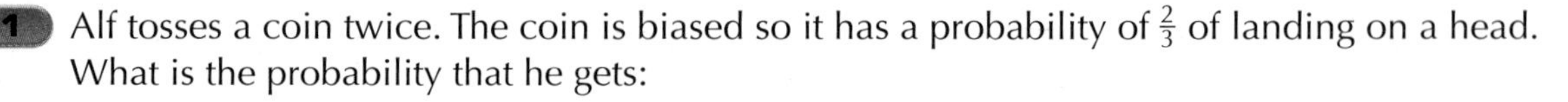

EXERCISE 11B

B

1 Alf tosses a coin twice. The coin is biased so it has a probability of $\frac{2}{3}$ of landing on a head. What is the probability that he gets:

a two heads

b a head and a tail (in any order)?

2 Bernice draws a card from a pack of cards, replaces it, shuffles the pack and then draws another card. What is the probability that the cards are:

a both Aces

b an Ace and a King (in any order)?

B

3 A dice is thrown twice. What is the probability that both scores are:

a even **b** one even and one odd (in any order)?

4 I throw a dice three times. What is the probability of getting three sixes?

5 A bag contains 15 white beads and 10 black beads. I take out a bead at random, replace it and take out another bead. What is the probability that:

a both beads are black

b one bead is black and the other white (in any order)?

FM **6** The probability that I am late for work on Monday is 0.4. The probability that I am late on Tuesday is 0.2. What is the probability of each of the following outcomes?

a I am late for work on Monday and Tuesday.

b I am late for work on Monday and on time on Tuesday.

c I am on time on both Monday and Tuesday.

A

FM **7** Ronda has to take a three-part language examination paper. The first part is speaking. She has a 0.7 chance of passing this part. The second part is listening. She has a 0.6 chance of passing this part. The third part is writing. She has a 0.8 chance of passing this part.

a If she passes all three parts, her father will give her £20. What is the probability that she gets the money?

b If she passes two parts only, she can resit the other part. What is the chance she will have to resit?

c If she fails all three parts, she will be thrown off the course. What is the chance she is thrown off the course?

FM **8** Roy regularly goes to Bristol by train.

The probability of the train arriving in Bristol late is 0.05.

The probability of it raining in Bristol is 0.8.

What is the probability of:

a Roy getting to Bristol on time and it not raining

b Roy travelling to Bristol five days in a row and not being late at all

c Roy travelling to Bristol three days is a row and it raining every day?

PS **9** What is the probability of rolling the same number on a dice five times in a row?

AU **10** Explain why picking a red card from a pack of cards and picking a King from a pack of cards are not independent events.

AU **11** A fair dice rolls 8 sixes in a row. Which of the following is the probability of a six on the 9th throw?

$0 \qquad \frac{1}{6} \qquad 1$

Explain your choice.

'At least' problems

In examination questions concerning combined events, it is common to ask for the probability of at least one of the events occurring. There are two ways to solve such problems.

- All possibilities can be written out, which takes a long time.
- Use P(at least one) = 1 – P(none)

The second option is much easier to work out and there is less chance of making a mistake.

EXAMPLE 5

A bag contains seven red and three black balls. A ball is taken out and replaced. This is repeated three times. What is the probability of getting:

a no red balls

b at least one red ball?

a P(no reds) = P(black, black, black) = $\frac{7}{10} \times \frac{7}{10} \times \frac{7}{10} = 0.343$

b P(at least one red) = 1 – P(no reds) = 1 – 0.343 = 0.657

Note that the answer to part **b** is 1 minus the answer to part **a**. Examination questions often build up answers in this manner.

EXERCISE 11C

1 A dice is thrown three times.

a What is the probability of not getting a 2?

b What is the probability of at least one 2?

2 Four coins are thrown. What is the probability of:

a four tails **b** at least one head?

FM **3** Adam, Bashir and Clem take a mathematics test. The probability that Adam passes is 0.6, the probability that Bashir passes is 0.9 and the probability that Clem passes is 0.7. What is the probability that:

a all three pass **b** Bashir and Adam pass but Clem does not

c all three fail **d** at least one passes?

4 A bag contains four red and six blue balls. A ball is taken out and replaced. Another ball is taken out. What is the probability that:

a both balls are red **b** both balls are blue **c** at least one is red?

A

5 **a** A dice is thrown three times. What is the probability of:

i three sixes **ii** no sixes **iii** at least one six?

b A dice is thrown four times. What is the probability of:

i four sixes **ii** no sixes **iii** at least one six?

c A dice is thrown five times. What is the probability of:

i five sixes **ii** no sixes **iii** at least one six?

d A dice is thrown n times. What is the probability of:

i n sixes **ii** no sixes **iii** at least one six?

FM 6 The probability that the school canteen serves chips on any day is $\frac{2}{3}$. In a week of five days, what is the probability that:

a chips are served every day

b chips are not served on any day

c chips are served on at least one day?

FM 7 The probability that Steve is late for work is $\frac{5}{6}$. The probability that Nigel is late for work is $\frac{9}{10}$. The probability that Gary is late for work is $\frac{1}{2}$. What is the probability that on a particular day:

a all three are late **b** none of them is late **c** at least one is late?

FM 8 A cricket test match is to be played in a coastal town of the West Indies. The test match lasts for five days.

In this town, at this time of year, the probability of rain on any day during the match is 0.35.

a What is the probability of no rain falling on any one day?

b What is the probability of no rain falling on two days in a row?

c On how many days during the cricket match could they expect rain?

d What is the probability of no rain falling during the cricket match?

e What is the probability of rain falling on at least one of the days of the cricket match?

PS 9 The probability of planting an orchid in Cardasica and it growing well is 0.6.

Kieron plants 10 orchids in Cardasica.

What is the probability that at least nine orchids will grow well?

AU 10 Jeff works for five days of the week in a call centre, cold calling people to sell them double glazing. Jeff's boss told him the probability of him making a sale each day. Explain how Jeff can work out the probability of making at least one sale in a week.

11.3

Independent events 2

This section will show you how to:
- use the connectors 'and' and 'or' in more advanced examples, to find the probability of combined events

Key words
and
independent events
or

More advanced use of 'and' and 'or'

You have already seen how certain probability problems related to **independent events** can be solved either by tree diagrams or by the use of the *and/or* method. Both methods are basically the same but the **and/or** method works better in the case of three events following one after another or in situations where the number of outcomes of one event is greater than two. This is simply because the tree diagram would get too large and involved.

EXAMPLE 6

Three cards are to be drawn from a pack of cards. Each card is to be replaced before the next one is drawn. What is the probability of drawing:

a three Kings
b exactly two Kings and one other card
c no Kings
d at least one King?

Let K be the event 'Drawing a King'. Let N be the event 'Not drawing a King'. Then:

a $P(KKK) = \frac{1}{13} \times \frac{1}{13} \times \frac{1}{13} = \frac{1}{2197}$

b P(exactly two Kings) = P(KKN) or P(KNK) or P(NKK)

$= (\frac{1}{13} \times \frac{1}{13} \times \frac{12}{13}) + (\frac{1}{13} \times \frac{12}{13} \times \frac{1}{13}) + (\frac{12}{13} \times \frac{1}{13} \times \frac{1}{13}) = \frac{36}{2197}$

c $P(\text{no Kings}) = P(NNN) = \frac{12}{13} \times \frac{12}{13} \times \frac{12}{13} = \frac{1728}{2197}$

d $P(\text{at least one King}) = 1 - P(\text{no Kings}) = 1 - \frac{1728}{2197} = \frac{469}{2197}$

Note that in part **b** the notation stands for the probability that the first card is a King, the second is a King and the third is not a King; or the first is a King, the second is not a King and the third is a King; or the first is not a King, the second is a King and the third is a King.

Note also that the probability of each component of part **b** is exactly the same. So you could have done the calculation as:

$3 \times \frac{1}{13} \times \frac{1}{13} \times \frac{12}{13} = \frac{36}{2197}$

Patterns of this kind often occur in probability.

EXERCISE 11D 

A

1 A bag contains three black balls and seven red balls. A ball is taken out and replaced. This is repeated twice. What is the probability that:

a all three are black **b** exactly two are black

c exactly one is black **d** none is black?

2 A dice is thrown four times. What is the probability that:

a four sixes are thrown **b** no sixes are thrown

c exactly one six is thrown?

FM **3** On my way to work I pass three sets of traffic lights. The probability that the first is green is $\frac{1}{2}$. The probability that the second is green is $\frac{1}{3}$. The probability that the third is green is $\frac{2}{3}$. What is the probability that:

a all three are green **b** exactly two are green **c** exactly one is green

d none are green **e** at least one is green?

FM **4** Alf is late for school with a probability of 0.9. Bert is late with a probability of 0.7. Chas is late with a probability of 0.6. On any particular day what is the probability of:

a exactly one of them being late **b** exactly two of them being late?

FM **5** Daisy takes four A-levels. The probability that she will pass English is 0.7. The probability that she will pass history is 0.6. The probability she will pass geography is 0.8. The probability that she will pass general studies is 0.9. What is the probability that:

a she passes all four subjects

b she passes exactly three subjects

c she passes at least three subjects?

FM **6** The driving test is in two parts, a written test and a practical test. It is known that 90% of people who take the written test pass and 60% of people who take the practical test pass. A person who passes the written test does not have to take it again. A person who fails the practical test does have to take it again.

a What is the probability that someone passes the written test?

b What is the probability that someone passes the practical test?

c What is the probability that someone passes both tests?

d What is the probability that someone passes the written test but takes two attempts to pass the practical test?

A

FM 7 Six out of 10 cars in Britain are made by foreign manufacturers. Three cars can be seen approaching in the distance.

a What is the probability that the first one is foreign?

b The first car is going so fast that its make could not be identified. What is the probability that the second car is foreign?

c What is the probability that exactly two of the three cars are foreign?

d Explain why, if the first car is foreign, the probability of the second car being foreign is still six out of 10.

8 Each day Mr Smith runs home from work. He has a choice of three routes: the road, the fields or the canal path. The road route is 4 miles, the fields route is 6 miles and the canal route is 5 miles. In a three-day period, what is the probability that Mr Smith runs a total distance of:

a exactly 17 miles
b exactly 13 miles
c exactly 15 miles
d over 17 miles?

9 A rock climber attempts a difficult route. There are three hard moves at points A, B and C in the climb. The climber has a probability of 0.6, 0.3 and 0.7 respectively of completing each of these moves. What is the probability that the climber:

a completes the climb
b fails at move A
c fails at move B
d fails at move C?

10 A car rally is being organised for the end of the year on 29 December, 30 December and 31 December.

If it snows on any of those days, the rally will finish and everyone must try to get home.

A long-range forecast gives the probability of snow on any one of these days as 0.25.

What is the probability that the rally will:

a last all three days
b only last two days
c only last one day
d not start?

PS 11 Evie's maths teacher told her that the probability of getting her mathematics homework correct is always the same and that, in any month of four homeworks, the chance of her getting at least one incorrect homework was 0.5904.

What is the probability of Evie getting her mathematics homework correct on any one night?

AU 12 James has been dealt two cards and knows that if he is now dealt a 10, Jack, Queen or a King he will win.

James thinks that the chance of his winning is now $\frac{16}{52}$.

Explain why he is wrong.

11.4 Conditional probability

This section will show you how to:

- work out the probability of combined events when the probabilities change after each event

Key words

conditional probability

The term **conditional probability** is used to describe the situation when the probability of an event is dependent on the outcome of another event. For instance, if a card is taken from a pack and not returned, then the probabilities for the next card drawn will be altered. The following example illustrates this situation.

EXAMPLE 7

A bag contains nine balls, of which five are white and four are black.

A ball is taken out and not replaced. Another is then taken out. If the first ball removed is black, what is the probability that:

a the second ball will be black

b both balls will be black?

When a black ball is removed, there are five white balls and three black balls left, reducing the total to eight.

Hence, when the second ball is taken out:

a P(second ball black) $= \frac{3}{8}$

b P(both balls black) $= \frac{4}{9} \times \frac{3}{8} = \frac{1}{6}$

EXERCISE 11E

FM **1** I put six CDs in my multi-player and put it on random play. Each CD has 10 tracks. Once a track is played, it is not played again.

a What is the chance that track 5 on CD 6 is the first one played?

b What is the maximum number of tracks that could be played before a track from CD 6 is played?

A

A

2 There are five white eggs and one brown egg in an egg box. Kate decides to make a two-egg omelette. She takes each egg from the box without looking at its colour.

a What is the probability that the first egg taken is brown?

b If the first egg taken is brown, what is the probability that the second egg taken will be brown?

c What is the probability that Kate gets an omelette made from:

i two white eggs **ii** one white and one brown egg **iii** two brown eggs?

A*

3 A box contains 10 red and 15 yellow balls. One is taken out and not replaced. Another is taken out.

a If the first ball taken out is red, what is the probability that the second ball is:

i red **ii** yellow?

b If the first ball taken out is yellow, what is the probability that the second ball is:

i red **ii** yellow?

4 A fruit bowl contains six Granny Smith apples and eight Golden Delicious apples. Kevin takes two apples at random.

a If the first apple is a Granny Smith, what is the probability that the second is:

i a Granny Smith **ii** a Golden Delicious?

b What is the probability that:

i both are Granny Smiths **ii** both are Golden Delicious?

5 Ann has a bargain box of tins. They are unlabelled but she knows that six tins contain soup and four contain peaches.

a She opens two tins. What is the probability that:

i they are both soup

ii they are both peaches?

b What is the probability that she has to open two tins before she gets a tin of peaches?

c What is the probability that she has to open three tins before she gets a tin of peaches?

d What is the probability that she will get a tin of soup if she opens five tins?

FM AU **6** One in three cars on British roads is made in Britain. A car comes down the road. It is a British-made car. John says that the probability of the next car being British made is one in two because a British-made car has just gone past. Explain why he is wrong.

7 A bag contains three black balls and seven red balls. A ball is taken out and not replaced. This is repeated twice. What is the probability that:

a all three are black **b** exactly two are black

c exactly one is black **d** none is black?

A*

8 On my way to work, I pass two sets of traffic lights. The probability that the first is green is $\frac{1}{3}$. If the first is green, the probability that the second is green is $\frac{1}{3}$. If the first is red, the probability that the second is green is $\frac{2}{3}$. What is the probability that:

a both are green **b** none are green **c** exactly one is green **d** at least one is green?

9 A hand of five cards is dealt. What is the probability that:

a all five are Spades **b** all five are the same suit

c they are four Aces and any other card **d** they are four of a kind and any other card?

FM 10 An engineering test is in two parts, a written test and a practical test. It is known that 90% of those who take the written test pass. When a person passes the written test, the probability that he or she will also pass the practical test is 60%. When a person fails the written test, the probability that he or she will pass the practical test is 20%.

a What is the probability that someone passes both tests?

b What is the probability that someone passes one test?

c What is the probability that someone fails both tests?

d What is the combined probability of the answers to parts **a**, **b** and **c**?

11 Each day Mr Smith runs home from work. He has a choice of three routes. The road, the fields or the canal path. On Monday, each route has an equal probability of being chosen. The route chosen on any day will not be picked the next day and so each of the other two routes has an equal probability of being chosen.

a Write down all the possible combinations so that Mr Smith runs home via the canal path on Wednesday (there are four of them).

b Calculate the probability that Mr Smith runs home via the canal path on Wednesday.

c Calculate the probability that Mr Smith runs home via the canal path on Tuesday.

d Using your results from parts **b** and **c**, write down the probability that Mr Smith runs home via the canal path on Thursday.

e Explain the answers to parts **b**, **c** and **d**.

FM 12 In the class box of calculators there are 10 calculators. Three of them are faulty.

What is the probability that:

a Dave takes the first and it is a good one

b Julie takes the second and it is a good one

c Andrew takes the third and it is faulty?

PS 13 What is the probability of being dealt four Aces in a row from a normal pack of cards?

AU 14 A bag contains some blue balls and some white balls, all the same size. Tony is asked to find the probability of taking out two balls of the same colour.

Explain to Tony how you would do this, explaining carefully the point where he is most likely to go wrong.

GRADE BOOSTER

B You can draw a tree diagram to work out the probability of combined events

A You can use *and/or* or a tree diagram to work out probabilities of specific outcomes of combined events

A* You can work out the probabilities of combined events when the probability of each event changes depending on the outcome of the previous event

What you should know now

- How to calculate theoretical probabilities of combined events

1 Matthew puts 3 red counters and 5 blue counters in a bag.

He takes at random a counter from the bag.

He writes down the colour of the counter.

He puts the counter in the bag again.

He then takes at random a second counter from the bag.

a Copy and complete the probability tree diagram. *(2 marks)*

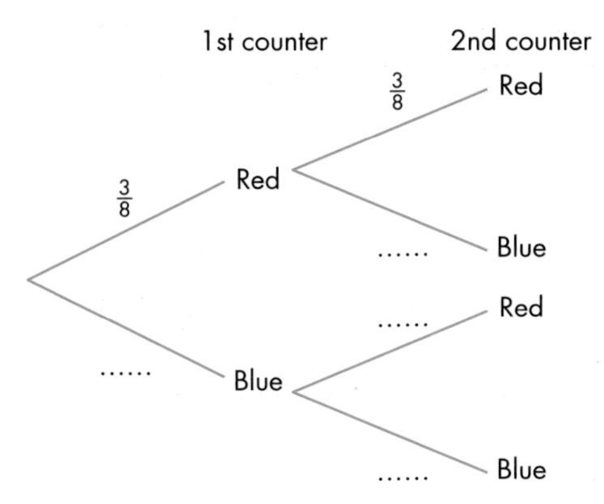

b Work out the probability that Matthew takes two red counters. *(2 marks)*

Edexcel A, November 2008, Paper 3 Higher, Question 21

2 Alan buys a box of 60 tulip bulbs with the promise that one-quarter are red, one-quarter are yellow, one-quarter are purple and one-quarter are white.

What is the probability that:

a the first one he takes out is a purple tulip

b the first two he takes out are the same colour

c the first three he takes out are a different colour?

3 There are 3 strawberry yoghurts, 2 peach yoghurts and 4 cherry yoghurts in a fridge.

Kate takes a yoghurt at random from the fridge.

She eats the yoghurt.

She then takes a second yoghurt at random from the fridge.

Work out the probability that both the yoghurts were the same flavour. *(4 marks)*

Edexcel B, March 2008, Paper 9 Higher, Question 5

4 Caroline cycles to school.

She passes through two sets of traffic lights.

The probability that she has to stop at the first set of traffic lights is $\frac{2}{5}$

If she has to stop at the first set of traffic lights, the probability that she has to stop at the second is $\frac{5}{6}$

If she does not have to stop at the first set of traffic lights, the probability that she has to stop at the second is $\frac{1}{2}$

Caroline cycles to school on the last day of term.

Work out the probability that she has to stop at only **one** set of traffic lights. *(4 marks)*

Edexcel B, November 2007, Paper 9 Higher, Question 6

5 There are 3 boys and 7 girls at a playgroup.

Mrs Gold selects two children at random.

a Copy and complete the probability tree diagram below. *(2 marks)*

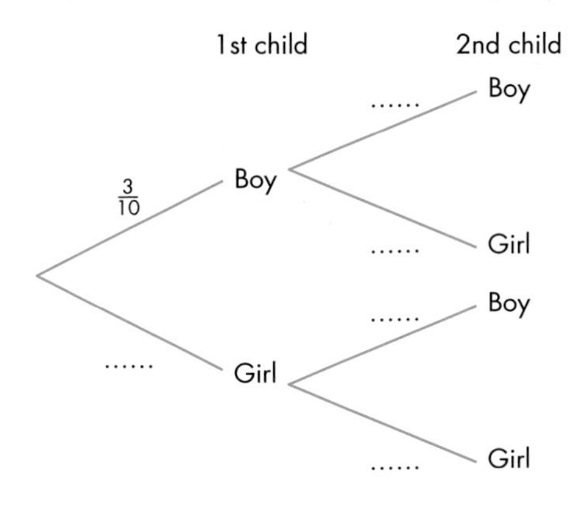

b Work out the probability that Mrs Gold selects two girls. *(2 marks)*

Edexcel B, June 2007, Paper 9 Higher, Question 4

6 William has two 10-sided spinners.

The spinners are equally likely to land on each of their sides.

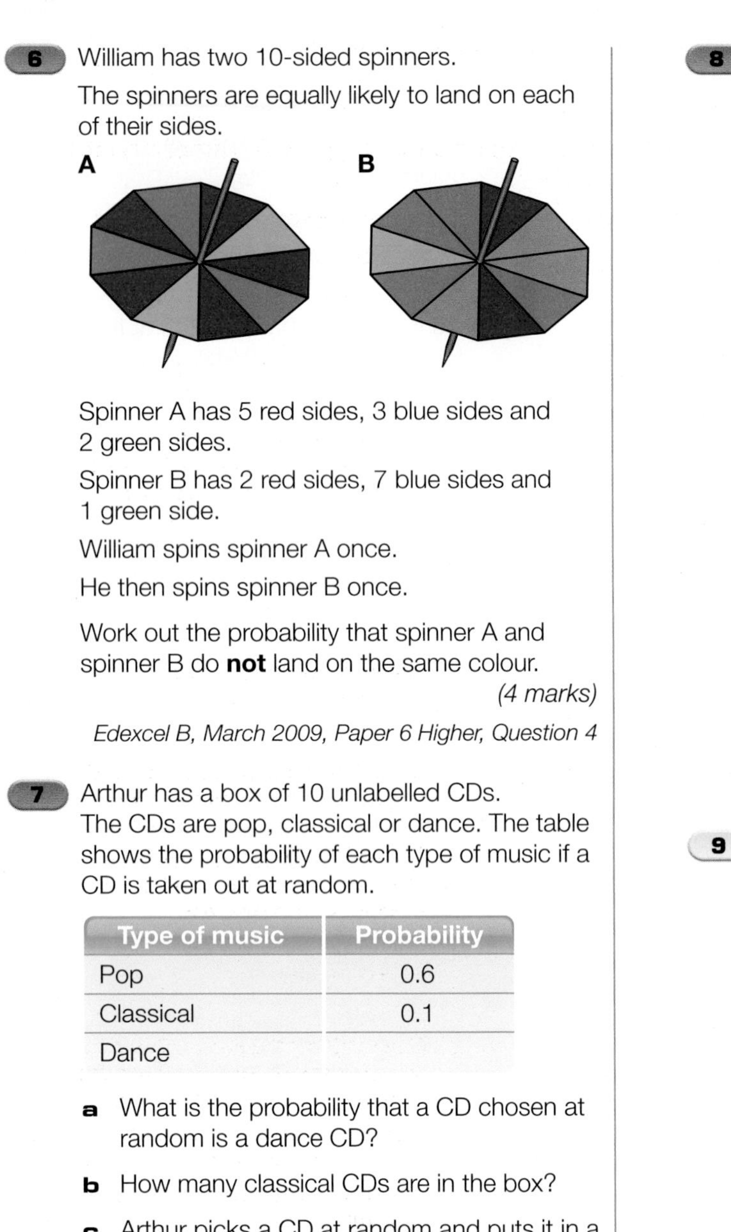

Spinner A has 5 red sides, 3 blue sides and 2 green sides.

Spinner B has 2 red sides, 7 blue sides and 1 green side.

William spins spinner A once.

He then spins spinner B once.

Work out the probability that spinner A and spinner B do **not** land on the same colour.

(4 marks)

Edexcel B, March 2009, Paper 6 Higher, Question 4

7 Arthur has a box of 10 unlabelled CDs. The CDs are pop, classical or dance. The table shows the probability of each type of music if a CD is taken out at random.

Type of music	Probability
Pop	0.6
Classical	0.1
Dance	

a What is the probability that a CD chosen at random is a dance CD?

b How many classical CDs are in the box?

c Arthur picks a CD at random and puts it in a 2-disc CD player. He then picks another CD at random and puts it in the player. Complete the tree diagram

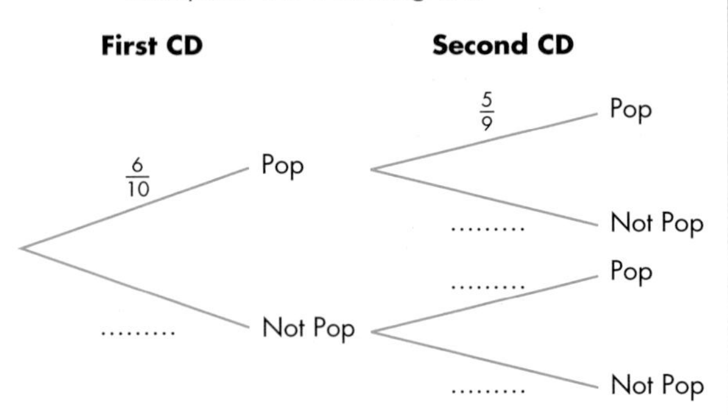

d What is the probability that neither of the CDs is pop?

8 At the end of a course, army cadets have to pass an exam to gain a certificate. The probability of passing the exam at the first attempt is 0.65.

Those who fail are allowed to re-sit.

The probability of passing the re-sit is 0.7.

No further attempts are allowed.

a i Complete the tree diagram.

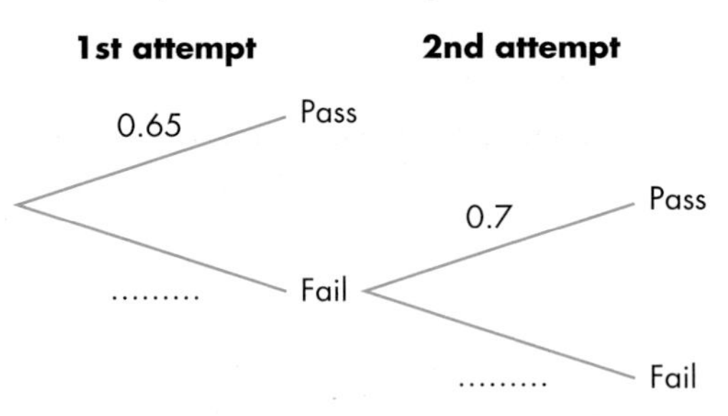

ii What is the probability that a cadet fails to gain a certificate after two attempts?

b Five cadets take the exam.

What is the probability that all of them gain a certificate?

9 A drinks machine uses cartridges to supply the drink. Billy has a job lot of eight cartridges which have lost their labels. He knows he has three teas and five coffees. He makes three drinks with the cartridges.

a What is the probability he gets three teas?

b What is the probability he gets exactly two coffees?

c What is the probability that he gets at least one coffee?

d Billy makes the three drinks and leaves the room. Betty comes in, tastes one of the drinks. She finds it is tea. What is the probability that the other two drinks are also tea?

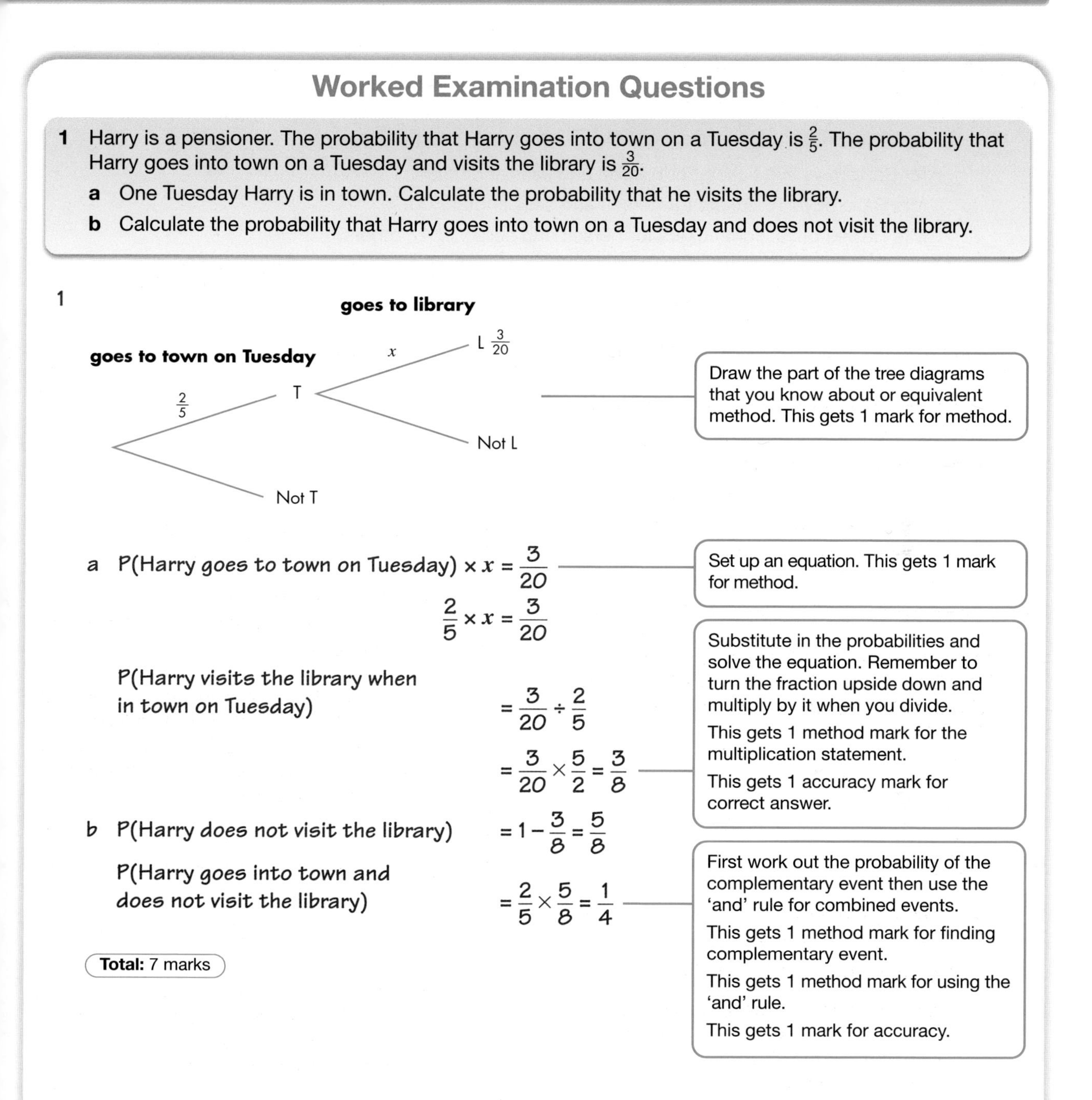

Worked Examination Questions

1 Harry is a pensioner. The probability that Harry goes into town on a Tuesday is $\frac{2}{5}$. The probability that Harry goes into town on a Tuesday and visits the library is $\frac{3}{20}$.

a One Tuesday Harry is in town. Calculate the probability that he visits the library.

b Calculate the probability that Harry goes into town on a Tuesday and does not visit the library.

1

Draw the part of the tree diagrams that you know about or equivalent method. This gets 1 mark for method.

a P(Harry goes to town on Tuesday) $\times x = \frac{3}{20}$

$$\frac{2}{5} \times x = \frac{3}{20}$$

Set up an equation. This gets 1 mark for method.

P(Harry visits the library when in town on Tuesday) $= \frac{3}{20} \div \frac{2}{5}$

$$= \frac{3}{20} \times \frac{5}{2} = \frac{3}{8}$$

Substitute in the probabilities and solve the equation. Remember to turn the fraction upside down and multiply by it when you divide.

This gets 1 method mark for the multiplication statement.

This gets 1 accuracy mark for correct answer.

b P(Harry does not visit the library) $= 1 - \frac{3}{8} = \frac{5}{8}$

P(Harry goes into town and does not visit the library) $= \frac{2}{5} \times \frac{5}{8} = \frac{1}{4}$

First work out the probability of the complementary event then use the 'and' rule for combined events.

This gets 1 method mark for finding complementary event.

This gets 1 method mark for using the 'and' rule.

This gets 1 mark for accuracy.

Total: 7 marks

Worked Examination Questions

FM **2** In a raffle 400 tickets have been sold. There is only one prize.

Mr Raza buys five tickets for himself and sells another 40.

Mrs Raza buys 10 tickets for herself and sells another 50.

Mrs Hewes buys eight tickets for herself and sells just 12 others.

a What is the probability of:

i Mr Raza winning the raffle

ii Mr Raza selling the winning ticket

iii Mr Raza either winning the raffle or selling the winning ticket?

b What is the probability of either Mr or Mrs Raza selling the winning ticket?

c What is the probability of Mrs Hewes not winning the lottery?

2 a i $\frac{5}{400}$ — This gets 1 mark.

ii $\frac{40}{400}$ — This gets 1 mark.

iii $\frac{(5+40)}{400} = \frac{45}{400}$ — This gets 1 mark for method of adding the probabilities and 1 mark for accuracy.

b $\frac{(40+50)}{400} = \frac{90}{400}$ — This gets 1 mark for method and 1 mark for accuracy.

c $1 - \frac{8}{400} = \frac{392}{400}$ — This gets 1 mark for method of subtracting from 1 and 1 mark for accuracy.

Total: 8 marks

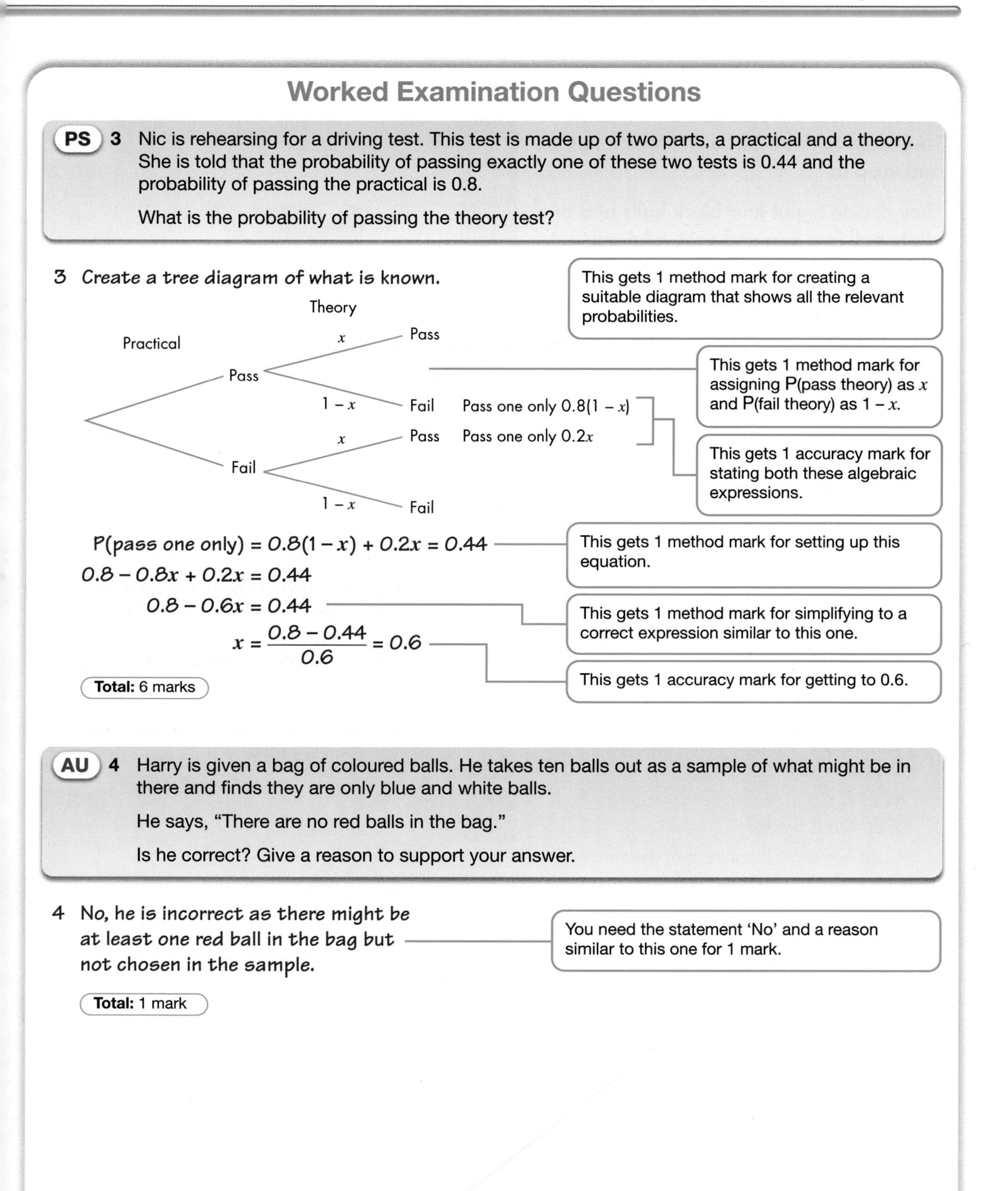

Worked Examination Questions

PS **3** Nic is rehearsing for a driving test. This test is made up of two parts, a practical and a theory. She is told that the probability of passing exactly one of these two tests is 0.44 and the probability of passing the practical is 0.8.

What is the probability of passing the theory test?

3 *Create a tree diagram of what is known.*

This gets 1 method mark for creating a suitable diagram that shows all the relevant probabilities.

This gets 1 method mark for assigning P(pass theory) as x and P(fail theory) as $1 - x$.

This gets 1 accuracy mark for stating both these algebraic expressions.

$$P(\text{pass one only}) = 0.8(1 - x) + 0.2x = 0.44$$

This gets 1 method mark for setting up this equation.

$$0.8 - 0.8x + 0.2x = 0.44$$

$$0.8 - 0.6x = 0.44$$

This gets 1 method mark for simplifying to a correct expression similar to this one.

$$x = \frac{0.8 - 0.44}{0.6} = 0.6$$

This gets 1 accuracy mark for getting to 0.6.

Total: 6 marks

AU **4** Harry is given a bag of coloured balls. He takes ten balls out as a sample of what might be in there and finds they are only blue and white balls.

He says, "There are no red balls in the bag."

Is he correct? Give a reason to support your answer.

4 *No, he is incorrect as there might be at least one red ball in the bag but not chosen in the sample.*

You need the statement 'No' and a reason similar to this one for 1 mark.

Total: 1 mark

11 Problem Solving
Sharing a trophy

A team of five people has won a trophy. They need to decide who will take it home and keep it.

They decide to put four black balls and one white ball in a bag and each take one out, without replacement. The person who picks the white ball keeps the trophy.

Your friend Pip is one of the team. She thinks that the method is not fair, because the person who goes first will have more chance of getting the white ball and keeping the trophy.

Getting started

1 Suppose there were just two balls, one black and one white.

What is the probability that the first ball taken out is white?

What is the probability that the first ball is black?

2 Now suppose there were two black balls and one white ball in the bag.

What is the probability that the first ball taken out is white?

What is the probability that the first ball is black?

Your task

1 What advice would you give your friend? You must explain your reasoning and give evidence to justify your advice.

2 Suppose there were 11 people in the team instead of 5. They use 10 black balls and one white one. What would your advice be this time? Justify your answer.

Extension

Your friend suggests an alternative method. Everyone takes a card from a pack in turn and the person with the highest card takes the trophy. Is this fair?

Why this chapter matters

Without algebra, humans would not have reached the moon and planes would not fly. Defining numbers with letters allows mathematicians to use formulae and solve the very complicated equations that are needed for today's technologies.

2000 BC The Babylonians discover that the ratio of the circumference of a circle to its diameter is approximately 3.125. This is the first time an approximation to π is used.

The Chinese made the first reference to negative numbers.

100 AD Heron of Alexandria did some work that led to the need for the square root of a negative number but this was dismissed for many centuries as something that was impossible.

1500 AD Italian mathematicians came across formulae that could only be solved if the square root of –1 was used.

1550 AD Rafael Bombelli was the first mathematician to introduce the notation $\sqrt{-1} = \mathrm{i}$.

1618 AD When doing work on logarithms, the Scottish mathematician John Napier published a list of 'natural logarithms' that were based on a number with a value of about 2.718, although this number was not quoted.

1680 AD A Swiss mathematician, Jacob Bernoulli, found the value 2.71828… which is the limit of $(1 + \frac{1}{n})^n$. (Try this on your calculator with a big number for n. If $n = 1000$, then $1.001^{1000} =$ 2.71692…)

1727 AD Another Swiss mathematician, Leonhard Euler, gave the value the letter e.

1748 AD Euler publishes his famous formula $\ln_e(\cos x + \mathrm{i}\sin x) = \mathrm{i}x$ which can also be expressed as $e^{\mathrm{i}\pi} = -1$.

Why is this so important? If you were an advanced alien culture you would need a unit of 1 as a basis for counting. You would need a symbol for zero. You would know about π, as circles are the same anywhere in the universe, and you would also know about e, as this number occurs naturally. Hence Euler's formula is literally universal and many people think that it proves that God exists.

Others think that it is just a coincidence that five of the most important numbers in mathematics, 0, 1, i, e and π are connected by such a neat formula.

Chapter

Algebra: Algebraic methods

1 Rearranging formulae

2 Changing the subject of a formula

3 Simultaneous equations

4 Solving problems using simultaneous equations

5 Linear and non-linear simultaneous equations

6 Algebraic fractions

7 Algebraic proof

The grades given in this chapter are target grades.

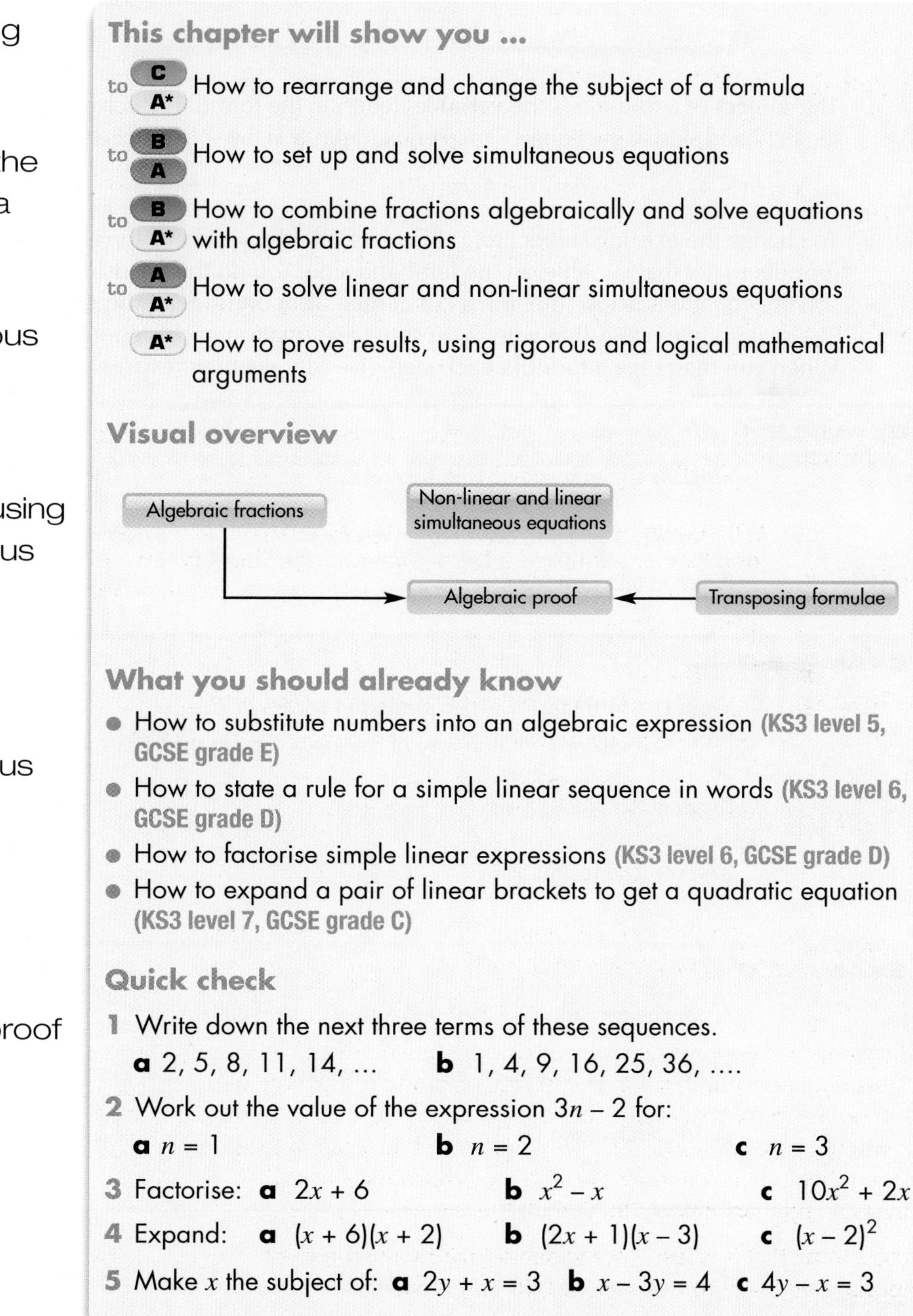

This chapter will show you ...

- C to A* How to rearrange and change the subject of a formula
- B to A How to set up and solve simultaneous equations
- B to A* How to combine fractions algebraically and solve equations with algebraic fractions
- A to A* How to solve linear and non-linear simultaneous equations
- A* How to prove results, using rigorous and logical mathematical arguments

Visual overview

What you should already know

- How to substitute numbers into an algebraic expression **(KS3 level 5, GCSE grade E)**
- How to state a rule for a simple linear sequence in words **(KS3 level 6, GCSE grade D)**
- How to factorise simple linear expressions **(KS3 level 6, GCSE grade D)**
- How to expand a pair of linear brackets to get a quadratic equation **(KS3 level 7, GCSE grade C)**

Quick check

1 Write down the next three terms of these sequences.

a 2, 5, 8, 11, 14, … **b** 1, 4, 9, 16, 25, 36, ….

2 Work out the value of the expression $3n - 2$ for:

a $n = 1$ **b** $n = 2$ **c** $n = 3$

3 Factorise: **a** $2x + 6$ **b** $x^2 - x$ **c** $10x^2 + 2x$

4 Expand: **a** $(x + 6)(x + 2)$ **b** $(2x + 1)(x - 3)$ **c** $(x - 2)^2$

5 Make x the subject of: **a** $2y + x = 3$ **b** $x - 3y = 4$ **c** $4y - x = 3$

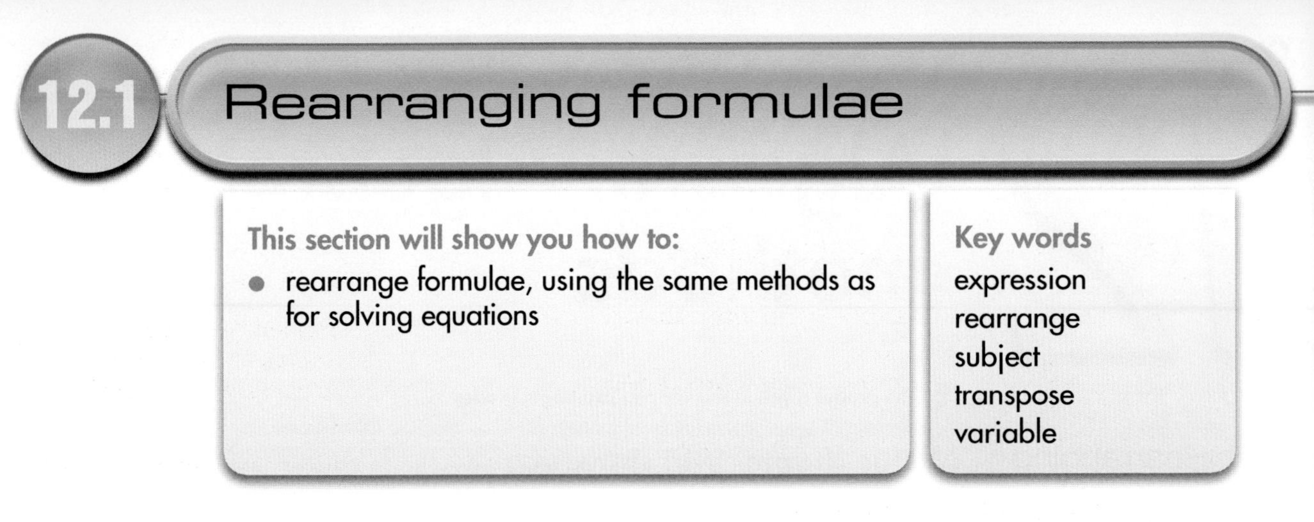

The **subject** of a formula is the **variable** (letter) in the formula which stands on its own, usually on the left-hand side of the equals sign. For example, x is the subject of each of the following equations.

$x = 5t + 4$ $\qquad$ $x = 4(2y - 7)$ $\qquad$ $x = \frac{1}{t}$

To change the existing subject to a different variable, you have to **rearrange** (**transpose**) the formula to get that variable on the left-hand side. You do this by using the same rules as for solving equations. Move the terms concerned from one side of the equals sign to the other. The main difference is that when you solve an equation each step gives a numerical value. When you rearrange a formula each step gives an algebraic **expression**.

EXAMPLE 1

Make m the subject of this formula. $\quad T = m - 3$

Move the 3 so that the m is on its own. $\quad T + 3 = m$

Reverse the formula. $\quad m = T + 3$

EXAMPLE 2

From the formula $P = 4t$, express t in terms of P.
(This is another common way of asking you to make t the subject.)

Divide both sides by 4: $\quad \frac{P}{4} = \frac{4t}{4}$

Reverse the formula: $\quad t = \frac{P}{4}$

EXAMPLE 3

From the formula $C = 2m^2 + 3$, make m the subject.

Move the 3 so that the $2m^2$ is on its own $\quad C - 3 = 2m^2$

Divide both sides by 2: $\quad \frac{C-3}{2} = \frac{2m^2}{2}$

Reverse the formula: $\quad m^2 = \frac{C-3}{2}$

Take the square root on both sides: $\quad m = \sqrt{\frac{C-3}{2}}$

EXERCISE 12A

HINTS AND TIPS

Remember about inverse operations, and the rule 'change sides, change signs'.

1 $T = 3k$ Make k the subject.

2 $X = y - 1$ Express y in terms of X.

3 $Q = \frac{p}{3}$ Express p in terms of Q.

4 $A = 4r + 9$ Make r the subject.

5 $W = 3n - 1$ Make n the subject.

6 $p = m + t$ **a** Make m the subject. **b** Make t the subject.

7 $g = \frac{m}{v}$ Make m the subject.

8 $t = m^2$ Make m the subject.

9 $C = 2\pi r$ Make r the subject.

10 $A = bh$ Make b the subject.

11 $P = 2l + 2w$ Make l the subject.

12 $m = p^2 + 2$ Make p the subject.

FM 13 The formula for converting degrees Fahrenheit to degrees Celsius is $C = \frac{5}{9}(F - 32)$.

a Show that when $F = -40$, C is also equal to -40.

b Find the value of C when $F = 68$.

c Use this flow diagram to establish the formula for converting degrees Celsius to degrees Fahrenheit.

°F → -32 → $\times 5$ → $\div 9$ → °C

FM 14 Kieran notices that the price of five cream buns is 75p more than the price of nine mince pies.
Let the price of a cream bun be x pence and the price of a mince pie be y pence.

a Express the cost of one mince pie, y, in terms of the price of a cream bun, x.

b If the price of a cream bun is 60p, how much is a mince pie?

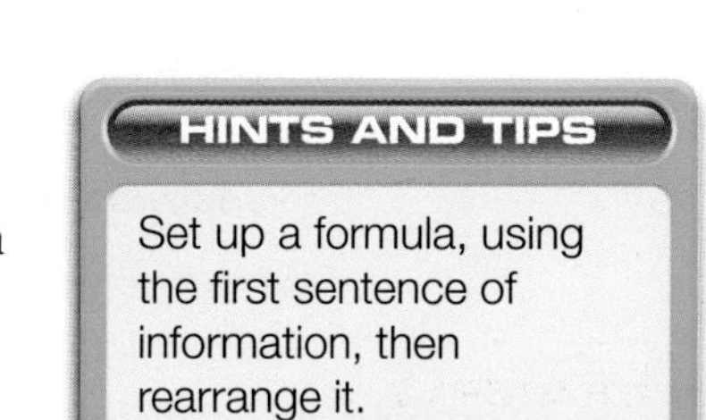

PS 15 Distance, speed and time are connected by the formula:

distance = speed × time.

A delivery driver drove 126 km in 1 hour and 45 minutes. On the return journey, he was held up at some road works so his average speed decreased by 9 km per hour.

How long was he held up at the road works?

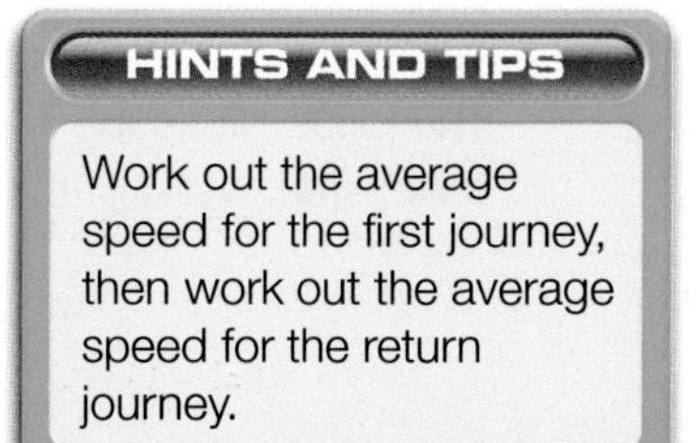

C

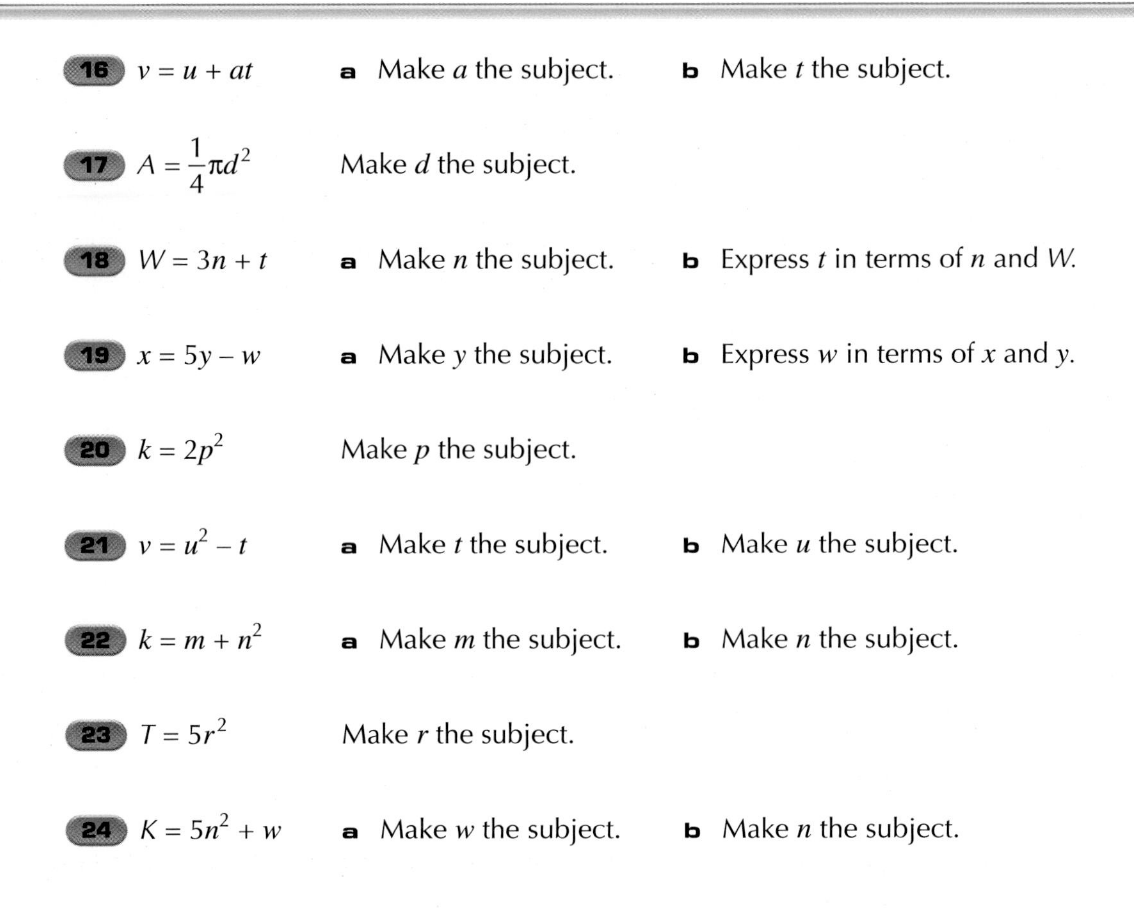

16 $v = u + at$ **a** Make a the subject. **b** Make t the subject.

17 $A = \frac{1}{4}\pi d^2$ Make d the subject.

18 $W = 3n + t$ **a** Make n the subject. **b** Express t in terms of n and W.

19 $x = 5y - w$ **a** Make y the subject. **b** Express w in terms of x and y.

20 $k = 2p^2$ Make p the subject.

21 $v = u^2 - t$ **a** Make t the subject. **b** Make u the subject.

22 $k = m + n^2$ **a** Make m the subject. **b** Make n the subject.

23 $T = 5r^2$ Make r the subject.

24 $K = 5n^2 + w$ **a** Make w the subject. **b** Make n the subject.

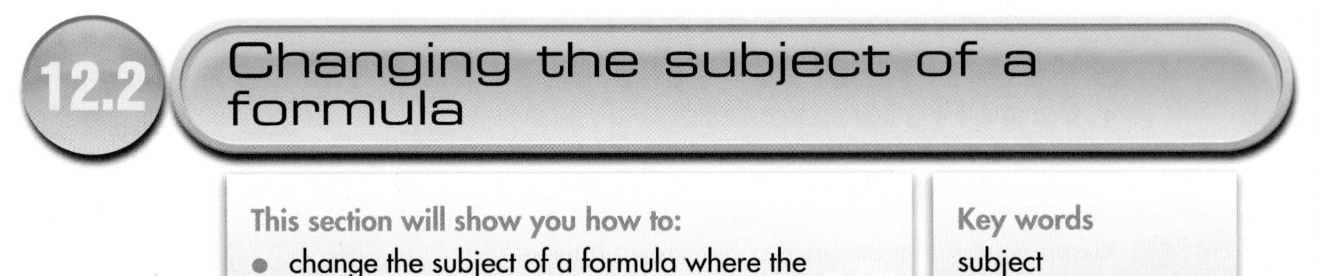

12.2 Changing the subject of a formula

This section will show you how to:
- change the subject of a formula where the subject occurs more than once

Key words
subject

You have just considered changing the **subject** of a formula in which the subject appears only once. This is like solving an equation but using letters. You have also solved equations in which the unknown appears on both sides of the equation. This requires you to collect the terms in the unknown (usually x) on one side and the numbers on the other.

You can do something similar, to rearrange formulae in which the subject appears more than once. The principle is the same. Collect all the subject terms on the same side and everything else on the other side. Most often, you then need to factorise the subject out of the resulting expression.

EXAMPLE 4

Make x the subject of this formula.

$$ax + b = cx + d$$

First, rearrange the formula to get all the x-terms on the left-hand side and all the other terms on the right-hand side. (The rule 'change sides – change signs' still applies.)

$$ax - cx = d - b$$

Factorise x out of the left-hand side to get:

$$x(a - c) = d - b$$

Divide by the expression in brackets, which gives:

$$x = \frac{d - b}{a - c}$$

EXAMPLE 5

Make p the subject of this formula.

$$5 = \frac{ap + b}{cp + d}$$

First, multiply both sides by the denominator of the algebraic fraction, which gives:

$$5(cp + d) = ap + b$$

Expand the brackets to get:

$$5cp + 5d = ap + b$$

Now continue as in Example 4:

$$5cp - ap = b - 5d$$

$$p(5c - a) = b - 5d$$

$$p = \frac{b - 5d}{5c - a}$$

EXERCISE 12B

In questions **1** to **10**, make the letter in brackets the subject of the formula.

1 $3(x + 2y) = 2(x - y)$ (x)

2 $3(x + 2y) = 2(x - y)$ (y)

3 $5 = \dfrac{a + b}{a - c}$ (a)

4 $p(a + b) = q(a - b)$ (a)

A

A

5 $p(a + b) = q(a - b) \quad (b)$

6 $A = 2\pi rh + \pi rk \quad (r)$

7 $v^2 = u^2 + av^2 \quad (v)$

8 $s(t - r) = 2r - 3 \quad (r)$

9 $s(t - r) = 2(r - 3) \quad (r)$

10 $R = \dfrac{x - 3}{x - 2} \quad (x)$

11 **a** The perimeter of the shape shown on the right is given by the formula $P = \pi r + 2kr$. Make r the subject of this formula.

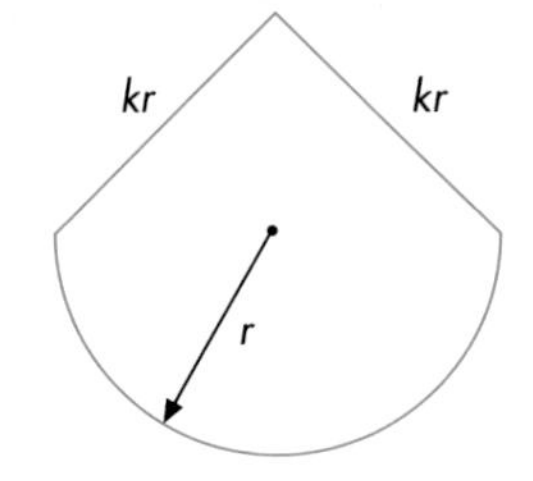

b The area of the same shape is given by $A = \frac{1}{2}[\pi r^2 + r^2\sqrt{(k^2 - 1)}]$
Make r the subject of this formula.

12 When £P is invested for Y years at a simple interest rate of R, the following formula gives the amount, A, at any time.

$$A = P + \frac{PRY}{100}$$

Make P the subject of this formula.

13 When two resistors with values a and b are connected in parallel, the total resistance is given by:

$$R = \frac{ab}{a + b}$$

a Make b the subject of the formula.

b Write the formula when a is the subject.

A*

AU 14 **a** Make x the subject of this formula.

$$y = \frac{x + 2}{x - 2}$$

b Show that the formula $y = 1 + \dfrac{4}{x - 2}$ can be rearranged to give:

$$x = 2 + \frac{4}{y - 1}$$

c Combine the right-hand sides of each formula in part **b** into single fractions and simplify as much as possible.

d What do you notice?

A*

15 The volume of the solid shown is given by:

$V = \frac{2}{3}\pi r^3 + \pi r^2 h$

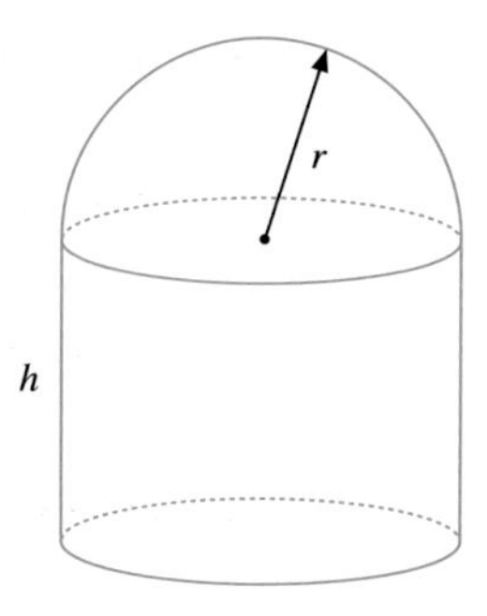

a Explain why it is not possible to make r the subject of this formula.

b Make π the subject.

c If $h = r$, can the formula be rearranged to make r the subject? If so, rearrange it to make r the subject.

16 Make x the subject of this formula.

$W = \frac{1}{2}z(x + y) + \frac{1}{2}y(x + z)$

PS 17 The following formulae in y can be rearranged to give the formulae in terms of x as shown.

$y = \dfrac{x + 1}{x + 2}$ gives $x = \dfrac{1 - 2y}{y - 1}$

$y = \dfrac{2x + 1}{x + 2}$ gives $x = \dfrac{1 - 2y}{y - 2}$

$y = \dfrac{3x + 2}{4x + 1}$ gives $x = \dfrac{2 - y}{4y - 3}$

$y = \dfrac{x + 5}{3x + 2}$ gives $x = \dfrac{5 - 2y}{3y - 1}$

Without rearranging the formula, write down $y = \dfrac{5x + 1}{2x + 3}$ as $x = \ldots$ and explain how you can do this without any algebra.

AU 18 A formula used in GCSE mathematics is the cosine rule, which relates the three sides of any triangle with the angle between two of the sides.

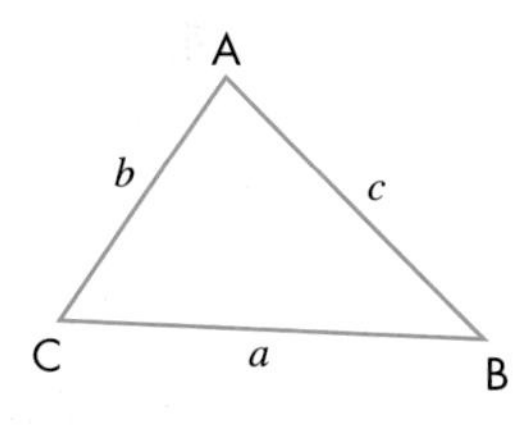

The formula is $a^2 = b^2 + c^2 - 2bc \cos A$.

This formula is known as an algebraic, cyclically symmetric formula.

That means that the various letters can be swapped with each other in a cycle, i.e. a becomes b, b becomes c and c becomes a.

a Write the formula so that it starts $b^2 = \ldots$

b When the formula is rearranged to make $\cos A$ the subject, then it becomes

$$\cos A = \frac{b^2 + c^2 - a^2}{2bc}$$

Write an equivalent formula that starts $\cos C = \ldots$

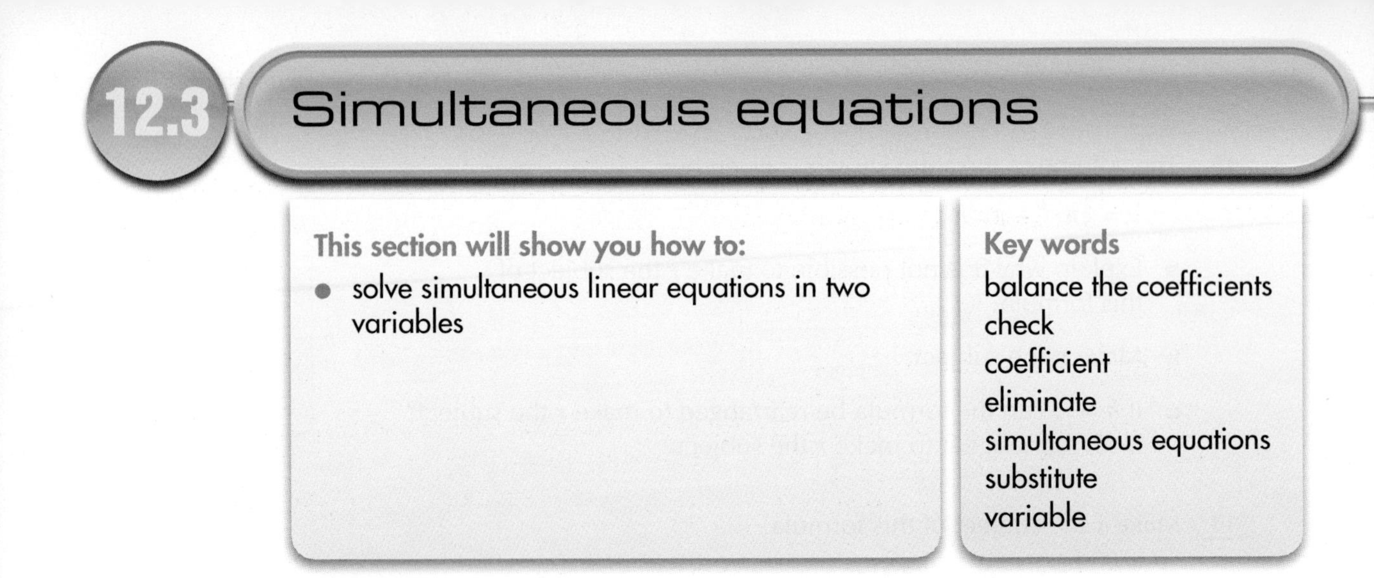

12.3 Simultaneous equations

This section will show you how to:

- solve simultaneous linear equations in two variables

Key words
balance the coefficients
check
coefficient
eliminate
simultaneous equations
substitute
variable

A pair of **simultaneous equations** is exactly that — two equations (usually linear) for which you want the same solution, and which you therefore *solve together*. For example,

$x + y = 10$ has many solutions:

$x = 2, y = 8$ $\quad$ $x = 4, y = 6$ $\quad$ $x = 5, y = 5 \ldots$

and $2x + y = 14$ has many solutions:

$x = 2, y = 10$ $\quad$ $x = 3, y = 8$ $\quad$ $x = 4, y = 6 \ldots$

But only *one* solution, $x = 4$ and $y = 6$, satisfies both equations at the same time.

Elimination method

Here, you solve simultaneous equations by the *elimination method*. There are six steps in this method. **Step 1** is to **balance the coefficients** of one of the **variables**. **Step 2** is to **eliminate** this variable by adding or subtracting the equations. **Step 3** is to solve the resulting linear equation in the other variable. **Step 4** is to **substitute** the value found back into one of the previous equations. **Step 5** is to solve the resulting equation. **Step 6** is to **check** that the two values found satisfy the original equations.

EXAMPLE 6

Solve the equations: $6x + y = 15$ and $4x + y = 11$

Label the equations so that the method can be clearly explained.

$6x + y = 15$ $\quad$ (1)

$4x + y = 11$ $\quad$ (2)

Step 1: Since the y-term in both equations has the same **coefficient** there is no need to balance them.

Step 2: Subtract one equation from the other. (Equation (1) minus equation (2) will give positive values.)

(1) – (2) $\quad$ $2x = 4$

Step 3: Solve this equation: $\quad$ $x = 2$

Step 4: Substitute $x = 2$ into one of the original equations. (Usually the one with smallest numbers involved.)

So substitute into: $4x + y = 11$

which gives: $8 + y = 11$

Step 5: Solve this equation: $y = 3$

Step 6: Test the solution in the original equations. So substitute $x = 2$ and $y = 3$ into $6x + y$, which gives $12 + 3 = 15$ and into $4x + y$, which gives $8 + 3 = 11$. These are correct, so you can confidently say the solution is $x = 2$ and $y = 3$.

EXAMPLE 7

Solve these equations.

$5x + y = 22$ (1)

$2x - y = 6$ (2)

Step 1: Both equations have the same y-coefficient but with *different* signs so there is no need to balance them.

Step 2: As the signs are different, *add* the two equations, to eliminate the y-terms.

(1) + (2) $7x = 28$

Step 3: Solve this equation: $x = 4$

Step 4: Substitute $x = 4$ into one of the original equations, $5x + y = 22$,

which gives: $20 + y = 22$

Step 5: Solve this equation: $y = 2$

Step 6: Test the solution by putting $x = 4$ and $y = 2$ into the original equations, $2x - y$, which gives $8 - 2 = 6$ and $5x + y$ which gives $20 + 2 = 22$. These are correct, so the solution is $x = 4$ and $y = 2$.

Substitution method

This is an alternative method. Which method you use depends very much on the coefficients of the variables and the way that the equations are written in the first place. There are five steps in the substitute method.

Step 1 is to rearrange one of the equations into the form $y = \dots$ or $x = \dots$.

Step 2 is to substitute the right-hand side of this equation into the other equation in place of the variable on the left-hand side.

Step 3 is to expand and solve this equation.

Step 4 is to substitute the value into the $y = \dots$ or $x = \dots$ equation.

Step 5 is to check that the values work in both original equations.

EXAMPLE 8

Solve the simultaneous equations: $y = 2x + 3$, $3x + 4y = 1$

Because the first equation is in the form $y = \dots$ it suggests that the substitution method should be used.

Again label the equations to help with explaining the method.

$y = 2x + 3$ (1)

$3x + 4y = 1$ (2)

Step 1: As equation (1) is in the form $y = \dots$ there is no need to rearrange an equation.

Step 2: Substitute the right-hand side of equation (1) into equation (2) for the variable y.

$$3x + 4(2x + 3) = 1$$

Step 3: Expand and solve the equation. $3x + 8x + 12 = 1$, $11x = -11$, $x = -1$

Step 4: Substitute $x = -1$ into $y = 2x + 3$: $y = -2 + 3 = 1$

Step 5: Test the values in $y = 2x + 3$ which gives $1 = -2 + 3$ and $3x + 4y = 1$, which gives $-3 + 4 = 1$. These are correct so the solution is $x = -1$ and $y = 1$.

EXERCISE 12C

B

1 Solve these simultaneous equations.

In question **1** parts **a** to **i** the coefficients of one of the variables are the same so there is no need to balance them. Subtract the equations when the identical terms have the same sign. Add the equations when the identical terms have opposite signs. In parts **j** to **l** use the substitution method.

a $4x + y = 17$
$2x + y = 9$

b $5x + 2y = 13$
$x + 2y = 9$

c $2x + y = 7$
$5x - y = 14$

d $3x + 2y = 11$
$2x - 2y = 14$

e $3x - 4y = 17$
$x - 4y = 3$

f $3x + 2y = 16$
$x - 2y = 4$

g $x + 3y = 9$
$x + y = 6$

h $2x + 5y = 16$
$2x + 3y = 8$

i $3x - y = 9$
$5x + y = 11$

j $2x + 5y = 37$
$y = 11 - 2x$

k $4x - 3y = 7$
$x = 13 - 3y$

l $4x - y = 17$
$x = 2 + y$

B

PS 2 In this sequence, the next term is found by multiplying the previous term by a and then adding b. a and b are positive whole numbers.

3 14 47 … …

a Explain why $3a + b = 14$

b Set up another equation in a and b.

c Solve the equations to solve for a and b.

d Work out the next two terms in the sequence.

Balancing coefficients in one equation only

You were able to solve all the pairs of equations in Exercise 12C, question **1** simply by adding or subtracting the equations in each pair, or just by substituting without rearranging. This does not always happen. The next examples show what to do when there are no identical terms to begin with, or when you need to rearrange.

EXAMPLE 9

Solve these equations. $3x + 2y = 18$ (1)

$2x - y = 5$ (2)

Step 1: Multiply equation (2) by 2. There are other ways to balance the coefficients but this is the easiest and leads to less work later. With practice, you will get used to which will be the best way to balance the coefficients.

$2 \times (2)$ $4x - 2y = 10$ (3)

Label this equation as number (3).

Be careful to multiply every term and not just the y-term, it sometimes helps to write:

$2 \times (2x - y = 5) \Rightarrow 4x - 2y = 10$ (3)

Step 2: As the signs of the y-terms are opposite, add the equations.

(1) + (3) $7x = 28$

Be careful to add the correct equations. This is why labelling them is useful.

Step 3: Solve this equation: $x = 4$

Step 4: Substitute $x = 4$ into any equation, say $2x - y = 5 \Rightarrow 8 - y = 5$

Step 5: Solve this equation: $y = 3$

Step 6: Check: (1), $3 \times 4 + 2 \times 3 = 18$ and (2), $2 \times 4 - 3 = 5$, which are correct so the solution is $x = 4$ and $y = 3$.

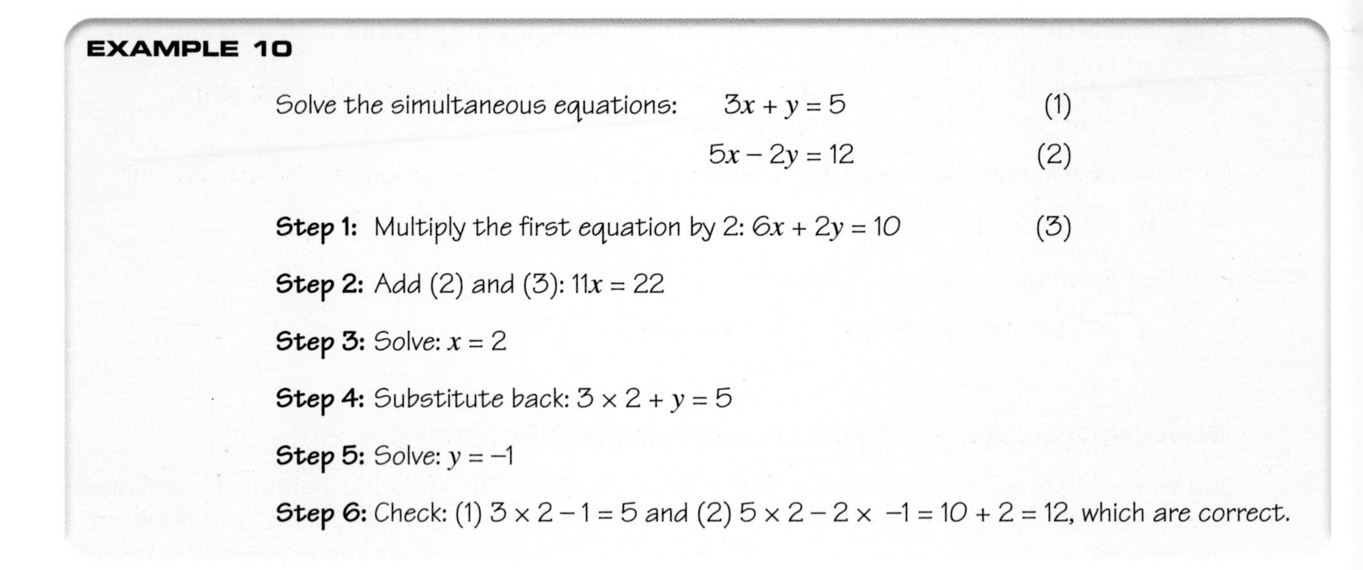

EXAMPLE 10

Solve the simultaneous equations: $3x + y = 5$ (1)

$5x - 2y = 12$ (2)

Step 1: Multiply the first equation by 2: $6x + 2y = 10$ (3)

Step 2: Add (2) and (3): $11x = 22$

Step 3: Solve: $x = 2$

Step 4: Substitute back: $3 \times 2 + y = 5$

Step 5: Solve: $y = -1$

Step 6: Check: (1) $3 \times 2 - 1 = 5$ and (2) $5 \times 2 - 2 \times -1 = 10 + 2 = 12$, which are correct.

EXERCISE 12D

B

1 Solve parts **a** to **c** by the substitution method and the rest by first changing one of the equations in each pair to obtain identical terms, and then adding or subtracting the equations to eliminate those terms.

a $5x + 2y = 4$
$4x - y = 11$

b $4x + 3y = 37$
$2x + y = 17$

c $x + 3y = 7$
$2x - y = 7$

d $2x + 3y = 19$
$6x + 2y = 22$

e $5x - 2y = 26$
$3x - y = 15$

f $10x - y = 3$
$3x + 2y = 17$

g $3x + 5y = 15$
$x + 3y = 7$

h $3x + 4y = 7$
$4x + 2y = 1$

i $5x - 2y = 24$
$3x + y = 21$

j $5x - 2y = 4$
$3x - 6y = 6$

k $2x + 3y = 13$
$4x + 7y = 31$

l $3x - 2y = 3$
$5x + 6y = 12$

AU **2** **a** Mary is solving the simultaneous equations $4x - 2y = 8$ and $2x - y = 4$.

She finds a solution of $x = 5$, $y = 6$ which works for both equations.

Explain why this is not a unique solution.

b Max is solving the simultaneous equations $6x + 2y = 9$ and $3x + y = 7$.

Why is it impossible to find a solution that works for both equations?

Balancing coefficients in both equations

There are also cases where *both* equations have to be changed to obtain identical terms. The next example shows you how this is done.

Note: The substitution method is not suitable for these types of equations as you end up with fractional terms.

EXAMPLE 11

Solve these equations. $4x + 3y = 27$ (1)

$5x - 2y = 5$ (2)

Both equations have to be changed to obtain identical terms in either x or y. However, you can see that if you make the y-coefficients the same, you will add the equations. This is always safer than subtraction, so this is obviously the better choice. We do this by multiplying the first equation by 2 (the y-coefficient of the other equation) and the second equation by 3 (the y-coefficient of the other equation).

Step 1: (1) × 2 or $2 \times (4x + 3y = 27)$ $\Rightarrow$ $8x + 6y = 54$ (3)

(2) × 3 or $3 \times (5x - 2y = 5)$ $\Rightarrow$ $15x - 6y = 15$ (4)

Label the new equations (3) and (4).

Step 2: Eliminate one of the variables: (3) + (4) $23x = 69$

Step 3: Solve the equation: $x = 3$

Step 4: Substitute into equation (1): $12 + 3y = 27$

Step 5: Solve the equation: $y = 5$

Step 6: Check: (1), $4 \times 3 + 3 \times 5 = 12 + 15 = 27$, and (2), $5 \times 3 - 2 \times 5 = 15 - 10 = 5$, which are correct so the solution is $x = 3$ and $y = 5$.

EXERCISE 12E

1 Solve the following simultaneous equations.

a $2x + 5y = 15$
$3x - 2y = 13$

b $2x + 3y = 30$
$5x + 7y = 71$

c $2x - 3y = 15$
$5x + 7y = 52$

d $3x - 2y = 15$
$2x - 3y = 5$

e $5x - 3y = 14$
$4x - 5y = 6$

f $3x + 2y = 28$
$2x + 7y = 47$

g $2x + y = 4$
$x - y = 5$

h $5x + 2y = 11$
$3x + 4y = 8$

i $x - 2y = 4$
$3x - y = -3$

j $3x + 2y = 2$
$2x + 6y = 13$

k $6x + 2y = 14$
$3x - 5y = 10$

l $2x + 4y = 15$
$x + 5y = 21$

m $3x - y = 5$
$x + 3y = -20$

n $3x - 4y = 4.5$
$2x + 2y = 10$

o $x - 5y = 15$
$3x - 7y = 17$

B

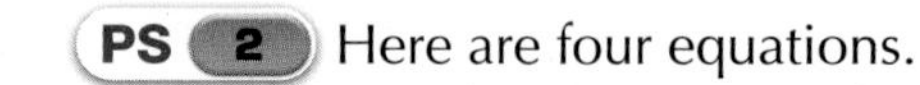

PS 2 Here are four equations.

A: $5x + 2y = 1$

B: $4x + y = 9$

C: $3x - y = 5$

D: $3x + 2y = 3$

Here are four sets of (x, y) values.

$(1, -2)$, $(-1, 3)$, $(2, 1)$, $(3, -3)$

Match each pair of (x, y) values to a pair of equations.

HINTS AND TIPS

You could solve each possible set of pairs but there are six to work out. Alternatively you can substitute values into the equations to see which work.

A

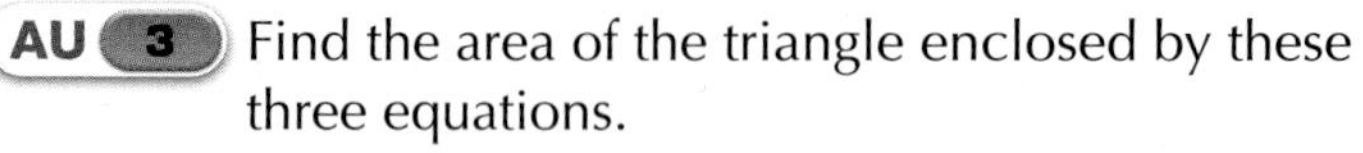

AU 3 Find the area of the triangle enclosed by these three equations.

$y - x = 2$ $\quad x + y = 6$ $\quad 3x + y = 6$

AU 4 Find the area of the triangle enclosed by these three equations.

$x - 2y = 6$ $\quad x + 2y = 6$ $\quad x + y = 3$

HINTS AND TIPS

Find the point of intersection of each pair of equations, plot the points on a grid and use any method to work out the area of the resulting triangle.

12.4 Solving problems using simultaneous equations

This section will show you how to:

- solve problems, using simultaneous linear equations in two variables

Key words

balance
check
coefficient
eliminate
simultaneous equations
substitute
variable

You are now going to meet a type of problem that has to be expressed as a pair of simultaneous equations so that it can be solved. The next example shows you how to tackle such a problem.

EXAMPLE 12

On holiday last year, I was talking over breakfast to two families about how much it cost them to go to the theatre. They couldn't remember how much was charged for each adult or each child, but they could both remember what they had paid altogether.

The Advani family, consisting of Mr and Mrs Advani with their daughter Rupa, paid £23.

The Shaw family, consisting of Mrs Shaw with her two children, Len and Sue, paid £17.50.

How much would I have to pay for my wife, my four children and myself?

Make a pair of simultaneous equations from the situation, as follows.

Let x be the cost of an adult ticket, and y be the cost of a child's ticket. Then

$2x + y = 23$ for the Advani family

and $x + 2y = 17.5$ for the Shaw family

Now solve these equations just as you have done in the previous examples, to obtain:

$x = £9.50$ and $y = £4$.

You can now find your cost, which will be $(2 \times £9.50) + (4 \times £4) = £35$.

EXERCISE 12F

Read each situation carefully, then make a pair of simultaneous equations in order to solve the problem.

PS 1 Amul and Kim have £10.70 between them. Amul has £3.70 more than Kim. Let x be the amount Amul has and y be the amount Kim has. Set up a pair of simultaneous equations. How much does each have?

FM 2 The two people in front of me at the Post Office were both buying stamps. One person bought 10 second-class and five first-class stamps at a total cost of £4.20. The other bought eight second-class and 10 first-class stamps at a total cost of £5.40.

- **a** Let x be the cost of a second-class stamp and y be the cost of a first-class stamp. Set up two simultaneous equations.
- **b** How much did I pay for three second-class and four first-class stamps?

3 At a local tea room I couldn't help noticing that at one table, where the customers had eaten six buns and had three teas, the bill came to £4.35. At another table, the customers had eaten 11 buns and had seven teas at a total cost of £8.80.

- **a** Let x be the cost of a bun and y be the cost of a cup of tea. Show the situation as a pair of simultaneous equations.
- **b** My family and I had five buns and six teas. What did it cost us?

PS 4 The sum of my son's age and my age this year is 72.

Six years ago my age was double that of my son.

Let my age now be x and my son's age now be y.

- **a** Explain why $x - 6 = 2(y - 6)$.
- **b** Find the values of x and y.

5 In a tea shop, three teas and five buns cost £8.10

In the same tea shop three teas and three buns cost £6.30

- **a** Using t to represent the cost of a tea and b to represent the cost of a bun, set up the above information as a pair of simultaneous equations.
- **b** How much will I pay for four teas and six buns?

B

A

6 Three chews and four bubblies cost 72p. Five chews and two bubblies cost 64p. What would three chews and five bubblies cost?

FM **7** On a nut-and-bolt production line, all the nuts had the same mass and all the bolts had the same mass. An order of 50 nuts and 60 bolts had a mass of 10.6 kg. An order of 40 nuts and 30 bolts had a mass of 6.5 kg. What should the mass of an order of 60 nuts and 50 bolts be?

FM **8** My local taxi company charges a fixed amount plus so much per mile. When I took a six mile journey the cost was £3.70. When I took a 10 mile journey the cost was £5.10. My next journey is going to be eight miles. How much will this cost?

FM **9** Two members of the same church went to the same shop to buy material to make Christingles. One bought 200 oranges and 220 candles at a cost of £65.60. The other bought 210 oranges and 200 candles at a cost of £63.30. They only needed 200 of each. How much should it have cost them?

FM **10** When you book Bingham Hall for a conference you pay a fixed booking fee plus a charge for each delegate. AQA booked a conference for 65 delegates and was charged £192.50. OCR booked a conference for 40 delegates and was charged £180. EDEXCEL wants to book for 70 delegates. How much will they be charged?

FM **11** My mother-in-law uses this formula to cook a turkey:

$$T = a + bW$$

where T is the cooking time (minutes), W is the weight of the turkey (kg) and a and b are constants. She says it takes 4 hours 30 minutes to cook a 12 kg turkey, and 3 hours 10 minutes to cook an 8 kg turkey. How long will it take to cook a 5 kg turkey?

FM **12** Four sacks of potatoes and two sacks of carrots weigh 188 pounds.

Five sacks of potatoes and one sack of carrots weigh 202 pounds.

Baz buys seven sacks of potatoes and eight sacks of carrots.

Will he be able to carry them in his trailer, which has a safe working load of 450 pounds?

HINTS AND TIPS

Set up two simultaneous equations using p and c for the weight of a sack of potatoes and carrots respectively.

FM **13** Five bags of bark chipping and four trays of pansies cost £24.50.

Three bags of bark chippings and five trays of pansies cost £19.25.

Camilla wants six bags of bark chippings and eight trays of pansies.

She has £30. Will she have enough money?

HINTS AND TIPS

Set up a pair of simultaneous equations using b and p for the cost of bark chippings and pansies and solve them.

AU **14** A teacher asks her class to solve these two simultaneous equations.

$$y = x + 4 \qquad (1)$$
$$2y - x = 10 \qquad (2)$$

Carmen says to Jeff, "Let's save time, you work out the x-value and I'll work out the y-value."
Jeff says, "Great idea."

This is Carmen's work.

$$y - x = 4 \qquad (3)$$
$$2y - x = 10 \qquad (2)$$
$$(2) - (3) \qquad 3y = 6$$
$$y = 2$$

This is Jeff's work.

Substitute (1) into (2)

$$2(x + 4) - x = 10$$
$$2x + 8 - x = 10$$
$$3x = 18$$
$$x = 6$$

When the teacher reads out the answer as "two, six" the students mark their work correct.
Explain all the mistakes that Carmen and Jeff have made.

12.5 Linear and non-linear simultaneous equations

This section will show you how to:
- solve linear and non-linear simultaneous equations

Key words
linear
non-linear
substitute

You have already seen the method of substitution for solving **linear** simultaneous equations (see page 321). Example 13 is a reminder.

EXAMPLE 13

Solve these simultaneous equations.

$$2x + 3y = 7 \qquad (1)$$
$$x - 4y = 9 \qquad (2)$$

First, rearrange equation (2) to obtain:

$$x = 9 + 4y$$

Substitute the expression for x into equation (1), which gives:

$$2(9 + 4y) + 3y = 7$$

Expand and solve this equation to obtain:

$$18 + 8y + 3y = 7$$
$$\Rightarrow 11y = -11$$
$$\Rightarrow y = -1$$

Now substitute for y into either equation (1) or (2) to find x. Using equation (1),

$$\Rightarrow 2x - 3 = 7$$
$$\Rightarrow x = 5$$

You can use a similar method when you need to solve a pair of equations, one of which is linear and the other of which is **non-linear**. But you must always **substitute** from the linear into the non-linear.

EXAMPLE 14

Solve these simultaneous equations.

$$x^2 + y^2 = 5$$
$$x + y = 3$$

Call the equations (1) and (2):

$$x^2 + y^2 = 5 \qquad (1)$$
$$x + y = 3 \qquad (2)$$

Rearrange equation (2) to obtain:

$$x = 3 - y$$

Substitute this into equation (1), which gives:

$$(3 - y)^2 + y^2 = 5$$

Expand and rearrange into the general form of the quadratic equation:

$$9 - 6y + y^2 + y^2 = 5$$
$$2y^2 - 6y + 4 = 0$$

Cancel by 2:

$$y^2 - 3y + 2 = 0$$

Factorise:

$$(y - 1)(y - 2) = 0$$
$$\Rightarrow y = 1 \text{ or } 2$$

Substitute for y in equation (2):

When $y = 1, x = 2$ and when $y = 2, x = 1$

Note that you should always give answers as a pair of values in x and y.

EXAMPLE 15

Find the solutions of the pair of simultaneous equations: $y = x^2 + x - 2$ and $y = 2x + 4$

This example is slightly different, as both equations are given in terms of y, so substituting for y gives:

$$2x + 4 = x^2 + x - 2$$

Rearranging into the general quadratic:

$$x^2 - x - 6 = 0$$

Factorising and solving gives:

$$(x + 2)(x - 3) = 0$$
$$x = -2 \text{ or } 3$$

Substituting back to find y:

When $x = -2, y = 0$

When $x = 3, y = 10$

So the solutions are $(-2, 0)$ and $(3, 10)$.

EXERCISE 12G

A A*

1 Solve these pairs of linear simultaneous equations using the substitution method.

a $2x + y = 9$
$x - 2y = 7$

b $3x - 2y = 10$
$4x + y = 17$

c $x - 2y = 10$
$2x + 3y = 13$

2 Solve these pairs of simultaneous equations.

a $xy = 2$
$y = x + 1$

b $xy = -4$
$2y = x + 6$

3 Solve these pairs of simultaneous equations.

a $x^2 + y^2 = 25$
$x + y = 7$

b $x^2 + y^2 = 9$
$y = x + 3$

c $x^2 + y^2 = 13$
$5y + x = 13$

4 Solve these pairs of simultaneous equations.

a $y = x^2 + 2x - 3$
$y = 2x + 1$

b $y = x^2 - 2x - 5$
$y = x - 1$

c $y = x^2 - 2x$
$y = 2x - 3$

5 Solve these pairs of simultaneous equations.

a $y = x^2 + 3x - 3$ and $y = x$

b $x^2 + y^2 = 13$ and $x + y = 1$

c $x^2 + y^2 = 5$ and $y = x + 1$

d $y = x^2 - 3x + 1$ and $y = 2x - 5$

e $y = x^2 - 3$ and $y = x + 3$

f $y = x^2 - 3x - 2$ and $y = 2x - 6$

g $x^2 + y^2 = 41$ and $y = x + 1$

AU 6 **a** Solve the simultaneous equations: $y = x^2 + 3x - 4$ and $y = 5x - 5$

b Which of the sketches below represents the graphs of the equations in part **a**?

Explain your choice.

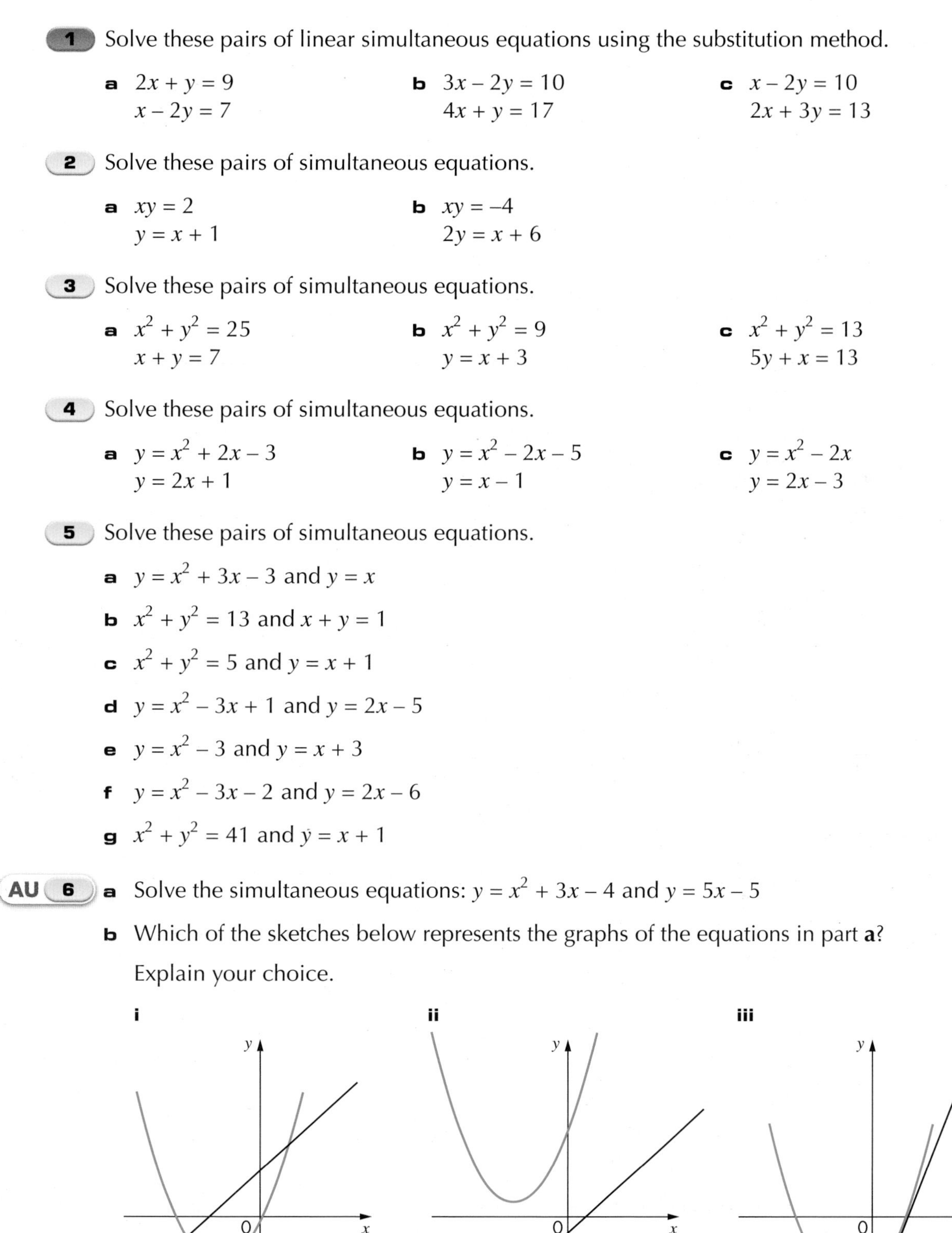

A*

PS 7 The simultaneous equations $x^2 + y^2 = 5$ and $y = 2x + 5$ only has one solution.

a Find the solution.

b Write down the intersection of each pair of graphs.

i $x^2 + y^2 = 5$ and $y = -2x + 5$

ii $x^2 + y^2 = 5$ and $y = -2x - 5$

iii $x^2 + y^2 = 5$ and $y = 2x - 5$

PS 8 Solve these pairs of simultaneous equations.

a $y = x^2 + x - 2$
$y = 5x - 6$

b $y = x^2 + 2x - 3$
$y = 4x - 4$

c What is the geometrical significance of the answers to parts **a** and **b**?

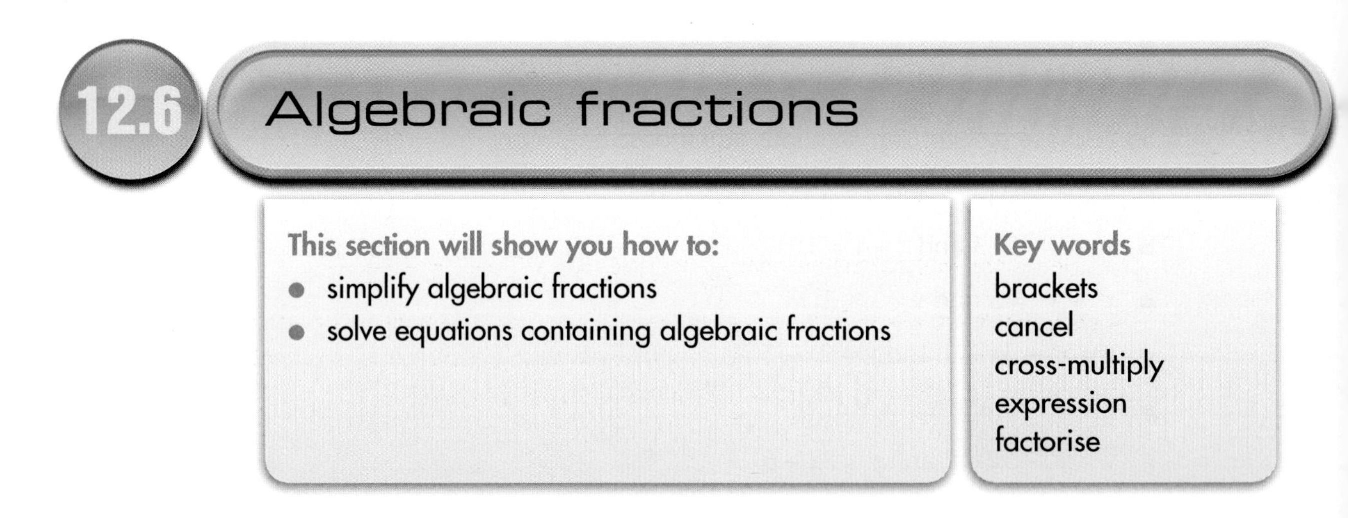

12.6 Algebraic fractions

This section will show you how to:
- simplify algebraic fractions
- solve equations containing algebraic fractions

Key words
brackets
cancel
cross-multiply
expression
factorise

The following four rules are used to work out the value of fractions.

Addition: $\dfrac{a}{b} + \dfrac{c}{d} = \dfrac{ad + bc}{bd}$

Subtraction: $\dfrac{a}{b} - \dfrac{c}{d} = \dfrac{ad - bc}{bd}$

Multiplication: $\dfrac{a}{b} \times \dfrac{c}{d} = \dfrac{ac}{bd}$

Division: $\dfrac{a}{b} \div \dfrac{c}{d} = \dfrac{ad}{bc}$

Note that a, b, c and d can be numbers, other letters or algebraic **expressions**. Remember:

- use **brackets**, if necessary
- **factorise** if you can
- **cancel** if you can.

EXAMPLE 16

Simplify **a** $\frac{1}{x} + \frac{x}{2y}$ **b** $\frac{2}{b} - \frac{a}{2b}$

a Using the addition rule: $\frac{1}{x} + \frac{x}{2x} = \frac{(1)(2y) + (x)(x)}{(x)(2y)} = \frac{2y + x^2}{2xy}$

b Using the subtraction rule: $\frac{2}{b} - \frac{a}{2b} = \frac{(2)(2b) - (a)(b)}{(b)(2b)} = \frac{4b - ab}{2b^2}$

$= \frac{\not{b}(4 - a)}{2b^{\not{2}}} = \frac{4 - a}{2b}$

Note: There are different ways of working out fraction calculations. Part **b** could have been done by making the denominator of each fraction the same.

$\frac{(2)2}{(2)b} - \frac{a}{2b} = \frac{4 - a}{2b}$

EXAMPLE 17

Simplify **a** $\frac{x}{3} \times \frac{x + 2}{x - 2}$ **b** $\frac{x}{3} \div \frac{2x}{7}$

a Using the multiplication rule: $\frac{x}{3} \times \frac{x + 2}{x - 2} = \frac{(x)(x + 2)}{(3)(x - 2)} = \frac{x^2 + 2x}{3x - 6}$

Remember that the line that separates the top from the bottom of an algebraic fraction acts as brackets as well as a division sign. Note that it is sometimes preferable to leave an algebraic fraction in a factorised form.

b Using the division rule: $\frac{x}{3} \div \frac{2x}{7} = \frac{(x)(7)}{(3)(2x)} = \frac{7}{6}$

EXAMPLE 18

Solve this equation. $\frac{x + 1}{3} - \frac{x - 3}{2} = 1$

Use the rule for combining fractions, and then **cross-multiply** to take the denominator of the left-hand side to the right-hand side.

$\frac{(2)(x + 1) - (3)(x - 3)}{(2)(3)} = 1$

$2(x + 1) - 3(x - 3) = 6 \ (= 1 \times 2 \times 3)$

Note the brackets. These will avoid problems with signs and help you to expand to get a linear equation.

$2x + 2 - 3x + 9 = 6 \Rightarrow -x = -5 \Rightarrow x = 5$

EXAMPLE 19

Solve this equation. $\frac{3}{x-1} - \frac{2}{x+1} = 1$

Use the rule for combining fractions, and cross-multiply to take the denominator of the left-hand side to the right-hand side, as in Example 13. Use brackets to help with expanding and to avoid problems with minus signs.

$3(x + 1) - 2(x - 1) = (x - 1)(x + 1)$

$3x + 3 - 2x + 2 = x^2 - 1$ (Right-hand side is the difference of two squares.)

Rearrange into the general quadratic form (see Chapter 6).

$x^2 - x - 6 = 0$

Factorise and solve $(x - 3)(x + 2) = 0 \Rightarrow x = 3$ or -2

Note that when your equation is rearranged into the quadratic form it should factorise. If it doesn't, then you have almost certainly made a mistake. If the question required an answer as a decimal or a surd it would say so.

EXAMPLE 20

Simplify this expression. $\frac{2x^2 + x - 3}{4x^2 - 9}$

Factorise the numerator and denominator: $\frac{(2x + 3)(x - 1)}{(2x + 3)(2x - 3)}$

Denominator is the difference of two squares.

Cancel any common factors: $\frac{\cancel{(2x+3)}(x - 1)}{\cancel{(2x+3)}(2x - 3)}$

If at this stage there isn't a common factor on top and bottom, you should check your factorisations.

The remaining fraction is the answer: $\frac{(x - 1)}{(2x - 3)}$

EXERCISE 12H

B

1 Simplify each of these.

a $\frac{x}{2} + \frac{x}{3}$ **b** $\frac{3x}{4} + \frac{x}{5}$ **c** $\frac{3x}{4} + \frac{2x}{5}$

d $\frac{x}{2} + \frac{y}{3}$ **e** $\frac{xy}{4} + \frac{2}{x}$ **f** $\frac{x + 1}{2} + \frac{x + 2}{3}$

g $\frac{2x + 1}{2} + \frac{3x + 1}{4}$ **h** $\frac{x}{5} + \frac{2x + 1}{3}$ **i** $\frac{x - 2}{2} + \frac{x - 3}{4}$

j $\frac{x - 4}{4} + \frac{2x - 3}{2}$

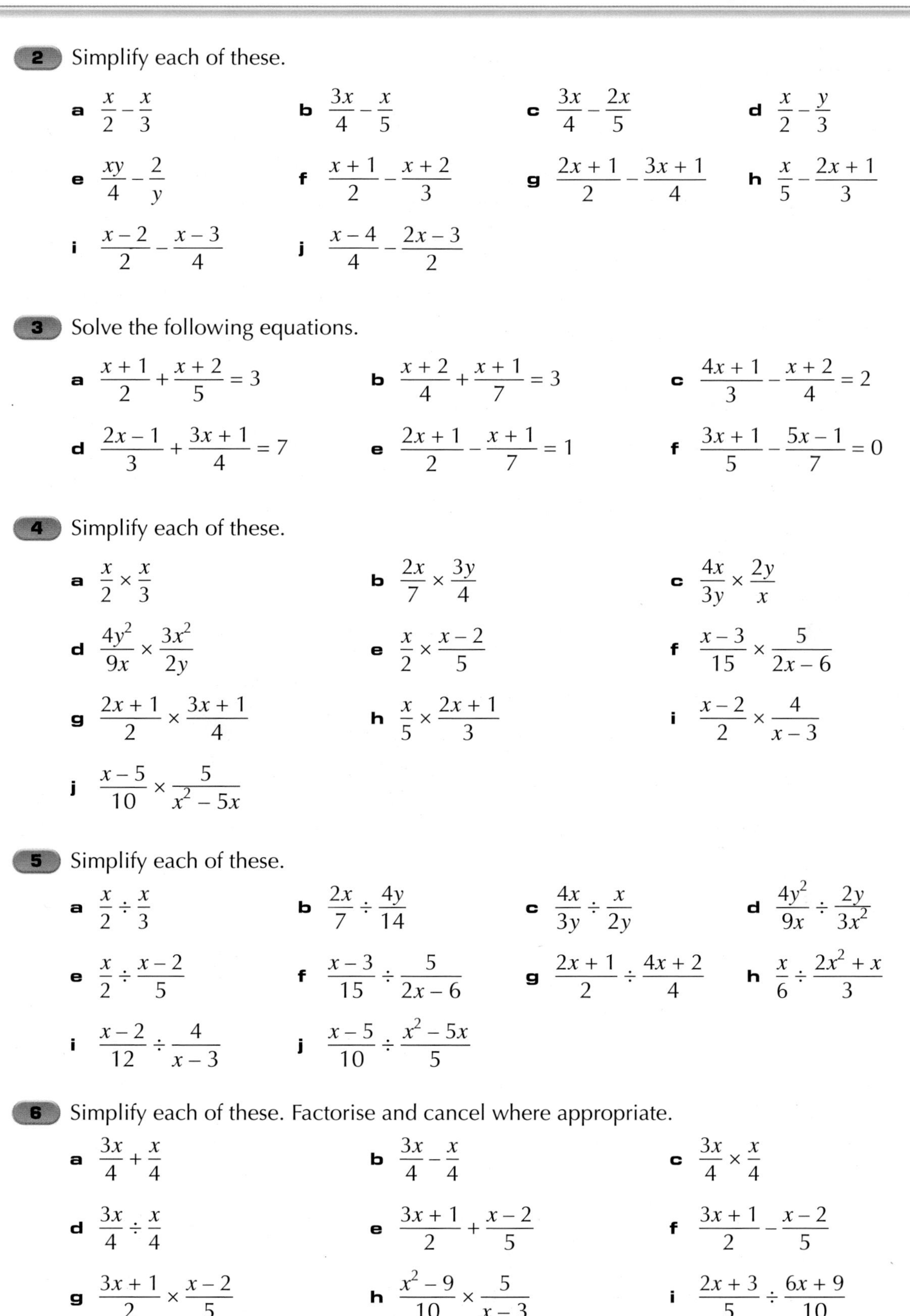

B

2 Simplify each of these.

a $\frac{x}{2} - \frac{x}{3}$ **b** $\frac{3x}{4} - \frac{x}{5}$ **c** $\frac{3x}{4} - \frac{2x}{5}$ **d** $\frac{x}{2} - \frac{y}{3}$

e $\frac{xy}{4} - \frac{2}{y}$ **f** $\frac{x+1}{2} - \frac{x+2}{3}$ **g** $\frac{2x+1}{2} - \frac{3x+1}{4}$ **h** $\frac{x}{5} - \frac{2x+1}{3}$

i $\frac{x-2}{2} - \frac{x-3}{4}$ **j** $\frac{x-4}{4} - \frac{2x-3}{2}$

3 Solve the following equations.

a $\frac{x+1}{2} + \frac{x+2}{5} = 3$ **b** $\frac{x+2}{4} + \frac{x+1}{7} = 3$ **c** $\frac{4x+1}{3} - \frac{x+2}{4} = 2$

d $\frac{2x-1}{3} + \frac{3x+1}{4} = 7$ **e** $\frac{2x+1}{2} - \frac{x+1}{7} = 1$ **f** $\frac{3x+1}{5} - \frac{5x-1}{7} = 0$

4 Simplify each of these.

a $\frac{x}{2} \times \frac{x}{3}$ **b** $\frac{2x}{7} \times \frac{3y}{4}$ **c** $\frac{4x}{3y} \times \frac{2y}{x}$

d $\frac{4y^2}{9x} \times \frac{3x^2}{2y}$ **e** $\frac{x}{2} \times \frac{x-2}{5}$ **f** $\frac{x-3}{15} \times \frac{5}{2x-6}$

g $\frac{2x+1}{2} \times \frac{3x+1}{4}$ **h** $\frac{x}{5} \times \frac{2x+1}{3}$ **i** $\frac{x-2}{2} \times \frac{4}{x-3}$

j $\frac{x-5}{10} \times \frac{5}{x^2-5x}$

A

5 Simplify each of these.

a $\frac{x}{2} \div \frac{x}{3}$ **b** $\frac{2x}{7} \div \frac{4y}{14}$ **c** $\frac{4x}{3y} \div \frac{x}{2y}$ **d** $\frac{4y^2}{9x} \div \frac{2y}{3x^2}$

e $\frac{x}{2} \div \frac{x-2}{5}$ **f** $\frac{x-3}{15} \div \frac{5}{2x-6}$ **g** $\frac{2x+1}{2} \div \frac{4x+2}{4}$ **h** $\frac{x}{6} \div \frac{2x^2+x}{3}$

i $\frac{x-2}{12} \div \frac{4}{x-3}$ **j** $\frac{x-5}{10} \div \frac{x^2-5x}{5}$

6 Simplify each of these. Factorise and cancel where appropriate.

a $\frac{3x}{4} + \frac{x}{4}$ **b** $\frac{3x}{4} - \frac{x}{4}$ **c** $\frac{3x}{4} \times \frac{x}{4}$

d $\frac{3x}{4} \div \frac{x}{4}$ **e** $\frac{3x+1}{2} + \frac{x-2}{5}$ **f** $\frac{3x+1}{2} - \frac{x-2}{5}$

g $\frac{3x+1}{2} \times \frac{x-2}{5}$ **h** $\frac{x^2-9}{10} \times \frac{5}{x-3}$ **i** $\frac{2x+3}{5} \div \frac{6x+9}{10}$

j $\frac{2x^2}{9} - \frac{2y^2}{3}$

A

7 Show that each algebraic fraction simplifies to the given expression.

a $\frac{2}{x+1} + \frac{5}{x+2} = 3$ simplifies to $3x^2 + 2x - 3 = 0$

b $\frac{4}{x-2} + \frac{7}{x+1} = 3$ simplifies to $3x^2 - 14x + 4 = 0$

c $\frac{3}{4x+1} - \frac{4}{x+2} = 2$ simplifies to $8x^2 + 31x + 2 = 0$

d $\frac{2}{2x-1} - \frac{6}{x+1} = 11$ simplifies to $22x^2 + 21x - 19 = 0$

e $\frac{3}{2x-1} - \frac{4}{3x-1} = 1$ simplifies to $x^2 - x = 0$

A*

PS 8 For homework a teacher asks her class to simplify the expression $\frac{x^2 - x - 2}{x^2 + x - 6}$.

This is Tom's answer:

$$\frac{\cancel{x^2} - x - \cancel{2}^{-1}}{\cancel{x^2} + x - \cancel{6}_{+3}}$$

$$= \frac{-x - 1}{x + 3} = \frac{x + 1}{x + 3}$$

When she marked the homework, the teacher was in a hurry and only checked the answer, which was correct.

Tom made several mistakes. What are they?

AU 9 An expression of the form $\frac{ax^2 + bx - c}{dx^2 - e}$ simplifies to $\frac{x-1}{2x-3}$.

What was the original expression?

10 Solve the following equations.

a $\frac{4}{x+1} + \frac{5}{x+2} = 2$ **b** $\frac{18}{4x-1} - \frac{1}{x+1} = 1$ **c** $\frac{2x-1}{2} - \frac{6}{x+1} = 1$

d $\frac{3}{2x-1} - \frac{4}{3x-1} = 1$

11 Simplify the following expressions.

a $\frac{x^2 + 2x - 3}{2x^2 + 7x + 3}$ **b** $\frac{4x^2 - 1}{2x^2 + 5x - 3}$ **c** $\frac{6x^2 + x - 2}{9x^2 - 4}$

d $\frac{4x^2 + x - 3}{4x^2 - 7x + 3}$ **e** $\frac{4x^2 - 25}{8x^2 - 22x + 5}$

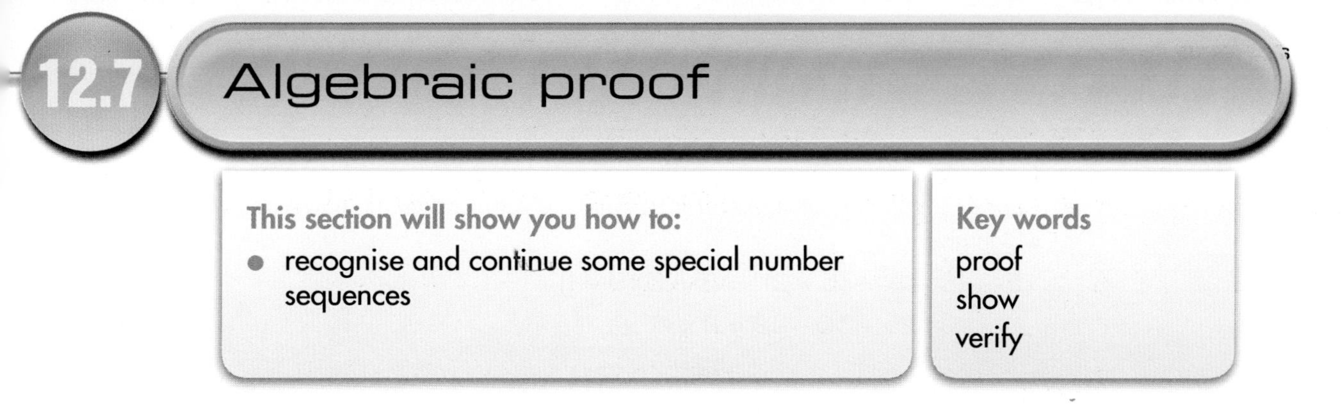

You will have met the fact that the sum of any two odd numbers is always an even number before, but can you prove it?

You can take any two odd numbers, add them together and get a number that divides exactly by 2. This does not prove the result, even if everyone in your class, or your school, or the whole of Britain, did this for a different pair of starting odd numbers. Unless you tried every pair of odd numbers (and there is an infinite number of them) you cannot be 100% certain this result is always true.

This is how to prove the result.

Let n be any whole number.

Whatever whole number is represented by n, $2n$ has to be even. So, $2n + 1$ represents any odd number.

Let one odd number be $2n + 1$, and let the other odd number be $2m + 1$.

The sum of these is:

$$\begin{aligned}(2n + 1) + (2m + 1) &= 2n + 2m + 1 + 1\\ &= 2n + 2m + 2\\ &= 2(n + m + 1), \text{ which must be even.}\end{aligned}$$

This proves the result, as n and m can be any numbers.

In an algebraic **proof**, every step must be shown clearly and the algebra must be done properly.

There are three levels of 'proof': **Verify** that ..., **Show** that ..., and Prove that ...

- At the lowest level (verification), all you have to do is substitute numbers into the result to show that it works.
- At the middle level, you have to show that both sides of the result are the same algebraically.
- At the highest level (proof), you have to manipulate the left-hand side of the result to become its right-hand side.

The following example demonstrates these three different procedures.

EXAMPLE 21

You are given that $n^2 + (n + 1)^2 - (n + 2)^2 = (n - 3)(n + 1)$.

a Verify that this result is true.

b Show that this result is true.

c Prove that this result is true.

EXAMPLE 21 continued

a Choose a number for n, say $n = 5$. Put this value into both sides of the expression, which gives:

$$5^2 + (5 + 1)^2 - (5 + 2)^2 = (5 - 3)(5 + 1)$$
$$25 + 36 - 49 = 2 \times 6$$
$$12 = 12$$

Hence, the result is true.

b Expand the LHS and RHS of the expression to get:

$$n^2 + n^2 + 2n + 1 - (n^2 + 4n + 4) = n^2 - 2n - 3$$
$$n^2 - 2n - 3 = n^2 - 2n - 3$$

That is, both sides are algebraically the same.

c Expand the LHS of the expression to get: $n^2 + n^2 + 2n + 1 - (n^2 + 4n + 4)$

Collect like terms, which gives $n^2 + n^2 - n^2 + 2n - 4n + 1 - 4 = n^2 - 2n - 3$

Factorise the collected result: $n^2 - 2n - 3 = (n - 3)(n + 1)$, which is the RHS of the original expression.

EXERCISE 12I

A*

AU 1 **a** Choose any odd number and any even number. Add these together. Is the result odd or even? Does this always work for any odd number and even number you choose?

b Let any odd number be represented by $2n + 1$. Let any even number be represented by $2m$, where m and n are integers. Prove that the sum of an odd number and an even number always gives an odd number.

AU 2 Prove the following results.

a The sum of two even numbers is even.

b The product of two even numbers is even.

c The product of an odd number and an even number is even.

d The product of two odd numbers is odd.

e The sum of four consecutive numbers is always even.

f Half the sum of four consecutive numbers is always odd.

AU 3 A Fibonacci sequence is formed by adding the previous two terms to get the next term. For example, starting with 3 and 4, the series is:

3, 4, 7, 11, 18, 29, 47, 76, 123, 199, …

a Continue the Fibonacci sequence 1, 1, 2, … up to 10 terms.

b Continue the Fibonacci sequence $a, b, a + b, a + 2b, 2a + 3b, \ldots$ up to 10 terms.

c Prove that the difference between the 8th term and the 5th term of any Fibonacci sequence is twice the 6th term.

AU **4** The nth term in the sequence of triangular numbers 1, 3, 6, 10, 15, 21, 28, … is given by $\frac{1}{2}n(n + 1)$.

a Show that the sum of the 11th and 12th terms is a perfect square.

b Explain why the $(n + 1)$th term of the triangular number sequence is given by $\frac{1}{2}(n + 1)(n + 2)$.

c Prove that the sum of any two consecutive triangular numbers is always a square number.

AU **5** The diagram shows part of a 10 × 10 'hundred square'.

12	13	14	15
22	23	24	25
32	33	34	35
42	43	44	45

a One 2 × 2 square is marked.

i Work out the difference between the product of the bottom-left and top-right values and the product of the top-left and bottom-right values:

$22 \times 13 - 12 \times 23$

ii Repeat this for any other 2 × 2 square of your choosing.

b Prove that this will always give an answer of 10 for any 2 × 2 square chosen.

c The diagram shows a calendar square (where the numbers are arranged in rows of seven).

Prove that you always get a value of 7 if you repeat the procedure in part **a i**.

1	2	3	4	5	6	7
8	9	10	11	12	13	14
15	16	17	18	19	20	21
22	23	24	25	26	27	28
29	30	31				

d Prove that in a number square that is arranged in rows of n numbers then the difference is always n if you repeat the procedure in part **a i**.

AU **6** Prove that if you add any two-digit number from the 9 times table to the reverse of itself (that is, swap the tens digit and units digit), the result will always be 99.

AU **7** Speed Cabs charges 45 pence per kilometre for each journey. Evans Taxis has a fixed charge of 90p plus 30p per kilometre.

a **i** Verify that Speed Cabs is cheaper for a journey of 5 km.

ii Verify that Evans Taxis is cheaper for a journey of 7 km.

b Show clearly why both companies charge the same for a journey of 6 km.

c Show that if Speed Cabs charges a pence per kilometre, and Evans Taxis has a fixed charge of £b plus a charge of c pence per kilometre, both companies charge the same for a journey of $\frac{4}{3x - 1}$ kilometres.

A*

AU 8 You are given that:

$$(a + b)^2 + (a - b)^2 = 2(a^2 + b^2)$$

a Verify that this result is true for $a = 3$ and $b = 4$.

b Show that the LHS is the same as the RHS.

c Prove that the LHS can be simplified to the RHS.

AU 9 Prove that: $(a + b)^2 - (a - b)^2 = 4ab$

AU 10 The rule for converting from degrees Fahrenheit to degrees Celsius is to subtract 32° and then to multiply by $\frac{5}{9}$.

Prove that the temperature that has the same value in both scales is –40°.

AU 11 The sum of the series $1 + 2 + 3 + 4 + \ldots + (n - 2) + (n - 1) + n$ is given by $\frac{1}{2}n(n + 1)$.

a Verify that this result is true for $n = 6$.

b Write down a simplified value, in terms of n, for the sum of these two series.

$$1 + 2 + 3 + \ldots + (n - 2) + (n - 1) + n$$

and $n + (n - 1) + (n - 2) + \ldots + 3 + 2 + 1$

c Prove that the sum of the first n integers is $\frac{1}{2}n(n + 1)$.

AU 12 The following is a 'think of a number' trick.

- Think of a number.
- Multiply it by 2.
- Add 10.
- Divide the result by 2.
- Subtract the original number.

The result is always 5.

a Verify that the trick works when you pick 7 as the original number.

b Prove why the trick always works.

AU 13 You are told that 'when two numbers have a difference of 2, the difference of their squares is twice the sum of the two numbers'.

a Verify that this is true for 5 and 7.

b Prove that the result is true.

c Prove that when two numbers have a difference of n, the difference of their squares is n times the sum of the two numbers.

AU 14 Four consecutive numbers are 4, 5, 6 and 7.

a Verify that their product plus 1 is a perfect square.

b Complete the multiplication square and use it to show that:

$$(n^2 - n - 1)^2 = n^4 - 2n^3 - n^2 + 2n + 1$$

	n^2	$-n$	-1
n^2	n^4		$-n^2$
$-n$		n^2	
-1			

c Let four consecutive numbers be $(n - 2)$, $(n - 1)$, n, $(n + 1)$.
Prove that the product of four consecutive numbers plus 1 is a perfect square.

AU 15 Here is another mathematical trick to try on a friend.

- Think of two single-digit numbers.
- Multiply one number (your choice) by 2.
- Add 5 to this answer.
- Multiply this answer by 5.
- Add the second number.
- Subtract 4.
- Ask your friend to state the final answer.
- Mentally subtract 21 from this answer.

The two digits you get are the two digits your friend first thought of.

Prove why this works.

EXERCISE 12J

You may not be able algebraically to prove all of these results. Some of them can be disproved by a counter-example. You should first try to verify each result, then attempt to prove it — or at least try to demonstrate that the result is probably true by trying lots of examples.

AU 1 T represents any triangular number. Prove the following.

a $8T + 1$ is always a square number.

b $9T + 1$ is always another triangular number.

AU 2 Lewis Carroll, who wrote *Alice in Wonderland*, was also a mathematician. In 1890, he suggested the following results.

a For any pair of numbers, x and y, if $x^2 + y^2$ is even, then $\frac{1}{2}(x^2 + y^2)$ is the sum of two squares.

b For any pair of numbers, x and y, $2(x^2 + y^2)$ is always the sum of two squares.

c Any number of which the square is the sum of two squares is itself the sum of two squares.

Can you prove these statements to be true or false?

AU 3 For all values of n, $n^2 - n + 41$ gives a prime number. True or false?

AU 4 Pythagoras' theorem says that for a right-angled triangle with two short sides a and b and a long side c, $a^2 + b^2 = c^2$. For any integer n, $2n$, $n^2 - 1$ and $n^2 + 1$ form three numbers that obey Pythagoras' theorem. Can you prove this?

AU 5 Waring's theorem states that: "Any whole number can be written as the sum of not more than four square numbers".

For example, $27 = 3^2 + 3^2 + 3^2$ and $23 = 3^2 + 3^2 + 2^2 + 1^2$

Is this always true?

A*

A*

AU **6** Take a three-digit multiple of 37, for example, $7 \times 37 = 259$. Write these digits in a cycle.

Take all possible three-digit numbers from the cycle, for example, 259, 592 and 925.

2 5 9

Divide each of these numbers by 37 to find that:

$259 = 7 \times 37 \quad 592 = 16 \times 37 \quad 925 = 25 \times 37$

Is this true for all three-digit multiples of 37?

Is it true for a five-digit multiple of 41?

AU **7** Prove that the sum of the squares of two consecutive integers is an odd number.

8 The difference of two squares is an identity, i.e.,

$$a^2 - b^2 \equiv (a + b)(a - b)$$

which means that it is true for all values of a and b whether they are numeric or algebraic.

Prove that $a^2 - b^2 \equiv (a + b)(a - b)$ is true when $a = 2x + 1$ and $b = x - 1$

9 The square of the sum of the first n consecutive whole numbers is equal to the sum of the cubes of the first n consecutive whole numbers.

a Verify that $(1 + 2 + 3 + 4)^2 = 1^3 + 2^3 + 3^3 + 4^3$

b The sum of the first n consecutive whole numbers is $\frac{1}{2}n(n + 1)$

Write down a formula for the sum of the cubes of the first n whole numbers.

c Test your formula for $n = 6$

GRADE BOOSTER

- **B** You can rearrange more complicated formulae where the subject may appear twice or as a power
- **B** You can solve linear equations involving algebraic fractions where the subject appears as the numerator
- **B** You can verify results by substituting numbers into them
- **A** You can rearrange a formula where the subject appears twice
- **A** You can combine fractions, using the four algebraic rules of addition, subtraction, multiplication and division
- **A** You can show that an algebraic statement is true, using both sides of the statement to justify your answer
- **A*** You can solve a quadratic equation obtained from algebraic fractions where the variable appears in the denominator
- **A*** You can simplify algebraic fractions by factorisation and cancellation
- **A*** You can solve a pair of simultaneous equations where one is linear and the other is non-linear
- **A*** You can prove algebraic results with rigorous and logical mathematical arguments

What you should know now

- How to manipulate algebraic fractions and solve equations resulting from the simplified fractions
- How to solve a pair of simultaneous equations where one is linear and one is non-linear
- How to rearrange a formula where the subject appears twice
- The meaning of the terms 'verify that', 'show that' and 'prove'
- How to prove some standard results in mathematics
- How to use your knowledge of proof to answer the questions throughout the book that are flagged with the proof icon

1

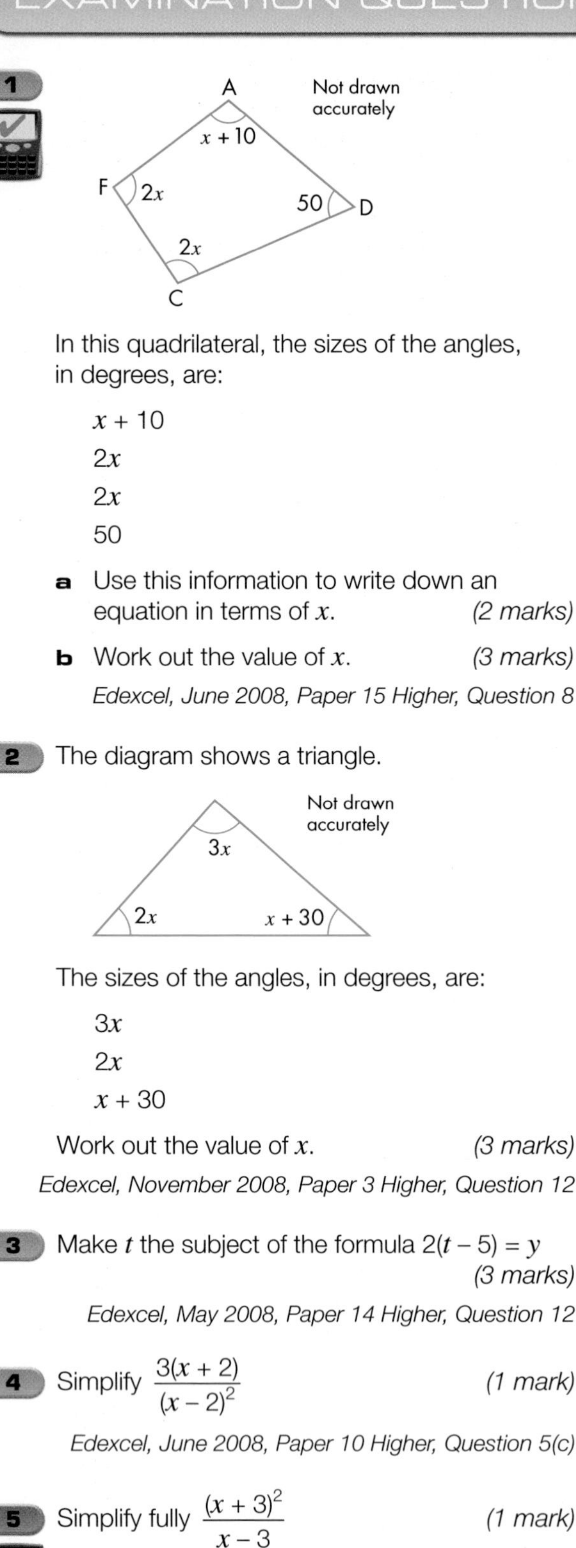

In this quadrilateral, the sizes of the angles, in degrees, are:

$x + 10$

$2x$

$2x$

50

a Use this information to write down an equation in terms of x. *(2 marks)*

b Work out the value of x. *(3 marks)*

Edexcel, June 2008, Paper 15 Higher, Question 8

2 The diagram shows a triangle.

Not drawn accurately

$3x$ $2x$ $x + 30$

The sizes of the angles, in degrees, are:

$3x$

$2x$

$x + 30$

Work out the value of x. *(3 marks)*

Edexcel, November 2008, Paper 3 Higher, Question 12

3 Make t the subject of the formula $2(t - 5) = y$ *(3 marks)*

Edexcel, May 2008, Paper 14 Higher, Question 12

4 Simplify $\dfrac{3(x + 2)}{(x - 2)^2}$ *(1 mark)*

Edexcel, June 2008, Paper 10 Higher, Question 5(c)

5 Simplify fully $\dfrac{(x + 3)^2}{x - 3}$ *(1 mark)*

Edexcel, March 2008, Paper 11 Higher, Question 5(e)

6 Simplify fully $\dfrac{x + 3}{4} + \dfrac{x - 5}{4}$ *(3 marks)*

Edexcel, November 2008, Paper 10 Higher, Question 8

7 Solve the simultaneous equations.

$2x + 3y = 0$

$x - 3y = 9$ *(3 marks*

Edexcel, May 2008, Paper 3 Higher, Question 1

8 $v^2 = u^2 + 2as$

$u = 6$

$a = 2.5$

$s = 9$

a Work out a value of v *(3 marks*

b Make s the subject of the formula

$v^2 = u^2 + 2as$ *(2 marks*

Edexcel, November 2008, Paper 3 Higher, Question 1

9 Simplify fully $\dfrac{3x + 6}{x^2 - 4}$ *(3 marks*

Edexcel, March 2007, Paper 11 Higher, Question

10 Simplify $\dfrac{x^2 + 5x + 6}{x + 2}$ *(3 marks*

Edexcel, June 2007, Paper 11 Higher, Question

11 Prove that $(n + 2)^2 - (n - 2)^2 = 8n$ for all values of n. *(2 marks*

Edexcel, June 2007, Paper 11 Higher, Question

12 Write $\dfrac{x}{x - 2} - \dfrac{3}{x(x - 2)}$

as a single fraction in its simplest form. *(2 marks*

Edexcel, June 2007, Paper 11 Higher, Question

13 Simplify fully $\dfrac{4a - 20}{a^2 - 25}$ *(3 marks*

Edexcel, November 2007, Paper 11 Higher, Question 1

14 Write as a single fraction in its simplest form

$\dfrac{4}{x + 5} + \dfrac{1}{x - 3}$ *(4 marks*

Edexcel, November 2007, Paper 11 Higher, Question

15 Simplify fully $\dfrac{x}{x - 3} - \dfrac{x}{x + 3}$ *(3 marks*

Edexcel, March 2008, Paper 11 Higher, Question 9(b

16 Simplify $\dfrac{p^2 - 9}{2p + 6}$ *(3 marks*

Edexcel, March 2008, Paper 10 Higher, Question

17 Simplify fully $\frac{6x^2 + 3x}{4x^2 - 1}$ *(3 marks)*

Edexcel, June 2008, Paper 10 Higher, Question 9

18 Simplify $\frac{x^2 + 2x + 1}{x^2 + 3x + 2}$ *(3 marks)*

Edexcel, November 2008, Paper 11 Higher, Question 10

19 Simplify fully $\frac{x^2 + x - 6}{x^2 - 7x + 10}$ *(3 marks)*

Edexcel, May 2008, Paper 3 Higher, Question 28

20 Solve the simultaneous equations

$x^2 + y^2 = 5$

$y = 3x + 1$ *(6 marks)*

Edexcel, November 2008, Paper 3 Higher, Question 28

21 Simplify $\frac{3(x - 2)}{x^2 - 7x + 10}$ *(2 marks)*

Edexcel, November 2008, Paper 10 Higher, Question 6(b)

22 Make b the subject of the formula

$a = \frac{2 - 7b}{b - 5}$ *(4 marks)*

Edexcel, May 2008, Paper 3 Higher, Question 22

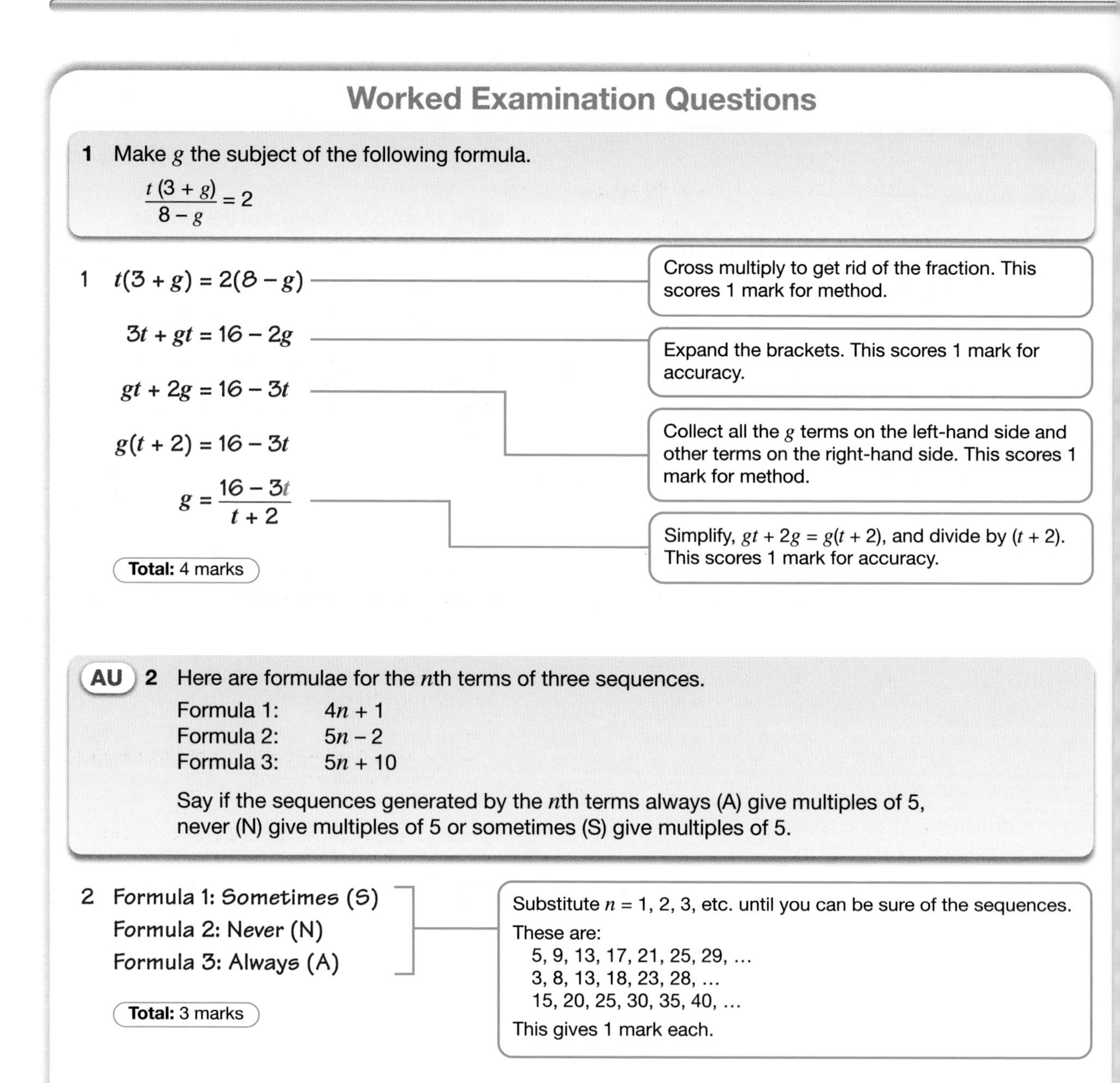

Worked Examination Questions

1 Make g the subject of the following formula.

$$\frac{t(3+g)}{8-g} = 2$$

1 $t(3 + g) = 2(8 - g)$ — Cross multiply to get rid of the fraction. This scores 1 mark for method.

$3t + gt = 16 - 2g$ — Expand the brackets. This scores 1 mark for accuracy.

$gt + 2g = 16 - 3t$ — Collect all the g terms on the left-hand side and other terms on the right-hand side. This scores 1 mark for method.

$g(t + 2) = 16 - 3t$

$g = \dfrac{16 - 3t}{t + 2}$ — Simplify, $gt + 2g = g(t + 2)$, and divide by $(t + 2)$. This scores 1 mark for accuracy.

Total: 4 marks

AU **2** Here are formulae for the nth terms of three sequences.

Formula 1: $4n + 1$
Formula 2: $5n - 2$
Formula 3: $5n + 10$

Say if the sequences generated by the nth terms always (A) give multiples of 5, never (N) give multiples of 5 or sometimes (S) give multiples of 5.

2 Formula 1: Sometimes (S)
Formula 2: Never (N)
Formula 3: Always (A)

Substitute $n = 1, 2, 3$, etc. until you can be sure of the sequences.
These are:
5, 9, 13, 17, 21, 25, 29, …
3, 8, 13, 18, 23, 28, …
15, 20, 25, 30, 35, 40, …
This gives 1 mark each.

Total: 3 marks

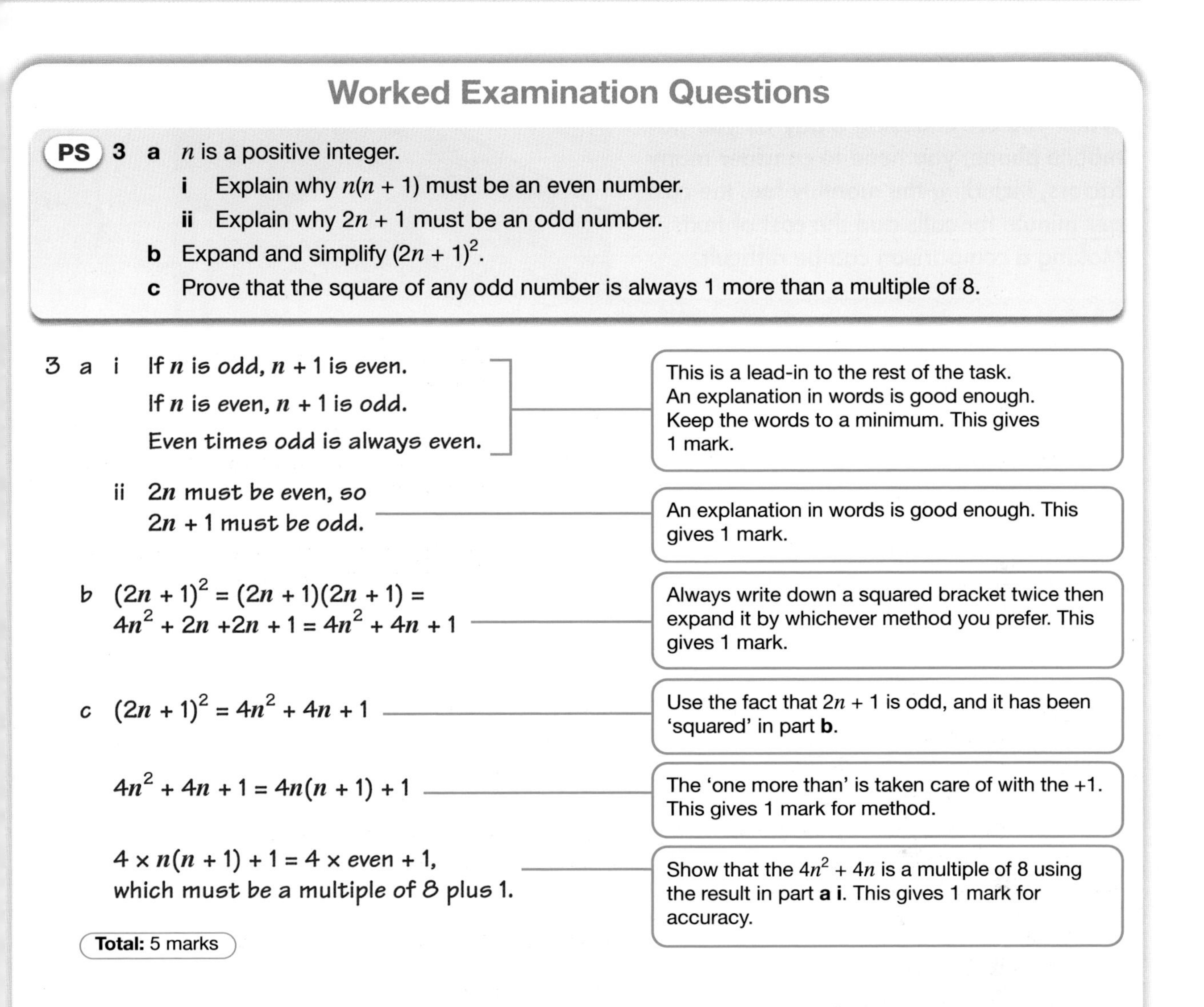

Worked Examination Questions

PS **3** **a** n is a positive integer.

i Explain why $n(n + 1)$ must be an even number.

ii Explain why $2n + 1$ must be an odd number.

b Expand and simplify $(2n + 1)^2$.

c Prove that the square of any odd number is always 1 more than a multiple of 8.

3 a i If n is odd, $n + 1$ is even.
If n is even, $n + 1$ is odd.
Even times odd is always even.

This is a lead-in to the rest of the task. An explanation in words is good enough. Keep the words to a minimum. This gives 1 mark.

ii $2n$ must be even, so $2n + 1$ must be odd.

An explanation in words is good enough. This gives 1 mark.

b $(2n + 1)^2 = (2n + 1)(2n + 1) =$
$4n^2 + 2n + 2n + 1 = 4n^2 + 4n + 1$

Always write down a squared bracket twice then expand it by whichever method you prefer. This gives 1 mark.

c $(2n + 1)^2 = 4n^2 + 4n + 1$

Use the fact that $2n + 1$ is odd, and it has been 'squared' in part **b**.

$4n^2 + 4n + 1 = 4n(n + 1) + 1$

The 'one more than' is taken care of with the +1. This gives 1 mark for method.

$4 \times n(n + 1) + 1 = 4 \times$ even $+ 1$,
which must be a multiple of 8 plus 1.

Show that the $4n^2 + 4n$ is a multiple of 8 using the result in part **a i**. This gives 1 mark for accuracy.

Total: 5 marks

12 Functional Maths
Choosing a mobile phone plan

When you are choosing a pay-as-you-go mobile phone, you need to consider many factors, including the monthly fee, the cost per minute for calls and the cost of texts. Making a comparison can be difficult.

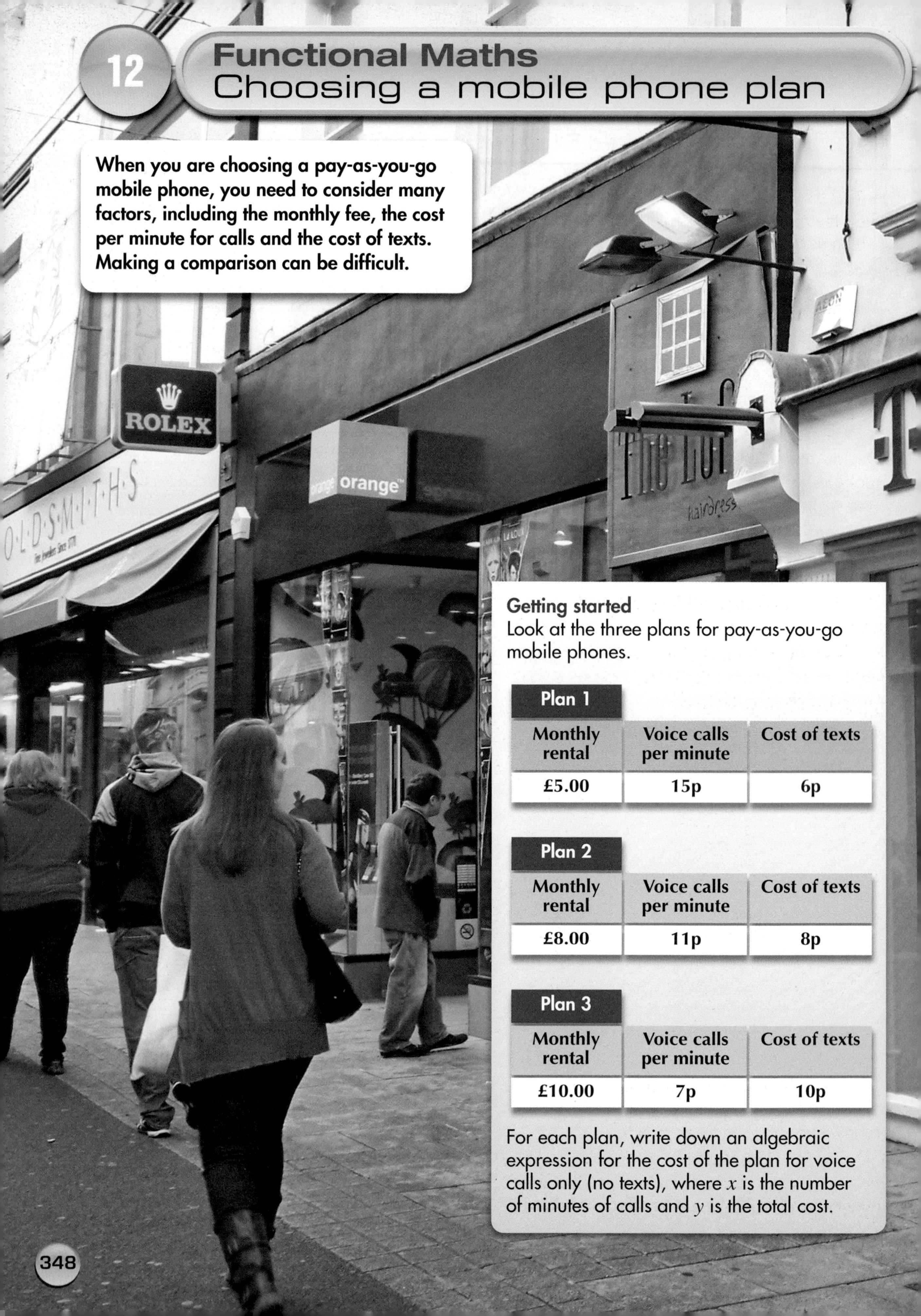

Getting started

Look at the three plans for pay-as-you-go mobile phones.

Plan 1

Monthly rental	Voice calls per minute	Cost of texts
£5.00	15p	6p

Plan 2

Monthly rental	Voice calls per minute	Cost of texts
£8.00	11p	8p

Plan 3

Monthly rental	Voice calls per minute	Cost of texts
£10.00	7p	10p

For each plan, write down an algebraic expression for the cost of the plan for voice calls only (no texts), where x is the number of minutes of calls and y is the total cost.

Your task

1 Compare the three mobile phone plans.

- Find the number of minutes of voice calls after which each plan becomes cheaper.

  Hint: Think about these questions.

  - After how many minutes does plan 2 become cheaper than plan 1?
  - After how many minutes does plan 3 become cheaper than plan 2?
  - After how many minutes does plan 3 become cheaper than plan 1?

- Decide which plan you would choose if you made 3 hours of voice calls per month and sent no texts.
- Write down an algebraic expression for the cost of the plan for 3 hours of voice calls and x texts, where y is the total cost.
- Work out the number of texts after which each plan becomes cheaper.

Using this information, decide which plan you would choose if you made 3 hours of voice calls per month and sent 250 texts.

Your task (continued)

2 Ask a few friends or relatives how many minutes of voice calls they make and how many texts they send per month.

Which of these plans would be best value for them? Write a short report for each of them, so that they can understand how you reached your conclusions.

Why this chapter matters

In many real-life situations, variables are connected by a rule or relationship. It may be that as one variable increases the other increases. Alternatively, it may be that as one variable increases the other decreases.

This chapter looks at how quantities vary when they are related in some way.

As this plant gets older it becomes taller.

As the storm increases the number of sunbathers decreases.

As this car gets older it is worth less (and eventually it is worthless!).

As more songs are downloaded, there is less money left on the voucher.

Try to think of other variables that are connected in this way.

Chapter 13

Number: Variation

The grades given in this chapter are target grades.

This chapter will show you ...

A how to solve problems where two variables are connected by a relationship that varies in direct or inverse proportion

Visual overview

Direct proportion → Inverse proportion

What you should already know

- Squares, square roots, cubes and cube roots of integers **(KS3 level 4–5, GCSE grade G–E)**
- How to substitute values into algebraic expressions **(KS3 level 5, GCSE grade E)**
- How to solve simple algebraic equations **(KS3 level 6, GCSE grade D)**

Quick check

1 Write down the value of each of the following.

a 5^2

b $\sqrt{81}$

c 3^3

d $\sqrt[3]{64}$

2 Calculate the value of y if $x = 4$.

a $y = 3x^2$

b $y = \dfrac{1}{\sqrt{x}}$

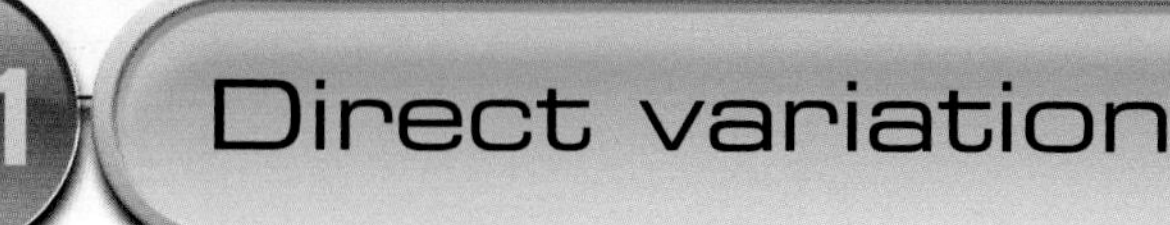

13.1 Direct variation

This section will show you how to:
- solve problems where two variables have a directly proportional relationship (direct variation)
- work out the constant of proportionality

Key words
constant of proportionality, k
direct proportion
direct variation

The term **direct variation** means the same as as **direct proportion**.

There is direct variation (or direct proportion) between two variables when one variable is a simple multiple of the other. That is, their ratio is a constant.

For example:

1 kilogram = 2.2 pounds	There is a multiplying factor of 2.2 between kilograms and pounds.
Area of a circle = πr^2	There is a multiplying factor of π between the area of a circle and the square of its radius.

An examination question involving direct variation usually requires you first to find this multiplying factor (called the **constant of proportionality**), then to use it to solve a problem.

The symbol for variation or proportion is $\propto$.

So the statement 'Pay is directly proportional to time' can be mathematically written as:

$pay \propto time$

which implies that:

$pay = k \times time$

where k is the constant of proportionality.

There are four steps to be followed when you are using proportionality to solve problems.

Step 1: Set up the statement, using the proportionality symbol (you may use symbols to represent the variables).

Step 2: Set up the equation, using a constant of proportionality.

Step 3: Use given information to work out the value of the constant of proportionality.

Step 4: Substitute the value of the constant of proportionality into the equation and use this equation to find unknown values.

EXAMPLE 1

The cost of an article is directly proportional to the time spent making it. An article taking 6 hours to make costs £30. Find:

a the cost of an article that takes 5 hours to make

b the length of time it takes to make an article costing £40.

Step 1: Let C be the cost of making an article and t the time it takes.

$C \propto t$

Step 2: Setting up the equation gives:

$C = kt$

where k is the constant of proportionality.

Note that you can 'replace' the proportionality sign $\propto$ with $= k$ to obtain the proportionality equation.

Step 3: Since $C = £30$ when $t = 6$ hours, then $30 = 6k$

$\Rightarrow \frac{30}{6} = k$

$\Rightarrow k = 5$

Step 4: So the formula is $C = 5t$.

a When $t = 5$ hours $\quad C = 5 \times 5 = 25$

So the cost is £25.

b When $C = £40 \quad 40 = 5 \times t$

$\Rightarrow \frac{40}{5} = t \Rightarrow t = 8$

So the time spent making the article is 8 hours.

EXERCISE 13A

For questions **1** to **4**, first find k, the constant of proportionality, and then the formula connecting the variables.

1 T is directly proportional to M. If $T = 20$ when $M = 4$, find:

a T when $M = 3$ **b** M when $T = 10$.

2 W is directly proportional to F. If $W = 45$ when $F = 3$, find:

a W when $F = 5$ **b** F when $W = 90$.

3 Q varies directly with P. If $Q = 100$ when $P = 2$, find:

a Q when $P = 3$ **b** P when $Q = 300$.

A

A

4 X varies directly with Y. If $X = 17.5$ when $Y = 7$, find:

a X when $Y = 9$

b Y when $X = 30$.

5 The distance covered by a train is directly proportional to the time taken for the journey. The train travels 105 miles in 3 hours.

a What distance will the train cover in 5 hours?

b How much time will it take for the train to cover 280 miles?

6 The cost of fuel delivered to your door is directly proportional to the weight received. When 250 kg is delivered, it costs £47.50.

a How much will it cost to have 350 kg delivered?

b How much would be delivered if the cost were £33.25?

FM **7** The number of children who can play safely in a playground is directly proportional to the area of the playground. A playground with an area of 210 m^2 is safe for 60 children.

a How many children can safely play in a playground of area 154 m^2?

b A playgroup has 24 children. What is the smallest playground area in which they could safely play?

8 The number of spaces in a car park is directly proportional to the area of the car park.

FM **a** A car park has 300 parking spaces in an area of 4500 m^2.

It is decided to increase the area of the car park by 500 m^2 to make extra spaces.

How many extra spaces will be made?

PS **b** The old part of the car park is redesigned so that the original area has 10% more parking spaces.

How many more spaces than originally will there be altogether if the number of spaces in the new area is directly proportional to the number in the redesigned car park?

AU **9** The number of passengers in a bus queue is directly proportional to the time that the person at the front of the queue has spent waiting.

Karen is the first to arrive at a bus stop. When she has been waiting 5 minutes the queue has 20 passengers.

A bus has room for 70 passengers.

How long had Karen been in the queue if the bus fills up from empty when it arrives and all passengers get on?

Direct proportions involving squares, cubes, square roots and cube roots

The process is the same as for a linear direct variation, as the next example shows.

EXAMPLE 2

The cost of a circular badge is directly proportional to the square of its radius. The cost of a badge with a radius of 2 cm is 68p. Find:

a the cost of a badge of radius 2.4 cm **b** the radius of a badge costing £1.53.

Step 1: Let C be the cost and r the radius of a badge.

$C \propto r^2$

Step 2: Setting up the equation gives:

$C = kr^2$

where k is the constant of proportionality.

Step 3: $C = 68\text{p}$ when $r = 2$ cm. So:

$68 = 4k$

$\Rightarrow \frac{68}{4} = k \Rightarrow k = 17$

Step 4: So the formula is $C = 17r^2$.

a When $r = 2.4$ cm $\quad C = 17 \times 2.4^2\text{p} = 97.92\text{p}$

Rounding gives the cost as 98p.

b When $C = 153\text{p}$ $\quad 153 = 17r^2$

$\Rightarrow \frac{153}{17} = 9 = r^2$

$\Rightarrow r = \sqrt{9} = 3$

Hence, the radius is 3 cm.

EXERCISE 13B

For questions **1** to **6**, first find k, the constant of proportionality, and then the formula connecting the variables.

1 T is directly proportional to x^2. If $T = 36$ when $x = 3$, find:

a T when $x = 5$ **b** x when $T = 400$.

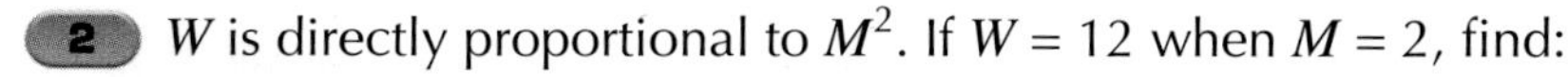

2 W is directly proportional to M^2. If $W = 12$ when $M = 2$, find:

a W when $M = 3$ **b** M when $W = 75$.

A

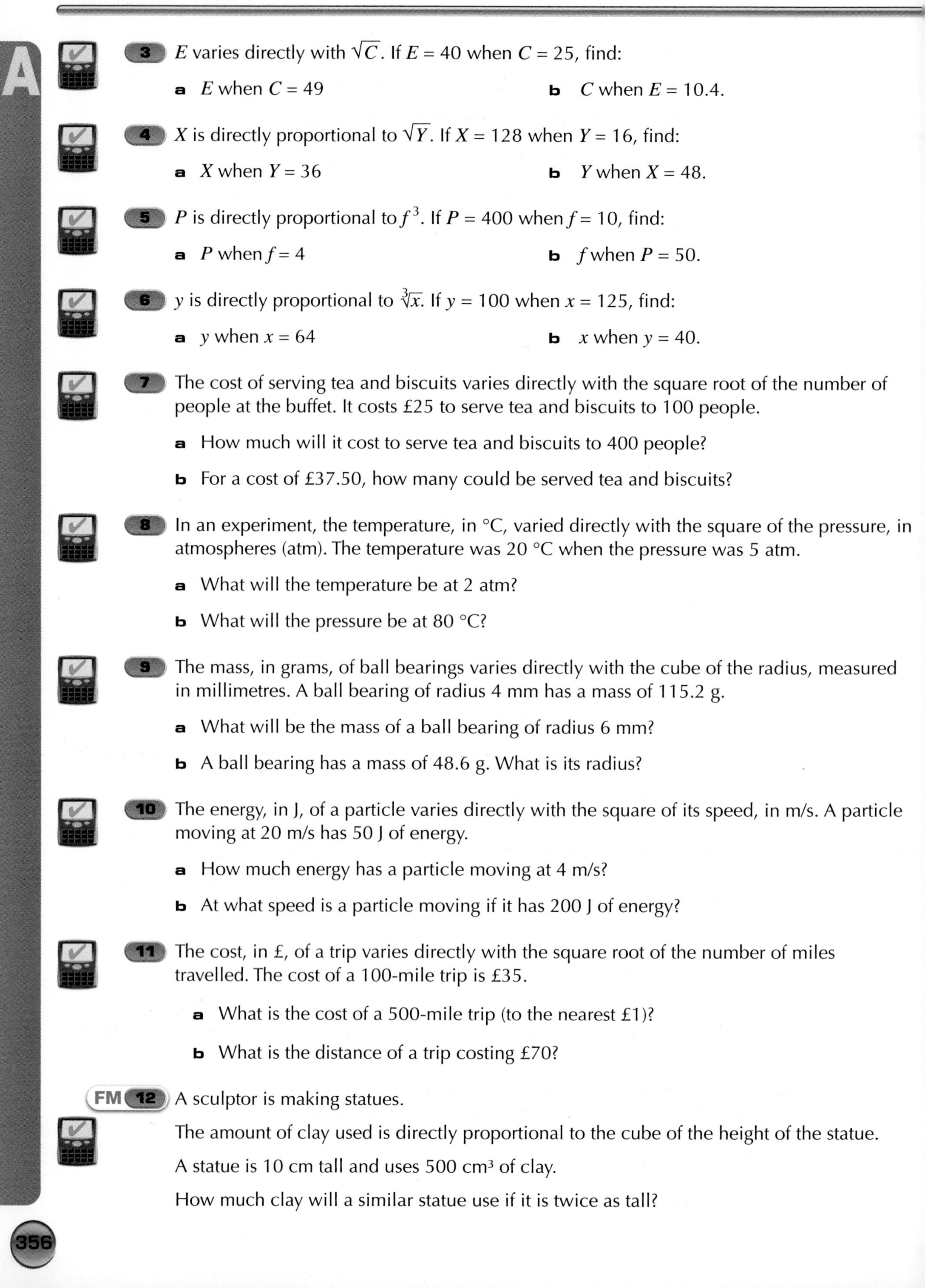

3 E varies directly with $\sqrt{C}$. If $E = 40$ when $C = 25$, find:

a E when $C = 49$ **b** C when $E = 10.4$.

4 X is directly proportional to $\sqrt{Y}$. If $X = 128$ when $Y = 16$, find:

a X when $Y = 36$ **b** Y when $X = 48$.

5 P is directly proportional to f^3. If $P = 400$ when $f = 10$, find:

a P when $f = 4$ **b** f when $P = 50$.

6 y is directly proportional to $\sqrt[3]{x}$. If $y = 100$ when $x = 125$, find:

a y when $x = 64$ **b** x when $y = 40$.

7 The cost of serving tea and biscuits varies directly with the square root of the number of people at the buffet. It costs £25 to serve tea and biscuits to 100 people.

a How much will it cost to serve tea and biscuits to 400 people?

b For a cost of £37.50, how many could be served tea and biscuits?

8 In an experiment, the temperature, in °C, varied directly with the square of the pressure, in atmospheres (atm). The temperature was 20 °C when the pressure was 5 atm.

a What will the temperature be at 2 atm?

b What will the pressure be at 80 °C?

9 The mass, in grams, of ball bearings varies directly with the cube of the radius, measured in millimetres. A ball bearing of radius 4 mm has a mass of 115.2 g.

a What will be the mass of a ball bearing of radius 6 mm?

b A ball bearing has a mass of 48.6 g. What is its radius?

10 The energy, in J, of a particle varies directly with the square of its speed, in m/s. A particle moving at 20 m/s has 50 J of energy.

a How much energy has a particle moving at 4 m/s?

b At what speed is a particle moving if it has 200 J of energy?

11 The cost, in £, of a trip varies directly with the square root of the number of miles travelled. The cost of a 100-mile trip is £35.

a What is the cost of a 500-mile trip (to the nearest £1)?

b What is the distance of a trip costing £70?

FM **12** A sculptor is making statues.

The amount of clay used is directly proportional to the cube of the height of the statue.

A statue is 10 cm tall and uses 500 cm³ of clay.

How much clay will a similar statue use if it is twice as tall?

FM **13** The cost of making different-sized machines is proportional to the time taken.

A small machine costs £100 and takes two hours to make.

How much will a large machine cost that takes 5 hours to build?

PS **14** The sketch graphs show each of these proportion statements.

a $y \propto x^2$ **b** $y \propto x$ **c** $y \propto \sqrt{x}$

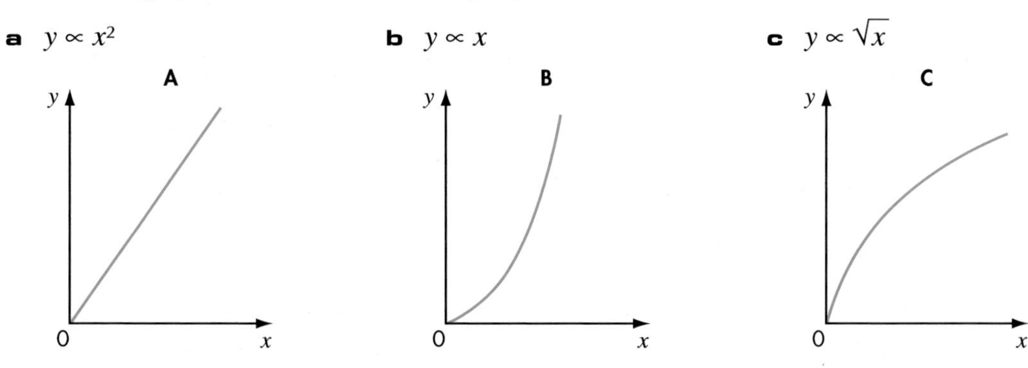

Match each statement to the correct sketch.

AU **15** Here are two tables.

Match each table to a graph in question **14**.

a

x	1	2	3
y	3	12	27

b

x	1	2	3
y	3	6	9

13.2 Inverse variation

This section will show you how to:

- solve problems where two variables have an inversely proportional relationship (inverse variation)
- work out the constant of proportionality

Key words

constant of proportionality, k
inverse proportion
inverse variation

The term **inverse variation** means the same as **inverse proportion**.

There is inverse variation between two variables when one variable is directly proportional to the *reciprocal* of the other. That is, the product of the two variables is constant. So, as one variable increases, the other decreases.

For example, the faster you travel over a given distance, the less time it takes. So there is an inverse variation between speed and time. Speed is inversely proportional to time.

$$S \propto \frac{1}{T} \text{ and so } S = \frac{k}{T}$$

which can be written as $ST = k$.

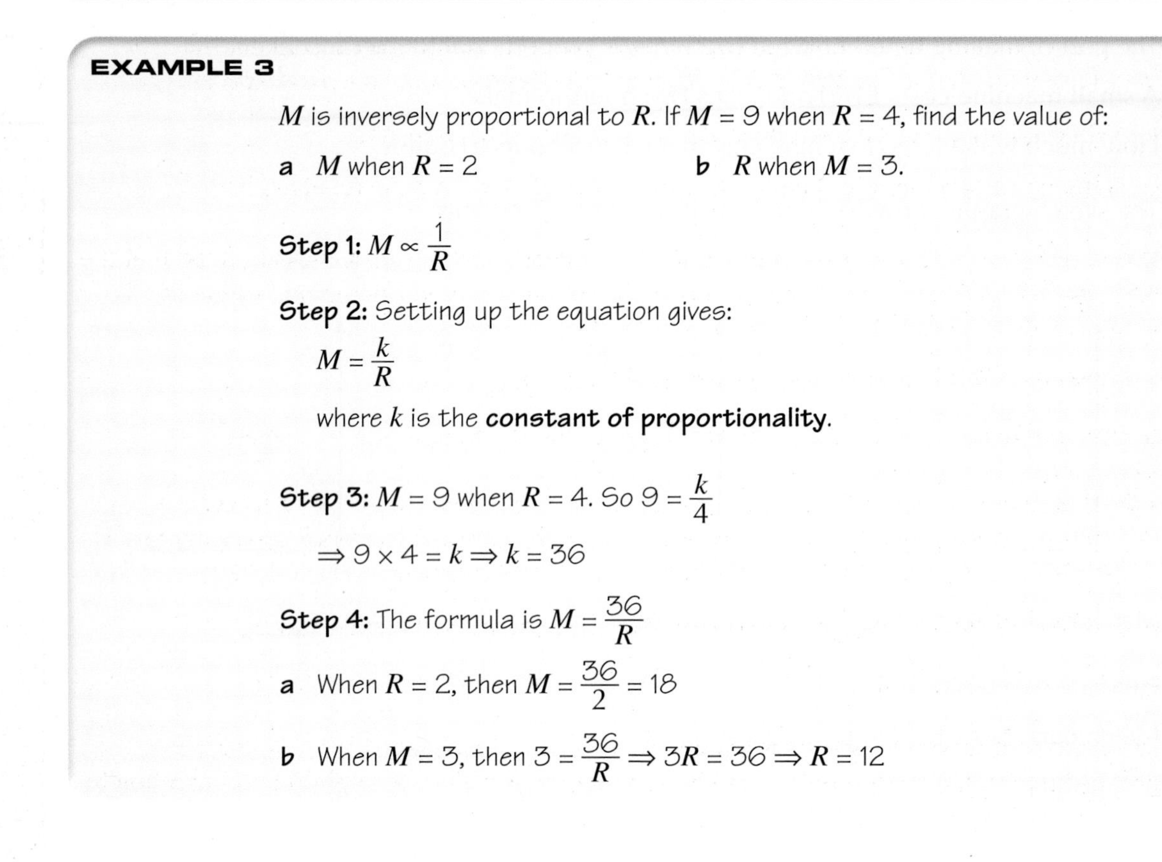

EXAMPLE 3

M is inversely proportional to R. If $M = 9$ when $R = 4$, find the value of:

a M when $R = 2$ **b** R when $M = 3$.

Step 1: $M \propto \frac{1}{R}$

Step 2: Setting up the equation gives:

$$M = \frac{k}{R}$$

where k is the **constant of proportionality**.

Step 3: $M = 9$ when $R = 4$. So $9 = \frac{k}{4}$

$\Rightarrow 9 \times 4 = k \Rightarrow k = 36$

Step 4: The formula is $M = \frac{36}{R}$

a When $R = 2$, then $M = \frac{36}{2} = 18$

b When $M = 3$, then $3 = \frac{36}{R} \Rightarrow 3R = 36 \Rightarrow R = 12$

EXERCISE 13C

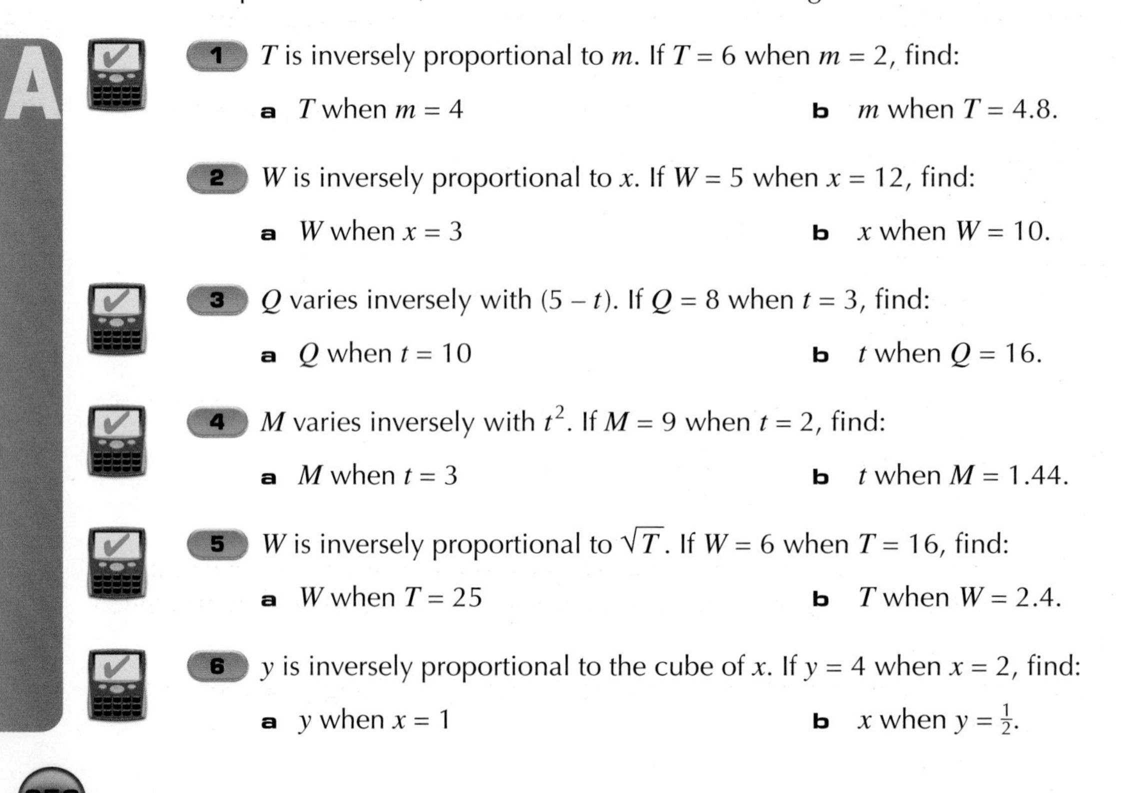

For questions **1** to **6**, first find the formula connecting the variables.

A

1 T is inversely proportional to m. If $T = 6$ when $m = 2$, find:

a T when $m = 4$ **b** m when $T = 4.8$.

2 W is inversely proportional to x. If $W = 5$ when $x = 12$, find:

a W when $x = 3$ **b** x when $W = 10$.

3 Q varies inversely with $(5 - t)$. If $Q = 8$ when $t = 3$, find:

a Q when $t = 10$ **b** t when $Q = 16$.

4 M varies inversely with t^2. If $M = 9$ when $t = 2$, find:

a M when $t = 3$ **b** t when $M = 1.44$.

5 W is inversely proportional to $\sqrt{T}$. If $W = 6$ when $T = 16$, find:

a W when $T = 25$ **b** T when $W = 2.4$.

6 y is inversely proportional to the cube of x. If $y = 4$ when $x = 2$, find:

a y when $x = 1$ **b** x when $y = \frac{1}{2}$.

A

7 The grant available to a section of society was inversely proportional to the number of people needing the grant. When 30 people needed a grant, they received £60 each.

a What would the grant have been if 120 people had needed one?

b If the grant had been £50 each, how many people would have received it?

8 While doing underwater tests in one part of an ocean, a team of scientists noticed that the temperature, in °C, was inversely proportional to the depth, in kilometres. When the temperature was 6 °C, the scientists were at a depth of 4 km.

a What would the temperature have been at a depth of 8 km?

b To what depth would they have had to go to find the temperature at 2 °C?

9 A new engine was being tested, but it had serious problems. The distance it went, in kilometres, without breaking down was inversely proportional to the square of its speed in metres per second (m/s). When the speed was 12 m/s, the engine lasted 3 km.

a Find the distance covered before a breakdown, when the speed is 15 m/s.

b On one test, the engine broke down after 6.75 km. What was the speed?

10 In a balloon it was noticed that the pressure, in atmospheres (atm), was inversely proportional to the square root of the height, in metres. When the balloon was at a height of 25 m, the pressure was 1.44 atm.

a What was the pressure at a height of 9 m?

b What would the height have been if the pressure was 0.72 atm?

FM 11 The amount of waste which a firm produces, measured in tonnes per hour, is inversely proportional to the square root of the area of the filter beds, in square metres (m^2). The firm produces 1.25 tonnes of waste per hour, with filter beds of size 0.16 m^2.

a The filter beds used to be only 0.01 m^2. How much waste did the firm produce then?

b How much waste could be produced if the filter beds were 0.75 m^2?

PS 12 Which statement is represented by the graph?
Give a reason for your answer.

A $y \propto x$ B $y \propto \frac{1}{x}$ C $y\sqrt{x}$

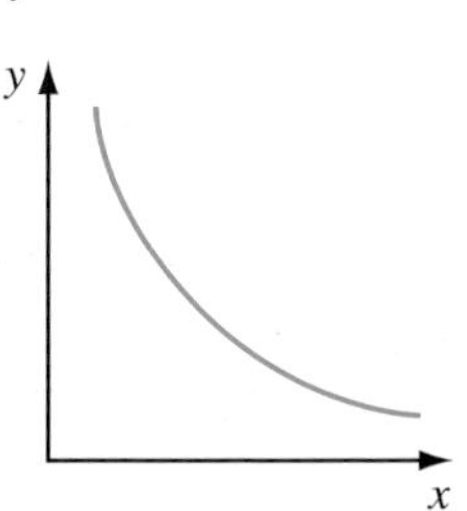

AU 13 In the table, y is inversely proportional to the cube root of x.

Complete the table, leaving your answers as fractions.

x	8	27	
y	1		$\frac{1}{2}$

FM 14 The fuel consumption, in miles per gallon (mpg) of a car is inversely proportional to its speed, in miles per hour (mph). When the car is travelling at 30 mph the fuel consumption is 60 mpg.

How much further would the car travel on 1 gallon of fuel by travelling at 60 mph instead of 70 mph on a motorway?

GRADE BOOSTER

A You can find formulae describing direct or inverse variation and use them to solve problems

What you should know now

- How to recognise direct and inverse variation
- What a constant of proportionality is, and how to find it
- How to find formulae describing inverse or direct variation
- How to solve problems involving direct or inverse variation

1 y is proportional to $\sqrt{x}$. Complete the table.

x	25		400
y	10	20	

2 The energy, E, of an object moving horizontally is directly proportional to the speed, v, of the object. When the speed is 10 m/s the energy is 40 000 Joules.

a Find an equation connecting E and v.

b Find the speed of the object when the energy is 14 400 Joules.

3 y is inversely proportional to the cube root of x. When $y = 8$, $x = \frac{1}{8}$.

a Find an expression for y in terms of x,

b Calculate

i the value of y when $x = \frac{1}{125}$

ii the value of x when $y = 2$.

4 The mass of a cube is directly proportional to the cube of its side. A cube with a side of 4 cm has a mass of 320 grams. Calculate the side length of a cube made of the same material with a mass of 36 450 grams.

5 y is directly proportional to the cube of x. When $y = 16$, $x = 3$. Find the value of y when $x = 6$.

6 d is directly proportional to the square of t. $d = 80$ when $t = 4$.

a Express d in terms of t.

b Work out the value of d when $t = 7$.

c Work out the positive value of t when $d = 45$. *(? marks)*

Edexcel, June 2005, Paper 5 Higher, Question 16

7 M is directly proportional to L^3.

When $L = 2$, $M = 160$

Find the value of M when $L = 3$ *(4 marks)*

Edexcel, May 2009, Paper 3, Question 21

8 Two variables, x and y, are known to be proportional to each other. When $x = 10$, $y = 25$.

Find the constant of proportionality, k, if:

a $y \propto x$

b $y \propto x^2$

c $y \propto \frac{1}{x}$

d $\sqrt{y} \propto \frac{1}{x}$

9 y is directly proportional to the cube root of x. When $x = 27$, $y = 6$.

a Find the value of y when $x = 125$.

b Find the value of x when $y = 3$.

10 The surface area, A, of a solid is directly proportional to the square of the depth, d. When $d = 6$, $A = 12\pi$.

a Find the value of A when $d = 12$. Give your answer in terms of π.

b Find the value of d when $A = 27\pi$.

11 q is inversely proportional to the square of t.

When $t = 4$, $q = 8.5$

a Find a formula for q in terms of t. *(3 marks)*

b Calculate the value of q when $t = 5$ *(1 mark)*

Edexcel, June 2008, Paper 4, Question 20

12 D is proportional to S^2.

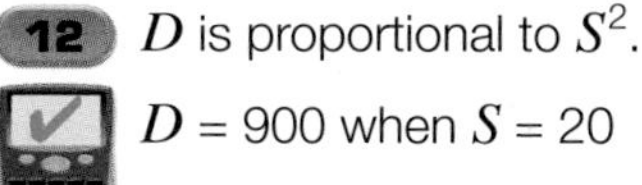

$D = 900$ when $S = 20$

Calculate the value of D when $S = 25$ *(4 marks)*

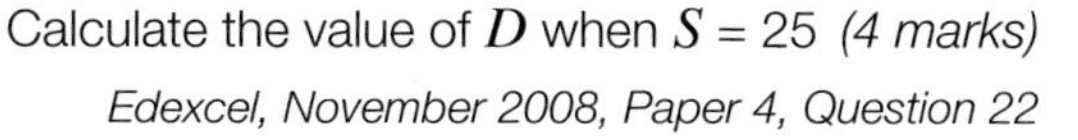

Edexcel, November 2008, Paper 4, Question 22

13 The frequency, f, of sound is inversely proportional to the wavelength, w. A sound with a frequency of 36 hertz has a wavelength of 20.25 metres.

Calculate the frequency when the frequency and the wavelength have the same numerical value.

14 t is proportional to m^3.

a When $m = 6$, $t = 324$. Find the value of t when $m = 10$.

Also, m is inversely proportional to the square root of w.

b When $t = 12$, $w = 25$. Find the value of w when $m = 4$.

15 P and Q are positive quantities. P is inversely proportional to Q^2. When $P = 160$, $Q = 20$. Find the value of P when $P = Q$.

Worked Examination Questions

1 y is inversely proportional to the square of x. When y is 40, $x = 5$.

a Find an equation connecting x and y.

b Find the value of y when $x = 10$.

1 a

$$y \propto \frac{1}{x^2}$$

$$y = \frac{k}{x^2}$$

First set up the proportionality relationship and replace the proportionality sign with $= k$.

This gets 1 method mark for stating first or second line or both.

$$40 = \frac{k}{25}$$

Substitute the given values of y and x into the proportionality equation to find the value of k.

This gets 1 accuracy mark for finding k.

$$\Rightarrow k = 40 \times 25 = 1000$$

$$y = \frac{1000}{x^2}$$

or $yx^2 = 1000$

Substitute the value of k to get the final equation connecting y and x.

This gets 1 mark for accuracy.

b When $x = 10$, $y = \dfrac{1000}{10^2} = \dfrac{1000}{100} = 10$

Substitute the value of x into the equation to find y.

This gets 1 method mark for substitution of $x = 10$ and 1 accuracy mark for correct answer.

Total: 5 marks

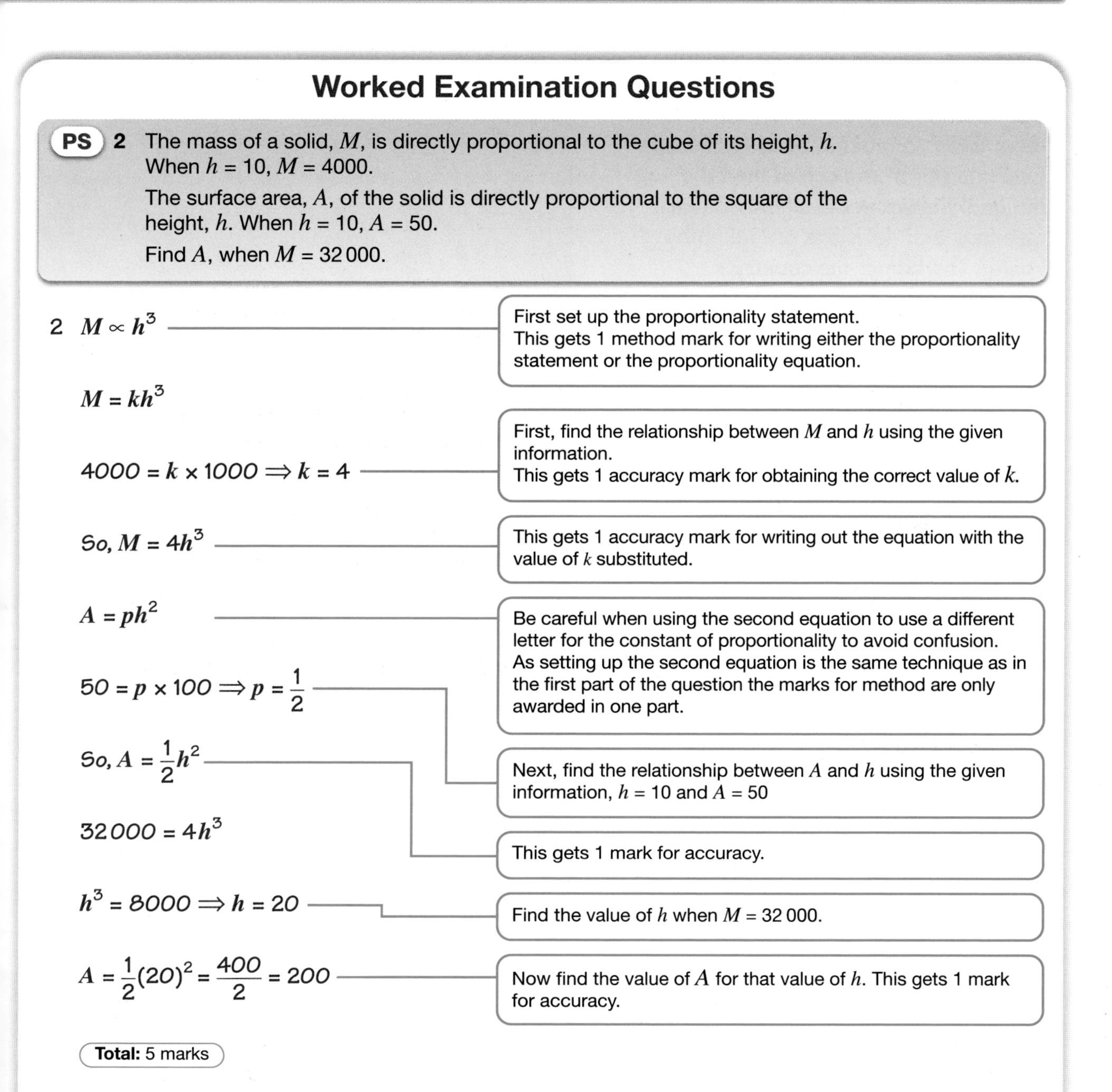

Worked Examination Questions

PS **2** The mass of a solid, M, is directly proportional to the cube of its height, h. When $h = 10$, $M = 4000$.

The surface area, A, of the solid is directly proportional to the square of the height, h. When $h = 10$, $A = 50$.

Find A, when $M = 32\,000$.

2 $M \propto h^3$ — First set up the proportionality statement. This gets 1 method mark for writing either the proportionality statement or the proportionality equation.

$M = kh^3$

$4000 = k \times 1000 \Rightarrow k = 4$ — First, find the relationship between M and h using the given information. This gets 1 accuracy mark for obtaining the correct value of k.

So, $M = 4h^3$ — This gets 1 accuracy mark for writing out the equation with the value of k substituted.

$A = ph^2$ — Be careful when using the second equation to use a different letter for the constant of proportionality to avoid confusion. As setting up the second equation is the same technique as in the first part of the question the marks for method are only awarded in one part.

$50 = p \times 100 \Rightarrow p = \frac{1}{2}$ — Next, find the relationship between A and h using the given information, $h = 10$ and $A = 50$

So, $A = \frac{1}{2}h^2$ — This gets 1 mark for accuracy.

$32\,000 = 4h^3$

$h^3 = 8000 \Rightarrow h = 20$ — Find the value of h when $M = 32\,000$.

$A = \frac{1}{2}(20)^2 = \frac{400}{2} = 200$ — Now find the value of A for that value of h. This gets 1 mark for accuracy.

Total: 5 marks

13 Functional Maths
Voting in the European Union

The Council of the European Union is the main decision-making body for Europe. One minister from each of the EU's national governments attends Council meetings and decisions are taken by voting. The bigger the country's population, the more votes it has, but numbers are currently weighted in favour

Getting started

In June 2007, Poland argued for a change to the rules for the voting in the Council of the European Union. The Polish suggested that each country's voting strength should be directly proportional to the square root of its population. This idea is known as **Pensore's rule.**

Let V be the voting strength (that is the number of votes a country gets) and P be the country's population.

- Write down a mathematical statement for Pensore's rule, using the symbol of variation, $\propto$.
- Write a proportionality equation for Pensore's rule, using a constant of proportionality, k.

Country	Population	Current number of votes in the Council of the European Union
UK	61 600 835	29
Poland	38 125 478	27
Romania	21 398 181	14
The Netherlands	16 518 199	13
Belgium	10 574 595	12
Sweden	9 290 113	10
Ireland	4 434 925	7
Luxembourg	472 569	4
Malta	408 009	3

Your task

1 Suppose that Pensore's rule was introduced and Poland gained an additional vote, making its voting strength 28. Your task is to determine how Pensore's rule would affect other countries' votes.

2 Now, suppose another member, such as the UK, proposed that each member's voting strength should be directly proportional to its country's population, making its voting strength 40. Your task is to determine how this suggested voting system would affect other countries' votes.

3 Imagine you are an advisor to the President of the European Union. How would you advise on Council voting? Write a letter setting out what you think of Poland's and the UK's suggestions, explaining the advantages and disadvantages. As an independent advisor, can you propose an alternative voting system that may be fairer to all countries?

Why this chapter matters

How accurate are we?

In real life it is not always sensible to use exact values. Sometimes it would be impossible to have exact measurements. People often round values without realising it. Rounding is done so that values are sensible.

Is it exactly 3 miles to Woodlaithes Village and exactly 4 miles to Rotherham?

Does this box contain exactly 750 g of flakes when full?

Was his time exactly 13.4 seconds?

Does the school have exactly 1500 students?

Imagine if people tried to use exact values all the time. Would life seem strange?

Number: Number and limits of accuracy

The grades given in this chapter are target grades.

This chapter will show you ...

C how to find the limits of numbers rounded to certain accuracies

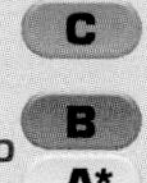

to **B** to **A*** how to use limits of accuracy in calculations

Visual overview

Limits of accuracy → Calculating with limits of accuracy

What you should already know

- How to round numbers to the nearest 10, 100 or 1000 **(KS3 level 4, GCSE grade G)**
- How to round numbers to a given number of decimal places **(KS3 level 5, GCSE grade F)**
- How to round numbers to a given number of significant figures **(KS3 level 6–7, GCSE grade E)**

Quick check

1 Round 6374 to:

a the nearest 10

b the nearest 100

c the nearest 1000.

2 Round 2.389 to:

a 1 decimal place

b 2 decimal places.

3 Round 47.28 to:

a 1 significant figure

b 3 significant figures.

14.1 Limits of accuracy

This section will show you how to:

- find the limits of accuracy of numbers that have been rounded to different degrees of accuracy

Key words
continuous data
discrete data
limits of accuracy
lower bound
upper bound

Any recorded measurements have usually been rounded.

The true value will be somewhere between the **lower bound** and the **upper bound**.

The lower and upper bounds are sometimes known as the **limits of accuracy**.

Discrete data

Discrete data can only take certain values within a given range; amounts of money and numbers of people are examples of discrete data.

EXAMPLE 1

A coach is carrying 50 people, to the nearest 10.

What are the minimum and maximum numbers of people on the coach?

45 is the lowest whole number that rounds to 50 to the nearest 10.
54 is the highest whole number that rounds to 50 to the nearest 10.

So minimum = 45 and maximum = 54

The limits are $45 \leqslant$ number of people $\leqslant 54$

Remember that you can only have a whole number of people.

Continuous data

Continuous data can take any value, within a given range; length and mass are examples of continuous data.

Upper and lower bounds

A journey of 26 miles measured to the nearest mile could actually be as long as 26.4999999… miles or as short as 25.5 miles. It could not be 26.5 miles, as this would round up to 27 miles. However, 26.499 999 9… is virtually the same as 26.5.

You overcome this difficulty by saying that 26.5 is the upper bound of the measured value and 25.5 is its lower bound. You can therefore write:

25.5 miles $\leqslant$ actual distance < 26.5 miles

which states that the actual distance is *greater than or equal to* 25.5 miles but *less than* 26.5 miles.

When stating the upper bound, follow the accepted practice, as demonstrated here, which eliminates the difficulties of using recurring decimals.

A mathematical peculiarity

Let:	$x = 0.999\,999\ldots$ (1)
Multiply by 10:	$10x = 9.999\,999\ldots$ (2)
Subtract (1) from (2):	$9x = 9$
Divide by 9:	$x = 1$
So, we have:	$0.\dot{9} = 1$

Hence, it is valid to give the upper bound without using recurring decimals.

EXAMPLE 2

A stick of wood measures 32 cm, to the nearest centimetre.

What are the lower and upper limits of the actual length of the stick?

The lower limit is 31.5 cm as this is the lowest value that rounds to 32 cm to the nearest centimetre.

The upper limit is 32.499 999 999... cm as this is the highest value that rounds to 32 cm to the nearest centimetre as 32.5 cm would round to 33 cm.

However, you say that 32.5 cm is the upper bound. So you write:

$31.5 \text{ cm} \leqslant \text{length of stick} < 32.5 \text{ cm}$

Note the use of the strict inequality ($<$) for the upper bound.

EXAMPLE 3

A time of 53.7 seconds is accurate to 1 decimal place.

What are the limits of accuracy?

The smallest possible value is 53.65 seconds.

The largest possible value is 53.749 999 999... but 53.75 seconds is the upper bound.

So the limits of accuracy are $53.65 \text{ seconds} \leqslant \text{time} < 53.75 \text{ seconds}$.

EXAMPLE 4

A skip has a mass of 220 kg measured to 3 significant figures. What are the limits of accuracy of the mass of the skip?

The smallest possible value is 219.5 kg.

The largest possible value is 220.499 999 99... kg but 220.5 kg is the upper bound.

So the limits of accuracy are $219.5 \text{ kg} \leqslant \text{mass of skip} < 220.5 \text{ kg}$.

EXERCISE 14A

C

1. Write down the limits of accuracy of the following.
   - **a** 7 cm measured to the nearest centimetre
   - **b** 120 g measured to the nearest 10 g
   - **c** 3400 km measured to the nearest 100 km
   - **d** 50 mph measured to the nearest miles per hour
   - **e** £6 given to the nearest pound
   - **f** 16.8 cm to the nearest tenth of a centimetre
   - **g** 16 kg to the nearest kilogram
   - **h** A football crowd of 14 500 given to the nearest 100
   - **i** 55 miles given to the nearest mile
   - **j** 55 miles given to the nearest 5 miles

B

2. Write down the limits of accuracy for each of the following values, which are rounded to the given degree of accuracy.
   - **a** 6 cm (1 significant figure)
   - **b** 17 kg (2 significant figures)
   - **c** 32 min (2 significant figures)
   - **d** 238 km (3 significant figures)
   - **e** 7.3 m (1 decimal place)
   - **f** 25.8 kg (1 decimal place)
   - **g** 3.4 h (1 decimal place)
   - **h** 87 g (2 significant figures)
   - **i** 4.23 mm (2 decimal places)
   - **j** 2.19 kg (2 decimal places)
   - **k** 12.67 min (2 decimal places)
   - **l** 25 m (2 significant figures)
   - **m** 40 cm (1 significant figure)
   - **n** 600 g (2 significant figures)
   - **o** 30 min (1 significant figure)
   - **p** 1000 m (2 significant figures)
   - **q** 4.0 m (1 decimal place)
   - **r** 7.04 kg (2 decimal places)
   - **s** 12.0 s (1 decimal place)
   - **t** 7.00 m (2 decimal places)

3. Write down the lower and upper bounds of each of these values, rounded to the accuracy stated.
   - **a** 8 m (1 significant figure)
   - **b** 26 kg (2 significant figures)
   - **c** 25 min (2 significant figures)
   - **d** 85 g (2 significant figures)
   - **e** 2.40 m (2 decimal places)
   - **f** 0.2 kg (1 decimal place)
   - **g** 0.06 s (2 decimal places)
   - **h** 300 g (1 significant figure)
   - **i** 0.7 m (1 decimal place)
   - **j** 366 d (3 significant figures)
   - **k** 170 weeks (2 significant figures)
   - **l** 210 g (2 significant figures)

B

PS **4** A bus has 53 seats of which 37 are occupied.

The driver estimates that at the next bus stop 20 people, to the nearest 10, will get on and no one will get off.

If he is correct, is it possible they will all get a seat?

FM **5** A chain is 30 m long, to the nearest metre.

A chain is needed to fasten a boat to a harbour wall, a distance that is also 30 m, to the nearest metre.

Which statement is definitely true? Explain your decision.

A: The chain will be long enough.

B: The chain will not be long enough.

C: It is impossible to tell whether or not the chain is long enough.

AU **6** A bag contains 2.5 kg of soil, to the nearest 100 g.

What is the least amount of soil in the bag?

Give your answer in kilograms and grams.

7 Billy has 40 identical marbles. Each marble has a mass of 65 g (to the nearest gram).

a What is the greatest possible mass of one marble?

b What is the least possible mass of one marble?

c What is the greatest possible mass of all the marbles?

d What is the least possible mass of all the marbles?

14.2 Problems involving limits of accuracy

This section will show you how to:
- combine limits of two or more variables together to solve problems

Key words
limits of accuracy
maximum
minimum

When rounded values are used for a calculation, the **minimum** and **maximum** possible exact values of the calculation can vary by large amounts.

There are four operations that can be performed on **limits of accuracy** — addition, subtraction, multiplication and division.

Addition and subtraction

Suppose you have two bags, each with the mass given to the nearest kilogram.

A mass 5 kg to nearest kg

B mass 9 kg to nearest kg

The limits for bag A are 4.5 kg ⩽ mass < 5.5 kg

The limits for bag B are 8.5 kg ⩽ mass < 9.5 kg

The minimum total mass of the two bags is 4.5 kg + 8.5 kg = 13 kg

The maximum total mass of the two bags is 5.5 kg + 9.5 kg = 15 kg

The minimum difference between the masses of the two bags is 8.5 kg – 5.5 kg = 3 kg

The maximum difference between the masses of the two bags is 9.5 kg – 4.5 kg = 5 kg

The table shows the combinations to give the minimum and maximum values for addition and subtraction of two numbers, a and b.

a and b lie within limits $a_{min} \leqslant a < a_{max}$ and $b_{min} \leqslant b < b_{max}$

Operation	Minimum	Maximum
Addition ($a + b$)	$a_{min} + b_{min}$	$a_{max} + b_{max}$
Subtraction ($a - b$)	$a_{min} - b_{max}$	$a_{max} - b_{min}$

Multiplication and division

Suppose a car is travelling at an average speed of 30 mph, to the nearest 5 mph, for 2 hours, to the nearest 30 minutes.

The limits for the average speed are:

27.5 mph ⩽ average speed < 32.5 mph

The limits for the time are:

1 hour 45 minutes (1.75 hours) ⩽ time < 2 hours 15 minutes (2.25 hours)

The minimum distance travelled = 27.5 × 1.75 = 48.125 miles

The maximum distance travelled = 32.5 × 2.25 = 73.125 miles

Suppose a lorry is travelling for 100 miles, to the nearest 10 miles, and takes 2 hours, to the nearest 30 minutes.

The limits for the distance are:

95 miles ⩽ distance < 105 miles

The limits for the time are the same as for the car.

The minimum average speed is $\frac{95}{2.25} = 42$ mph

The maximum average speed is $\frac{105}{1.75} = 60$ mph

The table shows the combinations to give the minimum and maximum values for multiplication and division of two numbers a and b.

a and b lie within limits $a_{min} \leqslant a < a_{max}$ and $b_{min} \leqslant b < b_{max}$

Operation	Minimum	Maximum
Multiplication ($a \times b$)	$a_{min} \times b_{min}$	$a_{max} \times b_{max}$
Division ($a \div b$)	$a_{min} \div b_{max}$	$a_{max} \div b_{min}$

When solving problems involving limits, write down all the limits for each value, then decide which combination to use to obtain the required solution.

When rounding answers, be careful to ensure your answers are within the acceptable range of the limits.

EXAMPLE 5

A rectangle has sides given as 6 cm by 15 cm, to the nearest centimetre.

Calculate the limits of accuracy of the area of the rectangle.

Write down the limits: 5.5 cm $\leqslant$ width < 6.5 cm, 14.5 cm $\leqslant$ length < 15.5 cm

For maximum area, multiply maximum width by maximum length, and for minimum area, multiply minimum width by minimum length.

The upper bound of the width is 6.5 cm and of the length is 15.5 cm. So the upper bound of the area of the rectangle is:

6.5 cm × 15.5 cm = 100.75 cm^2

The lower bound of the width is 5.5 cm and of the length is 14.5 cm. So the lower bound of the area of the rectangle is:

5.5 cm × 14.5 cm = 79.75 cm^2

Therefore, the limits of accuracy for the area of the rectangle are:

79.75 cm^2 $\leqslant$ area < 100.75 cm^2

EXAMPLE 6

The distance from Barnsley to Sheffield is 15 miles, to the nearest mile. The time Jeff took to drive between Barnsley and Sheffield was 40 minutes, to the nearest 10 minutes.

Calculate the upper limit of Jeff's average speed.

Write down the limits: 14.5 miles $\leqslant$ distance < 15.5 miles, 35 mins $\leqslant$ time < 45 mins

speed = distance ÷ time

To get the maximum speed you need the maximum distance ÷ minimum time.

15.5 miles ÷ 35 mins = 0.443 (3 significant figures) miles per minute

0.443 mph × 60 = 26.6 mph

The upper limit of Jeff's average speed = 26.6 mph

EXERCISE 14B

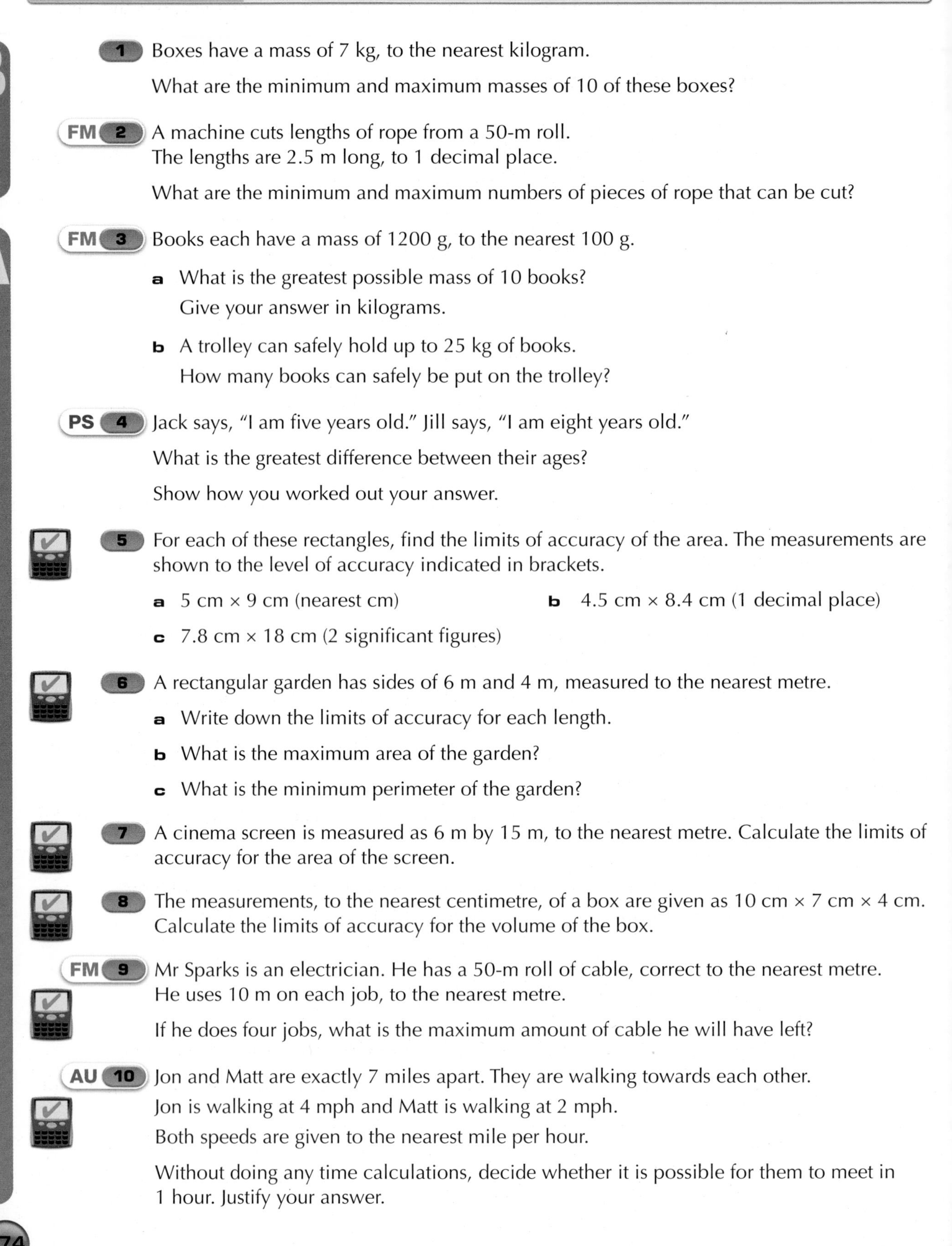

B

A

1 Boxes have a mass of 7 kg, to the nearest kilogram.

What are the minimum and maximum masses of 10 of these boxes?

FM **2** A machine cuts lengths of rope from a 50-m roll.
The lengths are 2.5 m long, to 1 decimal place.

What are the minimum and maximum numbers of pieces of rope that can be cut?

FM **3** Books each have a mass of 1200 g, to the nearest 100 g.

a What is the greatest possible mass of 10 books?
Give your answer in kilograms.

b A trolley can safely hold up to 25 kg of books.
How many books can safely be put on the trolley?

PS **4** Jack says, "I am five years old." Jill says, "I am eight years old."

What is the greatest difference between their ages?

Show how you worked out your answer.

5 For each of these rectangles, find the limits of accuracy of the area. The measurements are shown to the level of accuracy indicated in brackets.

a 5 cm × 9 cm (nearest cm)

b 4.5 cm × 8.4 cm (1 decimal place)

c 7.8 cm × 18 cm (2 significant figures)

6 A rectangular garden has sides of 6 m and 4 m, measured to the nearest metre.

a Write down the limits of accuracy for each length.

b What is the maximum area of the garden?

c What is the minimum perimeter of the garden?

7 A cinema screen is measured as 6 m by 15 m, to the nearest metre. Calculate the limits of accuracy for the area of the screen.

8 The measurements, to the nearest centimetre, of a box are given as 10 cm × 7 cm × 4 cm. Calculate the limits of accuracy for the volume of the box.

FM **9** Mr Sparks is an electrician. He has a 50-m roll of cable, correct to the nearest metre.
He uses 10 m on each job, to the nearest metre.

If he does four jobs, what is the maximum amount of cable he will have left?

AU **10** Jon and Matt are exactly 7 miles apart. They are walking towards each other.
Jon is walking at 4 mph and Matt is walking at 2 mph.
Both speeds are given to the nearest mile per hour.

Without doing any time calculations, decide whether it is possible for them to meet in 1 hour. Justify your answer.

11 The area of a rectangular field is given as 350 m^2, to the nearest 10 m^2. One length is given as 16 m, to the nearest metre. Find the limits of accuracy for the other length of the field.

A*

12 In triangle ABC, AB = 9 cm, BC = 7 cm, and $\angle ABC = 37°$. All the measurements are given to the nearest unit. Calculate the limits of accuracy for the area of the triangle.

13 The price of pure gold is £18.25 per gram. The density of gold is 19.3 g/cm^3. (Assume these figures are exact.) A solid gold bar in the shape of a cuboid has sides 4.6 cm, 2.2 cm and 6.6 cm. These measurements are made to the nearest 0.1 cm.

a **i** What are the limits of accuracy for the volume of this gold bar?

ii What are the upper and lower limits of the cost of this bar?

The gold bar was weighed and given a mass of 1296 g, to the nearest gram.

b What are the upper and lower limits for the cost of the bar now?

c Explain why the price ranges are so different.

FM 14 A stopwatch records the time for the winner of a 100-metre race as 14.7 seconds, measured to the nearest one-tenth of a second.

a What are the greatest and least possible times for the winner?

b The length of the 100-metre track is correct to the nearest 1 m. What are the greatest and least possible lengths of the track?

c What is the fastest possible average speed of the winner, with a time of 14.7 seconds in the 100-metre race?

15 A cube has a side measured as 8 cm, to the nearest millimetre. What is the greatest percentage error of the following?

a The calculated area of one face **b** The calculated volume of the cube

16 A cube has a volume of 40 cm^3, to the nearest cm^3. Find the range of possible values of the side length of the cube.

17 A cube has a volume of 200 cm^3, to the nearest 10 cm^3. Find the limits of accuracy of the side length of the cube.

18 A model car travels 40 m, measured to one significant figure, at a speed of 2 m/s, measured to one significant figure. Between what limits does the time taken lie?

PS 19 Here is a formula for calculating the tension, T newtons, in some coloured springs.

$$T = \frac{20x}{l}$$

x is the length that the spring is extended. l is the unstretched length of the spring.

If x and l are accurate to one decimal place, decide which colour of spring, if any, has the greater tension.

Red spring: $x = 3.4$ cm and $l = 5.3$ cm
Green spring: $x = 1.5$ cm and $l = 2.4$ cm
Blue spring: $x = 0.5$ cm and $l = 0.9$ cm

GRADE BOOSTER

B You can find measures of accuracy for numbers given to whole-number accuracies

A You can find measures of accuracy for numbers given to decimal-place or significant-figure accuracies

A* You can calculate the limits of compound measures

What you should know now

- How to find the limits of numbers given to various accuracies
- How to find the limits of compound measures by combining the appropriate limits of the variables involved

1 A school has 1850 pupils to the nearest 10.

a What is the least number of pupils at the school?

b What is the greatest number of pupils at the school?

2 The longest river in Britain is the River Severn. It is 220 miles long to the nearest 10 miles. What is the least length it could be?

3 The length of a line is 63 centimetres, correct to the nearest centimetre.

a Write down the **least** possible length of the line. *(1 mark)*

b Write down the **greatest** possible length of the line. *(1 mark)*

Edexcel, May 2009, Paper 3, Question 11

4 Bob measures the length of his book.

The length of the book is 22 cm correct to the nearest centimetre.

a Write down the maximum possible length it could be. *(1 mark)*

b Write down the minimum possible length it could be. *(1 mark)*

Edexcel, June 2007, Paper 11, Question 4

5 A ball is thrown vertically upwards with a speed V metres per second.

The height, H metres, to which it rises is given by $H = \dfrac{V^2}{2g}$

where g m/s^2 is the acceleration due to gravity.

$V = 24.4$ correct to 3 significant figures.

$g = 9.8$ correct to 2 significant figures.

a Write down the lower bound of g. *(1 mark)*

b Calculate the upper bound of H.

Give your answer correct to 3 significant figures. *(2 marks)*

Edexcel, November 2008, Paper 4, Question 23

6  Katy drove for 238 miles, correct to the nearest mile.

She used 27.3 litres of petrol, to the nearest tenth of a litre.

$$\text{Petrol consumption} = \frac{\text{Number of miles travelled}}{\text{Number of litres of petrol used}}$$

Work out the upper bound for the petrol consumption for Katy's journey.

Give your answer correct to 2 decimal places. *(3 marks)*

Edexcel, June 2008, Paper 4, Question 22

7 Jerry measures a piece of wood as 60 cm correct to the nearest centimetre.

a Write down the minimum possible length of the piece of wood.

b Write down the maximum possible length of the piece of wood.

Edexcel, March 2005, Paper 13 Higher, Question 1

8 The base of a triangle is 10 cm measured to the nearest centimetre. The area of the triangle is 100 cm^2 measured to the nearest square centimetre. Calculate the least and greatest values of the height of the triangle.

9 $x = 1.8$ measured to 1 decimal place,
$y = 4.0$ measured to 2 significant figures,
$z = 2.56$ measured to 3 significant figures.

Calculate the upper limit of $\dfrac{x^2 + y}{z}$.

10 A girl runs 60 metres in a time of 8.0 seconds. The distance is measured to the nearest metre and the time is measured to 2 significant figures.

What is the least possible speed?

11 **a** Calculate the length of the diagonal x in this cube of side 3 m.

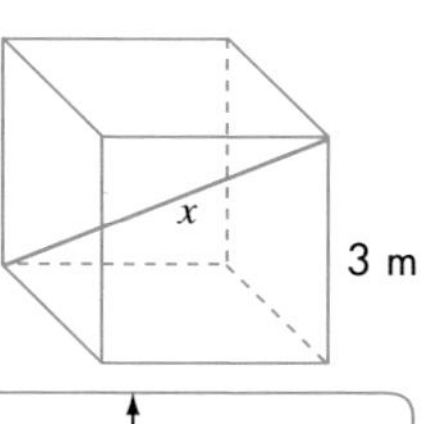

b A man is carrying a pole of length 5 m down a long corridor. The pole is measured to the nearest centimetre. At the end of the corridor is a right-angled corner. The corridor is 3 m wide and 3 m high, both measurements correct to the nearest 10 cm. Will the pole be certain to get round the corner?

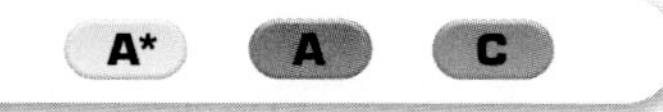

Worked Examination Questions

1 The magnification of a lens is given by the formula

$$m = \frac{u}{v}$$

In an experiment, u is measured as 8.5 cm and v is measured as 14.0 cm, both correct to the nearest 0.1 cm. Find the least possible value of m. You must show full details of your calculation.

1 $8.45 \leq u < 8.55$

Write down the limits of both variables. This gets 1 mark each.

$13.95 \leq v < 14.05$

As the calculation is a division the least value will be given by least $u \div$ greatest v.

This gets 1 mark for method.

Least m = least $u \div$ greatest $v = 8.45 \div 14.05$

$= 0.6014234875$

This gets 1 mark for accuracy.

$= 0.601$ or 0.6

It is good practice to round to a suitable degree of accuracy.

Total: 4 marks

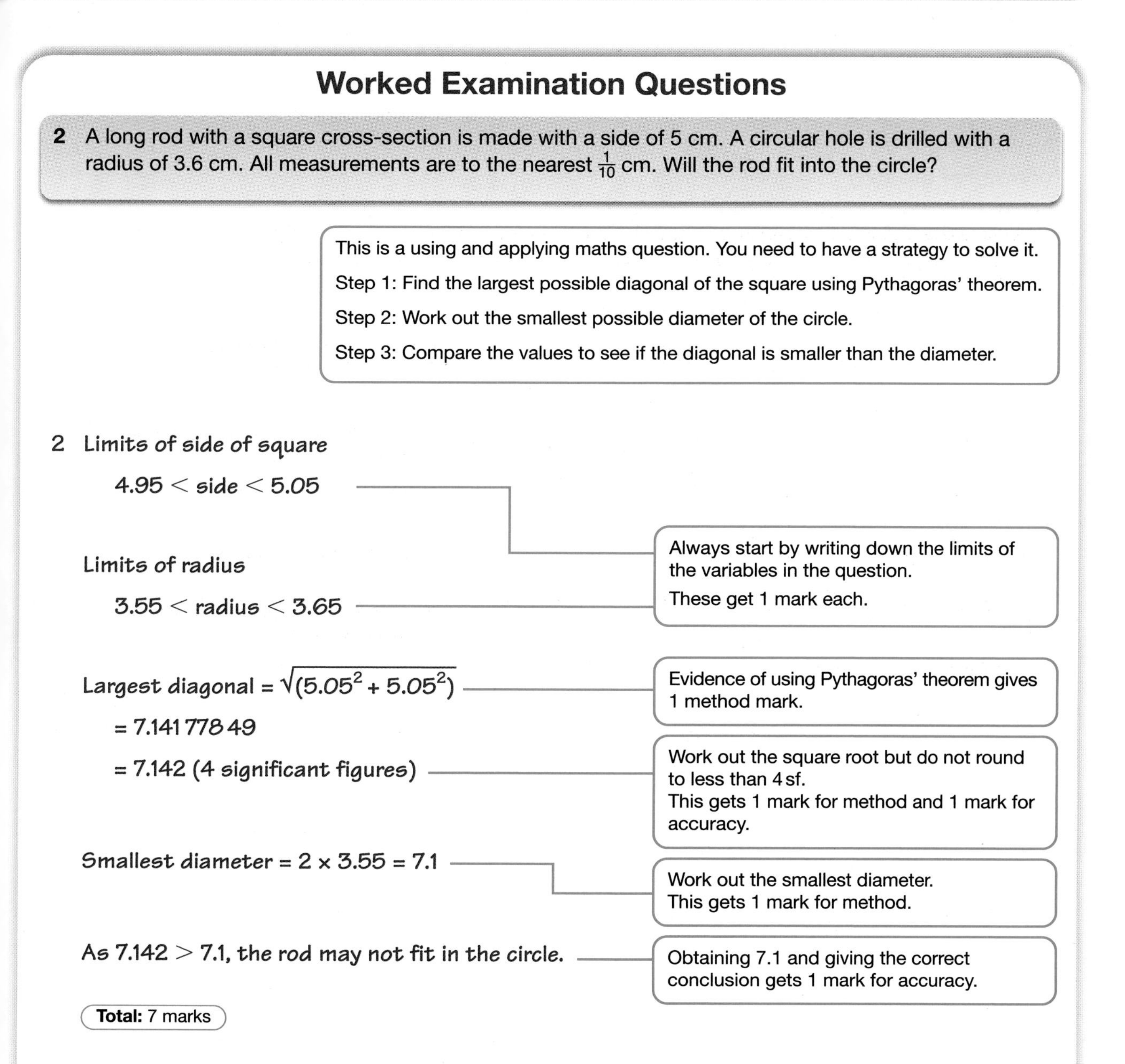

Worked Examination Questions

2 A long rod with a square cross-section is made with a side of 5 cm. A circular hole is drilled with a radius of 3.6 cm. All measurements are to the nearest $\frac{1}{10}$ cm. Will the rod fit into the circle?

This is a using and applying maths question. You need to have a strategy to solve it.

Step 1: Find the largest possible diagonal of the square using Pythagoras' theorem.

Step 2: Work out the smallest possible diameter of the circle.

Step 3: Compare the values to see if the diagonal is smaller than the diameter.

2 Limits of side of square

$4.95 < \text{side} < 5.05$

Limits of radius

$3.55 < \text{radius} < 3.65$

Always start by writing down the limits of the variables in the question.
These get 1 mark each.

Largest diagonal = $\sqrt{(5.05^2 + 5.05^2)}$

= 7.141 778 49

= 7.142 (4 significant figures)

Evidence of using Pythagoras' theorem gives 1 method mark.

Work out the square root but do not round to less than 4 sf.
This gets 1 mark for method and 1 mark for accuracy.

Smallest diameter = $2 \times 3.55 = 7.1$

Work out the smallest diameter.
This gets 1 mark for method.

As $7.142 > 7.1$, the rod may not fit in the circle.

Obtaining 7.1 and giving the correct conclusion gets 1 mark for accuracy.

Total: 7 marks

Functional Maths
Buying a house

Estate agents regularly measure the dimensions of rooms. For this they use an electronic signal reader, which makes measurements to a stated accuracy.

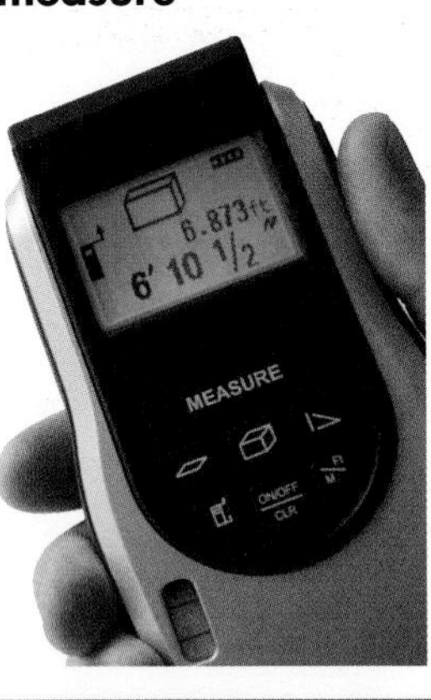

Getting started

Below are three measurements, given in metres.

- 4 m
- 109 m
- 80 m

If each of these measurements is an approximation and only expressed to the nearest metre, what is the smallest number each measurement could be? If each measurement was actually measured to the nearest centimetre, to be more accurate, what is the largest number each measurement could be?

Discuss the difference between these two answers.

Your task

An estate agent has measured the rooms in a house. The figures are shown in the plan opposite.

The measurements are given correct to 10 cm.

You are interested in buying the house but want to know its measurements more accurately as this could affect the price that you offer for it.

- Work out the maximum and minimum area of the floor space in the house, writing down and explaining your calculations so that you can inform your family later.
- Based on your calculations, work out the amount of flooring that will be needed for each room. You must also work out the maximum and minimum cost (inclusive of VAT) of your flooring options for each room so that you can factor the additional cost into your final price for the house.

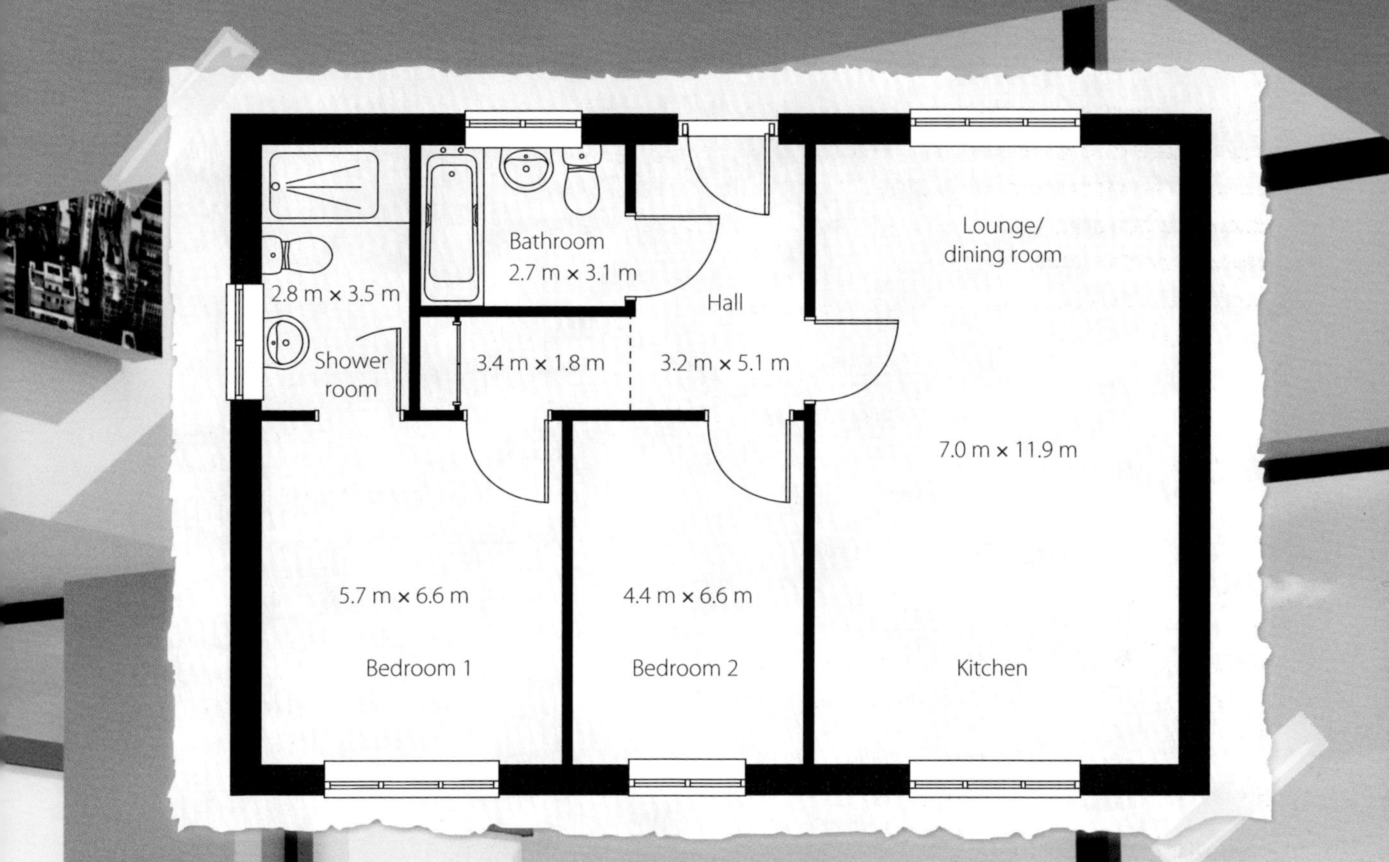

Here are the prices quoted by several flooring specialists in your area. The prices given exclude VAT (at 17.5%).

Flooring	Cost (per m^2)
Carpet	£11.58–£28.99
Laminate	£8.31–£16.50
Solid wood	£42.98–£52.98
Vinyl	£6.00–£10.00

Flooring	Size (mm)/ number per pack	Cost (per pack)
Stone tiles	305 × 305 / 5	£19.98–£22.98
Ceramic floor tiles	333 × 333 / 10	£15.58–£17.58

Why this chapter matters

Vectors are used to represent any quantity that has both magnitude and direction. The velocity of a speeding car may be described in terms of its direction and its speed. The speed is the magnitude, but when it has direction is becomes the velocity – a vector.

To understand how a force acts on an object, you need to know the magnitude of the force and the direction in which it moves – the two bits of information that define a vector.

And when you watch the weather report, you are told which way the wind will blow tomorrow, and how strongly – again, a direction and a magnitude together making up a vector.

Vectors are used to describe many quantities in science, such as displacement, acceleration and momentum.

But are vectors used in real life? Yes! Here are some examples.

In the 1950s, a group of talented Brazilian footballers invented the **swerving free kick**. By kicking the ball in just the right place, they managed to make it curl around the wall of defending players and, quite often, go straight into the back of the net. When a ball is in flight, it is acted upon by various forces. Some of these depend on the way the ball is spinning. The forces at work here can be described by vectors.

Formula One teams always employ physicists and mathematicians to help them build the perfect racing car. **Aerodynamics** is the study of how the air moves. Since vectors describe movements and forces, they are used as the basis of a car's design.

Pilots have to consider wind speed and direction when they plan to land an aircraft at an airport. Vectors are an integral part of the computerised landing system.

The science of aerodynamics is used in the design of aircraft; vectors play a key role in the design of wings, where an upward force or lift is needed to enable the aircraft to fly.

Meteorologists or weather forecasters use vectors to map out weather patterns. Wind speeds can be represented by vectors of different lengths to indicate the intensity of the wind.

Vectors are used extensively in computer graphics. Software designed to give the viewer the impression that an object or person is moving around a scene makes extensive use of the mathematics of vectors.

Chapter

Geometry: Vectors

Properties of vectors

Vectors in geometry

Geometric proof

The grades given in this chapter are target grades.

This chapter will show you ...

- **A** how to add and subtract vectors
- **A** to **A*** the properties of vectors
- **A*** how to use vectors to solve geometrical problems
- **A*** how to prove geometric results using rigorous and logical mathematical arguments

Visual overview

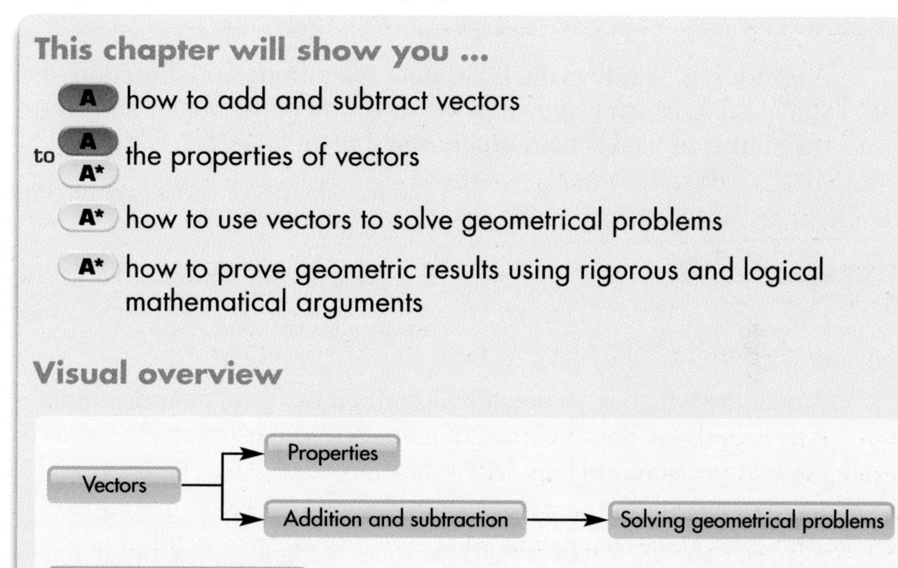

What you should already know

- Vectors are used to describe translations **(KS3 level 7, GCSE grade C)**

Quick check

Use column vectors to describe these translations.

a A to C
b B to D
c C to D
d D to E

15.1 Properties of vectors

This section will show you how to:

- add and subtract vectors

Key words
direction
magnitude
vector

A vector is a quantity which has both **magnitude** and **direction**. It can be represented by a straight line which is drawn in the direction of the vector and whose length represents the magnitude of the vector. Usually, the line includes an arrowhead.

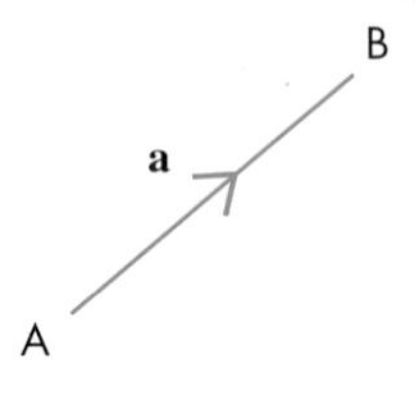

The translation or movement from A to B is represented by the vector **a**.

a is always printed in bold type, but is written as $\underline{a}$.

a can also be written as $\overrightarrow{AB}$.

A quantity which is completely described by its magnitude, and has no direction associated with it, is called a scalar. The mass of a bus (10 tonnes) is an example of a scalar. Another example is a linear measure, such as 25.4 mm.

Multiplying a vector by a number (scalar) alters its magnitude (length) but not its direction. For example, the vector 2**a** is twice as long as the vector **a**, but in the same direction.

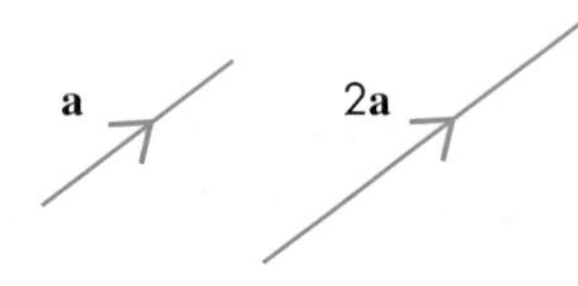

A negative vector, for example **–b**, has the same magnitude as the vector **b**, but is in the opposite direction.

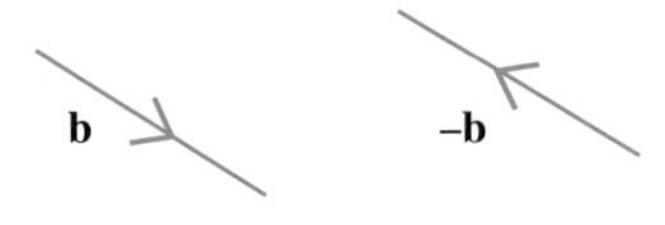

Addition and subtraction of vectors

Take two non-parallel vectors **a** and **b**, then **a** + **b** is defined to be the translation of **a** followed by the translation of **b**. This can easily be seen on a vector diagram.

Similarly, **a** – **b** is defined to be the translation of **a** followed by the translation of –**b**.

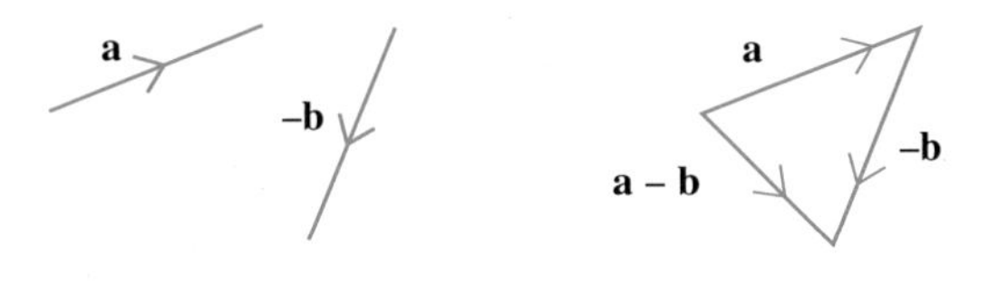

Look at the parallelogram grid below. **a** and **b** are two independent vectors that form the basis of this grid. It is possible to define the position, with reference to O, of any point on this grid by a vector expressed in terms of **a** and **b**. Such a vector is called a position vector.

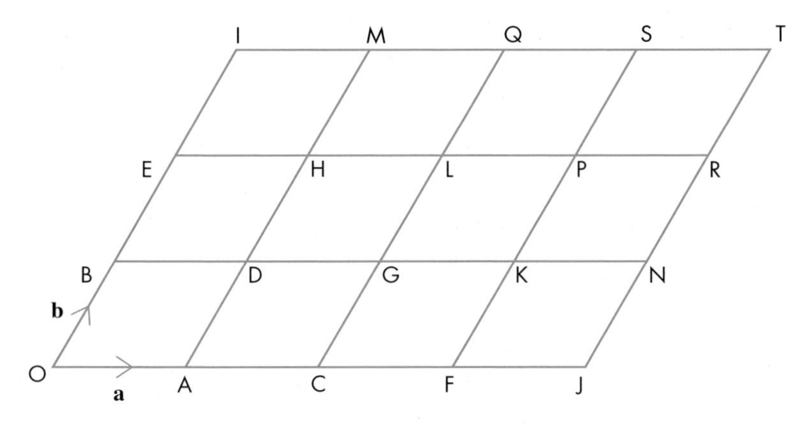

For example, the position vector of K is $\overrightarrow{OK}$ or $\mathbf{k} = 3\mathbf{a} + \mathbf{b}$, the position vector of E is $\overrightarrow{OE}$ or $\mathbf{e} = 2\mathbf{b}$. The vector $\overrightarrow{HT} = 3\mathbf{a} + \mathbf{b}$, the vector $\overrightarrow{PN} = \mathbf{a} - \mathbf{b}$, the vector $\overrightarrow{MK} = 2\mathbf{a} - 2\mathbf{b}$, and the vector $\overrightarrow{TP} = -\mathbf{a} - \mathbf{b}$.

Note $\overrightarrow{OK}$ and $\overrightarrow{HT}$ are called equal vectors because they have exactly the same length and are in the same direction. $\overrightarrow{MK}$ and $\overrightarrow{PN}$ are parallel vectors but $\overrightarrow{MK}$ is twice the magnitude of $\overrightarrow{PN}$.

EXAMPLE 1

a Using the grid above, write down the following vectors in terms of **a** and **b**.

i $\overrightarrow{BH}$ ii $\overrightarrow{HP}$ iii $\overrightarrow{GT}$

iv $\overrightarrow{TI}$ v $\overrightarrow{FH}$ vi $\overrightarrow{BQ}$

b What is the relationship between the following vectors?

i $\overrightarrow{BH}$ and $\overrightarrow{GT}$ ii $\overrightarrow{BQ}$ and $\overrightarrow{GT}$ iii $\overrightarrow{HP}$ and $\overrightarrow{TI}$

c Show that B, H and Q lie on the same straight line.

a i $\mathbf{a} + \mathbf{b}$ ii $2\mathbf{a}$ iii $2\mathbf{a} + 2\mathbf{b}$ iv $-4\mathbf{a}$ v $-2\mathbf{a} + 2\mathbf{b}$ vi $2\mathbf{a} + 2\mathbf{b}$

b i $\overrightarrow{BH}$ and $\overrightarrow{GT}$ are parallel and $\overrightarrow{GT}$ is twice the length of $\overrightarrow{BH}$.

ii $\overrightarrow{BQ}$ and $\overrightarrow{GT}$ are equal.

iii $\overrightarrow{HP}$ and $\overrightarrow{TI}$ are in opposite directions and $\overrightarrow{TI}$ is twice the length of $\overrightarrow{HP}$.

c $\overrightarrow{BH}$ and $\overrightarrow{BQ}$ are parallel and start at the same point B. Therefore, B, H and Q must lie on the same straight line.

EXAMPLE 2

Use a vector diagram to show that $\mathbf{a} + \mathbf{b} = \mathbf{b} + \mathbf{a}$.

Take two independent vectors $\mathbf{a}$ *and* $\mathbf{b}$:

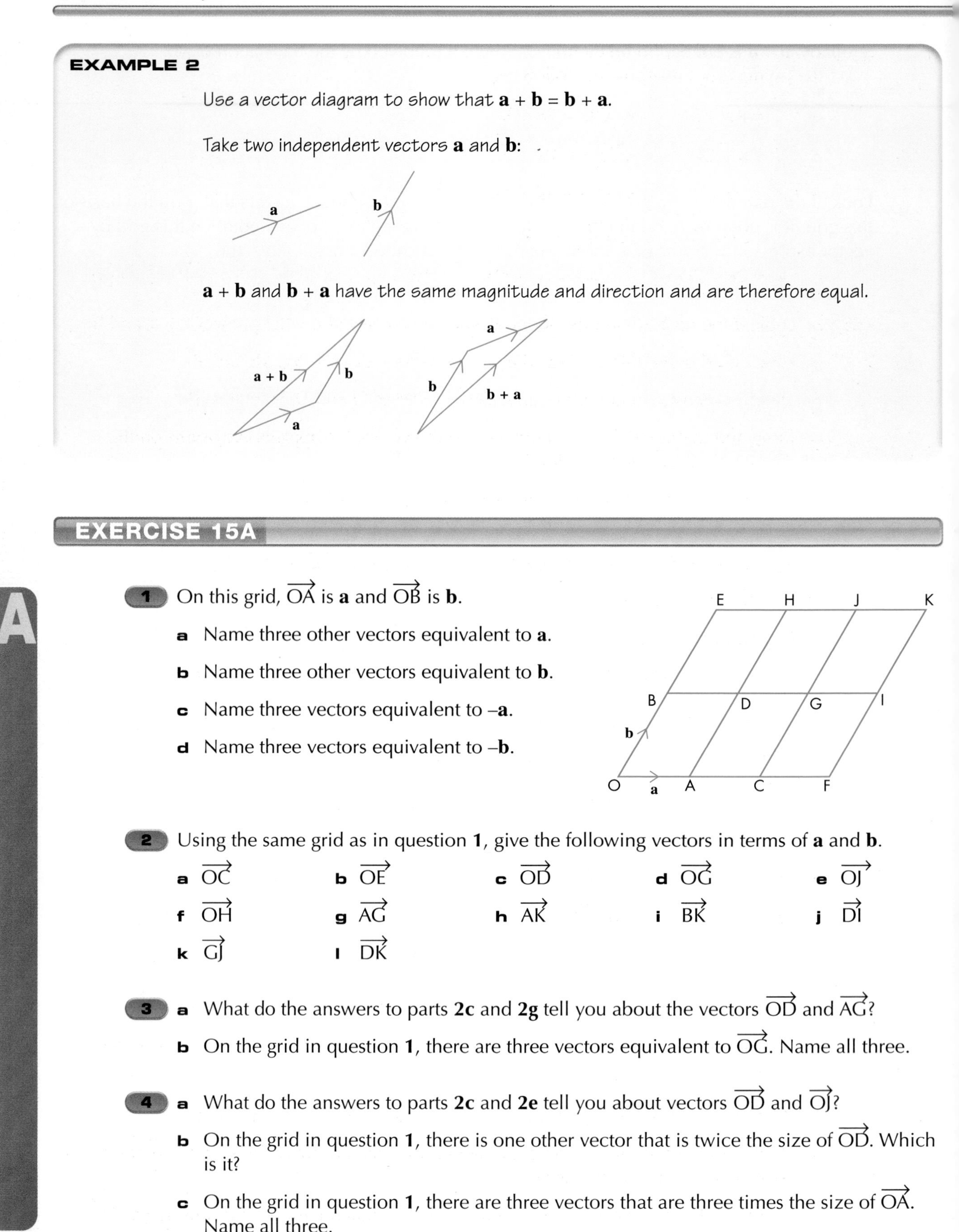

$\mathbf{a} + \mathbf{b}$ *and* $\mathbf{b} + \mathbf{a}$ *have the same magnitude and direction and are therefore equal.*

EXERCISE 15A

1 On this grid, $\overrightarrow{OA}$ is $\mathbf{a}$ and $\overrightarrow{OB}$ is $\mathbf{b}$.

- **a** Name three other vectors equivalent to $\mathbf{a}$.
- **b** Name three other vectors equivalent to $\mathbf{b}$.
- **c** Name three vectors equivalent to $-\mathbf{a}$.
- **d** Name three vectors equivalent to $-\mathbf{b}$.

2 Using the same grid as in question **1**, give the following vectors in terms of $\mathbf{a}$ and $\mathbf{b}$.

a $\overrightarrow{OC}$ **b** $\overrightarrow{OE}$ **c** $\overrightarrow{OD}$ **d** $\overrightarrow{OG}$ **e** $\overrightarrow{OJ}$

f $\overrightarrow{OH}$ **g** $\overrightarrow{AG}$ **h** $\overrightarrow{AK}$ **i** $\overrightarrow{BK}$ **j** $\overrightarrow{DI}$

k $\overrightarrow{GJ}$ **l** $\overrightarrow{DK}$

3 **a** What do the answers to parts **2c** and **2g** tell you about the vectors $\overrightarrow{OD}$ and $\overrightarrow{AG}$?

b On the grid in question **1**, there are three vectors equivalent to $\overrightarrow{OG}$. Name all three.

4 **a** What do the answers to parts **2c** and **2e** tell you about vectors $\overrightarrow{OD}$ and $\overrightarrow{OJ}$?

b On the grid in question **1**, there is one other vector that is twice the size of $\overrightarrow{OD}$. Which is it?

c On the grid in question **1**, there are three vectors that are three times the size of $\overrightarrow{OA}$. Name all three.

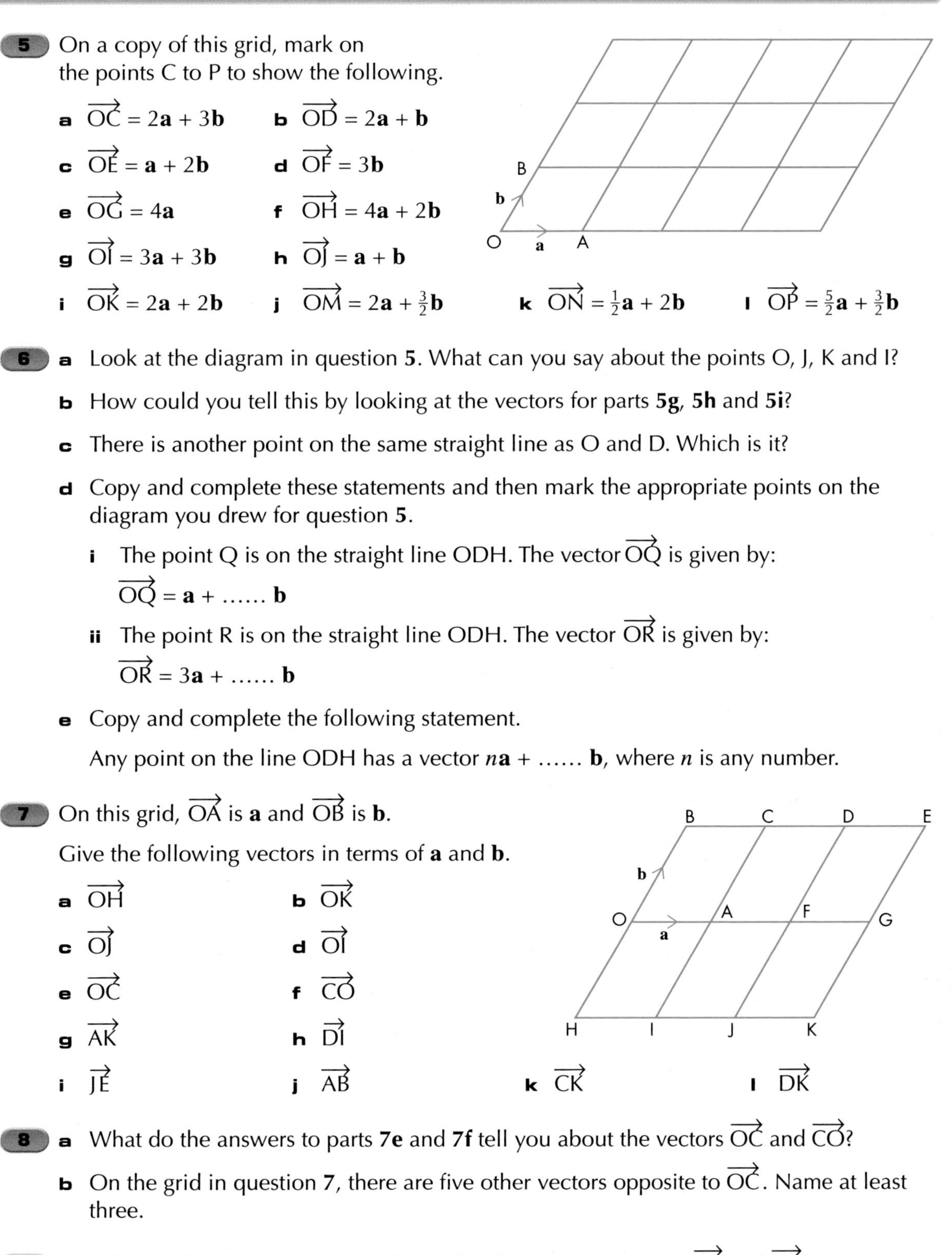

5 On a copy of this grid, mark on the points C to P to show the following.

a $\overrightarrow{OC} = 2\mathbf{a} + 3\mathbf{b}$ **b** $\overrightarrow{OD} = 2\mathbf{a} + \mathbf{b}$

c $\overrightarrow{OE} = \mathbf{a} + 2\mathbf{b}$ **d** $\overrightarrow{OF} = 3\mathbf{b}$

e $\overrightarrow{OG} = 4\mathbf{a}$ **f** $\overrightarrow{OH} = 4\mathbf{a} + 2\mathbf{b}$

g $\overrightarrow{OI} = 3\mathbf{a} + 3\mathbf{b}$ **h** $\overrightarrow{OJ} = \mathbf{a} + \mathbf{b}$

i $\overrightarrow{OK} = 2\mathbf{a} + 2\mathbf{b}$ **j** $\overrightarrow{OM} = 2\mathbf{a} + \frac{3}{2}\mathbf{b}$ **k** $\overrightarrow{ON} = \frac{1}{2}\mathbf{a} + 2\mathbf{b}$ **l** $\overrightarrow{OP} = \frac{5}{2}\mathbf{a} + \frac{3}{2}\mathbf{b}$

6 **a** Look at the diagram in question **5**. What can you say about the points O, J, K and I?

b How could you tell this by looking at the vectors for parts **5g**, **5h** and **5i**?

c There is another point on the same straight line as O and D. Which is it?

d Copy and complete these statements and then mark the appropriate points on the diagram you drew for question **5**.

i The point Q is on the straight line ODH. The vector $\overrightarrow{OQ}$ is given by:

$\overrightarrow{OQ} = \mathbf{a} + \ldots\ldots\ \mathbf{b}$

ii The point R is on the straight line ODH. The vector $\overrightarrow{OR}$ is given by:

$\overrightarrow{OR} = 3\mathbf{a} + \ldots\ldots\ \mathbf{b}$

e Copy and complete the following statement.

Any point on the line ODH has a vector $n\mathbf{a} + \ldots\ldots\ \mathbf{b}$, where n is any number.

7 On this grid, $\overrightarrow{OA}$ is **a** and $\overrightarrow{OB}$ is **b**.

Give the following vectors in terms of **a** and **b**.

a $\overrightarrow{OH}$ **b** $\overrightarrow{OK}$

c $\overrightarrow{OJ}$ **d** $\overrightarrow{OI}$

e $\overrightarrow{OC}$ **f** $\overrightarrow{CO}$

g $\overrightarrow{AK}$ **h** $\overrightarrow{DI}$

i $\overrightarrow{JE}$ **j** $\overrightarrow{AB}$ **k** $\overrightarrow{CK}$ **l** $\overrightarrow{DK}$

8 **a** What do the answers to parts **7e** and **7f** tell you about the vectors $\overrightarrow{OC}$ and $\overrightarrow{CO}$?

b On the grid in question **7**, there are five other vectors opposite to $\overrightarrow{OC}$. Name at least three.

9 **a** What do the answers to parts **7j** and **7k** tell you about vectors $\overrightarrow{AB}$ and $\overrightarrow{CK}$?

b On the grid in question **7**, there are two vectors that are twice the size of $\overrightarrow{AB}$ and in the opposite direction. Which is it?

c On the grid in question **7**, there are three vectors that are three times the size of $\overrightarrow{OA}$ and in the opposite direction. Name all three.

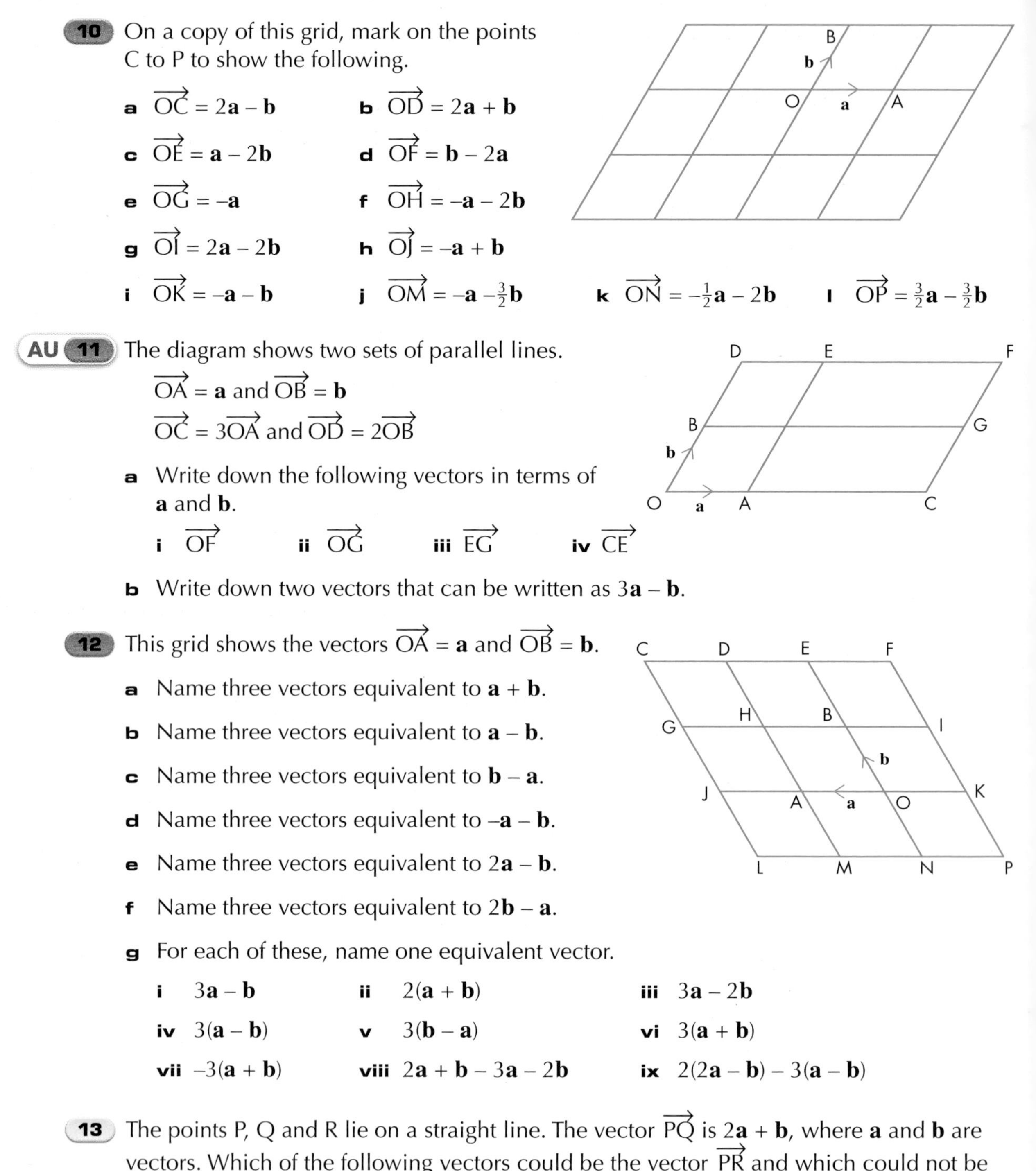

10 On a copy of this grid, mark on the points C to P to show the following.

a $\overrightarrow{OC} = 2\mathbf{a} - \mathbf{b}$ **b** $\overrightarrow{OD} = 2\mathbf{a} + \mathbf{b}$

c $\overrightarrow{OE} = \mathbf{a} - 2\mathbf{b}$ **d** $\overrightarrow{OF} = \mathbf{b} - 2\mathbf{a}$

e $\overrightarrow{OG} = -\mathbf{a}$ **f** $\overrightarrow{OH} = -\mathbf{a} - 2\mathbf{b}$

g $\overrightarrow{OI} = 2\mathbf{a} - 2\mathbf{b}$ **h** $\overrightarrow{OJ} = -\mathbf{a} + \mathbf{b}$

i $\overrightarrow{OK} = -\mathbf{a} - \mathbf{b}$ **j** $\overrightarrow{OM} = -\mathbf{a} - \frac{3}{2}\mathbf{b}$ **k** $\overrightarrow{ON} = -\frac{1}{2}\mathbf{a} - 2\mathbf{b}$ **l** $\overrightarrow{OP} = \frac{3}{2}\mathbf{a} - \frac{3}{2}\mathbf{b}$

AU 11 The diagram shows two sets of parallel lines.

$\overrightarrow{OA} = \mathbf{a}$ and $\overrightarrow{OB} = \mathbf{b}$

$\overrightarrow{OC} = 3\overrightarrow{OA}$ and $\overrightarrow{OD} = 2\overrightarrow{OB}$

a Write down the following vectors in terms of **a** and **b**.

i $\overrightarrow{OF}$ **ii** $\overrightarrow{OG}$ **iii** $\overrightarrow{EG}$ **iv** $\overrightarrow{CE}$

b Write down two vectors that can be written as $3\mathbf{a} - \mathbf{b}$.

12 This grid shows the vectors $\overrightarrow{OA} = \mathbf{a}$ and $\overrightarrow{OB} = \mathbf{b}$.

a Name three vectors equivalent to $\mathbf{a} + \mathbf{b}$.

b Name three vectors equivalent to $\mathbf{a} - \mathbf{b}$.

c Name three vectors equivalent to $\mathbf{b} - \mathbf{a}$.

d Name three vectors equivalent to $-\mathbf{a} - \mathbf{b}$.

e Name three vectors equivalent to $2\mathbf{a} - \mathbf{b}$.

f Name three vectors equivalent to $2\mathbf{b} - \mathbf{a}$.

g For each of these, name one equivalent vector.

i $3\mathbf{a} - \mathbf{b}$ **ii** $2(\mathbf{a} + \mathbf{b})$ **iii** $3\mathbf{a} - 2\mathbf{b}$

iv $3(\mathbf{a} - \mathbf{b})$ **v** $3(\mathbf{b} - \mathbf{a})$ **vi** $3(\mathbf{a} + \mathbf{b})$

vii $-3(\mathbf{a} + \mathbf{b})$ **viii** $2\mathbf{a} + \mathbf{b} - 3\mathbf{a} - 2\mathbf{b}$ **ix** $2(2\mathbf{a} - \mathbf{b}) - 3(\mathbf{a} - \mathbf{b})$

13 The points P, Q and R lie on a straight line. The vector $\overrightarrow{PQ}$ is $2\mathbf{a} + \mathbf{b}$, where **a** and **b** are vectors. Which of the following vectors could be the vector $\overrightarrow{PR}$ and which could not be the vector $\overrightarrow{PR}$ (two of each).

a $2\mathbf{a} + 2\mathbf{b}$ **b** $4\mathbf{a} + 2\mathbf{b}$ **c** $2\mathbf{a} - \mathbf{b}$ **d** $-6\mathbf{a} - 3\mathbf{b}$

14 The points P, Q and R lie on a straight line. The vector $\overrightarrow{PQ}$ is $3\mathbf{a} - \mathbf{b}$, where **a** and **b** are vectors.

a Write down any other vector that could represent $\overrightarrow{PR}$.

b How can you tell from the vector $\overrightarrow{PS}$ that S lies on the same straight line as P, Q and R?

15 Use a vector diagram to prove that $\mathbf{a} + (\mathbf{b} + \mathbf{c}) = (\mathbf{a} + \mathbf{b}) + \mathbf{c}$.

AU 16 OABC is a quadrilateral.

P, Q, R and S are the midpoints of OA, AB, BC and OC respectively.

$\overrightarrow{OA} = 2\mathbf{a}$, $\overrightarrow{OB} = 2\mathbf{b}$ and $\overrightarrow{OC} = 2\mathbf{c}$

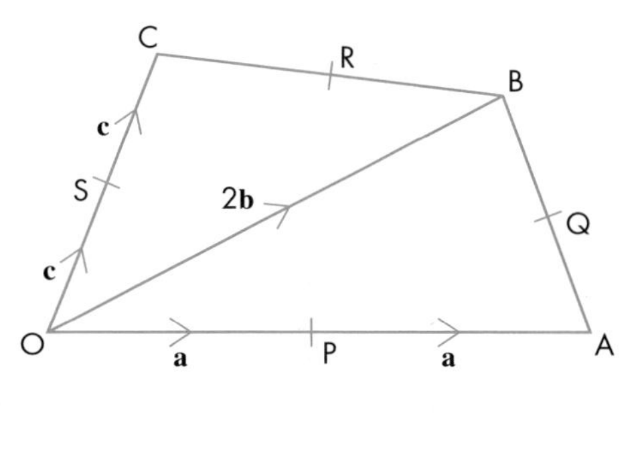

a Find the following vectors in terms of **a**, **b** and **c**.

Give your answers in their simplest form.

i $\overrightarrow{AB}$ **ii** $\overrightarrow{SP}$ **iii** $\overrightarrow{BC}$ **iv** $\overrightarrow{PR}$

b Use vectors to prove that PQRS is a parallelogram.

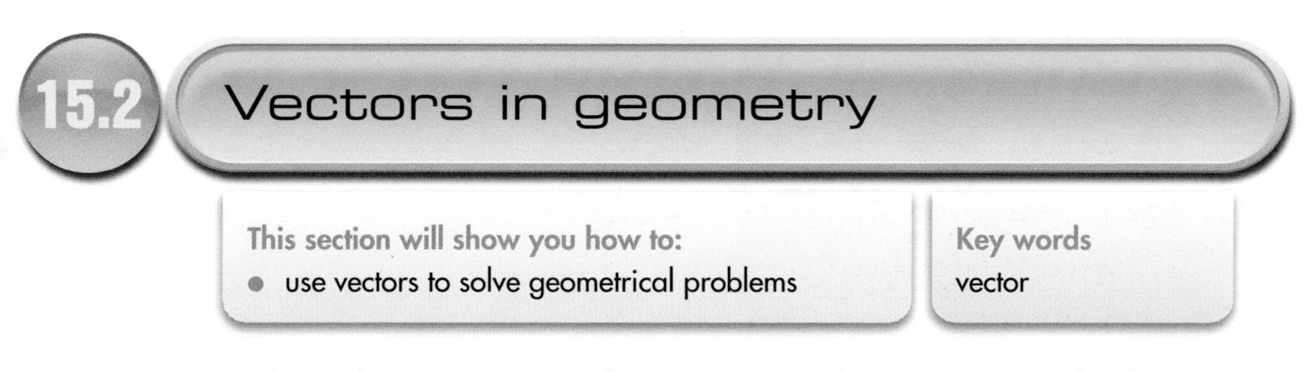

15.2 Vectors in geometry

This section will show you how to:

- use vectors to solve geometrical problems

Key words

vector

Vectors can be used to prove many results in geometry, as the following examples show.

EXAMPLE 3

In the diagram, $\overrightarrow{OA} = \mathbf{a}$, $\overrightarrow{OB} = \mathbf{b}$, and $\overrightarrow{BC} = 1.5\mathbf{a}$. M is the midpoint of BC, N is the midpoint of AC and P is the midpoint of OB.

a Find these vectors in terms of **a** and **b**.

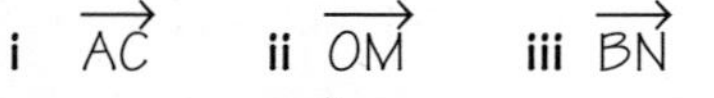

i $\overrightarrow{AC}$ ii $\overrightarrow{OM}$ iii $\overrightarrow{BN}$

b Prove that $\overrightarrow{PN}$ is parallel to $\overrightarrow{OA}$.

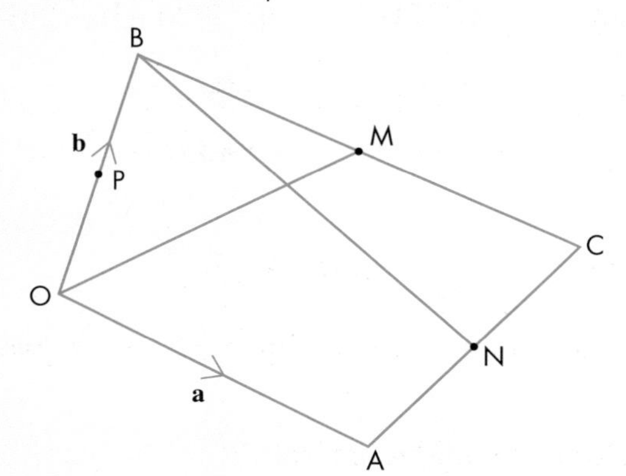

EXAMPLE 3 (continued)

a **i** You have to get from A to C in terms of vectors that you know.

$$\overrightarrow{AC} = \overrightarrow{AO} + \overrightarrow{OB} + \overrightarrow{BC}$$

Now $\overrightarrow{AO} = -\overrightarrow{OA}$, so you can write,

$$\overrightarrow{AC} = -\mathbf{a} + \mathbf{b} + \tfrac{3}{2}\mathbf{a}$$

$$= \tfrac{1}{2}\mathbf{a} + \mathbf{b}$$

Note that the letters 'connect up' as we go from A to C, and that the negative of a vector represented by any pair of letters is formed by reversing the letters.

ii In the same way:

$$\overrightarrow{OM} = \overrightarrow{OB} + \overrightarrow{BM} = \overrightarrow{OB} + \tfrac{1}{2}\overrightarrow{BC}$$

$$= \mathbf{b} + \tfrac{1}{2}(\tfrac{3}{2}\mathbf{a})$$

$$\overrightarrow{OM} = \tfrac{3}{4}\mathbf{a} + \mathbf{b}$$

iii $\overrightarrow{BN} = \overrightarrow{BC} + \overrightarrow{CN} = \overrightarrow{BC} - \tfrac{1}{2}\overrightarrow{AC}$

$$= \tfrac{3}{2}\mathbf{a} - \tfrac{1}{2}(\tfrac{1}{2}\mathbf{a} + \mathbf{b})$$

$$= \tfrac{3}{2}\mathbf{a} - \tfrac{1}{4}\mathbf{a} - \tfrac{1}{2}\mathbf{b}$$

$$= \tfrac{5}{4}\mathbf{a} - \tfrac{1}{2}\mathbf{b}$$

Note that if you did this as $\overrightarrow{BN} = \overrightarrow{BO} + \overrightarrow{OA} + \overrightarrow{AN}$, you would get the same result.

b $\overrightarrow{PN} = \overrightarrow{PO} + \overrightarrow{OA} + \overrightarrow{AN}$

$$= \tfrac{1}{2}(-\mathbf{b}) + \mathbf{a} + \tfrac{1}{2}(\tfrac{1}{2}\mathbf{a} + \mathbf{b})$$

$$= -\tfrac{1}{2}\mathbf{b} + \mathbf{a} + \tfrac{1}{4}\mathbf{a} + \tfrac{1}{2}\mathbf{b}$$

$$= \tfrac{5}{4}\mathbf{a}$$

$\overrightarrow{PN}$ is a multiple of $\mathbf{a}$ only, so must be parallel to $\overrightarrow{OA}$.

EXAMPLE 4

OACB is a parallelogram. $\overrightarrow{OA}$ is represented by the vector $\mathbf{a}$. $\overrightarrow{OB}$ is represented by the vector $\mathbf{b}$. P is a point $\tfrac{2}{3}$ the distance from O to C, and M is the midpoint of AC. Show that B, P and M lie on the same straight line.

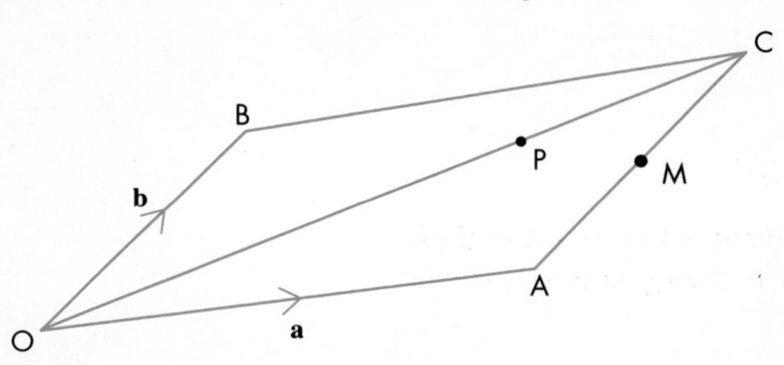

$\overrightarrow{OC} = \overrightarrow{OA} + \overrightarrow{AC} = \mathbf{a} + \mathbf{b}$

$\overrightarrow{OP} = \frac{2}{3}\overrightarrow{OC} = \frac{2}{3}\mathbf{a} + \frac{2}{3}\mathbf{b}$

$\overrightarrow{OM} = \overrightarrow{OA} + \overrightarrow{AM} = \overrightarrow{OA} + \frac{1}{2}\overrightarrow{AC} = \mathbf{a} + \frac{1}{2}\mathbf{b}$

$\overrightarrow{BP} = \overrightarrow{BO} + \overrightarrow{OP} = -\mathbf{b} + \frac{2}{3}\mathbf{a} + \frac{2}{3}\mathbf{b} = \frac{2}{3}\mathbf{a} - \frac{1}{3}\mathbf{b} = \frac{1}{3}(2\mathbf{a} - \mathbf{b})$

$\overrightarrow{BM} = \overrightarrow{BO} + \overrightarrow{OM} = -\mathbf{b} + \mathbf{a} + \frac{1}{2}\mathbf{b} = \mathbf{a} - \frac{1}{2}\mathbf{b} = \frac{1}{2}(2\mathbf{a} - \mathbf{b})$

Therefore, $\overrightarrow{BM}$ is a multiple of $\overrightarrow{BP}$ ($\overrightarrow{BM} = \frac{3}{2}\overrightarrow{BP}$).

Therefore, $\overrightarrow{BP}$ and $\overrightarrow{BM}$ are parallel and as they have a common point, B, they must lie on the same straight line.

EXERCISE 15B

A*

1 The diagram shows the vectors $\overrightarrow{OA} = \mathbf{a}$ and $\overrightarrow{OB} = \mathbf{b}$. M is the midpoint of AB.

a **i** Work out the vector $\overrightarrow{AB}$.

ii Work out the vector $\overrightarrow{AM}$.

iii Explain why $\overrightarrow{OM} = \overrightarrow{OA} + \overrightarrow{AM}$.

iv Using your answers to parts **ii** and **iii**, work out $\overrightarrow{OM}$ in terms of **a** and **b**.

b **i** Work out the vector $\overrightarrow{BA}$.

ii Work out the vector $\overrightarrow{BM}$.

iii Explain why $\overrightarrow{OM} = \overrightarrow{OB} + \overrightarrow{BM}$.

iv Using your answers to parts **ii** and **iii**, work out $\overrightarrow{OM}$ in terms of **a** and **b**.

c Copy the diagram and show on it the vector $\overrightarrow{OC}$ which is equal to **a** + **b**.

d Describe in geometrical terms the position of M in relation to O, A, B and C.

2 The diagram shows the vectors $\overrightarrow{OA} = \mathbf{a}$ and $\overrightarrow{OC} = -\mathbf{b}$. N is the midpoint of AC.

a **i** Work out the vector $\overrightarrow{AC}$.

ii Work out the vector $\overrightarrow{AN}$.

iii Explain why $\overrightarrow{ON} = \overrightarrow{OA} + \overrightarrow{AN}$.

iv Using your answers to parts **ii** and **iii**, work out $\overrightarrow{ON}$ in terms of **a** and **b**.

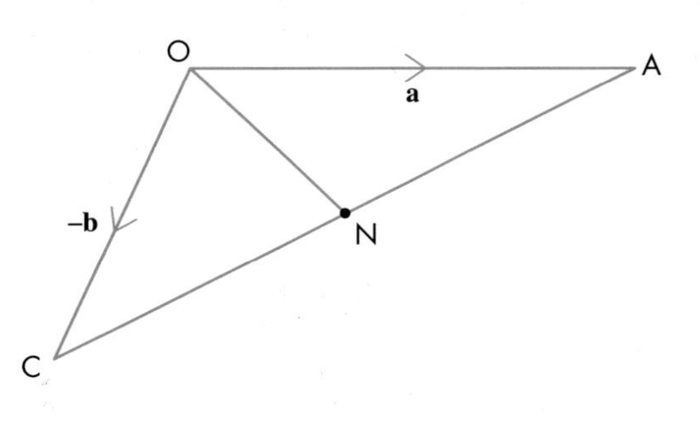

A*

b **i** Work out the vector $\overrightarrow{CA}$.

ii Work out the vector $\overrightarrow{CN}$.

iii Explain why $\overrightarrow{ON} = \overrightarrow{OC} + \overrightarrow{CN}$.

iv Using your answers to parts **ii** and **iii**, work out $\overrightarrow{ON}$ in terms of **a** and **b**.

c Copy the diagram above and show on it the vector $\overrightarrow{OD}$ which is equal to **a** – **b**.

d Describe in geometrical terms the position of N in relation to O, A, C and D.

3 The diagram shows the vectors $\overrightarrow{OA} = \mathbf{a}$ and $\overrightarrow{OB} = \mathbf{b}$.
The point C divides the line AB in the ratio 1:2.

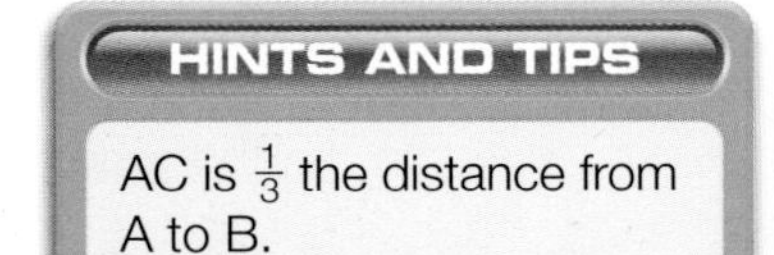

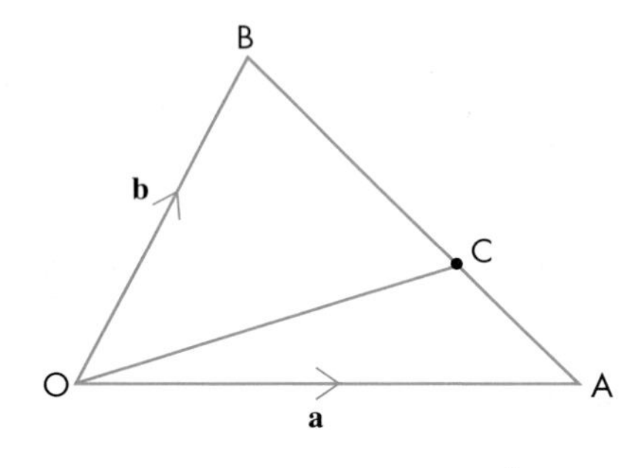

a **i** Work out the vector $\overrightarrow{AB}$. **ii** Work out the vector $\overrightarrow{AC}$.

iii Work out the vector $\overrightarrow{OC}$ in terms of **a** and **b**.

b If C now divides the line AB in the ratio 1:3, write down the vector that represents $\overrightarrow{OC}$.

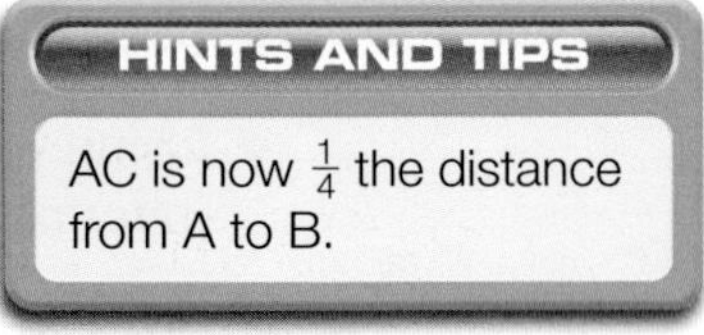

4 The diagram shows the vectors $\overrightarrow{OA} = \mathbf{a}$ and $\overrightarrow{OB} = \mathbf{b}$.

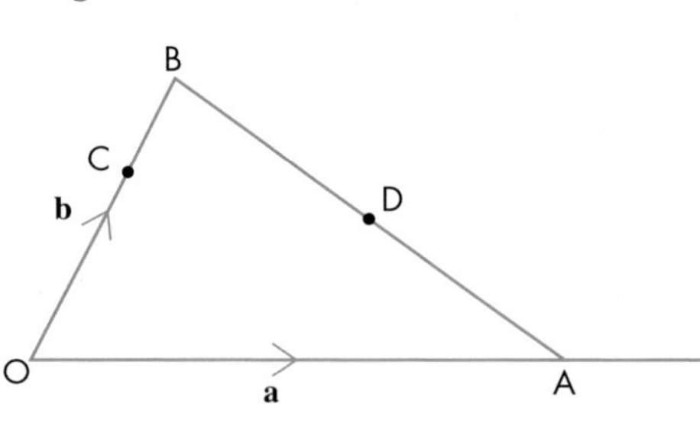

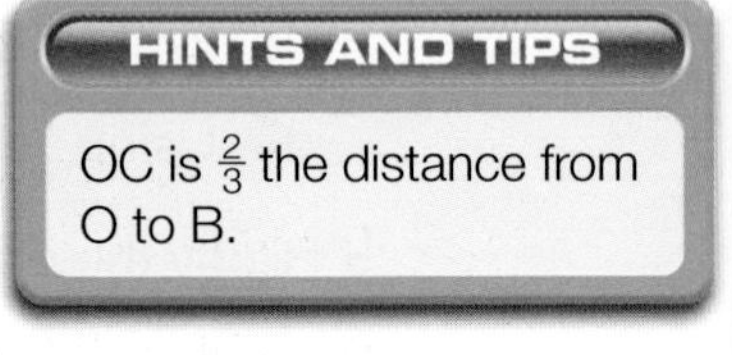

The point C divides OB in the ratio 2:1. The point E is such that $\overrightarrow{OE} = 2\overrightarrow{OA}$. D is the midpoint of AB.

a Write down (or work out) these vectors in terms of **a** and **b**.

i $\overrightarrow{OC}$ **ii** $\overrightarrow{OD}$ **iii** $\overrightarrow{CO}$

b The vector $\overrightarrow{CD}$ can be written as $\overrightarrow{CD} = \overrightarrow{CO} + \overrightarrow{OD}$. Use this fact to work out $\overrightarrow{CD}$ in terms of **a** and **b**.

c Write down a similar rule to that in part **b** for the vector $\overrightarrow{DE}$. Use this rule to work out $\overrightarrow{DE}$ in terms of **a** and **b**.

d Explain why C, D and E lie on the same straight line.

5 ABCDEF is a regular hexagon. $\overrightarrow{AB}$ is represented by the vector $\mathbf{a}$ and $\overrightarrow{BC}$ by the vector $\mathbf{b}$.

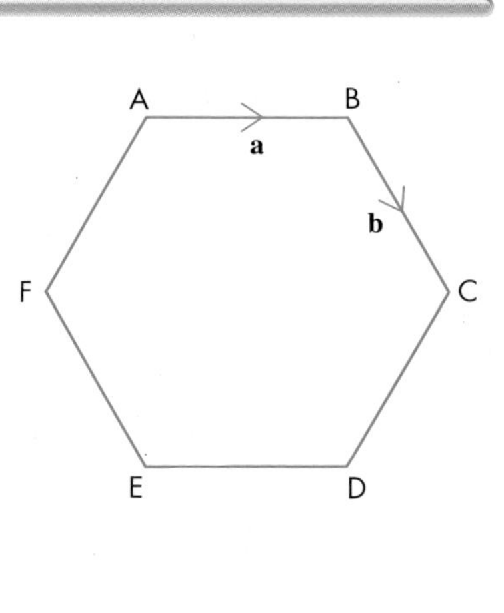

a By means of a diagram, or otherwise, explain why $\overrightarrow{CD} = \mathbf{b} - \mathbf{a}$.

b Express these vectors in terms of $\mathbf{a}$ and $\mathbf{b}$.

i $\overrightarrow{DE}$ **ii** $\overrightarrow{EF}$ **iii** $\overrightarrow{FA}$

c Work out the answer to:

$\overrightarrow{AB} + \overrightarrow{BC} + \overrightarrow{CD} + \overrightarrow{DE} + \overrightarrow{EF} + \overrightarrow{FA}$

Explain your answer.

d Express these vectors in terms of $\mathbf{a}$ and $\mathbf{b}$.

i $\overrightarrow{AD}$ **ii** $\overrightarrow{BE}$ **iii** $\overrightarrow{CF}$ **iv** $\overrightarrow{AE}$ **v** $\overrightarrow{DF}$

6 ABCDEFGH is a regular octagon. $\overrightarrow{AB}$ is represented by the vector $\mathbf{a}$, and $\overrightarrow{BC}$ by the vector $\mathbf{b}$.

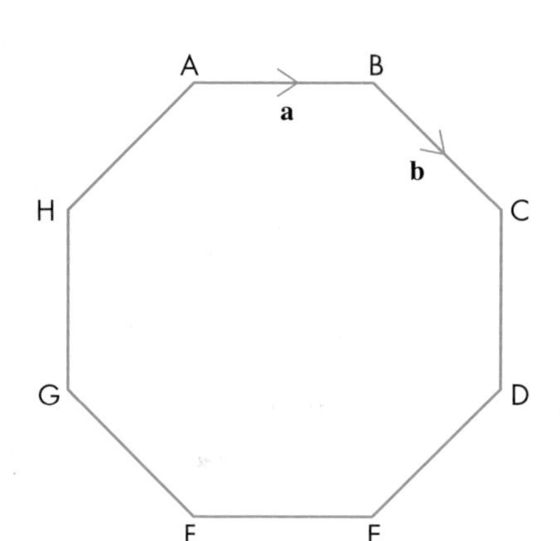

a By means of a diagram, or otherwise, explain why $\overrightarrow{CD} = \sqrt{2}\mathbf{b} - \mathbf{a}$.

b By means of a diagram, or otherwise, explain why $\overrightarrow{DE} = \mathbf{b} - \sqrt{2}\mathbf{a}$.

c Express the following vectors in terms of $\mathbf{a}$ and $\mathbf{b}$.

i $\overrightarrow{EF}$ **ii** $\overrightarrow{FG}$ **iii** $\overrightarrow{GH}$ **iv** $\overrightarrow{HA}$

v $\overrightarrow{HC}$ **vi** $\overrightarrow{AD}$ **vii** $\overrightarrow{BE}$ **viii** $\overrightarrow{BF}$

7 In the quadrilateral OABC, M, N, P and Q are the midpoints of the sides as shown. $\overrightarrow{OA}$ is represented by the vector $\mathbf{a}$, and $\overrightarrow{OC}$ by the vector $\mathbf{c}$. The diagonal $\overrightarrow{OB}$ is represented by the vector $\mathbf{b}$.

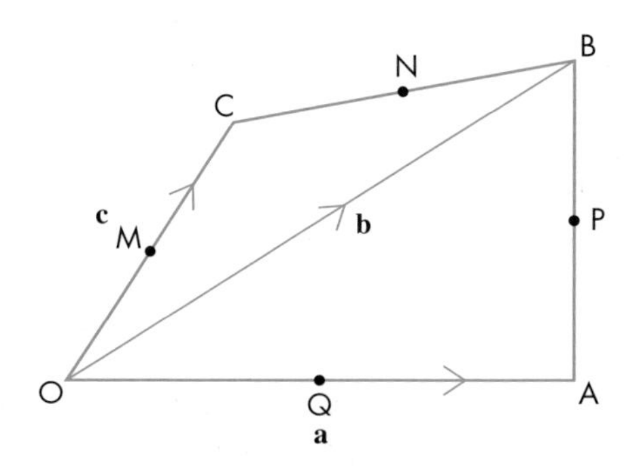

a Express these vectors in terms of $\mathbf{a}$, $\mathbf{b}$ and $\mathbf{c}$.

i $\overrightarrow{AB}$ **ii** $\overrightarrow{AP}$ **iii** $\overrightarrow{OP}$

Give your answers as simply as possible.

b **i** Express the vector $\overrightarrow{ON}$ in terms of $\mathbf{b}$ and $\mathbf{c}$.

ii Hence express the vector $\overrightarrow{PN}$ in terms of $\mathbf{a}$ and $\mathbf{c}$.

c **i** Express the vector $\overrightarrow{QM}$ in terms of $\mathbf{a}$ and $\mathbf{c}$.

ii What relationship is there between $\overrightarrow{PN}$ and $\overrightarrow{QM}$?

iii What sort of quadrilateral is PNMQ?

d Prove that $\overrightarrow{AC} = 2\overrightarrow{QM}$.

8 L, M, N, P, Q, R are the midpoints of the line segments, as shown.

$\overrightarrow{OA} = \mathbf{a}$, $\overrightarrow{OB} = \mathbf{b}$ and $\overrightarrow{OC} = \mathbf{c}$

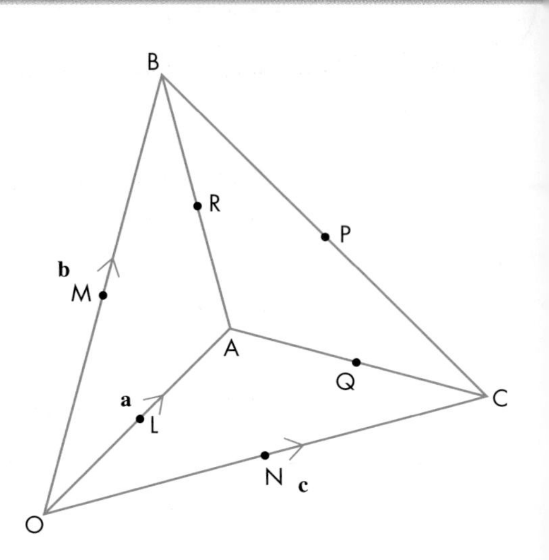

a Express these vectors in terms of **a** and **c**.

i $\overrightarrow{OL}$ **ii** $\overrightarrow{AC}$

iii $\overrightarrow{OQ}$ **iv** $\overrightarrow{LQ}$

b Express these vectors in terms of **a** and **b**.

i $\overrightarrow{LM}$ **ii** $\overrightarrow{QP}$

c Prove that the quadrilateral LMPQ is a parallelogram.

d Find two other sets of four points that form parallelograms.

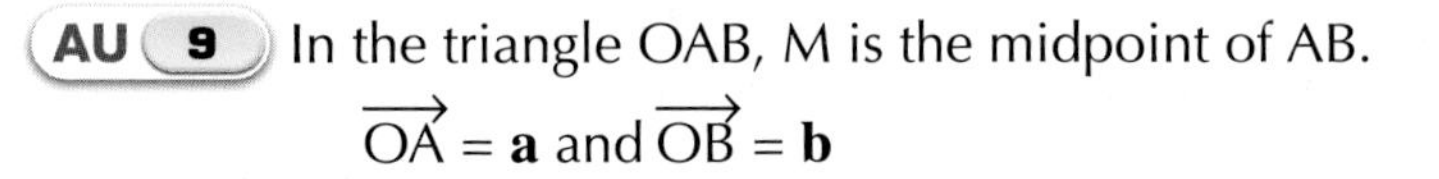

AU 9 In the triangle OAB, M is the midpoint of AB.

$\overrightarrow{OA} = \mathbf{a}$ and $\overrightarrow{OB} = \mathbf{b}$

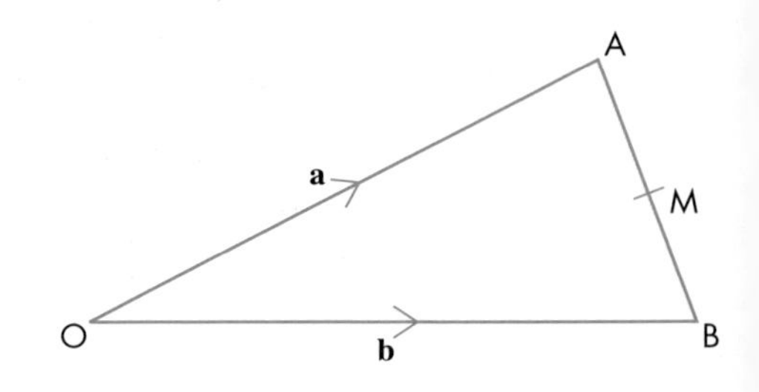

a Find $\overrightarrow{AM}$ in terms of **a** and **b**.

Give your answer in its simplest form.

b $\overrightarrow{OC} = \mathbf{a} + \mathbf{b}$

The length of OA is equal to the length of OB.

i Write down the name of the shape OACB.

ii Write down one fact about the points O, M and C.

Give a reason for your answer.

AU 10 ABCD is a trapezium with AB parallel to DC.

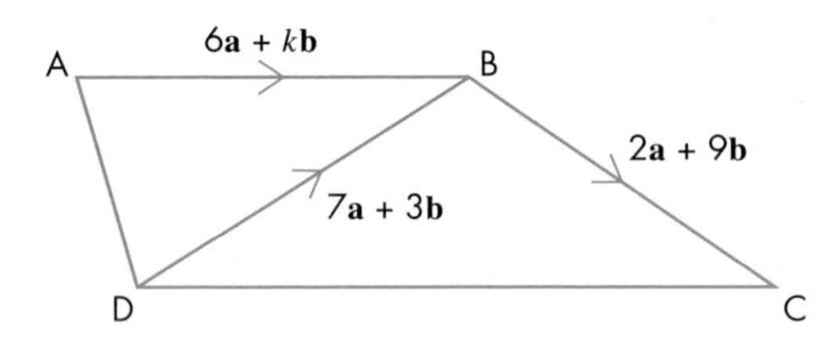

$\overrightarrow{AB} = 6\mathbf{a} + k\mathbf{b}$, $\overrightarrow{BC} = 2\mathbf{a} + 9\mathbf{b}$ and $\overrightarrow{DB} = 7\mathbf{a} + 3\mathbf{b}$, where k is a number.

Work out the value of k.

15.3 Geometric proof

This section will show you how to:
- understand the difference between a proof and a demonstration

Key words
demonstration
proof
prove

You should already know these.

- The angle sum of the interior angles in a triangle (180°)
- The circle theorems
- Pythagoras' theorem

Can you **prove** them?

For a mathematical **proof**, you must proceed in logical steps, establishing a series of mathematical statements by using facts that are already known to be true.

Below are three standard proofs: *the sum of the interior angles of a triangle is 180°*, *Pythagoras' theorem* and *congruency*. Read through them, following the arguments carefully. Make sure you understand each step in the process.

Proof that the sum of the interior angles of a triangle is 180°

One of your earlier activities in geometry may have been to draw a triangle, to cut off its corners and to stick them down to *show that* they make a straight line.

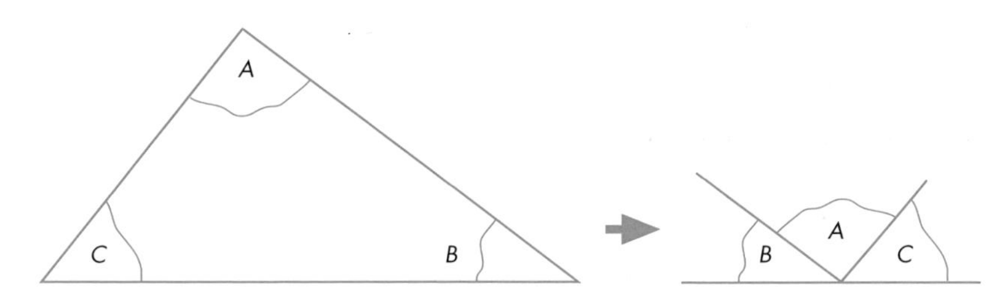

Does this prove that the interior angles make 180° or were you just lucky and picked a triangle that worked? Was the fact that everyone else in the class managed to pick a triangle that worked also a lucky coincidence?

Of course not! But this was a **demonstration**, not a proof. You would have to show that this method worked for *all* possible triangles (there is an infinite number!) to say that you have proved this result.

Your proof must establish that the result is true for *all* triangles.

Look at the following proof.

Start with triangle ABC with angles α, β and γ (figure **i**).

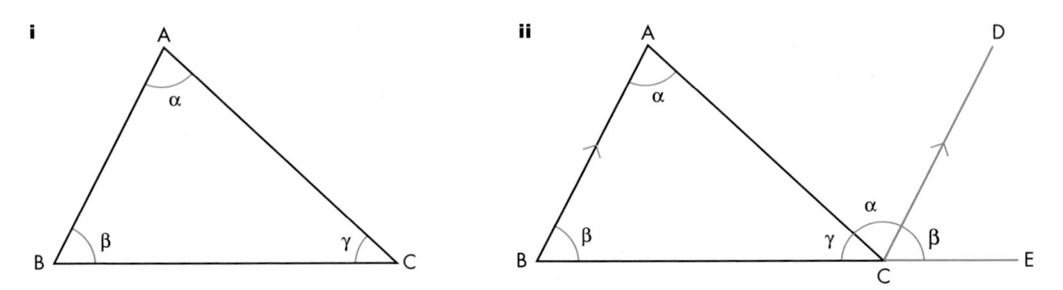

On figure **i** draw a line CD parallel to side AB and extend BC to E, to give figure **ii**.

Since AB is parallel to CD:

$\angle ACD = \angle BAC = \alpha$ (alternate angles) $\angle DCE = \angle ABC = \beta$ (corresponding angles)

BCE is a straight line, so $\gamma + \alpha + \beta = 180°$. Therefore the interior angles of a triangle = 180°.

This proof assumes that alternate angles are equal and that corresponding angles are equal. Strictly speaking, we should prove these results, but we have to accept certain results as true. These are based on Euclid's axioms from which all geometric proofs are derived.

Proof of Pythagoras' theorem

Draw a square of side c inside a square of side $(a + b)$, as shown.

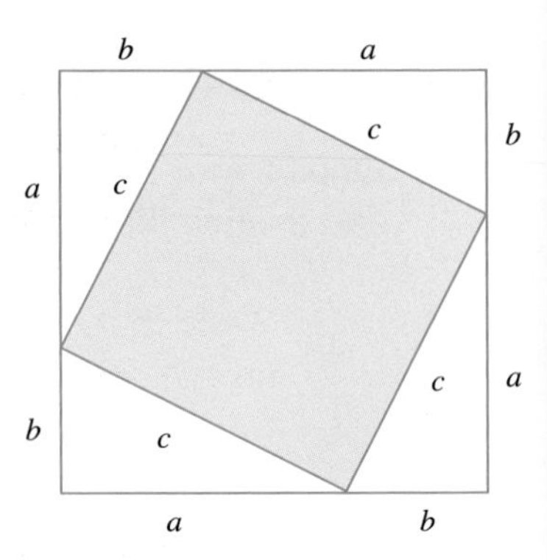

The area of the exterior square is $(a + b)^2 = a^2 + 2ab + b^2$.

The area of each small triangle around the shaded square is $\frac{1}{2}ab$.

The total area of all four triangles is $4 \times \frac{1}{2}ab = 2ab$.

Subtracting the total area of the four triangles from the area of the large square gives the area of the shaded square:

$$a^2 + 2ab + b^2 - 2ab = a^2 + b^2$$

But the area of the shaded square is c^2, so

$$c^2 = a^2 + b^2$$

which is Pythagoras' theorem.

Congruency proof

There are four conditions to prove congruency. These are commonly known as SSS (three sides the same), SAS (two sides and the included angle the same), ASA (or AAS) (two angles and one side the same) and RHS (right-angled triangle, hypotenuse, and one short side the same). **Note:** AAA (three angles the same) is not a condition for congruency.

When you prove a result, you must explain or justify every statement or line. Proofs have to be rigorous and logical.

EXAMPLE 5

ABCD is a parallelogram. X is the point where the diagonals meet.

Prove that triangles *AXB* and *CXD* are congruent.

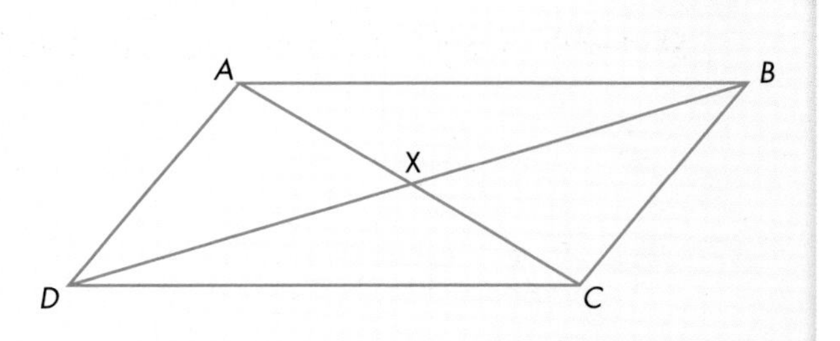

$\angle BAX = \angle DCX$ (alternate angles)

$\angle ABX = \angle CDX$ (alternate angles)

$AB = CD$ (opposite sides in a parallelogram)

Hence ΔAXB is congruent to ΔCXD (ASA).

Note that you could have used $\angle AXB = \angle CXD$ (vertically opposite angles) as the second line but whichever approach is used you *must* give a reason for each statement.

EXERCISE 15C

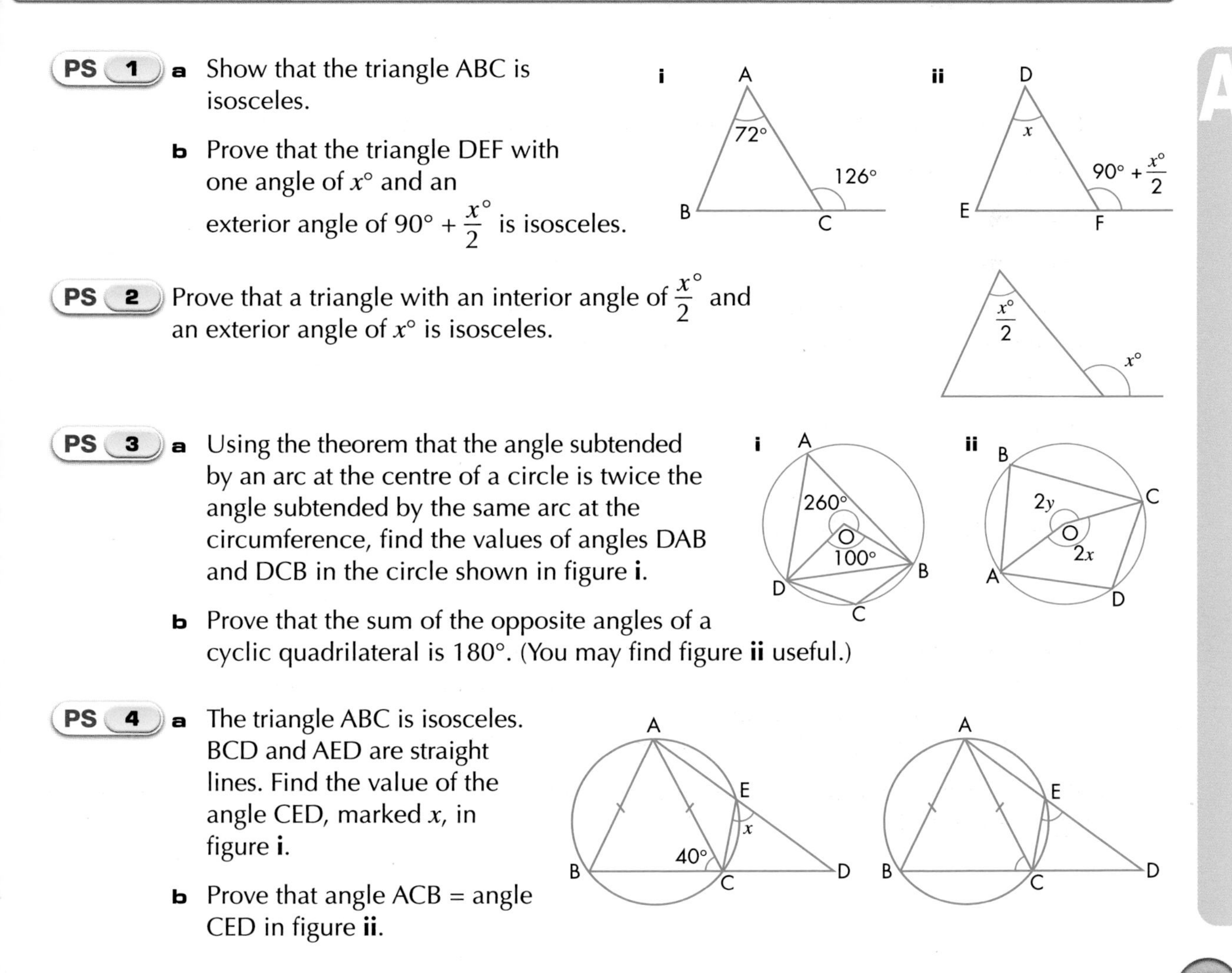

PS 1 **a** Show that the triangle ABC is isosceles.

b Prove that the triangle DEF with one angle of $x°$ and an exterior angle of $90° + \frac{x°}{2}$ is isosceles.

PS 2 Prove that a triangle with an interior angle of $\frac{x°}{2}$ and an exterior angle of $x°$ is isosceles.

PS 3 **a** Using the theorem that the angle subtended by an arc at the centre of a circle is twice the angle subtended by the same arc at the circumference, find the values of angles DAB and DCB in the circle shown in figure **i**.

b Prove that the sum of the opposite angles of a cyclic quadrilateral is 180°. (You may find figure **ii** useful.)

PS 4 **a** The triangle ABC is isosceles. BCD and AED are straight lines. Find the value of the angle CED, marked x, in figure **i**.

b Prove that angle ACB = angle CED in figure **ii**.

A*

PS 5 PQRS is a parallelogram. Prove that triangles PQS and RQS are congruent.

PS 6 OB is a radius of a circle, centre O. C is the point where the perpendicular bisector of OB meets the circumference. Prove that triangle OBC is equilateral.

PS 7 The grid is made up of identical parallelograms.
In the grid, $\overrightarrow{OA} = \mathbf{a}$ and $\overrightarrow{OB} = \mathbf{b}$.
Prove that AB is parallel to EF.

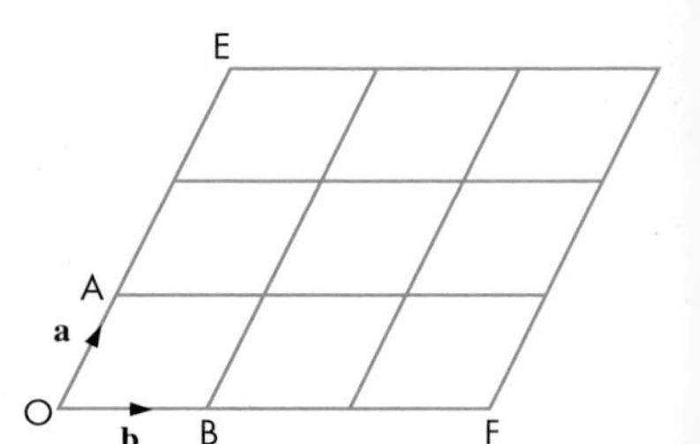

PS 8 **a** Prove the alternate segment theorem.

b Two circles touch internally at T. The common tangent at T is drawn. Two lines TAB and TXY are drawn from T. Prove that AX is parallel to BY.

PS 9 Two circles touch externally at T. A line ATB is drawn through T. The common tangent at T and the tangents at A and B meet at P and Q. Prove that PB is parallel to AQ.

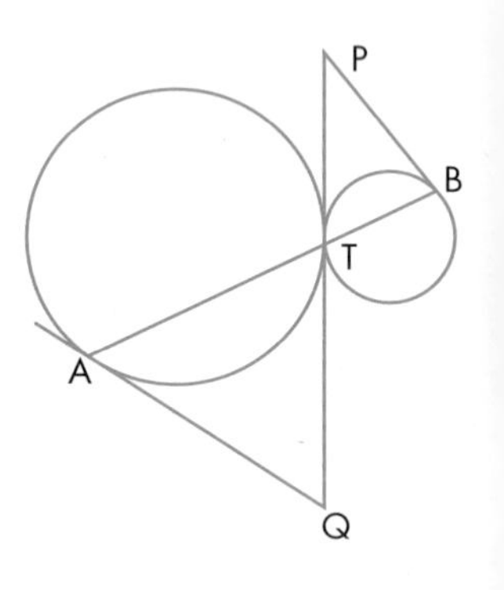

10 **a** and **b** are vectors.

$\overrightarrow{XY} = \mathbf{a} + \mathbf{b}$ $\quad\quad$ $\overrightarrow{YZ} = 2\mathbf{a} + \mathbf{b}$ $\quad\quad$ $\overrightarrow{ZW} = \mathbf{a} + 2\mathbf{b}$

a Show that $\overrightarrow{YW}$ is parallel to $\overrightarrow{XY}$.

b Write down the ratio YW : XY.

c What do your answers to **a** and **b** tell you about the points X, Y and W?

d O is the origin.
A, B and C are three points such that:

$\overrightarrow{OA} = \begin{pmatrix} 6 \\ 2 \end{pmatrix}$ $\quad\quad$ $\overrightarrow{OB} = \begin{pmatrix} 1 \\ 1 \end{pmatrix}$ $\quad\quad$ $\overrightarrow{OC} = \begin{pmatrix} 2 \\ -4 \end{pmatrix}$

Prove that angle ABC is a right angle.

GRADE BOOSTER

A You can solve problems, using addition and subtraction of vectors

A* You can solve complex geometrical problems, using vectors

A* You can use proof in geometrical problems

What you should know now

- How to add and subtract vectors
- How to apply vector methods to solve geometrical problems

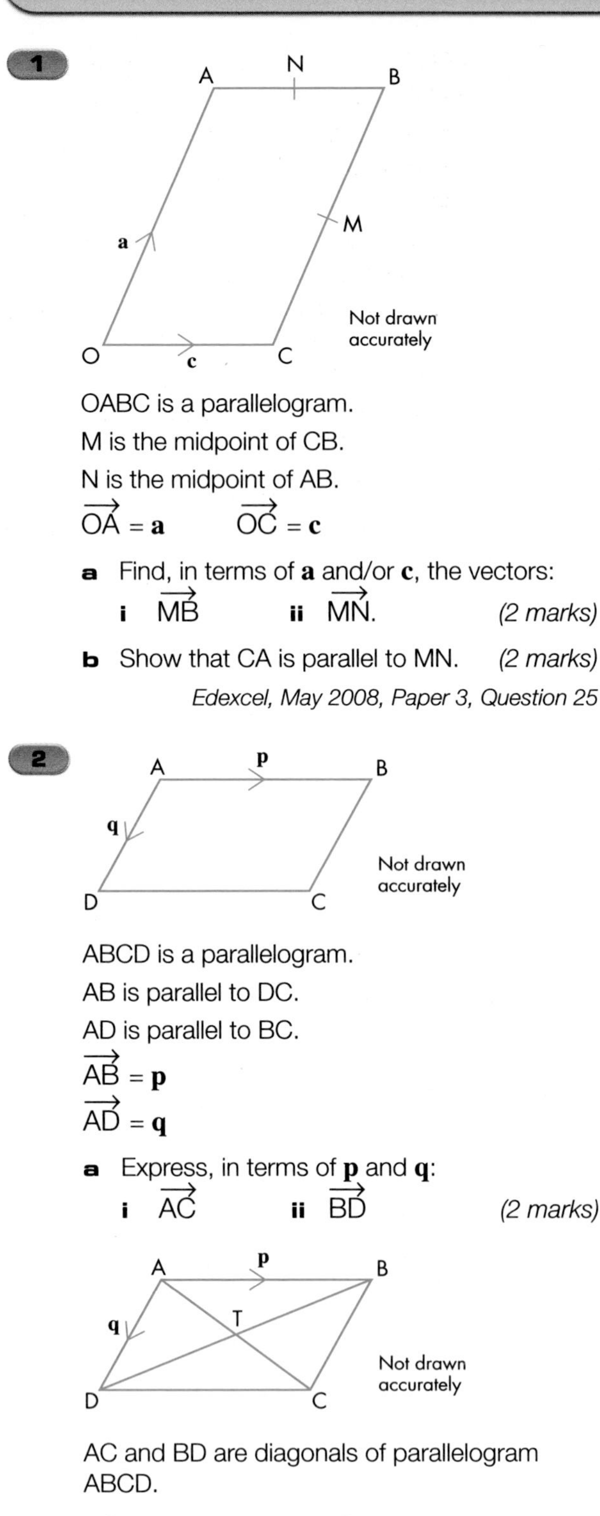

1

Not drawn accurately

OABC is a parallelogram.

M is the midpoint of CB.

N is the midpoint of AB.

$\overrightarrow{OA} = \mathbf{a}$ $\quad \overrightarrow{OC} = \mathbf{c}$

a Find, in terms of $\mathbf{a}$ and/or $\mathbf{c}$, the vectors:

i $\overrightarrow{MB}$ **ii** $\overrightarrow{MN}$. *(2 marks)*

b Show that CA is parallel to MN. *(2 marks)*

Edexcel, May 2008, Paper 3, Question 25

2

Not drawn accurately

ABCD is a parallelogram.

AB is parallel to DC.

AD is parallel to BC.

$\overrightarrow{AB} = \mathbf{p}$

$\overrightarrow{AD} = \mathbf{q}$

a Express, in terms of $\mathbf{p}$ and $\mathbf{q}$:

i $\overrightarrow{AC}$ **ii** $\overrightarrow{BD}$ *(2 marks)*

Not drawn accurately

AC and BD are diagonals of parallelogram ABCD.

AC and BD intersect at T.

b Express AT in terms of $\mathbf{p}$ and $\mathbf{q}$. *(1 mark)*

Edexcel, June 2006, Paper 6 Higher, Question 13

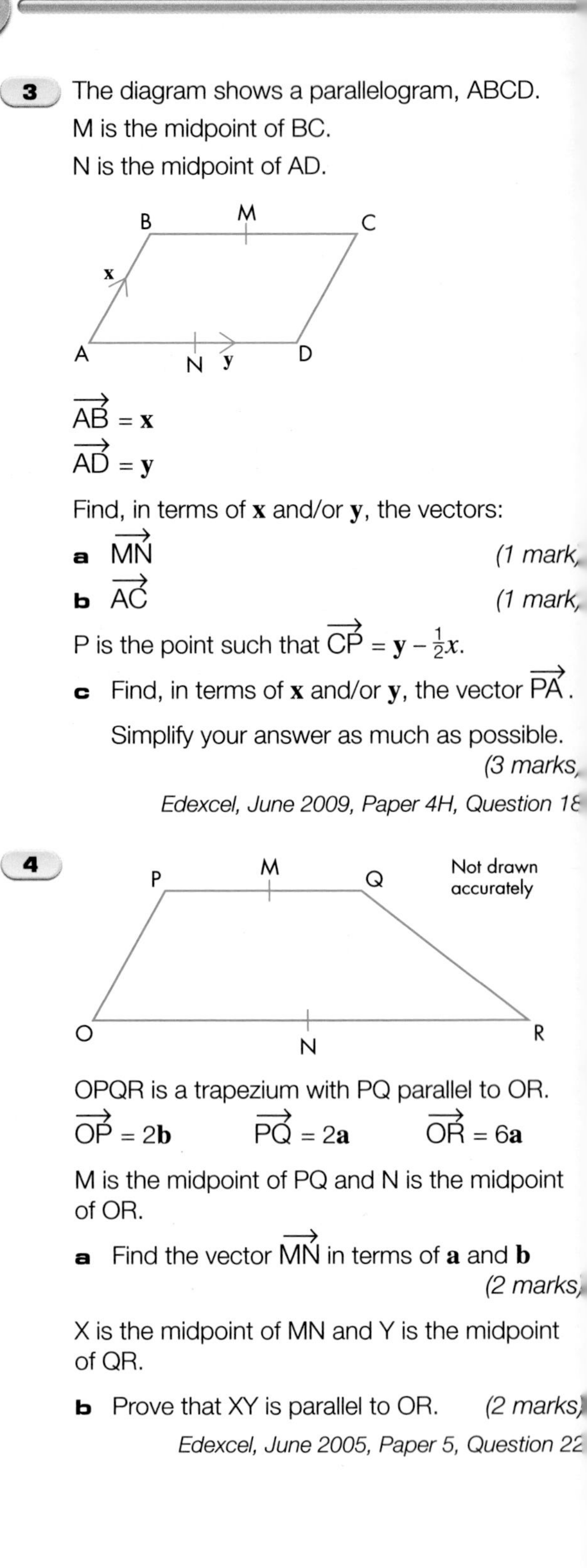

3 The diagram shows a parallelogram, ABCD.

M is the midpoint of BC.

N is the midpoint of AD.

$\overrightarrow{AB} = \mathbf{x}$

$\overrightarrow{AD} = \mathbf{y}$

Find, in terms of $\mathbf{x}$ and/or $\mathbf{y}$, the vectors:

a $\overrightarrow{MN}$ *(1 mark)*

b $\overrightarrow{AC}$ *(1 mark)*

P is the point such that $\overrightarrow{CP} = \mathbf{y} - \frac{1}{2}x$.

c Find, in terms of $\mathbf{x}$ and/or $\mathbf{y}$, the vector $\overrightarrow{PA}$.

Simplify your answer as much as possible. *(3 marks)*

Edexcel, June 2009, Paper 4H, Question 18

4

Not drawn accurately

OPQR is a trapezium with PQ parallel to OR.

$\overrightarrow{OP} = 2\mathbf{b}$ $\quad \overrightarrow{PQ} = 2\mathbf{a}$ $\quad \overrightarrow{OR} = 6\mathbf{a}$

M is the midpoint of PQ and N is the midpoint of OR.

a Find the vector $\overrightarrow{MN}$ in terms of $\mathbf{a}$ and $\mathbf{b}$ *(2 marks)*

X is the midpoint of MN and Y is the midpoint of QR.

b Prove that XY is parallel to OR. *(2 marks)*

Edexcel, June 2005, Paper 5, Question 22

5

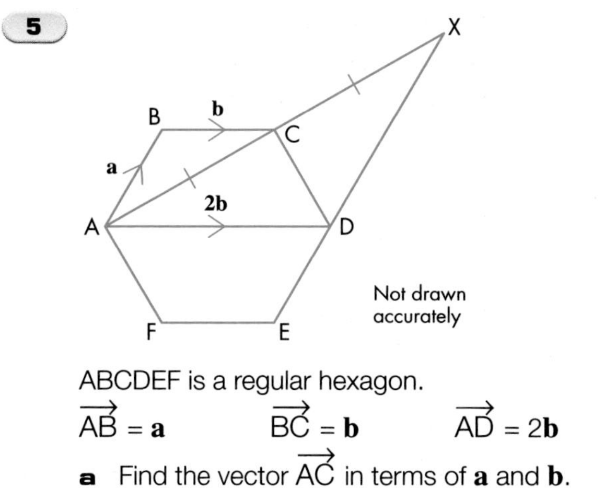

ABCDEF is a regular hexagon.

$\overrightarrow{AB} = \mathbf{a}$ $\overrightarrow{BC} = \mathbf{b}$ $\overrightarrow{AD} = 2\mathbf{b}$

a Find the vector $\overrightarrow{AC}$ in terms of **a** and **b**. *(1 mark)*

$\overrightarrow{AC} = \overrightarrow{CX}$

b Prove that AB is parallel to DX. *(3 marks)*

Edexcel, June 2007, Paper 5, Question 22

6 ABCD is a square.

BEC and DCF are equilateral triangles.

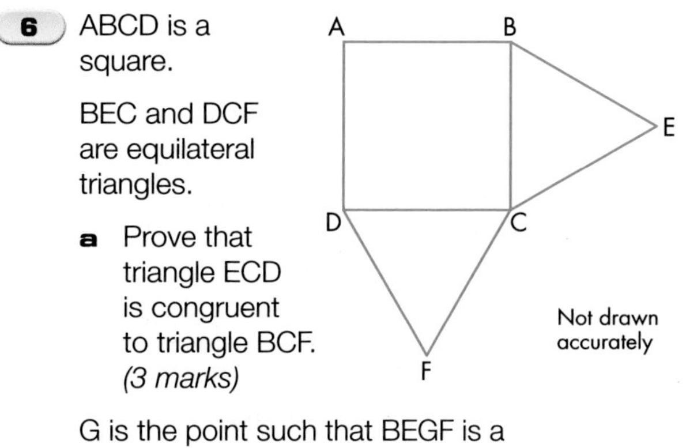

a Prove that triangle ECD is congruent to triangle BCF. *(3 marks)*

G is the point such that BEGF is a parallelogram.

b Prove that ED = EG. *(2 marks)*

Edexcel, June 2006, Paper 5, Question 21

Worked Examination Questions

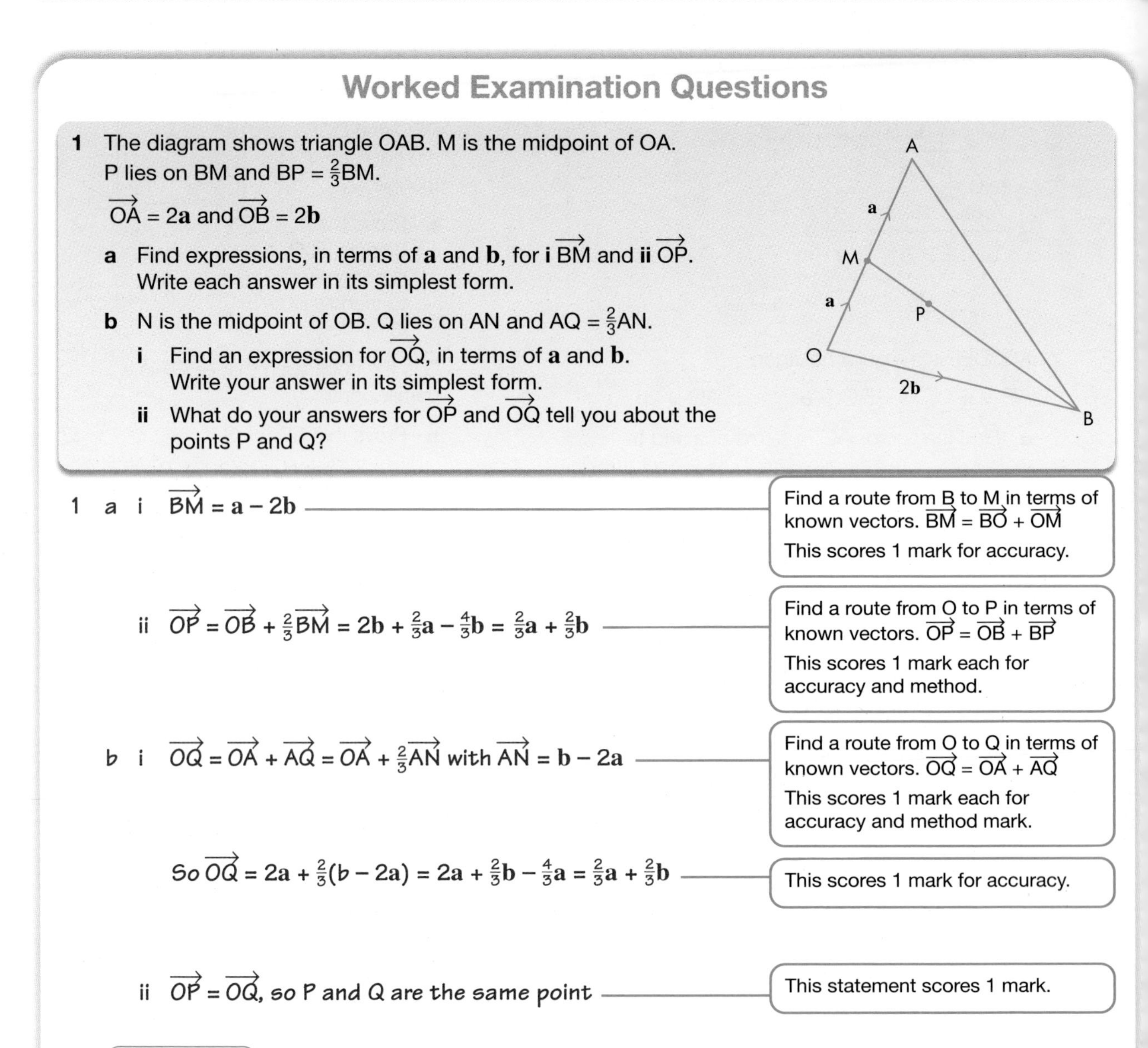

1 The diagram shows triangle OAB. M is the midpoint of OA.
P lies on BM and BP = $\frac{2}{3}$BM.

$\overrightarrow{OA} = 2\mathbf{a}$ and $\overrightarrow{OB} = 2\mathbf{b}$

a Find expressions, in terms of **a** and **b**, for **i** $\overrightarrow{BM}$ and **ii** $\overrightarrow{OP}$. Write each answer in its simplest form.

b N is the midpoint of OB. Q lies on AN and AQ = $\frac{2}{3}$AN.

i Find an expression for $\overrightarrow{OQ}$, in terms of **a** and **b**. Write your answer in its simplest form.

ii What do your answers for $\overrightarrow{OP}$ and $\overrightarrow{OQ}$ tell you about the points P and Q?

1 a i $\overrightarrow{BM} = \mathbf{a} - 2\mathbf{b}$

Find a route from B to M in terms of known vectors. $\overrightarrow{BM} = \overrightarrow{BO} + \overrightarrow{OM}$
This scores 1 mark for accuracy.

ii $\overrightarrow{OP} = \overrightarrow{OB} + \frac{2}{3}\overrightarrow{BM} = 2\mathbf{b} + \frac{2}{3}\mathbf{a} - \frac{4}{3}\mathbf{b} = \frac{2}{3}\mathbf{a} + \frac{2}{3}\mathbf{b}$

Find a route from O to P in terms of known vectors. $\overrightarrow{OP} = \overrightarrow{OB} + \overrightarrow{BP}$
This scores 1 mark each for accuracy and method.

b i $\overrightarrow{OQ} = \overrightarrow{OA} + \overrightarrow{AQ} = \overrightarrow{OA} + \frac{2}{3}\overrightarrow{AN}$ with $\overrightarrow{AN} = \mathbf{b} - 2\mathbf{a}$

Find a route from O to Q in terms of known vectors. $\overrightarrow{OQ} = \overrightarrow{OA} + \overrightarrow{AQ}$
This scores 1 mark each for accuracy and method mark.

So $\overrightarrow{OQ} = 2\mathbf{a} + \frac{2}{3}(\mathbf{b} - 2\mathbf{a}) = 2\mathbf{a} + \frac{2}{3}\mathbf{b} - \frac{4}{3}\mathbf{a} = \frac{2}{3}\mathbf{a} + \frac{2}{3}\mathbf{b}$

This scores 1 mark for accuracy.

ii $\overrightarrow{OP} = \overrightarrow{OQ}$, so P and Q are the same point

This statement scores 1 mark.

Total: 7 marks

Worked Examination Questions

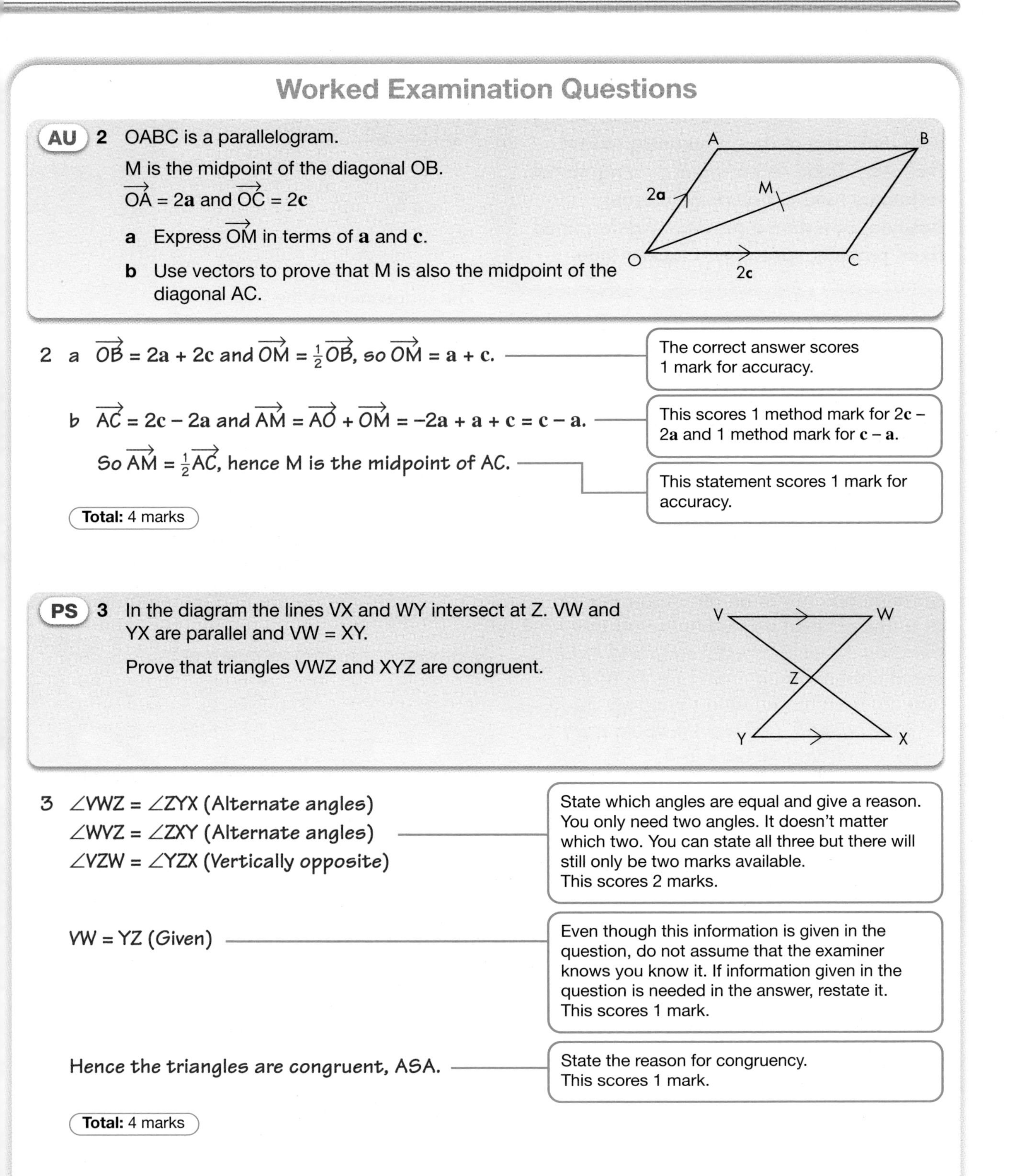

AU **2** OABC is a parallelogram.

M is the midpoint of the diagonal OB.

$\overrightarrow{OA} = 2\mathbf{a}$ and $\overrightarrow{OC} = 2\mathbf{c}$

a Express $\overrightarrow{OM}$ in terms of $\mathbf{a}$ and $\mathbf{c}$.

b Use vectors to prove that M is also the midpoint of the diagonal AC.

2 a $\overrightarrow{OB} = 2\mathbf{a} + 2\mathbf{c}$ and $\overrightarrow{OM} = \frac{1}{2}\overrightarrow{OB}$, so $\overrightarrow{OM} = \mathbf{a} + \mathbf{c}$. — The correct answer scores 1 mark for accuracy.

b $\overrightarrow{AC} = 2\mathbf{c} - 2\mathbf{a}$ and $\overrightarrow{AM} = \overrightarrow{AO} + \overrightarrow{OM} = -2\mathbf{a} + \mathbf{a} + \mathbf{c} = \mathbf{c} - \mathbf{a}$. — This scores 1 method mark for $2\mathbf{c} - 2\mathbf{a}$ and 1 method mark for $\mathbf{c} - \mathbf{a}$.

So $\overrightarrow{AM} = \frac{1}{2}\overrightarrow{AC}$, hence M is the midpoint of AC. — This statement scores 1 mark for accuracy.

Total: 4 marks

PS **3** In the diagram the lines VX and WY intersect at Z. VW and YX are parallel and VW = XY.

Prove that triangles VWZ and XYZ are congruent.

3 ∠VWZ = ∠ZYX (Alternate angles)
∠WVZ = ∠ZXY (Alternate angles)
∠VZW = ∠YZX (Vertically opposite)

— State which angles are equal and give a reason. You only need two angles. It doesn't matter which two. You can state all three but there will still only be two marks available. This scores 2 marks.

VW = YZ (Given) — Even though this information is given in the question, do not assume that the examiner knows you know it. If information given in the question is needed in the answer, restate it. This scores 1 mark.

Hence the triangles are congruent, ASA. — State the reason for congruency. This scores 1 mark.

Total: 4 marks

15 Functional Maths
Navigational techniques of ants

Scientists have discovered that some desert ants make use of dead reckoning to find their way. Dead reckoning is a navigational technique used to determine current position, based on a previously determined fixed position, speed and elapsed time.

Getting started

Scientists conducted an experiment to prove that Tunisian desert ants navigate using dead reckoning rather than other means such as laying down scents.

They conducted the following experiment:

The Tunisian desert ant set out from its nest at A. As soon as it found food at O, the scientists moved it to an alternative position at B. The ant then headed in exactly the direction it should have taken to find its nest, had it been returning from O to A, as if it had not been moved, and so ending up at C. If the ant had used scent it would have gone from B straight back to A.

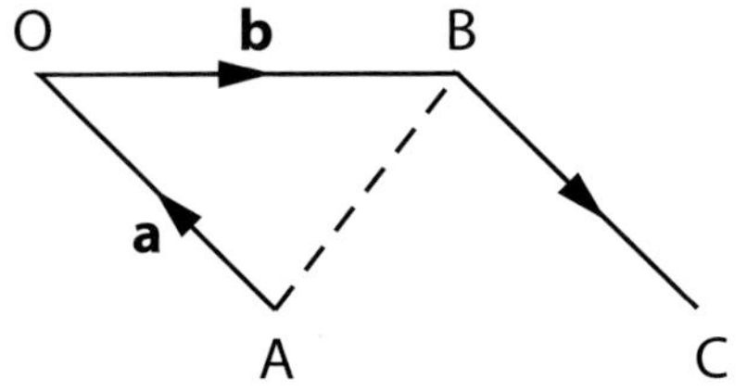

The diagram uses the vectors $\overrightarrow{OA} = -\mathbf{a}$ and $\overrightarrow{OB} = \mathbf{b}$.

Describe in terms of **a** and **b**:

- the route the ant took, from its nest to its end point after it had been moved, if it was using dead reckoning
- the route the ant would have taken from its nest to its end point, if it had navigated by scent rather than by dead reckoning.

Your task

Look at the diagram below. An ant's nest is at O. Each time the ant sets off it finds food at a different location: D, E, F, G, H, I or J.

1 Suggest possible vectors for its route to food at each of these points, writing them in terms of **a** and **b**.

2 Now imagine you are a scientist researching colonies of ants in the jungle and the desert. The position of the ants' nest, in each case, is again described as O. Choose one location for food at K, L, M, N or P. As soon as an ant finds food, you move it to point C.

Use vectors to describe the route the ant takes:

- from its nest to the food
- then when it is moved to C
- back to its nest, assuming it is an ant that lives in the jungle and relies on scent
- to its end point, assuming it is a desert ant and relies on dead reckoning.

Write a scientific report, with three sections, describing the method, results and conclusion, and using a vector diagram to explain your theory.

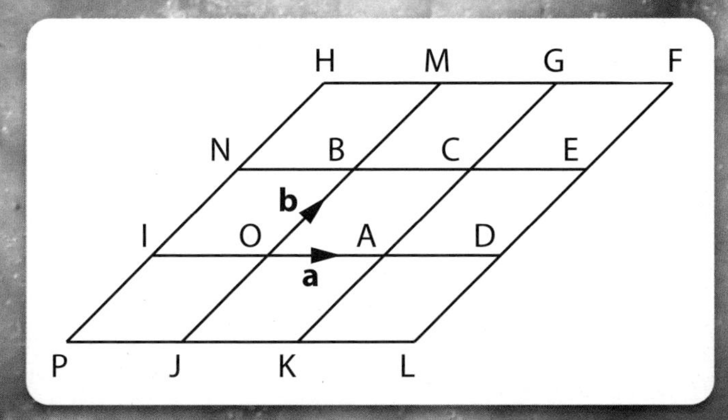

Why this chapter matters

By transforming graphs into other shapes, it is possible to change a circle into an aerofoil. This enables aeroplane engineers to do much simpler calculations when designing wings. This chapter will not help you to design aeroplane wings, but everyone has to start somewhere!

So far you have met four transformations:

- Translation
- Rotation
- Reflection
- Enlargement

Can you remember how to describe each of these?

Reflection

Original

Enlargement

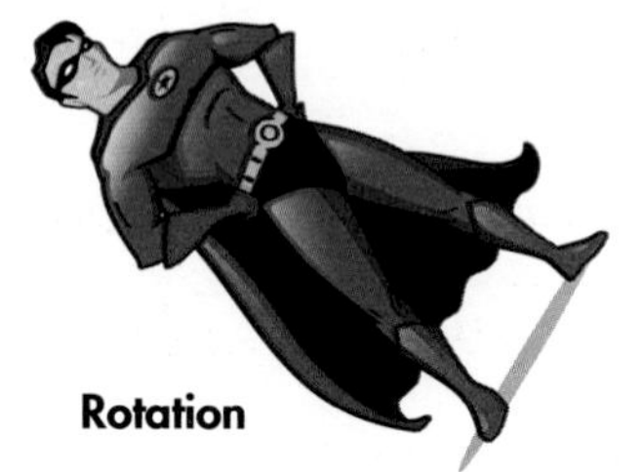

Rotation

Translation

There are many other transformations, but one that you will need here is the 'stretch'. It is exactly what it says it is.

The stretch can be in any direction, but in GCSE mathematics the stretch will be in the x- and y-directions.

A stretch is defined by a direction and a scale factor.

Original

Stretch scale factor 1.5 in the y-direction

Stretch scale factor 0.5 in the x-direction

Stretch scale factor 3 in the x-direction and 0.5 in the y-direction

Algebra: Transformation of graphs

The grades given in this chapter are target grades.

1 Transformations of the graph $y = f(x)$

This chapter will show you ...

A* how to transform a graph

A* how to recognise the relationships between graphs and their equations

Visual overview

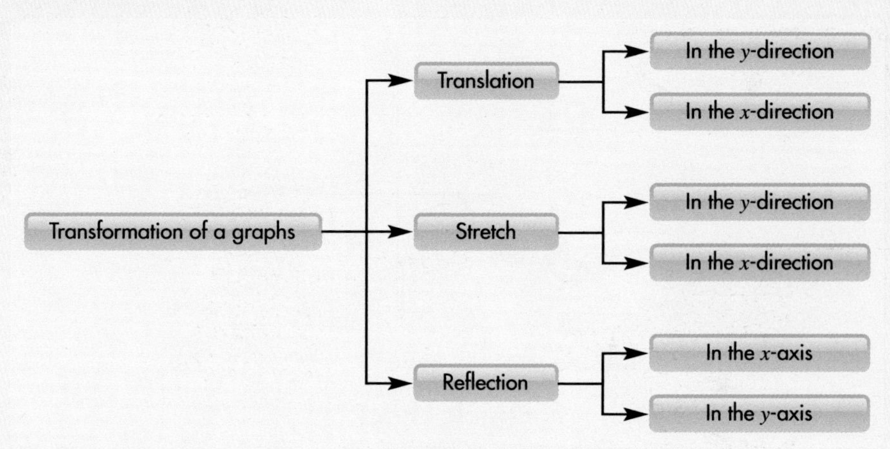

What you should already know

- How to transform a shape by a translation and a reflection **(KS3 level 6, GCSE grade D)**
- A translation is described by a column vector **(KS3 level 7, GCSE grade C)**
- A reflection is described by a mirror line **(KS3 level 5, GCSE grade E)**

continued

- The graphs of $y = x^2$, $y = x^3$, $y = \frac{1}{x}$, $y = \sin x$, $y = \cos x$ and $y = \tan x$

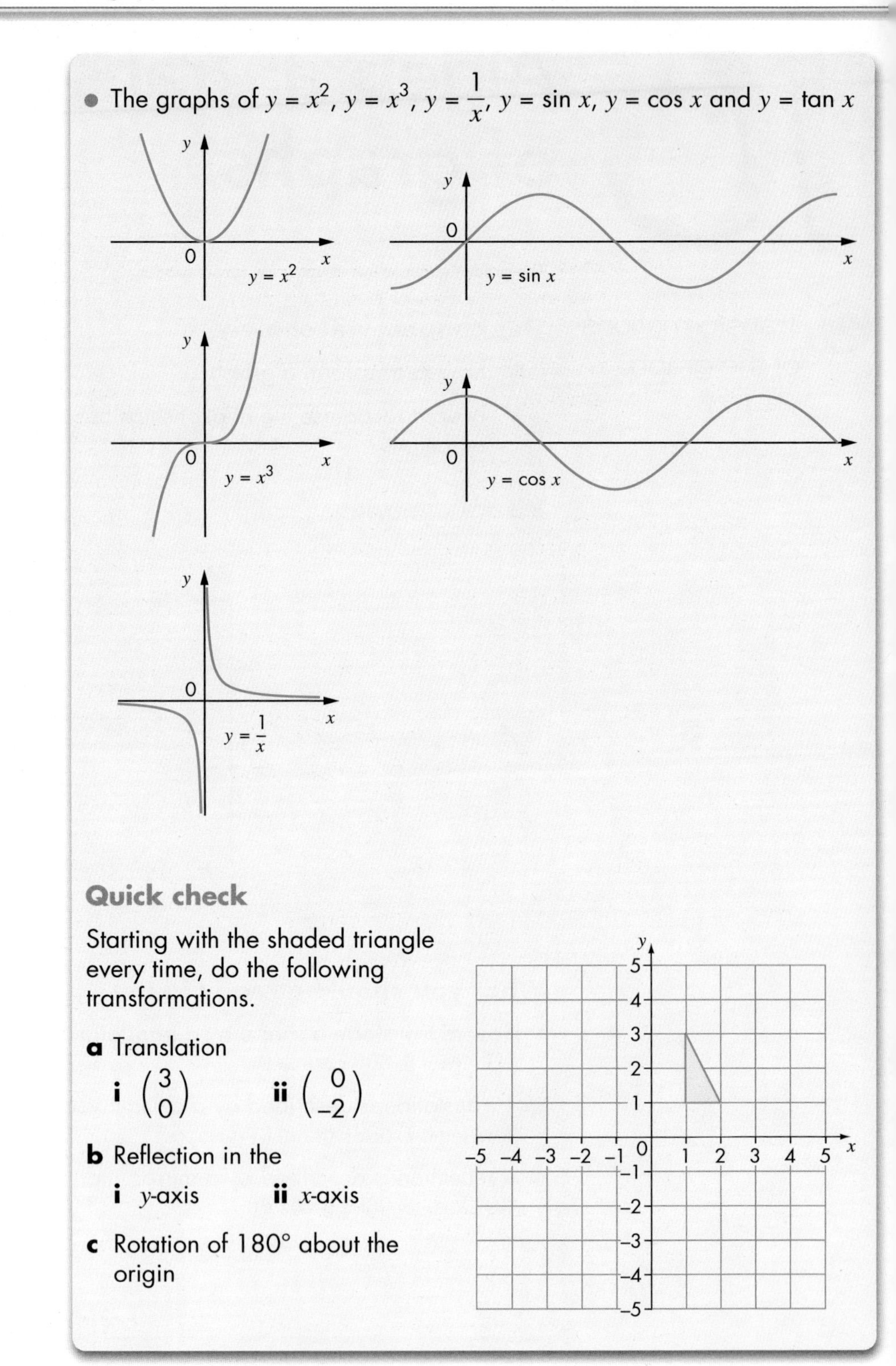

Quick check

Starting with the shaded triangle every time, do the following transformations.

a Translation

i $\begin{pmatrix} 3 \\ 0 \end{pmatrix}$ **ii** $\begin{pmatrix} 0 \\ -2 \end{pmatrix}$

b Reflection in the

i y-axis **ii** x-axis

c Rotation of 180° about the origin

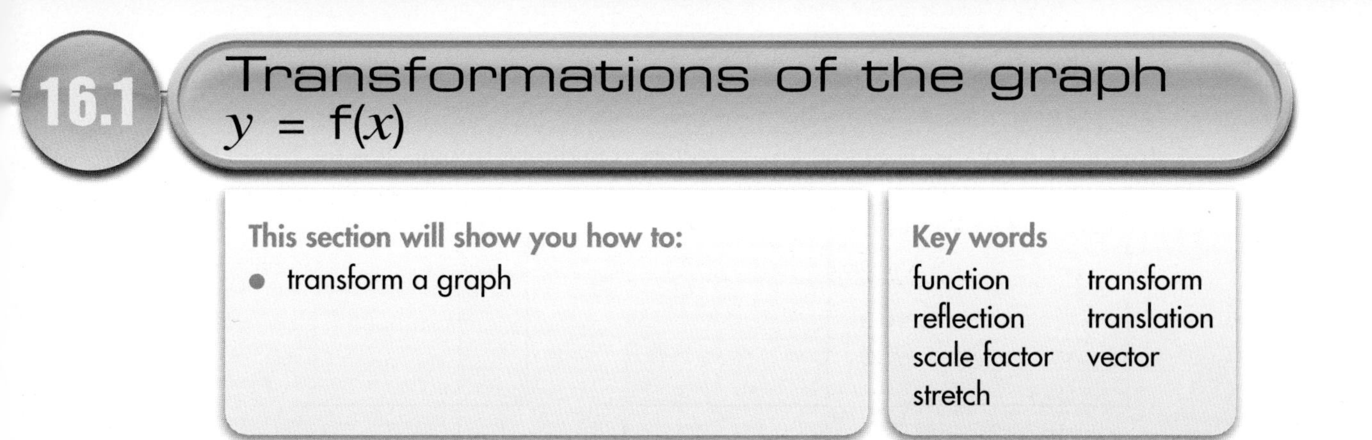

16.1 Transformations of the graph $y = f(x)$

This section will show you how to:
- transform a graph

Key words
function, transform, reflection, translation, scale factor, vector, stretch

The notation f(x) is used to represent a **function** of x. A function of x is any algebraic expression in which x is the only variable. Examples of functions are: $f(x) = x + 3$, $f(x) = 5x$, $f(x) = 2x - 7$, $f(x) = x^2$, $f(x) = x^3 + 2x - 1$, $f(x) = \sin x$ and $f(x) = \frac{1}{x}$.

On this and the next page are six general statements or rules about **transforming** graphs.

This work is much easier to understand if you can use to a graphics calculator or a graph-drawing computer program.

The graph on the right represents any function $y = f(x)$.

Rule 1 The graph of $y = f(x) + a$ is a **translation** of the graph of $y = f(x)$ by a **vector** $\begin{pmatrix} 0 \\ a \end{pmatrix}$.

Rule 2 The graph of $y = f(x - a)$ is a translation of the graph of $y = f(x)$ by a vector $\begin{pmatrix} a \\ 0 \end{pmatrix}$.

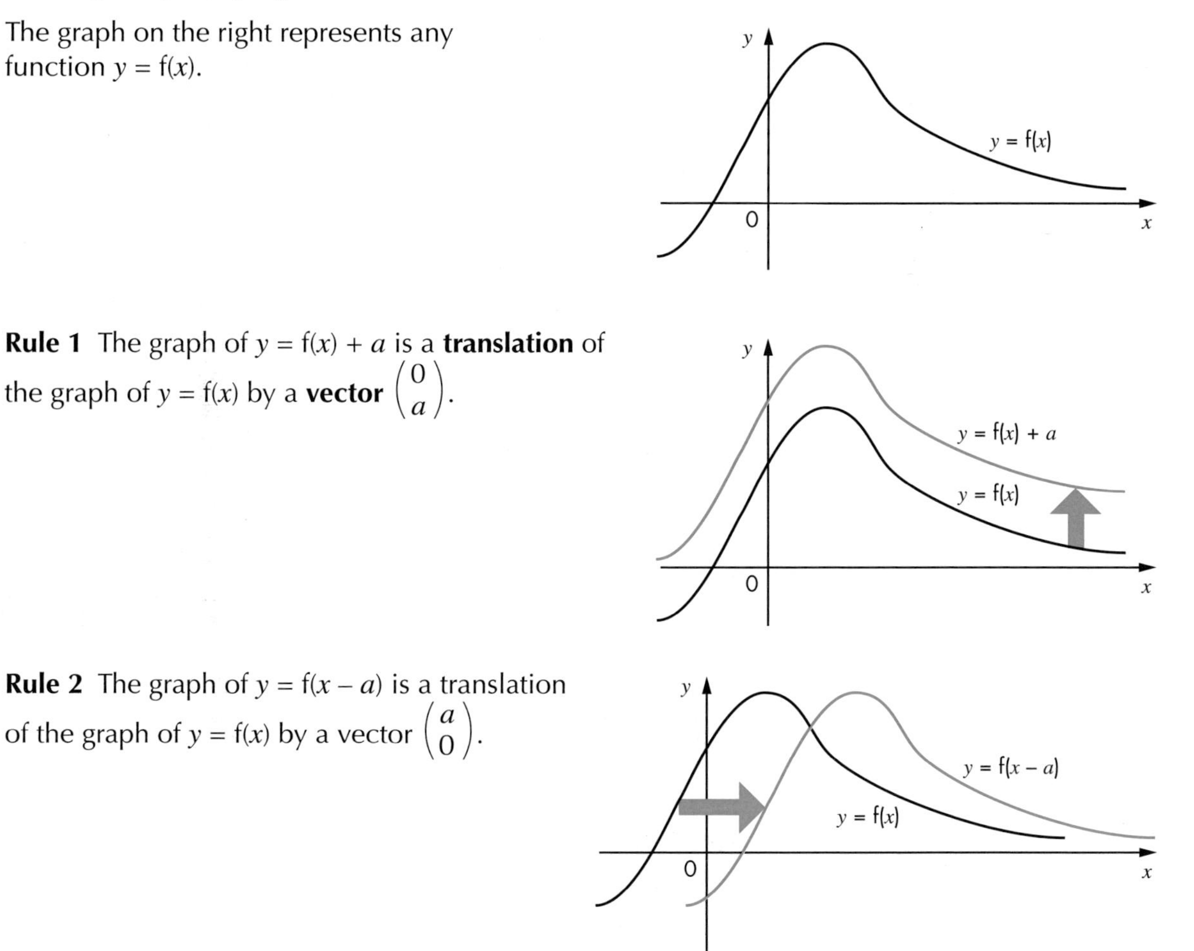

Note: The sign in front of a in the bracket is negative, but the translation is in the positive direction. $f(x + a)$ would translate $f(x)$ by the vector $\begin{pmatrix} -a \\ 0 \end{pmatrix}$.

A **stretch** is an enlargement that takes place in one direction only. It is described by a **scale factor** and the direction of the stretch.

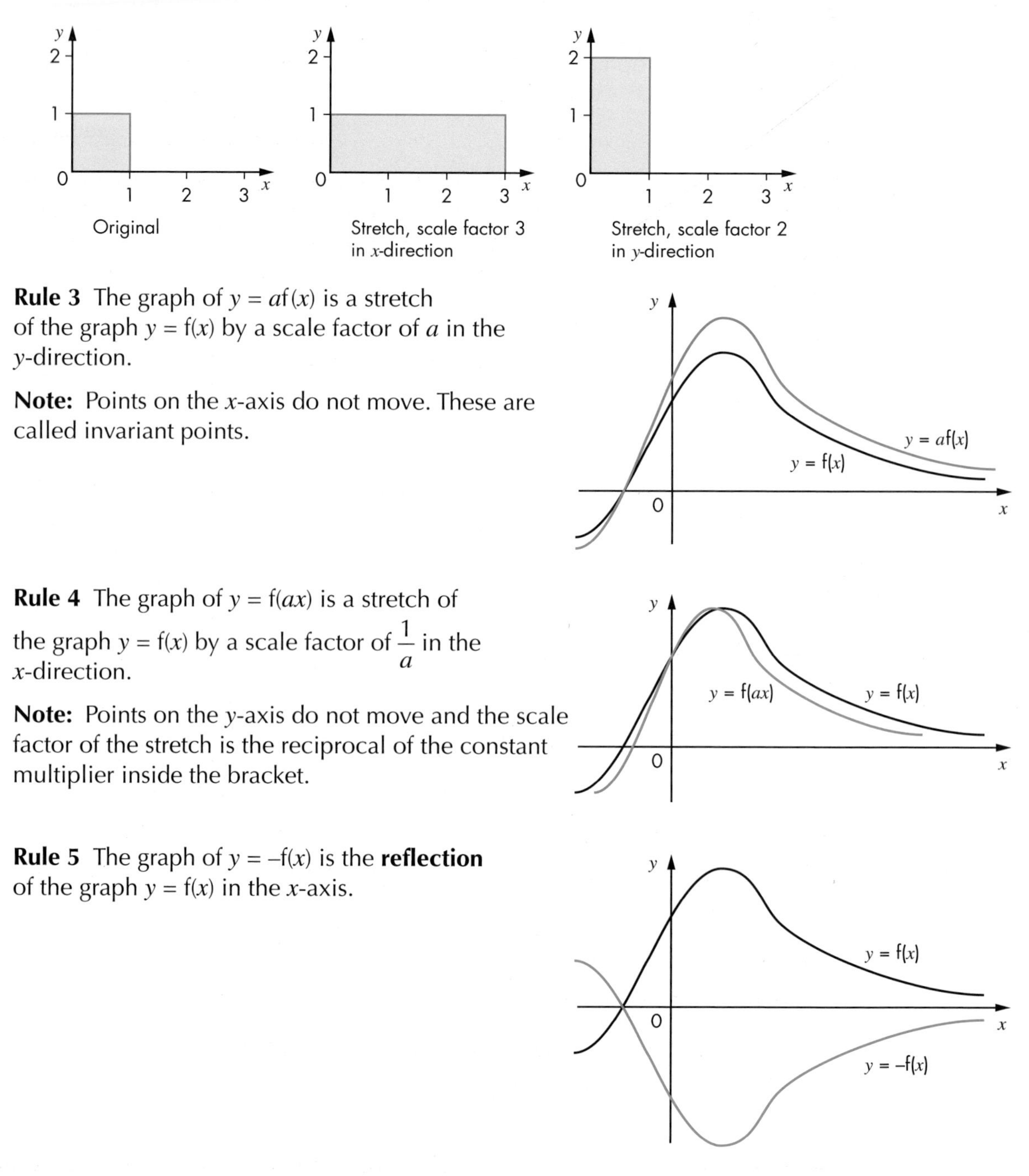

Rule 3 The graph of $y = af(x)$ is a stretch of the graph $y = f(x)$ by a scale factor of a in the y-direction.

Note: Points on the x-axis do not move. These are called invariant points.

Rule 4 The graph of $y = f(ax)$ is a stretch of the graph $y = f(x)$ by a scale factor of $\frac{1}{a}$ in the x-direction.

Note: Points on the y-axis do not move and the scale factor of the stretch is the reciprocal of the constant multiplier inside the bracket.

Rule 5 The graph of $y = -f(x)$ is the **reflection** of the graph $y = f(x)$ in the x-axis.

Rule 6 The graph of $y = f(-x)$ is the reflection of the graph $y = f(x)$ in the y-axis.

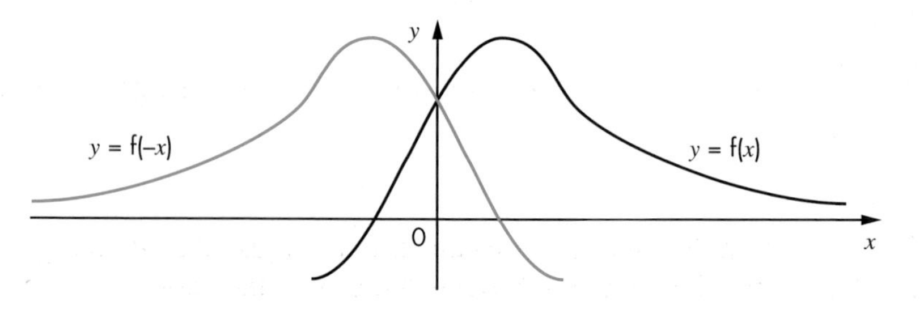

EXAMPLE 1

Sketch the following graphs.

a $y = x^2$ **b** $y = 5x^2$ **c** $y = x^2 - 5$

d $y = -x^2$ **e** $y = (x - 5)^2$ **f** $y = 2x^2 + 3$

Describe the transformation(s) that change(s) graph **a** to each of the other graphs.

Graph **a** is the basic graph to which you will apply the rules to make the necessary transformations: graph **b** uses Rule 3, graph **c** uses Rule 1, graph **d** uses Rule 5, graph **e** uses Rule 2, and graph **f** uses Rules 3 and 1.

The graphs are:

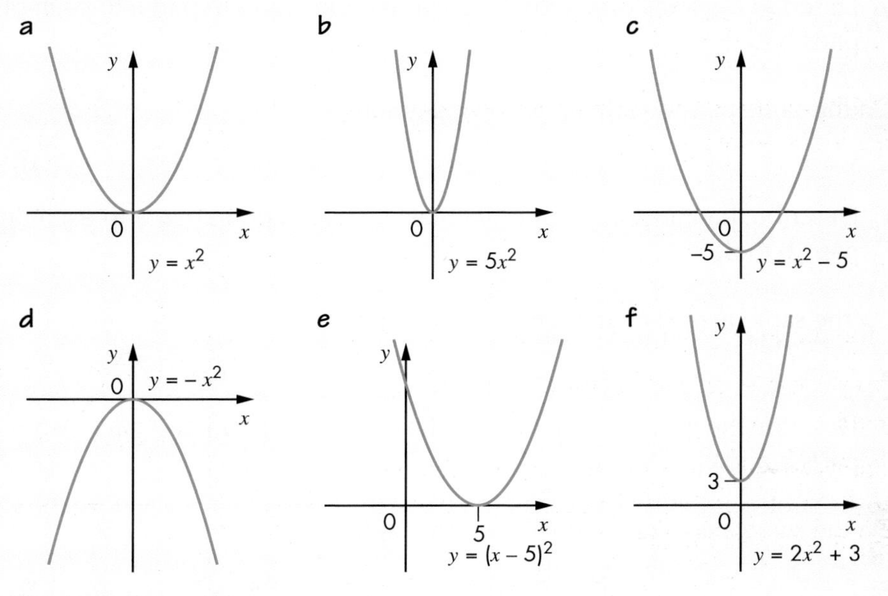

The transformations are:

graph **b** is a stretch of scale factor 5 in the y-direction,

graph **c** is a translation of $\begin{pmatrix} 0 \\ -5 \end{pmatrix}$,

graph **d** is a reflection in the x-axis,

graph **e** is a translation of $\begin{pmatrix} 5 \\ 0 \end{pmatrix}$,

graph **f** is a stretch of scale factor 2 in the y-direction, followed by a translation of $\begin{pmatrix} 0 \\ 3 \end{pmatrix}$.

Note that two of the transformations cause problems because they seem to do the opposite of what is expected. These are:

$y = f(x + a)$ (Rule 2)

The translation is $\begin{pmatrix} -a \\ 0 \end{pmatrix}$, so the sign of the constant inside the bracket changes in the vector (see part **e** in Example 1).

$y = f(ax)$ (Rule 4)

This does not look like a stretch. It actually closes the graph up. Just like an enlargement can make something smaller, a stretch can make it squeeze closer to the axes.

EXERCISE 16A

A*

1 On the same axes sketch the following graphs.

a $y = x^2$ **b** $y = 3x^2$ **c** $y = \frac{1}{2}x^2$ **d** $y = 10x^2$

e Describe the transformation(s) that take(s) the graph in part **a** to each of the graphs in parts **b** to **d**.

2 On the same axes sketch the following graphs.

a $y = x^2$ **b** $y = x^2 + 3$ **c** $y = x^2 - 1$ **d** $y = 2x^2 + 1$

e Describe the transformation(s) that take(s) the graph in part **a** to each of the graphs in parts **b** to **d**.

3 On the same axes sketch the following graphs.

a $y = x^2$ **b** $y = (x + 3)^2$ **c** $y = (x - 1)^2$ **d** $y = 2(x - 2)^2$

e Describe the transformation(s) that take(s) the graph in part **a** to each of the graphs in parts **b** to **d**.

4 On the same axes sketch the following graphs.

a $y = x^2$ **b** $y = (x + 3)^2 - 1$ **c** $y = 4(x - 1)^2 + 3$

d Describe the transformation(s) that take(s) the graph in part **a** to each of the graphs in parts **b** and **c**.

5 On the same axes sketch the following graphs.

a $y = x^2$ **b** $y = -x^2 + 3$ **c** $y = -3x^2$ **d** $y = -2x^2 + 1$

e Describe the transformation(s) that take(s) the graph in part **a** to each of the graphs in parts **b** to **d**.

6 On the same axes sketch the following graphs.

a $y = \sin x$ **b** $y = 2\sin x$ **c** $y = \frac{1}{2}\sin x$ **d** $y = 10\sin x$

e Describe the transformation(s) that take(s) the graph in part **a** to each of the graphs in parts **b** to **d**.

7 On the same axes sketch the following graphs.

a $y = \sin x$ **b** $y = \sin 3x$ **c** $y = \sin \frac{x}{2}$ **d** $y = 5\sin 2x$

e Describe the transformation(s) that take(s) the graph in part **a** to each of the graphs in parts **b** to **d**.

8 On the same axes sketch the following graphs.

a $y = \sin x$ **b** $y = \sin (x + 90°)$ **c** $y = \sin (x - 45°)$ **d** $y = 2\sin (x - 90°)$

e Describe the transformation(s) that take(s) the graph in part **a** to the graphs in parts **b** to **d**.

A*

9 On the same axes sketch the following graphs.

a $y = \sin x$ **b** $y = \sin x + 2$ **c** $y = \sin x - 3$ **d** $y = 2\sin x + 1$

e Describe the transformation(s) that take(s) the graph in part **a** to the graphs parts **b** to **d**.

10 On the same axes sketch the following graphs.

a $y = \sin x$ **b** $y = -\sin x$ **c** $y = \sin(-x)$ **d** $y = -\sin(-x)$

e Describe the transformation(s) that take(s) the graph in part **a** to the graphs in parts **b** to **d**.

11 On the same axes sketch the following graphs.

a $y = \cos x$ **b** $y = 2\cos x$ **c** $y = \cos(x - 60°)$ **d** $y = \cos x + 2$

e Describe the transformation(s) that take(s) the graph in part **a** to the graphs in parts **b** to **d**.

AU 12 Which of the equations below represents the graph shown?

A: $y = \sin x$

B: $y = \cos(x - 90°)$

C: $y = -\sin(-x)$

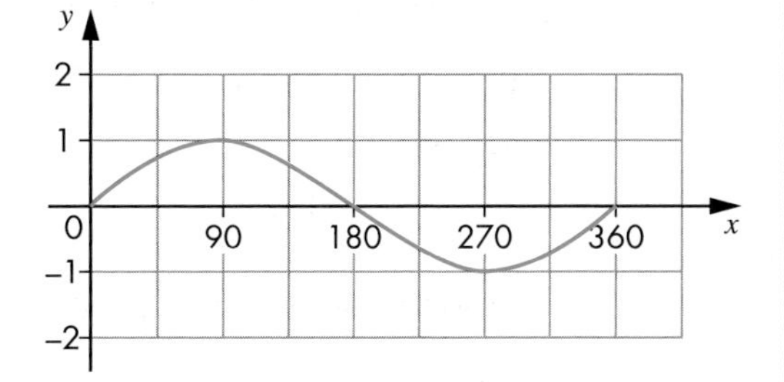

13 **a** Describe the transformations of the graph of $y = x^2$ needed to obtain these graphs.

i $y = 4x^2$ **ii** $y = 9x^2$ **iii** $y = 16x^2$

b Describe the transformations of the graph of $y = x^2$ needed to obtain these graphs.

i $y = (2x)^2$ **ii** $y = (3x)^2$ **iii** $y = (4x)^2$

c Describe two different transformations that take the graph of $y = x^2$ to the graph of $y = (ax)^2$, where a is a positive number.

14 On the right is a sketch of the function $y = f(x)$. Use this to sketch the following.

a $y = f(x) + 2$ **b** $y = 2f(x)$ **c** $y = f(x - 3)$

d $y = -f(x)$ **e** $y = 2f(x) + 3$ **f** $y = -f(x) - 2$

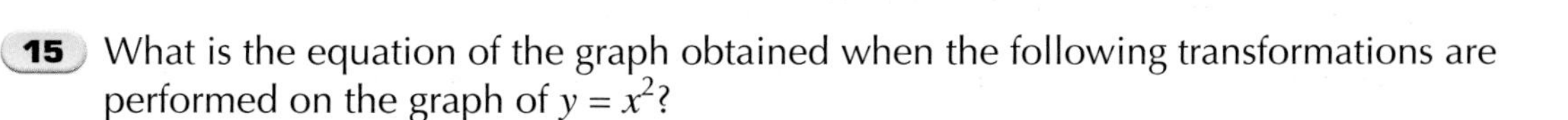

15 What is the equation of the graph obtained when the following transformations are performed on the graph of $y = x^2$?

a Stretch by a factor of 5 in the y-direction

b Translation of $\begin{pmatrix} 0 \\ 7 \end{pmatrix}$ **c** Translation of $\begin{pmatrix} -3 \\ 0 \end{pmatrix}$ **d** Translation of $\begin{pmatrix} -2 \\ -3 \end{pmatrix}$

e Stretch by a factor of 3 in the y-direction followed by a translation of $\begin{pmatrix} 0 \\ 4 \end{pmatrix}$

f Reflection in the x-axis, followed by a stretch, scale factor 3, in the y-direction

A*

16 What is the equation of the graph obtained when the following transformations are performed on the graph of $y = \cos x$?

a Stretch by a factor of 6 in the y-direction

b Translation of $\begin{pmatrix} 0 \\ 3 \end{pmatrix}$

c Translation of $\begin{pmatrix} -30 \\ 0 \end{pmatrix}$

d Translation of $\begin{pmatrix} 45 \\ -2 \end{pmatrix}$

e Stretch by a factor of 3 in the y-direction followed by a translation of $\begin{pmatrix} 0 \\ -2 \end{pmatrix}$

17 **a** Sketch the graph $y = x^3$.

b Use your sketch in part **a** to draw the graphs obtained after $y = x^3$ is transformed as follows.

i Reflection in the x-axis

ii Translation of $\begin{pmatrix} 0 \\ -2 \end{pmatrix}$

iii Stretch by a scale factor of 3 in the y-direction

iv Translation of $\begin{pmatrix} -2 \\ 0 \end{pmatrix}$

c Give the equation of each of the graphs obtained in part **b**.

18 **a** Sketch the graph of $y = \frac{1}{x}$.

b Use your sketch in part **a** to draw the graphs obtained after $y = \frac{1}{x}$ is transformed as follows.

i Translation of $\begin{pmatrix} 0 \\ 4 \end{pmatrix}$

ii Translation of $\begin{pmatrix} 4 \\ 0 \end{pmatrix}$

iii Stretch, scale factor 3 in the y-direction

iv Stretch, scale factor $\frac{1}{2}$ in the x-direction

c Give the equation of each of the graphs obtained in part **b**.

PS 19 A teacher asked her class to apply the following transformations to the function $f(x) = x^2$.

a $f(-x)$

b $-f(x)$

Martyn said that they must be the same as $-x^2 = x^2$.

Is Martyn correct? Explain your answer.

20 The graphs below are all transformations of $y = x^2$. Two points through which each graph passes are indicated. Use this information to work out the equation of each graph.

a y, x, 0, (0, 2), (1, 3)

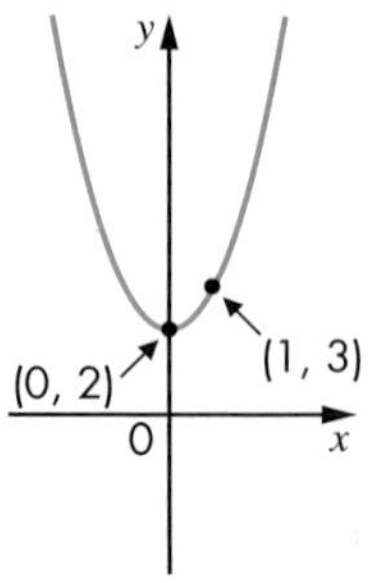

b y, x, 0, (1, 1), (2, 0)

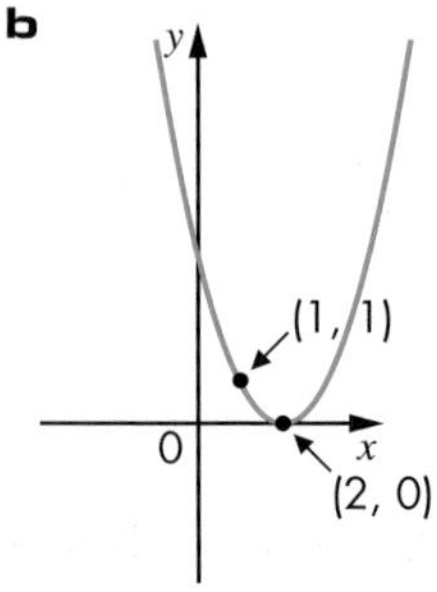

c y, x, 0, (2, 8), (0, 0)

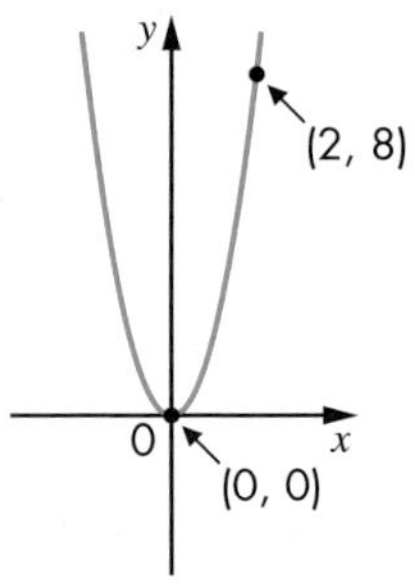

d y, x, 0, (0, 4), (2, 0)

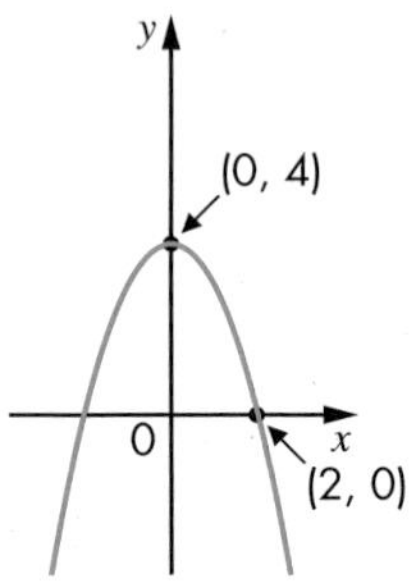

21 The graphs below are all transformations of $y = \sin x$. Two points through which each graph passes are indicated. Use this information to work out the equation of each graph.

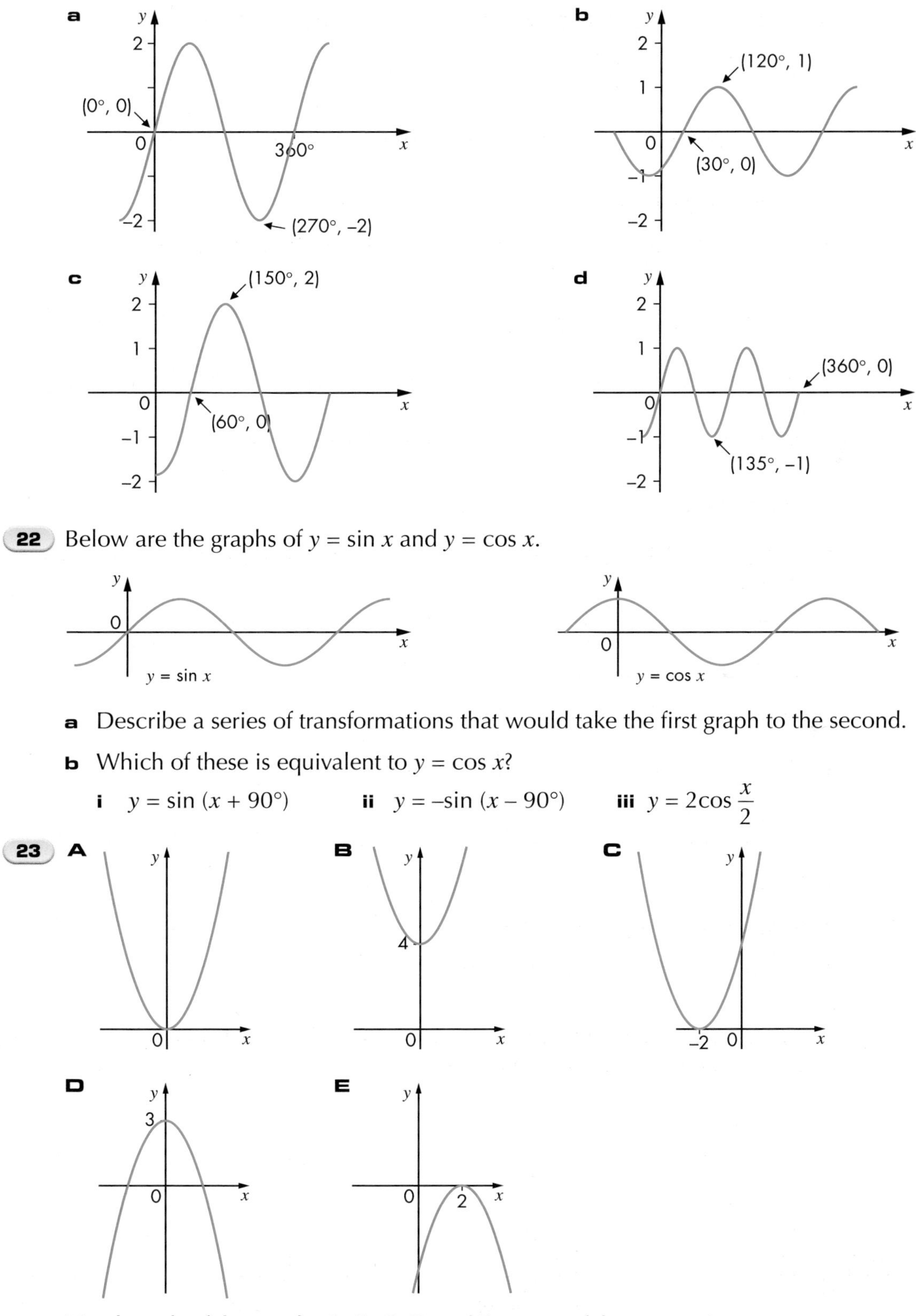

22 Below are the graphs of $y = \sin x$ and $y = \cos x$.

a Describe a series of transformations that would take the first graph to the second.

b Which of these is equivalent to $y = \cos x$?

i $y = \sin (x + 90°)$ **ii** $y = -\sin (x - 90°)$ **iii** $y = 2\cos \frac{x}{2}$

23 **A** **B** **C** **D** **E**

Match each of the graphs **A**, **B**, **C**, **D** and **E** to one of these equations.

i $y = x^2$ **ii** $y = -x^2 + 3$ **iii** $y = -(x - 2)^2$ **iv** $y = (x + 2)^2$ **v** $y = x^2 + 4$

GRADE BOOSTER

A* You can transform the graph of a given function

A* You can identify the equation of a function from its graph, which has been formed by a transformation on a known function

What you should know now

- How to sketch the graphs of functions such as $y = \mathrm{f}(ax)$ and $y = \mathrm{f}(x + a)$ from the known graph of $y = \mathrm{f}(x)$
- How to describe from their graphs the transformation of one function into another
- How to identify equations from the graphs of transformations of known graphs

1 The diagram shows part of the curve with equation $y = f(x)$.

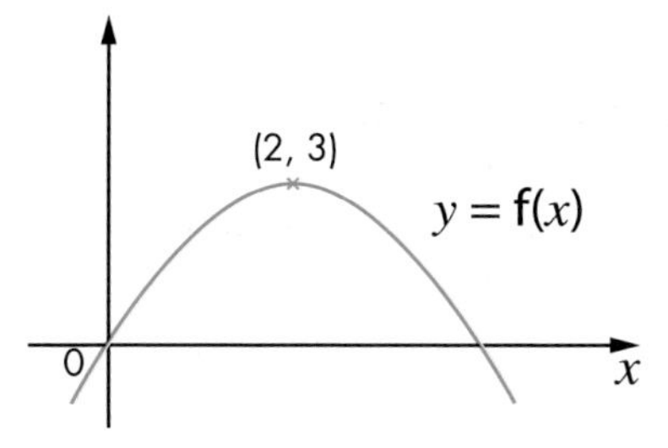

The coordinates of the maximum point of this curve are (2, 3).

Write down the coordinates of the maximum point of the curve with equation:

a $y = f(x - 2)$ *(1 mark)*

b $y = 2f(x)$ *(1 mark)*

Edexcel, May 2008, Paper 3 Higher, Question 27

2 The graph of $y = \sin x°$ for $0 \leqslant x \leqslant 360$ is drawn below.

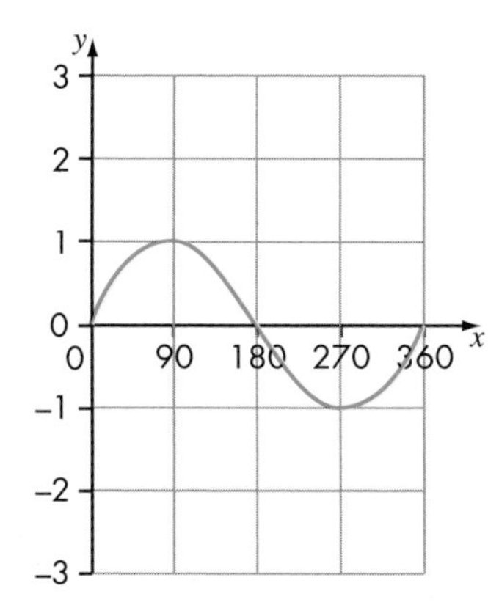

On copies of the same axes draw sketches of the following graphs.

a $y = 2 \sin x°$

b $y = \sin 2x°$

c $y = \sin x° - 2$

3 The graph of $y = f(x)$ is shown below.

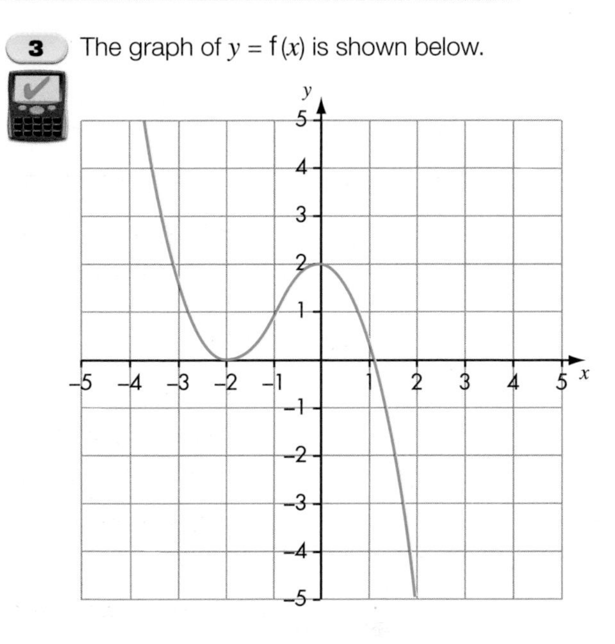

On copies of the grid, sketch the following graphs:

a $y = f(x + 1)$

b $y = 2f(x)$

Edexcel, June 2005, Paper 6 Higher, Question 20

4 The sketch to the right is of the graph of $y = x^2$

Copy these axes and sketch the following graphs.

a $y = x^2 + 2$

b $y = (x - 2)^2$

c $y = \frac{1}{2}x^2$

AQA, November 2004, Paper 1 Higher, Question 21

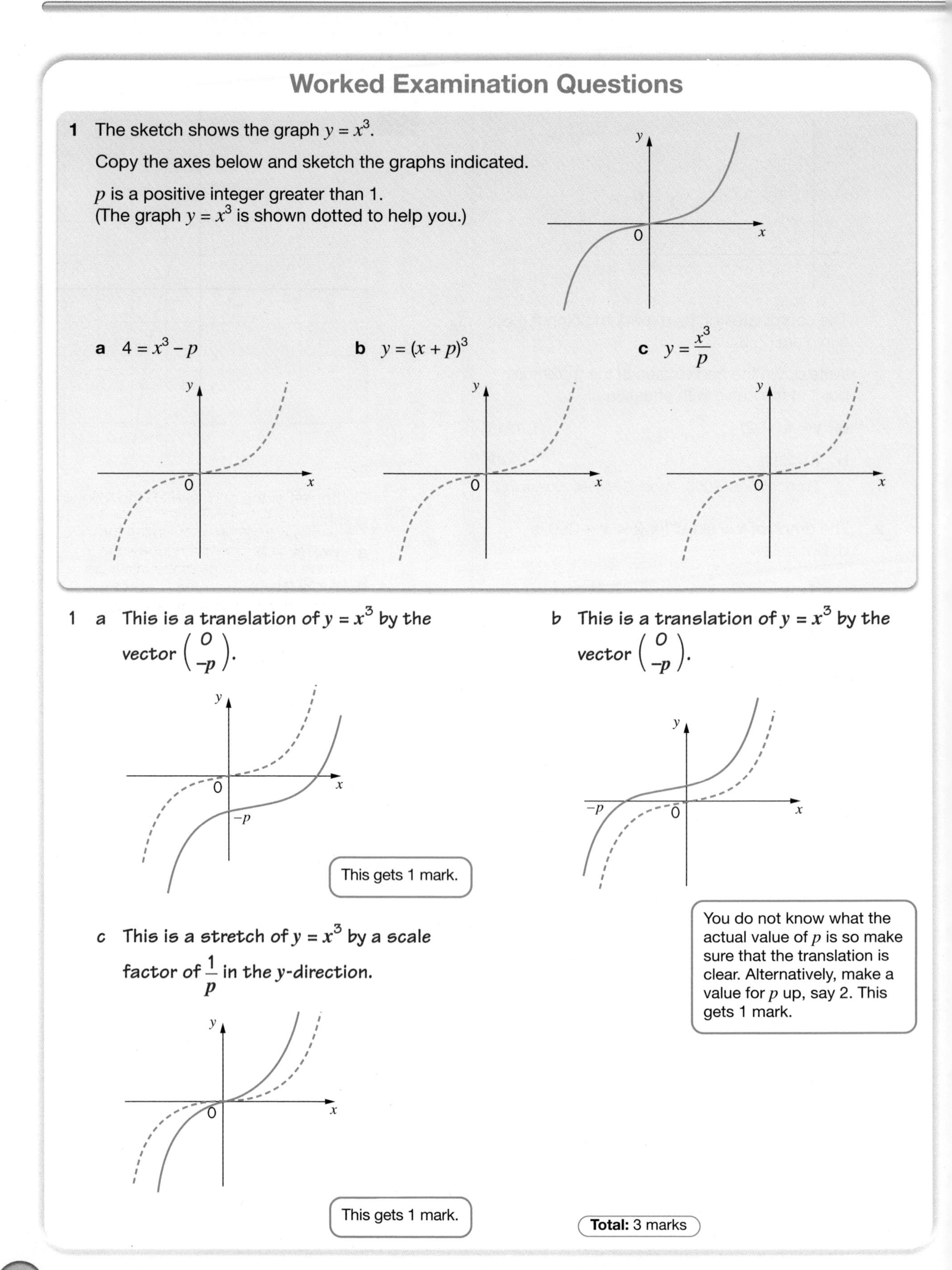

Worked Examination Questions

1 The sketch shows the graph $y = x^3$.

Copy the axes below and sketch the graphs indicated.

p is a positive integer greater than 1.
(The graph $y = x^3$ is shown dotted to help you.)

a $4 = x^3 - p$ **b** $y = (x + p)^3$ **c** $y = \frac{x^3}{p}$

1 a This is a translation of $y = x^3$ by the vector $\begin{pmatrix} 0 \\ -p \end{pmatrix}$.

This gets 1 mark.

b This is a translation of $y = x^3$ by the vector $\begin{pmatrix} 0 \\ -p \end{pmatrix}$.

You do not know what the actual value of p is so make sure that the translation is clear. Alternatively, make a value for p up, say 2. This gets 1 mark.

c This is a stretch of $y = x^3$ by a scale factor of $\frac{1}{p}$ in the y-direction.

This gets 1 mark.

Total: 3 marks

Worked Examination Questions

PS **2** The sketch shows the graph of $y = x^2 - 6x + 5$.

The minimum point of the graph is (3, –4).

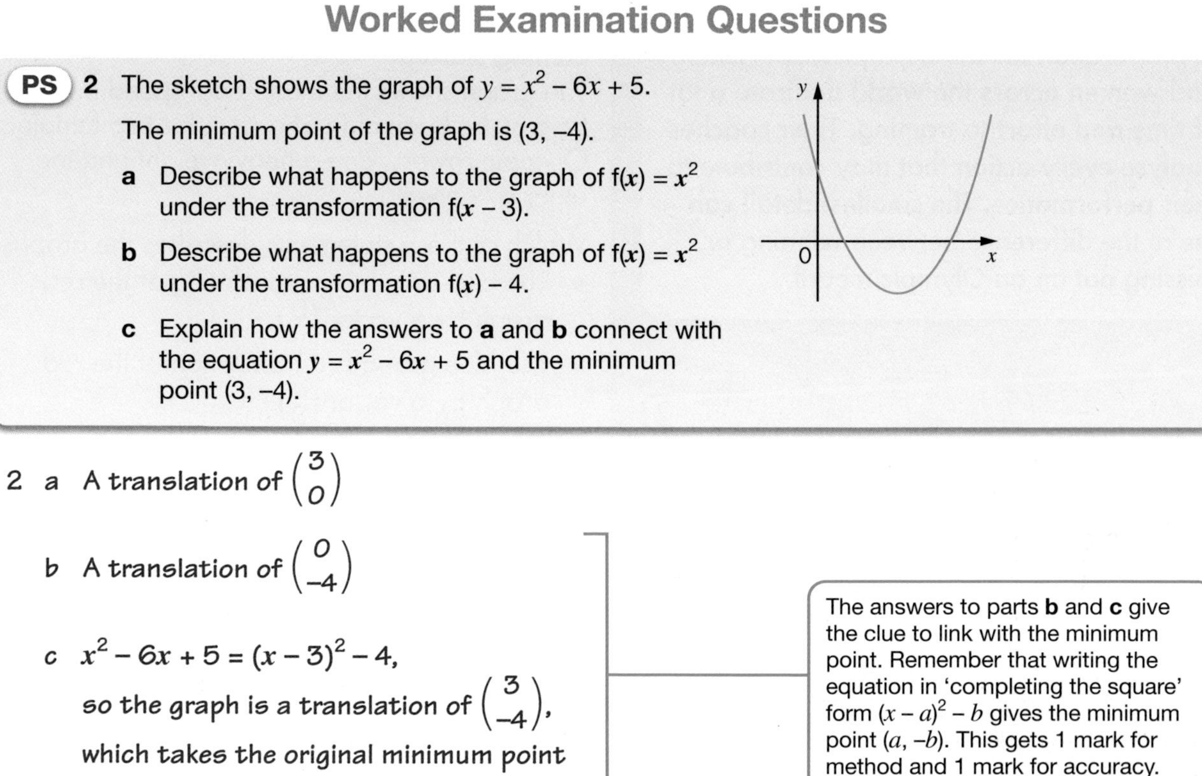

a Describe what happens to the graph of $f(x) = x^2$ under the transformation $f(x - 3)$.

b Describe what happens to the graph of $f(x) = x^2$ under the transformation $f(x) - 4$.

c Explain how the answers to **a** and **b** connect with the equation $y = x^2 - 6x + 5$ and the minimum point (3, –4).

2 a A translation of $\begin{pmatrix} 3 \\ 0 \end{pmatrix}$

b A translation of $\begin{pmatrix} 0 \\ -4 \end{pmatrix}$

c $x^2 - 6x + 5 = (x - 3)^2 - 4$,

so the graph is a translation of $\begin{pmatrix} 3 \\ -4 \end{pmatrix}$,

which takes the original minimum point (0, 0) to (3, –4).

The answers to parts **b** and **c** give the clue to link with the minimum point. Remember that writing the equation in 'completing the square' form $(x - a)^2 - b$ gives the minimum point $(a, -b)$. This gets 1 mark for method and 1 mark for accuracy.

Total: 2 marks

16 Functional Maths
Diving at the Olympics

In preparation for the Olympics, sportsmen and women across the world dedicate a lot of time and effort to training. Their coaches analyse every action that may contribute to their performance. The smallest detail can make the difference between winning or missing out on an Olympic medal.

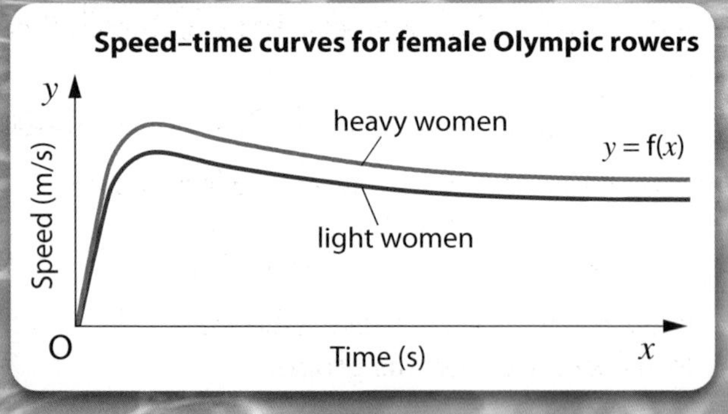

Getting started

The graph shows the mean boat speed against time at the beginning of a race for two female Olympic rowers: one a heavyweight and the other a lightweight.

Which of these statements describes the graphs?

- The blue graph is a translation of the red graph by a vector $\binom{0}{a}$.
- The blue graph is a translation of the red graph by a vector $\binom{a}{0}$.
- The blue graph is a stretch of the red graph in the vertical or y-direction.
- The blue graph is a reflection of the red graph in the horizontal or x-axis.

Dive 1

y

O 1 x

Your task

An Olympic diver begins her training for the day by performing a warm-up dive from the side of the pool. Her coach analyses the path of the dive and describes it as

$$y = -x^2 + x \qquad \text{(dive 1)}$$

where y represents the distance from the surface of the water and x represents the distance from the side of the pool. The graph to the left shows how this looks.

The diver performs a second dive from the side of the pool. Its graph is like the one for dive 1 but with a stretch in the y-direction (dive 2).

Then the diver climbs to the first diving platform, arches her back with arms outstretched and performs a swan dive (dive 3). The graph for this third dive is a transformation of the graph for dive 2.

Finally, the diver is joined by a team-mate to practise for the synchronised diving event. They dive, in unison, from two adjacent platforms along the side of the pool, both at the height of the second diving platform. The coach calls this dive 4.

Read the description of each dive above carefully.

Then write an appropriate equation and an accurate description for each of the transformations for dive 2 and dive 3.

Now consider dive 4. Describe the transformation for this graph.

Write the coach's end-of-day diving team report, sketching some graphs to show the shapes of the dives.

Answers to Chapter 1

Quick check

1 a 137 b 65 c 161 d 42
e 6.5 f 4.6 g 13.5 h 1.3

2 a i $\frac{12}{5}$ ii $\frac{13}{4}$ iii $\frac{16}{9}$
b i $1\frac{5}{6}$ ii $2\frac{1}{3}$ iii $3\frac{2}{7}$

3 a 14 b 20 c 1 d 7

4 a $1\frac{5}{12}$ b $\frac{17}{35}$ c $\frac{11}{20}$ d $\frac{14}{15}$

1.1 Basic calculations and using brackets

Exercise 1A

1 a 144 b 108

2 a 12.54 b 27.45

3 a 26.7 b 24.5 c 145.3 d 1.5

4 Sovereign is 102.48p per litre, Bridge is 102.73p per litre so Sovereign is cheaper.

5 Abby 1.247, Bobby 2.942, Col 5.333, Donna 6.538
Col is correct.

6 $31 \times 3600 \div 1610 = 69.31677 \approx 70$

7 a 167.552 b 196.48

8 a 2.77 b 6

9 a 497.952 b 110.978625

1.2 Adding and subtracting fractions with a calculator

Exercise 1B

1 a $6\frac{11}{20}$ b $8\frac{8}{15}$ c $16\frac{1}{4}$ d $11\frac{147}{200}$
e $7\frac{43}{80}$ f $11\frac{63}{80}$ g $3\frac{11}{30}$ h $2\frac{29}{48}$
i $3\frac{17}{96}$ j $7\frac{167}{240}$ k $7\frac{61}{80}$ l $4\frac{277}{396}$

2 $\frac{1}{12}$

3 a $12\frac{1}{4}$ b $3\frac{1}{4}$

4 Use the fraction key to input $\frac{3}{25}$, then key in + and then use the fraction key again to input $\frac{7}{10}$.

[fraction] 3 ▼ 2 5 ▶ + [fraction] 7 ▼ 1 0 ▶ =

5 $\frac{47}{120}$

6 a $-\frac{77}{1591}$ b Answer negative

7 a $\frac{223}{224}$ b $\frac{97}{1248}$ c $-\frac{97}{273}$
d One negative and one positive so $\frac{5}{7} > \frac{14}{39} > \frac{9}{32}$.

8 a Answers will vary
b Yes, always true, unless fractions are equivalent then answer is also equivalent.

9 $18\frac{11}{12}$ cm

10 $\frac{5}{12}$ (anticlockwise) or $\frac{7}{12}$ (clockwise)

1.3 Multiplying and dividing fractions with a calculator

Exercise 1C

1 a $\frac{3}{5}$ b $\frac{7}{12}$ c $\frac{9}{25}$ d $\frac{27}{200}$
e $\frac{21}{320}$ f $\frac{27}{128}$ g $5\frac{2}{5}$ h $5\frac{1}{7}$
i $2\frac{1}{16}$ j $\frac{27}{40}$ k $3\frac{9}{32}$ l $\frac{11}{18}$

2 $\frac{1}{6}$ m^2

3 15

4 a $\frac{27}{64}$ b $\frac{27}{64}$

5 a $\frac{4}{5}$ b $\frac{4}{5}$ c $\frac{16}{21}$ d $\frac{16}{21}$

6 a $8\frac{11}{20}$ b $18\frac{1}{60}$ c $65\frac{91}{100}$ d $22\frac{1}{8}$
e $7\frac{173}{320}$ f $52\frac{59}{160}$ g $2\frac{17}{185}$ h $2\frac{22}{103}$
i $1\frac{305}{496}$ j $5\frac{17}{65}$ k $7\frac{881}{4512}$ l $5\frac{547}{1215}$

7 $18\frac{5}{12}$ m^2

8 $3\frac{11}{32}$ cm^3

9 $90\frac{5}{8}$ miles

10 3

11 3

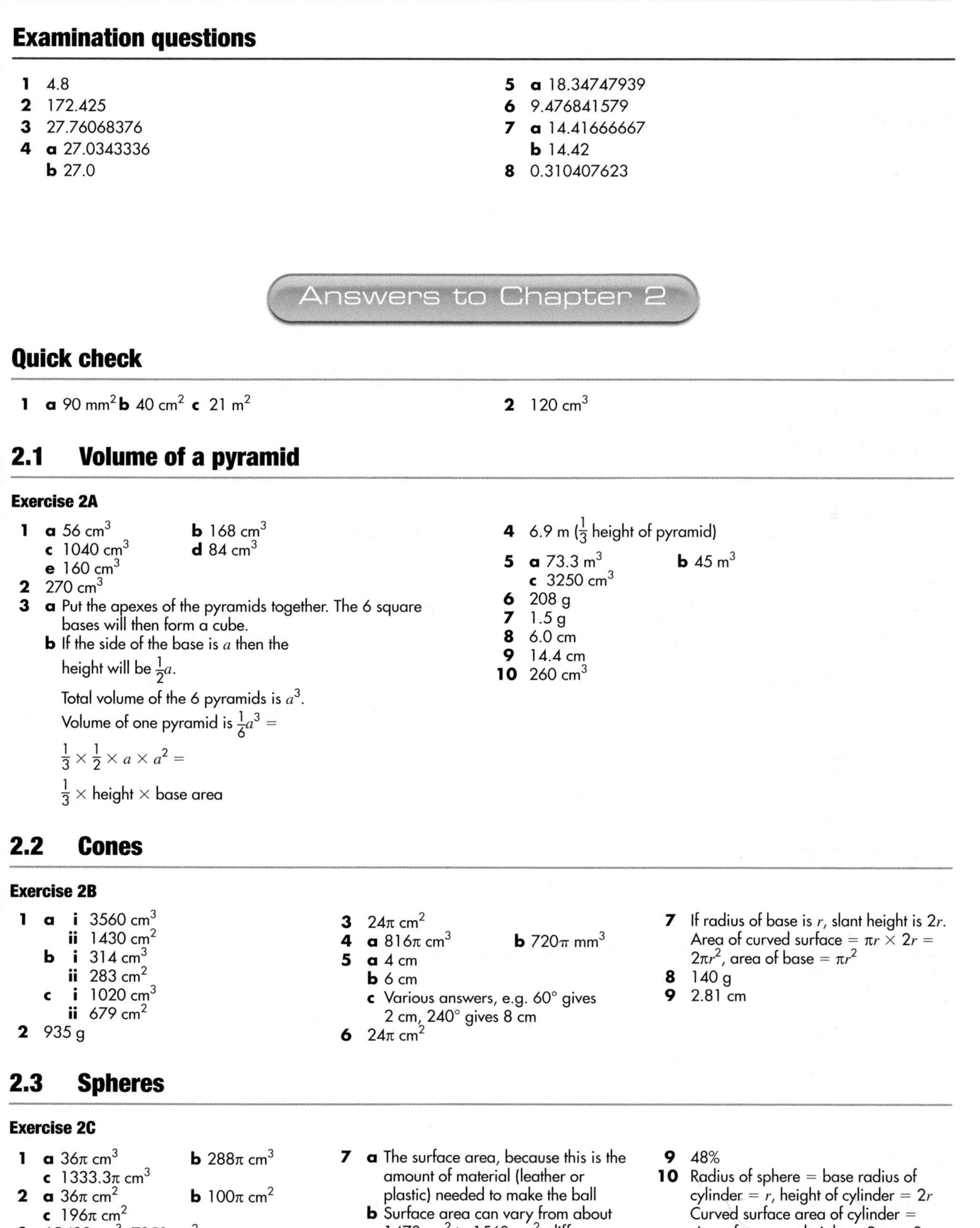

Examination questions

1 4.8
2 172.425
3 27.76068376
4 **a** 27.0343336
b 27.0
5 **a** 18.34747939
6 9.476841579
7 **a** 14.41666667
b 14.42
8 0.310407623

Answers to Chapter 2

Quick check

1 **a** 90 mm^2 **b** 40 cm^2 **c** 21 m^2
2 120 cm^3

2.1 Volume of a pyramid

Exercise 2A

1 **a** 56 cm^3 **b** 168 cm^3
c 1040 cm^3 **d** 84 cm^3
e 160 cm^3
2 270 cm^3
3 **a** Put the apexes of the pyramids together. The 6 square bases will then form a cube.
b If the side of the base is a then the height will be $\frac{1}{2}a$.
Total volume of the 6 pyramids is a^3.
Volume of one pyramid is $\frac{1}{6}a^3 =$
$\frac{1}{3} \times \frac{1}{2} \times a \times a^2 =$
$\frac{1}{3} \times$ height $\times$ base area
4 6.9 m ($\frac{1}{3}$ height of pyramid)
5 **a** 73.3 m^3 **b** 45 m^3
c 3250 cm^3
6 208 g
7 1.5 g
8 6.0 cm
9 14.4 cm
10 260 cm^3

2.2 Cones

Exercise 2B

1 **a** **i** 3560 cm^3
ii 1430 cm^2
b **i** 314 cm^3
ii 283 cm^2
c **i** 1020 cm^3
ii 679 cm^2
2 935 g
3 24π cm^2
4 **a** 816π cm^3 **b** 720π mm^3
5 **a** 4 cm
b 6 cm
c Various answers, e.g. 60° gives 2 cm, 240° gives 8 cm
6 24π cm^2
7 If radius of base is r, slant height is $2r$. Area of curved surface $= \pi r \times 2r = 2\pi r^2$, area of base $= \pi r^2$
8 140 g
9 2.81 cm

2.3 Spheres

Exercise 2C

1 **a** 36π cm^3 **b** 288π cm^3
c 1333.3π cm^3
2 **a** 36π cm^2 **b** 100π cm^2
c 196π cm^2
3 65 400 cm^3, 7850 cm^2
4 **a** 1960 cm^2 **b** 8180 cm^3
5 125 cm
6 6231
7 **a** The surface area, because this is the amount of material (leather or plastic) needed to make the ball
b Surface area can vary from about 1470 cm^2 to 1560 cm^2, difference of about 90 cm^2. This seems surprisingly large.
8 7.8 cm
9 48%
10 Radius of sphere = base radius of cylinder $= r$, height of cylinder $= 2r$
Curved surface area of cylinder = circumference $\times$ height $= 2\pi r \times 2r = 4\pi r^2$ = surface area of sphere

Examination questions

1 96π cm^2
2 $6x$
3 **a** 905 m^3
b 4.92 m

Quick check

1 5.3
2 246.5
3 0.6
4 2.8
5 16.1
6 0.7

3.1 Trigonometric ratios

Exercise 3A

1 **a** 0.682 **b** 0.829 **c** 0.922 **d** 1 **e** 0.707 **f** 0.342 **g** 0.375 **h** 0
2 **a** 0.731 **b** 0.559 **c** 0.388 **d** 0 **e** 0.707 **f** 0.940 **g** 0.927 **h** 1
3 45°
4 **a** **i** 0.574 **ii** 0.574
b **i** 0.208 **ii** 0.208
c **i** 0.391 **ii** 0.391
d i and ii are the same
e **i** sin 15° is the same as cos 75°
ii cos 82° is the same as sin 8°
iii sin x is the same as cos $(90° - x)$
5 **a** 0.933 **b** 1.48 **c** 2.38 **d** Infinite **e** 1 **f** 0.364 **g** 0.404 **h** 0
6 **a** 0.956 **b** 0.899 **c** 2.16 **d** 0.999 **e** 0.819 **f** 0.577 **g** 0.469 **h** 0.996
7 Has values > 1
8 **a** 4.53 **b** 4.46 **c** 6 **d** 0
9 **a** 10.7 **b** 5.40 **c** Infinite **d** 0
10 **a** 3.56 **b** 8.96 **c** 28.4 **d** 8.91
11 **a** 5.61 **b** 11.3 **c** 6 **d** 10
12 **a** 1.46 **b** 7.77 **c** 0.087 **d** 7.15
13 **a** 7.73 **b** 48.6 **c** 2.28 **d** 15.2
14 **a** 29.9 **b** 44.8 **c** 20.3 **d** 2.38
15 **a** $\frac{4}{5}, \frac{3}{5}, \frac{4}{3}$
b $\frac{5}{13}, \frac{12}{13}, \frac{5}{12}$
c $\frac{7}{25}, \frac{24}{25}, \frac{7}{24}$

3.2 Calculating angles

Exercise 3B

1 **a** 30.0° **b** 51.7° **c** 39.8° **d** 61.3° **e** 87.4° **f** 45.0°
2 **a** 60.0° **b** 50.2° **c** 2.6° **d** 45.0 **e** 78.5° **f** 45.6°
3 **a** 31.0° **b** 20.8° **c** 41.8° **d** 46.4° **e** 69.5° **f** 77.1°
4 **a** 53.1° **b** 41.8° **c** 44.4° **d** 56.4° **e** 2.4° **f** 22.6°
5 **a** 36.9° **b** 48.2° **c** 45.6° **d** 33.6° **e** 87.6° **f** 67.4°
6 **a** 31.0° **b** 37.9° **c** 15.9° **d** 60.9° **e** 57.5° **f** 50.2°
7 Error message, largest value 1, smallest value −1
8 **a** **i** 17.5° **ii** 72.5° **iii** 90°
b Yes

3.3 Using the sine and cosine functions

Exercise 3C

1 **a** 17.5° **b** 22.0° **c** 32.2°
2 **a** 5.29 cm **b** 5.75 cm **c** 13.2 cm
3 **a** 4.57 cm **b** 6.86 cm **c** 100 cm
4 **a** 5.12 cm **b** 9.77 cm **c** 11.7 cm **d** 15.5 cm
5 **a** 47.2° **b** 5.42 cm **c** 13.7 cm **d** 38.0°
6 **a** 6 **b** 15 **c** 30

Exercise 3D

1 **a** 51.3° **b** 75.5° **c** 51.3°
2 **a** 6.47 cm **b** 32.6 cm **c** 137 cm
3 **a** 7.32 cm **b** 39.1 cm **c** 135 cm
4 **a** 5.35 cm **b** 14.8 cm **c** 12.0 cm **d** 8.62 cm
5 **a** 5.59 cm **b** 46.6° **c** 9.91 cm **d** 40.1°
6 **a** 10 **b** 39 **c** 2.5

3.4 Using the tangent function

Exercise 3E

1 **a** 33.7° **b** 36.9° **c** 52.1°
2 **a** 5.09 cm **b** 30.4 cm
c 1120 cm
3 **a** 8.24 cm **b** 62.0 cm
c 72.8 cm
4 **a** 9.02 cm **b** 7.51 cm
c 7.14 cm **d** 8.90 cm
5 **a** 13.7 cm **b** 48.4°
c 7.03 cm **d** 41.2°
6 12, 12, 2

3.5 Which ratio to use

Exercise 3F

1 **a** 12.6 **b** 59.6 **c** 74.7
d 16.0 **e** 67.9 **f** 20.1
2 **a** 44.4° **b** 39.8° **c** 44.4°
d 49.5° **e** 58.7° **f** 38.7°
3 **a** 67.4° **b** 11.3 **c** 134
d 28.1° **e** 39.7 **f** 263
g 50.2° **h** 51.3° **i** 138
j 22.8
4 **a** Sides of right-hand triangle are sine and cosine
b Pythagoras' theorem
c Students should check the formulae

3.6 Solving problems using trigonometry 1

Exercise 3G

1 65°
2 The safe limits are between 1.04 m and 2.05 m. The ladder will reach between 5.64 m and 5.91 m up the wall.
3 44°
4 6.82 m
5 31°
6 **a** 25°
b 2.10 m
c Thickness of wood has been ignored
7 **a** 20° **b** 4.78 m
8 She would calculate 100 tan 23°. The answer is about 42.4 m
9 21.1 m
10 One way is stand opposite a feature, such as a tree, on the opposite bank, move a measured distance, x, along your bank and measure the angle, θ, between your bank and the feature. Width of river is $x \tan \theta$. This of course requires measuring equipment! An alternative is to walk along the bank until the angle is 45° (if that is possible). This angle is easily found by folding a sheet of paper. This way an angle measurer is not required.

Exercise 3H

1 10.1 km
2 22°
3 429 m
4 **a** 156 m
b No. the new angle of depression is $\tan^{-1}\left(\frac{200}{312}\right) = 33°$ and half of 52° is 26°
5 **a** 222 m **b** 42°
6 **a** 21.5 m **b** 17.8 m
7 13.4 m
8 19°
9 The angle is 16° so Cara is not quite correct.

3.7 Solving problems using trigonometry 2

Exercise 3I

1 **a** 73.4 km **b** 15.6 km
2 **a** 14.7 miles **b** 8.5 miles
3 120°
4 **a** 59.4 km **b** 8.4 km
5 **a** 15.9 km **b** 24.1 km
c 31.2 km **d** 052°
6 2.28 km
7 235°
8 **a** 66.2 km **b** 11.7 km
c 13.1 km **d** 170°
9 48.4 km, 100°

Exercise 3J

1 **a** 5.79 cm **b** 48.2°
c 7.42 cm **d** 81.6 cm
2 9.86 m
3 **a** 36.4 cm^2 **b** 115 cm^2
c 90.6 cm^2 **d** 160 cm^2
4 473 cm^2

Examination questions

1 **a** 33.7°
b 17.7 cm
2 34.8°
3 21.8°
4 **a** 30.7°
b 121°

Answers to Chapter 4

Quick check

1 $a = 50°$

2 $b = 140°$

3 $c = d = 65°$

4.1 Circle theorems

Exercise 4A

1 **a** 56° **b** 62° **c** 105° **d** 55° **e** 45° **f** 30° **g** 60° **h** 145°

2 **a** 55° **b** 52° **c** 50° **d** 24° **e** 39° **f** 80° **g** 34° **h** 30°

3 **a** 41° **b** 49° **c** 41°

4 **a** 72° **b** 37° **c** 72°

5 ∠AZY = 40° (angles in a triangle), $a = 50°$ (angle in a semicircle = 90°)

6 **a** $x = y = 40°$
b $x = 131°, y = 111°$
c $x = 134°, y = 23°$
d $x = 32°, y = 19°$
e $x = 59°, y = 121°$
f $x = 155°, y = 12.5°$

7 68°

8 ∠ABC = 180° − x (angles on a line), ∠AOC = 360° − $2x$ (angle at centre is twice angle at circumference), reflex ∠AOC = 360° − (360° − $2x$) = $2x$ (angles at a point)

9 **a** x
b $2x$
c ∠ABC = $(x + y)$ and ∠AOC = $2(x + y)$

4.2 Cyclic quadrilaterals

Exercise 4B

1 **a** $a = 50°, b = 95°$
b $c = 92°, x = 90°$
c $d = 110°, e = 110°, f = 70°$
d $g = 105°, h = 99°$
e $j = 89°, k = 89°, l = 91°$
f $m = 120°, n = 40°$
g $p = 44°, q = 68°$
h $x = 40°, y = 34°$

2 **a** $x = 26°, y = 128°$
b $x = 48°, y = 78°$
c $x = 133°, y = 47°$
d $x = 36°, y = 72°$
e $x = 55°, y = 125°$
f $x = 35°$
g $x = 48°, y = 45°$
h $x = 66°, y = 52°$

3 **a** $x = 49°, y = 49°$
b $x = 70°, y = 20°$
c $x = 80°, y = 100°$
d $x = 100°, y = 75°$

4 **a** $x = 50°, y = 62°$
b $x = 92°, y = 88°$
c $x = 93°, y = 42°$
d $x = 55°, y = 75°$

5 **a** $x = 95°, y = 138°$
b $x = 14°, y = 62°$
c $x = 32°, y = 48°$
d 52°

6 **a** 71° **b** 125.5° **c** 54.5°

7 **a** $x + 2x - 30° = 180°$ (opposite angles in a cyclic quadrilateral), so $3x - 30° = 180°$
b $x = 70°$, so $2x - 30° = 110°$ ∠DOB = 140° (angle at centre equals twice angle at circumference), $y = 80°$ (angles in a quadrilateral)

8 **a** x
b $360° - 2x$
c ∠ADC = $\frac{1}{2}$ reflex ∠AOC = $180° - x$, so ∠ADC + ∠ABC = 180°

9 Let ∠AED = x, then ∠ABC = x (opposite angles are equal in a parallelogram), ∠ADC = 180° − x (opposite angles in a cyclic quadrilateral), so ∠ADE = x (angles on a line)

4.3 Tangents and chords

Exercise 4C

1 **a** 38° **b** 110° **c** 15° **d** 45°

2 **a** 6 cm **b** 10.8 cm **c** 3.21 cm **d** 8 cm

3 **a** $x = 12°, y = 156°$
b $x = 100°, y = 50°$
c $x = 62°, y = 28°$
d $x = 30°, y = 60°$

4 **a** 62° **b** 66° **c** 19° **d** 20°

5 19.5 cm

6 5.77 cm

7 ∠OCD = 58° (triangle OCD is isosceles), ∠OCB = 90° (tangent/radius theorem), so ∠DCB = 32°, hence triangle BCD is isosceles (2 equal angles)

8 **a** ∠AOB = $\cos^{-1}\frac{OA}{OB} = \cos^{-1}\frac{OC}{OB}$ = ∠COB
b As ∠AOB = ∠COB, so ∠ABO = ∠CBO, so OB bisects ∠ABC

4.4 Alternate segment theorem

Exercise 4D

1 **a** $a = 65°, b = 75°, c = 40°$
b $d = 79°, e = 58°, f = 43°$
c $g = 41°, h = 76°, i = 76°$
d $k = 80°, m = 52°, n = 80°$

2 **a** $a = 75°, b = 75°, c = 75°, d = 30°$
b $a = 47°, b = 86°, c = 86°, d = 47°$
c $a = 53°, b = 53°$
d $a = 55°$

3 **a** 36° **b** 70°

4 **a** $x = 25°$
b $x = 46°, y = 69°, z = 65°$
c $x = 38°, y = 70°, z = 20°$
d $x = 48°, y = 42°$

5 ∠ACB = 64° (angle in alternate segment), ∠ACX = 116° (angles on a line), ∠CAX = 32° (angles in a triangle), so triangle ACX is isosceles (two equal angles)

6 ∠AXY = 69° (tangents equal and so triangle AXY is isosceles), ∠XZY = 69° (alternate segment), ∠XYZ = 55° (angles in a triangle)

7 **a** $2x$
b $90° - x$
c OPT = 90°, so angles APT = x = angle PBA

Examination questions

1 **a** **i** 90°
ii angle in a semicircle
b **i** 65°
ii angle at centre = twice angle at circumference

2 **a** **i** 140°
ii angle at centre = twice angle at circumference
b **i** 110°
ii opposite angles in a cyclic quadrilateral add up to 180°

3 **a** 90° **b** 140°
c angle OQP = 20° (isosceles triangle), angle PQT = 90° − 20° or PQT = 70° (isosceles triangle)

4 **a** 30°
b angle ABC = 75° (angle ABO = 90°, radius/tangent), triangle ABC is isosceles (equal tangents)

5 46°

6 angle OSR = $90° - x$ (radius meets tangent at 90°)
angle ORS = $90° - x$ (triangle OSR is isosceles)
angle ROS = $180° - (90° - x) - (90° - x) = 2x$

Answers to Chapter 5

Quick check

1 **a** 0.6 **b** 0.44 **c** 0.375

2 **a** $\frac{17}{100}$ **b** $\frac{16}{25}$ **c** $\frac{429}{500}$

3 **a** $\frac{13}{15}$ **b** $\frac{9}{40}$

4 **a** 5 **b** 4

5.1 Powers (indices)

Exercise 5A

1 **a** 2^4 **b** 3^5 **c** 7^2 **d** 5^3
e 10^7 **f** 6^4 **g** 4^1 **h** 1^7
i 0.5^4 **j** 100^3

2 **a** $3 \times 3 \times 3 \times 3$
b $9 \times 9 \times 9$
c 6×6
d $10 \times 10 \times 10 \times 10 \times 10$
e $2 \times 2 \times 2 \times 2 \times 2 \times 2 \times 2 \times 2 \times 2 \times 2$
f 8
g $0.1 \times 0.1 \times 0.1$
h 2.5×2.5
i $0.7 \times 0.7 \times 0.7$
j 1000×1000

3 **a** 16 **b** 243
c 49 **d** 125
e 10 000 000 **f** 1296
g 4 **h** 1
i 0.0625 **j** 1 000 000

4 **a** 81 **b** 729
c 36 **d** 100 000
e 1024 **f** 8
g 0.001 **h** 6.25
i 0.343 **j** 1 000 000

5 125 m^3

6 **b** 10^2 **c** 2^3 **d** 5^2

7 **a** 1 **b** 4 **c** 1
d 1 **e** 1

8 Any power of 1 is equal to 1.

9 10^6

10 10^6

11 **a** 1 **b** −1 **c** 1
d 1 **e** −1

12 **a** 1 **b** −1 **c** −1
d 1 **e** 1

13 2^{24}, 4^{12}, 8^8, 16^6

Exercise 5B

1 **a** $\frac{1}{5^3}$ **b** $\frac{1}{6}$ **c** $\frac{1}{10^5}$ **d** $\frac{1}{3^2}$
e $\frac{1}{8^2}$ **f** $\frac{1}{9}$ **g** $\frac{1}{w^2}$ **h** $\frac{1}{t}$
i $\frac{1}{x^m}$ **j** $\frac{4}{m^3}$

2 **a** 3^{-2} **b** 5^{-1} **c** 10^{-3} **d** m^{-1}
e t^{-n}

3 **a** **i** 2^4 **ii** 2^{-1} **iii** 2^{-4} **iv** -2^3
b **i** 10^3 **ii** 10^{-1} **iii** 10^{-2} **iv** 10^6
c **i** 5^3 **ii** 5^{-1} **iii** 5^{-2} **iv** 5^{-4}
d **i** 3^2 **ii** 3^{-3} **iii** 3^{-4} **iv** -3^5

4 **a** $\frac{5}{x^3}$ **b** $\frac{6}{t}$ **c** $\frac{7}{m^2}$ **d** $\frac{4}{q^4}$
e $\frac{10}{y^5}$ **f** $\frac{1}{2x^3}$ **g** $\frac{1}{2m}$ **h** $\frac{3}{4t^4}$
i $\frac{4}{5y^3}$ **j** $\frac{7}{8x^5}$

5 **a** $7x^{-3}$ **b** $10p^{-1}$ **c** $5t^{-2}$
d $8m^{-5}$ **e** $3y^{-1}$

6 **a** **i** 25 **ii** $\frac{1}{125}$ **iii** $\frac{4}{5}$
b **i** 64 **ii** $\frac{1}{16}$ **iii** $\frac{5}{256}$
c **i** 8 **ii** $\frac{1}{32}$ **iii** $\frac{9}{2}$ or $4\frac{1}{2}$
d **i** 1 000 000 **ii** $\frac{1}{1000}$ **iii** $\frac{1}{4}$

7 24 (32 − 8)
8 $x = 8$ and $y = 4$ (or $x = y = 1$)

9 $\frac{1}{2097152}$

10 **a** x^{-5}, x^0, x^5 **b** x^5, x^0, x^{-5} **c** x^5, x^{-5}, x^0

Exercise 5C

1 **a** 5^4 **b** 5^3 **c** 5^2 **d** 5^3 **e** 5^{-5}
2 **a** 6^3 **b** 6^0 **c** 6^6 **d** 6^{-7} **e** 6^2
3 **a** a^3 **b** a^5 **c** a^7 **d** a^4 **e** a^2 **f** a^1
4 **a** Any two values such that $x + y = 10$
b Any two values such that $x - y = 10$

5 **a** 4^6 **b** 4^{15} **c** 4^6 **d** 4^{-6} **e** 4^6 **f** 4^0
6 **a** $6a^5$ **b** $9a^2$ **c** $8a^6$ **d** $-6a^4$ **e** $8a^8$ **f** $-10a^{-3}$
7 **a** $3a$ **b** $4a^3$ **c** $3a^4$ **d** $6a^{-1}$ **e** $4a^7$ **f** $5a^{-4}$
8 **a** $8a^5b^4$ **b** $10a^3b$ **c** $30a^{-2}b^{-2}$ **d** $2ab^3$ **e** $8a^{-5}b^7$

9 **a** $3a^3b^2$ **b** $3a^2c^4$ **c** $8a^2b^2c^3$
10 **a** Possible answer: $6x^2 \times 2y^5$ and $3xy \times 4xy^4$
b Possible answer: $24x^2y^7 \div 2y^2$ and $12x^6y^8 \div x^4y^3$
11 12 ($a = 2$, $b = 1$, $c = 3$)
12 $1 = \frac{a^x}{a^x} = a^x \div a^x = a^{x-x} = a^0$

Exercise 5D

1 **a** 5 **b** 10 **c** 8 **d** 9 **e** 25 **f** 3 **g** 4 **h** 10 **i** 5 **j** 8 **k** 12 **l** 20 **m** 5 **n** 3 **o** 10 **p** 3 **q** 2 **r** 2 **s** 6 **t** 6 **u** $\frac{1}{4}$ **v** $\frac{1}{2}$ **w** $\frac{1}{3}$ **x** $\frac{1}{5}$ **y** $\frac{1}{10}$

2 **a** $\frac{5}{6}$ **b** $1\frac{2}{3}$ **c** $\frac{8}{9}$ **d** $1\frac{4}{5}$ **e** $\frac{5}{8}$ **f** $\frac{3}{5}$ **g** $\frac{1}{4}$ **h** $2\frac{1}{2}$ **i** $\frac{4}{5}$ **j** $1\frac{1}{7}$
3 $(x^{\frac{1}{n}})^n = x^{\frac{n}{n}} = x^1 = x$, and
$(\sqrt[n]{x})^n = \sqrt[n]{x} \times \sqrt[n]{x} \ldots n$ times $= x$,
so $x^{\frac{1}{n}} = \sqrt[n]{x}$

4 $64^{-\frac{1}{2}} = \frac{1}{8}$, others are both $\frac{1}{2}$
5 Possible answer: The negative power gives the reciprocal, so $27^{-\frac{1}{3}} = \frac{1}{27^{\frac{1}{3}}}$
The power one-third means cube root, so you need the cube root of 27 which is 3, so $27^{\frac{1}{3}} = 3$ and $\frac{1}{27^{\frac{1}{3}}} = \frac{1}{3}$
6 Possible answer: $x = 1$ and $y = 1$, $x = 8$ and $y = \frac{1}{64}$

Exercise 5E

1 **a** 16 **b** 25 **c** 216 **d** 81
2 **a** $t^{\frac{2}{3}}$ **b** $m^{\frac{3}{4}}$ **c** $k^{\frac{2}{5}}$ **d** $x^{\frac{3}{2}}$
3 **a** 4 **b** 9 **c** 64 **d** 3125
4 **a** $\frac{1}{5}$ **b** $\frac{1}{6}$ **c** $\frac{1}{2}$ **d** $\frac{1}{3}$ **e** $\frac{1}{4}$ **f** $\frac{1}{2}$ **g** $\frac{1}{2}$ **h** $\frac{1}{3}$

5 **a** $\frac{1}{125}$ **b** $\frac{1}{216}$ **c** $\frac{1}{8}$ **d** $\frac{1}{27}$ **e** $\frac{1}{256}$ **f** $\frac{1}{4}$ **g** $\frac{1}{4}$ **h** $\frac{1}{9}$
6 **a** $\frac{1}{100000}$ **b** $\frac{1}{12}$ **c** $\frac{1}{25}$ **d** $\frac{1}{27}$ **e** $\frac{1}{32}$ **f** $\frac{1}{32}$ **g** $\frac{1}{81}$ **h** $\frac{1}{13}$
7 $8^{-\frac{2}{3}} = \frac{1}{4}$, others are both $\frac{1}{8}$

8 Possible answer: The negative power gives the reciprocal, so $27^{-\frac{2}{3}} = \frac{1}{27^{\frac{2}{3}}}$
The power one-third means cube root, so we need the cube root of 27 which is 3 and the power 2 means square, so $3^2 = 9$, so $27^{\frac{2}{3}} = 9$ and $\frac{1}{27^{\frac{2}{3}}} = \frac{1}{9}$

5.2 Standard form

Exercise 5F

1 **a** 31 **b** 310 **c** 3100 **d** 31 000
2 **a** 65 **b** 650 **c** 6500 **d** 65000
3 **a** 0.31 **b** 0.031 **c** 0.0031 **d** 0.000 31
4 **a** 0.65 **b** 0.065 **c** 0.0065 **d** 0.000 65
5 **a** 250 **b** 34.5 **c** 4670 **d** 346 **e** 207.89 **f** 56780 **g** 246 **h** 0.76 **i** 999 000 **j** 23 456 **k** 98 765.4 **l** 43 230 000 **m** 345.78 **n** 6000 **o** 56.7 **p** 560 045
6 **a** 0.025 **b** 0.345 **c** 0.004 67 **d** 3.46 **e** 0.207 89 **f** 0.056 78 **g** 0.0246 **h** 0.0076 **i** 0.000 000 999 **j** 2.3456 **k** 0.098 7654 **l** 0.000 043 23 **m** 0.000 000 034 578 **n** 0.000 000 000 06 **o** 0.000 000 567 **p** 0.005 600 45
7 **a** 230 **b** 578 900 **c** 4790 **d** 57 000 000 **e** 216 **f** 10 500 **g** 0.000 32 **h** 9870
8 **a**, **b** and **c**
9 Venus, because power 24 means more digits in the answer.
10 6

Exercise 5G

1 **a** 0.31 **b** 0.031 **c** 0.0031 **d** 0.000 31
2 **a** 0.65 **b** 0.065 **c** 0.0065 **d** 0.000 65
3 **a** $9999999999 \times 10^{99}$
b $0.000000001 \times 10^{-99}$ (depending on number of digits displayed)
4 **a** 31 **b** 310 **c** 3100 **d** 31 000
5 **a** 65 **b** 650 **c** 6500 **d** 65 000
6 **a** 250 **b** 34.5 **c** 0.004 67 **d** 34.6 **e** 0.020 789 **f** 5678 **g** 246 **h** 7600 **i** 897 000 **j** 0.008 65 **k** 60 000 000 **l** 0.000 567
7 **a** 2.5×10^2 **b** 3.45×10^{-1} **c** 4.67×10^4 **d** 3.4×10^9 **e** 2.078×10^{10} **f** 5.678×10^{-4} **g** 2.46×10^3 **h** 7.6×10^{-2} **i** 7.6×10^{-4} **j** 9.99×10^{-1} **k** 2.3456×10^2 **l** 9.87654×10^1 **m** 6×10^{-4} **n** 5.67×10^{-3} **o** 5.60045×10^1
8 2.7797×10^4
9 2.81581×10^5, 3×10^1, 1.382101×10^6
10 1.298×10^7, 2.997×10^9, 9.3×10^4
11 100
12 36 miles

Exercise 5H

1 **a** 5.67×10^3 **b** 6×10^2 **c** 3.46×10^{-1} **d** 7×10^{-4} **e** 5.6×10^2 **f** 6×10^5 **g** 7×10^3 **h** 1.6 **i** 2.3×10^7 **j** 3×10^{-6} **k** 2.56×10^6 **l** 4.8×10^2 **m** 1.12×10^2 **n** 6×10^{-1} **o** 2.8×10^6

2 **a** 1.08×10^8 **b** 4.8×10^6 **c** 1.2×10^9 **d** 1.08 **e** 6.4×10^2 **f** 1.2×10^1 **g** 2.88 **h** 2.5×10^7 **i** 8×10^{-6}

3 **a** 1.1×10^8 **b** 6.1×10^6 **c** 1.6×10^9 **d** 3.9×10^{-2} **e** 9.6×10^8 **f** 4.6×10^{-7} **g** 2.1×10^3 **h** 3.6×10^7 **i** 1.5×10^2 **j** 3.5×10^9 **k** 1.6×10^4

4 **a** 2.7×10 **b** 1.6×10^{-2} **c** 2×10^{-1} **d** 4×10^{-8} **e** 2×10^5 **f** 6×10^{-2} **g** 2×10^{-5} **h** 5×10^2 **i** 2×10

5 **a** 5.4×10 **b** 2.9×10^{-3} **c** 1.1 **d** 6.3×10^{-10} **e** 2.8×10^2 **f** 5.5×10^{-2} **g** 4.9×10^2 **h** 8.6×10^6

6 2×10^{13}, 1×10^{-10}, mass $= 2 \times 10^3$ g (2 kg)

7 **a** (2^{63}) 9.2×10^{18} grains **b** $2^{64} - 1 = 1.8 \times 10^{19}$

8 **a** 6×10^7 sq miles **b** 30%

9 1.5×10^7 sq miles

10 5×10^4

11 2.3×10^5

12 455 070 000 kg or 455 070 tonnes

13 80 000 000 (80 million)

14 **a** 2.048×10^6 **b** 4.816×10^6

15 250

16 9.41×10^4

17 Any value from 1.00000001×10^8 to 1×10^9 (excluding 1×10^9), i.e. any value of the form $a \times 10^8$ where $1 \leqslant a < 10$

5.3 Rational numbers and reciprocals

Exercise 5I

1 **a** 0.5 **b** $0.\dot{3}$ **c** 0.25 **d** 0.2 **e** $0.1\dot{6}$ **f** $0.\dot{1}4285\dot{7}$ **g** 0.125 **h** $0.\dot{1}$ **i** 0.1 **j** $0.\dot{0}7692\dot{3}$

2 **a** $\frac{4}{7} = 0.571\,428\,5\ldots$

$\frac{5}{7} = 0.714\,285\,7\ldots$

$\frac{6}{7} = 0.857\,142\,8\ldots$

b They all contain the same pattern of digits, starting at a different point in the pattern.

3 $0.\dot{1}$, $0.\dot{2}$, $0.\dot{3}$, etc. Digit in decimal fraction same as numerator.

4 $0.\dot{0}\dot{9}$, $0.\dot{1}\dot{8}$, $0.\dot{2}\dot{7}$, etc. Recurring digits follow the nine times table.

5 0.444 ..., 0.454 ..., 0.428 ..., 0.409 ..., 0.432 ..., 0.461 ...;

$\frac{9}{22}, \frac{3}{7}, \frac{16}{37}, \frac{4}{9}, \frac{5}{11}, \frac{6}{13}$

6 $\frac{38}{120}, \frac{35}{120}, \frac{36}{120}, \frac{48}{120}, \frac{50}{120}$

$\frac{7}{24}, \frac{3}{10}, \frac{19}{60}, \frac{2}{5}, \frac{5}{12}$

7 **a** $\frac{1}{8}$ **b** $\frac{17}{50}$ **c** $\frac{29}{40}$ **d** $\frac{5}{16}$ **e** $\frac{89}{100}$ **f** $\frac{1}{20}$ **g** $2\frac{7}{20}$ **h** $\frac{7}{32}$

8 **a** $0.08\dot{3}$ **b** 0.0625 **c** 0.05 **d** 0.04 **e** 0.02

9 **a** $\frac{4}{3}$ **b** $\frac{6}{5}$ **c** $\frac{5}{2}$ **d** $\frac{10}{7}$ **e** $\frac{20}{11}$ **f** $\frac{15}{4}$

10 **a** 0.75, $1.\dot{3}$; $0.8\dot{3}$, 1.2; 0.4, 2.5; 0.7, $1.\dot{4}2857\dot{1}$; 0.55, $1.\dot{8}\dot{1}$; $0.2\dot{6}$, 3.75

b Not always true, e.g. reciprocal of 0.4 $(\frac{2}{5})$ is $\frac{5}{2} = 2.5$

11 $1 \div 0$ is infinite, so there is no finite answer.

12 **a** 10

b 2

c The reciprocal of a reciprocal is always the original number.

13 The reciprocal of x is greater than the reciprocal of y. For example, $2 < 10$, reciprocal of 2 is 0.5, reciprocal of 10 is 0.1, and $0.5 > 0.1$

14 Possible answer: $-\frac{1}{2} \times -2 = 1$, $-\frac{1}{3} \times -3 = 1$

15 **a** 24.24242 ... **b** 24 **c** $\frac{24}{99} = \frac{8}{33}$

16 **a** $\frac{8}{9}$ **b** $\frac{34}{99}$ **c** $\frac{5}{11}$ **d** $\frac{21}{37}$ **e** $\frac{4}{9}$ **f** $\frac{2}{45}$ **g** $\frac{13}{90}$ **h** $\frac{1}{22}$ **i** $2\frac{7}{9}$ **j** $7\frac{7}{11}$ **k** $3\frac{1}{3}$ **l** $2\frac{2}{33}$

17 **a** true **b** true **c** recurring

18 **a** $\frac{9}{9}$ **b** $\frac{45}{90} = \frac{1}{2} = 0.5$

5.4 Surds

Exercise 5J

1 **a** $\sqrt{6}$ **b** $\sqrt{15}$ **c** 2 **d** 4 **e** $2\sqrt{10}$ **f** 3 **g** $2\sqrt{3}$ **h** $\sqrt{21}$ **i** $\sqrt{14}$ **j** 6 **k** 6 **l** $\sqrt{30}$

2 **a** 2 **b** $\sqrt{5}$ **c** $\sqrt{6}$ **d** $\sqrt{3}$ **e** $\sqrt{5}$ **f** 1 **g** $\sqrt{3}$ **h** $\sqrt{7}$ **i** 2 **j** $\sqrt{6}$ **k** 1 **l** 3

3 **a** $2\sqrt{3}$ **b** 15 **c** $4\sqrt{2}$ **d** $4\sqrt{3}$ **e** $8\sqrt{5}$ **f** $3\sqrt{3}$ **g** 24 **h** $3\sqrt{7}$ **i** $2\sqrt{7}$ **j** $6\sqrt{5}$ **k** $6\sqrt{3}$ **l** 30

4 **a** $\sqrt{3}$ **b** 1 **c** $2\sqrt{2}$ **d** $\sqrt{2}$ **e** $\sqrt{5}$ **f** $\sqrt{3}$ **g** $\sqrt{2}$ **h** $\sqrt{7}$ **i** $\sqrt{7}$ **j** $2\sqrt{3}$ **k** $2\sqrt{3}$ **l** 1

5 **a** a **b** 1 **c** $\sqrt{a}$

6 **a** $3\sqrt{2}$ **b** $2\sqrt{6}$ **c** $2\sqrt{3}$ **d** $5\sqrt{2}$ **e** $2\sqrt{2}$ **f** $3\sqrt{3}$ **g** $4\sqrt{3}$ **h** $5\sqrt{3}$ **i** $3\sqrt{5}$ **j** $3\sqrt{7}$ **k** $4\sqrt{2}$ **l** $10\sqrt{2}$ **m** $10\sqrt{10}$ **n** $5\sqrt{10}$ **o** $7\sqrt{2}$ **p** $9\sqrt{3}$

7 **a** 36 **b** $16\sqrt{30}$ **c** 54 **d** 32 **e** $48\sqrt{6}$ **f** $48\sqrt{6}$ **g** $18\sqrt{15}$ **h** 84 **i** 64 **j** 100 **k** 50 **l** 56

8 **a** $20\sqrt{6}$ **b** $6\sqrt{15}$ **c** 24 **d** 16 **e** $12\sqrt{10}$ **f** 18 **g** $20\sqrt{3}$ **h** $10\sqrt{21}$ **i** $6\sqrt{14}$ **j** 36 **k** 24 **l** $12\sqrt{30}$

9 **a** 6 **b** $3\sqrt{5}$ **c** $6\sqrt{6}$ **d** $2\sqrt{3}$ **e** $4\sqrt{5}$ **f** 5 **g** $7\sqrt{3}$ **h** $2\sqrt{7}$ **i** 6 **j** $2\sqrt{7}$ **k** 5 **l** 24

10 **a** $2\sqrt{3}$ **b** 4 **c** $6\sqrt{2}$ **d** $4\sqrt{2}$ **e** $6\sqrt{5}$ **f** $24\sqrt{3}$ **g** $3\sqrt{2}$ **h** $\sqrt{7}$ **i** $10\sqrt{7}$ **j** $8\sqrt{3}$ **k** $10\sqrt{3}$ **l** 6

11 **a** abc **b** $\frac{a}{c}$ **c** $c\sqrt{b}$

12 **a** 20 **b** 24 **c** 10 **d** 24 **e** 3 **f** 6

13 **a** $\frac{3}{4}$ **b** $8\frac{1}{3}$ **c** $\frac{5}{16}$ **d** 12 **e** 2

14 **a** False **b** False

15 Possible answer: $\sqrt{3} \times 2\sqrt{3}$ (= 6)

Exercise 5K

1 Expand the brackets each time.
2 **a** $2\sqrt{3} - 3$ **b** $3\sqrt{2} - 8$ **c** $10 + 4\sqrt{5}$
d $12\sqrt{7} - 42$ **e** $15\sqrt{2} - 24$ **f** $9 - \sqrt{3}$
3 **a** $2\sqrt{3}$ **b** $1 + \sqrt{5}$ **c** $-1 - \sqrt{2}$
d $\sqrt{7} - 30$ **e** -41 **f** $7 + 3\sqrt{6}$
g $9 + 4\sqrt{5}$ **h** $3 - 2\sqrt{2}$ **i** $11 + 6\sqrt{2}$
4 **a** $3\sqrt{2}$ cm **b** $2\sqrt{3}$ cm **c** $2\sqrt{10}$ cm
5 **a** $\sqrt{3} - 1$ cm^2 **b** $2\sqrt{5} + 5\sqrt{2}$ cm^2 **c** $2\sqrt{3} + 18$ cm^2
6 **a** $\frac{\sqrt{3}}{3}$ **b** $\frac{\sqrt{2}}{2}$ **c** $\frac{\sqrt{5}}{5}$ **d** $\frac{\sqrt{3}}{6}$ **e** $\sqrt{3}$ **f** $\frac{5\sqrt{2}}{2}$
g $\frac{3}{2}$ **h** $\frac{5\sqrt{2}}{2}$ **i** $\frac{\sqrt{21}}{3}$ **j** $\frac{\sqrt{2} + 2}{2}$ **k** $\frac{2\sqrt{3} - 3}{3}$
l $\frac{5\sqrt{3} + 6}{3}$
7 **a** **i** 1 **ii** -4 **iii** 2
iv 17 **v** -44
b They become whole numbers. Difference of two squares makes the 'middle terms' (and surds) disappear.
8 **a** Possible answer: $\sqrt{2}$ and $\sqrt{2}$ or $\sqrt{2}$ and $\sqrt{8}$
b Possible answer: $\sqrt{2}$ and $\sqrt{3}$
9 **a** Possible answer: $\sqrt{2}$ and $\sqrt{2}$ or $\sqrt{8}$ and $\sqrt{2}$
b Possible answer: $\sqrt{3}$ and $\sqrt{2}$
10 Possible answer: $80^2 = 6400$, so $80 = \sqrt{6400}$ and $10\sqrt{70} = \sqrt{7000}$
Since $6400 < 7000$, there is not enough cable.
11 $9 + 6\sqrt{2} + 2 - (1 - 2\sqrt{8} + 8) = 11 - 9 + 6\sqrt{2} + 4\sqrt{2}$
$= 2 + 10\sqrt{2}$
12 $x^2 - y^2 = (1 + \sqrt{2})^2 - (1 - \sqrt{8})^2 =$
$1 + 2\sqrt{2} + 2 - (1 - 2\sqrt{8} + 8) =$
$3 - 9 + 2\sqrt{2} + 4\sqrt{2} = -6 + 6\sqrt{2}$
$(x + y)(x - y) = (2 - \sqrt{2})(3\sqrt{2}) = 6\sqrt{2} - 6$

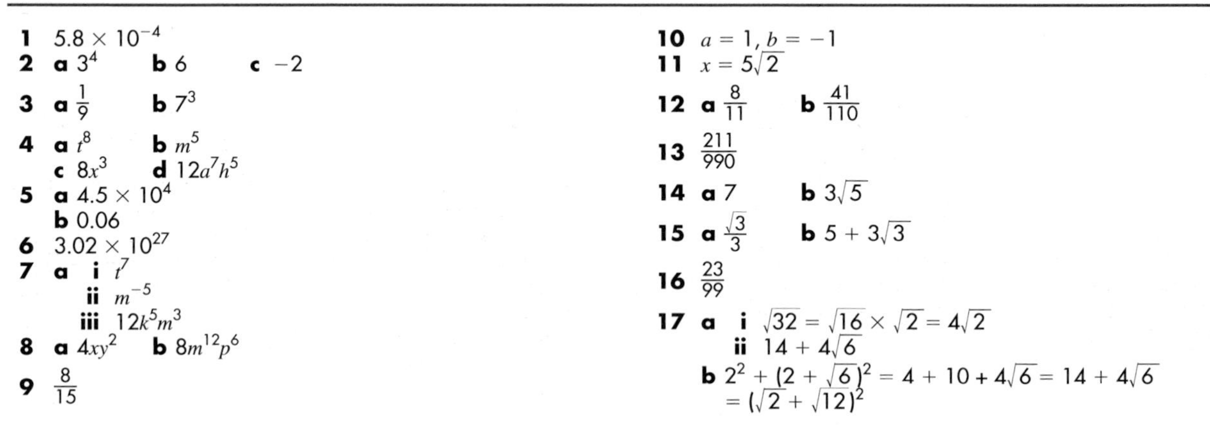

Examination questions

1 5.8×10^{-4}
2 **a** 3^4 **b** 6 **c** -2
3 **a** $\frac{1}{9}$ **b** 7^3
4 **a** t^8 **b** m^5
c $8x^3$ **d** $12a^7h^5$
5 **a** 4.5×10^4
b 0.06
6 3.02×10^{27}
7 **a** **i** t^7
ii m^{-5}
iii $12k^5m^3$
8 **a** $4xy^2$ **b** $8m^{12}p^6$
9 $\frac{8}{15}$
10 $a = 1, b = -1$
11 $x = 5\sqrt{2}$
12 **a** $\frac{8}{11}$ **b** $\frac{41}{110}$
13 $\frac{211}{990}$
14 **a** 7 **b** $3\sqrt{5}$
15 **a** $\frac{\sqrt{3}}{3}$ **b** $5 + 3\sqrt{3}$
16 $\frac{23}{99}$
17 **a** **i** $\sqrt{32} = \sqrt{16} \times \sqrt{2} = 4\sqrt{2}$
ii $14 + 4\sqrt{6}$
b $2^2 + (2 + \sqrt{6})^2 = 4 + 10 + 4\sqrt{6} = 14 + 4\sqrt{6}$
$= (\sqrt{2} + \sqrt{12})^2$

Quick check

1 **a** $-3x$ **b** $2x$ **c** $-3x$ **d** $6m^2$ **e** $-6x^2$ **f** $-12p^2$ **2** **a** -6 **b** $-\frac{1}{2}$ **c** $\frac{2}{3}$

6.1 Expanding brackets

Exercise 6A

1 $x^2 + 5x + 6$
2 $t^2 + 7t + 12$
3 $w^2 + 4w + 3$
4 $m^2 + 6m + 5$
5 $k^2 + 8k + 15$
6 $a^2 + 5a + 4$
7 $x^2 + 2x - 8$
8 $t^2 + 2t - 15$
9 $w^2 + 2w - 3$
10 $f^2 - f - 6$
11 $g^2 - 3g - 4$
12 $y^2 + y - 12$
13 $x^2 + x - 12$
14 $p^2 - p - 2$
15 $k^2 - 2k - 8$
16 $y^2 + 3y - 10$
17 $a^2 + 2a - 3$
18 $x^2 - 9$
19 $t^2 - 25$
20 $m^2 - 16$
21 $t^2 - 4$
22 $y^2 - 64$
23 $p^2 - 1$
24 $25 - x^2$
25 $49 - g^2$
26 $x^2 - 36$
27 $(x + 2)$ and $(x + 3)$
28 **a** B: 1; $(x - 2)$; $1(x - 2)$
C: 1; 2; 1×2
D: $(x - 1)$; 2; $2(x - 1)$
b $(x - 2) + 2 + 2(x - 1)$
$= 3x - 2$
c Area A $= (x - 1)(x - 2)$
= area of square minus areas (B + C + D)
$= x^2 - (3x - 2)$
$= x^2 - 3x + 2$
29 **a** $x^2 - 9$
b **i** 9991
ii 39991

Exercise 6B

1 $6x^2 + 11x + 3$
2 $12y^2 + 17y + 6$
3 $6t^2 + 17t + 5$
4 $8t^2 + 2t - 3$
5 $10m^2 - 11m - 6$
6 $12k^2 - 11k - 15$
7 $6p^2 + 11p - 10$
8 $10w^2 + 19w + 6$
9 $6a^2 - 7a - 3$
10 $8r^2 - 10r + 3$
11 $15g^2 - 16g + 4$
12 $12d^2 + 5d - 2$
13 $8p^2 + 26p + 15$
14 $6t^2 + 7t + 2$
15 $6p^2 + 11p + 4$
16 $6 - 7t - 10t^2$
17 $12 + n - 6n^2$
18 $6f^2 - 5f - 6$
19 $12 + 7q - 10q^2$
20 $3 - 7p - 6p^2$
21 $4 + 10t - 6t^2$
22 **a** $x^2 + 2x + 1$ **b** $x^2 - 2x + 1$ **c** $x^2 - 1$ **d** $p + q = (x + 1 + x - 1) = 2x$
$(p + q)^2 = (2x)^2 = 4x^2$
$p^2 + 2pq + q^2 = x^2 + 2x + 1 + 2(x^2 - 1) + x^2 - 2x + 1$
$= 4x^2 + 2x - 2x + 2 - 2 = 4x^2$
23 **a** $(3x - 2)(2x + 1) = 6x^2 - x - 2$, $(2x - 1)(2x - 1) = 4x^2 - 4x + 1$, $(6x - 3)(x + 1) = 6x^2 + 3x - 3$, $(3x + 2)(2x + 1) = 6x^2 + 7x + 2$ **b** Multiply the x terms to match the x^2 term and/or multiply the constant terms to get the constant term in the answer.

Exercise 6C

1 $4x^2 - 1$
2 $9t^2 - 4$
3 $25y^2 - 9$
4 $16m^2 - 9$
5 $4k^2 - 9$
6 $16h^2 - 1$
7 $4 - 9x^2$
8 $25 - 4t^2$
9 $36 - 25y^2$
10 $a^2 - b^2$
11 $9t^2 - k^2$
12 $4m^2 - 9p^2$
13 $25k^2 - g^2$
14 $a^2b^2 - c^2d^2$
15 $a^4 - b^4$
16 **a** $a^2 - b^2$ **b** Dimensions: $a + b$ by $a - b$; Area: $a^2 - b^2$ **c** Areas are the same, so $a^2 - b^2 = (a + b) \times (a - b)$
17 First shaded area is $(2k)^2 - 1^2 = 4k^2 - 1$
Second shaded area is $(2k + 1)(2k - 1) = 4k^2 - 1$

Exercise 6D

1 $x^2 + 10x + 25$
2 $m^2 + 8m + 16$
3 $t^2 + 12t + 36$
4 $p^2 + 6p + 9$
5 $m^2 - 6m + 9$
6 $t^2 - 10t + 25$
7 $m^2 - 8m + 16$
8 $k^2 - 14k + 49$
9 $9x^2 + 6x + 1$
10 $16t^2 + 24t + 9$
11 $25y^2 + 20y + 4$
12 $4m^2 + 12m + 9$
13 $16t^2 - 24t + 9$
14 $9x^2 - 12x + 4$
15 $25t^2 - 20t + 4$
16 $25r^2 - 60r + 36$
17 $x^2 + 2xy + y^2$
18 $m^2 - 2mn + n^2$
19 $4t^2 + 4ty + y^2$
20 $m^2 - 6mn + 9n^2$
21 $x^2 + 4x$
22 $x^2 - 10x$
23 $x^2 + 12x$
24 $x^2 - 4x$
25 **a** Bernice has just squared the first term and the second term. She hasn't written down the brackets twice. **b** Pete has written down the brackets twice but has worked out $(3x)^2$ as $3x^2$ and not $9x^2$. **c** $9x^2 + 6x + 1$
26 Whole square is $(2x)^2 = 4x^2$.
Three areas are $2x - 1$, $2x - 1$ and 1.
$4x^2 - (2x - 1 + 2x - 1 + 1) = 4x^2 - (4x - 1) = 4x^2 - 4x + 1$

6.2 Quadratic factorisation

Exercise 6E

1 $(x + 2)(x + 3)$
2 $(t + 1)(t + 4)$
3 $(m + 2)(m + 5)$
4 $(k + 4)(k + 6)$
5 $(p + 2)(p + 12)$
6 $(r + 3)(r + 6)$
7 $(w + 2)(w + 9)$
8 $(x + 3)(x + 4)$
9 $(a + 2)(a + 6)$
10 $(k + 3)(k + 7)$
11 $(f + 1)(f + 21)$
12 $(b + 8)(b + 12)$
13 $(t - 2)(t - 3)$
14 $(d - 4)(d - 1)$
15 $(g - 2)(g - 5)$
16 $(x - 3)(x - 12)$
17 $(c - 2)(c - 16)$
18 $(t - 4)(t - 9)$
19 $(y - 4)(y - 12)$
20 $(j - 6)(j - 8)$
21 $(p - 3)(p - 5)$
22 $(y + 6)(y - 1)$
23 $(t + 4)(t - 2)$
24 $(x + 5)(x - 2)$
25 $(m + 2)(m - 6)$
26 $(r + 1)(r - 7)$
27 $(n + 3)(n - 6)$
28 $(m + 4)(m - 11)$
29 $(w + 4)(w - 6)$
30 $(t + 9)(t - 10)$
31 $(h + 8)(h - 9)$
32 $(t + 7)(t - 9)$
33 $(d + 1)^2$
34 $(y + 10)^2$
35 $(t - 4)^2$
36 $(m - 9)^2$
37 $(x - 12)^2$
38 $(d + 3)(d - 4)$
39 $(t + 4)(t - 5)$
40 $(q + 7)(q - 8)$
41 $(x + 2)(x + 3)$, giving areas of $2x$ and $3x$, or $(x + 1)(x + 6)$, giving areas of x and $6x$.
42 **a** $x^2 + (a + b)x + ab$ **b** **i** $p + q = 7$ **ii** $pq = 12$ **c** 7 can only be 1×7 and $1 + 7 \neq 12$

Exercise 6F

1 $(x + 3)(x - 3)$
2 $(t + 5)(t - 5)$
3 $(m + 4)(m - 4)$
4 $(3 + x)(3 - x)$
5 $(7 + t)(7 - t)$
6 $(k + 10)(k - 10)$
7 $(2 + y)(2 - y)$
8 $(x + 8)(x - 8)$
9 $(t + 9)(t - 9)$
10 **a** x^2 **b** **i** $(x - 2)$ **ii** $(x + 2)$ **iii** x^2 **iv** 4 **c** $A + B - C = x^2 - 4$, which is the area of D, which is $(x + 2)(x - 2)$.
11 **a** $x^2 + 4x + 4 - (x^2 + 2x + 1) = 2x + 3$ **b** $(a + b)(a - b)$ **c** $(x + 2 + x + 1)(x + 2 - x - 1) = (2x + 3)(1) = 2x + 3$ **d** The answers are the same. **e** $(x + 1 + x - 1)(x + 1 - x + 1) = (2x)(2) = 4x$
12 $(x + y)(x - y)$
13 $(x + 2y)(x - 2y)$
14 $(x + 3y)(x - 3y)$
15 $(3x + 1)(3x - 1)$
16 $(4x + 3)(4x - 3)$
17 $(5x + 8)(5x - 8)$
18 $(2x + 3y)(2x - 3y)$
19 $(3t + 2w)(3t - 2w)$
20 $(4y + 5x)(4y - 5x)$

Exercise 6G

1 $(2x + 1)(x + 2)$
2 $(7x + 1)(x + 1)$
3 $(4x + 7)(x - 1)$
4 $(3t + 2)(8t + 1)$
5 $(3t + 1)(5t - 1)$
6 $(4x - 1)^2$
7 $3(y + 7)(2y - 3)$
8 $4(y + 6)(y - 4)$
9 $(2x + 3)(4x - 1)$
10 $(2t + 1)(3t + 5)$
11 $(x - 6)(3x + 2)$
12 $(x - 5)(7x - 2)$
13 $4x + 1$ and $3x + 2$
14 **a** All the terms in the quadratic have a common factor of 6.
b $6(x + 2)(x + 3)$. This has the highest common factor taken out.

6.3 Solving quadratic equations by factorisation

Exercise 6H

1 −2, −5
2 −3, −1
3 −6, −4
4 −3, 2
5 −1, 3
6 −4, 5
7 1, −2
8 2, −5
9 7, −4
10 3, 2
11 1, 5
12 4, 3
13 −4, −1
14 −9, −2
15 2, 4
16 3, 5
17 −2, 5
18 −3, 5
19 −6, 2
20 −6, 3
21 −1, 2
22 −2
23 −5
24 4
25 −2, −6
26 7
27 **a** $x(x - 3) = 550$, $x^2 - 3x - 550 = 0$
b $(x - 25)(x + 22) = 0$, $x = 25$
28 $x(x + 40) = 48000$, $x^2 + 40x - 48000 = 0$, $(x + 240)(x - 200) = 0$
Fence is $2 \times 200 + 2 \times 240 = 880$ m.
29 −6, −4
30 2, 16
31 −6, 4
32 −9, 6
33 −10, 3
34 −4, 11
35 −8, 9
36 8, 9
37 1
38 Mario was correct. Sylvan did not make it into a standard quadratic and only factorised the x terms. She also incorrectly solved the equation $x - 3 = 4$.

Exercise 6I

1 **a** $\frac{1}{3}, -3$ **b** $1\frac{1}{3}, -\frac{1}{2}$
c $-\frac{1}{5}, 2$ **d** $-2\frac{1}{2}, 3\frac{1}{2}$
e $-\frac{1}{6}, -\frac{1}{3}$ **f** $\frac{2}{3}, 4$
g $\frac{1}{2}, -3$ **h** $2\frac{1}{2}, -1\frac{1}{6}$
i $-1\frac{2}{3}, 1\frac{2}{5}$ **j** $1\frac{3}{4}, 1\frac{2}{7}$
k $\frac{2}{3}, \frac{1}{8}$ **l** $\pm\frac{1}{4}$
m $-2\frac{1}{4}, 0$ **n** $\pm 1\frac{2}{5}$
o $-\frac{1}{3}, 3$
2 **a** 7, −6 **b** $-2\frac{1}{2}, 1\frac{1}{2}$
c 7, −6 **d** $-1, \frac{11}{13}$
e 3, −2 **f** $-\frac{2}{5}, \frac{1}{2}$
g $-\frac{1}{3}, -\frac{1}{2}$ **h** $\frac{1}{5}, -2$
i 4 **j** $-2, \frac{1}{8}$
k $-\frac{1}{3}, 0$ **l** ±5
m $-1\frac{2}{3}$ **n** $\pm 3\frac{1}{2}$
o $-2\frac{1}{2}, 3$
3 **a** Both only have one solution: $x = 1$.
b B is a linear equation, but A and C are quadratic equations.
4 **a** $(5x - 1)^2 = (2x + 3)^2 + (x + 1)^2$, when expanded and collected into the general quadratics, gives the required equation.
b $(10x + 3)(2x - 3)$, $x = 1.5$; area $= 7.5\text{ cm}^2$.

6.4 Solving a quadratic equation by the quadratic formula

Exercise 6J

1 1.77, −2.27
2 −0.23, −1.43
3 3.70, −2.70
4 0.29, −0.69
5 −0.19, −1.53
6 −1.23, −2.43
7 −0.41, −1.84
8 −1.39, −2.27
9 1.37, −4.37
10 2.18, 0.15
11 −0.39, −5.11
12 0.44, −1.69
13 1.64, 0.61
14 0.36, −0.79
15 1.89, 0.11
16 13
17 $x^2 - 3x - 7 = 0$
18 Terry gets $x = \frac{4 + \sqrt{0}}{8}$ and June gets $(2x - 1)^2 = 0$ which only give one solution $x = \frac{1}{2}$

6.5 Solving a quadratic equation by completing the square

Exercise 6K

1 **a** $(x + 2)^2 - 4$ **b** $(x + 7)^2 - 49$ **c** $(x - 3)^2 - 9$
d $(x + 3)^2 - 9$ **e** $(x - 2)^2 - 4$ **f** $(x - 5)^2 - 25$
g $(x + 10)^2 - 100$ **h** $(x + 5)^2 - 25$ **i** $(x + 4)^2 - 16$
j $(x - 1)^2 - 1$ **k** $(x + 1)^2 - 1$
2 **a** $(x + 2)^2 - 5$ **b** $(x + 7)^2 - 54$ **c** $(x - 3)^2 - 6$
d $(x + 3)^2 - 2$ **e** $(x - 2)^2 - 5$ **f** $(x + 3)^2 - 6$
g $(x - 5)^2 - 30$ **h** $(x + 10)^2 - 101$ **i** $(x + 4)^2 - 22$
j $(x + 1)^2 - 2$ **k** $(x - 1)^2 - 8$ **l** $(x + 1)^2 - 10$
3 **a** $-2 \pm \sqrt{5}$ **b** $-7 \pm 3\sqrt{6}$ **c** $3 \pm \sqrt{6}$
d $-3 \pm \sqrt{2}$ **e** $2 \pm \sqrt{5}$ **f** $-3 \pm \sqrt{6}$
g $5 \pm \sqrt{30}$ **h** $-10 \pm \sqrt{101}$ **i** $-4 \pm \sqrt{22}$
j $-1 \pm \sqrt{2}$ **k** $1 \pm 2\sqrt{2}$ **l** $-1 \pm \sqrt{10}$
4 **a** 1.45, −3.45 **b** 5.32, −1.32 **c** −4.16, 2.16
5 Check or correct proof.
6 $p = -14$, $q = -3$
7 **a** 3rd, 1st, 4th and 2nd – in that order
b Rewrite $x^2 - 4x - 3 = 0$ as $(x - 2)^2 - 7 = 0$, take −7 over the equals sign, square root both sides, take −2 over the equals sign
c **i** $x = -3 \pm \sqrt{2}$ **ii** $x = 2 \pm \sqrt{7}$
8 H, C, B, E, D, J, A, F, G, I

6.6 Problems involving quadratic equations

Exercise 6L

1 52, two
2 65, two
3 24, two
4 85, two
5 145, two
6 68, two
7 −35, none
8 −23, none
9 41, two
10 40, two
11 −135, none
12 37, two
13 $x^2 + 3x - 1 = 0$; $x^2 - 3x - 1 = 0$; $x^2 + x - 3 = 0$; $x^2 - x - 3 = 0$

Exercise 6M

1 15 m, 20 m
2 29
3 6.54, 0.46
4 5, 0.5
5 16 m by 14 m
6 48 km/h
7 45, 47
8 2.54 m, 3.54 m
9 30 km/h
10 10p
11 1.25, 0.8
12 10
13 5 h
14 0.75 m
15 Area = 22.75, width = 3.5 m

Examination questions

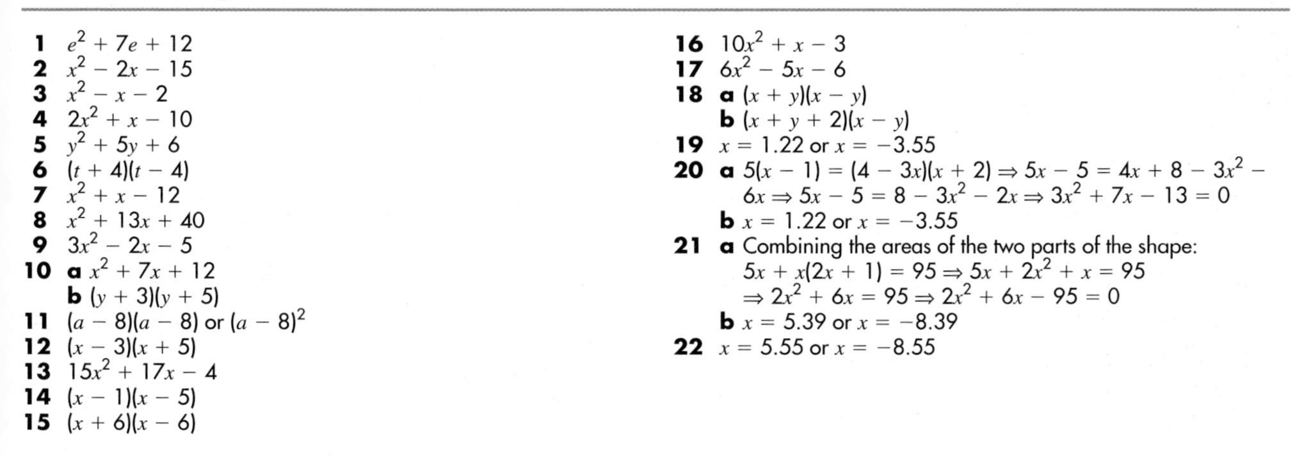

1 $e^2 + 7e + 12$
2 $x^2 - 2x - 15$
3 $x^2 - x - 2$
4 $2x^2 + x - 10$
5 $y^2 + 5y + 6$
6 $(t + 4)(t - 4)$
7 $x^2 + x - 12$
8 $x^2 + 13x + 40$
9 $3x^2 - 2x - 5$
10 **a** $x^2 + 7x + 12$
b $(y + 3)(y + 5)$
11 $(a - 8)(a - 8)$ or $(a - 8)^2$
12 $(x - 3)(x + 5)$
13 $15x^2 + 17x - 4$
14 $(x - 1)(x - 5)$
15 $(x + 6)(x - 6)$
16 $10x^2 + x - 3$
17 $6x^2 - 5x - 6$
18 **a** $(x + y)(x - y)$
b $(x + y + 2)(x - y)$
19 $x = 1.22$ or $x = -3.55$
20 **a** $5(x - 1) = (4 - 3x)(x + 2) \Rightarrow 5x - 5 = 4x + 8 - 3x^2 - 6x \Rightarrow 5x - 5 = 8 - 3x^2 - 2x \Rightarrow 3x^2 + 7x - 13 = 0$
b $x = 1.22$ or $x = -3.55$
21 **a** Combining the areas of the two parts of the shape: $5x + x(2x + 1) = 95 \Rightarrow 5x + 2x^2 + x = 95 \Rightarrow 2x^2 + 6x = 95 \Rightarrow 2x^2 + 6x - 95 = 0$
b $x = 5.39$ or $x = -8.39$
22 $x = 5.55$ or $x = -8.55$

Quick check

1 **a** 3 : 4 **b** 4 : 5 **c** 4 : 1 **d** 3 : 2
2 **a** 14 **b** 2.5 **c** 8 **d** 6

7.1 Similar triangles

Exercise 7A

1 **a** 2 **b** 3
2 **a** Yes, 4
b No, corresponding sides have different ratios.
3 **a** PQR is an enlargement of ABC
b 1 : 3
c Angle R
d BA
4 **a** Sides in same ratio
b Angle P **c** PR
5 **a** Same angles **b** Angle Q
c AR
6 **a** 8 cm
b 7.5 cm
c $x = 6.67$ cm, $y = 13.5$ cm
d $x = 24$ cm, $y = 13$ cm
e AB = 10 cm, PQ = 6 cm
f 4.2 cm
7 **a** Sides in same ratio
b 1 : 3 **c** 13 cm **d** 39 cm
8 5.2 m
9 Corresponding sides are not in the same ratio, 12 : 15 ≠ 16 : 19.
10 DE = 17.5 cm; AC : EC = BA : DE, 5 : 12.5 = 7 : DE, DE = 7 × 12.5 ÷ 5 = 17.5 cm

Exercise 7B

1 a ABC and ADE; 9 cm
b ABC and ADE; 12 cm
2 a 5 cm
b 5 cm
c $x = 60$ cm, $y = 75$ cm
d $x = 45$ cm, $y = 60$ cm
e DC = 10 cm, EB = 8 cm
3 82 m
4 220 feet
5 15 m
6 3.3 m
7 1.8 m
8 13.5 cm
9 c

Exercise 7C

1 5 cm
2 6 cm
3 10 cm
4 $x = 6$ cm, $y = 7.5$ cm
5 $x = 15$ cm, $y = 21$ cm
6 $x = 3$ cm, $y = 2.4$ cm

7.2 Areas and volumes of similar shapes

Exercise 7D

1 a 4 : 25 b 8 : 125
2 a 16 : 49 b 64 : 343
3

Linear scale factor	Linear ratio	Linear fraction	Area scale factor	Volume scale factor
2	1 : 2	$\frac{2}{1}$	4	8
3	1 : 3	$\frac{3}{1}$	9	27
$\frac{1}{4}$	4 : 1	$\frac{1}{4}$	$\frac{1}{16}$	$\frac{1}{64}$
5	1 : 5	$\frac{5}{1}$	25	125
$\frac{1}{10}$	10 : 1	$\frac{1}{10}$	$\frac{1}{100}$	$\frac{1}{1000}$

4 135 cm^2
5 a 56 cm^2
b 126 cm^2
6 a 48 m^2
b 3 m^2
7 a 2400 cm^3
b 8100 cm^3
8 4 litres
9 1.38 m^3
10 £6
11 4 cm
12 8×60p = £4.80 so it is better value
13 a 3 : 4
b 9 : 16
c 27 : 64
14 $720 \div 8 = 90$ cm^3

Exercise 7E

1 6.2 cm, 10.1 cm
2 4.26 cm, 6.74 cm
3 9.56 cm
4 3.38 m
5 8.39 cm
6 26.5 cm
7 16.9 cm
8 a 4.33 cm, 7.81 cm
b 143 g, 839 g
9 53.8 kg
10 1.73 kg
11 8.8 cm
12 7.9 cm and 12.6 cm
13 b

Examination questions

1 a 12 cm b 5 cm
2 a 39 cm b 30 cm
3 a 2 cm b 5.25 cm
4 a 8 cm b 96 cm^3
5 a 12 cm b 2700π cm^3

Quick check

1 8.60 cm
2 13.0 cm
3 21.6°
4 8.40 cm

8.1 Some 2D problems

Exercise 8A

1 13.1 cm
2 73.7°
3 9.81 cm
4 33.5 m
5 a 10.0 cm b 11.5° c 4.69 cm
6 63.0°
7 774 m
8 a $\sqrt{2}$ cm
b i $\frac{\sqrt{2}}{2}$ (an answer of $\frac{1}{\sqrt{2}}$ would also be accepted)
ii $\frac{\sqrt{2}}{2}$ iii 1
9 14.1°

8.2 Some 3D problems

Exercise 8B

1 25.1°
2 **a** 58.6° **b** 20.5 cm
c 2049 cm^3 **d** 64.0°
3 **a** 3.46 m **b** 75.5°
c 73.2° **d** 60.3 m^2
e £1507.50 (using rounded figure from **d**)
4 **a** 24.0° **b** 48.0°
c 13.5 cm **d** 16.6°
5 **a** 3.46 m **b** 70.5°
6 For example, the length of the diagonal of the base is $\sqrt{b^2 + c^2}$ and taking this as the base of the triangle with the height of the edge, then the hypotenuse is $\sqrt{(a^2 + (\sqrt{b^2 + c^2})^2)} = \sqrt{a^2 + b^2 + c^2}$
7 It is 44.6°; use triangle XDM where M is the midpoint of BD; triangle DXB is isosceles, as X is over the point where the diagonals of the base cross; the length of DB is $\sqrt{656}$, the cosine of the required angle is $0.5\sqrt{656} \div 18$.

8.3 Trigonometric ratios of angles between 90° and 360°

Exercise 8C

1 **a** 36.9°, 143.1° **b** 53.1°, 126.9°
c 48.6°, 131.4° **d** 224.4°, 315.6°
e 194.5°, 345.5° **f** 198.7°, 341.3°
g 190.1°, 349.9° **h** 234.5°, 305.5°
i 28.1°, 151.9° **j** 185.6°, 354.4°
k 33.6°, 146.4° **l** 210°, 330°
2 Sin 234°, as the others all have the same numerical value.
3 **a** 438° or 78° + $360n$° or 102 + $360n$
b −282° or 78° − $360n$° or 102 − $360n$
c Line symmetry about ±$90n$° where n is an odd integer.
Rotational symmetry about ±$180n$° where n is an integer.

Exercise 8D

1 **a** 53.1°, 306.9° **b** 54.5°, 305.5°
c 62.7°, 297.3° **d** 54.9°, 305.1°
e 79.3°, 280.7° **f** 143.1°, 216.9°
g 104.5°, 255.5° **h** 100.1°, 259.9°
i 111.2°, 248.8° **j** 166.9°, 193.1°
k 78.7°, 281.3° **l** 44.4°, 315.6°
2 Cos 58°, as the others are negative.
3 **a** 492° or 132° + $360n$° or 228 + $360n$
b −228° or 132° − $360n$° or 228 − $360n$
c Line symmetry about ±$180n$° where n is an integer.
Rotational symmetry about ±$90n$° where n is an odd integer.

Exercise 8E

1 **a** 0.707 **b** −1 (−0.9998)
c −0.819 **d** 0.731
2 **a** −0.629 **b** −0.875
c −0.087 **d** 0.999
3 **a** 21.2°, 158.8° **b** 209.1°, 330.9°
c 50.1°, 309.9° **d** 150.0°, 210.0°
e 60.9°, 119.1° **f** 29.1°, 330.9°
4 30°, 150°
5 −0.755
6 **a** 1.41 **b** −1.37
c −0.0367 **d** −0.138
e 1.41 **f** −0.492
7 True
8 **a** Cos 65° **b** Cos 40°
9 Possible Answers:
a 10°, 130° **b** 12.7°, 59.3°
10 38.2°, 141.8°

Exercise 8F

1 **a** 14.5°, 194.5° **b** 38.1°, 218.1°
c 50.0°, 230.0° **d** 61.9°, 241.9°
e 68.6°, 248.6° **f** 160.3°, 340.3°
g 147.6°, 327.6° **h** 135.4°,315.4°
i 120.9°, 300.9° **j** 105.2°, 285.2°
k 54.4°, 234.4° **l** 42.2°, 222.2°
m 160.5°, 340.5° **n** 130.9°, 310.9°
o 76.5°, 256.5° **p** 116.0°, 296.0°
q 174.4°, 354.4° **r** 44.9°, 224.9°
s 50.4°, 230.4° **t** 111.8°, 291.8°
2 Tan 235°, as the others have a numerical value of 1 (1 or −1)
3 **a** 425° or 65° + $180n$°, $n \geq 2$
b −115° or 65° − $180n$°
c No Line symmetry
Rotational symmetry about ±$180n$° where n is an integer.

8.4 Solving any triangle

Exercise 8G

1 **a** 3.64 m
b 8.05 cm
c 19.4 cm
2 **a** 46.6° **b** 68.0° **c** 36.2°
3 50.3°, 129.7°
4 2.88 cm, 20.9 cm
5 **a** **i** 30° **ii** 40°
b 19.4 m
6 36.5 m
7 22.2 m
8 3.47 m
9 767 m
10 26.8 km/h
11 64.6 km
12 Check students' answers.
13 134°
14 Check that proof is valid.

Exercise 8H

1 **a** 7.71 m **b** 29.1 cm
c 27.4 cm
2 **a** 76.2°
b 125.1°
c 90°
d Right-angled triangle
3 5.16 cm
4 65.5 cm
5 **a** 10.7 cm **b** 41.7°
c 38.3° **d** 6.69 cm
e 54.4 cm^2
6 72.3°
7 25.4 cm, 38.6 cm
8 58.4 km at 092.5°
9 21.8°
10 **a** 82.8° **b** 8.89 cm
11 42.5 km
12 Check students' answers.
13 111°; the largest angle is opposite the longest side

Exercise 8I

1 **a** 8.60 m **b** 90° **c** 27.2 cm **d** 26.9° **e** 41.0° **f** 62.4 cm **g** 90.0° **h** 866 cm **i** 86.6 cm
2 7 cm
3 11.1 km
4 19.9 knots
5 **a** 27.8 miles **b** 262°
6 **a** $A = 90°$; this is Pythagoras' theorem **b** A is acute **c** A is obtuse
7 142 m

8.5 Trigonometric ratios in surd form

Exercise 8J

1 $\frac{3}{5}$
2 $\frac{\sqrt{10}}{5}$
3 $\sqrt{19}$, $\sin x = \frac{\sqrt{6}}{\sqrt{19}}$, $\cos x = \frac{\sqrt{13}}{\sqrt{19}}$, $\tan x = \frac{\sqrt{6}}{\sqrt{13}}$
4 **a** $\sqrt{157}$ **b** $\sin A = \frac{6}{\sqrt{157}}$, $\cos A = \frac{11}{\sqrt{157}}$
5 $9\sqrt{3}$ cm^2
6 400 cm^2
7 $3\sqrt{6}$ cm

8.6 Using sine to find the area of a triangle

Exercise 8K

1 **a** 24.0 cm^2 **b** 26.7 cm^2 **c** 243 cm^2 **d** 21 097 cm^2 **e** 1224 cm^2
2 4.26 cm
3 **a** 42.3° **b** 49.6°
4 103 cm^2
5 2033 cm^2
6 21.0 cm^2
7 **a** 33.2° **b** 25.3 cm^2
8 Check that proof is valid.
9 **a** $\frac{1}{\sqrt{2}}$ **b** 21 cm^2
10 726 cm^2
11 $\frac{a^2\sqrt{3}}{4}$
12 c

Examination questions

1 85.5°
2 53.8 cm^2
3 5.89 cm
4 **a** **i** $\angle ABM = 60°$, BM = 1m $\cos 60° = \frac{BM}{AB} = \frac{1}{2}$
ii $\frac{\sqrt{3}}{2}$
b $(\frac{1}{2})^2 + (\sqrt{\frac{3}{2}})^2 = \frac{1}{4} + \frac{3}{4} = 1$
5 21.7 m

Answers to Chapter 9

Quick check

1

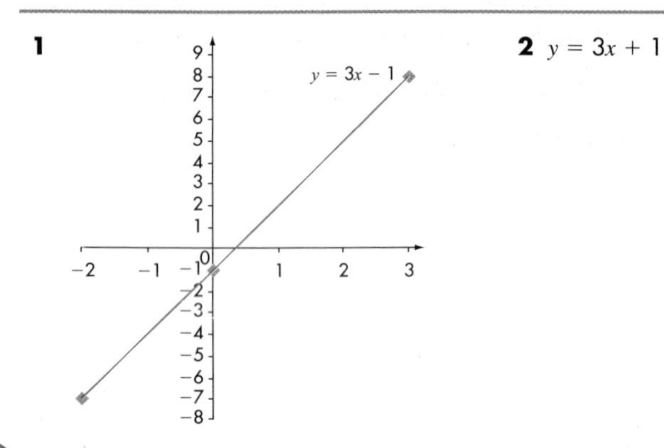

2 $y = 3x + 1$

9.1 Uses of graphs

Exercise 9A

1 **a** 32° F
b $\frac{9}{5}$ (Take gradient at C = 10° and 30°)
c $F = \frac{9}{5}C + 32$

2 **a** 0.07 (Take gradient at U = 0 and 500)
b £10
c C = £(10 + 0.07U) or Charge = £10 + 7p/unit

3 **a** \$1900 – \$1400 = \$500
b **i** \$7500 **ii** £3783

4

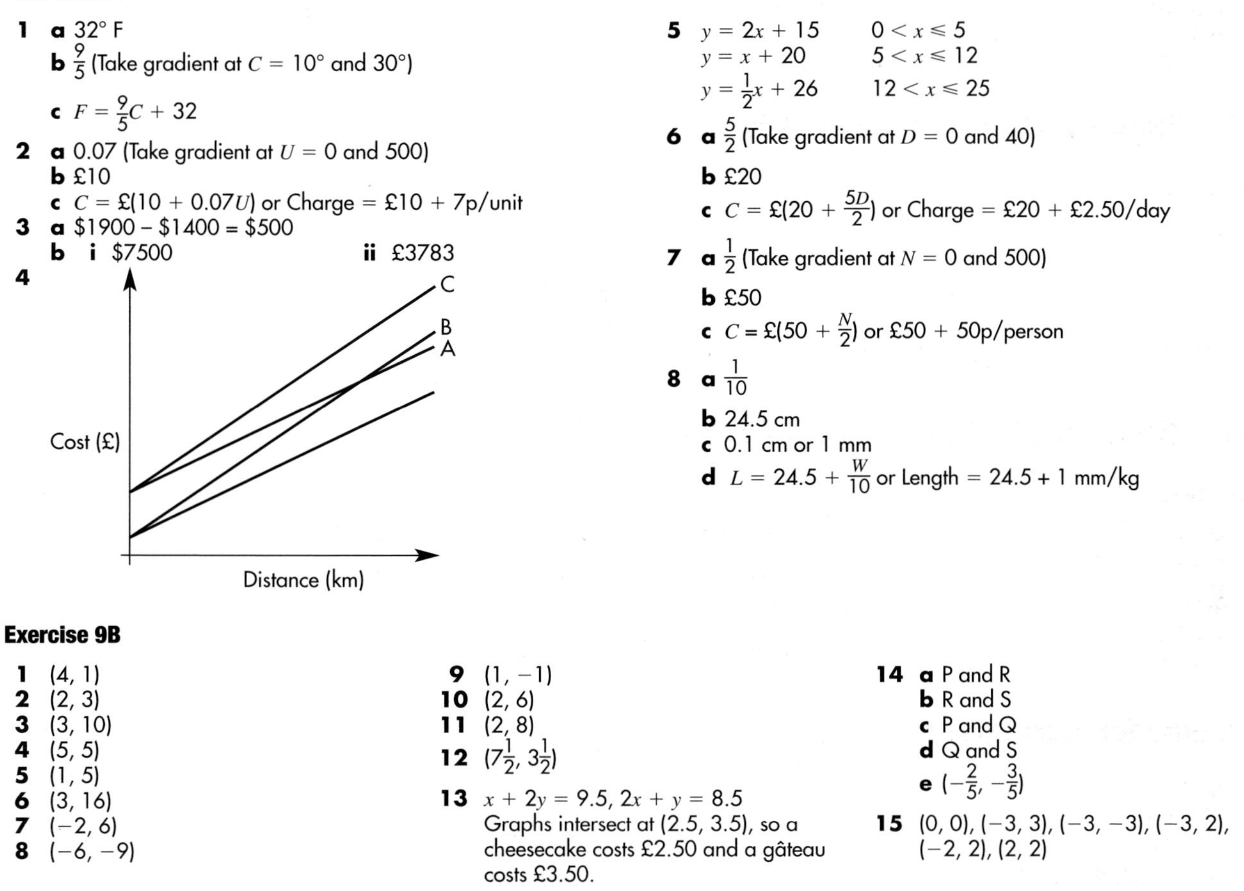

5 $y = 2x + 15$ $0 < x \leqslant 5$
$y = x + 20$ $5 < x \leqslant 12$
$y = \frac{1}{2}x + 26$ $12 < x \leqslant 25$

6 **a** $\frac{5}{2}$ (Take gradient at D = 0 and 40)
b £20
c C = £(20 + $\frac{5D}{2}$) or Charge = £20 + £2.50/day

7 **a** $\frac{1}{2}$ (Take gradient at N = 0 and 500)
b £50
c C = £(50 + $\frac{N}{2}$) or £50 + 50p/person

8 **a** $\frac{1}{10}$
b 24.5 cm
c 0.1 cm or 1 mm
d $L = 24.5 + \frac{W}{10}$ or Length = 24.5 + 1 mm/kg

Exercise 9B

1 (4, 1)
2 (2, 3)
3 (3, 10)
4 (5, 5)
5 (1, 5)
6 (3, 16)
7 (−2, 6)
8 (−6, −9)
9 (1, −1)
10 (2, 6)
11 (2, 8)
12 $(7\frac{1}{2}, 3\frac{1}{2})$
13 $x + 2y = 9.5$, $2x + y = 8.5$
Graphs intersect at (2.5, 3.5), so a cheesecake costs £2.50 and a gâteau costs £3.50.
14 **a** P and R
b R and S
c P and Q
d Q and S
e $(-\frac{2}{5}, -\frac{3}{5})$
15 (0, 0), (−3, 3), (−3, −3), (−3, 2), (−2, 2), (2, 2)

9.2 Parallel and perpendicular lines

Exercise 9C

1 **a** Line parallel to the original, intersecting the charge axis at a higher point
b Line passing through the same intersecting point on the charge axis, with a steeper gradient

2 **a** Line A does not pass through (0, 1).
b Line C is perpendicular to the other two.
c (i)

3 **a** $-\frac{1}{2}$ **b** $\frac{1}{3}$ **c** $-\frac{1}{5}$ **d** 1 **e** −2 **f** −4
g 3 **h** $\frac{3}{2}$ **i** $-\frac{2}{3}$ **j** $-\frac{1}{10}$ **k** $\frac{1}{6}$ **l** $-\frac{3}{4}$

4 **a** $y = -\frac{1}{2}x - 1$ **b** $y = \frac{1}{3}x + 1$ **c** $y = -x + 2$
d $y = x + 2$ **e** $y = -2x + 3$ **f** $y = -4x - 3$
g $y = 3x$ **h** $y = 1.5x - 5$

5 **a** $y = 4x + 1$ **b** $y = \frac{1}{2}x - 2$ **c** $y = -x + 3$

6 **a** $y = -\frac{1}{3}x - 1$ **b** $y = 3x + 5$ **c** $y = -x + 1$

7 **a** $y = -x + 14$ **b** $y = x + 2$

8 $y = 2x + 6$

9 $y = -\frac{1}{4}x + 2$

10 **a** (0, −20) **b** $y = -\frac{1}{5}x + 6$ **c** $y = 5x - 20$

11 $y = -\frac{1}{2}x + 5$

12 **a** $y = 3x - 6$
b Bisector of AB is $y = -2x + 9$, bisector of AC is $y = \frac{1}{2}x + \frac{3}{2}$, solving these equations shows the lines intersect at (3, 3).
c (3, 3) lies on $y = 3x - 6$ because (3 × 3) − 6 = 3

13 $y = 2x + \frac{9}{2}$

9.3 Other graphs

Exercise 9D

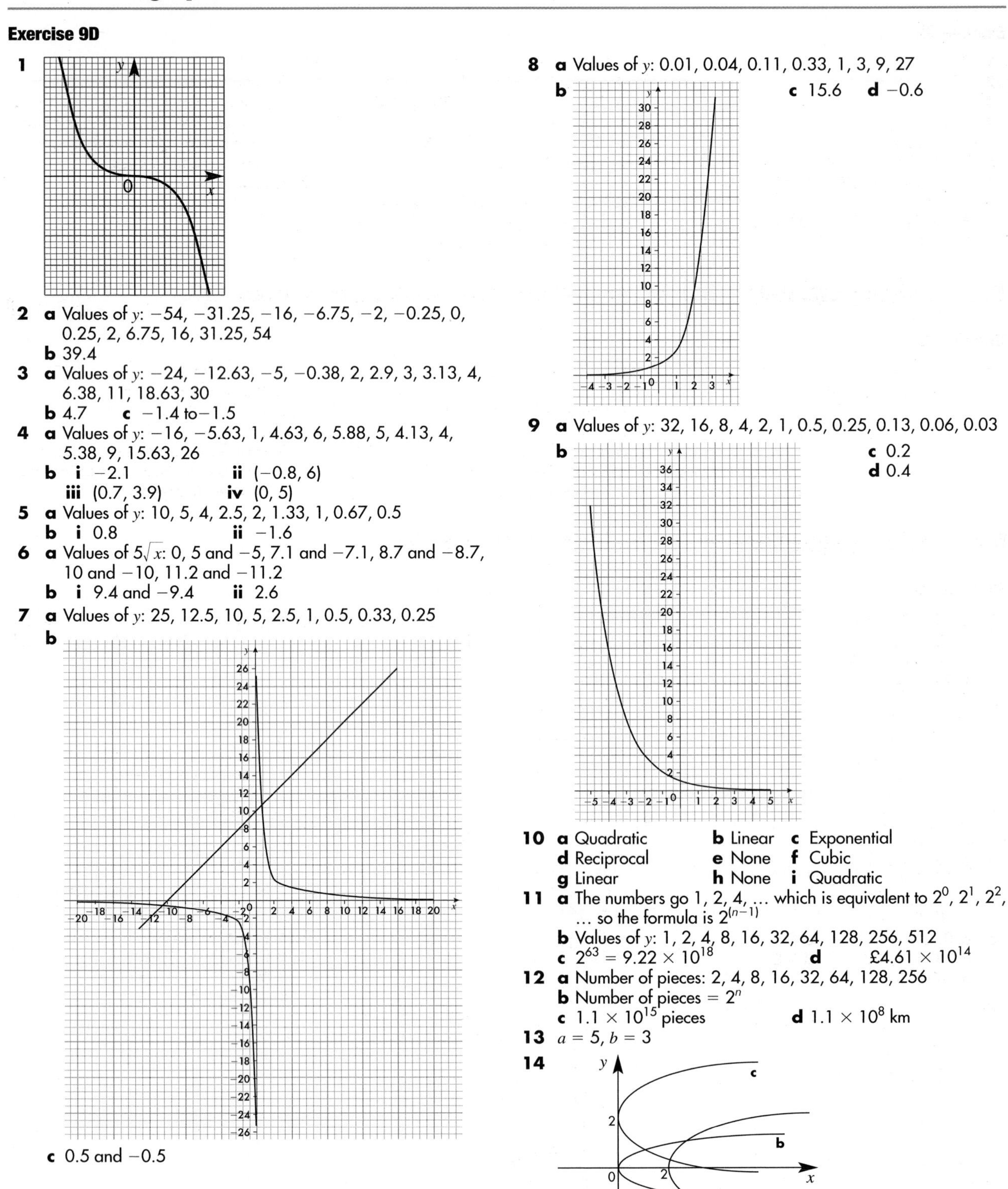

1

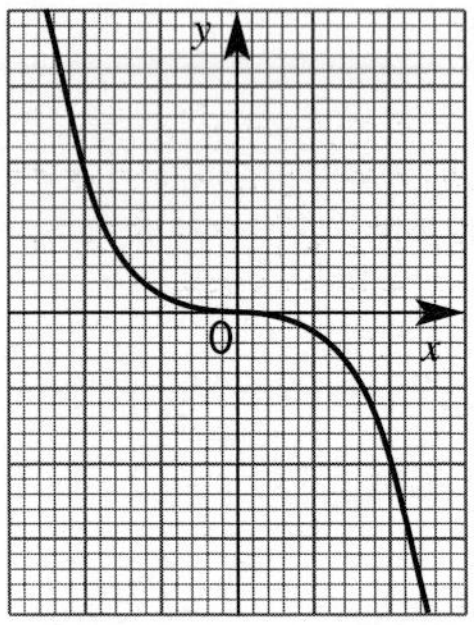

2 **a** Values of y: −54, −31.25, −16, −6.75, −2, −0.25, 0, 0.25, 2, 6.75, 16, 31.25, 54
b 39.4

3 **a** Values of y: −24, −12.63, −5, −0.38, 2, 2.9, 3, 3.13, 4, 6.38, 11, 18.63, 30
b 4.7 **c** −1.4 to −1.5

4 **a** Values of y: −16, −5.63, 1, 4.63, 6, 5.88, 5, 4.13, 4, 5.38, 9, 15.63, 26
b **i** −2.1 **ii** (−0.8, 6)
iii (0.7, 3.9) **iv** (0, 5)

5 **a** Values of y: 10, 5, 4, 2.5, 2, 1.33, 1, 0.67, 0.5
b **i** 0.8 **ii** −1.6

6 **a** Values of $5\sqrt{x}$: 0, 5 and −5, 7.1 and −7.1, 8.7 and −8.7, 10 and −10, 11.2 and −11.2
b **i** 9.4 and −9.4 **ii** 2.6

7 **a** Values of y: 25, 12.5, 10, 5, 2.5, 1, 0.5, 0.33, 0.25
b

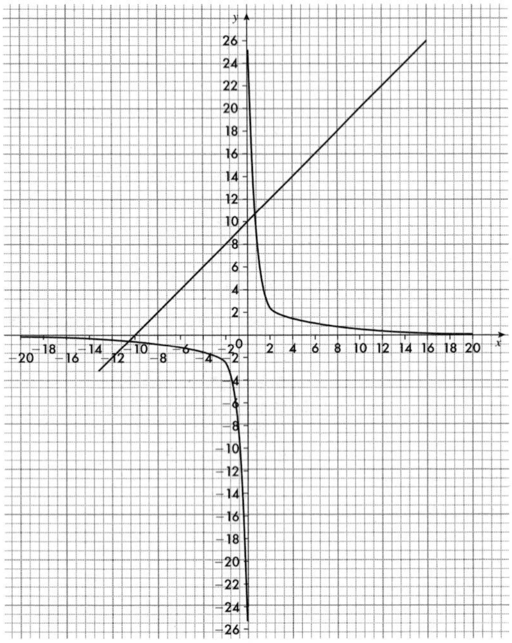

c 0.5 and −0.5

8 **a** Values of y: 0.01, 0.04, 0.11, 0.33, 1, 3, 9, 27
b

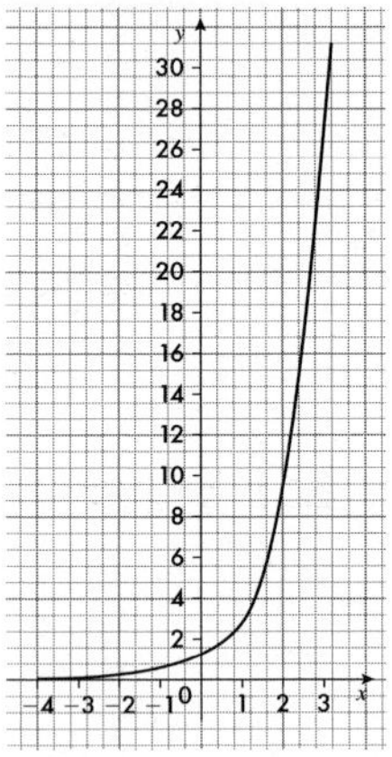

c 15.6 **d** −0.6

9 **a** Values of y: 32, 16, 8, 4, 2, 1, 0.5, 0.25, 0.13, 0.06, 0.03
b

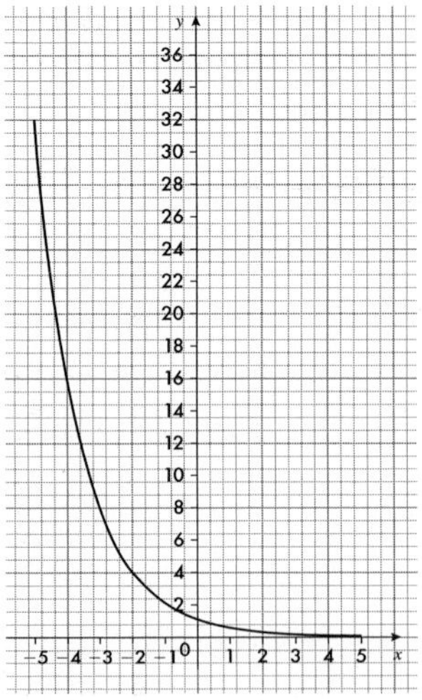

c 0.2
d 0.4

10 **a** Quadratic **b** Linear **c** Exponential
d Reciprocal **e** None **f** Cubic
g Linear **h** None **i** Quadratic

11 **a** The numbers go 1, 2, 4, … which is equivalent to 2^0, 2^1, 2^2, … so the formula is $2^{(n-1)}$
b Values of y: 1, 2, 4, 8, 16, 32, 64, 128, 256, 512
c $2^{63} = 9.22 \times 10^{18}$ **d** £4.61×10^{14}

12 **a** Number of pieces: 2, 4, 8, 16, 32, 64, 128, 256
b Number of pieces $= 2^n$
c 1.1×10^{15} pieces **d** 1.1×10^8 km

13 $a = 5$, $b = 3$

14

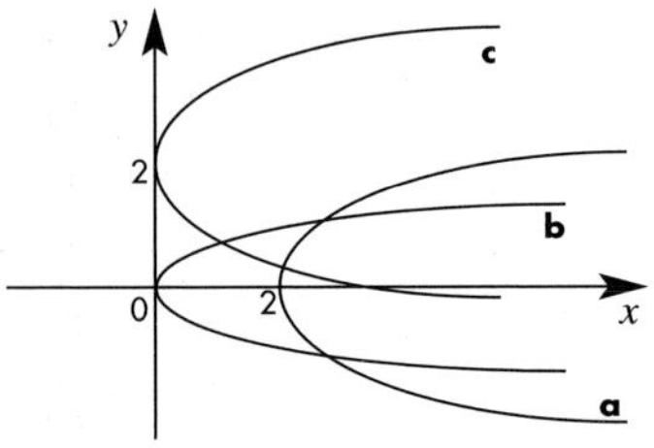

9.4 The circular function graph

Exercise 9E

1 115°
2 327°
3 324°
4 195°
5 210°, 330°
6 135°, 225°
7 **a** Say 32°, sin 32° = 0.53, cos 58° = 0.53
b Say 70°, sin 70° = 0.94, cos 20° = 0.94
c Sin x = cos (90 − x)°
d Cos x = sin (90 − x)°
8 **a** 64° **b** 206°, 334°
c 116°, 244°
9 **a** −0.385 **b** 113°
10 **a** 0.822 **b** 55.3
c No
d The calculator has given the value of the acute angle but the angle 124.7° has the same positive sign.
11 **a** 1.1307
b Error
c If you tried to draw this triangle accurately then you would see that the line that is 12 long does not intersect with the base.
12 **a to e** All true

9.5 Solving equations, one linear and one non-linear, with graphs

Exercise 9F

1 **a** (0.7, 0.7), (−2.7, −2.7)
b (6, 12), (−1, −2)
c (4, −3), (−3, 4)
d (0.8, 1.8), (−1.8, −0.8)
e (4.6, 8.2), (0.5, 0)
f (3, 6), (−2, 1)
g (4.8, 6.6), (0.2, −2.6)
h (2.6, 1.6), (−1.6, −2.6)
2 **a** (1, 0)
b Only one intersection point
c $x^2 + x(3 - 5) + (-4 + 5) = 0$
d $(x - 1)^2 = 0 \Rightarrow x = 1$
e Only one solution as line is a tangent to curve.
3 **a** There is no solution.
b The graphs do not intersect.
c $x^2 + x + 4 = 0$
d $b^2 - 4ac = -15$
e No solution as the discriminant is negative and there is no square root of a negative number.

9.6 Solving equations by the method of intersection

Exercise 9G

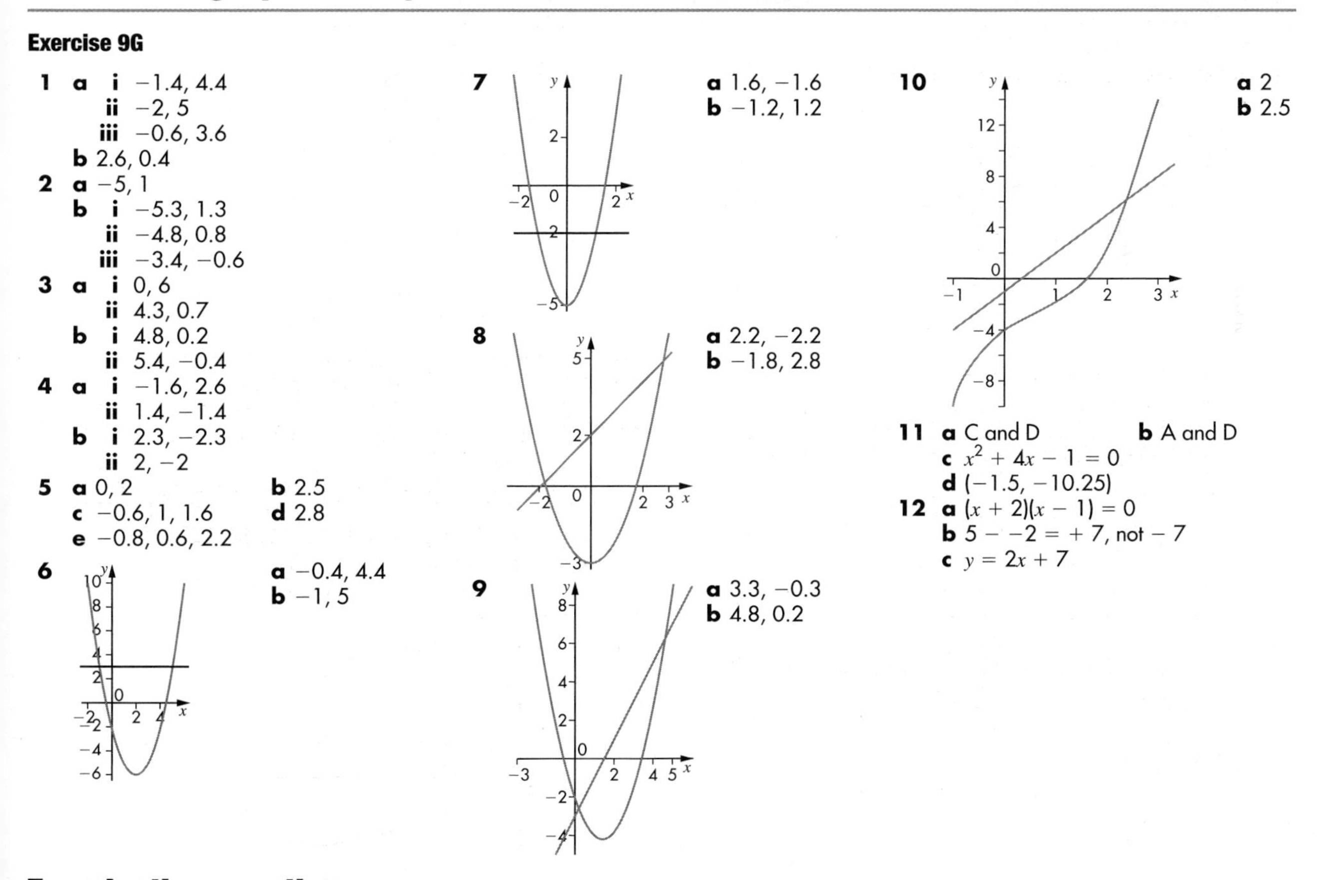

1 **a** **i** −1.4, 4.4
ii −2, 5
iii −0.6, 3.6
b 2.6, 0.4
2 **a** −5, 1
b **i** −5.3, 1.3
ii −4.8, 0.8
iii −3.4, −0.6
3 **a** **i** 0, 6
ii 4.3, 0.7
b **i** 4.8, 0.2
ii 5.4, −0.4
4 **a** **i** −1.6, 2.6
ii 1.4, −1.4
b **i** 2.3, −2.3
ii 2, −2
5 **a** 0, 2 **b** 2.5
c −0.6, 1, 1.6 **d** 2.8
e −0.8, 0.6, 2.2
6 **a** −0.4, 4.4
b −1, 5
7 **a** 1.6, −1.6
b −1.2, 1.2
8 **a** 2.2, −2.2
b −1.8, 2.8
9 **a** 3.3, −0.3
b 4.8, 0.2
10 **a** 2
b 2.5
11 **a** C and D **b** A and D
c $x^2 + 4x - 1 = 0$
d (−1.5, −10.25)
12 **a** $(x + 2)(x - 1) = 0$
b 5 − −2 = + 7, not − 7
c $y = 2x + 7$

Examination questions

1 $a = 5, k = 1.4$
2 0.25

Answers to Chapter 10

Quick check

1 29.0

10.1 Cumulative frequency diagrams

Exercise 10A

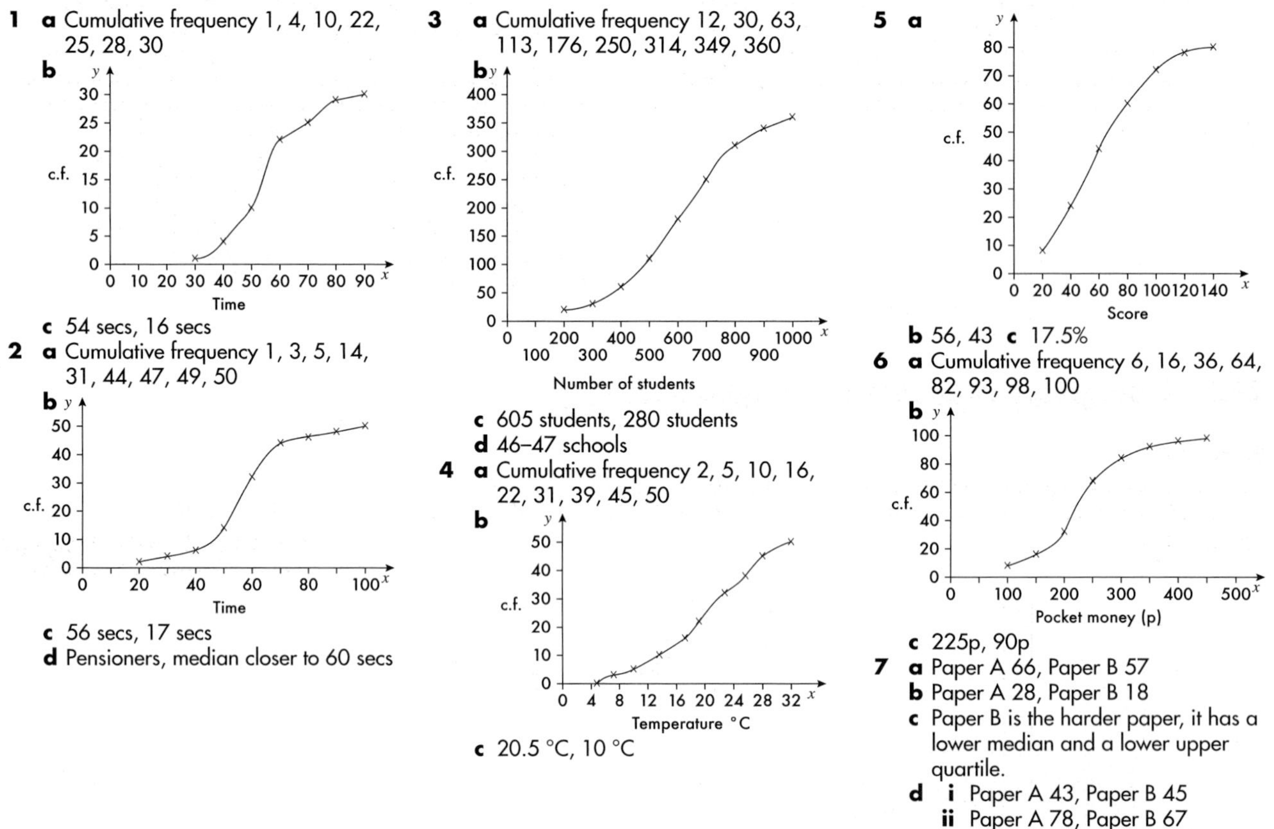

1 **a** Cumulative frequency 1, 4, 10, 22, 25, 28, 30

b

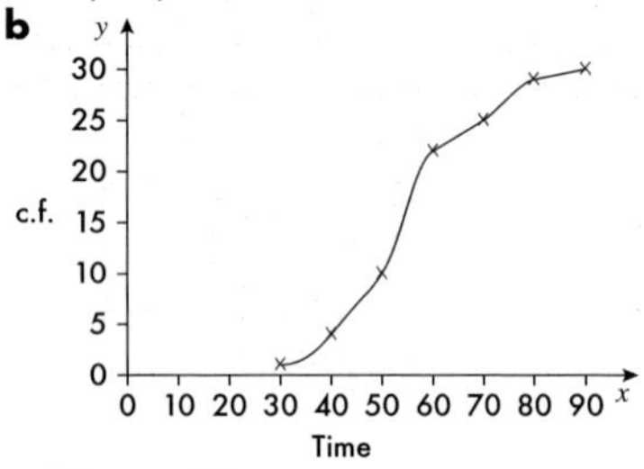

c 54 secs, 16 secs

2 **a** Cumulative frequency 1, 3, 5, 14, 31, 44, 47, 49, 50

b

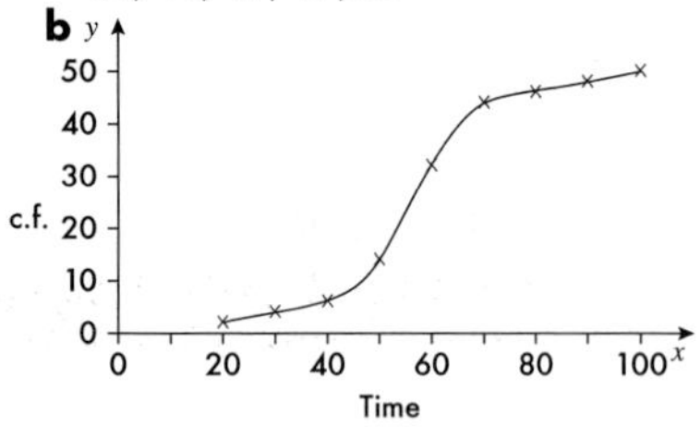

c 56 secs, 17 secs

d Pensioners, median closer to 60 secs

3 **a** Cumulative frequency 12, 30, 63, 113, 176, 250, 314, 349, 360

b

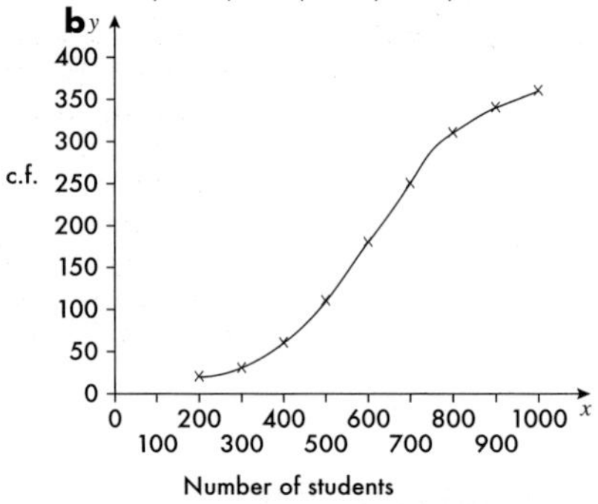

c 605 students, 280 students

d 46–47 schools

4 **a** Cumulative frequency 2, 5, 10, 16, 22, 31, 39, 45, 50

b

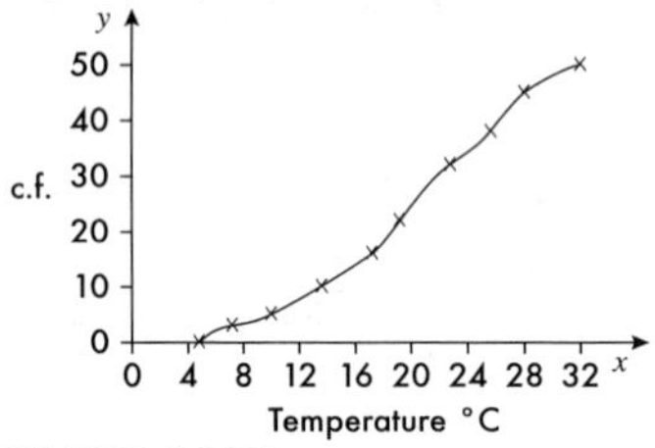

c 20.5 °C, 10 °C

5 **a**

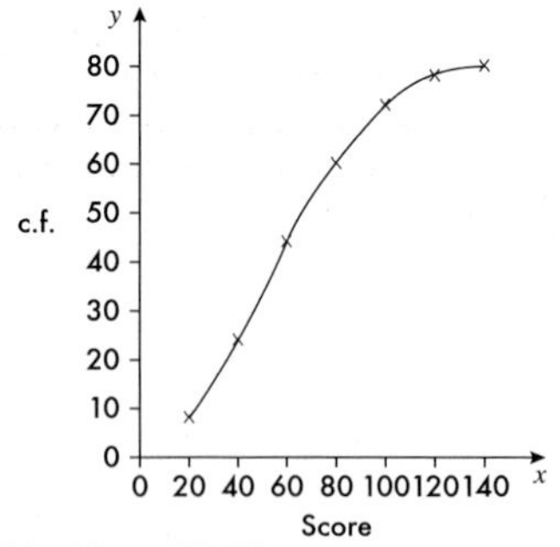

b 56, 43 **c** 17.5%

6 **a** Cumulative frequency 6, 16, 36, 64, 82, 93, 98, 100

b

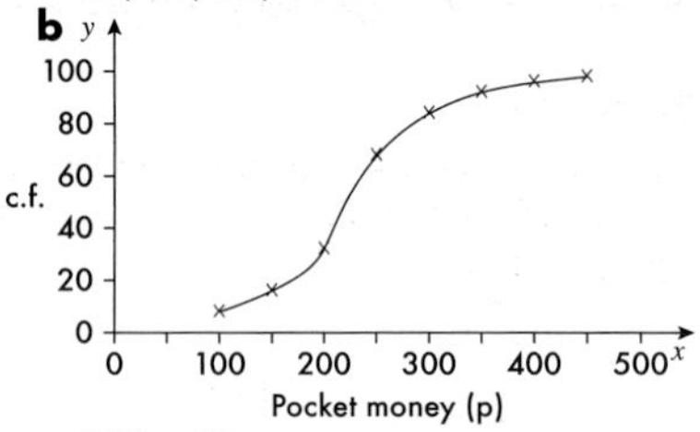

c 225p, 90p

7 **a** Paper A 66, Paper B 57

b Paper A 28, Paper B 18

c Paper B is the harder paper, it has a lower median and a lower upper quartile.

d **i** Paper A 43, Paper B 45

ii Paper A 78, Paper B 67

8 9.25 minutes (create a grouped frequency chart)

9 Find the top 10% on the cumulative frequency scale, read along to the graph and read down to the marks. The mark seen will be the minimum mark needed for this top grade.

10.2 Box plots

Exercise 10B

1 **a**

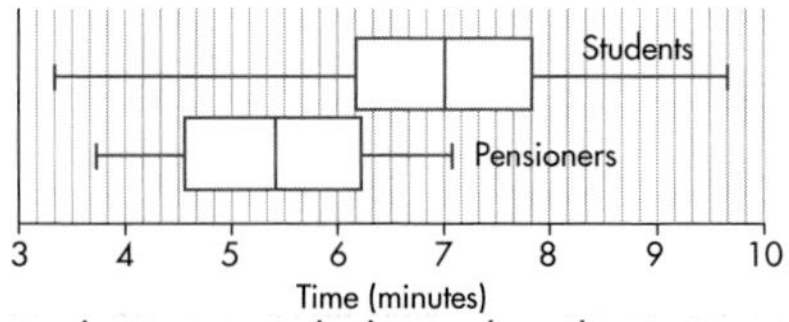

b Students are much slower than the pensioners. Both distributions have the same interquartile range, but students' median and upper quartiles are 1 minute, 35 seconds higher. The fastest person to complete the calculations was a student, but so was the slowest.

2 **a**

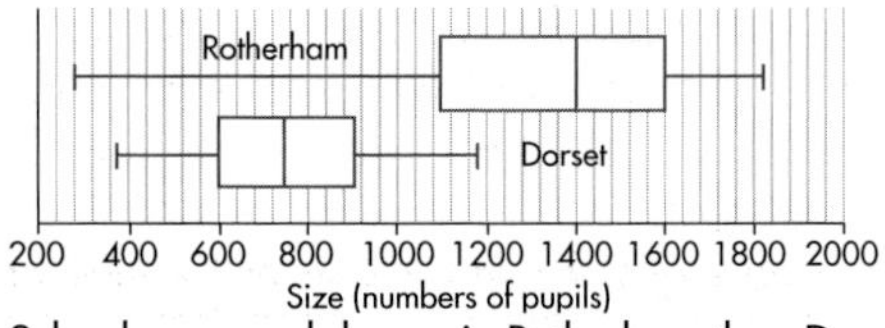

b Schools are much larger in Rotherham than Dorset. The Dorset distribution is symmetrical, but the Rotherham distribution is negatively skewed – so most Rotherham schools are large.

3 **a** The resorts have similar median temperatures, but Resort B has a much wider temperature range, where the greatest extremes of temperature are recorded.
b Resort A is probably a better choice as the weather seems more consistent.

4 **a**

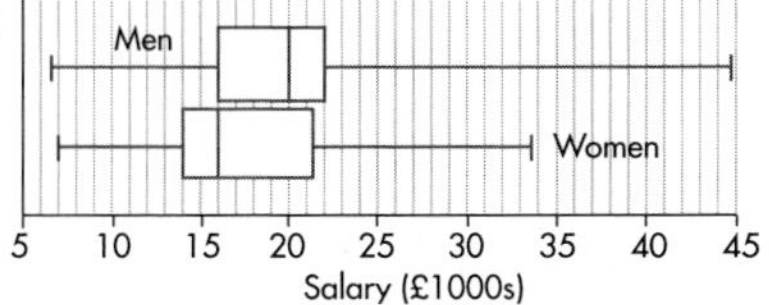

b Both distributions have a similar interquartile range, and there is little difference between the upper quartile values. Men have a wider range of salaries, but the higher men's median and the fact that the men's distribution is negatively skewed and the women's distribution is positively skewed indicates that men are better paid than women.

5 **a**

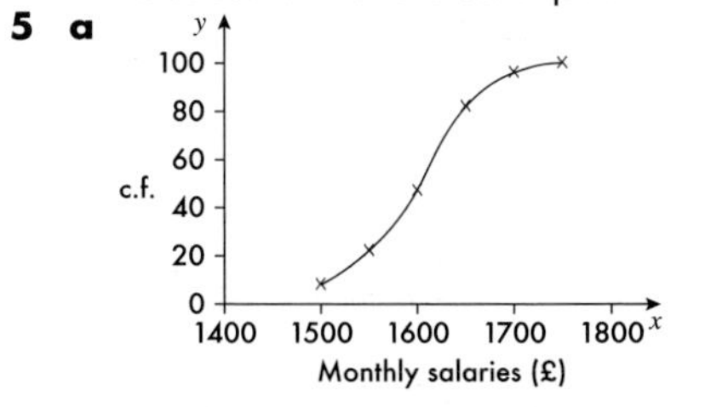

b £1605, £85
c **i**

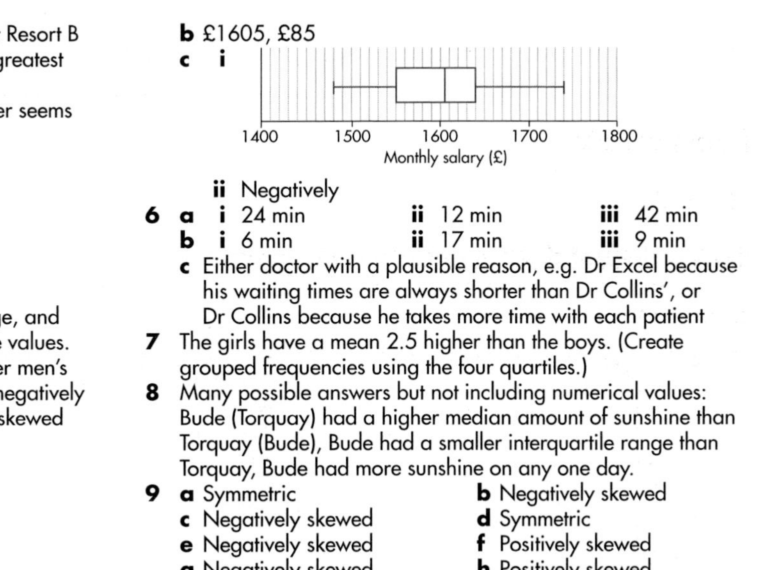

ii Negatively

6 **a** **i** 24 min **ii** 12 min **iii** 42 min
b **i** 6 min **ii** 17 min **iii** 9 min
c Either doctor with a plausible reason, e.g. Dr Excel because his waiting times are always shorter than Dr Collins', or Dr Collins because he takes more time with each patient

7 The girls have a mean 2.5 higher than the boys. (Create grouped frequencies using the four quartiles.)

8 Many possible answers but not including numerical values: Bude (Torquay) had a higher median amount of sunshine than Torquay (Bude), Bude had a smaller interquartile range than Torquay, Bude had more sunshine on any one day.

9 **a** Symmetric **b** Negatively skewed
c Negatively skewed **d** Symmetric
e Negatively skewed **f** Positively skewed
g Negatively skewed **h** Positively skewed
i Positively skewed **j** Symmetric

10 A and X, B and Y, C and W, D and Z

10.3 Histograms with bars of unequal width

Exercise 10C

1 The respective frequency densities on which each histogram should be based are:
a 2.5, 6.5, 6, 2, 1, 1.5 **b** 4, 27, 15, 3 **c** 17, 18, 12, 6.67 **d** 0.4, 1.2, 2.8, 1 **e** 9, 21, 13.5, 9

2 **a** **b**

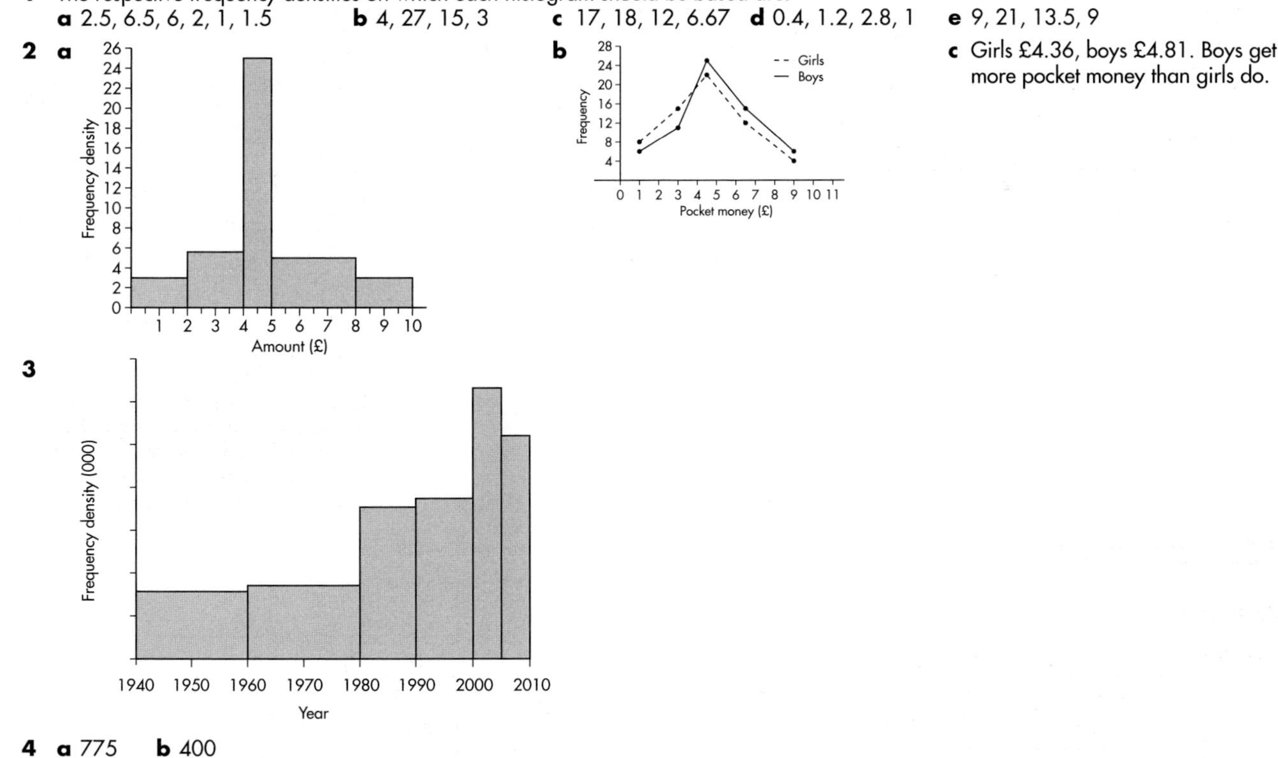

c Girls £4.36, boys £4.81. Boys get more pocket money than girls do.

3

4 **a** 775 **b** 400

5 Divide the frequency of the class interval by the width of the class interval.

6 **a** **i**

Age, y (years)	$9 < y \leq 10$	$10 < y \leq 12$	$12 < y \leq 14$	$14 < y \leq 17$	$17 < y \leq 19$	$19 < y \leq 20$
Frequency	4	12	8	9	5	1

ii 10–12 **iii** 13
iv 11, 16, 5 **v** 13.4

b **i**

Temperature, t (°C)	$10 < t \leq 11$	$11 < t \leq 12$	$12 < t \leq 14$	$14 < t \leq 16$	$16 < t \leq 19$	$19 < t \leq 21$
Frequency	15	15	50	40	45	15

ii 12–14°C **iii** 14.5°C
iv 12 or 13, 17, 5 or 4 (with units) **v** 14.8°C

c **i**

Weight, w (kg)	$50 < w \leq 70$	$70 < w \leq 90$	$90 < w \leq 100$	$100 < w \leq 120$	$120 < w \leq 170$
Frequency	160	200	120	120	200

ii 70–90 kg and 120–170 kg **iii** 93.33 kg
iv 74 kg, 120 kg, 46 kg **v** 99.0 kg

7 **a** 7.33 hours **b** 8.44 hours **c** 7 hours

8 **a**

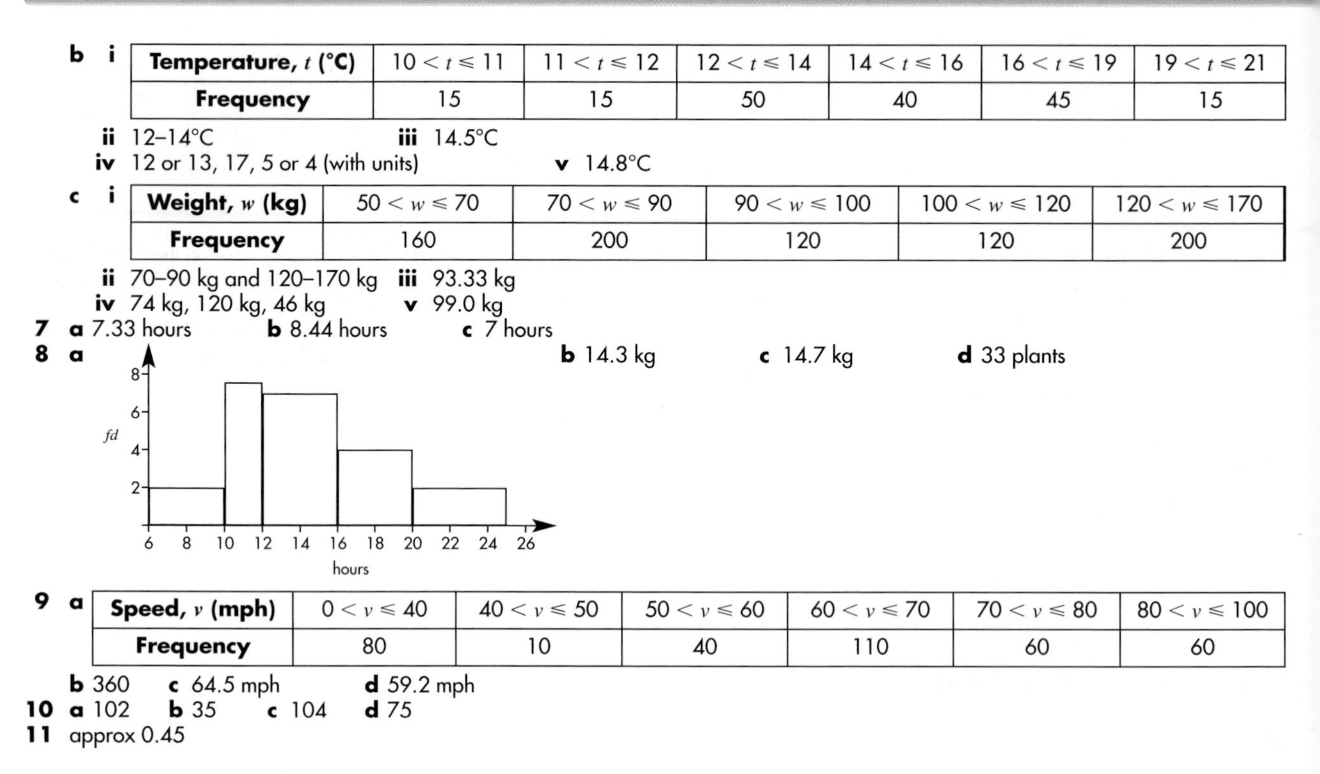

b 14.3 kg **c** 14.7 kg **d** 33 plants

9 **a**

Speed, v (mph)	$0 < v \leq 40$	$40 < v \leq 50$	$50 < v \leq 60$	$60 < v \leq 70$	$70 < v \leq 80$	$80 < v \leq 100$
Frequency	80	10	40	110	60	60

b 360 **c** 64.5 mph **d** 59.2 mph

10 **a** 102 **b** 35 **c** 104 **d** 75

11 approx 0.45

Examination questions

1 **a** **i** 12 goals
ii 5 goals
b Check students' box plots

2 **a** Cumulative frequencies: 4, 12, 23, 39, 48, 55, 60
b Check students' box plots for best fit
c 35

3 **a** Check students' box plots
b Boys have a larger range than the girls. Boys and girls both have the same interquartile range.

4 John's garage has a larger range of prices than Peter's. Peter's prices have a smaller interquartile range than John's.

5 £3.06

6 Find the top 15% on the cumulative frequency scale, read along to the graph and read down to the marks. The mark seen will be the minimum mark needed for this top grade.

7 **a** missing frequencies: 18, 20 and 12
b Check students' histogram (0.008, 0.002)

8 **a** missing frequencies: 25 and 26
b Check students' histogram

9 Check students' histogram

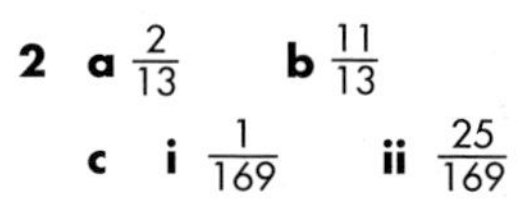

Quick check

1 **a** Perhaps around 0.6 **b** Very close to 0.1 **c** Very close to 0
d 1 **e** 1

11.1 Tree diagrams

Exercise 11A

1 **a** $\frac{1}{4}$ **b** $\frac{1}{2}$ **c** $\frac{3}{4}$

2 **a** $\frac{2}{13}$ **b** $\frac{11}{13}$
c **i** $\frac{1}{169}$ **ii** $\frac{25}{169}$

3 **a** $\frac{2}{3}$ **b** $\frac{1}{2}$

c

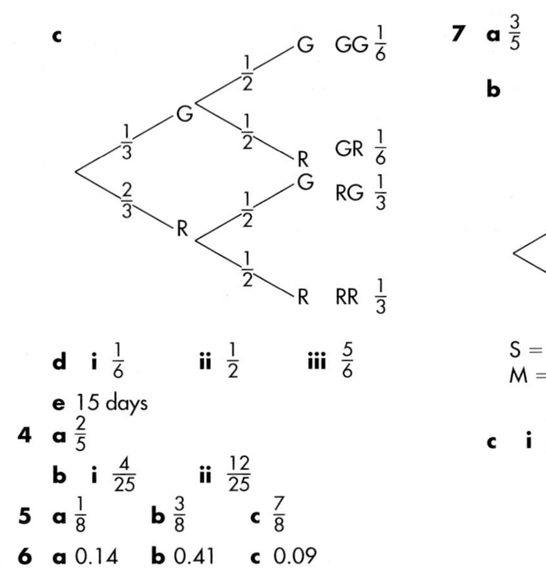

d i $\frac{1}{6}$ ii $\frac{1}{2}$ iii $\frac{5}{6}$

e 15 days

4 a $\frac{2}{5}$

b i $\frac{4}{25}$ ii $\frac{12}{25}$

5 a $\frac{1}{8}$ b $\frac{3}{8}$ c $\frac{7}{8}$

6 a 0.14 b 0.41 c 0.09

7 a $\frac{3}{5}$

b

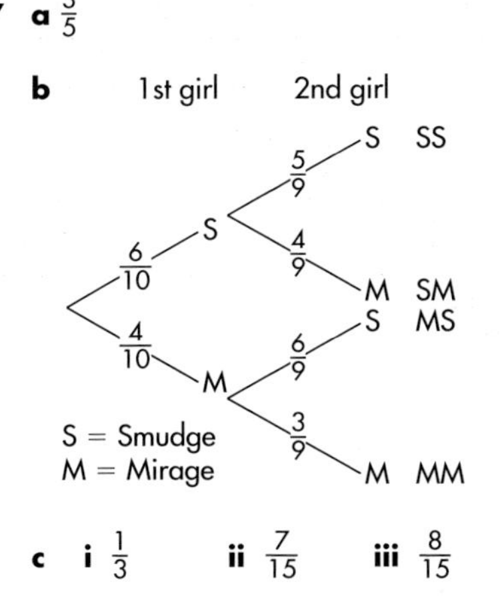

c i $\frac{1}{3}$ ii $\frac{7}{15}$ iii $\frac{8}{15}$

8 a 1 b 1

c

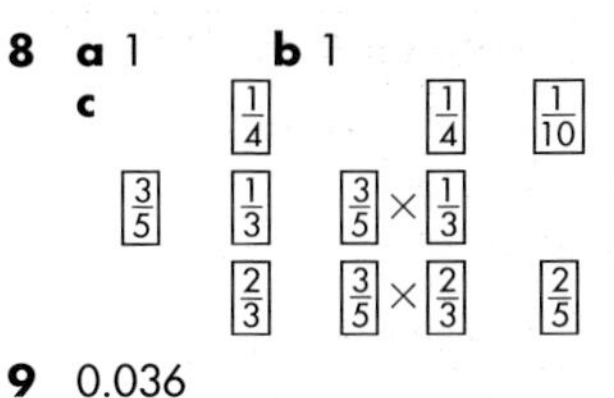

9 0.036

10 It will help to show all the 27 different possible events and which ones give the three different coloured sweets, then the branches will help you to work out the chance of each.

11.2 Independent events 1

Exercise 11B

1 a $\frac{4}{9}$ b $\frac{4}{9}$

2 a $\frac{1}{169}$ b $\frac{2}{169}$

3 a $\frac{1}{4}$ b $\frac{1}{2}$

4 $\frac{1}{216}$

5 a $\frac{4}{25}$ b $\frac{12}{25}$

6 a 0.08 b 0.32 c 0.48

7 a 0.336 b 0.452 c 0.024

8 a 0.19 b 0.77 c 0.512

9 0.000 77

10 Check students' understanding

11 $\frac{1}{6}$ as each throw is independent of any previous throws.

Exercise 11C

1 a $\frac{125}{216}$ (0.579) b $\frac{91}{216}$ (0.421)

2 a $\frac{1}{16}$ b $\frac{15}{16}$

3 a 0.378 b 0.162 c 0.012 d 0.988

4 a $\frac{4}{25}$ b $\frac{9}{25}$ c $\frac{16}{25}$

5 a i $\frac{1}{216}$ (0.005)

ii $\frac{125}{216}$ (0.579)

iii $\frac{91}{216}$ (0.421)

b i $\frac{1}{1296}$ (0.000 77)

ii $\frac{625}{1296}$ (0.482)

iii $\frac{671}{1296}$ (0.518)

c i $\frac{1}{7776}$ (0.000 13)

ii $\frac{3125}{7776}$ (0.402)

iii $\frac{4651}{7776}$ (0.598)

d i $\frac{1}{6^n}$ ii $\frac{5^n}{6^n}$ iii $1 - \frac{5^n}{6^n}$

6 a $\frac{32}{243}$ (0.132) b $\frac{1}{243}$ (0.004)

c $\frac{242}{243}$ (0.996)

7 a $\frac{3}{8}$ b $\frac{1}{120}$ c $\frac{119}{120}$

8 a 0.65 b 0.4225 c 2 days d 0.116 e 0.884

9 $10 \times 0.6^9 \times 0.4 + 0.6^{10} = 0.046$

10 Here P(S) = probability of a sale on one day, then probability of at least one sale in week = $(1 - [1 - P(S)]^5)$.

11.3 Independent events 2

Exercise 11D

1 a $\frac{27}{1000}$ b $\frac{189}{1000}$ c $\frac{441}{1000}$

d $\frac{343}{1000}$

2 a $\frac{1}{1296}$ (0.000 77) b $\frac{625}{1296}$ (0.482)

c $\frac{125}{324}$

3 a $\frac{1}{9}$ b $\frac{7}{18}$ c $\frac{7}{18}$

d $\frac{1}{9}$ e $\frac{8}{9}$

4 a 0.154 b 0.456

5 a 0.3024 b 0.4404

c 0.7428 (P(3 or 4))

6 a 0.9 b 0.6 c 0.54 d 0.216

7 a 0.6 b 0.6 c 0.432 d Independent events

8 a $\frac{1}{9}$ b $\frac{1}{9}$ c $\frac{7}{27}$ d $\frac{1}{27}$

9 a 0.126 b 0.4 c 0.42 d 0.054

10 a 0.42

b $0.75^2 \times 0.25 = 0.14$

c $0.75 \times 0.25 = 0.19$

d 0.25

11 0.8

12 He may already have a 10 or Jack or Queen or King in his hand, in which case the probability fraction will have a different numerator and, as he already has two cards, the denominator should be 50.

11.4 Conditional probability

Exercise 11E

1 **a** $\frac{1}{60}$ **b** 50

2 **a** $\frac{1}{6}$ **b** 0

c **i** $\frac{2}{3}$ **ii** $\frac{1}{3}$ **iii** 0

3 **a** **i** $\frac{3}{8}$ **ii** $\frac{5}{8}$

b **i** $\frac{5}{12}$ **ii** $\frac{7}{12}$

4 **a** **i** $\frac{5}{13}$ **ii** $\frac{8}{13}$

b **i** $\frac{15}{91}$ **ii** $\frac{4}{13}$

5 **a** **i** $\frac{1}{3}$ **ii** $\frac{2}{15}$

b $\frac{4}{15}$ **c** $\frac{1}{6}$ **d** 1

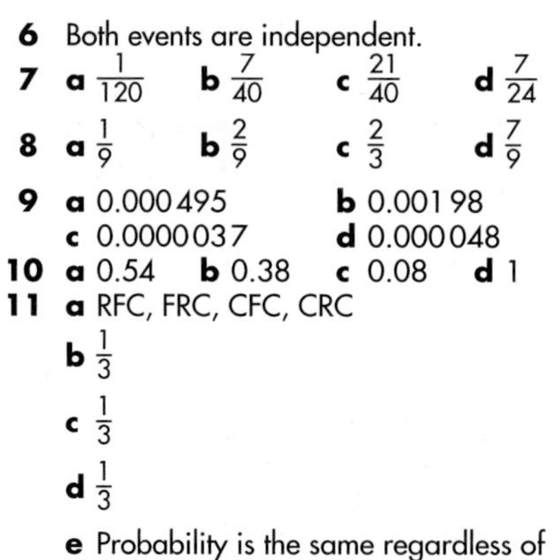

6 Both events are independent.

7 **a** $\frac{1}{120}$ **b** $\frac{7}{40}$ **c** $\frac{21}{40}$ **d** $\frac{7}{24}$

8 **a** $\frac{1}{9}$ **b** $\frac{2}{9}$ **c** $\frac{2}{3}$ **d** $\frac{7}{9}$

9 **a** 0.000 495 **b** 0.001 98
c 0.000 0037 **d** 0.000 048

10 **a** 0.54 **b** 0.38 **c** 0.08 **d** 1

11 **a** RFC, FRC, CFC, CRC

b $\frac{1}{3}$

c $\frac{1}{3}$

d $\frac{1}{3}$

e Probability is the same regardless of which day he chooses

12 **a** 0.7 **b** 0.667 **c** 0.375

13 $\frac{1}{270725}$ or 0.000 003 694

14 Find P(B) and also P(W). Then find P(B) × P(B second) remembering numerator and denominator will each be one less, and P(W) × P(W second) again remembering the numerator and denominator will each be one less. Finally add together these probabilities.

Examination questions

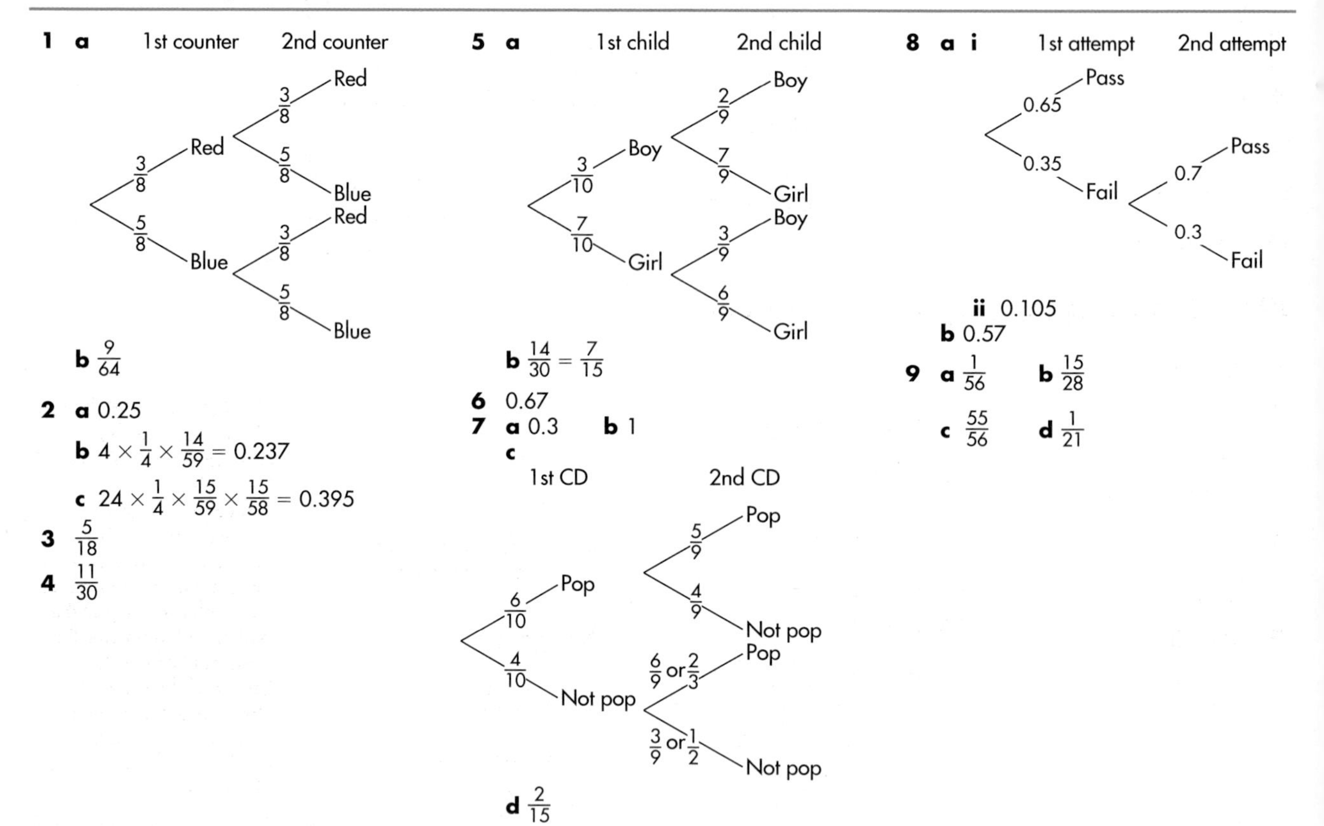

1 **a**

b $\frac{9}{64}$

2 **a** 0.25

b $4 \times \frac{1}{4} \times \frac{14}{59} = 0.237$

c $24 \times \frac{1}{4} \times \frac{15}{59} \times \frac{15}{58} = 0.395$

3 $\frac{5}{18}$

4 $\frac{11}{30}$

5 **a**

b $\frac{14}{30} = \frac{7}{15}$

6 0.67

7 **a** 0.3 **b** 1

c

d $\frac{2}{15}$

8 **a** **i**

ii 0.105

b 0.57

9 **a** $\frac{1}{56}$ **b** $\frac{15}{28}$

c $\frac{55}{56}$ **d** $\frac{1}{21}$

Answers to Chapter 12

Quick check

1 **a** 17, 20, 23 **b** 49, 64, 81
2 **a** 1 **b** 4 **c** 7
3 **a** $2(x + 3)$ **b** $x(x - 1)$ **c** $2x(5x + 1)$
4 **a** $x^2 + 8x + 12$ **b** $2x^2 - 5x - 3$ **c** $x^2 - 4x + 4$
5 **a** $x = 3 - 2y$ **b** $x = 4 + 3y$ **c** $x = 4y - 3$

12.1 Rearranging formulae

Exercise 12A

1 $k = \frac{T}{3}$
2 $y = X + 1$
3 $p = 3Q$
4 $r = \frac{A - 9}{4}$
5 $n = \frac{W + 1}{3}$
6 **a** $m = p - t$
b $t = p - m$
7 $m = gv$
8 $m = \sqrt{t}$
9 $r = \frac{C}{2\pi}$
10 $b = \frac{A}{h}$
11 $l = \frac{P - 2w}{2}$
12 $p = \sqrt{m - 2}$
13 **a** $-40 - 32 = -72$, $-72 \div 9 = -8$, $5 \times -8 = -40$
b $68 - 32 = 36$, $36 \div 9 = 4$, $4 \times 5 = 20$
c $F = \frac{9}{5}C + 32$
14 **a** $5x = 9y + 75$, $y = \frac{5x - 75}{9}$
b 25p
15 Average speeds: outward journey = 72 kph, return journey = 63 kph, taking 2 hours. He was held up for 15 minutes.
16 **a** $a = \frac{v - u}{t}$
b $t = \frac{v - u}{a}$
17 $d = \sqrt{\frac{4A}{\pi}}$
18 **a** $n = \frac{W - t}{3}$
b $t = W - 3n$
19 **a** $y = \frac{x + w}{5}$
b $w = 5y - x$
20 $p = \sqrt{\frac{k}{2}}$
21 **a** $t = u^2 - v$
b $u = \sqrt{v + t}$
22 **a** $m = k - n^2$
b $n = \sqrt{k - m}$
23 $r = \sqrt{\frac{T}{5}}$
24 **a** $w = K - 5n^2$
b $n = \sqrt{\frac{K - w}{w}}$

12.2 Changing the subject of a formula

Exercise 12B

1 $-8y$
2 $-\frac{x}{8}$
3 $\frac{b + 5c}{4}$
4 $\frac{b(q + p)}{q - p}$
5 $\frac{a(q - p)}{q + p}$
6 $\frac{A}{\pi(2h + k)}$
7 $\frac{u}{\sqrt{(1 - a)}}$
8 $\frac{3 + st}{2 + s}$
9 $\frac{6 + st}{2 + s}$
10 $\frac{2R - 3}{R - 1}$
11 **a** $\frac{P}{\pi + 2k}$
b $\sqrt{\frac{2A}{\pi + \sqrt{(k^2 - 1)}}}$
12 $\frac{100A}{100 + RY}$
13 **a** $b = \frac{Ra}{a - R}$
b $a = \frac{Rb}{b - R}$
14 **a** $\frac{2 + 2y}{y - 1}$
b $y - 1 = \frac{4}{x - 2}$, $(x - 2)(y - 1) = 4$, $x - 2 = \frac{4}{y - 1}$, $x = 2 + \frac{4}{y - 1}$
c $y = 1 + \frac{4}{x - 2} = \frac{x - 2 + 4}{x - 2} = \frac{x + 2}{x - 2}$ and $x = \frac{2 + 2y}{y - 1}$
d Same formulae as in **a**
15 **a** Cannot factorise the expression
b $\frac{3V}{r^2(2r + 3h)}$
c Yes, $\sqrt[3]{\frac{3V}{5\pi}}$
16 $x = \frac{2W - 2zy}{z + y}$
17 $x = \frac{1 - 3y}{2y - 5}$

The first number at the top of the answer is the constant term on the top of the original.
The coefficient of y at the top of the answer is the negative constant term on the bottom of the original.
The coefficient of y at the bottom of the answer the coefficient of x on the bottom of the original.
The constant term on the bottom is negative the coefficient of x on the top of the original.

18 **a** $b^2 = c^2 + a^2 - 2ac \cos B$
b $\cos C = \frac{a^2 + b^2 - c^2}{2ab}$

12.3 Simultaneous equations

Exercise 12C

1 **a** $x = 4, y = 1$ **b** $x = 1, y = 4$ **c** $x = 3, y = 1$ **d** $x = 5, y = -2$ **e** $x = 7, y = 1$ **f** $x = 5, y = \frac{1}{2}$ **g** $x = 4\frac{1}{2}, y = 1\frac{1}{2}$ **h** $x = -2, y = 4$ **i** $x = 2\frac{1}{2}, y = -1\frac{1}{2}$ **j** $x = 2\frac{1}{4}, y = 6\frac{1}{2}$ **k** $x = 4, y = 3$ **l** $x = 5, y = 3$

2 **a** 3 is the first term. The next term is $3 \times a + b$, which equals 14.
b $14a + b = 47$ **c** $a = 3, b = 5$ **d** 146, 443

Exercise 12D

1 **a** $x = 2, y = -3$ **b** $x = 7, y = 3$ **c** $x = 4, y = 1$ **d** $x = 2, y = 5$ **e** $x = 4, y = -3$ **f** $x = 1, y = 7$ **g** $x = 2\frac{1}{2}, y = 1\frac{1}{2}$ **h** $x = -1, y = 2\frac{1}{2}$ **i** $x = 6, y = 3$ **j** $x = \frac{1}{2}, y = -3$ **k** $x = -1, y = 5$ **l** $x = 1\frac{1}{2}, y = \frac{3}{4}$

2 **a** They are the same equation. Divide the first by 2 and it is the second, so they have an infinite number of solutions.
b Double the second equation to get $6x + 2y = 14$ and subtract to get $9 = 14$. The left-hand sides are the same if the second is doubled so they cannot have different values.

Exercise 12E

1 **a** $x = 5, y = 1$ **b** $x = 3, y = 8$ **c** $x = 9, y = 1$ **d** $x = 7, y = 3$ **e** $x = 4, y = 2$ **f** $x = 6, y = 5$ **g** $x = 3, y = -2$ **h** $x = 2, y = \frac{1}{2}$ **i** $x = -2, y = -3$ **j** $x = -1, y = 2\frac{1}{2}$ **k** $x = 2\frac{1}{2}, y = -\frac{1}{2}$ **l** $x = -1\frac{1}{2}, y = 4\frac{1}{2}$ **m** $x = -\frac{1}{2}, y = -6\frac{1}{2}$ **n** $x = 3\frac{1}{2}, y = 1\frac{1}{2}$ **o** $x = -2\frac{1}{2}, y = -3\frac{1}{2}$

2 (1, −2) is the solution to equations A and C; (−1, 3) is the solution to equations A and D; (2, 1) is the solution to B and C; (3, −3) is the solution to B and D.

3 Intersection points are (0, 6), (1, 3) and (2, 4). Area is 2 square units.

4 Intersection points are (0, 3), (6, 0) and (4, −1). Area is 6 square units.

12.4 Solving problems using simultaneous equations

Exercise 12F

1 Amul £7.20, Kim £3.50

2 **a** $10x + 5y = 420$, $8x + 10y = 540$
b £2.11

3 **a** $6x + 3y = 435$, $11x + 7y = 880$
b £5.55

4 **a** My age minus 6 equals 2 × (my son's age minus 6)
b $x = 46$ and $y = 26$

5 **a** $3t + 5b = 810$, $3t + 3b = 630$
b £10.20

6 84p

7 10.3 kg

8 £4.40

9 £62

10 £195

11 2 hr 10 min

12 $p = 36$, $c = 22$. Total weight for Baz is 428 pounds so he can carry the load safely on his trailer.

13 $b = £3.50$, $p = £1.75$. Camilla needs £35 so she will not have enough money.

14 When Carmen worked out (2) − (3), she should have got $y = 6$
When Jeff rearranged $2x + 8 - x = 10$, he should have got $x = 2$
They also misunderstood 'two, six' as this means $x = 2$ and $y = 6$, not the other way round.

12.5 Linear and non-linear simultaneous equations

Exercise 12G

1 **a** (5, −1)
b (4, 1)
c (8, −1)

2 **a** (1, 2) and (−2, −1)
b (−4, 1) and (−2, 2)

3 **a** (3, 4) and (4, 3)
b (0, 3) and (−3, 0)
c (3, 2) and (−2, 3)

4 **a** (2, 5) and (−2, −3)
b (−1, −2) and (4, 3)
c (3, 3) and (1, −1)

5 **a** (−3, −3), (1, 1)
b (3, −2), (−2, 3)
c (−2, −1), (1, 2)
d (2, −1), (3, 1)
e (−2, 1), (3, 6)
f (1, −4), (4, 2)
g (4, 5), (−5, −4)

6 **a** (1, 0)
b **iii** as the straight line just touches the curve

7 **a** (−2, 1)
b **i** (2, 1)
ii (−2, −1)
iii (2, −1)

8 **a** (2, 4)
b (1, 0)
c The line is a tangent to the curve.

12.6 Algebraic fractions

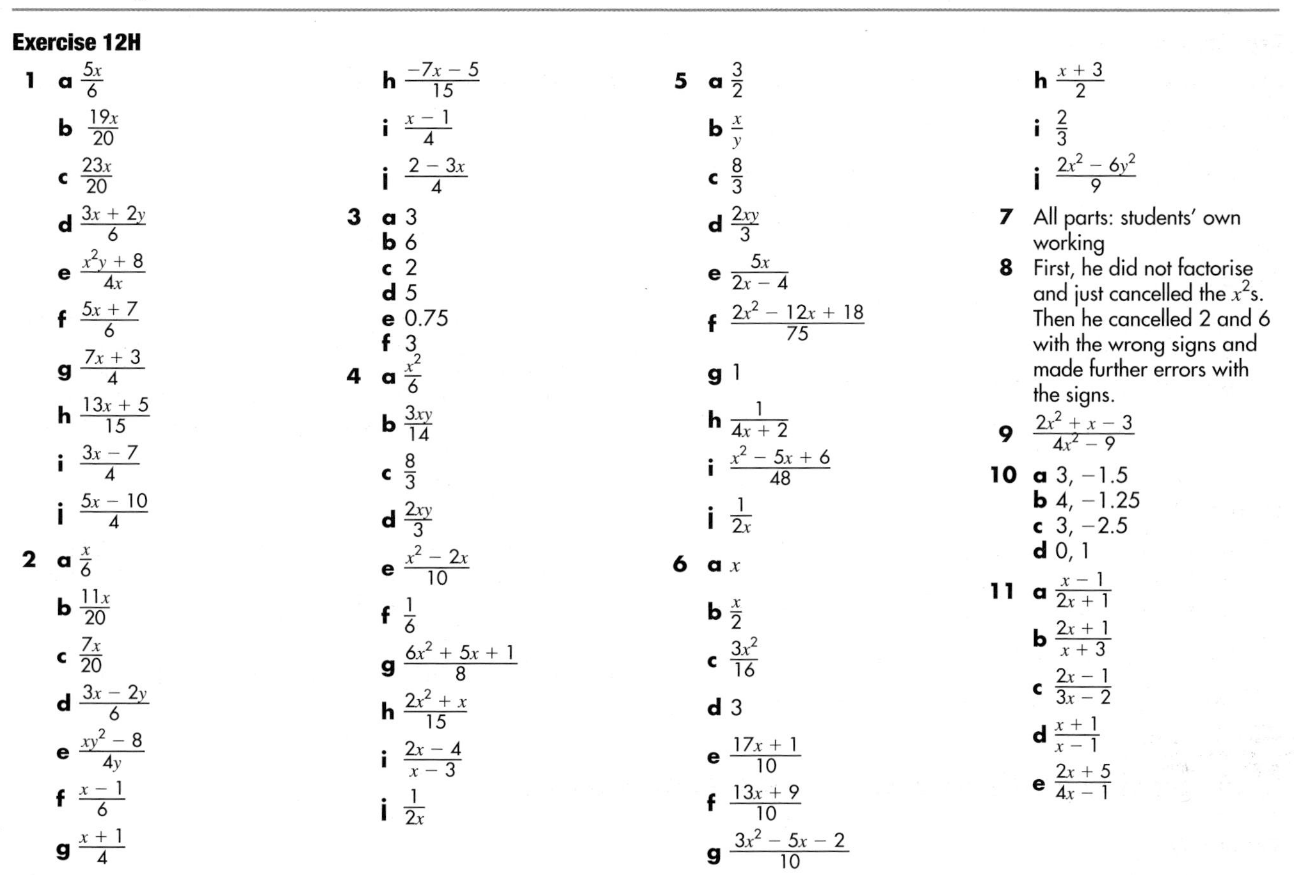

Exercise 12H

1 **a** $\frac{5x}{6}$
b $\frac{19x}{20}$
c $\frac{23x}{20}$
d $\frac{3x + 2y}{6}$
e $\frac{x^2y + 8}{4x}$
f $\frac{5x + 7}{6}$
g $\frac{7x + 3}{4}$
h $\frac{13x + 5}{15}$
i $\frac{3x - 7}{4}$
j $\frac{5x - 10}{4}$

2 **a** $\frac{x}{6}$
b $\frac{11x}{20}$
c $\frac{7x}{20}$
d $\frac{3x - 2y}{6}$
e $\frac{xy^2 - 8}{4y}$
f $\frac{x - 1}{6}$
g $\frac{x + 1}{4}$
h $\frac{-7x - 5}{15}$
i $\frac{x - 1}{4}$
j $\frac{2 - 3x}{4}$

3 **a** 3
b 6
c 2
d 5
e 0.75
f 3

4 **a** $\frac{x^2}{6}$
b $\frac{3xy}{14}$
c $\frac{8}{3}$
d $\frac{2xy}{3}$
e $\frac{x^2 - 2x}{10}$
f $\frac{1}{6}$
g $\frac{6x^2 + 5x + 1}{8}$
h $\frac{2x^2 + x}{15}$
i $\frac{2x - 4}{x - 3}$
j $\frac{1}{2x}$

5 **a** $\frac{3}{2}$
b $\frac{x}{y}$
c $\frac{8}{3}$
d $\frac{2xy}{3}$
e $\frac{5x}{2x - 4}$
f $\frac{2x^2 - 12x + 18}{75}$
g 1
h $\frac{1}{4x + 2}$
i $\frac{x^2 - 5x + 6}{48}$
j $\frac{1}{2x}$

6 **a** x
b $\frac{x}{2}$
c $\frac{3x^2}{16}$
d 3
e $\frac{17x + 1}{10}$
f $\frac{13x + 9}{10}$
g $\frac{3x^2 - 5x - 2}{10}$
h $\frac{x + 3}{2}$
i $\frac{2}{3}$
j $\frac{2x^2 - 6y^2}{9}$

7 All parts: students' own working

8 First, he did not factorise and just cancelled the x^2s. Then he cancelled 2 and 6 with the wrong signs and made further errors with the signs.

9 $\frac{2x^2 + x - 3}{4x^2 - 9}$

10 **a** 3, −1.5
b 4, −1.25
c 3, −2.5
d 0, 1

11 **a** $\frac{x - 1}{2x + 1}$
b $\frac{2x + 1}{x + 3}$
c $\frac{2x - 1}{3x - 2}$
d $\frac{x + 1}{x - 1}$
e $\frac{2x + 5}{4x - 1}$

12.7 Algebraic proof

Exercise 12I

1 **a** Odd, yes
b $(2n + 1) + 2m = 2(n + m) + 1$, which is odd
2 Check students' proofs
3 **a** 3, 5, 8, 13, 21, 34, 55
b $3a + 5b$, $5a + 8b$, $8a + 13b$, $13a + 21b$, $21a + 34b$
c $(8a + 13b) - (2a + 3b) = 6a + 10b = 2(3a + 5b)$
4 Check students' proofs
5 **a** **i** 10
b–d Check students' proofs
6–15 Check students' proofs

Exercise 12J

All answers student's own proof

Examination questions

1 **a** $5x + 60 = 360°$
b $5x = 300$, $x = 60°$
2 $6x + 30 = 180°$, $6x = 150°$, $x = 25°$
3 $t = \frac{1}{2}(y + 10)$ or equivalent
4 $\frac{3}{x + 2}$
5 $x + 3$
6 $\frac{x - 1}{2}$
7 $x = 3$, $y = -2$
8 **a** 9 or −9
b $s = \frac{v^2 - u^2}{2a}$
9 $\frac{3}{x - 2}$
10 $x + 3$
11 $(n + 2)^2 - (n - 2)^2 = n^2 + 4n + 4 - (n^2 - 4n + 4)$
12 $\frac{x^2 - 3}{x(x - 2)}$
13 $\frac{4}{a + 5}$
14 $\frac{5x - 7}{(x + 5)(x - 3)}$
15 $\frac{6x}{(x - 3)(x + 3)}$
16 $\frac{p - 3}{2}$

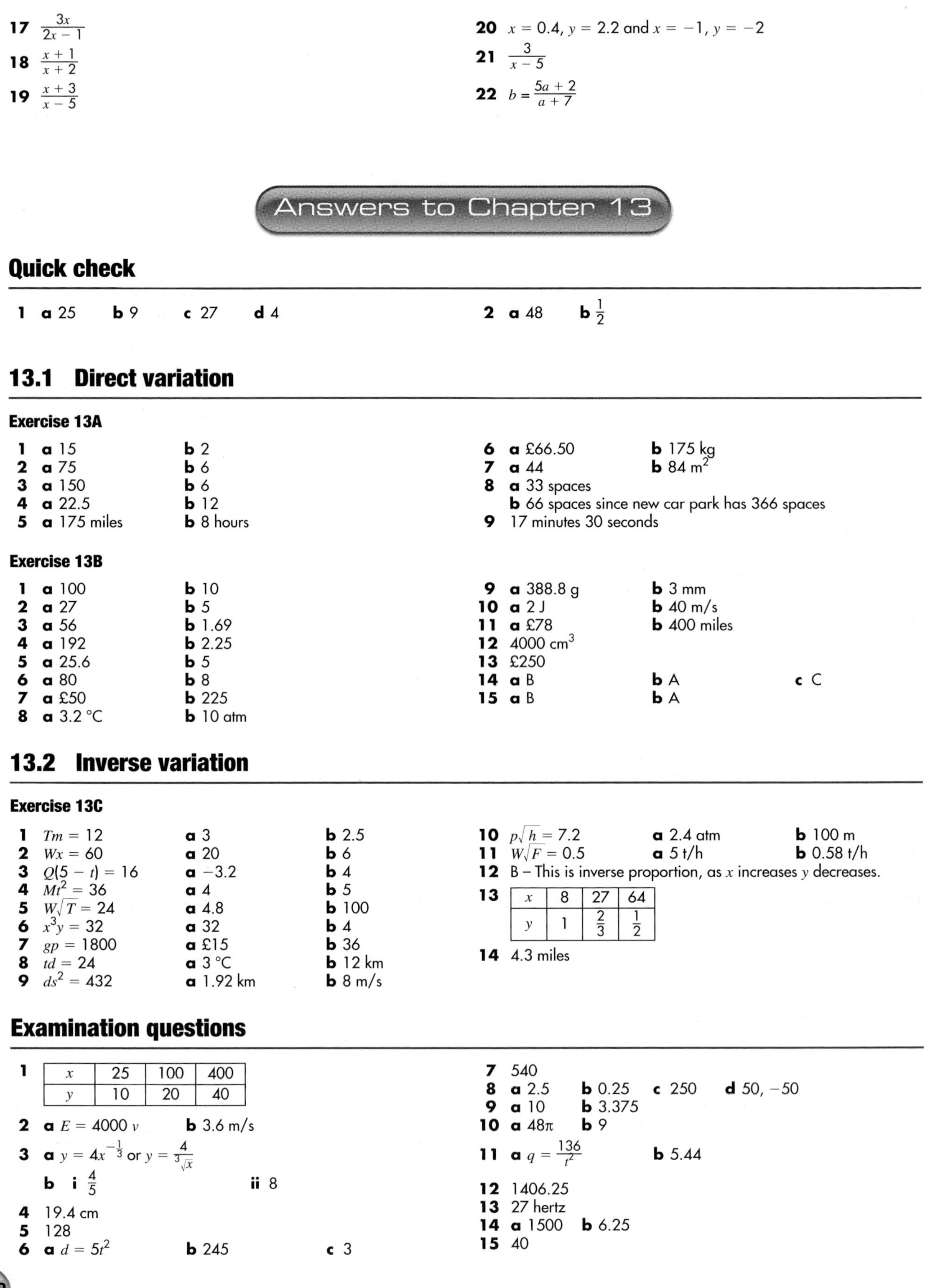

17 $\frac{3x}{2x-1}$

18 $\frac{x+1}{x+2}$

19 $\frac{x+3}{x-5}$

20 $x = 0.4$, $y = 2.2$ and $x = -1$, $y = -2$

21 $\frac{3}{x-5}$

22 $b = \frac{5a+2}{a+7}$

Answers to Chapter 13

Quick check

1 **a** 25 **b** 9 **c** 27 **d** 4

2 **a** 48 **b** $\frac{1}{2}$

13.1 Direct variation

Exercise 13A

1 **a** 15 **b** 2
2 **a** 75 **b** 6
3 **a** 150 **b** 6
4 **a** 22.5 **b** 12
5 **a** 175 miles **b** 8 hours
6 **a** £66.50 **b** 175 kg
7 **a** 44 **b** 84 m^2
8 **a** 33 spaces
b 66 spaces since new car park has 366 spaces
9 17 minutes 30 seconds

Exercise 13B

1 **a** 100 **b** 10
2 **a** 27 **b** 5
3 **a** 56 **b** 1.69
4 **a** 192 **b** 2.25
5 **a** 25.6 **b** 5
6 **a** 80 **b** 8
7 **a** £50 **b** 225
8 **a** 3.2 °C **b** 10 atm
9 **a** 388.8 g **b** 3 mm
10 **a** 2 J **b** 40 m/s
11 **a** £78 **b** 400 miles
12 4000 cm^3
13 £250
14 **a** B **b** A **c** C
15 **a** B **b** A

13.2 Inverse variation

Exercise 13C

1 $Tm = 12$ **a** 3 **b** 2.5
2 $Wx = 60$ **a** 20 **b** 6
3 $Q(5 - t) = 16$ **a** -3.2 **b** 4
4 $Mt^2 = 36$ **a** 4 **b** 5
5 $W\sqrt{T} = 24$ **a** 4.8 **b** 100
6 $x^3y = 32$ **a** 32 **b** 4
7 $gp = 1800$ **a** £15 **b** 36
8 $td = 24$ **a** 3 °C **b** 12 km
9 $ds^2 = 432$ **a** 1.92 km **b** 8 m/s
10 $p\sqrt{h} = 7.2$ **a** 2.4 atm **b** 100 m
11 $W\sqrt{F} = 0.5$ **a** 5 t/h **b** 0.58 t/h
12 B – This is inverse proportion, as x increases y decreases.
13

x	8	27	64
y	1	$\frac{2}{3}$	$\frac{1}{2}$

14 4.3 miles

Examination questions

1

x	25	100	400
y	10	20	40

2 **a** $E = 4000\,v$ **b** 3.6 m/s
3 **a** $y = 4x^{-\frac{1}{3}}$ or $y = \frac{4}{\sqrt[3]{x}}$
b **i** $\frac{4}{5}$ **ii** 8
4 19.4 cm
5 128
6 **a** $d = 5t^2$ **b** 245 **c** 3
7 540
8 **a** 2.5 **b** 0.25 **c** 250 **d** 50, -50
9 **a** 10 **b** 3.375
10 **a** 48π **b** 9
11 **a** $q = \frac{136}{t^2}$ **b** 5.44
12 1406.25
13 27 hertz
14 **a** 1500 **b** 6.25
15 40

Answers to Chapter 14

Quick check

1 **a** 6370 **b** 6400 **c** 6000
2 **a** 2.4 **b** 2.39
3 **a** 50 **b** 47.3

14.1 Limits of accuracy

Exercise 14A

1 **a** $6.5 \leq 7 < 7.5$ **b** $115 \leq 120 < 125$
c $3350 \leq 3400 < 3450$ **d** $49.5 \leq 50 < 50.5$
e $5.50 \leq 6 < 6.50$ **f** $16.75 \leq 16.8 < 16.85$
g $15.5 \leq 16 < 16.5$ **h** $14\,450 \leq 14\,500 < 14\,550$
i $54.5 \leq 55 < 55.5$ **j** $52.5 \leq 55 < 57.5$
2 **a** $5.5 \leq 6 < 6.5$ **b** $16.5 \leq 17 < 17.5$
c $31.5 \leq 32 < 32.5$ **d** $237.5 \leq 238 < 238.5$
e $7.25 \leq 7.3 < 7.35$ **f** $25.75 \leq 25.8 < 25.85$
g $3.35 \leq 3.4 < 3.45$ **h** $86.5 \leq 87 < 87.5$
i $4.225 \leq 4.23 < 4.235$ **j** $2.185 \leq 2.19 < 2.195$
k $12.665 \leq 12.67 < 12.675$ **l** $24.5 \leq 25 < 25.5$
m $35 \leq 40 < 45$ **n** $595 \leq 600 < 605$
o $25 \leq 30 < 35$ **p** $995 \leq 1000 < 1050$
q $3.95 \leq 4.0 < 4.05$ **r** $7.035 \leq 7.04 < 7.045$
s $11.95 \leq 12.0 < 12.05$ **t** $6.995 \leq 7.00 < 7.005$

3 **a** 7.5, 8.5 **b** 25.5, 26.5
c 24.5, 25.5 **d** 84.5, 85.5
e 2.395, 2.405 **f** 0.15, 0.25
g 0.055, 0.065 **h** 250, 350
i 0.65, 0.75 **j** 365.5, 366.5
k 165, 175 **l** 205, 215
4 There are 16 empty seats and the number getting on the bus is from 15 to 24 so it is possible if 15 or 16 get on.
5 C: The chain and distance are both any value between 29.5 and 30.5 metres, so there is no way of knowing if the chain is longer or shorter than the distance.
6 2 kg 450 grams
7 **a** <65.5 g **b** 64.5 g
c <2620 g **d** 2580 g

14.2 Problems involving limits of accuracy

Exercise 14B

1 Minimum 65 kg, maximum 75 kg
2 Minimum is 19, maximum is 20
3 **a** 12.5 kg **b** 20
4 3 years 364 days (Jack is on his fifth birthday; Jill is 9 years old tomorrow)
5 **a** $38.25\text{ cm}^2 \leq \text{area} < 52.25\text{ cm}^2$
b $37.1575\text{ cm}^2 \leq \text{area} < 38.4475\text{ cm}^2$
c $135.625\text{ cm}^2 \leq \text{area} < 145.225\text{ cm}^2$
6 **a** $5.5\text{ m} \leq \text{length} < 6.5\text{ m}$, $3.5\text{ m} \leq \text{width} < 4.5\text{ m}$
b 29.25 m^2
c 18 m
7 $79.75\text{ m}^2 \leq \text{area} < 100.75\text{ m}^2$
8 $216.125\text{ m}^3 \leq \text{volume} < 354.375\text{ m}^3$
9 12.5 metres
10 Yes, because they could be walking at 4.5 mph and 2.5 mph meaning that they would cover 4.5 miles + 2.5 miles = 7 miles in 1 hour
11 $20.9\text{ m} \leq \text{length} < 22.9\text{ m}$ (3 sf)
12 $16.4\text{ cm}^2 \leq \text{area} < 21.7\text{ cm}^2$ (3 sf)

13 **a** **i** $64.1\text{ cm}^3 \leq \text{volume} < 69.6\text{ cm}^3$ (3 sf)
ii £22 578 $\leq$ price $<$ £24 515 (nearest £ using rounded figures from ai)
b 23 643 $\leq$ price $<$ £23 661 (nearest £)
c Errors in length compounded by being used 3 times in **a**, but errors in weight only used once in **b**
14 **a** $14.65\text{ s} \leq \text{time} < 14.75\text{ s}$
b $99.5\text{ m} \leq \text{length} < 100.5\text{ m}$
c 6.86 m/s (3 sf)
15 **a** +1.25% (3 sf)
b +1.89% (3 sf)
16 $3.41\text{ cm} \leq \text{length} < 3.43\text{ cm}$ (3 sf)
17 $5.80\text{ cm} \leq \text{length} < 5.90\text{ cm}$ (3 sf)
18 $14\text{ s} \leq \text{time} < 30\text{ s}$
19 Cannot be certain as limits of accuracy for all three springs overlap:
Red: 12.5 newtons to 13.1 newtons
Green: 11.8 newtons to 13.2 newtons
Blue: 9.5 newtons to 12.9 newtons

Examination questions

1 **a** 1845 **b** 1854
2 215 miles
3 **a** 62.5 cm **b** 63.5 cm
4 **a** 22.5 cm **b** 21.5 cm
5 **a** 9.75 m/s^2 **b** 30.7 metres
6 8.75 miles per litre
7 **a** 59.5 cm **b** 60.5 cm
8 18.95 cm and 21.16 cm
9 2.92
10 7.39 m/s
11 **a** 5.196 m **b** Yes (min corridor 5.11 m, max pole 5.005 m)

Answers to Chapter 15

Quick check

a $\begin{pmatrix}1\\3\end{pmatrix}$ **b** $\begin{pmatrix}3\\0\end{pmatrix}$ **c** $\begin{pmatrix}2\\-1\end{pmatrix}$ **d** $\begin{pmatrix}-1\\-2\end{pmatrix}$

15.1 Properties of vectors

Exercise 15A

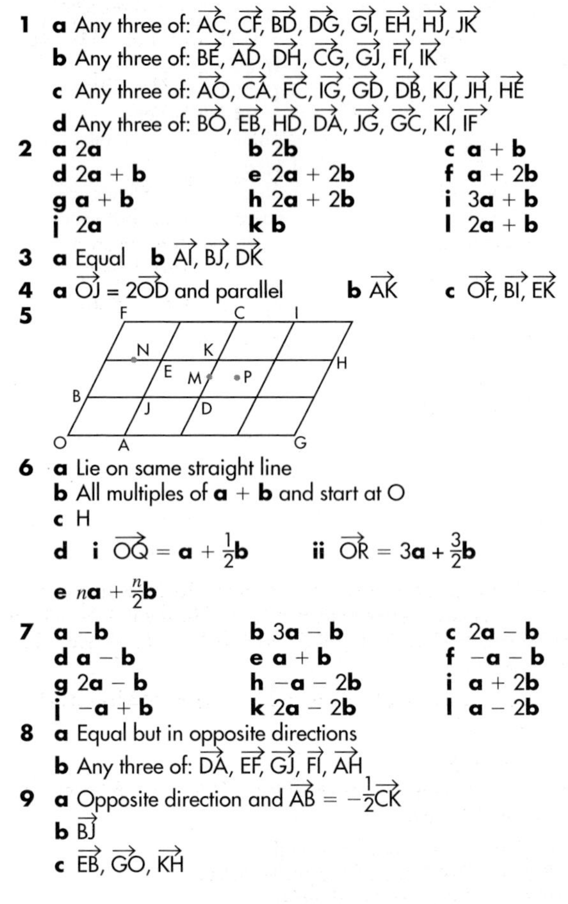

1 **a** Any three of: $\overrightarrow{AC}$, $\overrightarrow{CF}$, $\overrightarrow{BD}$, $\overrightarrow{DG}$, $\overrightarrow{GI}$, $\overrightarrow{EH}$, $\overrightarrow{HJ}$, $\overrightarrow{JK}$
b Any three of: $\overrightarrow{BE}$, $\overrightarrow{AD}$, $\overrightarrow{DH}$, $\overrightarrow{CG}$, $\overrightarrow{GJ}$, $\overrightarrow{FI}$, $\overrightarrow{IK}$
c Any three of: $\overrightarrow{AO}$, $\overrightarrow{CA}$, $\overrightarrow{FC}$, $\overrightarrow{IG}$, $\overrightarrow{GD}$, $\overrightarrow{DB}$, $\overrightarrow{KJ}$, $\overrightarrow{JH}$, $\overrightarrow{HE}$
d Any three of: $\overrightarrow{BO}$, $\overrightarrow{EB}$, $\overrightarrow{HD}$, $\overrightarrow{DA}$, $\overrightarrow{JG}$, $\overrightarrow{GC}$, $\overrightarrow{KI}$, $\overrightarrow{IF}$

2 **a** $2\mathbf{a}$ **b** $2\mathbf{b}$ **c** $\mathbf{a}+\mathbf{b}$
d $2\mathbf{a}+\mathbf{b}$ **e** $2\mathbf{a}+2\mathbf{b}$ **f** $\mathbf{a}+2\mathbf{b}$
g $\mathbf{a}+\mathbf{b}$ **h** $2\mathbf{a}+2\mathbf{b}$ **i** $3\mathbf{a}+\mathbf{b}$
j $2\mathbf{a}$ **k** $\mathbf{b}$ **l** $2\mathbf{a}+\mathbf{b}$

3 **a** Equal **b** $\overrightarrow{AI}$, $\overrightarrow{BJ}$, $\overrightarrow{DK}$

4 **a** $\overrightarrow{OJ} = 2\overrightarrow{OD}$ and parallel **b** $\overrightarrow{AK}$ **c** $\overrightarrow{OF}$, $\overrightarrow{BI}$, $\overrightarrow{EK}$

5

6 **a** Lie on same straight line
b All multiples of $\mathbf{a}+\mathbf{b}$ and start at O
c H
d **i** $\overrightarrow{OQ} = \mathbf{a} + \frac{1}{2}\mathbf{b}$ **ii** $\overrightarrow{OR} = 3\mathbf{a} + \frac{3}{2}\mathbf{b}$
e $n\mathbf{a} + \frac{n}{2}\mathbf{b}$

7 **a** $-\mathbf{b}$ **b** $3\mathbf{a}-\mathbf{b}$ **c** $2\mathbf{a}-\mathbf{b}$
d $\mathbf{a}-\mathbf{b}$ **e** $\mathbf{a}+\mathbf{b}$ **f** $-\mathbf{a}-\mathbf{b}$
g $2\mathbf{a}-\mathbf{b}$ **h** $-\mathbf{a}-2\mathbf{b}$ **i** $\mathbf{a}+2\mathbf{b}$
j $-\mathbf{a}+\mathbf{b}$ **k** $2\mathbf{a}-2\mathbf{b}$ **l** $\mathbf{a}-2\mathbf{b}$

8 **a** Equal but in opposite directions
b Any three of: $\overrightarrow{DA}$, $\overrightarrow{EF}$, $\overrightarrow{GJ}$, $\overrightarrow{FI}$, $\overrightarrow{AH}$

9 **a** Opposite direction and $\overrightarrow{AB} = -\frac{1}{2}\overrightarrow{CK}$
b $\overrightarrow{BJ}$
c $\overrightarrow{EB}$, $\overrightarrow{GO}$, $\overrightarrow{KH}$

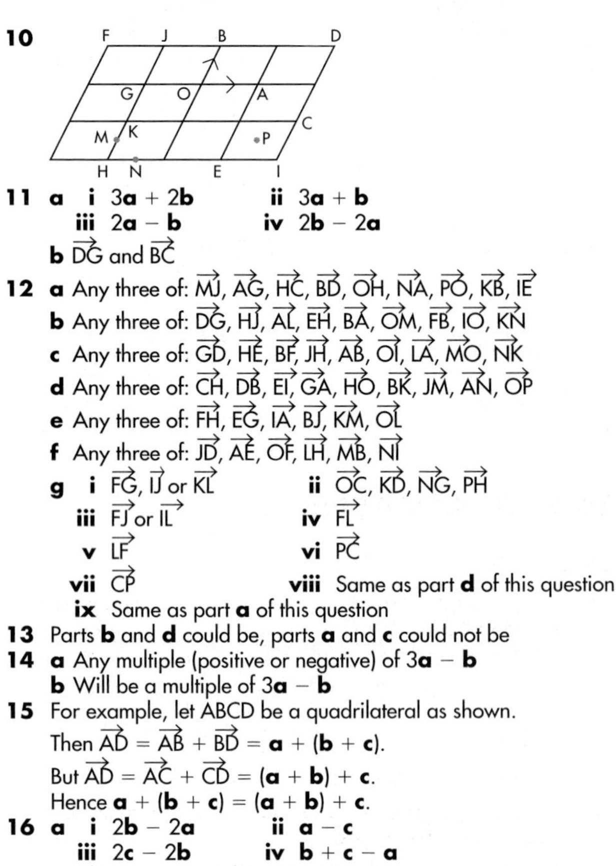

10

11 **a** **i** $3\mathbf{a}+2\mathbf{b}$ **ii** $3\mathbf{a}+\mathbf{b}$
iii $2\mathbf{a}-\mathbf{b}$ **iv** $2\mathbf{b}-2\mathbf{a}$
b $\overrightarrow{DG}$ and $\overrightarrow{BC}$

12 **a** Any three of: $\overrightarrow{MJ}$, $\overrightarrow{AG}$, $\overrightarrow{HC}$, $\overrightarrow{BD}$, $\overrightarrow{OH}$, $\overrightarrow{NA}$, $\overrightarrow{PO}$, $\overrightarrow{KB}$, $\overrightarrow{IE}$
b Any three of: $\overrightarrow{DG}$, $\overrightarrow{HJ}$, $\overrightarrow{AL}$, $\overrightarrow{EH}$, $\overrightarrow{BA}$, $\overrightarrow{OM}$, $\overrightarrow{FB}$, $\overrightarrow{IO}$, $\overrightarrow{KN}$
c Any three of: $\overrightarrow{GD}$, $\overrightarrow{HE}$, $\overrightarrow{BF}$, $\overrightarrow{JH}$, $\overrightarrow{AB}$, $\overrightarrow{OI}$, $\overrightarrow{LA}$, $\overrightarrow{MO}$, $\overrightarrow{NK}$
d Any three of: $\overrightarrow{CH}$, $\overrightarrow{DB}$, $\overrightarrow{EI}$, $\overrightarrow{GA}$, $\overrightarrow{HO}$, $\overrightarrow{BK}$, $\overrightarrow{JM}$, $\overrightarrow{AN}$, $\overrightarrow{OP}$
e Any three of: $\overrightarrow{FH}$, $\overrightarrow{EG}$, $\overrightarrow{IA}$, $\overrightarrow{BJ}$, $\overrightarrow{KM}$, $\overrightarrow{OL}$
f Any three of: $\overrightarrow{JD}$, $\overrightarrow{AE}$, $\overrightarrow{OF}$, $\overrightarrow{LH}$, $\overrightarrow{MB}$, $\overrightarrow{NI}$
g **i** $\overrightarrow{FG}$, $\overrightarrow{IJ}$ or $\overrightarrow{KL}$ **ii** $\overrightarrow{OC}$, $\overrightarrow{KD}$, $\overrightarrow{NG}$, $\overrightarrow{PH}$
iii $\overrightarrow{FJ}$ or $\overrightarrow{IL}$ **iv** $\overrightarrow{FL}$
v $\overrightarrow{LF}$ **vi** $\overrightarrow{PC}$
vii $\overrightarrow{CP}$ **viii** Same as part **d** of this question
ix Same as part **a** of this question

13 Parts **b** and **d** could be, parts **a** and **c** could not be

14 **a** Any multiple (positive or negative) of $3\mathbf{a}-\mathbf{b}$
b Will be a multiple of $3\mathbf{a}-\mathbf{b}$

15 For example, let ABCD be a quadrilateral as shown.
Then $\overrightarrow{AD} = \overrightarrow{AB} + \overrightarrow{BD} = \mathbf{a} + (\mathbf{b}+\mathbf{c})$.
But $\overrightarrow{AD} = \overrightarrow{AC} + \overrightarrow{CD} = (\mathbf{a}+\mathbf{b}) + \mathbf{c}$.
Hence $\mathbf{a} + (\mathbf{b}+\mathbf{c}) = (\mathbf{a}+\mathbf{b}) + \mathbf{c}$.

16 **a** **i** $2\mathbf{b}-2\mathbf{a}$ **ii** $\mathbf{a}-\mathbf{c}$
iii $2\mathbf{c}-2\mathbf{b}$ **iv** $\mathbf{b}+\mathbf{c}-\mathbf{a}$
b $\overrightarrow{RQ} = \mathbf{a}-\mathbf{c} = \overrightarrow{SP}$, so two opposite sides are equal and parallel, hence PQRS is a parallelogram

15.2 Vectors in geometry

Exercise 15B

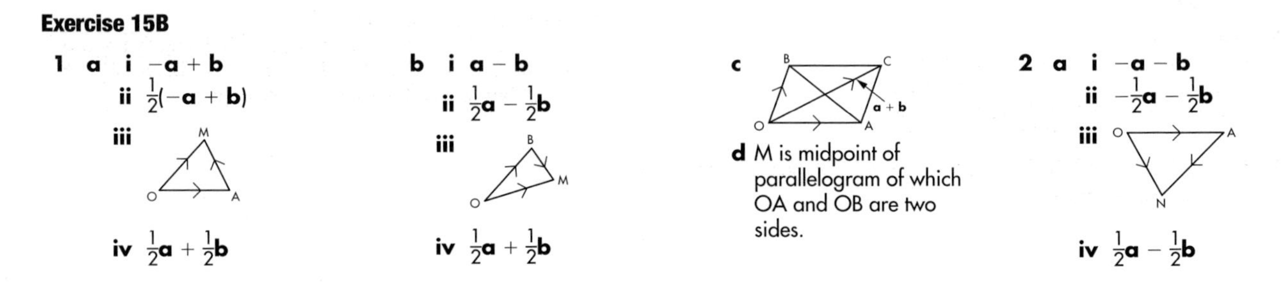

1 **a** **i** $-\mathbf{a}+\mathbf{b}$
ii $\frac{1}{2}(-\mathbf{a}+\mathbf{b})$
iii

iv $\frac{1}{2}\mathbf{a} + \frac{1}{2}\mathbf{b}$

b **i** $\mathbf{a}-\mathbf{b}$
ii $\frac{1}{2}\mathbf{a} - \frac{1}{2}\mathbf{b}$
iii

iv $\frac{1}{2}\mathbf{a} + \frac{1}{2}\mathbf{b}$

c

d M is midpoint of parallelogram of which OA and OB are two sides.

2 **a** **i** $-\mathbf{a}-\mathbf{b}$
ii $-\frac{1}{2}\mathbf{a} - \frac{1}{2}\mathbf{b}$
iii

iv $\frac{1}{2}\mathbf{a} - \frac{1}{2}\mathbf{b}$

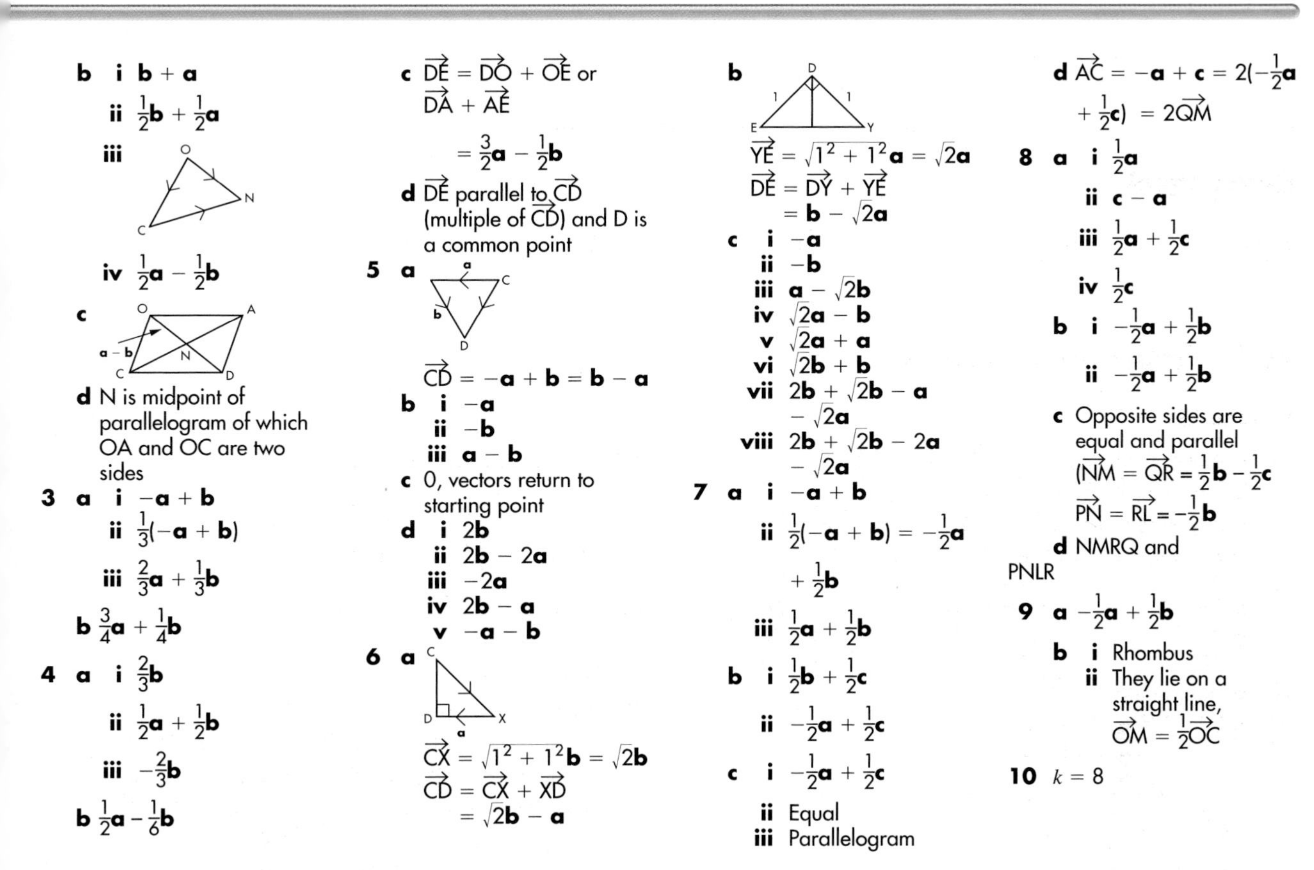

b **i** $\mathbf{b} + \mathbf{a}$
ii $\frac{1}{2}\mathbf{b} + \frac{1}{2}\mathbf{a}$
iii
iv $\frac{1}{2}\mathbf{a} - \frac{1}{2}\mathbf{b}$
c
d N is midpoint of parallelogram of which OA and OC are two sides

3 **a** **i** $-\mathbf{a} + \mathbf{b}$
ii $\frac{1}{3}(-\mathbf{a} + \mathbf{b})$
iii $\frac{2}{3}\mathbf{a} + \frac{1}{3}\mathbf{b}$
b $\frac{3}{4}\mathbf{a} + \frac{1}{4}\mathbf{b}$

4 **a** **i** $\frac{2}{3}\mathbf{b}$
ii $\frac{1}{2}\mathbf{a} + \frac{1}{2}\mathbf{b}$
iii $-\frac{2}{3}\mathbf{b}$
b $\frac{1}{2}\mathbf{a} - \frac{1}{6}\mathbf{b}$

c $\overrightarrow{DE} = \overrightarrow{DO} + \overrightarrow{OE}$ or $\overrightarrow{DA} + \overrightarrow{AE}$
$= \frac{3}{2}\mathbf{a} - \frac{1}{2}\mathbf{b}$
d $\overrightarrow{DE}$ parallel to $\overrightarrow{CD}$ (multiple of $\overrightarrow{CD}$) and D is a common point

5 **a**
$\overrightarrow{CD} = -\mathbf{a} + \mathbf{b} = \mathbf{b} - \mathbf{a}$
b **i** $-\mathbf{a}$
ii $-\mathbf{b}$
iii $\mathbf{a} - \mathbf{b}$
c 0, vectors return to starting point
d **i** $2\mathbf{b}$
ii $2\mathbf{b} - 2\mathbf{a}$
iii $-2\mathbf{a}$
iv $2\mathbf{b} - \mathbf{a}$
v $-\mathbf{a} - \mathbf{b}$

6 **a**
$\overrightarrow{CX} = \sqrt{1^2 + 1^2}\mathbf{b} = \sqrt{2}\mathbf{b}$
$\overrightarrow{CD} = \overrightarrow{CX} + \overrightarrow{XD}$
$= \sqrt{2}\mathbf{b} - \mathbf{a}$
b
$\overrightarrow{YE} = \sqrt{1^2 + 1^2}\mathbf{a} = \sqrt{2}\mathbf{a}$
$\overrightarrow{DE} = \overrightarrow{DY} + \overrightarrow{YE}$
$= \mathbf{b} - \sqrt{2}\mathbf{a}$
c **i** $-\mathbf{a}$
ii $-\mathbf{b}$
iii $\mathbf{a} - \sqrt{2}\mathbf{b}$
iv $\sqrt{2}\mathbf{a} - \mathbf{b}$
v $\sqrt{2}\mathbf{a} + \mathbf{a}$
vi $\sqrt{2}\mathbf{b} + \mathbf{b}$
vii $2\mathbf{b} + \sqrt{2}\mathbf{b} - \mathbf{a} - \sqrt{2}\mathbf{a}$
viii $2\mathbf{b} + \sqrt{2}\mathbf{b} - 2\mathbf{a} - \sqrt{2}\mathbf{a}$

7 **a** **i** $-\mathbf{a} + \mathbf{b}$
ii $\frac{1}{2}(-\mathbf{a} + \mathbf{b}) = -\frac{1}{2}\mathbf{a} + \frac{1}{2}\mathbf{b}$
iii $\frac{1}{2}\mathbf{a} + \frac{1}{2}\mathbf{b}$
b **i** $\frac{1}{2}\mathbf{b} + \frac{1}{2}\mathbf{c}$
ii $-\frac{1}{2}\mathbf{a} + \frac{1}{2}\mathbf{c}$
c **i** $-\frac{1}{2}\mathbf{a} + \frac{1}{2}\mathbf{c}$
ii Equal
iii Parallelogram
d $\overrightarrow{AC} = -\mathbf{a} + \mathbf{c} = 2(-\frac{1}{2}\mathbf{a} + \frac{1}{2}\mathbf{c}) = 2\overrightarrow{QM}$

8 **a** **i** $\frac{1}{2}\mathbf{a}$
ii $\mathbf{c} - \mathbf{a}$
iii $\frac{1}{2}\mathbf{a} + \frac{1}{2}\mathbf{c}$
iv $\frac{1}{2}\mathbf{c}$
b **i** $-\frac{1}{2}\mathbf{a} + \frac{1}{2}\mathbf{b}$
ii $-\frac{1}{2}\mathbf{a} + \frac{1}{2}\mathbf{b}$
c Opposite sides are equal and parallel ($\overrightarrow{NM} = \overrightarrow{QR} = \frac{1}{2}\mathbf{b} - \frac{1}{2}\mathbf{c}$
$\overrightarrow{PN} = \overrightarrow{RL} = -\frac{1}{2}\mathbf{b}$
d NMRQ and PNLR

9 **a** $-\frac{1}{2}\mathbf{a} + \frac{1}{2}\mathbf{b}$
b **i** Rhombus
ii They lie on a straight line, $\overrightarrow{OM} = \frac{1}{2}\overrightarrow{OC}$

10 $k = 8$

15.3 Geometric proof

Exercise 15C

1 **a** Angles ACB = 180 – 126 = 54° (angles in a straight line);
Angle ABC = 180 – 72 – 54 = 54° (angles in a Δ);
Angle ABC = Angle ACB hence ΔABC is isosceles
b $\angle DFE = 180° - (90° + \frac{x}{2}°) = 90° - \frac{x}{2}°$
$\angle DEF = 180° - x° - (90° - \frac{x}{2}°) = 90° - \frac{x}{2}°$
∠DFE = ∠DEF hence DEF is isosceles.

2 The exterior angle of a triangle is equal to the sum of the opposite 2 interior angles.
$x° = \frac{x}{2}° + \frac{x}{2}°$, hence the triangle is isosceles.

3 **a** ∠DAB = 50°, ∠DCB = 130°
b ∠AOC = $2x°$, hence ∠ABC = $x°$
reflex angle AOC = $2y°$, hence ∠ADC = $y°$
But $2x° + 2y° = 360°$ (angles around a point)
hence $2(x° + y°) = 360°$
giving $x° + y° = 180°$

4 **a** $x = 40°$
b ∠CED + ∠AEC = 180° (angles on a straight line)
∠ABC + ∠AEC = 180° (cyclic quadrilateral)
But ∠ABC = ∠ACB (isosceles triangle)
Hence ∠ACB = ∠CED

5 PS = QR, RS = PQ, both triangles share side QS hence by SSS triangles are congruent.

6 Join O to C and C to B to form a triangle. Let X be the point where the perpendicular bisector meets OB. Then OX = BX. By pythagoras OC = CB. But OC = OB (both radii). Hence OBC is equilateral.

7 $\overrightarrow{AB} = \mathbf{b} - \mathbf{a}$, $\overrightarrow{EF} = 3\mathbf{b} - 3\mathbf{a} = 3(\mathbf{b} - \mathbf{a}) = 3\overrightarrow{AB}$ hence $\overrightarrow{AB}$ is parallel to $\overrightarrow{EF}$.

8 **a** Check students' proofs
b By the alternate segment theorem ∠TXA = ∠TYB hence AX is parallel to BY.

9 ∠QAT = ∠QTA (isosceles triangle)
∠PTB = ∠QTA (vertically opposite angles)
∠PTB = ∠PBT (isosceles triangle)
Hence ∠PBT = ∠QAT and PB is parallel to AQ.

10 **a** $\overrightarrow{YW} = \overrightarrow{YZ} + \overrightarrow{ZW} = 2\mathbf{a} + \mathbf{b} + \mathbf{a} + 2\mathbf{b} = 3\mathbf{a} + 3\mathbf{b}$
$= 3(\mathbf{a} + \mathbf{b}) = 3\overrightarrow{XY}$
b 3 : 1
c They lie on a straight line.
d Points are A(6, 2), B(1, 1) and C(2, −4). Using Pythagoras' theorem, $AB^2 = 26$, $BC^2 = 26$ and $AC^2 = 52$ so $AB^2 + BC^2 = AC^2$ hence ∠ABC must be a right angle.

Examination questions

1 a i $\frac{1}{2}\mathbf{a}$ **ii** $\frac{1}{2}\mathbf{a} - \frac{1}{2}\mathbf{c}$
b $\overrightarrow{CA} = \mathbf{a} - \mathbf{c} = 2\overrightarrow{MN}$, so parallel

2 a i $\mathbf{p} + \mathbf{q}$ **ii** $\mathbf{q} - \mathbf{p}$
b $\frac{1}{2}\mathbf{p} + \frac{1}{2}\mathbf{q}$

3 a $-\mathbf{x}$ **b** $\mathbf{x} + \mathbf{y}$
c $-\frac{1}{2}\mathbf{x} - \mathbf{y}$

4 a $2\mathbf{a} - 2\mathbf{b}$
b $\overrightarrow{XY} = 2\mathbf{a}$ and $\overrightarrow{OR} = 6\mathbf{a}$, so parallel

5 a $\mathbf{a} + \mathbf{b}$
b $\overrightarrow{DX} = 2\mathbf{a}$ and $\overrightarrow{AB} = \mathbf{a}$, so parallel

6 a CD = BC, CE = CF, angle DCE = angle BCF, so congruent (SAS)
b ED = BF from **a** and BF = EG as parallelogram, so ED = EG

Answers to Chapter 16

Quick check

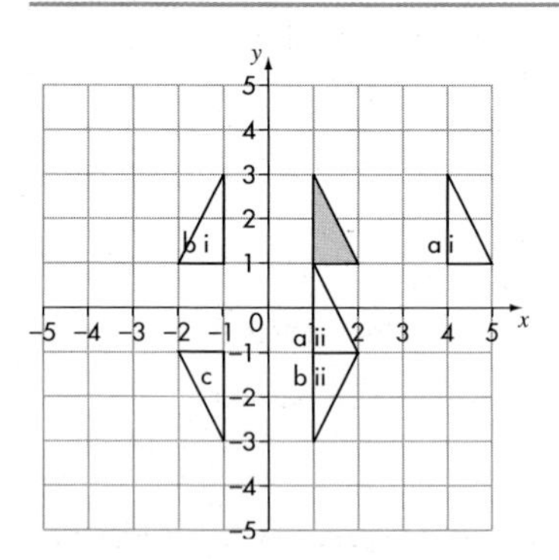

16.1 Transformations of the graph $y = f(x)$

Exercise 16A

1 a–d

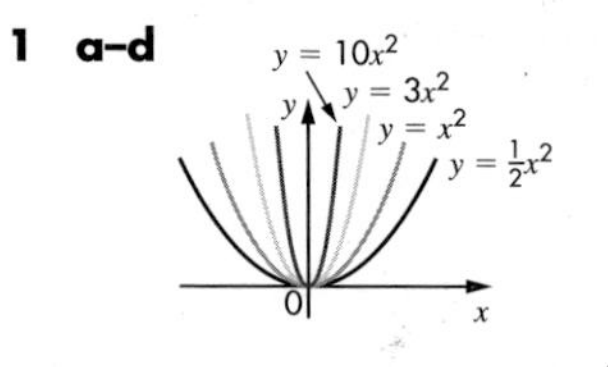

e Stretch sf in y-direction: 3, $\frac{1}{2}$, 10

2 a–d

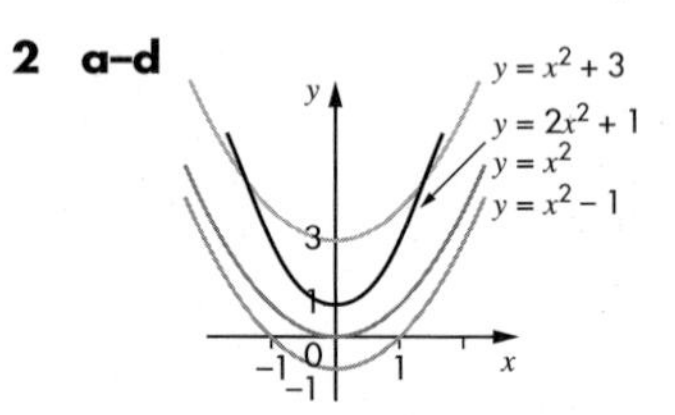

e b Translation $\begin{pmatrix}0\\3\end{pmatrix}$

c Translation $\begin{pmatrix}0\\-1\end{pmatrix}$

d Stretch sf 2 in y-direction, followed by translation $\begin{pmatrix}0\\1\end{pmatrix}$

3 a–d

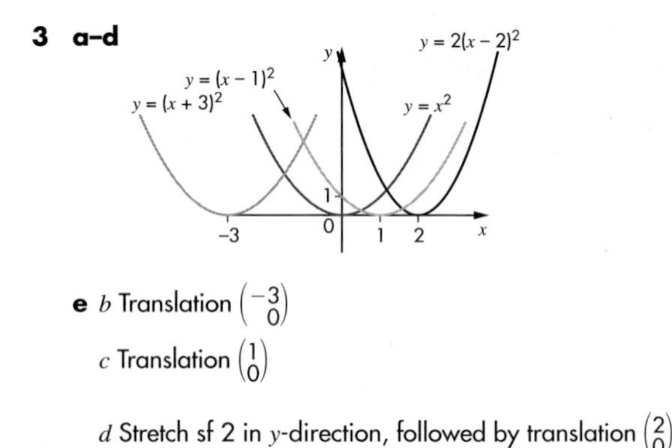

e b Translation $\begin{pmatrix}-3\\0\end{pmatrix}$

c Translation $\begin{pmatrix}1\\0\end{pmatrix}$

d Stretch sf 2 in y-direction, followed by translation $\begin{pmatrix}2\\0\end{pmatrix}$

4 a–c

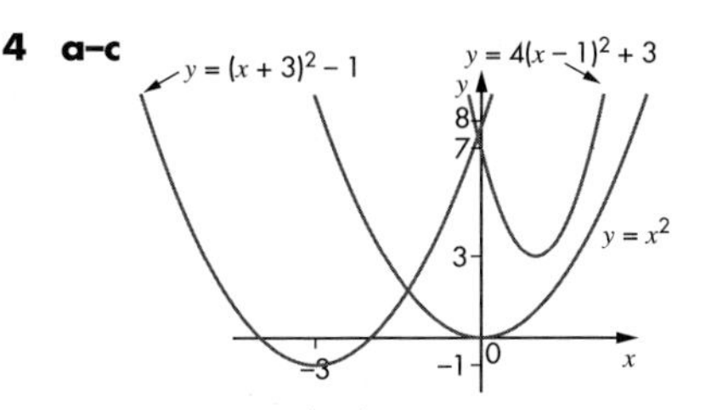

d *b* Translation $\begin{pmatrix}-3\\-1\end{pmatrix}$

c Translation $\begin{pmatrix}1\\3\end{pmatrix}$ followed by stretch sf 4 in y-direction

5 a–d

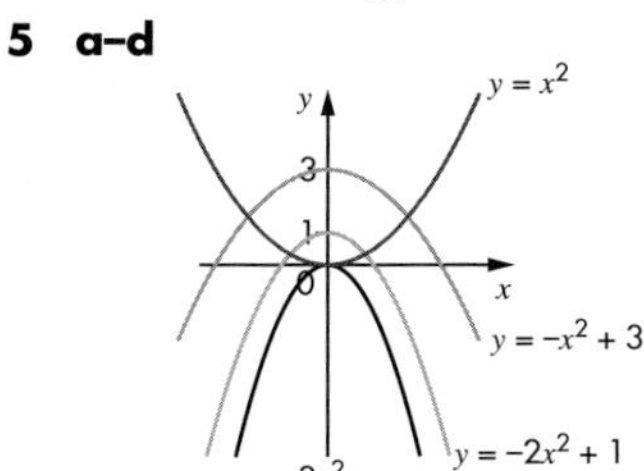

e *b* Reflection in x-axis, followed by translation $\begin{pmatrix}0\\3\end{pmatrix}$

c Reflection in the x-axis, followed by stretch sf 3 in y-direction

d Reflection in x-axis, followed by stretch sf 2 in y-direction and translation $\begin{pmatrix}0\\1\end{pmatrix}$

6 a–d

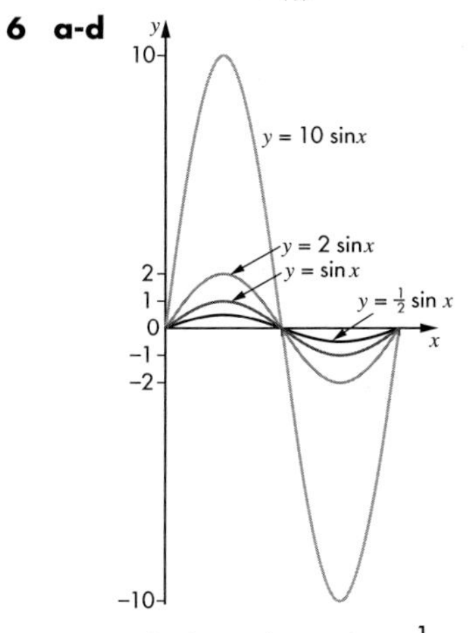

e Stretch sf in y-direction: 2, $\frac{1}{2}$, 10

7 a–d

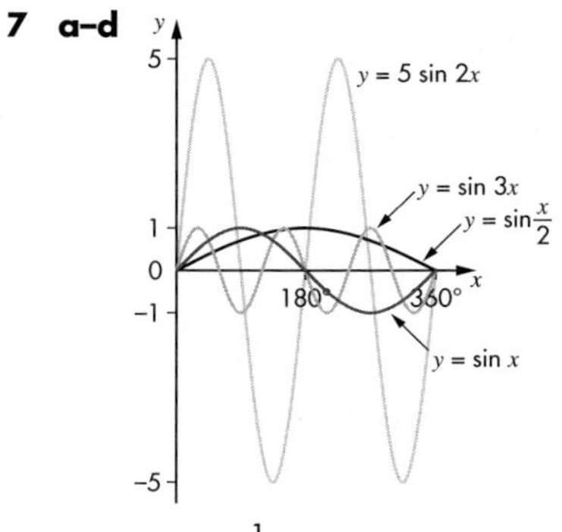

e *b* Stretch sf $\frac{1}{3}$ in x-direction

c Stretch sf 2 in x-direction

d Stretch sf 5 in y-direction, followed by stretch sf $\frac{1}{2}$ in x-direction

8 a–d

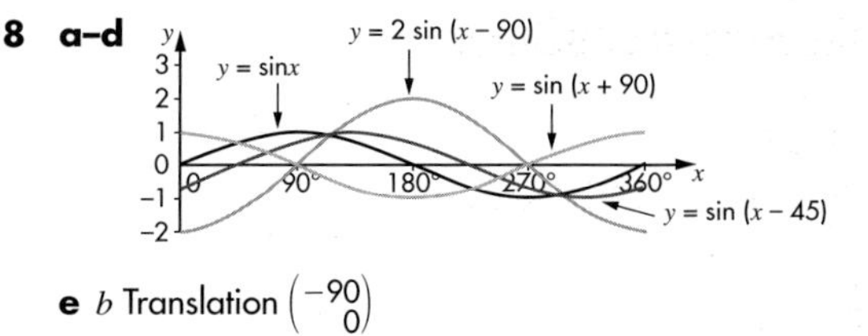

e *b* Translation $\begin{pmatrix}-90\\0\end{pmatrix}$

c Translation $\begin{pmatrix}40\\0\end{pmatrix}$

d Stretch sf 2 in y-direction followed by translation $\begin{pmatrix}90\\0\end{pmatrix}$

9 a–d

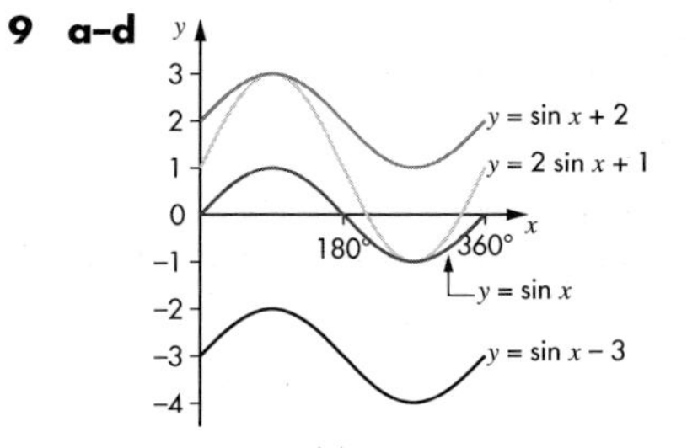

e *b* Translation $\begin{pmatrix}0\\2\end{pmatrix}$

c Translation $\begin{pmatrix}0\\-3\end{pmatrix}$

d Stretch sf 2 in y-direction followed by translation $\begin{pmatrix}0\\1\end{pmatrix}$

10 a–d

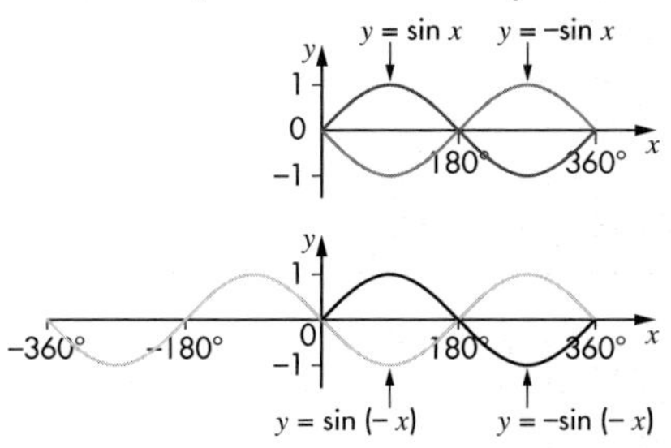

e *b* Reflection in x-axis

c Reflection in y-axis

d This leaves the graph in the same place and is the identity transformation

11 a–d

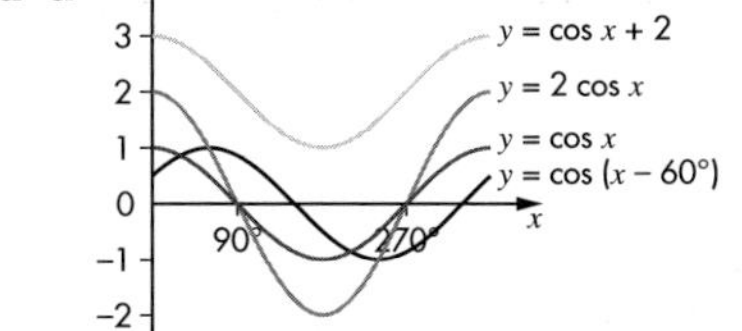

e *b* Stretch sf 2 in y-direction

c Translation $\begin{pmatrix}60\\0\end{pmatrix}$

d Translation $\begin{pmatrix}0\\2\end{pmatrix}$

12 All of them.

13 a **i** Stretch sf 4 in y-direction

ii Stretch sf 9 in y-direction

iii Stretch sf 16 in y-direction

b **i** Stretch sf $\frac{1}{2}$ in x-direction

ii Stretch sf $\frac{1}{3}$ in x-direction

iii Stretch sf $\frac{1}{4}$ in x-direction

c Stretch sf a^2 in y-direction, or stretch sf $\frac{1}{a}$ in x-direction

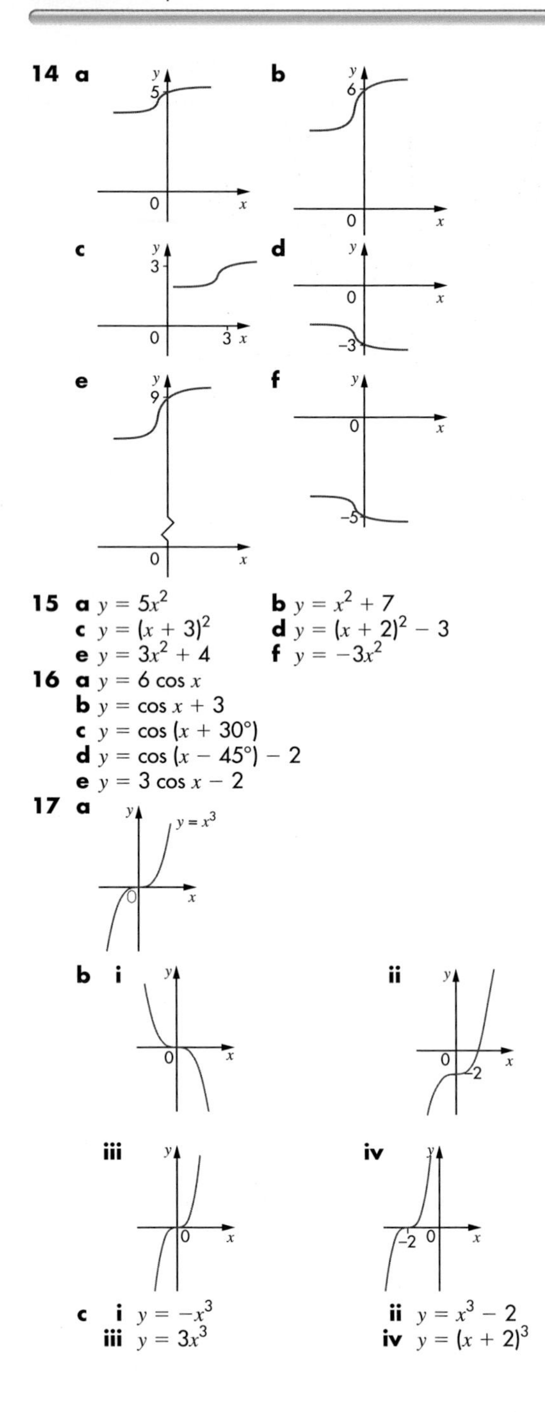

14 **a** **b** **c** **d** **e** **f**

15 **a** $y = 5x^2$ **b** $y = x^2 + 7$
c $y = (x + 3)^2$ **d** $y = (x + 2)^2 - 3$
e $y = 3x^2 + 4$ **f** $y = -3x^2$

16 **a** $y = 6 \cos x$
b $y = \cos x + 3$
c $y = \cos (x + 30°)$
d $y = \cos (x - 45°) - 2$
e $y = 3 \cos x - 2$

17 **a**

b **i** **ii** **iii** **iv**

c **i** $y = -x^3$ **ii** $y = x^3 - 2$
iii $y = 3x^3$ **iv** $y = (x + 2)^3$

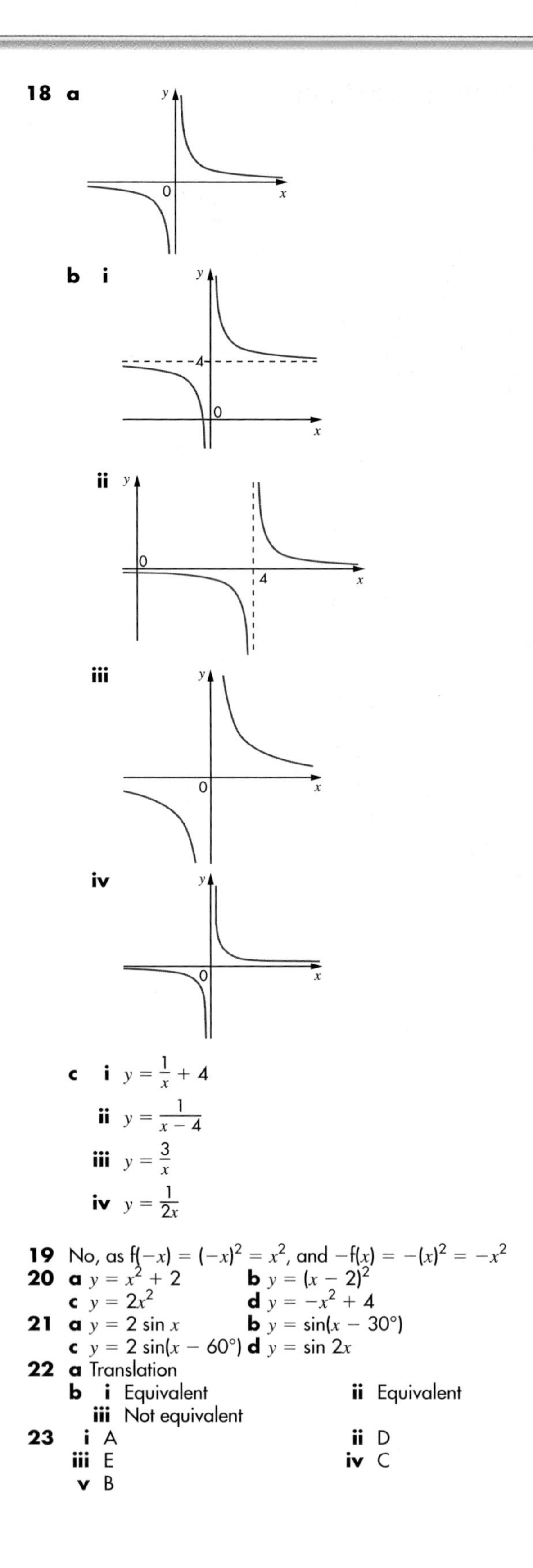

18 **a**

b **i** **ii** **iii** **iv**

c **i** $y = \frac{1}{x} + 4$
ii $y = \frac{1}{x - 4}$
iii $y = \frac{3}{x}$
iv $y = \frac{1}{2x}$

19 No, as $f(-x) = (-x)^2 = x^2$, and $-f(x) = -(x)^2 = -x^2$

20 **a** $y = x^2 + 2$ **b** $y = (x - 2)^2$
c $y = 2x^2$ **d** $y = -x^2 + 4$

21 **a** $y = 2 \sin x$ **b** $y = \sin(x - 30°)$
c $y = 2 \sin(x - 60°)$ **d** $y = \sin 2x$

22 **a** Translation
b **i** Equivalent **ii** Equivalent
iii Not equivalent

23 **i** A **ii** D
iii E **iv** C
v B

Examination questions

1 **a** (4, 3) **b** (2, 6)

2

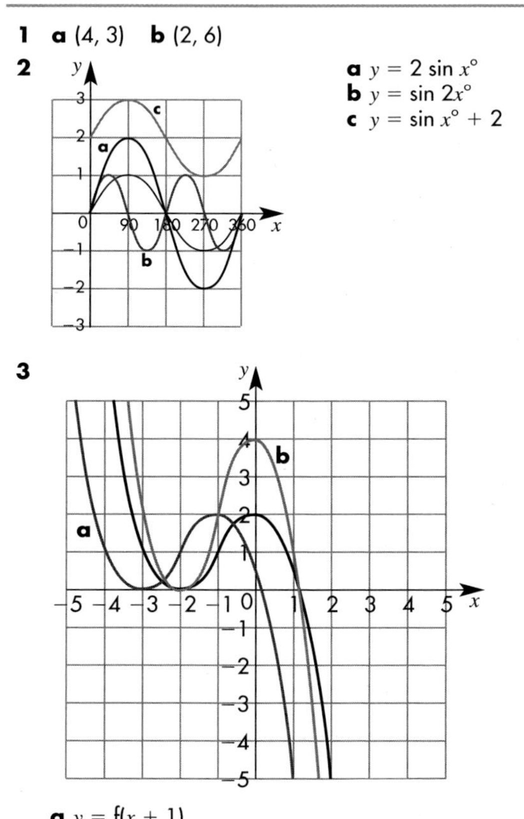

a $y = 2 \sin x°$
b $y = \sin 2x°$
c $y = \sin x° + 2$

3

a $y = \text{f}(x + 1)$
b $y = 2\text{f}(x)$

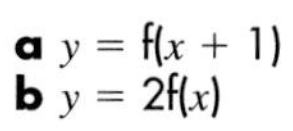

4

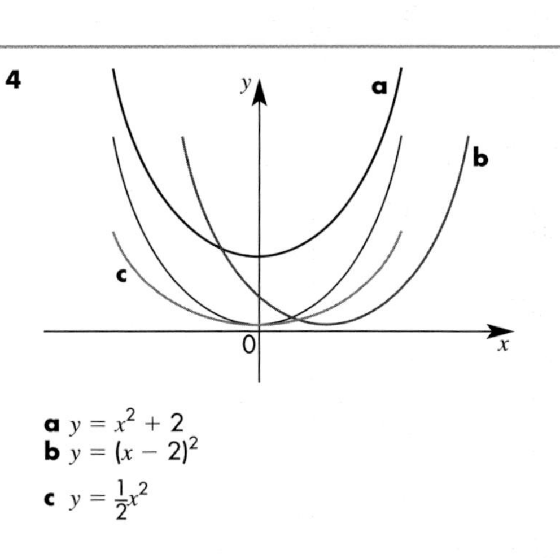

a $y = x^2 + 2$
b $y = (x - 2)^2$
c $y = \frac{1}{2}x^2$

GLOSSARY

adjacent side In a triangle, rectangle or square, the side adjacent to (next to) the angle or side being worked on.

alternate segment The 'other' segment. In a circle divided by a chord, the alternate segment lies on the other side of the chord.

and In solving probability problems, if two events occur, A and B, the probability of them both occurring is P(A) × P(B).

angle of depression The angle you have to turn downwards from looking along the horizontal to look at the ground or sea from the top of a tower, tree, or cliff, etc.

angle of elevation The angle you have to turn upwards from looking along the horizontal to look at the top of a tree, cliff, flagpole, etc.

apex The highest vertex in the given orientation of a polygon such as triangle or a 3D shape such as a pyramid or cone.

arc A curve forming part of the circumference of a circle.

area Measurement of the flat space a shape occupies. Usually measured in square units or hectares. (See also *surface area*.)

area ratio The ratio of the areas of two similar shapes is equal to the ratio of the squares of their corresponding lengths. The area ratio, or area scale factor, is the square of the length ratio.

area scale factor See *area ratio*.

area sine rule The area of a triangle is given by $\frac{1}{2}ab \sin C$.

asymptote A straight line whose perpendicular distance from a curve decreases to zero as the distance from the origin increases without limit.

balance Equality on either side of an equation.

balance the coefficients The first step to solving simultaneous equations is to make the coefficients of one of the variables the same. This is called balancing the coefficients.

bearing The direction relative to a fixed point.

box plot Or box-and-whisker plot. This is a way of displaying data for comparison. It requires five pieces of data; lowest value, the lower quartile (Q1), the median (Q2), the upper quartile (Q3), and the highest value.

box-and-whisker plot A way of displaying data for comparison. It requires five pieces of data; lowest value, the lower quartile (Q1), the median (Q2), the upper quartile (Q3) and the highest value.

brackets The symbols '(' and ')' which are used to separate part of an expression. This may be for clarity or to indicate a part to be worked out individually. When a number and/or value is placed immediately before an expression or value inside a pair of brackets, the two are to be multiplied together. For example, $6a(5b + c) = 30ab + 6ac$.

cancel A fraction can be simplified to an equivalent fraction by dividing the numerator and denominator by a common factor. This is called cancelling.

check Calculations can be checked by carrying out the inverse operation. Solutions to equations can be checked by substituting values of the variable(s).

chord A line joining two points on the circumference of a circle.

circle A circle is the path of a point that is always equidistant from another point (the centre).

circular function A function with a repeating set of values, such as sine and cosine, where the values cycle through from 0 to 1 to 0 to –1 to 0, and repeat.

circumference The outline of a circle. The distance all the way around this outline.

class interval The size or spread of the measurement defining a class. For example, heights could be grouped in 1 cm or 10 cm class intervals.

coefficient The number in front of an unknown quantity (the letter) in an algebraic term. For example, in $8x$, 8 is the coefficient of x.

combined events Two or more events (independent or mutually exclusive) that may occur during a trial.

completing the square A method of solving quadratic equations which involves rewriting the equation $x^2 + px + q$ in the form $(x + a)^2 + b$.

conditional probability This describes the situation when the probability of an event is dependent upon the outcome of another event. For example, the probability of the colour of a second ball drawn from a bag is conditional to the colour of the first ball drawn from the bag – if the first ball is not replaced.

constant of proportionality, k This describes the situation when the probability of an event is dependent upon the outcome of another event. For example, the probability of the colour of a second ball drawn from a bag is conditional to the colour of the first ball drawn from the bag – if the first ball is not replaced.

constant term A term in an algebraic expression that does not change because it does not contain a variable, the number term. For example, in $6x^2 + 5x + 7$, 7 is the constant term.

continuous data Data that can be measured rather than counted, such as weight and height. It can have an infinite number of values within a range.

cosine The ratio of the adjacent side to the hypotenuse in a right-angled triangle.

cosine rule A formula used to find the lengths of sides or the size of an angle in a triangle. $a^2 = b^2 + c^2 - 2bc \cos A$.

cross-multiply A method for taking the denominator from one side of an equals sign to the numerator on the other side of the equals sign. You are actually multiplying both sides of the equation by the denominator, but it cancels out on the original side. (Take care that you multiply all the terms by the denominator.)

cube A 3D shape with six identical square faces.

cubic A cubic expression or equation contains an 'x^3' term.

cumulative frequency Can be found by adding each frequency to the sum of all preceding frequencies.

cumulative frequency diagram A graph where the value of the variable is plotted against the total frequency so far. The last plot will show the total frequency.

cyclic Repeating indefinitely in both directions.

cyclic quadrilateral A quadrilateral whose vertices lie on a circle.

demonstration Logically presented proof of how a theory generates a certain result.

diameter A straight line across a circle, from circumference to circumference and passing through the centre. It is the longest chord of a circle and two radii long. (See also *radius*.)

difference of two squares The result $a^2 - b^2 = (a - b)(a + b)$ is called the difference of two squares.

direct proportion Two values or measurements may vary in direct proportion. That is, if one increases, then so does the other.

direct variation Another name for *direct proportion*.

direction The way something is facing or pointing. Direction can be described using the compass points (north, south, south-east, etc.) or using bearings (the clockwise angle turned from facing north). The direction of a vector is given by the angle it makes with a line of reference. Two vectors that differ only in magnitude will be parallel.

discrete data Data that is counted, rather than measured, such as favourite colour or a measurement that occurs in integer values, such as a number of days. It can only have specific values within a range.

discriminant The quantity $(b^2 - 4ac)$ in the quadratic formula is called the discriminant.

dispersion Another name for the interquartile range.

eliminate To remove a quantity such as a variable from an equation.

exact values An alternative name for surds.

exponential function Non-linear equations that have the form $y = kx$, where k is a positive number.

expression Collection of symbols representing a number. These can include numbers, variables (x, y, etc.), operations (+, ×, etc.), functions (squaring, cosine, etc.), but there will be no equals sign (=).

factorisation (noun) Finding one or more factors of a given number or expression. (verb: factorise)

factorise See *factorisation*.

factors A whole number that divides exactly into a given number. For example, the factors of 16 are 1, 2, 4, 8, 16.

formula (plural: formulae) An equation that enables you to convert or find a measurement from another known measurement or measurements. For example, the conversion formula from the Fahrenheit scale of temperature to the more common Celsius scale is $C = \frac{5}{9}(F - 32)$ where C is the temperature on the Celsius scale and F is the temperature on the Fahrenheit scale.

frustum The base of a cone or pyramid. The shape obtained by removing the top of a cone or pyramid.

function A function of x is any algebraic expression in which x is the only variable. This is often represented by the function notation $f(x)$ or 'function of x'.

highest value The largest value in a range of values.

hypotenuse The longest side of a right-angled triangle. The side opposite the right angle.

included angle The angle between two lines or sides of a polygon.

independent events Two events are independent if the occurrence of one has no influence on the occurrence of the other. For example, getting heads when tossing a coin has no influence over scoring a 2 with a dice. Missing a bus and getting to school on time are not independent events.

index (plural: indices) A power or exponent. For example, in the expression 3^4, 4 is the index, power or exponent.

indices See *index*.

interquartile range The difference between the values of the lower and upper quartiles.

inverse Inverse operations cancel each other out or reverse the effect of each other. The inverse of a number is the reciprocal of that number.

inverse cosine The reverse of the cosine function. It tells you the value of the angle with that cosine. $\text{Cos}^{-1}A$.

inverse proportion Two quantities vary in inverse proportion when one quantity is directly proportional to the reciprocal of the other. As one quantity increases, the other decreases.

inverse sine The reverse of the sine function. It tells you the value of the angle with that sine. $\text{Sin}^{-1}A$.

inverse variation Another name for inverse proportion.

isosceles triangle A triangle with two sides that are equal. It also has two equal angles.

length ratio The ratio of lengths in similar figures.

limit of accuracy No measurement is entirely accurate. The accuracy depends on the tool used to measure it. The value of every measurement will be rounded to within certain limits. For example you can probably measure with a ruler to the nearest half-centimetre. Any measurement you take could be inaccurate by up to half a centimetre. This is your limit of accuracy. (See also *lower bound* and *upper bound*.)

line symmetry Symmetry that uses a line to divide a figure or shape into halves, such that both halves are an exact mirror image of each other. Many shapes have more than one line of symmetry.

linear Forming a line.

linear scale factor Also called the *scale factor* or *length ratio*. The ratio of corresponding lengths in two similar shapes is constant.

lower bound The lower limit of a measurement. (See also *limit of accuracy*.)

lower quartile The value of the item at one-quarter of the total frequency. It is the $\frac{n}{4}$th value.

lowest value The smallest value in a range of values.

magnitude Size. Magnitude is always a positive value.

maximum The greatest value of something. The turning point or point at which the graph of a parabola $y = -ax^2 + bx + c$ is at its highest.

median The middle value of a sample of data that is arranged in order. For example, the sample 3, 2, 6, 2, 2, 3, 7, 4 may be arranged in order as follows 2, 2, 2, 3, 3, 4, 6, 7. The median is the fourth value, which is 3. If there is an even number of values the median is the mean of the two middle values. For example, 2, 3, 6, 7, 8, 9, has a median of 6.5.

minimum The smallest value of something. The turning point or point at which the graph of a parabola $y = ax^2 + bx + c$ is at its lowest.

negative reciprocal The reciprocal multiplied by –1. The gradients of perpendicular lines are the negative reciprocal of each other.

non-linear An expression or equation that does not form a straight line. The highest power of x is greater than 1.

opposite side In a triangle, rectangle or square, the side opposite (or, on the other side of) the angle or side being worked on.

or In solving probability problems, if two events occur, A or B, the probability of either of them occurring is P(A) + P(B).

parallel Straight lines that are always the same distance apart.

perpendicular At right angles. Two perpendicular lines are at right angles to each other. A line or plane can also be perpendicular to another plane.

point of contact The point where a tangent touches a circle.

polygon A closed shape with three or more straight sides.

power When a number or expression is multiplied by itself, the power is how many 'lots' are multiplied together. For example, $2^3 = 2 \times 2 \times 2$ (note that it is not the number of times that 2 is multiplied by itself; it is one more than the number of times it is multiplied by itself.) The name given to the symbol to indicate this, such as 2. (See also *square* and *cube*.)

probability space diagram A diagram or table showing all the possible outcomes of an event.

proof An argument that establishes a fact about numbers or geometry for all cases. Showing that the fact is true for specific cases is a demonstration.

prove The process of explaining a proof. (See also *proof*.)

pyramid A polyhedron on a triangular (see triangle), square, or polygonal (see polygon) base, with triangular faces meeting at a vertex. The volume, V, of a pyramid of base area A and perpendicular height h, is given by the formula: $V = \frac{1}{3}A \times h$.

Pythagoras' theorem The theorem states that the square on the hypotenuse of a right-angled triangle is equal to the sum of the squares on the other two sides.

quadratic expression An expression involving an x^2 term.

quadratic formula A formula for solving quadratic equations. The solution of the equation $ax^2 + bx + c = 0$ is given by: $x = \frac{-b \pm \sqrt{b^2 - 4ac}}{2a}$

quartile One of four equal parts, formed when data is divided.

radius (plural: radii) The distance from the centre of a circle to its circumference.

ratio The ratio of A to B is a number found by dividing A by B. It is written as A : B. For example, the ratio of 1 m to 1 cm is written as 1 m : 1 cm = 100 : 1. Notice that the two quantities must both be in the same units if they are to be compared in this way.

rational number A rational number is a number that can be written as a fraction. For example, $\frac{1}{4}$ or $\frac{10}{3}$.

rationalise To make into a ratio. Removing a surd from a denominator (by multiplying the numerator and denominator by that surd) is called rationalising the denominator.

rearrange See rearrangement.

reciprocal The reciprocal of any number is 1 divided by the number. The effect of finding the reciprocal of a fraction is to turn it upside down. The reciprocal of 3 is $\frac{1}{3}$. The reciprocal of $\frac{1}{4}$ is 4. The reciprocal of $\frac{10}{3}$ is $\frac{3}{10}$.

recurring decimal A decimal number that repeats forever with a repeating pattern of digits.

reflection The image formed after being reflected. The process of reflecting an object.

rotational symmetry A shape which can be turned about a point so that it coincides exactly with its original position at least twice in a complete rotation.

rule An alternative name for a formula.

scale factor The ratio by which a length or other measurement is increased or deceased.

segment A part of a circle between a chord and the circumference.

semicircle Half a circle.

show The middle level in an algebraic proof. You are required to show that both sides of the result are the same algebraically. (See also *proof* and *verify*.)

similar The same shape but a different size.

similar triangles Triangles with the same size angles. The lengths of the sides of the triangles are different but vary in a constant proportion.

simultaneous equations Two or more equations that are true at the same time.

sine The ratio of the opposite side to the hypotenuse in a right-angled triangle.

sine rule In a triangle, the ratio of the length of a side to the sine of the opposite angle is constant, hence $\frac{a}{\text{Sin}A} = \frac{b}{\text{Sin}B} = \frac{c}{\text{Sin}C}$.

slant height The distance along the sloping edge of a cone or pyramid.

soluble Something that can be solved.

solve Finding the value or values of a variable (x) that satisfy the given equation or problem.

sphere A three-dimensional round body. All points of its surface are equidistant from its centre.

square
1. A polygon with four equal sides and all the interior angles equal to 90°.
2. The result of multiplying a number by itself. For example, 5^2 or 5 squared is equal to $5 \times 5 = 25$.

square root The square of a square root of number gives you the number. The square root of 9 (or $\sqrt{9}$) is 3, $3 \times 3 = 3^2 = 9$.

standard form Also called standard index form. Standard form is a way of writing very large and very small numbers using powers of 10. A number is written as a value between 1 and 10 multiplied by a power of 10. That is, $a \times 10^n$ where $1 \leqslant a < 10$, and n is a whole number.

standard index form See *standard form*.

stretch An enlargement that takes place in one direction only. It is described by a scale factor and the direction of the stretch.

subject The subject of a formula is the letter on its own on one side of the equals sign. For example, t is the subject of this formula: $t = 3f + 7$.

substitute When a letter in an equation, expression or formula is replaced by a number, we have substituted the number for the letter. For example, if $a = b + 2x$, and we know $b = 9$ and $x = 6$, we can write $a = 9 + 2 \times 6$. So $a = 9 + 12 = 17$.

subtended Standing on. An angle made by two radii at the centre of a circle is the angle subtended by the arc which joins the points on the circumference at the ends of the radii.

surd A number written as $\sqrt{x}$. For example, $\sqrt{7}$.

surd form The square root sign is left in the final expression when $\sqrt{x}$ is an irrational number.

surface area The area of the surface of a 3D shape, such as a prism. The area of a net will be the same as the surface area of the shape.

tangent
1. A straight line that touches the circumference of a circle at one point only.
2. The ratio of the opposite side to the adjacent side in a right-angled triangle.

terminating decimal A terminating decimal can be written down exactly. $\frac{33}{100}$ can be written as 0.33, but $\frac{3}{10}$ is 0.3333... with the 3s recurring forever.

three-figure bearing The angle of a bearing is given with three digits. The angle is less than 100°, a zero (or zeros) is placed in front, such as 045° for north-east.

transform To change.

translation A transformation in which all points of a plane figure are moved by the same amount and in the same direction.

transpose To rearrange a formula.

tree diagram A diagram to show all the possible outcomes of combined events.

triangle A three-sided polygon. The interior angles add up to 180°. Triangles may be classified as:

1. scalene – no sides of the triangle are equal in length (and no angles are equal)
2. equilateral – all the sides of the triangles are equal in length (and all the angles are equal)
3. isosceles – two of the sides of the triangle are equal in length (and two angles are equal)
4. a right-angled triangle has an interior angle equal to 90°.

trigonometry The branch of mathematics that shows how to explain and calculate the relationships between the sides and angles of triangles.

upper bound The higher limit of a measurement. (See also *limit of accuracy*.)

upper quartile The value of the item at three-quarters of the total frequency. It is the $\frac{3n}{4}$th value.

variable A quantity that can have many values. These values may be discrete or continuous. They are often represented by x and y in an expression. (See also *expression*.)

vector A quantity with magnitude and direction. (See also *magnitude* and *direction*.)

verify The first of the three levels in an algebraic proof. Each step must be shown clearly and at this level you need to substitute numbers into the result to show how it works. (See also *proof* and *show*.)

vertical height The perpendicular height from the base to the apex of a triangle, cone, or pyramid.

volume The amount of space occupied by a substance or object or enclosed within a container.

volume ratio The ratio of the volumes of two similar shapes is equal to the ratio of the cubes of their corresponding lengths. The volume ratio, or volume scale factor, is the cube of the length ratio.

volume scale factor See *volume ratio*.

INDEX